C0-AZS-537

Ecological Studies

Analysis and Synthesis

Volume 91

Ecological Studies

Volume 75
The Grazing Land Ecosystems of the African Sahel (1989)
By H.N. Le Houérou

Volume 76
Vascular Plants as Epiphytes: Evolution and Ecophysiology (1989)
Edited by U. Lüttge

Volume 77
Air Pollution and Forest Decline: A Study of Spruce *(Picea abies)* on Acid Soils (1989)
Edited by E.-D. Schulze, O.L. Lange, and R. Oren

Volume 78
Agroecology: Researching the Ecological Basis for Sustainable Agriculture (1990)
Edited by S.R. Gliessman

Volume 79
Remote Sensing of Biosphere Functioning (1990)
Edited by R.J. Hobbs and H.A. Mooney

Volume 80
Plant Biology of the Basin and Range (1990)
Edited by C.B. Osmond, L.F. Pitelka, and G.M. Hidy

Volume 81
Nitrogen in Terrestrial Ecosystems (1990)
By C.O. Tamm

Volume 82
Quantitative Methods in Landscape Ecology (1990)
Edited by M.G. Turner and R.H. Gardner

Volume 83
The Rivers of Florida (1990)
Edited by R.J. Livingston

Volume 84
Fire in the Tropical Biota: Ecosystem Processes and Global Challenges (1990)
Edited by J.G. Goldammer

Volume 85
The Mosaic-Cycle Concept of Ecosystems (1991)
Edited by H. Remmert

Volume 86
Ecological Heterogeneity (1991)
Edited by J. Kolasa and S.T.A. Pickett

Volume 87
Horses and Grasses: The Nutritional Ecology of Equids, and Their Impact on the Camargue (1991)
By P. Duncan

Volume 88
Pinnipeds and El Nino: Responses to Environmental Stress (1991)
Edited by F. Trillmich and K.A. Ono

Volume 89
Plantago: A Multidisciplinary Study (1992)
Edited by P.J.C. Kuiper and M. Bos

Volume 90
Biogeochemistry of a Subalpine Ecosystem: Loch Vale Watershed (1992)
Edited by Jill Baron

Volume 91
Atmospheric Deposition and Forest Nutrient Cycling
Edited by D.W. Johnson and S.E. Lindberg

Dale W. Johnson
Steven E. Lindberg
Editors

Atmospheric Deposition and Forest Nutrient Cycling

A Synthesis of the Integrated Forest Study

With 236 Illustrations

Springer-Verlag
New York Berlin Heidelberg London Paris
Tokyo Hong Kong Barcelona Budapest

Dale W. Johnson
Desert Research Institute
University of Nevada System
Biological Sciences Center
Reno, NV 89506 USA
and
Range, Wildlife and Forestry
College of Agriculture
University of Nevada, Reno
Reno, NV 89512 USA

Steven E. Lindberg
Oak Ridge National Laboratory
Environmental Sciences Division
Oak Ridge, TN 37831 USA

Library of Congress Cataloging-in-Publication Data
Atmospheric deposition and forest nutrient cycling : a synthesis of the integrated forest study / Dale W. Johnson, Steven E. Lindberg, editors.
p. cm. — (Ecological studies)
Includes bibliographical references and index.
ISBN 0-387-97632-9 (alk. paper)
1. Forest ecology. 2. Acid deposition—Environmental aspects. 3. Mineral cycle (Biogeochemistry) I. Johnson, D.W. (Dale W.), 1946– . II. Lindberg, Steven E. III. Series.
QH541.5.F6A85 1992
581.5′2642—dc20 91-20122

Printed on acid-free paper.

Production managed by Henry Krell; Manufacturing supervised by Jacqui Ashri.
Typeset by Impressions, Inc., Ann Arbor, MI.
Printed and bound by Braun-Brumfield, Ann Arbor, MI.
Printed in the United States of America.

9 8 7 6 5 4 3 2 1

ISBN 0-387-97632-9 Springer-Verlag New York Berlin Heidelberg
ISBN 3-540-97632-9 Springer-Verlag Berlin Heidelberg New York

Preface

During the past decade, there has been considerable interest in the effects of atmospheric deposition on forest ecosystems. The basis for this interest includes general concern about effects of acidic deposition, or acid rain, on natural ecosystems and the reports of forest decline in Europe and eastern North America, plus the suggestion that air pollutants might be responsible.

To determine what relationships might exist between atmospheric deposition and forest health, a number of major research programs were initiated in the early 1980s. These included the Forest Effects Program of the U.S. National Acid Precipitation Assessment Program (NAPAP), the programs of Germany and other European countries, and several large projects funded by the Electric Power Research Institute (EPRI). This volume describes the results of one of the latter projects, the Integrated Forest Study (IFS).

The Integrated Forest Study was a $15 million project to evaluate the effects of atmospheric deposition on nutrient cycling in forest ecosystems. The research involved monitoring deposition and nutrient cycling at 17 forested sites in the northwestern, northeastern, and southeastern United States and in Canada and Norway. These sites represent a range of conditions in climate, air quality, soils, and vegetation, which facilitated the testing of hypotheses regarding the effects of atmospheric sulfur and nitrogen deposition on forest nutrient cycles. In addition to field measurements of atmospheric deposition and nutrient cycling, the IFS included a component of experimental research, incorporating both laboratory and field studies, to investigate selected atmospheric and soil processes in greater detail.

Several features of the IFS make it unusual and significant both as an applied research project to evaluate the effects of pollution on ecosystems and as a basic investigation of forest nutrient cycling. As an applied study, the IFS is unusual in its focus on understanding the ecosystem-level processes controlling the nutritional status of forests. Only by evaluating the entire ecosystem nutrient budget, including all inputs, outputs, and internal fluxes, has it been possible to develop a complete understanding of the status of individual nutrients and to identify whether any significant changes in status are occurring.

As a more basic investigation of ecosystem nutrient cycling, the IFS is noteworthy for its incorporation of state-of-the-art methods for measuring dry and cloud water deposition in addition to wet deposition. Another important feature of the study is the integration of a series of experimental tasks (e.g., on soil weathering, foliar leaching, organic and inorganic sulfur immobilization) with the deposition and nutrient cycling monitoring tasks. But perhaps the most significant aspect of the study is the fact that deposition and nutrient cycling were monitored at a large number of field sites during essentially the same time period using the same instruments and protocols. Most previous analyses of nutrient cycling patterns across different ecosystems have involved comparisons of results from independent studies that used different protocols. As a consequence of the integrated nature of the IFS, we can be far more confident that any dissimilarities in measured patterns of nutrient cycling across sites reflect real differences rather than artifacts of different methodologies or measurement periods.

We believe that this volume, and the results of the IFS, will be of lasting value in several ways. First, the results address the current issue of acid rain by providing the basis for evaluating how acidic deposition does affect the mineral nutrient status of forest ecosystems. Most important, the study places this effect in perspective, showing how it varies as a function of other ecosystem properties and how it compares to other processes that influence nutrient balances. Second, the comparative data set on deposition and nutrient cycling at the 17 forest sites, as well as the analyses of these data, will provide valuable new insights into the processes and regulation of forest nutrient cycling. These data are now available to be examined and analyzed further by other scientists. Finally, an important product of the project is the model, NuCM, that is described in Chapter 14. This is a more comprehensive and detailed model of forest nutrient cycling than heretofore has been available. It should be useful to both managers and basic researchers for evaluating the factors that control nutrient cycling in forest systems.

L.F. Pitelke
D.W. Johnson
S.E. Lindberg

Acknowledgments

The IFS was funded primarily by the Electric Power Research Institute (EPRI), but many other organizations and government agencies either funded parts of the project or otherwise cooperated in helping to ensure the success of this unusual and complex research effort. The Empire State Electric Energy Research Corporation and Southern Company Services, Inc., and Forestry Canada each funded individual sites, and the National Acid Precipitation Assessment Program funded two sites. The Norwegian Forest Research Institute, the Norwegian Forest Research Institute for Air Research, the U.S. Forest Service, and the U.S. National Park Service cooperated by providing personnel time or facilities at particular sites. The support provided by these organizations reflects both a commitment to understanding the effects of atmospheric deposition on ecosystems and a recognition of the value of a large integrated project that addressed the issue in a comprehensive manner.

The project also was enhanced by annual meetings with a large number of project reviewers and advisors. We would like to especially acknowledge Bill Reiners, Bill McFee, Dave Grigal, Don Gatz, and Marvin Weseley for their advice and support as project reviewers. These individuals, representing universities, research institutes, funding organizations, and utilities, provided valuable advice and encouragement. Ultimately, however, the key to the success of the IFS was the large group of researchers and advisors who were able to work together in a relationship that remained harmonious and productive through the course of the project.

Many people, some of whom are not listed as authors of this volume, deserve our thanks for helping make this project a success. In the realm of management, we must first and foremost express our great appreciation to Dr. Louis Pitelka, our project manager at EPRI, for his steady hand in keeping the Integrated Forest Study (IFS) viable and strong for so many years. We wish to thank Dr. John Huckabee of EPRI for his crucial role in the initial formation and organization of the project; without his input and encouragement at this crucial early stage, the IFS would never have begun. We also wish to acknowledge the late Ernest Bondietti for his role as Oak Ridge National Laboratory (ORNL) manager of the project in its earliest stages, and to Dr. Robert I. Van Hook of ORNL for his crucial role in making midcourse changes in management structure.

Numerous technical personnel conducted most of the actual field and laboratory work of the IFS project, leaving the principal investigators to reap the scientific benefits. They include: Donald E. Todd, Jr., Jim Owens, Wil Petty, Kris Dearstone, and Ruth Montgomery of Oak Ridge National Laboratory; David Silsbee of the Uplands Research Center, Great Smoky Mountains National Park; Paul Hazlett, Wayne Johns, Dieter Ropke, Don Kurylo, Jo Ramakers, and Linda Irwin of Forestry Canada, Ontario Region; Bob Gonyea, Bert Hasselberg, Mike Johnson, Dick Hinshaw, Jay Kuhn, Jacquie Fenning, Gordon Wolfe, Tony Basabe, and Jana Compton at the University of Washington; Ute Valentine and Paul Conklin, Duke University; Magne Huse, Norwegian Forest Research Institute; and Lee Reynolds, James Buchannan, Bryant Cunningham, Jim Deal, Lisa Leatherman, Bob McCollum, Barbara Reynolds, Mary Lou Rollins, and David White from Coweeta Hydrologic Laboratory.

The framework used in the H^+ analysis was developed in discussions with a variety of colleagues, most notably Dan Richter, Dave Valentine, Phil Sollins, and Dale Johnson.

Thanks go to Judith Tarplee who assisted in the field and in the laboratory and to Nicole Ruderman, Lisa Hu, Beth Van Schaack, John Figurelli, Veronica Blette and Jill Newton for their assistance with sample preparation, analytical procedures, and preparation of the manuscript. Special gratitude is extended to Dianne Keller whose Master's thesis research at Colgate University contributed greatly to the completion and overall goals of this project, and for her valuable comments and assistance in preparing this manuscript.

We would like to express our appreciation to Dr. David Grigal and Dr. Michael Unsworth for their constructively critical reviews of this book.

Finally, we wish to thank Susan Sawatzky of the Desert Research Institute and Opal Grooms of Oak Ridge National Laboratory for their tireless (and often thankless) efforts in word processing this volume.

This research was sponsored by the Electric Power Research Institute under Contract TP-2621–1 and the Atmospheric and Climate Research Division, Office of Health and Environmental Research, U.S. Department of

Energy, under Contract No. DE-AC05–84OR21400 with Martin Marietta Energy Systems, Inc., and under contract RP-2621–3 with the Desert Research Institute, University of Nevada System.

This volume is Publication No. 3668, Environmental Sciences Division, ORNL.

Contents

Contributors

D. Aamlid	Norwegian Forest Research Institute, Postbox 61 Aas–NLH, Norway
G. Abrahamsen	Department of Forest Soils, Agricultural University of Norway, N-1432 Aas-NLH, Norway
E. Allen	Department of Environmental and Engineering Science, University of Florida, Gainesville, FL 32601, USA
A. Autry	Department of Microbiology, University of Georgia, Athens, GA 30602, USA
Richard April	Department of Geology, Colgate University, Hamilton, NY 13345, USA
J. Battles	School of Natural Resources, Cornell University, Ithaca, NY 14850, USA

Dan Binkley	Forest and Wood Sciences, Colorado State University, Ft. Collins, CO 80523, USA
Ernie Bondietti	Environmental Sciences Division, Oak Ridge National Laboratory, Oak Ridge, TN 37831-6034, USA
Dale Cole	College of Forest Resources AR-10, University of Washington, Seattle, WA 98195, USA
Paul Conklin	School of Forestry and Environmental Studies, Duke University, Durham, NC 27706, USA
K. Dai	School of Forestry and Environmental Studies, Duke University, Durham, NC 27706, USA
R. Edmonds	College of Forest Resources AR-10, University of Washington, Seattle, WA 98195, USA
John Fitzgerald	Department of Microbiology, University of Georgia, Athens, GA 30602, USA
Neil Foster	Canadian Forestry Service, Sault Ste. Marie, Ontario, Canada P6A 5M7
Andrew J. Friedland	Environmental Studies Program, Dartmouth College, Hanover, NH 03755, USA
Steven Gherini	Tetra-Tech, Inc., Lafayette, CA 94549, USA
Paul Hanson	Environmental Sciences Division, Oak Ridge National Laboratory, Oak Ridge, TN 37831-6034, USA
Robert B. Harrison	College of Forest Resources, AR-10, University of Washington, Seattle, WA 98195, USA

P. Homann — Department of Forest Science, Oregon State University, Corvallis, OR 97331, USA

R. Hudson — Tetra-Tech, Inc., Lafayette, CA 94549, USA

T. Huntington — Department of Geology, University of Pennsylvania, Philadelphia, PA 19104, USA

Art Johnson — Department of Geology, University of Pennsylvania, Philadelphia, PA 19104, USA

Dale W. Johnson — Biological Sciences Center, Desert Research Institute, Reno, NV 89506, USA

Einar Joranger — Norwegian Institute for Air Research, N-2001 Lillestrom, Norway

Ken Knoerr — School of Forestry and Environmental Studies, Duke University, Durham, NC 27706, USA

Steven E. Lindberg — Environmental Sciences Division, Oak Ridge National Laboratory, Oak Ridge, TN 37831-6034, USA

S. Liu — Tetra-Tech, Inc., Lafayette, CA 94549, USA

Gary M. Lovett — Institute of Ecosystem Studies, Cary Arboretum, Millbrook, NY 12545, USA

S. McLaughlin — Environmental Sciences Division, Oak Ridge National Laboratory, Oak Ridge, TN 37831-6034, USA

Eric Miller — State University of New York at Albany, Atmospheric Sciences Research Center, Wilmington, NY 12997, USA

Myron Mitchell	Department of Environmental and Forest Biology, College of Env. Science and Forestry, State University of New York, Syracuse, NY 13210, USA
R. Munson	Tetra-Tech, Inc., Lafayette, CA 94549, USA
Robert M. Newton	Department of Geology, Smith College, Northampton, MA 01063, USA
Louis F. Pitelka	Ecological Studies Program, Electric Power Research Institute, Palo Alto, CA 94303, USA
Larry H. Ragsdale	Human and Natural Ecology Program, Biology Department, Emory University, Atlanta, GA 30322, USA
D. Raynal	Department of Environmental and Forest Biology, College of Env. Science and Forestry, State University of New York, Syracuse, NY 13210, USA
Dan D. Richter	School of Forestry and Environmental Studies, Duke University, Durham, NC 27706, USA
M. Ross-Todd	Environmental Sciences Division, Oak Ridge National Laboratory, Oak Ridge, TN 37831-6034, USA
Douglas A. Schaefer	Department of Civil Engineering, Syracuse University, Syracuse, NY 13244-1190, USA
Jim Shepard	Department of Environmental Forest and Biology, College of Environmental Sciences & Forestry, State University of New York, Syracuse, NY 13210, USA

David Silsbee	National Park Service, Great Smoky Mountains National Park, Gatlinburg, TN 37738, USA
A. Stevens	Norwegian Forest Research Institute, Postbox 61 Aas-NLH, Norway
G. Strimbeck	Department of Geology, University of Pennsylvania, Philadelphia, PA 19104, USA
Arne O. Stuanes	Norwegian Forest Research Institute, Postbox 61 Aas-NLH, Norway
K. Summers	Tetra-Tech, Inc., Lafayette, CA 94549, USA
Wayne T. Swank	Coweeta Hydrologic Laboratory, U.S. Forest Service, Otto, NC 28763, USA
George E. Taylor, Jr.	Environmental Sciences Division, Oak Ridge National Laboratory, Oak Ridge, TN 37831-6034, USA
Helga Van Miegroet	Environmental Sciences Division, Oak Ridge National Laboratory, Oak Ridge, TN 37831-6034, USA
D. Vann	Department of Geology, University of Pennsylvania, Philadelphia, PA 19104, USA
K. Venn	Norwegian Forest Research Institute, Postbox 61 Aas-NLH, Norway
J. Vose	Coweeta Hydrologic Laboratory, U.S. Forest Service, Otto, NC 28763, USA
K. Wilkinson	Tetra-Tech, Inc., Lafayette, CA 94549, USA
Y. Zhang	Department of Environmental and Forest Biology, College of Env. Science and Forestry, State University of New York, Syracuse, NY 13210, USA

1. Introduction

D.W. Johnson, S.E. Lindberg, and L.F. Pitelka

Early conceptualizations of forest nutrient cycles depicted atmospheric deposition as an input to the system and, by implication, considered it to be beneficial (Cole et al. 1968; Curlin 1970; Duvigneaud and Denaeyer-De-Smet 1970). These studies took an ecosystem perspective, and detailed process-level research was left for the future. With the widespread reporting of acid deposition in the late 1960s, atmospheric deposition began to be viewed as a cause of nutrient loss rather than gain.

Nearly two decades ago, the Swedish Report to the United Nations forecast forest growth declines of approximately 1.5% per year as a result of accelerated Ca^{2+} leaching caused by acid deposition (Engstrom et al. 1971). These estimates were based on estimated atmospheric H^+ inputs, exchangeable Ca^{2+} pools, and a general correlation between site index and soil exchangeable Ca^{2+}. The implicit assumption was that the soil was a homogeneous, unvegetated block with certain exchangeable cation reserves which would become more acid, given sufficiently large acid inputs over a sufficient length of time.

Our knowledge of the effects that acid deposition and natural processes have on soil change and forest fertility has improved enormously since these crude and incomplete calculations were made, as has our ablilty to measure atmospheric inputs. However, the original concern over forest declines as the result of soil acidification and nutrient deficiency remains very much with us today. Many hypotheses for recent forest declines invoke soil acidification, declines in site Ca, Mg, or K status, and more recently, Al toxicity as major or contributing causes (Ulrich 1983; Shortle and Smith 1988; Binkley et al. 1989).

To understand the effects of atmospheric deposition on nutrient gains and losses, acid deposition scientists initially took a process-level approach, focusing primarily on atmospheric deposition, canopy interactions, decomposition, and soil leaching (e.g., Overrein 1972; Wood and Bormann 1974; Johnson and Cole 1977). A general model of forest nutrient cycling was well established by the time the acid rain issue emerged (e.g., Cole et al. 1968; Curlin 1970), but this model tended to concentrate on aboveground processes such as litterfall, uptake, translocation, etc., and was not widely used in early assessments of acid deposition effects.

Reuss (1976), Sollins et al. (1980), and Ulrich (1980) began to introduce an ecosystem-level perspective to the acid deposition–soil acidification issue by noting that acid deposition is merely an increment to a variety of internal acidification processes (carbonic, organic, and nitric acid production within the soil; humus formation; and plant uptake). These authors proposed that internal acid-generating processes could be described and quantified by constructing "hydrogen ion budgets," a concept later elaborated by Binkley and Richter (1987).

Many initial analyses of the role of atmospheric deposition in forest nutrient cycling were based on input estimates from collections of bulk deposition (e.g., Cole and Rapp 1981). The degree of error in quantifying both deposition and the fate of deposited ions in forests from bulk deposition is now understood (NRC 1983) and has been shown to be significant (Lindberg et al. 1986; Richter and Lindberg 1988). Recent developments in both wet and dry deposition methods allows these measurements to be made in the context of extensive nutrient cycling studies (Hicks et al. 1986).

The primary objective of the Integrated Forest Study (IFS) was to analyze the effects of atmospheric deposition on a variety of forest ecosystems from a nutrient cycling perspective (Figure 1.1). To achieve this objective, the IFS combined the long-established techniques used to analyze nutrient cycling in forest ecosystems with the more recent process-level approach taken to assess the effects of acid deposition on ion flux. The IFS also included a thorough analysis of air chemistry, hydrology, meteorology, and canopy characteristics necessary for accurate estimates of total atmospheric deposition. For the first time these data have been collected simultaneously with nutrient cycling data at numerous sites across a wide range of forest types, geography, and climate (Figure 1.2). Modeling provides a logical mechanism for the integration of such a large database and assessment of atmospheric deposition effects on nutrient cycles in so many diverse ecosystems. Thus, a PC-based Nutrient Cycling Model (NuCM) was developed as a part of the IFS project with input from all principal investigators during its construction.

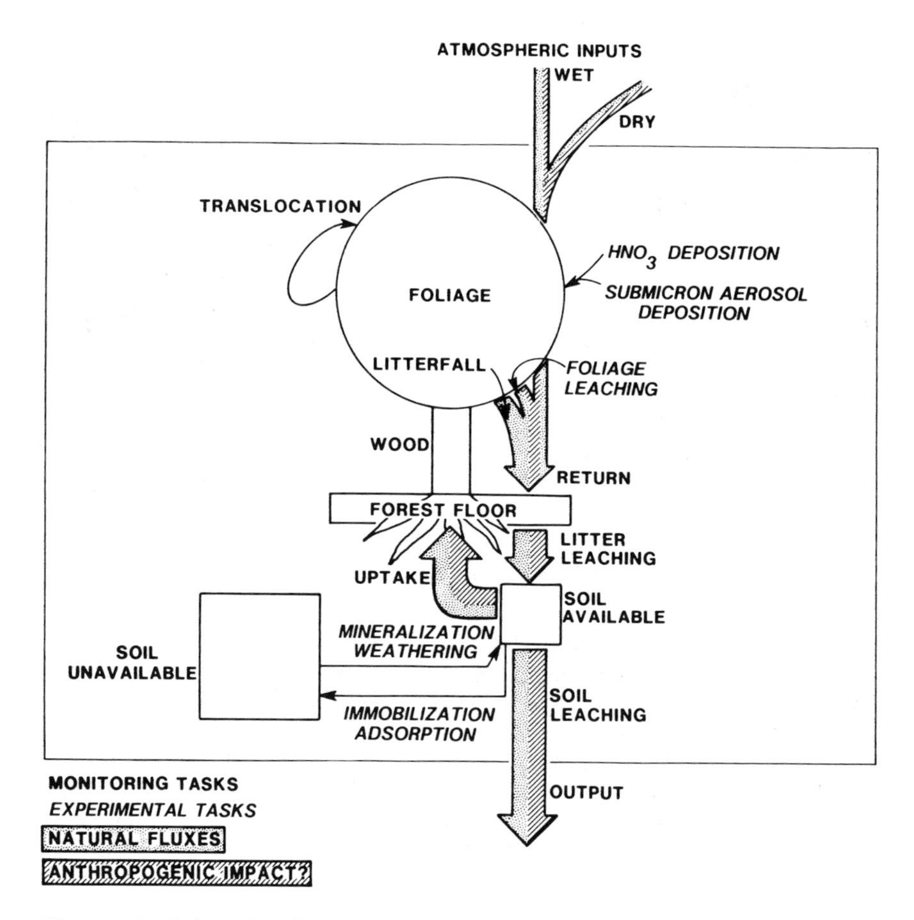

Figure 1.1. Schematic of the nutrient cycles quantified during the Integrated Forest Study (IFS).

In addition to the basic site monitoring tasks and modeling described, the IFS project included several specific experimental tasks designed to examine important deposition and nutrient cycling processes in greater detail (Figure 1.3). These tasks were divided into those concerning atmosphere/canopy interactions and those on soil chemical interactions. The atmosphere/canopy interactions tasks included the following field and laboratory experiments: (1) field studies of the factors influencing throughfall fluxes and canopy leaching of ions including controlled leaching following exposure to ozone and H^+; (2) chamber studies of the mechanisms of NO_x and HNO_3 dry deposition to plant surfaces; and (3) development of canopy-specific transfer efficiencies for dry deposition of submicron aerosols for each IFS site using naturally radioactive aerosol tracers. The results of the first task are sum-

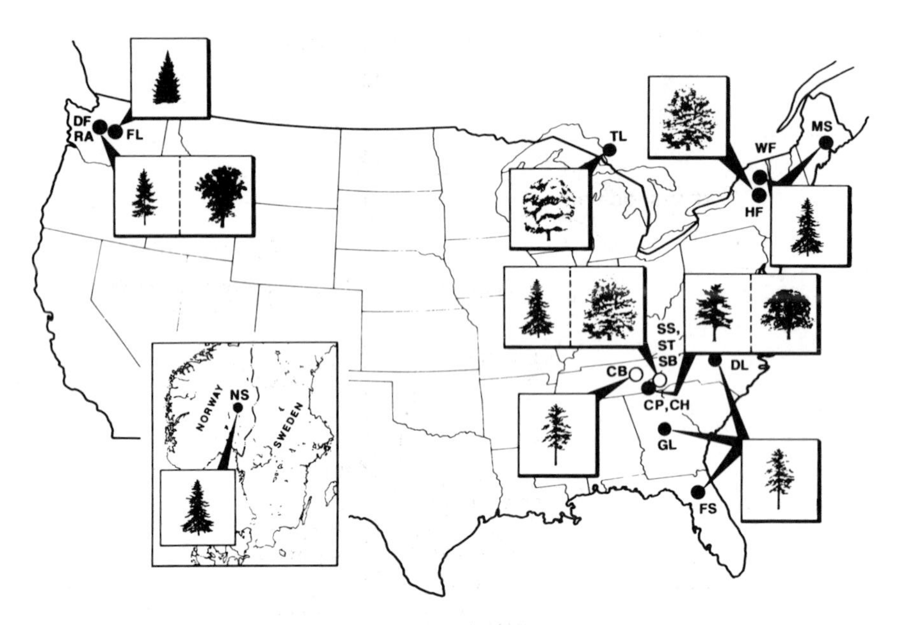

Figure 1.2. Sites studied during the Integrated Forest Study (IFS). Codes: CH, Coweeta Hydrologic Laboratory, southern hardwoods; CP, Coweeta Hydrologic Laboratory, white pine; DF, Thompson Forest, Douglas fir; DL, Duke Forest, loblolly pine; FL, Findley Lake, Pacific silver fir; FS, Florida, slash pine; GL, B.F. Grant Forest, loblolly pine; HF, Huntington Forest, mixed hardwood; LP, Oak Ridge, loblolly pine; MS, Howland, Maine, spruce; NS, Nordmoen, Norway, Norway spruce; RA, Thompson Forest, red alder; SB, Great Smoky Mountains, beech; SS, Great Smoky Mountains, spruce (Becking site); ST, Great Smoky Mountains, spruce (Tower site); TL, Turkey Lakes Watershed, mixed hardwood; WF, Whiteface Mountain, spruce/fir.

marized in Chapter 8, and those of the second task are in Chapter 6. Because of the untimely death of one investigator (E.A. Bondietti), the results of the third task have not been summarized here, but the factors were computed and used throughout our studies (see the Appendix) and the method has been published elsewhere (Bondietti et al. 1984).

Within the soil chemical interactions category, four specific tasks were identified and completed during the IFS study: sulfur saturation, nitrogen saturation, weathering/mineralogy, and cation exchange/aluminum release. The first of these tasks was divided into two subtasks: inorganic chemical reactions causing sulfur to be retained by soils, and organic sulfur retention in soils (see Chapter 5). Both these tasks were laboratory studies conducted on soils from all the IFS sites. However, nitrogen retention processes are not easily studied in a laboratory environment. Thus, the nitrogen saturation task was originally designed to be conducted at the one site—red alder, Thompson Forest, Washington—that we knew to be nitrogen saturated at

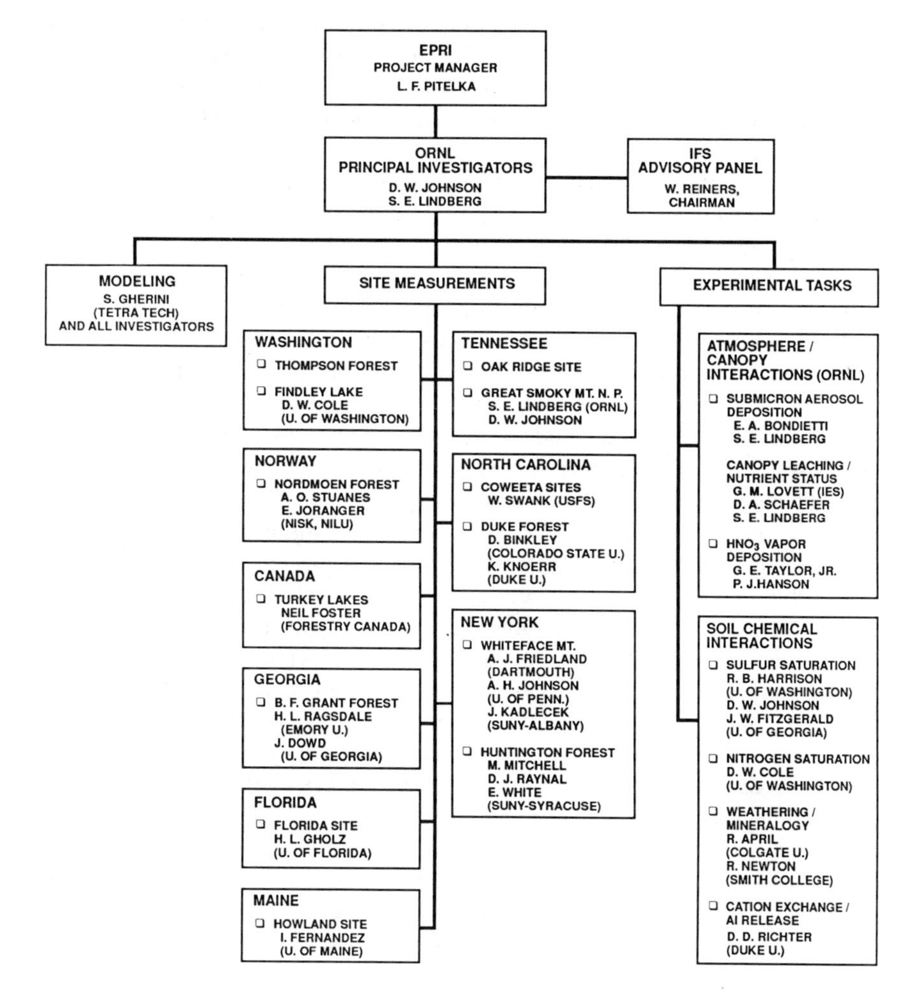

Figure 1.3. Tasks included in the Integrated Forest Study (IFS).

the beginning of the study. This was designed to reduce high nitrogen inputs via fixation by removing the nitrogen fixers, and recovery from nitrogen saturation and acidification could be monitored. The results of this experiment were then combined with the results of an experiment involving artificial irrigation with sulfuric acid and subsequent recovery in Norway and summarized in Chapter 12.

The third and fourth soil chemical interaction tasks are concerned with cation cycling. The soil weathering/mineralogy task was designed to characterize the important primary and secondary minerals at each of the IFS sites and to evaluate the potential for base cation replenishment of the soils at these sites by weathering. The fourth task involved laboratory studies of the cation exchange processes controlling both base cation leaching and Al mobilization in soils from all the IFS sites. Results of this task are summarized in Chapter 9, with complementary studies of cation composition of wood in tree rings as potential indices of historical patterns of soil and soils solution chemistry.

Some other studies conducted within the IFS project and summarized within this volume were not set forth as specific experimental tasks at the onset, but nevertheless contributed to our overall knowledge of these forest ecosystems. These include modeling and validation studies of water balances (Chapter 3), measurements of tropospheric ozone (Chapter 4), and studies of N, P, and S interactions in forest ecosystems (Chapter 7).

Chapter 13 places the IFS results in a larger perspective by briefly summarizing acid deposition effects and their potential role in forest health in several forest types represented in the IFS project. This chapter gives brief overviews of the situation in eastern spruce-fir, eastern hardwood, and southern pine forests in North America, and a very brief overview of air pollution and forest ecosystems in Europe, with special emphasis on the situation in Norway where the single European IFS site was located.

In achieving its primary objective, the IFS project has developed a database on forest nutrient cycles in many diverse ecosystems. The development of this database was, in fact, a major secondary objective of the IFS project. Recognizing the value of baseline nutrient cycling data from the international Biological Program (IBP) (summarized by Cole and Rapp 1981) in understanding basic patterns of forest nutrient cycling (Vitousek 1982) and the effects of perturbations (Johnson and Richter 1983), the IFS project was designed to provide the scientific community with an additional nutrient cycling database with better quality control and atmospheric deposition data than previously available. It is the hope of all investigators in this project that the data sets produced here will provide a base for many future analyses, hypotheses, and research programs on forest ecosystems.

References

Binkley D., Richter D. 1987. Nutrient cycles and H^+ budgets of forest ecosystems. Adv. Ecol. Res. 16:1–51.

Binkley D., Driscoll C.T., Allen H.L., Schoenberger P., McAvoy D. 1989. Acid Deposition and Forest Soils. Springer-Verlag, New York.

Bondietti E.A., Hoffman P.O., Larsen I.L. 1984. Air-to-vegetation transfer rates of natural submicron aerosols. J. Environ. Radioact. 1:5–27.

Cole D.W., Rapp M. 1981. Elemental cycling in forest ecosystems. In: Reichle D.E., (ed.) Dynamic Properties of Forest Ecosystems. Cambridge University Press, London, pp. 341–409.

Cole D.W., Gessel S.P., Dice S.F. 1968. Distribution and cycling of nitrogen, phosphorus, potassium, calcium in a second-growth Douglas-fir forest. In: Young H.E., (ed.) Primary Production and Mineral Cycling in Natural Ecosystems. University of Maine Press, Orono, pp. 197–213.

Curlin J.W. 1970. Nutrient cycling as a factor in site productivity and forest fertilization. In: Youngberg C.T., Davey C.R., (eds.) Tree Growth and Forest Soils. Oregon State University Press, Corvallis, pp. 313–326.

Duvigneaud P., Denaeyer-DeSmet S. 1970. Biological cycling of minerals in temperate deciduous forests. In: Reichle D.E., (ed.) Analysis of Forest Ecosystems. Springer-Verlag, New York. pp. 199–255.

Engstrom A., Backstrand G., Stenram H. (eds.) 1971. Air Pollution across National Boundaries: The Impact on the Environment of Sulfur in Air and Precipitation. Report No. 93 Ministry for Foreign Affairs/Ministry for Agriculture, Stockholm, Sweden.

Hicks B.B., Wesley M.L., Lindberg S.E., Bromberg S.M., (eds.) 1986. Proceedings of the Dry Deposition Workshop of the National Acid Precipitation Assessment Program, March 25–27, 1986. NOAA/ATDD, Oak Ridge, Tennessee.

Johnson D.W., Cole D.W. 1977. Sulfate mobility in an outwash soil in western Washington. Water Air Soil Pollut. 7:489–495.

Johnson D.W., Richter D.D. 1983. Effects of atmospheric deposition on forest nutrient cycles. TAPPI Journal 67:82–85.

Lindberg S.E., Lovett G.M., Richter D.D., Johnson D.W. 1986. Atmospheric deposition and canopy interactions of major ions in a forest. Science 231:141–145.

National Research Council (NRC). 1983. Atmospheric Processes in Eastern North America. National Academy of Sciences Press, Washington, D.C.

Overrein L.N. 1972. Sulfur pollution patterns observed: leaching of calcium in forest soils determined. AMBIO 1:145–147.

Reuss J.O. 1976. Chemical/biological relationships relevant to ecological effects of acid rainfall. Water Air Soil Pollut. 7:461–478.

Richter D.D., Lindberg S.E. 1988. Wet deposition estimates from long-term bulk and event wet-only samples of incident precipitation and throughfall. J. Environ. Qual. 17:619–622.

Shortle W.C., Smith K.T. 1988. Aluminum-induced calcium deficiency syndrome in declining red spruce. Science 240:1017–1018.

Sollins P., Grier C.C., McCorison F.M., Cromack K. Jr., Fogel R., Fredriksen R.L. 1980. The internal element cycles of an old-growth Douglas-fir ecosystem in western Oregon. Ecol. Monogr. 50:261–285.

Ulrich B. 1983. Soil acidity and its relation to acid deposition. In: Ulrich B., Pankrath J. (eds.) Effects of Accumulation of Air Pollutants in Ecosystems. D. Reidel, New York, NY pp. 127–146.

Ulrich B. 1980. Production and consumption of hydrogen ions in the ecosphere. In: Hutchinson T.C., Havas M. (eds.) Effects of Acid Precipitation on Terrestrial Ecosystems. Plenum, New York, pp. 255–282.

Vitousek P.M. 1982. Nutrient cycling and nutrient use efficiency. Am. Nat. 119:553–572.

Wood T., Bormann F.H. 1974. The effects of an artificial acid mist upon the growth of *Betula alleghenienisis*. Environ. Pollut. 7:259–268.

2. Background on Research Sites and Methods

S.E. Lindberg, D.W. Johnson, and E.A. Bondietti

Site Descriptions

This chapter briefly summarizes characteristics of the field sites participating in the Integrated Forest Study (IFS). The information presented here was abstracted from a published report (Bondietti 1990), which presents greater detail about activities and characteristics at each of the research sites including a complete description of the meteorological towers for atmospheric sampling. Tables 2.1 and 2.2 summarize selected site data and the site code abbreviations used throughout this volume, and Figure 1.2 shows site locations.

Great Smoky Mountains Beech and Spruce-Fir

Three sites (one beech and two red spruce) were located near Clingmans Dome at elevations between 1650 and 1800 m in the Great Smoky Mountains National Park in eastern Tennessee–western North Carolina. The beech site (SB) was located on a southerly slope at an elevation of approximately 1600 m, 1 km west of Newfound Gap on the road to Clingmans Dome. The Becking (spruce) (SS) site was so named because it was located on one of the vegetation inventory plots established by Rudolf Becking in 1976. It was located at an elevation of 1800 m on a southwesterly slope on a spruce ridge west of Nolan Divide near Clingman's Dome. The Tower (spruce) (ST) site was located at an elevation of 1740 m on a southerly slope on Nolan Divide near a spur road branching off the main road to Clingmans Dome. The site was so named because the meteorological tower for the detailed IFS atmospheric deposition measurements was located there.

Vegetation

At the Beech site, overstory vegetation consisted primarily of beech (*Fagus grandifolia*) with occasional buckeye (*Aesculus octandra*) and red spruce (*Picea rubens*). Understory was dominated by carex (*Carex eastivalis*), angelica (*Angelica triquinata*), stinging nettle (*Urtica* spp.), white snakeroot (*Eupatorium rugosum*), blackberry (*Rubus canadensis*), goldenrod (*Solidago casesia*), and various ferns. At the Becking site, overstory vegetation consisted primarily of old-growth (200- to 300-year-old) red spruce with occasional Fraser fir (*Abies fraseri*), many of which were suffering from infestations of the balsam woolly adelgid. The understory was dominated by patches of regenerating Fraser fir ranging from 4 to 12 cm in diameter and undergoing significant mortality. Other understory species of note were blackberry, vaccinium (*Vaccinium erythrocarpum*), witch hobble (*Viburnum alnifolium*), sorbus (*Sorbus americana*), oxalis (*Oxalis acetosella*), various ferns, and abundant mosses. At the Tower site, overstory vegetation was dominated by vigorous, old-growth (200–300 years old) red spruce with an occasional yellow birch (*Betula alleghanensis*). Understory consisted of patches of Fraser fir and occasional red spruce regeneration, blackberry, witch hobble, vaccinium, sorbus, and various ferns.

Geology and Soils

At the Beech site, soils were classified as Umbric Dystrochrepts derived from the Anakeesta formation. The soil profile consisted of a thin, patchy litter layer (Oi, 1 cm; Oa, <1 cm), a 20-cm-thick A horizon of dark reddish-brown loam, a 12-cm-thick BA horizon of brown loam, a 23-cm-thick Bw horizon of strong brown clay loam, a 15-cm-thick Cb horizon of brown, shaley sand loam, and a C horizon of undetermined thickness (but at least 30 cm) of brown, very shaley sandy loam. At the Becking site, the soils were classified as Umbric Dystrochrepts derived from Thunderhead sandstone. The landscape was characterized by frequent large sandstone boulders on top of and within the soil profile. The profile consisted of a 3-cm-thick Oe + Oi horizon of leaves and needles in various stages of decay, a 6-cm-thick Oa horizon of mucky humus, a 15-cm-thick A horizon of dark brown, bouldery, mucky silt loam, a 22-cm-thick Bw horizon of yellowish-brown, bouldery loam, a 15-cm-thick BC horizon of yellowish-brown, very bouldery loam, and a 16-cm-thick C horizon of pale brown, extremely bouldery, sandy loam overlying sandstone bedrock. At the Tower site the soils were also Umbric Dystrochrepts derived from Thunderhead sandstone. The profile consisted of 4-cm-thick Oi + Oe horizon of needles and leaves in var-

Table 2.1. Summary of Geographic and Climatic Information for Field Research Sites Participating in the Integrated Forest Study

Site Code	Forest Type	Location	Latitude, Longitude	Elevation (m)	Mean Annual Precipitation[a] (cm)	Snow, %	Institution
CH[b]	Southern hardwoods	Coweeta Hydrologic Laboratory, NC	35°03′N. 83°27′W.	725	138	<3	Coweeta Hydrologic Laboratory
CP	White pine	Coweeta Hydrologic Laboratory, NC	35°03′N. 83°27′W.	725	144	<3	Coweeta Hydrologic Laboratory
DF	Douglas fir	Thompson Forest, Washington	47°23′N. 121°56′W.	220	114	<5	University of Washington
DL	Loblolly pine	Duke Forest, North Carolina	36°12′N. 79°17′W.	213	113	<2	Duke University, Colorado State University
FL[b]	Fir, hemlock	Findley Lake, Washington	47°04′N. 121°25′W.	1130	270	80	University of Washington
FS	Slash pine	Gainesville, Florida	29°44′N. 82°30′W.	38	112	0	University of Florida
GL	Loblolly pine	B.F. Grant Forest, Georgia	33°22′N. 83°27′W.	145	87	<1	Emory University, University of Georgia
HF	Northern hardwoods	Huntington Forest, New York	43°59′N. 74°14′W.	530	97	47	State University of New York at Syracuse

LP	Loblolly pine	Oak Ridge, Tennessee	35°54′N. 84°20′W.	300	114	3	Oak Ridge National Laboratory
MS	Red spruce	Howland, Maine	45°10′N. 68°40′W.	65	79	38	University of Maine
NS	Norway spruce	Nordmoen, Norway	60°16′N. 11°06′E.	200	107	29	Norwegian Forest Research Institute
RA	Red alder	Thompson Forest, Washington	47°23′N. 121°56′W.	220	114	<5	University of Washington
SB[b]	American beech	Great Smoky Mountains National Park	35°37′N. 83°26′W.	1600	151	10	Oak Ridge National Laboratory
SS[b]	Red spruce	Great Smoky Mountains National Park	35°34′N. 83°29′W.	1800	151	10	Oak Ridge National Laboratory
ST	Red spruce	Great Smoky Mountains National Park	35°34′N. 83°28′W.	1740	203	10	Oak Ridge National Laboratory
TL	Northern hardwoods	Turkey Lakes, Ontario	47°03′N. 84°25′W.	350	121	27	Forestry Canada, Ontario
WF	Red spruce, balsam fir, white birch	Whiteface Mt., New York	44°22′N. 73°54′W.	1000	115	19	Dartmouth College University of Pennsylvania, State University of New York at Albany

[a] During the study period.
[b] Bulk deposition only.

Table 2.2. Summary of Biomass and Geological Information for Field Research Sites Participating in the Integrated Forest Study

Site	Stand Age (yr)	Basal Area (m^2/ha)	Stems (number/ha)	LAI[a] (m^2/m^2)	Parent Material	Soil Type	Comments
CH	Uneven	40.6	—	—	Metamorphic	Typic Hapludult	Natural second growth
CP	30	39.6	950	17**	Metamorphic	Typic Hapludult	Plantation
DF	55	50	1100	5.3**	Glacial till	Dystric entic Durocrept	Plantation
DL	23	32.4	760	6.5**	Igneous rock	Typic Hapludult	Old field, plantation
FL	180	74	510	—	Volcanic ash over andesite	Typic cryohumod	Old growth
FS	22	24.3	1,056	5.1**	Marine sands	Haplaquods	Plantation, unthinned
GL	17	309	17,700	11**	Gneiss	Typic Hapludult	Plantation
HF	100	27.5	643	6.5*	Glacial till	Typic Haplorthod	Unmanaged
LP	35	26.7	430	3.6**	Alluvium	Fluventic Dystrochrept	Plantation
MS	85	45	2,200	5.7**	Basal till	Haplorthods	Commercial forest
NS	40	31.9	2,080	10.5**	Outwash sand	Typic Udipsamment	Plantation
RA	55	36	800	4.5*	Glacial till	Dystric entic Durocrept	Unmanaged
SB[b]	Unknown	24.0	670	—	Shale	Umbric Dystochrept	Old growth
SS	~250	42.6	440	—	Sandstone	Umbric Dystrochrept	Old growth
ST	~250	54.1	360	9.0**	Sandstone	Umbric Dystrochrept	Old growth
TL	135	28.6	682	5.3*	Glacial till	Haplorthods	Natural second growth
WF	Virgin	26.8	1,949	8.2**	Anorthosite	Typic Cryohumods	Old growth

[a] Growing season mean; *, one-sided; **, all-sided.
[b] For site codes, see Table 2.1.

ious stages of decay, a 4-cm-thick Oa horizon of mucky humus, an 8-cm-thick dark, reddish-brown, mucky loam A horizon, a 27-cm-thick dark brown, sandy loam Bw horizon, a 35-cm-thick dark, yellowish-brown loamy CB horizon, and an olive-brown, loamy sand C horizon at least 20 cm-thick overlying sandstone bedrock.

Oak Ridge Loblolly Pine

The Oak Ridge Loblolly Pine (LP) site was located on the U.S. Department of Energy Reservation, National Environmental Research Park, near Oak Ridge, Tennessee.

Vegetation

Overstory vegetation was loblolly pine (*Pinus taeda*). Understory consisted of very occasional red maple (*Acer rubrum*), yellow poplar (*Liriodendron tulipifera*), black cherry (*Prunus serotina*), and dogwood (*Cornus florida*). Ground cover consisted of extensive grass (*Eulalia viminea*) with patches of blackberry (*Rubus* spp.) and honeysuckle (*Lonicera japonica*).

Geology and Soils

The site was located on an alluvial soil derived from shale on one of the upper terraces of the Clinch River. Parent material is alluvium, consisting mainly of shale with some sandstone and siltstone from the Rome and Conasauga formations. The soil was tentatively identified as Fluventic Dystrochrept and was characterized by a thin (1–2 cm) Oi + Oe horizon with dense grass cover, a 15-cm-thick dark, yellowish-brown Ap horizon of gravelly silty clay loam, a 15-cm-thick yellowish-brown B/A horizon of gravelly silty clay loam, a 15-cm-thick dark, yellowish-brown Bw horizon of gravelly silty clay loam, and a dark yellowish-brown BC horizon of gravelly silty clay of undetermined thickness.

Whiteface Mountain Spruce-Fir

Whiteface Mountain (summit elevation, 1483 m) is located near Wilmington, New York, in the northeast section of the Adirondack Mountains. The study site (WF) was located in a narrow west-northwest-trending drainage basin on the flank of Esther Mountain (summit elevation, 1292 m), the eastern peak of the Whiteface massif. A restricted access road to the summit of Whiteface cuts northeast-southwest across the basin at an elevation of approximately 960 m. The IFS nutrient cycling activities took place between 970 and 1100 m elevation on a north-facing slope of the drainage basin. Airflow at the site was dominated by westerly and northwesterly upslope winds.

Vegetation

Vegetation zones on Esther Mountain include the northern hardwood forest at the base, hardwood-fir-spruce transition, fir-spruce, and fir krummholz toward the summit. Balsam fir [*Abies balsamea* (L.) Mill.], red spruce (*Picea rubens* Sarg.), and white birch [*Betula papyrifera* var. *cordifolia* (Reg.) Fernald] dominated the study area. Mountain ash (*Sorbus americana* Marsh) was also present. The species composition represents a typical forest at a 1000-m elevation in the Adirondacks; the study area has not been disturbed by fire or logging in recent history, and can be considered, in general, mature and overmature forest.

Geology and Soils

The bedrock on Esther Mountain is primarily a Precambrian anorthosite. Soils are Typic Cryohumods or Typic Cryorthods developed on anorthosite colluvium and basal till of local origin. The average depth to bedrock is approximately 60 cm. Areas of well-developed spodosols to 1 m deep and organic mats with little or no mineral soil development overlying anorthosite boulders are equally common.

Duke Forest Loblolly Pine

The Duke Forest Loblolly Pine (DL) site was located in the Dailey Tract of the Duke Forest, about 60 km west of Durham, North Carolina (about 10 km north of Mebane, North Carolina) in Alamance County. Before acquisition by Duke University in the 1960s, the land was used for farming various crops; no records were kept of soil treatments.

Vegetation

One-year-old loblolly pine seedlings were planted by machine in rows in 1966–1967 in an old field. Furrows are clear but gentle, with about 5 cm of relief between ridge and trough. Planting density was about 1200 stems ha^{-1}. The stand was thinned by rows (one row removed from each four rows, plus selected stems in other rows) in early 1984, and the canopy was at maximum biomass at the initiation of research. Site index was 17.7 m at 25 years, which is generally a very good site for the Piedmont region of North Carolina. Understory vegetation was negligible.

Geology and Soils

The Piedmont region is characterized by old, highly weathered landscapes, and land use over the past few centuries has strongly affected the topsoil in most areas. Soil series and associations vary on a small scale (from 1 to 100 ha), primarily because of a patchwork of parent materials. The Durham soil series (Typic Hapludult) at the Dailey Tract formed in residual parent material of highly weathered felsic igneous rocks; other common parent materials in the region include mafic igneous rocks and mixtures of felsic and mafic rocks.

Thompson Forest Douglas Fir and Red Alder

The IFS Douglas Fir (DF) and Red Alder (RA) sites were located at the Thompson Research Center, located in the Cedar River Watershed 56 km southeast of Seattle, Washington. It lies at an elevation of 220 m on the west side of the Cascade Mountains.

Vegetation

Douglas fir and red alder control plots are located near each other. The Douglas fir stand was planted in 1931 after a series of wildfires, following logging of the original old-growth forest between 1910 and 1920. The stand contained approximately 1100 stems ha^{-1} and had a basal area of 50 m^2 ha^{-1}. It was classified as a low site III with a site index of 106 for Douglas fir at age 50. The understory vegetation consisted mainly of salal (*Gaultheria shallon* Pursh.), Oregon grape [*Berberis nervosa* (Pursh.) Nutt.], and bracken fern (*Pteridium aquilinum* Kuhn var. *pubescens* Underw.). There were also several species of mosses, predominantly *Hylocomium* spp. and *Eurynchium oreganum* (Sull.) Jaeg. The adjacent red alder forest established naturally in the burnt area a few years later. Stand density was generally lower than in the Douglas fir stand, with only 800 stems ha^{-1} and a basal area of 36 m^2 ha^{-1}. At age 50 this pioneer species was approaching the end of its rotation, and as a consequence the canopy was rapidly opening up. The understory was more prominent and consisted mainly of a dense growth of sword fern [*Polystichum munitum* (Kaulf.) Presl.] and bracken fern intermixed with some Oregon grape.

Geology and Soil

The soil underlying both alder and Douglas fir belongs to the Alderwood series (Dystric Entic Durochrept). It is of glacial origin, developed from ablation till overlying compacted basal till and has a gravelly, sandy loam texture. The presence of the basal till layer caused drainage to be restricted in the lower parts of the soil profile, and a shallow water table was not uncommon during the winter months.

Findley Lake Pacific Silver Fir

The Findley Lake (FL) site lies within the Cedar River Watershed, approximately 65 km southeast of Seattle, Washington. It was located on the west slopes of the Cascade Mountains at an elevation of 1130 m.

Vegetation

The study area lies within the *Abies amabilis* vegetation zone. The stand was dominated by approximately 180-year-old fir. Associated species were *Tsuga mertensiana* (Bong.) Carr. and *Tsuga heterophylla* (Raf.) Sarg. Major shrubs included *Vaccinium membranaceum* Dougl. ex Hook., *V. ovalifolium* Smith, *V. alaskaense* How., *Sorbus sitchensis* Romemer, and *Alnum sinuata* (Reg.) Rydb. Low shrubs and herbs included *Cornus canadensis* L., *Xerophyllum tenax* (Pursh.) Nutt., *Clintonia uniflora* (Shult.) Kunth, *Achlys triphylla* (Smith) D.C., *Rubus pedatus* J.E. Smith, and *Viola sempervirens* Greene.

Geology and Soils

Study area parent material is primarily fractured andesite overlain with andesite moraines 1 to 3 m thick. Volcanic ash layers 5 to 25+ cm thick overlie these materials. The soil in this mature Pacific silver fir forest was a Spodosol, which has tentatively been classified as a Typic Cryohumod (Chickamin Series).

Huntington Forest Mixed Hardwood

The Huntington Forest (HF) is a 6066-ha property of the State University of New York, College of Environmental Science and Forestry (S.U.N.Y.-C.E.S.F.) located in western Essex County and eastern Hamilton County within the Adirondack State Park of New York. The mixed northern hardwood site was located at an elevation of 530 m and about 200 m southeast of an access road to the Arbutus Lake complex.

Vegetation

This area was typical of much of the Adirondack region, which was heavily cut 75 years ago and has regenerated into a northern deciduous forest dominated by *Acer saccharum Marsh* (sugar maple), *Fagus grandifolia* Ehrh. (American beech), *Acer rubrum* L. (red maple), *Betula lutea* Michx. F. (yellow birch), *Prunus serotina* Erhrh. (black cherry), and *Picea rubens* Sarg. (red spruce). The herbaceous layer was dominated by *Dryopteris austriaca* (Jacq.) *Woynar* (shield fern), while *Lycopodium annotinum* L. (stiff clubmoss), and *Dennstaedtia punctilobula* (Michx.) Moore (hay-scented fern) are also common.

Geology and Soils

The underlying bedrock of the Huntington Forest, like that of most of the Adirondacks, was characterized by Precambrian rocks mostly of the hornblende-granitic gneiss type and occasionally by smaller amounts of metamorphosed sediments from the Grenville Formation, including impure marble, quartzites, amphibolites, and schists. Within the Huntington Forest, 46 soil taxa have been described as belonging to the Histosol, Inceptisol, and Spodosol orders, with the latter being most common. The soil at the IFS site was a Becket, bouldery, fine sandy, loam, a Typic Haplorthod. Similar soils are found throughout the Adirondack Region and other areas of the Northeast United States. The soil overlies a bedrock of gneiss, is shallow (<1 m depth), and contains a hardpan derived from the parent material of glacial till.

Coweeta Hydrologic Laboratory White Pine and Southern Hardwoods

The Coweeta study sites for the IFS included a 30-year-old white pine (CP) plantation located on Watershed Number 1 and an uneven-aged, mixed hardwood stand (CH) located on Watershed 2. Both plots are at approximately 720 m elevation with a slope of about 40% to 45%. Within the Coweeta basin, elevations range from about 680 to 1600 m; the plots thus are at low elevations, only slightly above nearby valley floors.

Vegetation

An essentially pure plantation of white pine (*Pinus strobus*) was established in 1957 following clear-cutting of the original hardwood forest on Watershed 1. Competing hardwood sprouts have been mostly removed by cutting and spraying. There was a thin, patchy understory of *Rhododendron maximum*. Watershed 2 was covered by native, mixed mesophytic hardwood forest typical of the region. It has served as a control watershed since 1934, and no human-caused disturbance has occurred there since 1923. The multistoried, uneven-aged stand was composed of *Quercus* spp. (50% of basal area), *Oxydendrum arboreum* (11%), *Rhododendron maximum* (11%), *Acer rubrum* (10%), *Carya* spp. (6%), *Cornus florida* (3%), and other species (9%).

Geology and Soils

Coweeta is situated on the deeply dissected eastern flanks of the Nantahala Mountain range, part of the Southern Blue Ridge Province. Both study plots are underlain by bedrock of the Tallulah Falls Formation, metamorphic rocks of upper Precambrian age. This formation is composed predominantly of biotite paragneiss and biotite schist, containing quartz, plagioclase, biotite, orthoglase, muscovite, and garnet. Within the formation are interlayers of pelitic schist and metasandstone.

With one exception, lysimeter pit soils are classified as mesic Typic Hapludults of the Fannin soils series. The exception, in the hardwood stand, was classified in the Cashiers soil series (mesic Umbric Dystrocrepts).

Turkey Lakes Watershed Mixed Hardwood

The overall study area, Turkey Lakes Watershed (TL), is located approximately 60 km north of Sault Ste. Marie, Ontario. Samples were collected from a 1-ha study site.

Vegetation

The site was established within an uneven-aged, sugar maple (*Acer saccharum* Marsh.) forest with a lesser component of yellow birch (*Betula alleghaniensis* Britton). In addition, there were scattered red maple (*Acer rubrum* L.), ironwood [*Ostrya virginiana* (Mill.) K. Koch], red oak (*Quercus rubra* L.), and white spruce [*Picea glauca* (Moench) Voss] in the overstory and understory; balsam fir [*Abies balsamea* (L.) Mill.] and eastern white cedar (*Thuja occidentalis* L.) were confined mainly to the understory.

Geology and Soils

The study site was located in moderately steep terrain at an elevation of 350 to 400 m. The soil consists of a stony, silty loam ablation till over a compacted basal till at a depth of 0.5 m. Soils are Orthic Humo-Ferric and Ferro-Humic Podzols (Spodosols) derived from basalt and granite.

Nordmoen Norway Spruce

The Norwegian site (NS) was located at Nordmoen Field Station (about 60 km north of Oslo) at an elevation of 200 m.

Vegetation

All six plots included in the site are located on the same flat plain. The mature vegetation was a mixture of Scots pine (*Pinus sylvestris* L.) and Norway spruce [*Picea abies* (L.) Karst.]. On drier soils, pine dominates. The dominating vascular plants were *Vaccinium myrtillus* L. and *Deschampsia flexuosa* (L.) Trin. (*Aira flexuosa* L.). Other vasculars were *Calluna vulgaris* (L.) Hull, *Vaccinium vitis-idaea* L., *Lula pilosa* (L.) Willd., *Diphasium complanatum* (L.) Rothm., *Lycopodium annotinum* L., *Maianthemum bifolium* (L.) F.W. Schm., *Melampyrum pratense* L., and *Trientalis europaea* L. The forest floor layer was dominated by moss species such as *Pleurozium schreberi* (Brid.) Mitt., *Hylocomium splendens* (Hedw.) Br Eur., *Ptilium crista-castrensis* (Hedw.) De Not., and *Dicranum* (Hedw.) spp.

Geology and Soils

The site was located on a flat plain of glaciofluvial deposits partly covered by a sheet of eolian sand. The deposits are about 60 m deep, overlying Precambrian and Permian crystalline bedrock. The soil in a nearby experimental plot was classified as a Typic Udipsamment.

B.F. Grant Forest Loblolly Pine

The loblolly pine (*Pinus taeda*) site was located in the B.F. Grant Memorial Forest (GL) in east-central Georgia near the town of Eatonton. The site was in the middle Piedmont of the southeastern United States, approximately 300 km from the Atlantic Ocean. The forest was owned and managed by the School of Forest Resources, University of Georgia, Athens, Georgia.

Vegetation

The stand was an approximately 2.5-ha remnant of a 17-year-old loblolly pine plantation that was decimated by the southern pine bark beetle in 1978. Loblolly pines were planted on 6 × 8 ft spacing and have not been thinned. Deciduous trees and vines were present throughout, a typical situation in planted pine stands in this region. Overstory vegetation was loblolly pine (*Pinus taeda*) with an occasional large white oak (*Quercus alba*) or red oak (*Quercus rubra*). In the subcanopy and understory, in order of prominence, were red maple (*Acer rubra*), tulip poplar (*Liriodendron tulpfera*), sweet gum (*Liquidambar styraciflua*), black cherry (*Prunus serotina*), and dogwood (*Cornus florida*).

Geology and Soils

The pine stand was located on an upland site with an elevation of 140 to 150 m. Soils at the site are primarily Typic Hapludults of the Cecil series. The A horizon of Piedmont soils was generally thin as a result of erosion from previous agricultural usage. The dominant minerals are biotite and mica gneiss. A typical profile has a sandy loam epidon 10 to 20 cm thick underlain by red clay subsoils. The organic horizon of the forest floor consists of decaying pine needles, some deciduous leaves, pine bark, pine branches, and pine boles of fallen trees. The organic litter layer, excluding branches and boles, was 1 to 5 cm thick. The A horizon (0–8 cm) was a brown, very friable, sandy loam soil of medium granular structure containing many fine roots. The E horizon (8–16 cm) was a yellowish-red, friable, sandy clay loam of weak medium granular structure containing many fine roots. The BE horizon (16–30 cm) was a red clay loam soil with fine subangular blocky structure and numerous medium roots. The deeper (30–81 cm) B horizons, Bt1 and Bt2, are red clays. The BC horizon (81+ cm) was a red clay loam with yellow mottles.

Howland, Maine Spruce-Fir

The Howland site (MS) was located in eastern central Maine within the 6300-ha International Paper Company Northern Experiment Forest. The topography of this site is relatively flat with an approximate elevation of 65 m. The Atlantic Ocean lies 60 miles to the south. Site establishment was funded by the Forest Response Program (Spruce-Fir Cooperative), U.S. Forest Service.

Vegetation

The major forest type that occurs at the study site was spruce-fir (*Picea rubens* and *Abies balsamea,* respectively), with a strong component of eastern hemlock (*Tsuga canadensis*). Additional species occurring with less frequency include northern white cedar (*Thuja occidentalis*), eastern white pine (*Pinus strobus*), and paper birch (*Betula papyrifera*). This forest type is typical of low-elevation, commercial softwood stands that occur over a large region in central Maine. The site has always existed as forest land. Mixed-stand composition has resulted from a history of partial cutting and high grading. An uneven-age management plan of single-tree selection over a 20-year cutting cycle was implemented before 1977, after which the area was dedicated as the International Paper Company's Northern Experiment Forest.

Geology and Soils

Bedrock geology was classified in the moderately metamorphosed Waterville and Vassalboro formations. The area was deglaciated between 14,000 and 13,000 years ago, leaving a layer of till that was primarily locally derived. A gravel esker passes within a few kilometers of the site. Three soil series, Plaisted, Skerry, and Westburg, have been identified within the vicinity of the selected study areas. The Plaisted series was classified as a coarse-loamy, mixed, frigid Typic Haplorthod; the Skerry series as a coarse-loamy, mixed, no-acid, frigid Aeric Haplaquod. These soils are all derived from basal till and range from well drained to poorly drained.

Florida Slash Pine

The Florida site (FS) was on a 60-ha plantation block of 20-year-old (in 1986) slash pine (*Pinus elliottii*) on land under a long-term commercial timber lease to Container Corporation of America. The site was 38 km northeast of Gainesville in Alachua County, Florida, established in 1986 under a coordinated project on carbon, water, and nutrient interactions (National Science Foundation Grant No. BSR 8516678). IFS work was funded by the Forest Response Program (Southern Commercial Forest Research Cooperative, USDA, Forest Service).

Geology and Soils

The topography of the area is flat. The predominant soil type is an Ultic Haplaquod (sandy, siliceous, thermic), although the relative development and depths of the subsurface spodic (organic) and argillic (clay) horizons are highly variable over the 60 ha. Soils are derived from marine deposits and are sandy and very low in nutrients; organic matter is concentrated in the forest floor and the spodic horizon. Because of the low relief, sandy soil texture, and underlying clay horizon, a water table is usually present above the 150-cm depth (average depth of 60 cm). There is little downward percolation of soil water and only slow drainage into lower surrounding wetlands. Flooded conditions can occur at any time depending on rainfall patterns.

Vegetation and Management History

Even-aged slash pine dominated the site. This species is indigenous to these sites (along with longleaf pine *P. palustris*). The understory also consists of indigenous species, with gallberry (*Ilex glabra*), saw palmetto (*Serenoa repens*), and wire grasses the most common. Site preparation after the stem-only harvest of the former plantation stands (the current stands are second rotation) consisted of chopping the residues, broadcast burning, and machine planting of seedlings. After establishment there was no further treatment of the stands.

Protocols

A very brief synopsis of the protocols and methods used for sample collection and for computation of estimates of field site deposition and nutrient cycling fluxes in the Integrated Forest Study is given here. Some flexibility in methods was necessary to accommodate individual site conditions. As a general rule, however, a set of specific protocols were followed, as described in detail in Lindberg et al. (1989).

Methods for Atmospheric Deposition Estimates to Forests

Field Measurements

At the IFS forests designated as intensive deposition measurement sites (see Table 2.1), data on atmospheric chemistry and wet deposition were collected on an event basis (Lindberg 1982) throughout the year. All measurements were taken at or adjacent to the study plot used for collection of nutrient cycling data (Figure 2.1). Wet deposition of all major ions (and weak acidity, Al, total N, and total P on subsets of selected samples) was collected with automatic collectors situated in nearby forest clearings on the basis of published criteria (NADP 1988). Fluxes in throughfall were sampled with the same methods using replicate automatic collectors below the canopy (n = 2–20, see Appendix; Lovett and Lindberg 1984). Methods for stemflow collection are described below.

Atmospheric chemistry measurements were also made on an event basis using samplers situated on towers that extended 5–10 m above the surround-

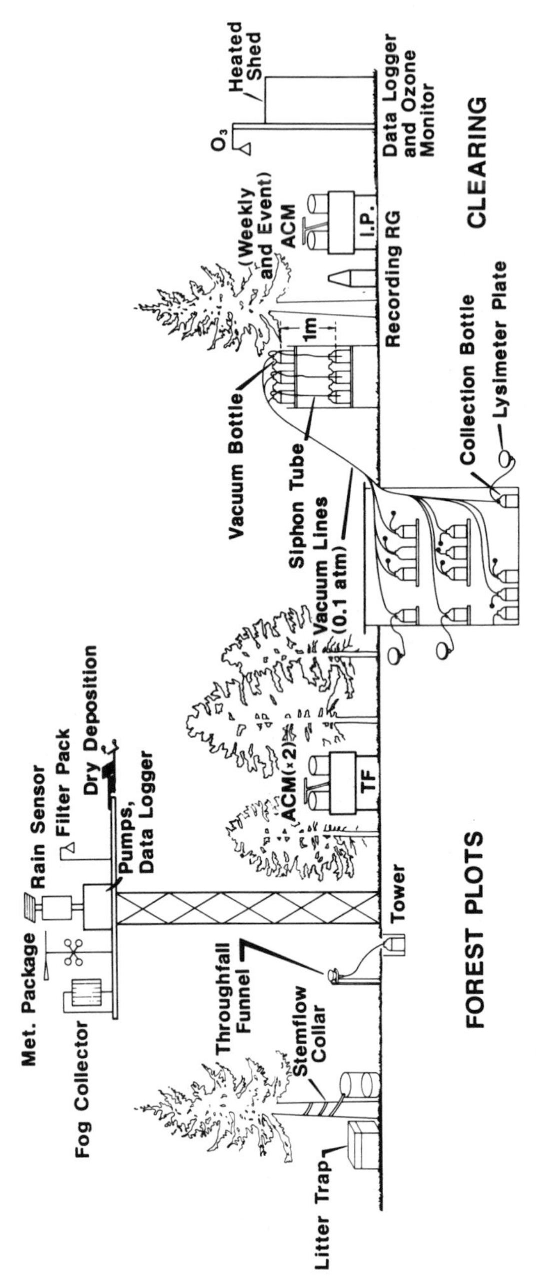

Figure 2.1. Schematic representation of equipment used in typical IFS site.

ing forest canopy; these measurements included: (1) major ions in aerosols collected on teflon filters and in coarse particles dry-deposited to inert plates (Lindberg and Lovett 1985); (2) SO_2 and HNO_3, using filter packs of three sequential filters (Teflon, nylon, and carbonate-treated cellulose; Lovett and Lindberg 1986; Matt et al. 1987), and (3) O_3 sampled continuously with ultraviolet (UV) absorption detectors located in the nearby clearings (see Chapter 4, this volume). Meteorological data were also collected continuously from the same towers and included temperature, relative humidity, solar radiation, wind speed and direction, surface wetness, and precipitation (Hicks et al. 1987) (see Figure 2.1). At fog-prone and high-elevation sites, cloud water was also collected on an event basis using passive string-type collectors (Lovett 1984). Complete details on all the atmospheric measurement methods used in the IFS have been published (Lindberg et al. 1989).

Computation of Deposition Fluxes

The seasonal and annual input of ions in rain, throughfall, and stemflow were calculated from the product of precipitation-weighted mean ion concentrations in each solution and the appropriate hydrologic flux. Depending on site accessibility, 50%–100% (median, 82%; see Appendix) of the wet deposition was sampled and analyzed. Hydrologic fluxes were based on continuous records of rain, throughfall, and stemflow (the latter two with extensive replicate-gage grids; Lindberg and Johnson 1989).

Dry deposition was calculated from the product of event mean air concentrations and appropriate deposition velocities (V_d). For vapors, V_d values were computed hourly using the inferential big-leaf model of Hicks et al. (1987) based on site-specific hourly meteorological measurements and canopy structure/physiology data. For aerosols, V_d values were derived from site-specific canopy structure data using the submicron radionuclide tracer methods of Bondietti et al. (1984). Event dry deposition was computed from the product of the event mean concentrations and V_d values. The annual dry fluxes were computed from time-weighted means of the event data and measurements of the total number of nonrain hours.

Dry deposition of coarse particles measured to inert surfaces must be scaled to the full forest canopy, which was done using scaling factors applicable to each forest. We developed a linear regression model to separate ion fluxes in net throughfall into their component parts of dry deposition washoff and foliar leaching based on event collections (Lovett and Lindberg 1984). We used this method to derive scaling factors for coarse particles based on the results of the regression model applied to calcium at each forest (Lindberg et al. 1988). Calcium in throughfall was used because it is an excellent coarse-particle tracer with no vapor-phase component. These factors were determined as annual means from regressions on all sampled wet events over the year, and were multiplied by the time-weighted mean measured fluxes of coarse particles to inert surfaces to estimate annual coarse-particle dry deposition.

Cloud water deposition was calculated from the product of the weighted mean ion concentrations in cloud water (weighted by liquid water content) and the total annual hydrologic flux of cloud water. Hydrologic fluxes were determined by a number of independent methods, allowing an estimate of the range of cloud water fluxes. These included application of the model of Lovett (1984) to meteorological and canopy structure data for each site, direct measurements of cloud water drip during selected events (extrapolated to a year, based on estimates of cloud immersion time), and various methods involving hydrologic fluxes in throughfall (Lovett 1988; Lindberg and Johnson 1989; Schaefer et al. 1990).

Nutrient Cycling

The basic unit of study was the plot, which ranged in area from approximately 0.05 to 0.10 ha, depending on site conditions and protocols for previous plot-level studies that may have been conducted at the site. At each site, organic matter, N, P, S, K, Ca, and Mg contents of vegetation, forest floor, and soils and nutrient fluxes via litterfall, throughfall, and soil solution were estimated for a minimum of 2 to 3 years. The equipment layout typically used in a site is depicted in Figure 2.1.

Vegetation Biomass and Nutrient Content

Aboveground vegetation biomass (foliage, branch, bole) was typically estimated from regression equations based on diameter at breast height (dbh, measured at 137 cm) or dbh and height. In some instances, foliage biomass was better estimated from litterfall, back-calculating for weight loss from foliage during senescence. Root biomass was obtained where possible, in most instances by random coring. Vegetation was analyzed for nutrients by component, and weighted average concentrations were multiplied by biomass to obtain nutrient content for each component.

Forest Floor and Soil Nutrient Content

Forest floor mass was determined by random destructive sampling within the plot. Soil mass and gravel contents were estimated from destructive measurements for bulk density (quantitative pit, core, or clod method, depending on soil texture and structure) and coarse-fragment (>2 mm sieving) content. Forest floor and the <2-mm soil fraction were analyzed for total nutrient content, and Oa and soil samples were also analyzed for extractable nutrients, extractable Al, and cation exchange capacity (by neutral NH_4Cl). Forest floor and soil nutrient contents were estimated by multiplying concentrations by weights for each component. Nutrient content of the >2-mm soil-size fraction was not normally estimated.

Litterfall Flux

Litterfall was collected in screened traps of known area, separated into components (needles, leaves, wood, reproductive parts), weighed, and analyzed for nutrient content.

Throughfall and Stemflow Flux

In addition to the wet-only event throughfall collections made as part of the atmospheric deposition studies, bulk throughfall was collected and analyzed at all sites. Stemflow was collected using typical collar methods.

Soil Solution Flux

Soil solutions were collected by means of tension lysimeters (set to 10–30 kPa tension, depending on local soil conditions). Tension was supplied to lysimeters either by the hanging column method (see Figure 2.1) or an electrical system with the vacuum pump and vacuum control. Solutions were collected at monthly intervals or less and analyzed for major cations and anions, pH, conductivity, total N, and total P. Volumes were recorded at each collection, and volume-weighted means for each constituent were multiplied by independent estimates of soil water flux (obtained from nearby gaged streams, modeling, Cl balance, or subtraction of evapotranspiration estimates from total precipitation).

Site Operation, Data Management, and Uncertainties

Responsibility for site operation throughout the IFS was in the hands of the principal investigators at each site. The separate experimental tasks, which were under way at various sites are indicated in Figure 1.3. Coordination of synthesis activities for the IFS project was shared by several individuals with expertise in the areas of either atmospheric science or forest ecosystems. These individuals were designated as synthesis group leaders for the major chemical components that have been studied at each IFS research site: sulfur, nitrogen, acidity, and base cations. Data from the principal investigators were supplied to these individuals by the IFS site researchers for summary in this volume.

The problem of assessing uncertainties in measured, calculated, and estimated quantities plagues all environmental research projects, and is often neglected. This is particularly true in nutrient cycling and deposition measurement projects, which rely on numerous assumptions and applications of unique methods. The approach to evaluating uncertainties in the IFS involved both qualitative and quantitative analyses. We attempted to identify the most important sources of potential bias or error and to then analyze the probable magnitude of these sources. Relative standard errors (RSE) were estimated from a combination of the measured standard errors in ion concentrations, analytical precision and accuracy, potential bias in certain physical measurements, and published sensitivity analyses of models used in the IFS. In addition, ranges of possible values based on "reasonable" physical bounds, where available, were used with these RSE values to derive overall estimates of uncertainty. The results of the qualitative and quantitative analyses have been published (Lindberg and Johnson 1989) and are discussed in the context of each ion within the major chapters of this book (see Chapters 5, 6, 8, and 11).

References

Bondietti E.A., Hoffman P.O., Larsen I.L. 1984. Air-to-vegetation transfer rates of natural submicron aerosols. J. Envir. Radioact. 1:5–27.

Bondietti E.A. 1990. Integrated forest study on effects of atmospheric deposition: site descriptions. ORNL/TM-11149, Oak Ridge National Laboratory, Tennessee.

Hicks B., Baldocchi D., Myers T., Hosker R., Matt D. 1987. A multiple resistance routine for dry deposition velocities. Water Air Soil Pollut. 36:311–330.

Lindberg S.E. 1982. Factors influencing trace metal, sulfate, and hydrogen ion concentrations in rain. Atmos. Environ. 16:1701–1709.

Lindberg S.E., Lovett G.M. 1985. Field measurements of particle dry deposition rates to foliage and inert surfaces in a forest canopy. Environ. Sci. Technol. 19:238–244.

Lindberg S.E., Johnson D.W. (eds.) 1989. 1988 Annual Report of the Integrated Forest Study. ORNL/TM 11121, Oak Ridge National Laboratory, Oak Ridge, Tennessee.

Lindberg S.E, Johnson D.W., Lovett G.M., Taylor G.E., Van Miegroet H., Owens J.G. 1989. Sampling and Analysis Protocols and Project Description for the Integrated Forest Study. ORNL/TM 11214 Oak Ridge National Laboratory, Oak Ridge, Tennessee.

Lindberg S.E., Lovett G.M., Schaefer D.A., Bredemeier M. 1988. Coarse aerosol deposition velocities and surface-to-canopy scaling factors from forest canopy throughfall. J. Aerosol Sci. 19:1187–1190.

Lovett G.M., Lindberg S.E. 1984. Dry deposition and canopy exchange in a mixed oak forest determined from analysis of throughfall, J. Appl. Ecol. 21:1013–1028.

Lovett G.M., Lindberg S.E. 1986. Dry deposition of nitrate to a deciduous forest. Biogeochemistry, 2:137–148.

Lovett G.M. 1988. A comparison of methods for estimating cloud water deposition to a New Hampshire subalpine forest. In Unsworth M., Fowler D. (eds.) Processes of Acidic Deposition in Mountainous Terrain. Kluwer Academic Publishers, London, pp. 309–320.

Lovett G.M. 1984. Rates and mechanisms of cloud water deposition to a balsam fir forest. Atmos. Environ. 18:361–367.

Matt D.R., McMillan R.T., Womack J.D., Hicks B.B. 1987. A comparison of estimated and measured SO_2 deposition velocities. Water Air Soil Pollut. 36:331–347.

NADP. 1988 Instruction Manual for NADP/NTN Site Operation. U.S. National Atmospheric Deposition Program, Coordination Office, Colorado State University, Fort Collins, Colorado.

Schaefer D.A., Lindberg S.E., Lovett G.M. 1990. Comparing cloud water flux estimates for a coniferous forest. Acidic deposition, its nature and impacts. In Conference Abstracts, Glasgow, U.K., September 1990.

3. Water Balances

J.M. Vose and W.T. Swank

Introduction

Purpose, Rationale, and Approach

Characterizing the nutrient cycles at the Integrated Forest Study (IFS) study sites required estimates of both nutrient inputs and nutrient outputs. To determine nutrient outputs, estimates of soil water flux had to be coupled with nutrient concentrations in soil lysimeters. However, only a few IFS sites had direct measures of soil water flux (e.g., gaged watersheds at Coweeta and Turkey Lakes), and even in these instances, water flux measurements were not directly comparable with water flux below the lowest lysimeter location. Thus, our purpose was to provide water flux estimates corresponding to lysimeter locations for the IFS sites. We used the hydrologic simulation model, PROSPER (Goldstein et al. 1974), to estimate evapotranspiration and soil water flux for several IFS study sites. A modeling approach was used because it was impractical to measure soil water flux at each site. While other hydrologic models were available (e.g., BROOK; Federer and Nash 1978), we chose PROSPER because it had been applied successfully in both hardwood and conifer forests at Coweeta (Swift et al. 1975; Huff and Swank 1985) and performed well in regional evapotranspiration assessments (USDA 1980).

Scope of Application to Various Sites

At least 1 year of evapotranspiration and water flux estimates were provided for the following sites and forest types:

Coweeta (mixed hardwood and white pine)
Florida (slash pine)
Norway (spruce)
Oak Ridge (Smokies spruce and Oak Ridge loblolly pine)
Washington (Douglas fir and red alder)
BF Grant (loblolly pine)
Turkey Lakes (mixed hardwood).

Three years of simulation were conducted on the Coweeta forests. No simulations were performed for Whiteface and Duke Forest, which lacked suitable input data, or for Huntington Forest, because PROSPER could not handle large water fluxes through shallow soil horizons.

Water flux estimates from PROSPER were provided to each IFS site on a monthly resolution. Use of the data was at the discretion of the principal site investigator. For example, at the Coweeta (hardwood and pine) and Oak Ridge (Smokies spruce) sites, water flux estimates were coupled directly with lysimeter chemistry to estimate nutrient outputs from the study sites. In other cases, site investigators had alternative measurements of soil water flux (e.g., other simulation models, chloride balance) and PROSPER was used for comparison, calibration, or validation of their estimates. By necessity, alternative water flux estimates were used by sites for years not simulated with PROSPER.

Site Characterizations

Precipitation

Descriptions of each study site are provided in detail in the first part of Chapter 2. In this section, we provide total precipitation for the period of simulation and compare these data to the long-term mean (Table 3.1). Precipitation during simulation periods at Coweeta was exceptionally low. The 1984–1985 simulation period, which had 143.3 cm precipitation, represented the beginning of a record-setting 4-year drought period (analyses of the first 3 years are reported in Swift et al. 1989), as precipitation was well below average for the 1985–1986 and 1986–1987 simulation periods as well. Other sites with below average precipitation for the simulation period were Florida (−13%), Oak Ridge loblolly (−14%), Oak Ridge Smokies spruce (−6%), and Washington (−22%); Norway (+13%) and Turkey Lakes (+3%) had above-average precipitation.

Table 3.1. Precipitation during Period of Simulation Relative to the Long-Term Average Precipitation.

Site	Precipitation (cm)		
	Simulation Period	Long-Term Average	% Deviation
B.F. Grant	96.3	not available	
Coweeta 5/84–4/85	143.3	178.8	−20%
5/85–4/86	118.9	178.8	−34%
5/86–4/87	150.0	178.8	−16%
Florida	113.1	130.0	−13%
Norway	100.8	88.0	+13%
Oak Ridge Loblolly	97.9	114.0	−14%
Great Smoky Mountains	187.2	200.0	−6%
Turkey Lakes	124.1	121.0	+3%
Washington	101.5	130.0	−22%

Model Description and Parameterization

PROSPER has been described in detail elsewhere (Goldstein and Mankin 1972; Goldstein et al. 1974; Huff and Swank 1985), thus, only a general description of the model is presented here. PROSPER is a phenomenological, one-dimensional model that links the atmosphere, vegetation, and soils. Plant and soil characteristics are combined in an evapotranspiration surface that is characterized by a "surface resistance" to water vapor loss. This resistance, which is analogous to the relationship between stomatal resistance and leaf water potential, is a function of the water potential of the evapotranspiration surface. Evapotranspiration is predicted by a combined energy balance–aerodynamic method that is a function of the surface resistance to vapor loss. PROSPER applies a water balance (through the use of electrical network equations) to the vegetation with the soil divided into layers. Hence, the flow of water within and between soil and plant is a function of soil conductivity, soil water potential, root characteristics for each soil layer, and surface water potential. Soil water flux during unsaturated soil conditions is governed by hydraulic conductivity, where hydraulic conductivity is estimated from the relationship between soil matric potential and moisture content using the procedure described in Luxmoore (1973). The version of PROSPER used in this study simulates evapotranspiration and soil water redistribution between soil layers on a daily time step.

PROSPER requires the following site-specific climatic data:

solar radiation (daily total)
precipitation (daily total)
windspeed (daytime mean)
air temperature (daytime mean)
vapor pressure (daytime mean).

With the exception of precipitation, these data are used to compute evaporative demand in the energy balance–aerodynamic equation (Swift et al. 1975).

Model parameter values were provided by investigators at each site; however, in some cases, site-specific parameter values were either unavailable or incomplete. In these instances, we used general values (e.g., a general estimate of surface resistance for conifers or hardwoods) or extrapolated the data to satisfy the range requirements for PROSPER. An example of a subset of the model parameters and actual parameter values for two IFS sites are listed in Table 3.2. For a complete list of model parameters, see Goldstein et al. (1974).

Simulation Results and Discussion

Annual Evapotranspiration and Soil Water Flux

Annual evapotranspiration (ET) predictions and soil water flux below the lowest soil layer (outflow) for each site are listed in Table 3.3. Predicted ET was generally lowest for cool and cloudy sites (i.e., Norway, Washington, Smokies, and Turkey Lakes) and highest in the warmer Southeastern sites (i.e., Coweeta, Florida, B.F. Grant). As demonstrated in previous applications, PROSPER also simulated ET differences due to vegetation type; for example, ET was 44 cm greater in the pine watershed at Coweeta as compared to the hardwood watershed in the 1984–1985 simulations. In addition, the difference in predicted outflow between watersheds (i.e., 30 cm) compares favorably with the 27-cm difference in water yield estimated by the paired watershed method of analysis for white pine on Watershed 1 (WS1) (Swank 1988). In the the drought period, the ET difference decreased to 29 and 20 cm for the 1985–1986 and 1986–1987 simulation periods, respectively. Thus, the simulated response to drought was more pronounced in the pine stand, which may be a result of greater water demand caused by the high leaf area as well as greater interception losses.

While the ET estimates appear reasonable, it is difficult to evaluate the accuracy of these predictions because true watershed values for ET are impossible to obtain. One approach is to compare drainage or outflow from the lowest soil layer of PROSPER to annual streamflow amounts with the assumption that the relationship

$$\text{streamflow} = \text{precipitation} - \text{ET}$$

provides an adequate assessment of the accuracy of the ET prediction. This has been done previously with PROSPER (Swift et al. 1975, Huff and Swank 1985); however, because PROSPER does not account for deep soil water flux processes that occur between the lowest soil layer and the stream, this

Table 3.2. Input Parameters and Parameter Values

Prosper Input Parameters	Units	Coweeta Hardwood	B.F. Grant Loblolly
Canopy and Aboveground			
Leaf area index–summer	m^2/m^2	5.0	9.0
Leaf area index–winter	m^2/m^2	0.5	5.0
Interception–winter	cm	0.15	0.20
Interception–summer	cm	0.35	0.30
Start leaf-out	Julian day	100	91
End leaf-out	Julian day	150	121
Start leaf-fall	Julian day	260	260
End leaf-fall	Julian day	306	306
Albedo–summer	%	22.0	12.0
Albedo–winter	%	16.0	10.0
Leaf length (mean)	cm	10.0	15.0
Max. resistance to diffusion of H_2O vapor–summer	sec/cm	50.0	50.0
Max. resistance to diffusion of H_2O vapor–winter	sec/cm	50.0	50.0
Min. resistance to diffusion of H_2O vapor–summer	sec/cm	2.5	3.5
Min. resistance to diffusion of H_2O vapor–winter	sec/cm	4.0	4.0
Ratio of convective to latent heat loss surface area–summer	cm^2/cm^2	2.0	1.5
Ratio of convective to latent heat loss surface area–winter	cm^2/cm^2	1.0	1.5
Soils and Belowground			
Thickness of each soil layer	cm	0–30;30–90;90–180	0–10;10–80;80–125
Field capacity by layer	cm^3/cm^3	0–30 0.29	0–10 0.15
		30–90 0.30	10–80 0.26
		90–180 0.28	80–125 0.37
Soil porosity by layer	cm^3/cm^3	0–30 0.610	0–10 0.40
		30–90 0.500	10–80 0.380
		90–180 0.499	80–125 0.420
Saturated hydraulic conductivity by layer	cm/day	0–30 1344	0–10 100
		30–90 168	10–80 10
		90–180 90	80–125 6
Root surface area and distribution by layer	m^2/m^2	0–30 0.20 (75%)	0–10 0.20 (35%)
		30–90 0.01 (25%)	10–80 0.01 (65%)

Table 3.3. Predicted Annual ET and Outflow below the Lowest Soil Layer

IFS Site	Species	Simulation Period	ET (cm)	Outflow (cm)
B.F. Grant	Loblolly	6/87 to 5/88	88.6	5.2
Coweeta	Mixed Hardwood	5/84 to 4/85	60.8	73.6
	White Pine	5/84 to 4/85	104.5	42.8
	Mixed Hardwood	5/85 to 4/86	64.7	55.9
	White Pine	5/85 to 5/86	93.7	34.5
	Mixed Hardwood	5/86 to 4/87	79.3	63.9
	White Pine	5/86 to 4/87	99.0	41.3
Florida	Slash	3/87 to 2/88	75.0	13.2
Norway	Spruce	1/87 to 12/88	59.0	43.6
Oak Ridge	Spruce (GSMP)	2/87 to 1/88	62.0	115.9
	Loblolly	1/87 to 12/87	89.4	18.4
Turkey Lakes	Mixed Hardwood	1/84 to 12/84	53.2	67.3
Washington	Red Alder	1/87 to 12/87	32.6	68.0
	Douglas Fir	1/87 to 12/87	60.2	41.0

type of comparison provides only a general assessment of model performance. Our watershed-level evaluations were restricted to gaged watersheds at Coweeta (WS1 and WS2) and a watershed (WS31) at Turkey Lakes. We also compared soil water fluxes with alternative measures determined by investigators at the other IFS sites.

Measured streamflow on Coweeta WS1 was 22.9 and 29.4 cm for the 1985–1986 and 1986–1987 simulation periods, respectively. By comparison, annual outflow estimates using PROSPER were 34.5 and 41.0 cm, respectively. Streamflow was also lower on Coweeta WS2 relative to outflow estimates using PROSPER. On WS2, streamflow was 31.4 and 47.8 cm for the 2 simulation years, respectively, while outflow estimates using PROSPER were 55.9 and 63.9 cm. Past applications of PROSPER at Coweeta have generally found good agreement between simulated outflow and streamflow (Swift et al. 1975; Huff and Swank 1985). However, these simulations were usually conducted during periods of average or above-average precipitation. Simulations conducted for 1984–1985, which represent the beginning of the dry period, resulted in excellent agreement between predicted outflow and measured streamflow on WS2 (i.e., WS2 outflow = 73.6, streamflow = 74.2), while on WS1, outflow was 25% lower than streamflow (WS1 outflow = 42.8, streamflow = 57.1). The 1985–1986 and 1986–1987 simulations were conducted during very dry periods (i.e., 16%–34% below average precipitation).

During these dry conditions, we hypothesize that outflow and streamflow do not agree (i.e., outflow > streamflow) because a large component of the outflow replenishes storage deficits of deeper soil layers and does not appear as surface discharge. PROSPER does not account for processes regulating deep soil storage or streamflow generation and hence cannot simulate streamflow when the linkages between drainage below the root zone and streamflow are less direct. This does not, however, affect the validity of linking outflow estimates with nutrient concentration data collected from lysimeters in the lower soil horizons. A similar comparison between predicted outflow and measured streamflow was performed on the Turkey Lakes watershed. In this case, measured annual (January 1984 to December 1984) streamflow was 67 cm and predicted outflow was 72 cm. The close agreement (i.e., 7%) between predicted outflow and measured streamflow provides at least circumstantial evidence of adequate model performance for predicting annual water fluxes.

Annual soil water flux was estimated by alternative approaches for the B.F. Grant, Florida, Norway, and Turkey Lakes IFS sites. We compared annual soil water flux below the deepest soil layer that was estimated by these alternative approaches with predictions obtained using PROSPER (Table 3.4). With the exception of the B.F. Grant site, we found excellent agreement between these site estimates and PROSPER estimates. Discrepancies between the B.F. Grant soil water flux estimates obtained with PROSPER and the soil water flux estimates obtained by WATFLO may be related to differences in ET estimates, as is discussed next.

Monthly Water Fluxes

Precipitation inputs and soil water outputs below the lowest soil layer for each IFS site are presented in Figure 3.1. Predicted ET and soil water flux through all soil layers are presented in Figure 3.2. Patterns shown in both figures reflect the influences of soil water retention characteristics and water use in plant transpiration. For example, in both Washington sites patterns of soil water flux through the soil horizons closely parallel precipitation and there is little storage within horizons. This is probably a result of low ET and the coarse texture of these soils, which results in high saturated and unsaturated conductivities (Table 3.5). In contrast, soils with greater water-

Table 3.4. Comparison of Annual Water Flux below the Deepest Soil Layer Estimated using PROSPER and Alternative Approaches Developed by IFS Study Sites

Site	Soil Depth (cm)	Prosper	Alternative/Method	
	flux in cm/yr			
B.F. Grant	120	5.8	22.2	hydrology model
Florida	120	13.2	13.1	hydrology model
Norway	50	41.3	39.2	Cl^- balance
Turkey Lakes	70	67.0	74	PPT-PET[1]

[1] Potential evapotranspiration (PET) predicted using the Thornthwaite (1948) model.

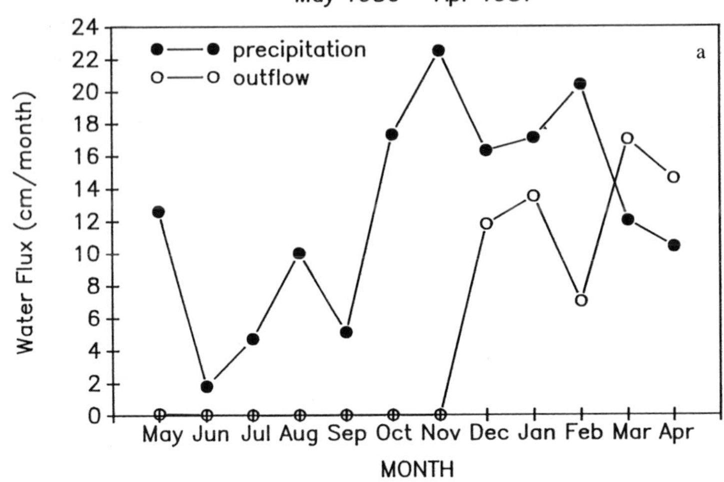

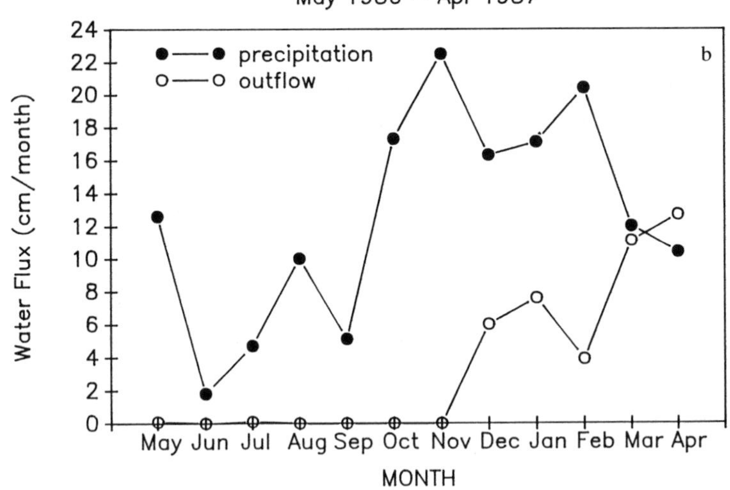

Figure 3.1. Monthly precipitation inputs (measured) and soil water flux outputs (outflow simulated with PROSPER) for IFS study sites. Although 3 years were simulated for the Coweeta sites, only 1 year of simulation is presented.

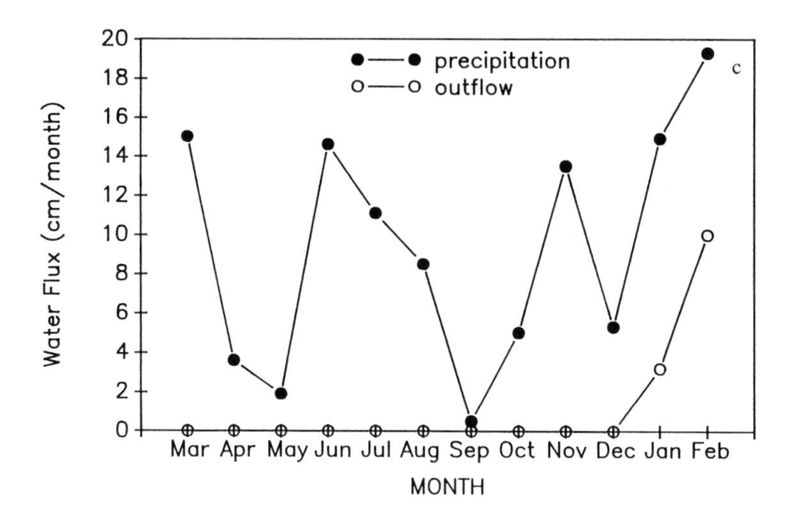

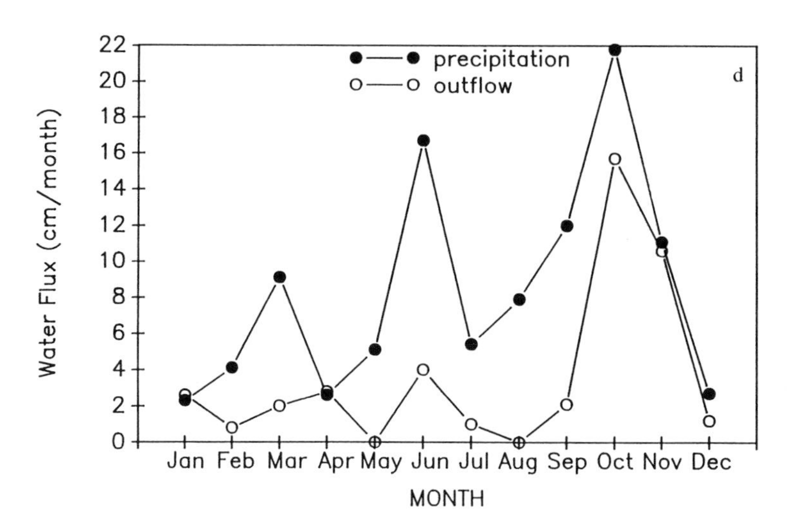

Figure 3.1. *Continued.*

holding capacity, such as the clay soils at B.F. Grant and Oak Ridge Loblolly, show a less direct relationship between precipitation inputs and outflow, probably a combined effect of greater soil retention and higher ET at the B.F. Grant and Oak Ridge sites. The influence of plant water use on soil water flux was evident because fluxes were highest when ET was low

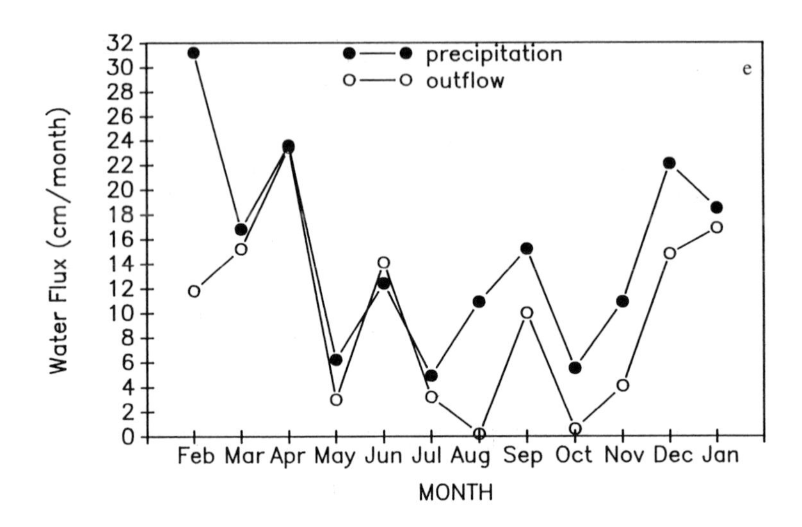

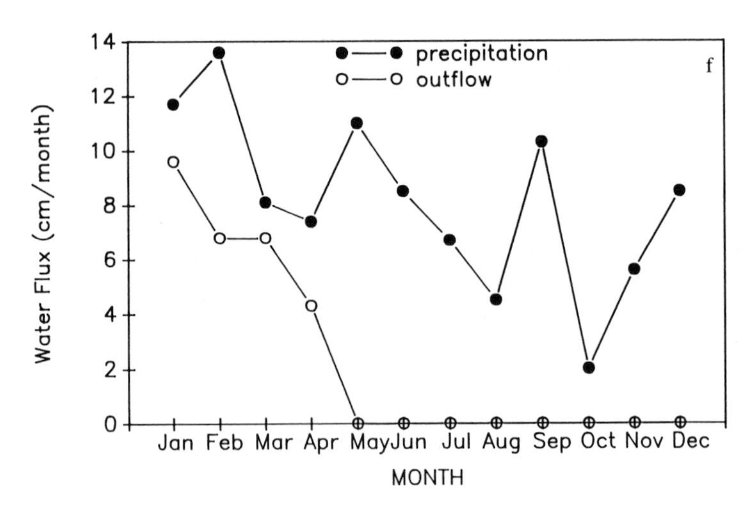

Figure 3.1. *Continued.*

regardless of soil water retention characteristics (e.g., during leafless conditions or periods of low energy input) and precipitation was high.

At the B.F. Grant Forest, monthly soil water fluxes were also determined using the WATFLO hydrologic model (data provided by J. Dowd). We compared soil water fluxes predicted with PROSPER with those predicted with WATFLO (Figure 3.3). Monthly soil water flux patterns were nearly iden-

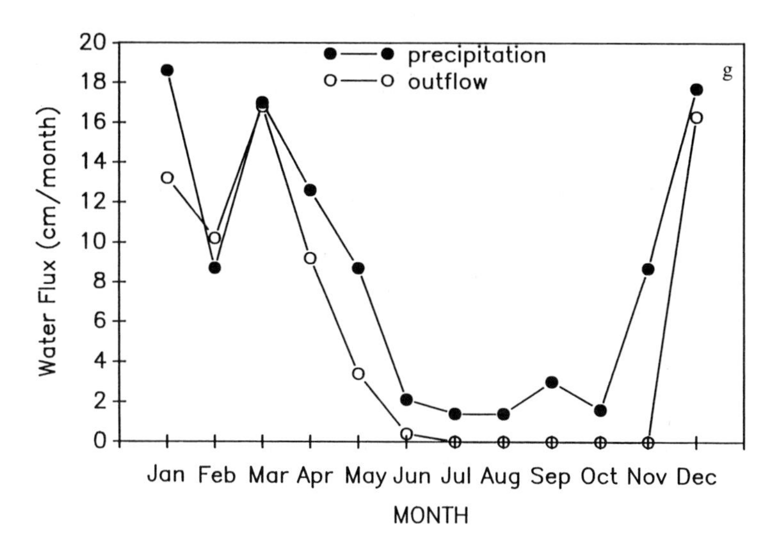

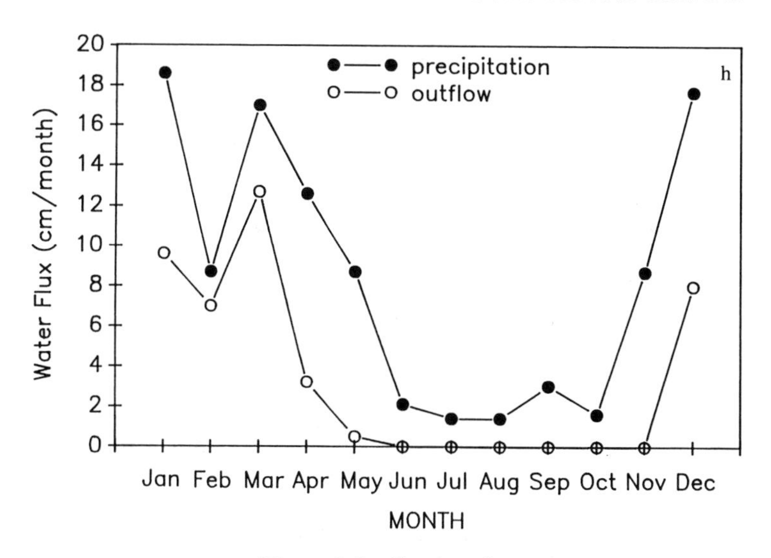

Figure 3.1. *Continued.*

tical between the models; however, PROSPER fluxes were generally greater than WATFLO at 10 cm and less than WATFLO at 120 cm. Differences between PROSPER and WATFLO in the upper soil layer may be at least partially related to the different soil depths used in the simulations. For example, soil water would be depleted by root uptake in the 10- to 20-cm

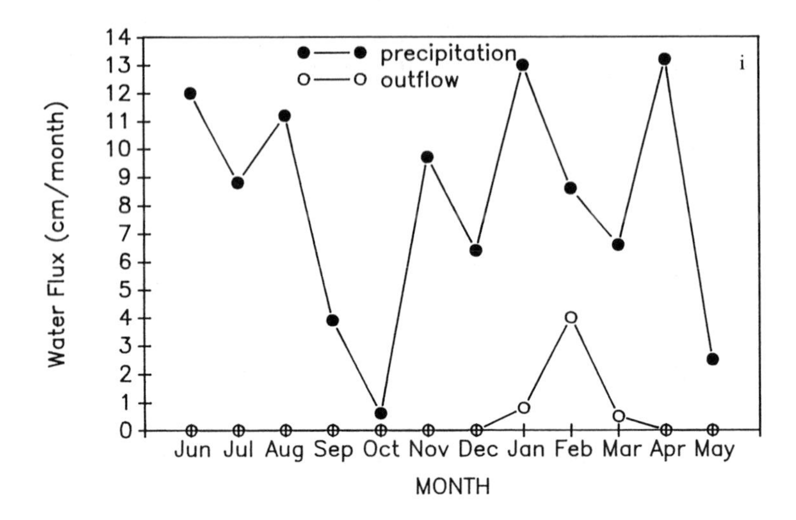

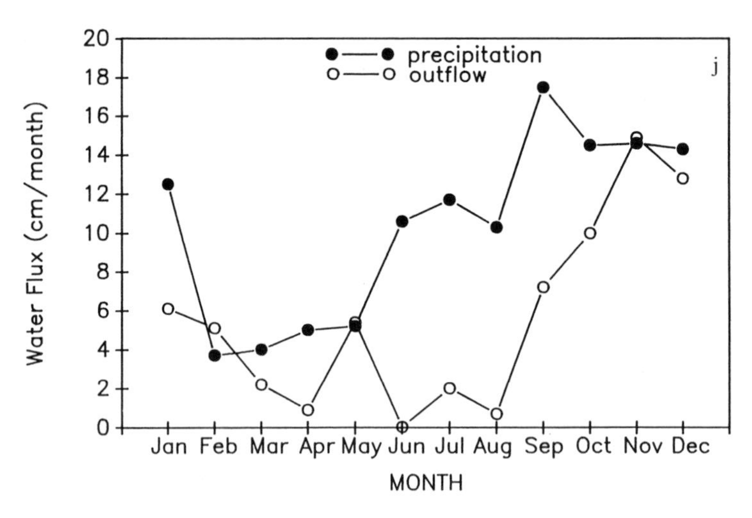

Figure 3.1. *Continued.*

depth zone, particularly during periods of high plant transpiration: differences at the 120-cm soil depth may be related to variation in the corresponding ET estimates. The ET component of the WATFLO simulation predicted considerably less ET than PROSPER (i.e., WATFLO ET = 56 cm; PROSPER ET = 89 cm). The WATFLO ET was estimated using the Thornthwaite (1948) model, which has been shown to underestimate ET.

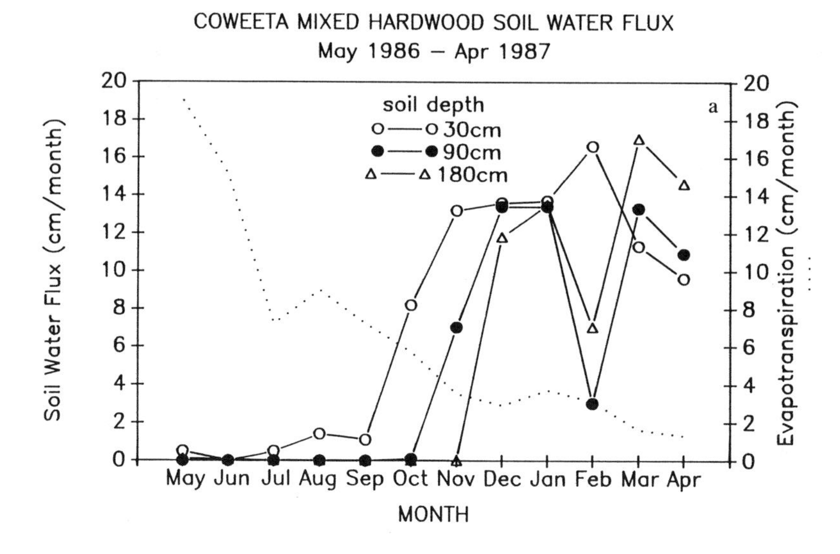

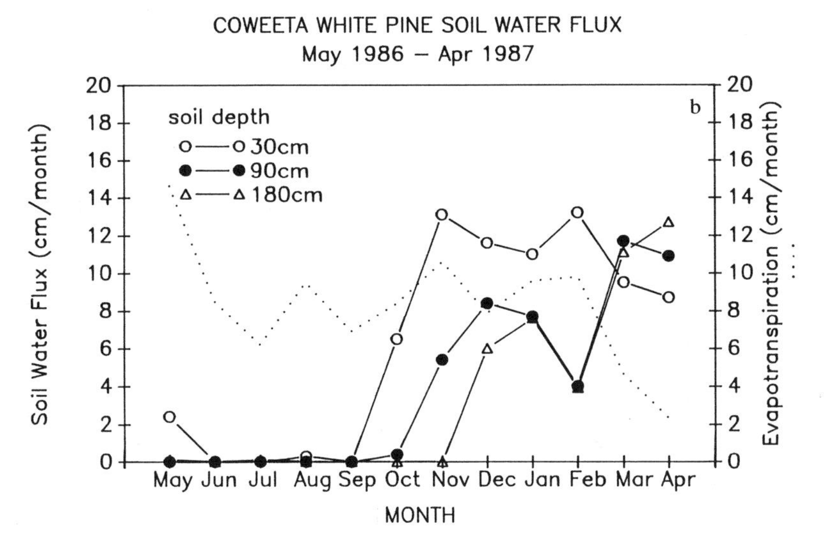

Figure 3.2. Monthly soil water flux by soil depth and monthly evapotranspiration for IFS sites. Although 3 years were simulated for the Coweeta sites, only 1 year of simulation is presented.

Hence, with PROSPER, an additional 33 cm of water was removed from the soil in evapotranspiration making less soil water available for outflow at 120 cm.

A similar comparison was made for the Turkey Lakes WS31, where we compared monthly outflow predicted with PROSPER with measured monthly

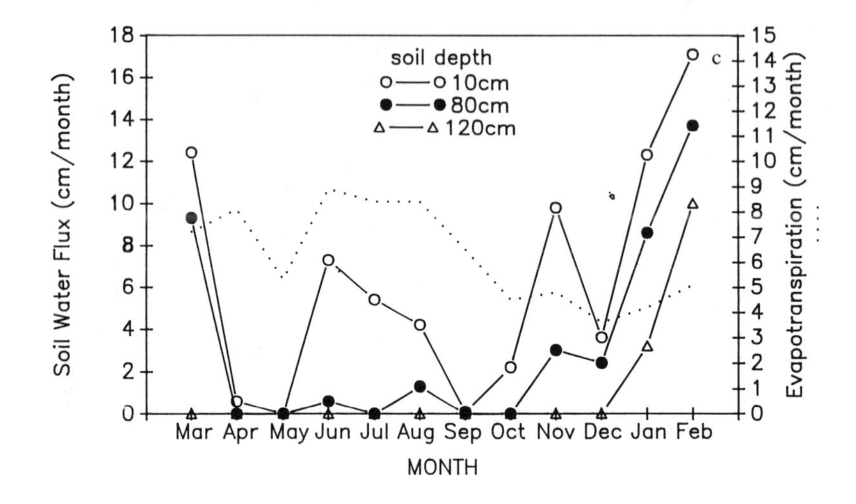

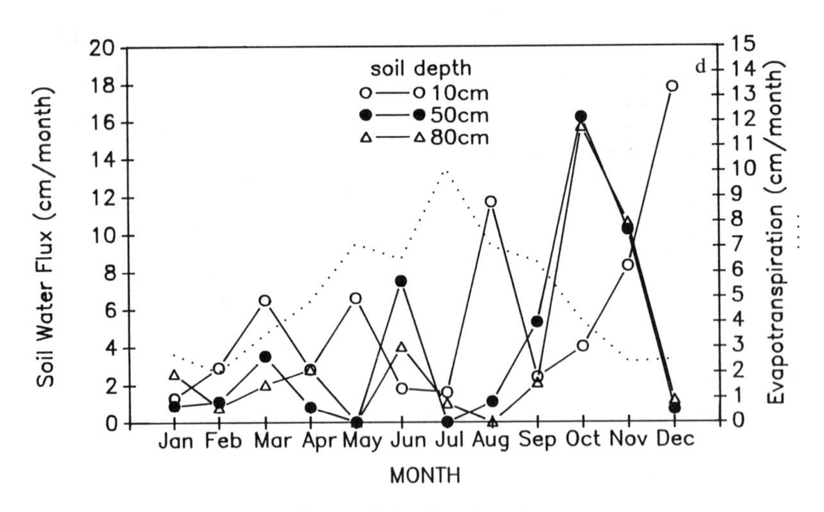

Figure 3.2. *Continued.*

streamflow (Figure 3.4). With the exception of the month of April, outflow and streamflow were in good agreement. The large difference between April streamflow and simulated outflow is related to snowmelt, and a quantitative discrepency could be expected because PROSPER does not consider snowmelt in the precipitation input component of the model. An agreement between monthly outflow and streamflow, as observed on Turkey Lakes WS31, would be less likely on watersheds that have deep soils (e.g., Coweeta)

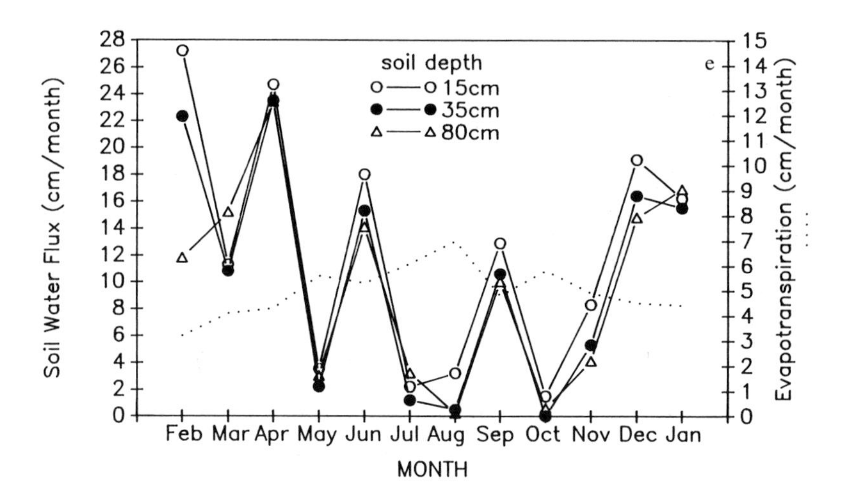

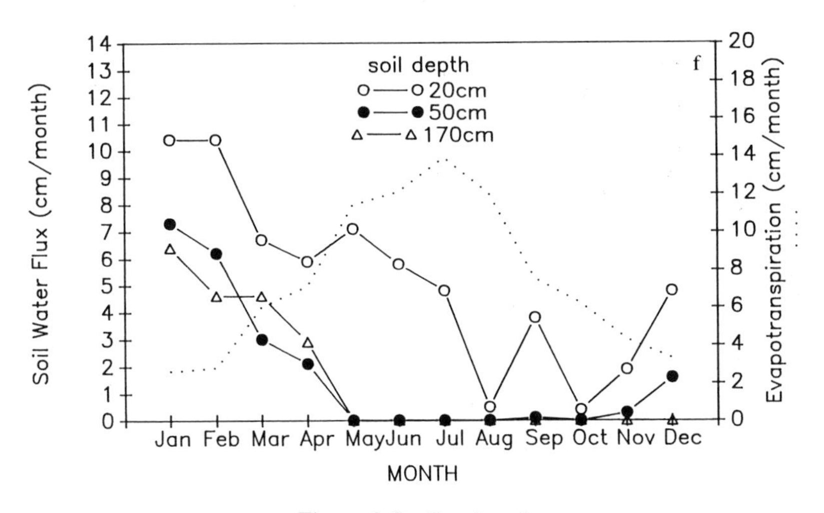

Figure 3.2. *Continued.*

because of a substantial lag between precipitation input and streamflow generation.

Validation of Coupled Water and Nutrient Fluxes

For the Coweeta sites, we compared annual nutrient fluxes for Cl^-, NO_3^--N, and K^+ estimated by combining water fluxes simulated with PROSPER and lysimeter concentration data with nutrient concentrations measured in

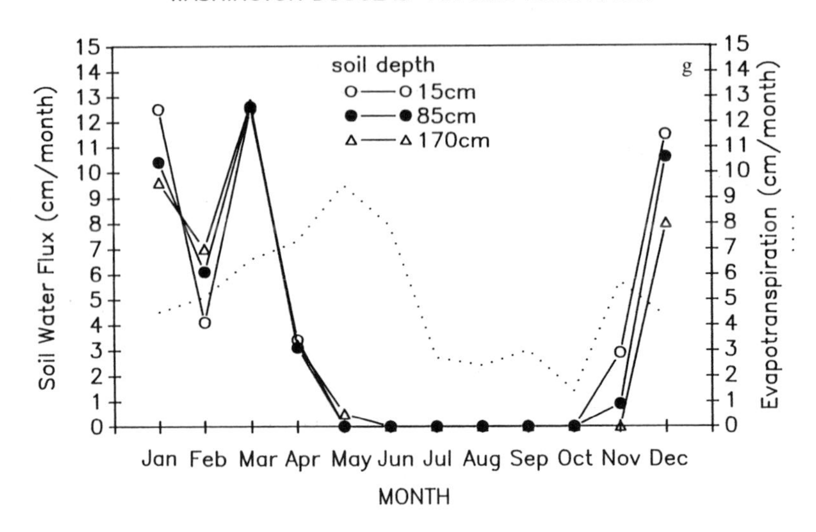

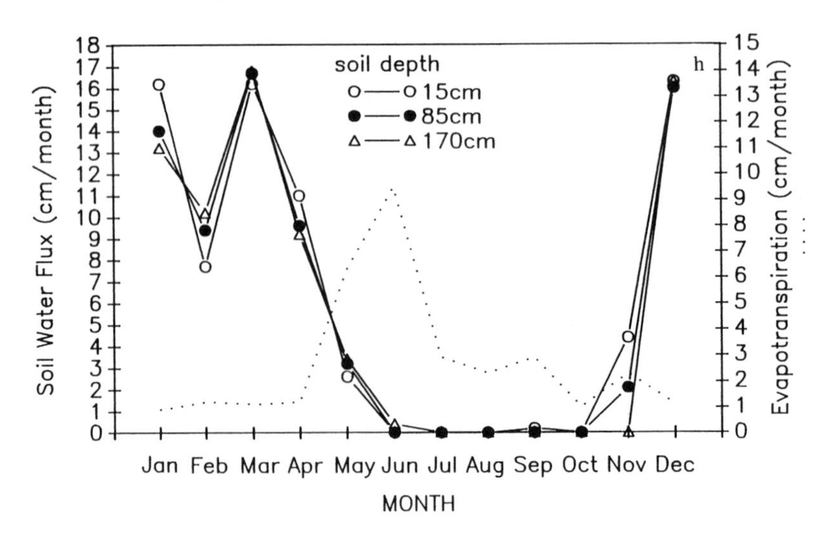

Figure 3.2. *Continued.*

streamwater (Table 3.6). Assuming Cl^- is a conservative tracer, fluxes from the B horizon were calculated over a 19-month period, normalized to annual values, and compared to measured long-term export for both the hardwood and pine watersheds. Modeled fluxes differed by 13% and 30% from measured exports in the Coweeta pine and hardwood watersheds, respectively. This indicates at least general agreement between Cl^- balance approaches and PROSPER simulation results. Stream exports of base cations should be

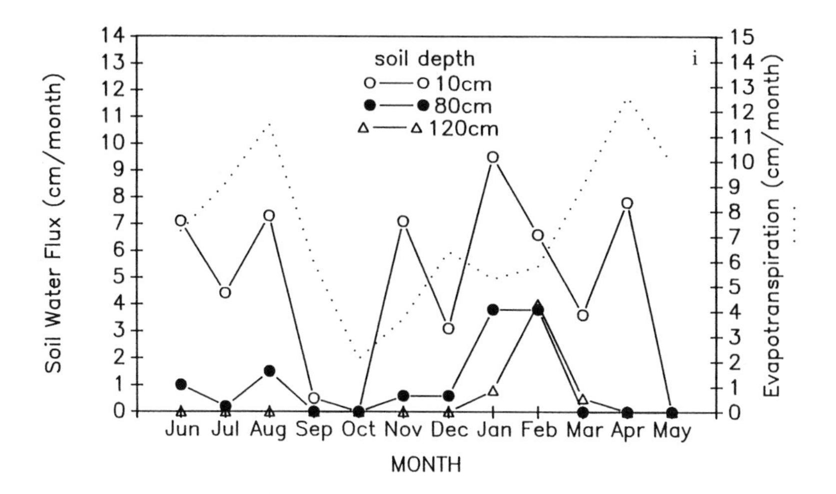

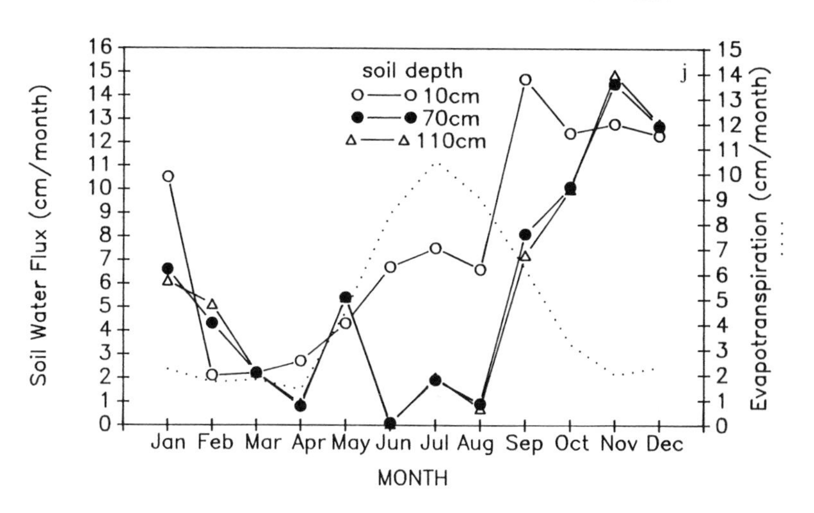

Figure 3.2. *Continued.*

higher than calculated values, as illustrated by K^+ in Table 3.6, from mineral weathering below the B horizon. Concentrations of NO_3^--N are low in both soil solutions and streams of both watersheds and fluxes are small, and thus there is not much significance to the discrepancies between predicted and measured NO_3-N.

Watershed Hydrologic Response

Although the IFS project was designed to assess the effects of acid deposition on nutrient cycles, findings could also have implications for consid-

Table 3.5. Saturated and Unsaturated Hydraulic Conductivity for Surface Soil Layers

Site	Saturated Hydraulic Conductivity	Calculated[1] Unsaturated Hydraulic Conductivity at Field Capacity
	cm/day	
Coweeta hardwood	1344	0.99
pine	1344	0.13
Florida	792	0.07
Norway	199	0.11
Oak Ridge GSMP	490	0.27
loblolly	144	0.26
Washington (both sites)	1116	0.34
B.F. Grant	100	0.08
Turkey Lakes	86	0.08

[1] Unsaturated hydraulic conductivity was calculated using the procedures described in Luxmoore (1973).

ering effects on surface water chemistry. The two most important hydrologic factors influencing the alteration of acidic rainfall as it moves through a catchment are contact time of water in the soil mantle and the flowpaths that the water follows (Peters and Driscoll 1987). The hydrologic response of forest land provides an integrated index of these factors and other variables such as topography and soil water storage and release characteristics. In the IFS project, studies were focused on relatively small areas, and the hydrologic response of watersheds thus may provide a method for inferring potential effects of solute responses to atmospheric deposition on surface or groundwater chemistry across ecosystems. Several different expressions of hydrologic response have been used (Hewlett and Hibbert 1967), and we selected the storm runoff/total runoff coefficient for our analysis because these data were most available across sites.

Our approach was to identify and obtain stream discharge records from forested watersheds representative of vegetation, topography, and soil in proximity of IFS study sites. Using these criteria, we were able to obtain discharge records for eight of the ecosystems studied in the IFS project (Table 3.7). In the case of Coweeta and Turkey Lakes, IFS plots were actually located on gaged watersheds. Good representation of study sites and quality discharge records were also available for Florida, Oak Ridge, and B.F. Grant. Shorter periods of record or poorer site representation were the case for Great Smoky Mountains and Duke. In the compilation of data, the original time-stage height-gaging records were analyzed (where needed) using translation, editing, and discharge programs developed at the Coweeta Hydrologic Laboratory (Hibbert and Cunningham 1967). Thus, procedures provided standardized methods for calculating quickflow (storm runoff) to facilitate comparison of hydrologic response across sites.

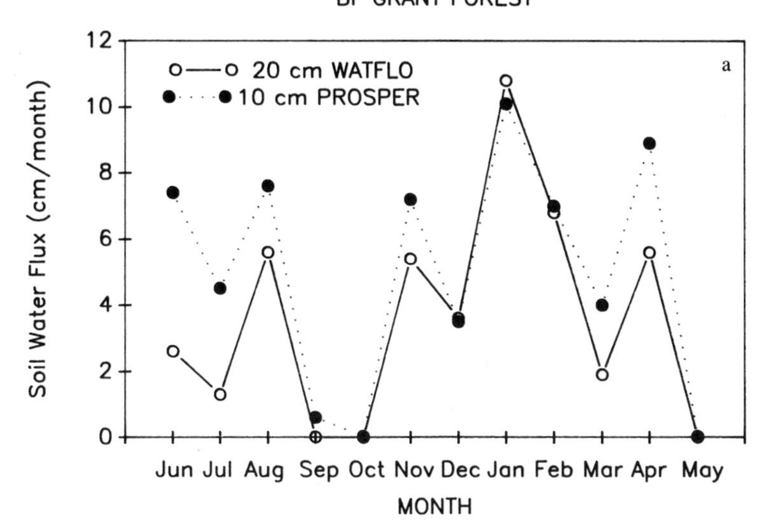

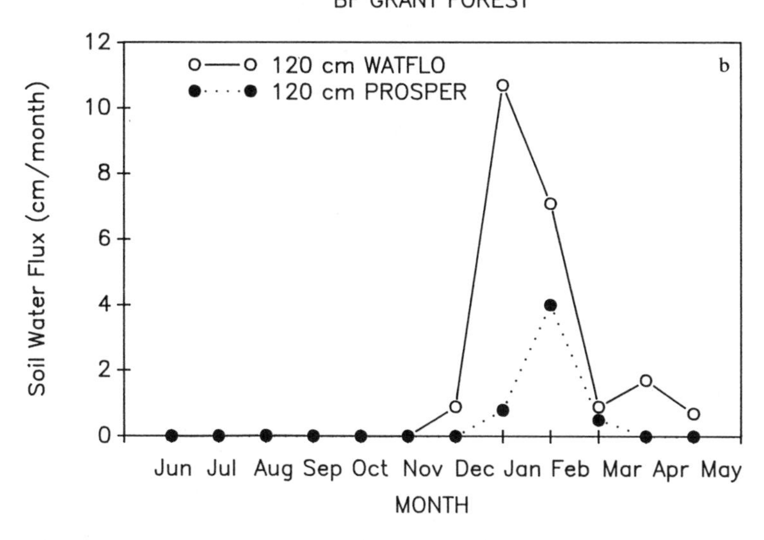

Figure 3.3. Comparison of soil water flux predicted with PROSPER and WATFLO for the B.F. Grant loblolly pine site.

Table 3.6. Comparison of Calculated and Measured Export (kg/ha/yr) of Select Ions for the Coweeta Hardwood and Pine Watersheds

	Cl^-		NO_3-N		K^+	
Watershed	Calculated	Measured	Calculated	Measured	Calculated	Measured
white pine	5.1	4.5	0.02	0.20	2.0	3.4
hardwood	4.3	6.2	0.02	0.02	1.6	4.7

The IFS sites used in this analyses represent a wide range of discharge conditions (Table 3.8), with annual flows ranging from 12 to 91 cm. Sites with high annual flows are associated with regions of high precipitation and moderate evapotranspiration, and sites with lower flows occur in regions of moderate precipitation and moderate to high evapotranspiration. Watersheds representative of the study sites also show a wide range of hydrologic responses. Quickflow or storm runoff accounts for only about 8% of the total annual flow at Coweeta but for more than 40% in Florida and B.F. Grant. Other sites are intermediate in their response and average about 25% of the total annual flow as storm runoff.

Factors contributing to this wide range of hydrologic responses include soil depth, texture, and structure, and to some extent, the form and distribution of precipitation. Soils of the IFS sites in Table 3.7 include clays, stony ablation till, sands, and loams with regolith depth ranging from 0.5 to 30 m. Regardless of causative factors, the indices of hydrologic response

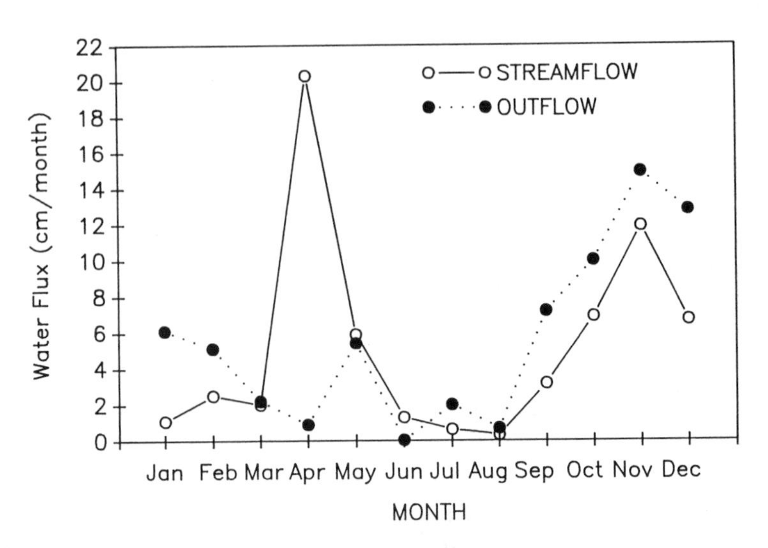

Figure 3.4. Comparison of predicted outflow with measured streamflow for the Turkey Lakes WS31.

Table 3.7. Summary of IFS Sites, Watershed Descriptor, and Source and Years of Data used in Hydrologic Response Analyses

IFS Site and Vegetation	Watershed Descriptor	Source of Data	Years
Coweeta, southern hardwoods	WS 2	Coweeta Hydrologic Laboratory, Southeastern Forest Exp. Station	1980–1987
Coweeta, white pine	WS 1	Coweeta Hydrologic Laboratory, Southeastern Forest Exp. Station	1980–1987
Turkey Lakes, northern hardwoods	WS 31	Dr. John Nicolson, Canadian Forest Service, Sault Ste. Marie, Ontario	1984–1986
Great Smoky Mountains, red spruce	Camel Hump	Dr. David Silsbee, GSMP, NPS	1979–1981
Florida, slash pine	Control, pine flatwoods	Riekerk (1989)	1978–1985
B.F. Grant, loblolly pine	B.F. Grant, WS 15	Dr. Wade Nutter, University of Georgia	1974–1980
Duke, loblolly pine	Hill Forest	Dr. Jim Gregory, North Carolina State University	1970
Oak Ridge, loblolly pine	Walker Branch	Luxmoore and Huff (1989)	1971–1975

Table 3.8. Discharge and Hydrologic Response of Watersheds Occupied by or Representative of Some IFS Study Sites

IFS Sites	Annual Flow	Quick Flow	Hydrologic Response % of Annual flow as Quickflow
	-------- cm	--------	
Coweeta hardwoods	72.2	5.8	8
Coweeta pine	49.0	3.4	7
Turkey Lakes	65.7	21.6	33
Great Smoky Mountains	66.0	17.7	27
Florida	26.0	10.4	40
B.F. Grant	29.3	15.8	54
Duke	12.3	2.3	19
Oak Ridge	91.0	22.0	24

suggest significant variability in soil water contact time across IFS sites and thus capability to buffer acidic precipitation. Wolock et al. (1989) derived indices of soil contact time and flowpath partitioning for 145 northeastern U.S. catchments used in the direct-delayed response project of the U.S. Environmental Protection Agency. They found that surface water alkalinity, and hence potential buffering capacity, was positively correlated with the index of soil contact time. As a general principle, IFS ecosystems with large hydrologic response indices could be expected to show less modification of catchment response to acidic precipitation than ecosystems with smaller indices. Of course, such interpretations must consider geochemical properties of soils, such as cation exchange capacity and base saturation, and biologically mediated transformations of ions, as discussed in other chapters of this book.

Summary

Although it is difficult to validate the ET and soil water flux predictions generated with PROSPER, simulations of ET and soil water flux for the IFS sites appeared quite reasonable and thus, provided good estimates of soil water flux that could then be linked with lysimeter nutrient concentration data. Evapotranspiration levels were within the range of values expected for these forests, and seasonal, geographical, and species-related ET differences appeared well characterized. It was also encouraging to note good agreement between simulation results obtained with PROSPER and alternative approaches utilized by the IFS sites. Analyses of hydrologic records for watersheds representative of eight IFS study sites show that the hydrologic response (storm runoff/total runoff) varies from 8% to more than 50%. The effects of nutrient cycles altered by acid precipitation on surface water chemistry are strongly related to hydrologic factors.

References

Federer C.A., Nash D. 1978. BROOK: A hydrologic simulation model for eastern forests. Research Paper 19, Water Resources Research Center, University of New Hampshire, Durham.

Goldstein R.A., Mankin J.B. 1972. Prosper: A model of atmosphere-plant-soil water flow. In Summer Computer Simulation Proceedings, pp. 1176–1181, San Diego, California.

Goldstein R.A., Mankin J.B., Luxmoore R.J. 1974. Documentation of PROSPER: A Model of Atmosphere-Soil-Plant Water Flow. EDFB/IBP-73/9, Oak Ridge National Laboratory, Oak Ridge, Tennessee.

Hewlett J.D., Hibbert A.R. 1967. Factors affecting the response of small watersheds to precipitation in humid areas. In Sopper W.E., Lull H.W. (eds.) Forest Hydrology. Symposium Publications Division, Pergamon Press, New York, pp. 275–290.

Hibbert A.R., Cunningham G.B. 1967. Streamflow data processing opportunities and application. In Sopper W.E., Lull H.W. (eds.) Forest Hydrology. Symposium Publications Division, Pergamon Press, New York, pp. 725–736.

Huff D.D., Swank W.T. 1985. Modelling changes in forest evapotranspiration. In Anderson M.G., Burt T.P. (eds.) Hydrological Forecasting. Wiley, New York, pp. 125–151.

Luxmoore R.J. 1973. Application of the Green and Corey Method for Computing Hydraulic Conductivity in Hydrologic Modeling. EDFB/IBP-74/4, Oak Ridge National Laboratory, Oak Ridge, Tennessee.

Luxmoore R.J., Huff D.D. 1989. Water. In Johnson D.W., VanHook R.I. (eds.) Analysis of Biogeochemical Cycling Processes in Walker Branch Watershed. Springer Advanced Texts in Life Sciences, Springer-Verlag, New York, pp. 164–196.

Peters N.E., Driscoll C.T. 1987. Hydrogeologic controls of surface-water chemistry in the Adirondack region of New York State. Biogeochemistry 3:163–180.
Riekerk H. 1989. Influence of silvicultural practices on the hydrology of pine flatwoods in Florida. Water Resour. Res. 24(4):713–719.
Swank W.T. 1988. Stream chemistry responses to disturbance. In Swank W.T., Crossley Jr. D.A. (eds.) Forest Hydrology and Ecology at Coweeta. Ecological Studies, Vol. 66, Springer-Verlag, New York, pp. 339–357.
Swift L.W., Jr., Swank W.T., Mankin J.B., Luxmoore R.J., Goldstein R.A. 1975. Simulation of evapotranspiration and drainage from mature and clear-cut forest and young pine plantations. Water Resour. Res. 11(5):667–673.
Swift L.W., Jr., Waide J.B., White D.L. 1989. Refinements in the Z-T method of extreme value analysis for small watersheds. In Sixth Conference on Applied Climatology, Charleston, South Carolina, March 7–10, 1989, pp. 60–65. American Meterological Society, Boston, Massachusetts.
Thornthwaite C.W. 1948. An approach toward the rational classification of climate. Geogr. Rev. 38:55–94.
U.S. Department of Agriculture (USDA), Forest Service. 1980. An Approach to Water Resources Evaluation of Non-Point Silvicultural Sources (A Procedural Handbook). EPA-600/8–80–12, Environmental Research Lab, Office of Research and Development, U.S. Environmental Protection Agency, Athens, Georgia.
Wolock D.M., Hornberger G.M., Beven K.J., Campbell W.G. 1989. The relationship of catchment topography and soil hydrologic characteristics to lake alkalinity in the northeastern United States. Water Resour. Res. 25(5):829–837.

4. Patterns of Tropospheric Ozone in Forested Landscapes of the Integrated Forest Study

G.E. Taylor, Jr., B.M. Ross-Todd, E. Allen, P. Conklin, R. Edmonds, E. Joranger, E. Miller, L. Ragsdale, J. Shepard, D. Silsbee, and W. Swank

Introduction

Interest in characterizing patterns of tropospheric ozone in continental landscapes has increased significantly in the last decade because of the documented effects of this trace gas on both human health in urban areas and vegetation in terrestrial ecosystems (EPA 1986). More recently, concern about the responsiveness of forest ecosystems to ozone (McLaughlin 1985; Johnson and Taylor 1989) has prompted similar analyses applicable to remote, forest landscapes (Evans et al. 1983; Taylor and Norby 1985; Meagher et al. 1987; Pinkerton and Lefohn 1987; Lefohn and Pinkerton 1988; Logan 1989). Few of these studies are in forested areas of greatest ecological concern, and many of the sites are not investigating the influence of air pollution stress on the long-term productivity and biogeochemistry of forest stands. Moreover, most sites were not originally selected with the intent of comparing patterns of ozone concentrations among a variety of forest stands that differ in climate, air quality, vegetation cover, and elevation. The Integrated Forest Study (IFS) is unusual in providing an array of sites with many of these contrasting characteristics.

Ozone is unique in many of its physiocochemical and toxicological properties, and consequently its analysis requires an unusual mix of statistical approaches for characterizing its patterns in space and time (Taylor and Norby 1985). These unique features are severalfold: (1) ozone precursors are both natural and anthropogenic (Altshuller 1988); (2) ozone concentrations in many locations are strongly dependent on regional meteorological variables, notably photochemistry and airshed dynamics (EPA 1986); (3) the residence time of ozone after deposition to forest canopies is measured on timeframes of seconds to minutes (Taylor and Norby 1985); and (4) ozone is nonconservative in terrestrial ecosystems (Taylor and Norby 1985). From an ecological and physiological perspective, the last two features are particularly unusual in comparison with other criteria pollutants (e.g., sulfur dioxide, nitrogen dioxide, heavy metals) in which the principles of deposition, ecosystem loading, and long-term biogeochemistry are applicable.

The objective of this chapter is to compare the patterns of ozone concentration among the sites of the Integrated Forest Study. This analysis, based on data collected in 1987 and 1988, addresses spatial and temporal patterns among sites rather than site-specific features. The temporal patterns of interest involve diurnal, monthly, seasonal, and annual time scales, whereas the spatial patterns focus on elevation, proximity to precursors, latitude, and regional meteorology conditions. Finally, the analysis compares patterns of ozone concentration across all sites with concurrent data on nitrogen and sulfur deposition.

Methodology

Site Description and Monitoring Protocol

A general description of each of the 11 core sites of the Integrated Forest Study is outlined in Chapter 2.1, (this volume). Of particular relevance to the study of ozone patterns are the gradients in latitude (30°–46°N), ranges of elevation (100–1250 m), forest cover types (conifer to mixed hardwood), scavenging efficiency of the biosphere (leaf area indexes of 4–11 m^2 m^{-2}), differences in proximity to anthropogenic sources of oxidant precursors, and variability in synoptic-scale meteorological conditions that determine airshed dynamics. Two of the highest elevation sites for ozone monitoring were ST (1250 m) and CP (1010 m) (see Chapter 2 for definition of codes). At both these locations, the site for ozone monitoring was at a different elevation than the heavily instrumented atmospheric deposition and nutrient cycling study (see Chapter 2), with ozone monitoring being lower at ST and higher at CP.

The protocol for monitoring ozone at each of the 11 principal forest sites of the Integrated Forest Study is described in Lindberg et al. (1989) and is based on the ambient air quality surveillance specifications of the U.S. Environmental Protection Agency. The most salient features of the protocol are the following: (1) common UV absorption methodology for ozone monitoring (Dasibi Model 1008AH), (2) continuous datalogging of 1-h means, (3) monitor operation in a temperature-controlled shed, (4) shed location in a clearing meeting the criteria for siting that minimized the influence of the surrounding forest canopy on ambient ozone levels, (5) air sampling through thermostatically heated Teflon lines equipped with particle filter cartridges, and (6) site-specific calibration of the monitors using National Bureau of Standards (NBS) traceable standards.

Statistics for Ozone Analysis

The basic statistic for all analyses was the 1-hour mean ozone concentration (ppb in vol vol^{-1}) for each hour of the day. Completeness requirements for subsequent analysis were set at 65% of the total possible observations for a 24-hour period.

The selection of statistics to address ozone patterns among sites was dictated by the diversity of ozone trends as exemplified in Figure 4.1. Irrespective of the quarter, the patterns were always one of two principal types: (1) diurnal cycle indicative of daytime photochemical processes controlling ozone formation similar to that observed in many urban areas (Guderian 1985; Altshuller 1988) and (2) diurnally less variable or flat profile (e.g., WF; Figure 4.1) reported in some high-elevation sites (Guderian 1985; Singh et al. 1978). The sites in the former pattern type were further classified as showing either a very pronounced diurnal cycle (absolute amplitude >40 ppb; e.g., LP; Figure 4.1) or cycles with comparatively less amplitude in concentration (<40 ppb) during a 24-hour period (e.g., NS; Figure 4.1).

Because of this diversity, multiple statistics were selected that were relevant in a context of either physiological ecology or atmospheric chemistry. The physiologically relevant statistics were the 7-hour daylight mean (0900–

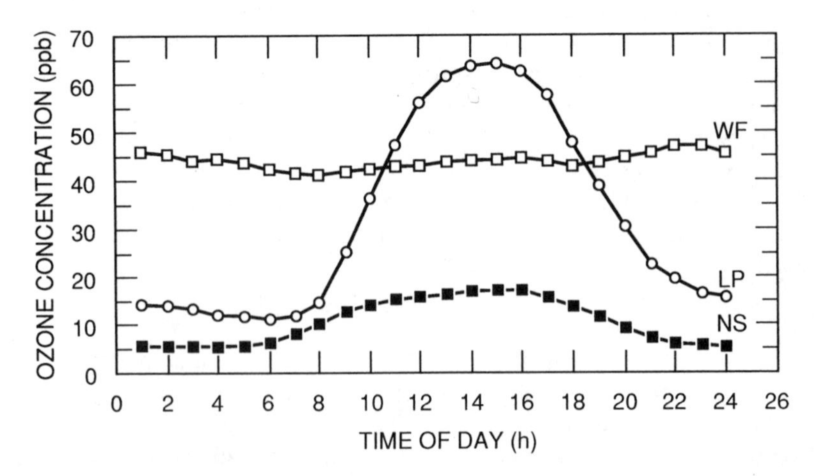

Figure 4.1. Representative diurnal variation in ozone concentration among sites of Integrated Forest Study. Abbreviations: WF, Whiteface, New York; LP, Oak Ridge, Tennessee; and NS, Nordmoen, Norway.

1600 hours, Local Standard Time), second highest 1-hour maximum, and the sum of hourly concentrations greater than or equal to 60 ppb (SUM60) and 80 ppb (SUM80). The 7-hour daylight mean has been the most common statistic used to develop dose–response relationships for a variety of agricultural crops (Heck et al. 1983; Rawlings et al. 1988), whereas the second highest 1-hour maximum is the basis for evaluating the degree of compliance with the ambient air quality standard for ozone in the United States. The last two statistics address only those periods in which the ozone concentration equaled or exceeded threshold values of 60 and 80 ppb, respectively. The rationale for these cumulative statistics is that they recognize the disproportionate role of episodically high ozone concentrations in governing the physiology and growth of terrestrial vegetation, and they appear to be particularly valuable in evaluating ozone air quality issues in nonurban areas (Hogsett et al. 1985, 1988). The traditional statistics from the perspective of atmospheric chemistry were the 24-hour mean and 12-hour daylight mean (0800–2000 hours Local Standard Time).

The daily statistics for each site were aggregated by month and quarter by standard annual meteorological season (i.e., the second or spring quarter is April–June, and third or summer quarter is July–September). The rationale for this aggregation by quarters is twofold. For most sites of the Integrated Forest Study, these two quarters encompass the principal months of the growing season. The second is demonstrated in Figure 4.2, in which the 4-year (1986–1989) mean ozone concentrations (24-hour mean) and 1-hour maximum by month are plotted for the Oak Ridge site (LP) (intermediate in longitude and latitude among the 11 sites), whose diurnal pattern is characteristic of many low elevations in North America. Annual minima for each statistic were consistently observed in the first 2 months of the first quarter and each month of the fourth quarter. Thus, the highest values for both the 24-hour mean and 1-hour maximum were clustered in the second and third quarters (see Figure 4.2).

Patterns of Ozone in the Integrated Forest Study

Temporal Patterns in Ozone Concentration Among the Low-Elevation Sites

The eight sites at elevations less than 530 m (HF, MS, LP, DF, GL, FS, DL, and NS) exhibited a diurnal pattern of ozone concentration similar to that observed in urban areas (Figure 4.3a–4.3d). This pattern was consistent in both the second and third quarters of 1987 and 1988 and was characterized by the following features: (1) invariant or declining predawn (0100–0800 hours) concentrations that were daily minima; (2) rapidly increasing concentrations at a rate of 5 to 20 ppb per hour in midmorning (0800–1100 hours); (3) site-specific, midday-to-late-afternoon periods of maximum ozone

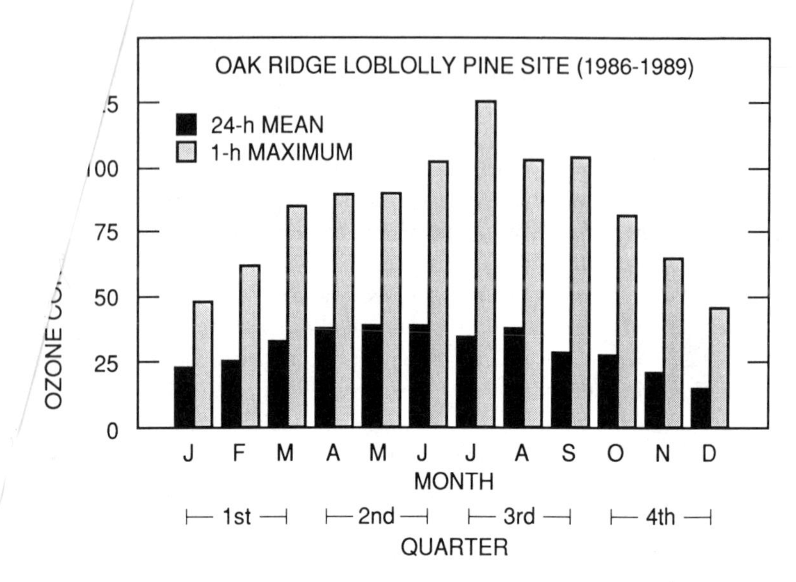

Figure 4.2. Monthly patterns in 24-h mean and 1-h maximum ozone concentrations at one low-elevation sites (LP, Oak Ridge, Tennessee) of Integrated Forest Study. Both peak and mean concentrations tended to be highest in the second and third quarters and least in the first and fourth quarters.

concentration from 1300 to 1800 hours; (4) for the majority of sites, a ratio of midday maximum to predawn minimum concentrations greater than or equal to 4; (5) quarterly maxima in the afternoon that varied by as much as a factor of 4 across all sites from ~20 ppb (NS in third quarter of 1987; Figure 4.3b) to ~80 ppb (GL and DL in second quarter of 1988; Figure 4.3c); and (5) slow decline in ozone concentration in the late evening until 2400 hours.

For most low-elevation sites, the rate of midmorning increase in ozone concentration was higher than the corresponding decline in the evening such that the diurnal distribution of ozone concentration tended to be skewed to the right (e.g., second quarter of 1988; Figure 4.3c). The ratio of predawn minima to late-afternoon maxima ranged from a value of ~2 (NS in third quarter of 1987 and 1988; Figure 4.3b and 4.3d, respectively) to ~14 (GL in third quarter of 1987; Figure 4.3b). Independent of the site, quarter, or year, the least variability in hourly ozone concentrations within or among days at each site was observed in the predawn hours, whereas the greatest variation was in the mid- to late afternoon. The predawn values of the DL site in the third quarter of 1987 were unusually high, more than twice that reported for any of the other low-elevation sites.

Differences in hourly ozone concentration among sites were least in predawn (±20 ppb), whereas variability in midafternoon often exceeded 40–50 ppb. Although there were sites in which the ranking between predawn

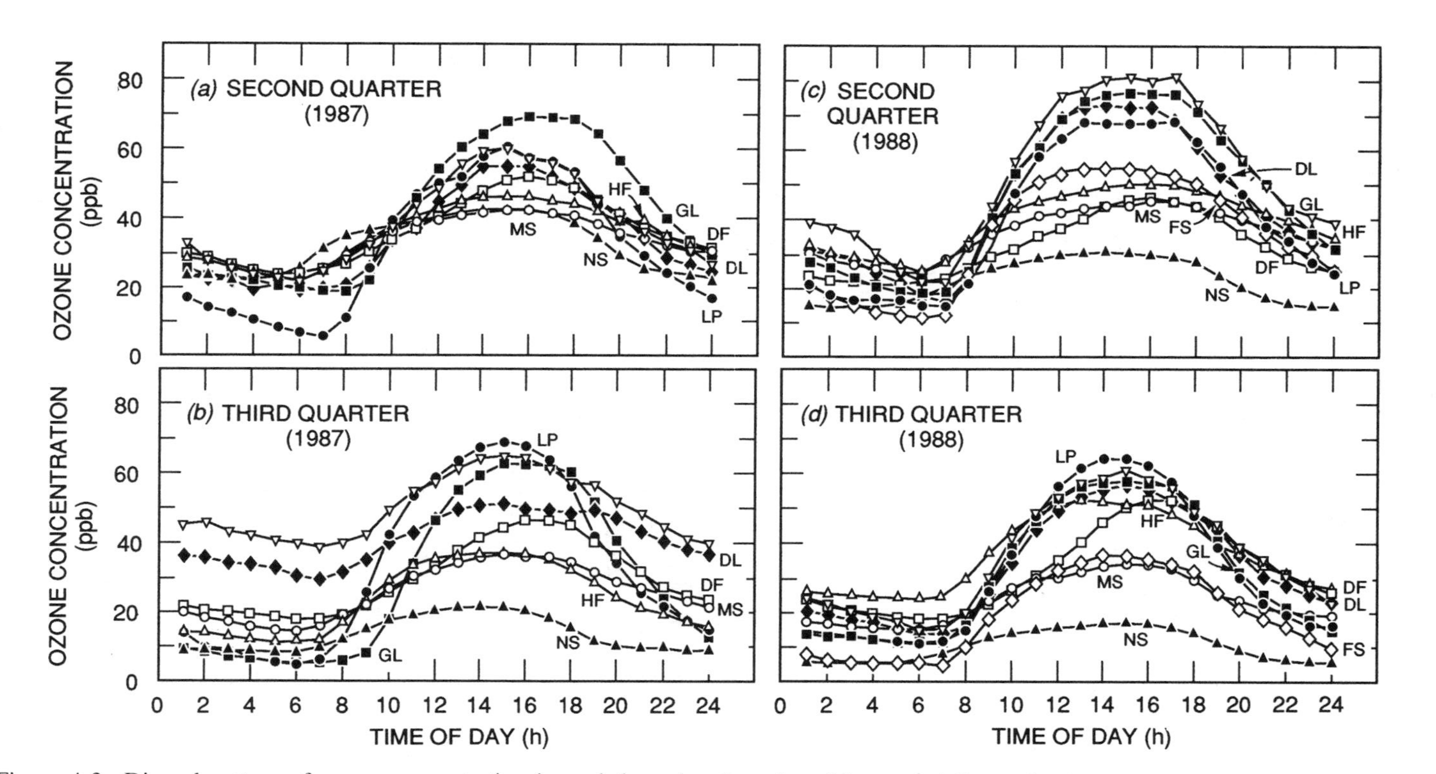

Figure 4.3. Diurnal pattern of ozone concentration in each low-elevation site of Integrated Forest Study. Data are aggregated by quarter for 1987 and 1988. Abbreviations: HF, Huntington, New York; LP, Oak Ridge, Tennessee; DL, Duke Forest, North Carolina; DF, Thompson Forest, Seattle, Washington; NS, Nordmoen, Norway; GL, Grant Forest, Macon, Georgia; MS, Maine Forest, Howland, Maine; Florida Forest, Gainesville, Florida.

minimum and midafternoon maximum were consistent (e.g., NS, FS), there were notable exceptions. For example, sites with the maximum midday concentrations in the third quarter of 1987 (GL and LP; Figure 4.3b) also had the lowest predawn minima and were nearly equivalent in their predawn values to those observed at NS.

Whereas the general pattern of a pronounced diurnal cycle in ozone concentration was common to all the lower elevation sites of the Integrated Forest Study, there were many features that were either site specific or region specific. For example, in those sites exhibiting the least amplitude in maximum to minimum concentrations (e.g., NS, MS, FS), the shape of the diurnal pattern was nearly symmetrical, with the rate of midmorning rise being equivalent to the corresponding rate of decline in the late afternoon to early evening. In each of these sites, the plateau in ozone concentration commonly was achieved early in midmorning (~1000 hours) and subsequently was extended through the late afternoon (~1800 hours), resulting in a midafternoon plateau that was broad rather than accentuated (e.g., Figure 4.3b–d).

Those sites with a very pronounced diurnal pattern (i.e., GL, LP, and DL) and in which the midafternoon maximum exceeded 40 ppb exhibited significant variability, best characterized by the amplitude in the midafternoon ozone peak. Unlike the sites with a less pronounced pattern, the diurnal patterns at GL, LP, and DL were asymmetric because the ratio of midmorning increase to evening decline approached a factor of 2. The maximum concentrations were typically abbreviated in duration, extending only from 1400 to 1600 hours, although the concentrations remained elevated above the predawn minima at most sites until late evening (2200 hours).

Although these observations indicate some common underlying mechanisms in ozone dynamics at multiple latitudes and longitudes, there were some site-specific features. For example, some sites showed marked peak values (e.g., LP in third quarter of 1987 and 1988), whereas others exhibited more of a plateau (e.g., GL in second and third quarters of 1988; DL in third quarter of 1987). The time of day during the highest concentrations was site dependent, with some occurring between 1200 and 1600 hours, whereas others extended well into the evening hours and approached 1800 to 2000 hours (e.g., DL in 1987 and 1988, Figure 4.3b and c; GL in third quarter of 1987, Figure 4.3b).

Differences and similarities among the lower elevation sites were reflected in many of the summary statistics (Table 4.1). The least diagnostic was the 24-hour mean, which only ranged from a minimum of 11 ppb (NS in third quarter of 1988) to maxima of 54 ppb (DL in second quarter of 1988). Within any given quarter, the absolute difference in maximum-to-minimum concentrations was only 10 ppb in the second quarter of 1987 and less than or equal to 32 ppb in each of the remaining three quarters (third in 1987 and second and third in 1988). The corresponding variability among the North American sites was less, being less than or equal to 28 ppb in all

Table 4.1. Summary Statistics for 11 Integrated Forest Study Sites. All statistical units are in ppb.

			High-Elevation Sites			Low-Elevation Sites							
Statistic	Year	Quarter	WF	ST	CP	HF	MS	LP	DF[a]	GL	FS	DL	NS
24-h (ppb)	1987	2	42	54	50	36	34	42	36	32	42	38	32
		3	45	53	47	24	26	29	30	33	29	52	14
	1988	2	49	71	61	40	36	40	32	47	35	54	22
		3	44	59	57	37	24	32	32	32	20	38	11
12-h (ppb)	1987	2	43	52	48	42	39	53	43	46	53	48	40
		3	44	51	44	32	32	44	36	52	44	59	18
	1988	2	50	70	59	46	41	57	39	63	48	69	28
		3	43	57	54	46	30	47	39	47	29	51	15
7-h (ppb)	1987	2	42	49	47	42	39	50	41	48	50	52	41
		3	43	49	42	33	31	41	34	54	41	50	20
	1988	2	49	68	59	46	41	58	37	64	51	75	29
		3	43	55	51	48	30	51	36	48	30	54	16
1-h max (ppb)	1987	2	104	99	85	88	69	112	103	99	—[b]	100	75
		3	114	95	95	76	76	105	94	102	—[b]	124	32
	1988	2	131	119	104	106	90	104	103	127	84	115	53
		3	119	120	100	91	71	122	140	116	70	141	30
SUM60 (10^3 ppb)	1987	2	13.2	57.1	32.4	9.8	1.9	39.5	10.7	26.1	—[b]	29.2	2.4
		3	30.1	34.3	24.1	5.4	3.8	24.3	10.3	31.3	—[b]	—[b]	0
	1988	2	33.5	126.3	81.6	19.2	8.1	26.4	8.1	53.1	23.4	—[b]	0
		3	22.6	74.7	63.6	18.6	1.7	19.7	13.5	24.1	1.9	52.9	0
SUM60 (10^3 ppb)	1987	2	2.5	10.9	2.6	0.9	0	13.5	3.6	5.1	—[b]	7.8	0
		3	11.8	8.8	2.4	0.2	0	9.0	2.1	10.3	—[b]	—[b]	0
	1988	2	13.9	61.2	18.5	6.1	2.9	9.8	2.3	21.9	0.5	—[b]	0
		3	10.4	22.2	19.8	2.7	0	7.7	6.7	7.4	0.1	23.4	0

[a] For the analysis of ozone data, the DF and RA sites were synonymous.
[b] Data were insufficient to calculate statistic.

quarters. The differences in 7-hour daylight means were more pronounced for each quarter, with differences per quarter approaching 40–45 ppb in three of four quarters. The second highest 1-hour maximum values were the most variable statistic, ranging by a factor of ~1.6 (69–112 ppb) in the second quarter of 1987 to a factor of ~5 (30–141 ppb) in the third quarter of 1988. The pattern of enhanced discrimination among sites in their ozone statistics with increasing attention to the periods of highest pollutant concentration (i.e., daylight 7-hour mean or 1-hour maximum versus 24-hour mean) was reflected in the cumulative threshold statistics of SUM60 and SUM80 (see Table 4.1).

With the exception of the second quarter of 1987, the minimum value for each cumulative statistic was consistently observed at NS and was zero with the exception of the second quarter of 1987. None of the low-elevation North American sites exhibited SUM60 values less than $1.7 \cdot 10^3$ ppb (MS in third quarter of 1988). The upper limit of the SUM60 statistic ranged from $39.5 \cdot 10^3$ ppb (LP in second quarter of 1987) to $53 \cdot 10^3$ ppb (GL in second quarter and DL in third quarter of 1988) and was consistently observed at one of three sites in southeastern North America (GL, LP, and DL). The absolute differences among sites in SUM60 were less than $22 \cdot 10^3$ ppb for all quarters. The lower limit of the SUM80 statistic was 0 (NS for each quarter of both years and MS in three of four quarters of both years). Several of the North American sites (i.e., HF, DF, and FS) reported SUM80 values across both quarters of 1987 and 1988 less than $3 \cdot 10^3$ ppb. The differences between quarters in these indices of episodically high ozone levels were not consistent across sites: for HF, MS, and LP, second-quarter values exceeded those of the third quarter across both years, whereas a pattern was not evident at either DF or GL.

Temporal Patterns among the Higher Elevation Sites

In contrast to the pronounced diurnal amplitude in ozone concentrations at most of the low-elevation sites, the diurnal concentrations at each of three higher elevation sites (elevation >900 m; CP, ST, and WF) were less variable, with most of the concentrations at each site being ±5 ppb of the site's 24-hour mean irrespective of the quarter or year (Figure 4.4a–4.4d). The less pronounced diurnal variation in ozone concentrations at the high-elevation sites was not a statistical artefact accountable to the aggregation procedure by month and quarter. However, this complacency was not observed across days. For example, in the second quarter of 1987 at both the southern (ST) and northern (WF) sites, daily 24-hour means ranged from 29–82 and 17–75 ppb, respectively; the interday variability in ozone concentrations was significantly greater than the corresponding intraday statistic, a contrasting pattern to that reported at all low-elevation sites. Thus, the commonly reported pattern of diurnal complacency in the ozone concentrations at high

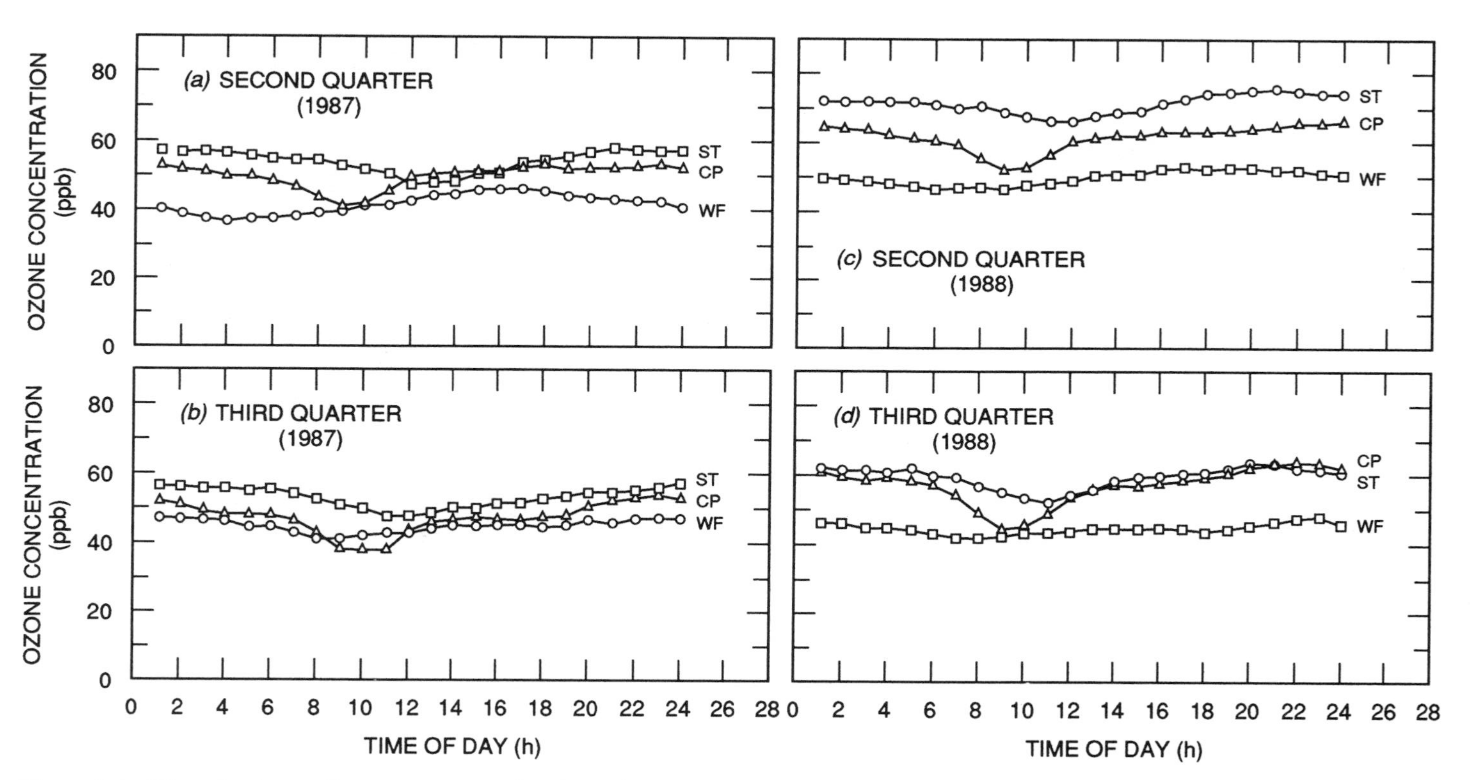

Figure 4.4. Diurnal pattern of ozone concentration in each high-elevation site of Integrated Forest Study. Data are aggregated by quarter for each year. Abbreviations: WF, Whiteface, New York; ST, Smoky Mountains National Park, North Carolina; CP, Coweeta, North Carolina.

elevations was solely applicable to an individual day; complacency was not a long-term feature of the ozone concentrations at high elevations.

Several additional features of the diurnal ozone patterns at high-elevation sites were reported. First was the similarity in absolute ozone concentration among the three sites in spite of their differences in geographical location. In most cases, the quarter's 1-hour means were within 10 ppb of one another over the entire diurnal period. Thus, the intersite variability in ozone concentration was much less than that reported at low elevations over the same spatial scale. Second was the persistent predawn-to-midday decline in ozone concentration at each site (Figure 4.4a–d). Many features of this phase were site- and quarter specific and related to the time of day, rate of decline, rate of recovery, duration, and amplitude. The most pronounced and consistent decline was that at CP, at which the minimum at each quarter was at least 10 ppb below the daily maximum. The morning decline in ozone concentration in both years was more pronounced in each of the more southern sites (CP and ST) in comparison with that reported at WF. The highest quarterly 1-hour mean concentrations occurred at ST (Figure 4.4a–d), with the exception of the third quarter of 1988 in which the maxima were nearly equivalent for both the southern sites (CP and ST). The concentrations at WF were consistently lower by 5–10 ppb.

The diurnal pattern of declining ozone concentration in midmorning is attributed to the updrafting of ozone-depleted air from the lower elevations with the daily convective turbulence created with solar radiation. The site-specific features of the decline phase are attributed to local topography, regional pollutant burdens, and the degree of nighttime ozone depletion in lower elevations (Guderian 1985; Altshuller 1988).

The temporal patterns of ozone on quarterly or annual time steps exhibited less definitive patterns. For both CP and ST, ozone concentrations were higher in the second versus the third quarter and markedly so for 1988. This pattern was similar to that reported for the most proximal low-elevation site (LP), which is less than 100 km to the west. At all sites in the second quarter of 1988 (Figure 4.4c), the 24-hour ozone concentrations exceeded by at least 20% the corresponding concentrations during the other quarters in both years. This pattern was also observed at the nearest low-elevation site (LP). The tendency for average ozone concentrations to be higher in the second versus the third quarter of the year is typical of many rural sites (Evans et al. 1983; Taylor and Norby 1985; Altshuller 1988) and is attributed to either stratospheric intrusions (Singh et al. 1978) or an increasing frequency of slow-moving, high-pressure systems that promote the formation of ozone (Altshuller 1988).

The summary statistics provide additional information regarding the differences and similarities among the three high-elevation sites (see Table 4.1). The 24-hour and 12-hour means indicated lower concentrations at WF in comparison with either CP or ST. The mean 12-hour and 24-hour ozone concentrations were consistently highest at the ST site across quarters and years. The 7-hour means showed less variation among sites, and without exception this daylight mean concentration was equal to (WF) or less than (ST and CP) the corresponding 12-hour and 24-hour means for the second and third quarters. As a consequence, the highest mean ozone values by quarter occurred in periods outside the daylight 0900–1600 hour window.

The second highest 1-hour maximum values were extremely variable spatially and temporally, and the pattern was not always comparable to that for the other statistics (see Table 4.1). For example, in three of the four quarters the highest maximum concentrations occurred at WF, which exhibited the lowest longer term means. Between the other two sites, the maximum concentrations at the ST sites were consistently greater than or equal to those at CP. The differences between years were substantial and consistent at all sites, with higher values in each quarter of 1988 versus 1987. Whereas these maximum values were higher on average than the combination of all low-elevation sites, the maximum for the two southern high-elevation sites (CP and ST) were not significantly different from that of the nearest low-elevation site (LP).

The cumulative SUM60 statistic differed by as much as a factor of 4 within a given quarter at the high-elevation sites, with the order of concentrations being ST greater than CP greater than WF (see Table 4.1). The pattern for the SUM80 statistic was less consistent among the three sites. With only one exception for both the SUM60 and SUM80 statistic, the values were higher in the second versus the third quarter of both years. These consistent differences between quarters indicate that periods of episodically high ozone levels at high-elevations in both northern and southern locations in eastern North America were more frequent in the earlier part of the growing season.

Spatial Patterns among Sites and Underlying Processes

The most pronounced spatial difference in ozone was that due to elevation, and the effect was manifest in both the diurnal ozone behavior as well as the various summary statistics. The unique feature of ozone's diurnal behavior among the high-elevation sites of the Integrated Forest Study relative to that reported previously (Singh et al. 1978; Guderian 1985) is the tendency for ozone to be depleted in midmorning. Moreover, this pattern contrasts with the corresponding rapid rise in ozone concentration in low-elevation sites within the same region. Ozone concentrations at both high and low elevations exhibited a diurnal behavior (albeit contrasting), but the daily fluctuations in concentration at low elevations tended to be several times greater than that at high-elevations. As already noted, the second major difference was the source of variation in ozone concentration in which the interday variability was far more significant at the low-elevation sites.

The dissimilarity in patterns of ozone clearly indicates pronounced differences in the mechanisms underlying ozone dynamics on a diurnal basis at high and low elevations. The behavior at low elevations is assumed to have been driven by photochemistry of ozone formation, precursor transport, and pollutant scavenging, in a fashion analogous to that documented for urban centers (Guderian 1985; EPA 1986; Seinfield 1989) as well as downwind regions (Altshuller 1988). The mechanism underlying the contrasting and less variable pattern at higher elevations is not fully resolved; candidate processes proposed include diminished ozone scavenging at high-elevations from nitric oxide and long-range transport and confluence of polluted air masses across a region into the high-elevations (EPA 1986). Guderian (1985) and Altshuller (1988) proposed that the reduced availability of foliar sites for ozone scavenging in high versus low elevations also contributed to the flat diurnal profile, but this hypothesis is not supported from the data in the Integrated Forest Study (see Chapter 3), which demonstrate high indices of leaf area at the higher elevations. Clearly, the atmospheric reservoir of the troposphere at these sites is uncoupled somewhat from the daytime photochemistry/scavenging in the surface boundary layer at the lower elevation sites.

The inverted midmorning reduction in ozone concentration at the Coweeta site across all quarters (see Figure 4.4a–d) is most likely driven by a somewhat isolated atmospheric reservoir in the lower levels of the free troposphere at this elevation, but one that is punctuated in the midmorning by updrafting of ozone-depleted air from the lower elevation's surface boundary layer.

The statistical data in Table 4.1 convey an array of patterns as a function of elevation, and the patterns differed between the northern (WF) and southern (CP and ST) mountain locations. In comparing the more northern locations (WF, HF, and MS), the values for the multiple ozone statistics at the high-elevation WF site were consistently greater than or equal to the values at HF and MS. This disparity was most pronounced for the 24-hour mean, 1-hour maximum, and cumulative statistics (SUM60 and SUM80). The pattern was still evident for the 12-hour and 7-hour means, although it was not as consistent over time and not as pronounced. The pattern in all the southern locations was more variable when comparing the two high-elevation sites (CP and ST) and the most proximal low-elevation sites (LP, GL, and DL). Specifically, for each of the multiple-hour statistics (24-hour, 12-hour, and 7-hour means), the values at the high-elevation sites were greater than or equal to those at the low-elevation sites. Conversely, the 1-hour maximum values at ST and CP were less than or equal to those at LP, DL, and GL. The cumulative statistics showed specific patterns with the SUM60 at high elevations being equal to or greater than that at low elevations whereas the converse was reported for SUM80. This indicates that cumulative high exposure levels of ozone in excess of 80 ppb are more frequent or prolonged at the lower elevation sites.

The significance of these comparisons must be evaluated in the context of physiological ecology of forest trees, particularly with regard to the statistics that are commonly endorsed as being the most robust in developing dose–response relationships (i.e., 7-hour mean, SUM60 and SUM80). Specifically, the tendency for the maximum quarterly means at high-elevations to occur outside the photoperiod indicates that the time period for the cumulative ozone statistics does not coincide with the period of maximum physiological activity of critical importance in governing ozone toxicity (Tingey and Taylor 1982). The SUM60 and SUM80 statistics were accumulated in low elevations during daylight hours, whereas the same statistic at high-elevations encompassed late-evening and early-morning hours. This is particularly important because stomatal conductance, the dominant factor governing ozone diffusion from the atmosphere to the physiologically sensitive sites in the leaf interior (Tingey and Taylor 1982), is likely to be lower in the evening and predawn hours for most species. However, it is important to recognize that stomatal conductance, even in the absence of light at high-elevations, may be enough to allow for ozone uptake at a rate sufficient to elicit injury (Winner et al. 1989).

Elevation was not the only spatial variable correlated with patterns in ozone concentration. Focusing only on sites in eastern North America (low and high-elevation), the correlation between latitude and the multiple ozone statistics revealed a number of associations (Figure 4.5). While negative correlations were documented for each of the summary statistics in both years, the strongest correlation coefficients were evident for the SUM60 (−0.74), 7-hour mean (−0.66), and 12-hour mean (−0.79) in 1987. The

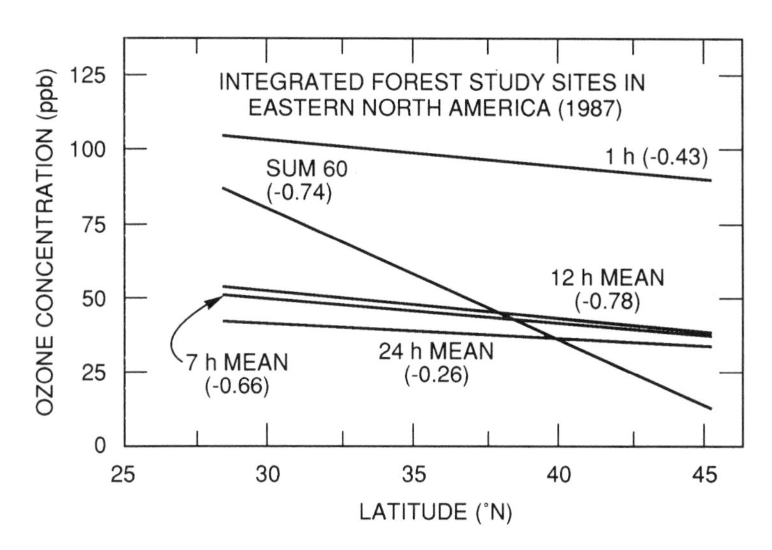

Figure 4.5. Relationship (linear regression analysis) between latitude and multiple 1987 ozone statistics in eastern North American sites of Integrated Forest Study, including both high and low elevations. Parenthetical data are correlation coefficients.

relationships for both the second highest 1-hour maximum and 24-hour means were also negative, but the coefficients were less than or equal to −0.43. Thus, with decreasing latitude from the north to south in eastern North America, all ozone statistics increased, with the greatest rate of increase being observed for SUM60 (Figure 4.5). Based on the associations in Figure 4.5, for each 10° shift in latitude from north to south the SUM60 and 1-hour mean ozone concentrations increased $43 \cdot 10^3$ and 10 ppb, respectively, whereas the corresponding increase in the 7-hour and 24-hour means were only 7 and 4 ppb, respectively. The underlying reason for a strong correlation with latitude lies in the associated pattern in solar radiation, which plays a dominant role in ozone formation.

Equally revealing in establishing spatial patterns in ozone among sites was the relationship between the multiple statistics and the historical frequency of air quality inversion events (Holzworth 1972). The inversion potential was simply defined as the sum of days during a 5-year period (1960–1964) in various regions of North America in which atmospheric conditions were indicative of inversions in the planetary boundary layer. Figure 4.6a shows the correlation between inversion potential and the ozone statistics for low-elevation sites in North America for 1987 only; similar patterns were documented for 1988 and with the inclusion of the high-elevation sites (for both years), although the associations were not as strong. In 1987, each of the statistics was positively correlated with inversion potential, with the strongest relationship being that for SUM60 (+0.80) and the second-highest 1-hour maximum (+0.78; Figure 4.6b). The most positive slope was that exhibited by the SUM60 statistic whereas the least was that for the 24-hour mean (Figure 4.6a). Given the relationship in Figure 4.6a, a doubling of the frequency of air quality inversion from 10 to 20 days would result in a $15 \cdot 10^3$ ppb and 10-ppb increase in the SUM60 and 1-hour maximum, respectively with far smaller increments approaching 2 ppb for both the 7-hour and 24-hour means.

Relationships among Statistics

The issue of which ozone statistic is the most appropriate to characterize ozone in forested landscapes is particularly important in light of the diversity of averaging times currently employed in the disciplines of atmospheric chemistry (24-hour and 12-hour means), plant sciences (7-hour daylight mean, SUM60 and SUM80), and human health (1-hour maximum). In urban areas, there is a strong correlation between the 1-hour maximum and longer term mean ozone statistics such that the maximum is a valid surrogate for addressing the effects of ozone on human health. Given the dissimilarity in ozone patterns between urban and rural landscapes (Guderian 1985; Altshuller 1988), the data from the Integrated Forest Study sites can provide a test of the hypothesis that the 1-hour maximum is not be a valid surrogate for characterizing ozone concentrations in more remote, rural landscapes.

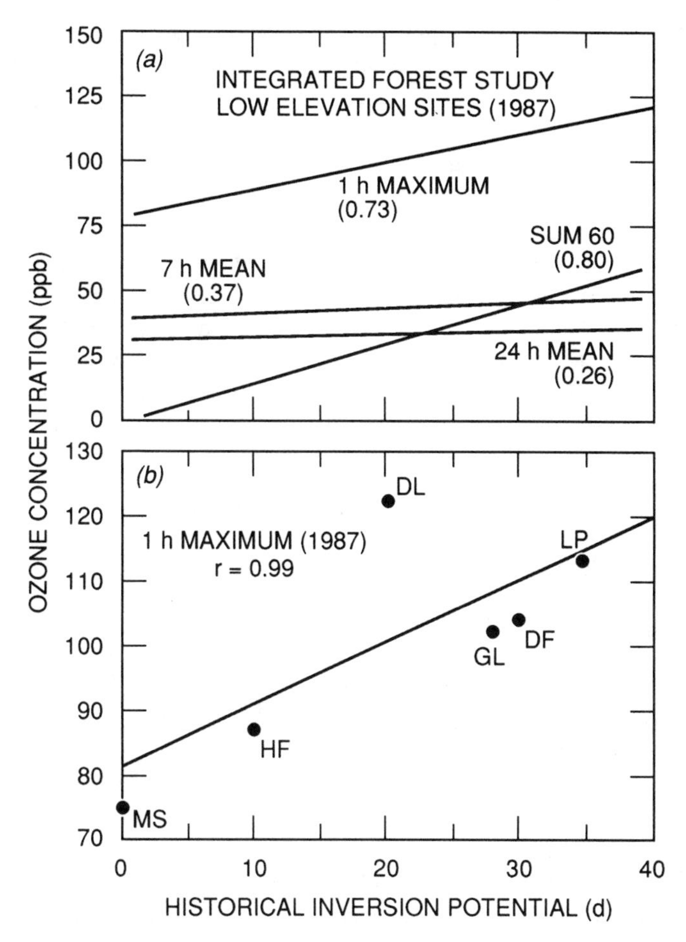

Figure 4.6. (a) Relationship (linear regression analysis) between multiple ozone statistics and historical frequency of atmospheric inversion frequency among North American low-elevation sites of Integrated Forest Study in 1987. Parenthetical data are correlation coefficients. (b) Scatter diagram is singular relationship (and corresponding linear regression line) for second-highest 1-h maximum value in 1987.

In both 1987 and 1988, the correlation coefficients among the statistics were all positive but ranged from very weak (0.26 for 24-hour mean and 1-hour maximum in 1987) to strong (0.98 for 7-hour and 12-hour mean in 1988). The strongest correlations were reported for the following pairs of statistics: 7-hour and 12-hour means (0.92), 24-hour mean and SUM60 (0.86), and SUM60 and SUM80 (0.84). Weak correlations (<0.50) were reported for all associations between the 1-hour maximum and most of the remaining statistics including the 24-hour mean (0.33), SUM60 (0.29), and SUM80

(0.39). Of the factors influencing the degree of correlation among the ozone statistics at the sites of the Integrated Forest Study (quarter, year, and elevation), interannual variation in ozone was far more important than either elevation or quarter.

The relationship among statistics indicated strong correlation of the cumulative exposure indices (SUM60 and SUM80) with the longer term averages commonly used in the disciplines of atmospheric chemistry (24-hour and 12-hour means) and plant sciences (7-hour daylight mean). Conversely, there was only a weak correlation between the index of peak concentration (second highest 1-hour maximum) and all other indices of ozone exposure. Thus, unlike that observed in urban areas, peak ozone concentrations are not appropriate surrogate statistics with which to characterize longer term exposure to ozone in forested landscapes.

Joint Occurrence of Ozone, Nitrogen, and Sulfur

Based on air quality data from intensively managed agricultural areas and associated dose-response relationships, the joint occurrence of multiple pollutants (i.e., sulfur dioxide, nitrogen dioxide, and ozone) is not regarded as a significant issue with respect to crop productivity in North America (Lefohn et al. 1987). One of the principal bases for this conclusion is the absence of cooccurrences in rural areas of ozone with either nitrogen dioxide or sulfur dioxide at levels sufficient to elicit changes in physiological function. The air quality data bases used in these analyses (Lefohn et al. 1987) were restricted to low-elevation sites and included a number of sites that were either urban or suburban (Lefohn et al. 1987).

This conclusion may not be valid in natural ecosystems because of cumulative effects of atmospheric deposition, either directly on physiological processes of forest trees or indirectly through changes in stand-level biogeochemistry. This argument is supported by the correlation of ozone concentrations among the sites of the Integrated Forest Study with data on dry deposition of sulfur (see Chapter 5) and nitrogen (see Chapter 6).

With respect to dry nitrogen deposition (mol nitrogen ha^{-1} yr^{-1}), principally in the form of nitric acid vapor, there was an extremely weak correlation between 1-hour maximum ozone concentration and dry nitrogen deposition (+0.28) but a substantially stronger correlation ($\geq$+0.60) for each of the remaining longer term statistics (e.g., 24-hour, 12-hour, and 7-hour means, and SUM60) (Figure 4.7). The slope was greatest for SUM60, indicating that for every additional 200 mol nitrogen ha^{-1} yr^{-1} in dry deposition, an increment in SUM60 of $18 \cdot 10^3$ ppb occurred. A similar pattern was evident for sulfur deposition (principally as sulfur dioxide and sulfate aerosol as eq SO_4^{-2} ha^{-1} yr^{-1}) but with a corresponding smaller increment of $5 \cdot 10^3$ ppb. Consequently, the forest landscapes with the highest loadings of sulfur and nitrogen via dry deposition tended to be the same forests with

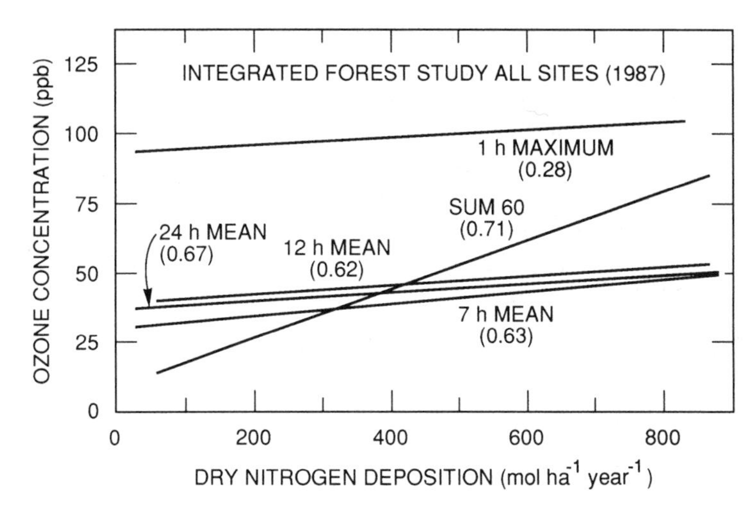

Figure 4.7. Relationships (linear regression analysis) between multiple ozone statistics and dry deposition of nitrogen (principally nitric acid vapor) among all sites of Integrated Forest Study. Parenthetical data are correlation coefficients.

the highest average ozone concentrations (e.g., ST, WF, LP, DL, and GL). Moreover, the most physiologically robust statistic of SUM60 exhibited the strongest correlation ($>+0.60$). These data indicate that joint occurrences of multiple pollutants in forested landscapes are important.

Physiological and Ecological Role of Ozone

The potential ecological significance of tropospheric ozone in the terrestrial ecosystems of the Integrated Forest Study is based on the documented effects of this gas on the physiology and growth of sensitive plant species in both natural and managed ecosystems (EPA 1986). Ambient levels of ozone are reported to influence the growth and productivity of major crop species in several regions of North America, including the Northeast and Southeast (Heck et al. 1988). Conversely, ozone effects on the productivity of forest trees are not as well documented. On the basis of controlled exposure studies, Reich and Amundson (1985) concluded that ambient levels of ozone were sufficient to inhibit net photosynthesis of several low elevation woody species, including *Pinus strobus* and *Acer saccharum,* and the study by Swank and Vose (in press) supports the sensitivity of *P. strobus*. Experimental studies of direct ozone effects under controlled field conditions lend credence to the hypothesis that some woody plants are responsive to ambient levels of the pollutant, with the most substantive effects reported for *Pinus taeda* in the Southeast (Adams et al. 1988; McLaughlin et al. 1988; Edwards et al., in press). Whereas corresponding studies on high-elevation species are limited, the data indicate that the physiology and growth of *Picea rubens* and *Abies fraseri* are not responsive to direct effects of ozone at near-ambient levels (Johnson and Taylor 1989).

Using boundary-layer analysis, Taylor and Norby (1985) proposed that sustained daylight ozone concentrations in excess of 50 ppb were sufficient to affect the growth of sensitive species in natural ecosystems. Based on this threshold, the daylight 7-hour mean ozone concentrations at WF, HF, MS, DF, and NS in either quarter of 1987 and 1988 (see Table 4.1) were insufficient to influence the growth of sensitive indigenous species. Many of the other locations exhibited at least one quarterly mean value exceeding this threshold, with the highest values at ST and CP (the higher elevation sites in the Southeast). At the lower elevation sites in North America, the pattern was less clear, but at least one of the quarterly 7-hour mean concentrations during the 2-year period exceeded the threshold for sensitive species at LP, GL, DL, and FS. Thus, the data on direct ozone effects in conjunction with the air quality measurements at the sites of the Integrated Forest Study indicated that ozone concentrations (particularly in the southern low-elevation sites) were high enough to influence the growth of sensitive species. In high-elevation sites, the ozone air quality was above the 50-ppb threshold for sensitive species, but the experimental data demonstrated that the most dominant species, *P. rubens* and *A. fraseri,* are resistant to direct ozone effects.

At high-elevation sites there is emerging evidence to suggest that indirect effects of ozone are more important than direct effects in governing the growth and physiology of high-elevation conifers (Johnson and Taylor 1989). The principal observation is that prolonged ozone exposure during the growing season inhibits early autumn hardening, thus promoting frost damage in ozone treated trees (Brown et al. 1987; Barnes and Davison 1988). Moreover, the ozone-induced delay in hardening is reported to be concentration dependent (Lucas et al. 1988). Preliminary data with *P. rubens* (R. Amundson, personal communication) indicated that carbohydrate metabolism is disrupted and foliar raffinose accumulation in the autumn is delayed with increasing ozone exposure; reduction in raffinose is correlated with higher incidences of bud mortality under natural conditions for some conifers. The same mechanism is proposed to affect the onset or degree of frost hardiness in *P. rubens*. The putative role of indirect ozone effects in high-elevation forest is particularly important in light of the marginal carbon budgets for many woody species in these ecosystems (McLaughlin et al., in press), potential long-term effects of anthropogenic pollutants on stand biogeochemistry, and the significance of regional climate in governing the growth of high-elevation forests (Johnson et al. 1988).

Summary

Patterns of ozone among the Integrated Forest Study sites indicate that the forests experienced very marked quantitative and qualitative differences in tropospheric ozone. The principal spatial factors underlying this variation were elevation, proximity to anthropogenic sources of oxidant precursors, regional-scale meteorological conditions, and airshed dynamics between the lower free troposphere and the surface boundary layer. The results do not support the hypothesis that patterns in ozone concentrations are always comparable across either large (e.g., $>10^5$ km^2) or small ($<2 \cdot 10^2$ km^2) geographical scales, particularly in areas where topographic features are significant.

Ozone concentrations were highly variable temporally on scales ranging from diurnal to interannual. The diurnal patterns were elevation dependent: at low elevations, the pattern was characteristic of daytime photochemical processes, showing a predawn minimum, rapidly increasing ozone concentration in midmorning, and an afternoon-to-early-evening peak or plateau in ozone concentration. Mean daylight values (7-hour mean) ranged from a low of 16–40 ppb at NS, to 30–50 ppb at HF, MS, DF, and FS, and to 40–75 ppb at LP, DL, and GL. At high elevations, the pattern was not an entirely flat or complacent profile but rather exhibited a consistent early-to-midmorning diminution in ozone concentration, followed by a resurgence in the mid- to late afternoon. Mean daylight values (7-hour means) in the second and third quarter ranged from 40–50 ppb at WF to 40–70 ppb at CP and ST. Although the diurnal pattern at high elevations exhibited less intraday variability than that at low elevations, the interday variability at high elevations was more pronounced than that at low elevations. Thus, a flat or complacent diurnal profile in ozone concentration at high elevations was not observed at any of the high-elevation sites of the Integrated Forest Study. Seasonal variation in ozone concentration was a significant source of variation, with the highest concentrations tending to occur in the second versus the third quarter. However, the most important temporal source of variation influencing ozone concentrations was that between years, with concentration tending to be greater in 1988, particularly for the high-elevation sites. The statistics reflecting episodes of ozone (SUM60 and SUM80) were ambivalent with respect to elevation.

The spatial patterns in ozone concentration were scale dependent. Over distances of less than 100 km, very pronounced differences in the diurnal behavior of ozone were reported. Among low-elevation sites in eastern North America, the daily maximum ozone concentration decreased with latitude, and the highest peak ozone concentrations within a quarter occurred in the more southerly locations. This correlation was attributed to the region's propensity to experience air quality inversion episodes. Across all sites, the correlations between the 1-hour maximum and all other indices of ozone were weak, indicating that forests having the highest long-term average ozone concentrations were not necessarily those that experienced the highest peak ozone levels. The sites with the highest multiple-hour mean ozone levels were positively correlated with dry deposition of both sulfur and nitrogen, suggesting that the cooccurrence of multiple pollutant loadings may be more important in forests than agricultural landscapes.

References

Adams M.B., Kelly J.M., Edwards N.T. 1988. Growth of *Pinus taeda* L. seedlings varies with family and ozone level. Water Air Soil Pollut. 38:137–150.

Altshuller, A.P. 1988. Meteorology—atmospheric chemistry and long range transport. In Heck W.W., Taylor O.C., Tingey D.T. (eds.) Assessment of Crop Loss from Air Pollutants. Elsevier Applied Science, New York, pp. 65–89.

Barnes J.D., Davison A.W. 1988. The influence of ozone on the winter hardiness of Norway spruce [*Picea abies* (L.) Karst.]. New Phytol. 108:159–166.

Brown K.A., Roberts T.M., Blank L.W. 1987. Interaction between ozone and cold sensitivity in Norway spruce: a factor contributing to the forest decline in central Europe? New Phytol. 105:149–155.

Edwards N.T., Taylor G.E., Jr., Adams M.B., Simmons G.L., Kelly J.M. Ozone, acidic rainfall, and soil Mg effects on growth and foliar pigments of *Pinus taeda* L. Tree Physiol. (in press).

EPA (U.S. Environmental Protection Agency). 1986. Air Quality Criteria for Ozone and Other Photochemical Oxidants, Vol III. EPA-600/8–84–020A, U.S. Environmental Protection Agency, Washington, D.C.

Evans G., Finkelstein P., Martin B., Possiel N., Graves M. 1983. Ozone measurements from a network of remote sites. J Air Pollut. Control Assoc. 33:291–296.

Guderian R. 1985. Air Pollution by Photochemical Oxidants. Ecological Studies 52. Springer-Verlag, New York.

Heck W.W., Taylor O.C., Tingey D.T. (eds.) 1988. Assessment of Crop Loss from Air Pollutants. Elsevier, New York.

Heck W.W., Adams R.M., Cure W.W., Heagle A.S., Heggestad H.E., Kohut R.J., Rawlings J.O., Taylor O.C. 1983. Assessing impacts of ozone on agricultural crops. I. Overview. Environ. Sci. Technol. 17:572A–581A.

Hogsett W.E., Tingey D.T., Holman S.R. 1985. A programmable exposure control system for determination of the effects of pollutant exposure regimes on plant growth. Atmos. Environ. 19:1135–1145.

Hogsett W.E., Tingey D.T., Lee E.H. 1988. Ozone exposure indices:concepts for development and evaluation of their use. In: Heck W.W., Taylor O.C., Tingey D.T. (eds.) Assessment of Crop Loss from Air Pollutants. Elsevier Applied Science, New York, pp. 107–138.

Holzworth G.C. 1972. Mixing Heights, Wind Speeds and Potential for Urban Air Pollution Throughout the Contiguous United States. Office of Air Programs Publication No. AP-11, U.S. Environmental Protection Agency, Washington, D.C.

Johnson A.H., Cook E.R., Siccama T.G. 1988. Climate and red spruce growth and decline in the Northern Appalachians. Proc. Nat. Acad. Sci. U.S.A. 85:5369–5373.

Johnson D.W., Taylor G.E., Jr. 1989. Role of air pollution in forest decline in eastern North America—a review update. Water Soil Air Pollut. 48:21–43.

Lefohn A.S., Pinkerton J.E. 1988. High resolution characterization of ozone data for sites located in forested areas of the United States. J Air Pollut. Control Assoc. 38:1504–1511.

Lefohn A.S., Davis C.E., Jones C.K., Tingey D.T., Hogsett W.E. 1987. Co-occurrence patterns of gaseous air pollutant pairs at different minimum concentrations in the United States. Atmos. Environ. 21:2435–2444.

Lindberg S.E., Johnson D.W., Lovett G.M., van Miegroet H., Taylor G.E., Jr., Owens J.G. 1989. Sampling and Analysis Protocols and Project Description for the Integrated Forest Study. TM-11214, Oak Ridge National Laboratory, Oak Ridge, Tennessee.

Logan J.A. 1989. Ozone in rural areas of the United States. J. Geophys. Res. 94:8511–8532.
Lucas P.W., Cottam D.A., Sheppard L.J., Francis B.J. 1988. Growth responses and delayed winter hardening in Stka spruce following summer exposure to ozone. New Phytol. 108:495–504.
McLaughlin S.B. 1985. Effects of air pollution on forests: a critical review. J Air Pollut. Control Assoc. 35:512–521.
McLaughlin S.B., Adams M.B., Edwards N.T., Hanson P.J., Layton P.A., O'Neill E.G., Roy W.K. 1988. Comparative sensitivity, mechanisms and whole-plant physiological implications of responses of loblolly pine genotypes to ozone and acid deposition. ORNL/TM-10777, Oak Ridge National Laboratory, Oak Ridge, Tennessee.
McLaughlin S.B., Anderson C.P., Edwards N.T., Roy W.K. Seasonal patterns of photosynthesis and respiration of red spruce saplings from two elevations in declining Southern Appalachian stands. Can. J. For. Res. (in press).
Meagher J.F., Lee N.T., Valente R.J., Parkhurst W.J. 1987. Rural ozone in the southeastern United States. Atmos. Environ. 28:60–70.
Pinkerton J.E., Lefohn A.S. 1987. The characterization of ozone data for sites located in forested areas of the eastern United States. J. Air Pollut. Control Assoc. 37:1005–1010.
Rawlings J.O., Lesser V.M., Dassel K.A. 1988. Statistical approaches to assessing crop losses. In: Heck W.W., Taylor O.C., Tingey D.T. (eds.) Assessment of Crop Loss from Air Pollutants. Elsevier, New York, pp. 389–416.
Reich P.B., Amundson R.G. 1985. Ambient levels of ozone reduce net photosynthesis in tree and crop species. Science 230:566–570.
Seinfeld J.H. 1989. Urban air pollution: state of the science. Science 243:745–752.
Singh H.B., Ludwig F.L., Johnson W.B. 1978. Tropospheric ozone: concentrations and variabilities in clean remote atmospheres. Atmos. Environ. 12:2185–2196.
Swank W.T., Vose J.M. Watershed scale responses to ozone events in a *Pinus strobus* L. plantation. Water Air Soil Pollut. (in press).
Taylor G.E. Jr., Norby R.J. 1985. The significance of elevated levels of ozone on natural ecosystem of North America. In: Lee S.D. (ed.) International Specialty Conference on Evaluation of the Scientific Basis for Ozone/Oxidant Standard. Air Pollution Control Association, Pittsburgh, Pennsylvania, pp. 152–175.
Tingey D.T., Taylor G.E., Jr. 1982. Variation in plant response to ozone: a conceptual model of physiological events. In: Unsworth M.H., Ormrod D.P. (eds.) Effects of Gaseous Air Pollutants on Agriculture and Horticulture. Butterworth, London, pp. 113–138.
Winner W.E., Lefohn A.E., Cotter I.S., Greitner C.S., Nellessen J., McEvoy L.R., Atkinson C.J., Moore L.D. 1989. Plant response to elevational gradients of ozone exposure in Virginia. Proc. Natl. Acad. Sci. U.S.A. 86:8826–8832.

5. Sulfur Chemistry, Deposition, and Cycling in Forests

Overview

M.J. Mitchell and S.E. Lindberg

Sulfur is an important element in the biogeochemistry of forest ecosystems because of its role as an essential plant nutrient (Duke and Reisenauer 1986; Lambert 1986; Lambert et al. 1976; Turner and Lambert 1980) and because of the contribution of SO_4^{2-} as a counter ion in altering the flux of other elements, especially acidic (H^+ and Al^{3+}) and basic (Ca^{2+}, Mg^{2+}, Na^+, and K^+) cations in soil solutions (Johnson et al. 1982; Reuss and Johnson 1986). The role of SO_4^{2}, as well as the other mobile anions (NO_3^-, Cl^-, and HCO_3^-), in affecting both pool sizes and fluxes of cations is detailed in Chapter 8 (under Base Cation Distribution and Cycling).

The combustion of fossil fuels, especially coal, has caused the input of sulfur to the atmosphere to increase. Consequently, greater inputs by both wet and dry deposition of sulfur to ecosystems have resulted, both near combustion sources and within broad geographical regions associated with industrialization and other anthropogenic activities (NRC 1983). Although the temporal pattern of sulfur inputs to the atmosphere varies among regions, there was a rapid increase in eastern North America from the late 1800s until the mid-1900s with a gradual small decline since the 1970s (Hussar 1986; Dignon and Hameed 1989). This increased amount of atmospheric sulfur as sulfuric acid and SO_2 has resulted in greater deposition of sulfur to terrestrial ecosystems and their associated surface waters (Shriner et al. 1980). Aspects of the atmospheric chemistry, transport, and deposition of sulfur oxides that influence these trends are discussed later in Chapter 5.

For those watersheds that are most sensitive to the accelerated inputs of this mineral acid, marked changes in water chemistry have included increased SO_4^{2-} concentrations (Henrikson 1980; Brakke et al. 1989), reductions of acid-neutralizing capacities (Schnoor 1984; Schindler 1988), and mobilization of toxic metals such as Al (Schofield and Trojnar 1980; Cronan et al. 1986). Because most sensitive watersheds are found within forested regions, the processing of sulfur within forest ecosystems is of great importance to understanding the acidification of surface waters. Moreover, these changes are directly linked to processes regulated by the surrounding terrestrial ecosystems, including flux of basic nutrient cations and mobilization of toxic monomeric Al species. It has been hypothesized that these changes may be contributing to the decline of some forests (Johnson and Siccama 1983; Klein and Perkins 1988), but there is little direct evidence to support this contention (Foster, in press). Further linkages between biogeochemical cycling of sulfur and forest health are discussed in Chapter 13 (Regional Issues).

A generalized diagram depicting the major pools and fluxes of sulfur within a forest ecosystem is shown in Figure 5.1. The pathways and fluxes are discussed for the various IFS sites and other selected forest ecosystems, with

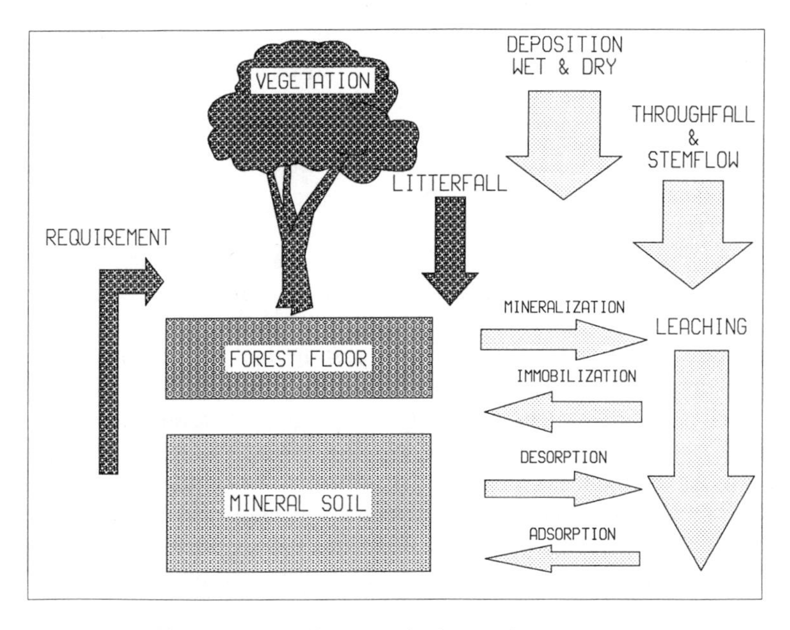

Figure 5.1. Sulfur cycle in forested ecosystems.

emphasis on similarities and differences associated with the sulfur cycling in these ecosystems and on the processes that influence these cycles.

Atmospheric Deposition and Canopy Interactions of Sulfur

S.E. Lindberg

Processes and Sources of Sulfur Deposition

Forest vegetation is an important receptor for airborne sulfur compounds produced from industrial activities because of the reactivity and scavenging efficiency of its large surface area canopy (Hosker and Lindberg 1982; Ulrich and Pankrath 1983). Plant leaves sorb gases having both beneficial and detrimental effects through a combination of active biological uptake and indiscriminant sorption (Murphy and Sigmon 1990). Foliage also provides a surface of interaction for substances that occur as aerosols or which have been dissolved in precipitation or cloud water. Scavenging of these materials is largely a physical process that is influenced by wind speed and turbulence, particle or droplet size, and canopy surface area (Lovett 1984; Davidson and Wu 1990).

The primary forms of airborne pollutant sulfur that are most important to deposition and its effects on forest systems are the oxides SO_2 and SO_4^{2-} (NRC 1983). Both are emitted directly from industrial sources, at a rate of approximately 15 Tg S yr^{-1} in North America, with SO_2 contributing more than 97% of the total (Tanner 1990). Airborne SO_2 undergoes several reactions, with the predominant gas-phase reaction being that with the hydroxyl radical to ultimately form H_2SO_4 (Calvert et al. 1985). Sulfuric acid is also formed by aqueous-phase oxidation in cloud water or precipitation, primarily by reaction with H_2O_2 (Calvert et al. 1985). Once formed, sulfuric acid may remain in solution where it can react with various compounds to form sulfate salts, or it can form fine-aerosol acid sulfate during cloud evaporation. This aerosol can, in turn, react to form various acidic, neutral, or basic sulfate compounds. SO_2 can also react directly with the surfaces of alkaline aerosols to form coarse-particle SO_4^{2-}. However, only about 20% of the SO_2 emitted in the United States is converted to SO_4^{2-} during its atmospheric lifetime (Tanner 1990), suggesting that SO_2 remains an important source of anthropogenic sulfur to the forest canopy.

Precipitation or wet deposition, dry deposition of SO_2 and aerosol SO_4^{2-}, and interception of cloud water or fog droplets containing SO_4^{2-} all contribute to the atmospheric deposition of sulfur. A schematic of these sources and their interactions with the forest canopy is shown in Figure 5.2. Dry-deposited SO_2 is the most likely species to have a significant biological interaction with internal foliar surfaces because it readily penetrates the stomata (Winner et al. 1985), which is less likely for both wet- and dry-deposited SO_4^{2-}. The relative importance of each deposition process, and the

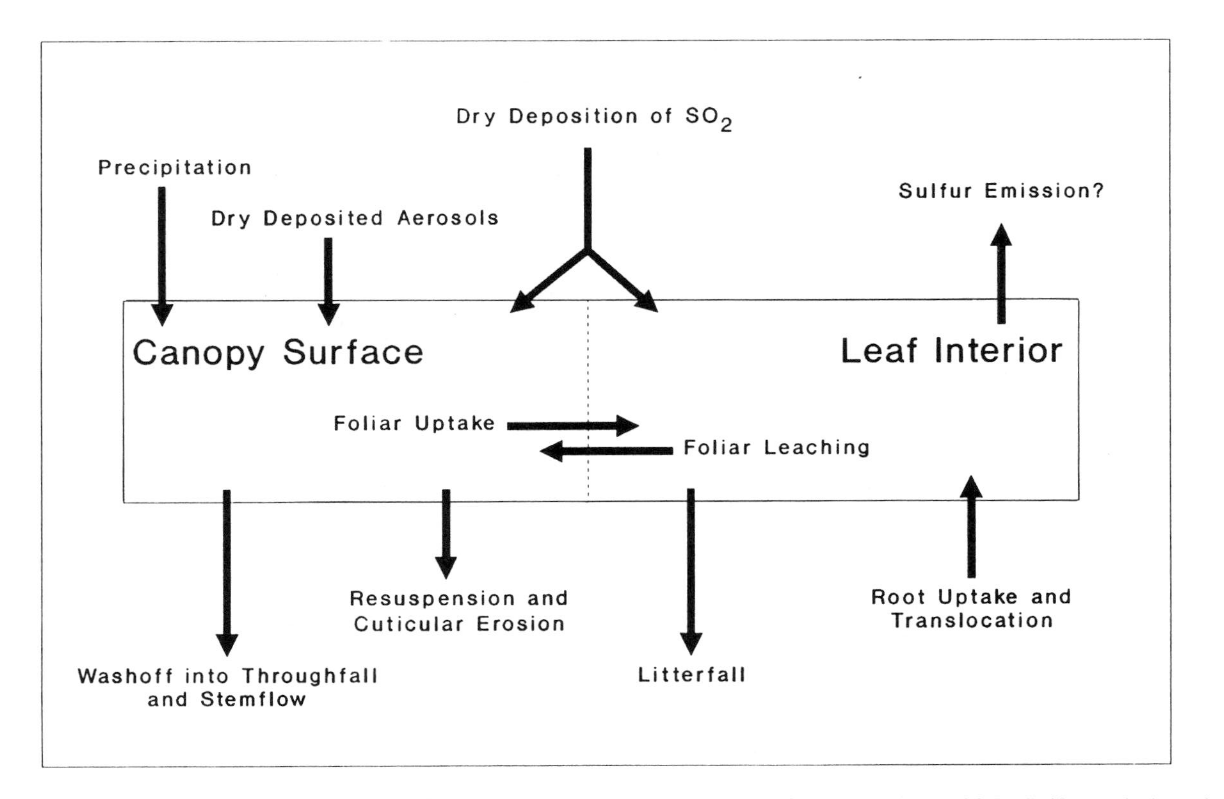

Figure 5.2. Schematic of atmospheric sulfur interactions with a forest canopy (adapted from Unsworth 1985). Sulfur emissions from trees, beyond the scope of the IFS, has been discussed in reference to one IFS site by Garten (1990).

degree of canopy interaction of airborne and deposited sulfur at the IFS forests, are described in this section through detailed comparisons of data from each site.

Atmospheric Deposition at IFS Forests

The Data Sets

Comparison of published studies of the deposition of airborne sulfur compounds to forests has been complicated by the wide range of sampling methods used, as reviewed in several recent books (Shriner et al. 1980; Winner et al. 1985; Legge and Krupa 1986; Hutchinson and Meema 1987; Unsworth and Fowler 1988; Johnson and Van Hook 1989; Adriano and Havas 1989; Lindberg et al. 1990a). In addition, the measurements have not consistently been accompanied by complete nutrient cycling studies. In this section, data from 13 of the 17 IFS nutrient cycling sites that made intensive deposition measurements (see Chapter 2 for site locations and codes) are compared through a series of figures ranking the sites from highest to lowest concentrations and fluxes. The data sets were collected between 1985 and 1989 using common protocols (see Chapter 2) and represent multiyear averages: 3 years for ST, LP, and CP (April 1986–March 1989); 2 years for DL, HF, WF, DF, RA, NS (mid-1986 to mid-1988), and GL (April 1987–March 1989); but only 1 year for MS and FS (mid-1988 to mid-1989). In addition, data from the TL site collected in 1984 as part of a related study were available for comparison. Data from the MS, FS, and TL sites must be considered cautiously in light of their more limited extent. Some aspects of the uncertainty in reported fluxes of sulfur at these and other sites are discussed below.

Trends in Airborne and Precipitation Sulfur

Total air concentrations of sulfur range from approximately 1.5 $\mu gS\ m^{-3}$ at the relatively unpolluted sites in Florida, Norway, and Washington (FS, NS, DF, RA) to about 8 in the mid-Southeast (LP, DL) (Figure 5.3). Between these values (~2–4 $\mu gS\ m^{-3}$) are several sites in the Southeast (GL, ST, CP) and Northeast (HF, MS, WF, TL). These patterns are not reflected in the precipitation data, which indicate that the highest concentrations (~40–50 $\mu eq\ L^{-1}$) occur at certain southeastern and northeastern United States sites (DL, LP, WF, GL) and at Norway, while the lowest (~15–25 $\mu eq\ L^{-1}$) occur at the Washington, Maine, and North Carolina (DF, RA, MS, CP) sites. Airborne sulfur at most of the IFS sites is dominated by SO_2, comprising an average of approximately 60% of the mass of S. The sites fall into two general categories, those for which SO_2 is about two-thirds or more of total airborne S (LP, DL, HF, GL, MS, TL, DF, RA), and those for which SO_2 is approximately less than half of airborne S (ST, CP, WF, FS, NS). As expected, fine-aerosol SO_4^{2-} is generally more important at

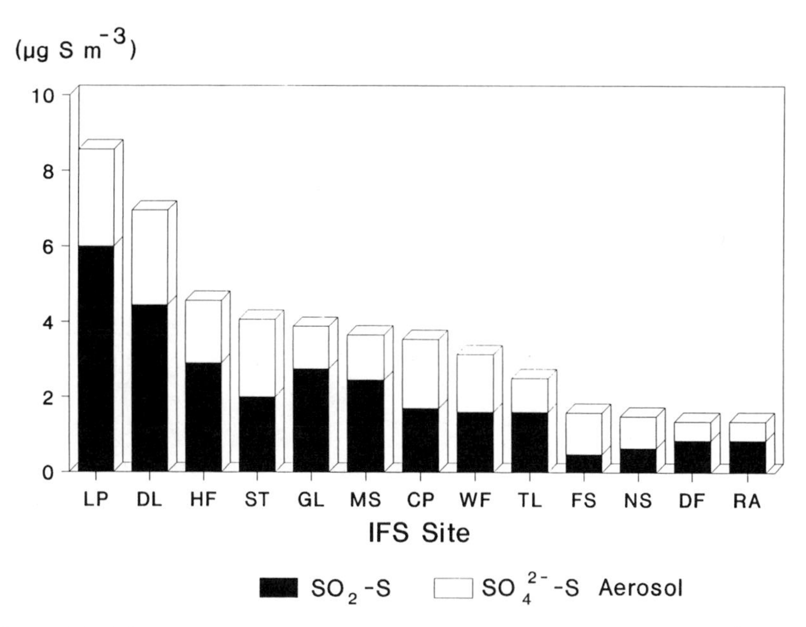

Figure 5.3. Mean atmospheric concentrations of SO_2 and fine-aerosol SO_4^{2-} across IFS sites for 1986–1989. Site codes are given in Chapter 2.1.

those sites that are more distant from local SO_2 emission sources. Fine-aerosol SO_4^{2-} is most strongly correlated with H^+ and NH_4^+, indicating a partially neutralized sulfuric acid aerosol. Sea salt SO_4^{2-} was an important component in precipitation (8%–16%) and aerosol SO_4^{2-} (~5%) only at the coastal sites in Washington and Florida (DF, RA, FS). At the near coastal sites (MS, DL, GL, and NS), sea salt comprised 2%–4% of precipitation SO_4^{2-} and 1%–2% of fine aerosol SO_4^{2-}. Sea salt was insignificant ($\ll$1%) at the remaining inland sites (LP, HF, WF, ST, CP). For more detailed analyses of precipitation chemistry at selected sites see Wolfe (1988) and Lindberg et al. (1990b).

Air concentrations of coarse-particle SO_4^{2-} are difficult to measure directly (Noll et al. 1985), but their dry deposition to inert surfaces can be quantified (Davidson et al. 1985). The pattern of the measured fluxes of coarse-particles among the IFS sites (Figure 5.4) is somewhat different from that for atmospheric S, and these fluxes are only weakly correlated with SO_2 ($r = 0.22$, $n = 12$). The coarse-particle fluxes are similar at 11 of 13 sites (~7–15 μeq m^{-2} day^{-1}), but much lower at CP and WF. These fluxes are influenced by windspeed and nearby coarse-particle sources. Fluxes at the coastal and near-coastal sites are significantly influenced by sea salt aerosols, which contribute up to 35% of the coarse SO_4^{2-} flux at FS, RA, DF, NS, GL, and DL. Fluxes corrected for sea salt SO_4^{2-} were somewhat more correlated with SO_2 ($r = 0.36$). Unlike the fine-aerosol SO_4^{2-}, ion equiv-

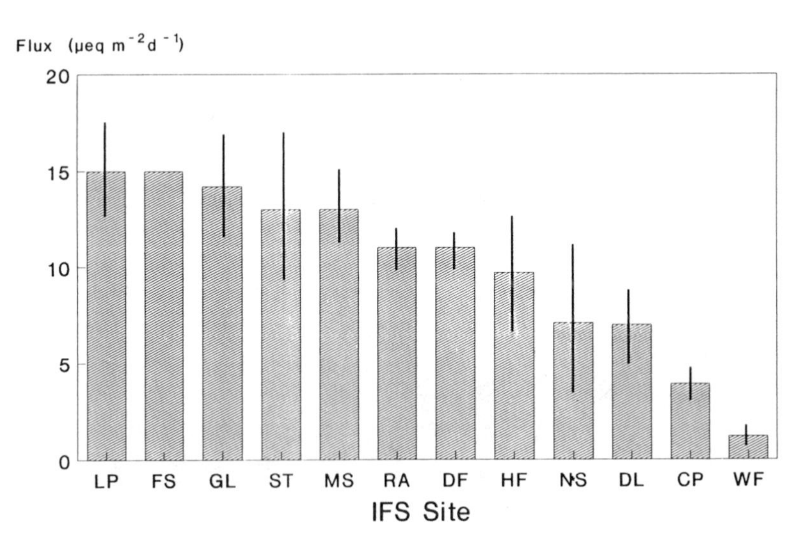

Figure 5.4. Mean (± 1 SE) dry deposition flux of SO_4^{2-} to inert deposition plates situated above forest canopies at each IFS site during 1986–1989.

alent ratios and correlations suggest that deposited coarse-particle SO_4^{2-} is primarily in the form of salts of Ca^{2+} and Mg^{2+}. It has been suggested that coarse-particle SO_4^{2-} is a result of the interaction between SO_2 and soil dust either before or during suspension in the air, reflecting another deposition pathway for SO_2 (Coe and Lindberg 1987; Butler 1988; Lindberg et al. 1990b).

Wet, Dry, and Cloud water Fluxes

The total atmospheric deposition of SO_4^{2-} to the IFS forests ranges over nearly an order of magnitude from approximately 300 to more than 2000 eq ha^{-1} yr^{-1} (Figure 5.5). The IFS sites fall into three general categories of sulfate loading: high (>1000 eq ha^{-1} yr^{-1}) that includes the mountain sites ST (1740 m elevation) and WF (1000 m) and the low-elevation southeastern United States sites that exhibited the highest concentrations of airborne sulfur (LP and DL); moderate (~500–800 eq ha^{-1} yr^{-1}), which includes the low-elevation northeastern and southeastern United States sites and Norway; and low (<350 eq ha^{-1} yr^{-1}), consisting of the two Washington sites. As expected, these fluxes were underestimated by bulk collectors operated at each site, with bulk to total ratios averaging 0.7 (range, 0.4–0.9). There are few, if any, directly comparable data in the literature. However, data from Hicks et al. (1989) and the United States National Atmospheric Deposition Program (NADP 1988) can be used to estimate total SO_4^{2-} deposition to low-elevation forests near the LP, HF, and GL sites during 1987. These fluxes (560–1100 eq ha^{-1} yr^{-1}) are within 20% of the values reported here for these sites, while deposition estimates for other mountain sites in

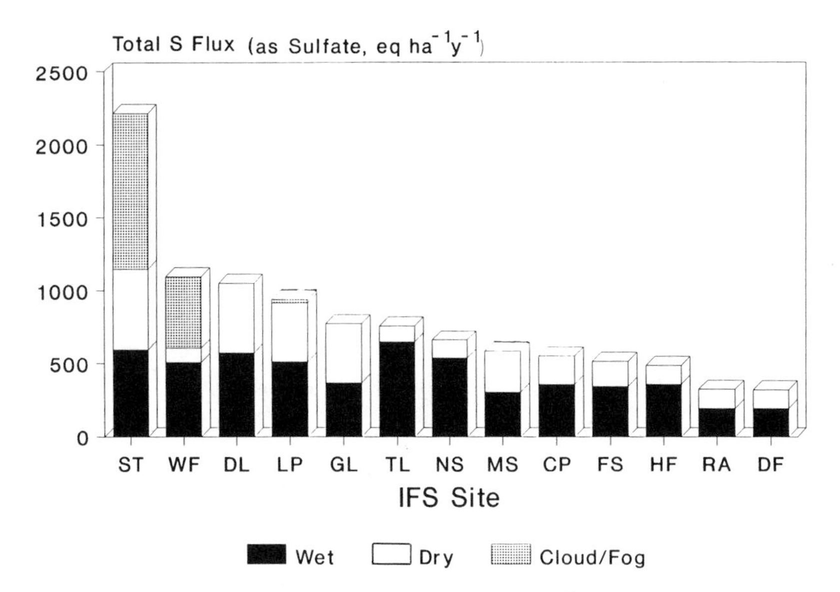

Figure 5.5. Total annual atmospheric deposition of SO_4^{2-} by each input process for IFS sites for 1986–1989.

the United States are comparable to or higher than those reported here for ST and WF (see review by Lovett and Kinsman, 1990). For example, Joslin et al. (1989) reported SO_4^{2-} fluxes of 2500–3100 eq ha^{-1} yr^{-1} at Whitetop Mt. in Virginia. In northeastern Germany, Bredemeier (1988) estimated that total sulfate fluxes to six low-elevation forests ranged from about 1800 to 5400 eq ha^{-1} yr^{-1}, suggesting that the only fluxes in the IFS network to approach those seen in north-central Europe are at the Smoky Mountains site (ST).

Wet deposition is less variable than total deposition across the IFS sites, varying by a factor of 3 from about 200 to 650 eq ha^{-1} yr^{-1}. The sites again fall into three general categories of wet deposition: high (~500–650 eq ha^{-1} yr^{-1}), moderate (340–370 eq ha^{-1} yr^{-1}), and low (~200 eq ha^{-1} yr^{-1}) (see Figure 5.5). However, this grouping of sites is not closely related to that described for total deposition, with the high and moderate groups both consisting of a mix of southeastern and northeastern sites, followed by the low-deposition Washington sites. This illustrates the degree to which atmospheric fluxes of sulfur are influenced by processes other than wet deposition alone. Wet deposition itself was more strongly influenced by the levels of sulfate in precipitation than by annual rainfall across the sites. In a linear regression, the annual weighted mean concentration of sulfate at each site explained about 70% of the variance in sulfate wet deposition while annual rainfall explained less than 10% of the variance.

Dry deposition was a significant input process at all IFS sites, regardless of total loading, and was most important at the drier Southeastern sites (see Figure 5.5). The relative contribution of dry deposition averaged about 30% of total input across the IFS, and ranged from about 10% to 15% at WF and TL to more than 40% at LP, DL, GL, and MS. The importance of dry deposition to forests has been previously quantified for sites in the eastern United States and Europe where the mean contribution by dry deposition is in the range of 40%–60% (Ulrich and Pankrath 1983; Grennfelt et al. 1985; Lindberg et al. 1986; Hicks et al. 1989). Dry to wet flux ratios, which may be useful for comparisons on large time and spatial scales (Hicks et al. 1989), are highly variable across the IFS (range, 0.19–1.1). However, these values illustrate regional differences, with the values at the southern sites (mean ratio, 0.78 $\pm$ 0.09) generally higher than those at the northern sites (mean ratio, 0.51 $\pm$ 0.12). To some extent this difference is the result of the record drought conditions in the Southeast, which affected all southern IFS sites between 1986 and 1988 (Cook et al. 1988), but also suggests real differences in the importance of dry deposition processes between these regions. The IFS data clearly indicate that application of a fixed dry to wet ratio to estimate dry deposition is subject to considerable uncertainty.

Dry-deposited SO_4^{2-} originates from three sources: impaction of fine-aerosol SO_4^{2-}, sedimentation plus interception of coarse-particle SO_4^{2-}, and uptake/sorption of SO_2 (for our calculations dry-deposited SO_2 is assumed to be quantitatively converted to SO_4^{2-} in the canopy; Garsed 1985). As was the case for the airborne concentrations, SO_2 was also the single most important source of dry-deposited sulfur to most of the forests (Figure 5.6). The sites fall into three general groups: those in which SO_2 clearly dominates total deposition, contributing more than 70% of SO_4^{2-} (WF, LP, DL, GL, HF); those where SO_2 and total aerosols are comparable (CP, RA, DF, MS, ST); and those where particles dominate (>70% of total; NS and FS). In general, both these latter two groups include sites more distant from local emissions or with an important sea salt aerosol SO_4^{2-} component.

Particle deposition is generally dominated by the coarse fraction, which contributed an average of about 70% of the particle flux. The importance of dry-deposited particles that are greater in size than those in the fine mode (~0.05–2 μm), which typically dominate airborne SO_4^{2-}, was proposed by Garland (1978) but has only recently been quantified (Davidson et al. 1982; Lindberg et al. 1986; Hicks et al. 1989). Supermicron particles (>1 μm) containing SO_4^{2-} are responsible for a disproportionate share of total SO_4^{2-} mass dry deposition onto both inert surfaces and natural vegetation, relative to submicron particles, because their deposition velocities are higher than those of submicron particles (Davidson et al. 1985). Microscopic examination of plant and inert surfaces has repeatedly confirmed the importance of this material to dry deposition (Davidson and Chu 1981; Fortmann 1982; Coe and Lindberg 1987). More detailed dry deposition data, including intercomparisons with different methods, have been published for the ST, LP,

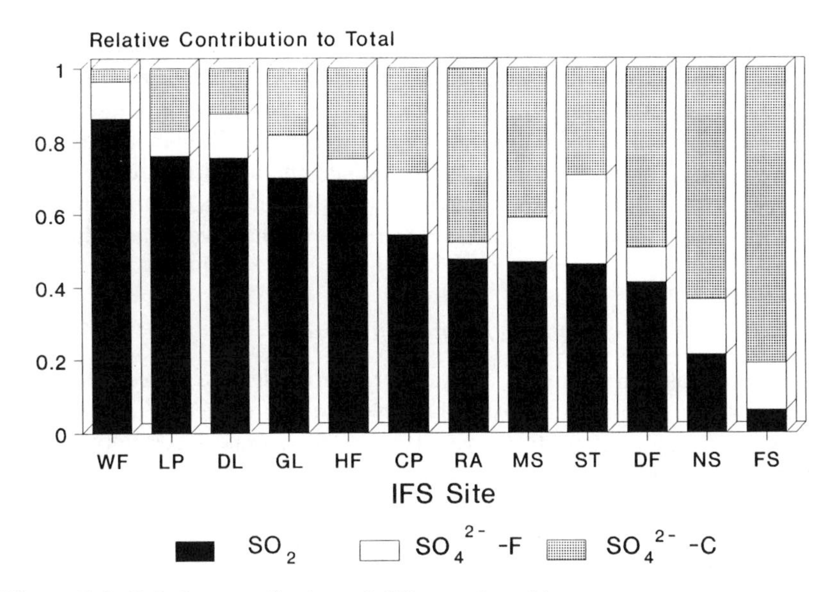

Figure 5.6. Relative contribution of different deposition processes to total dry deposition of SO_4^{2-} for IFS sites for 1986–1989. SO_4^{2-}-F represents fine-aerosol sulfate, and SO_4^{2-}-C, coarse-particle sulfate.

HF, TL, and CP sites (Lindberg et al. 1988a, 1988b; Hicks et al. 1989; Shepard et al. 1989; Lindberg et al. 1990b; Vose and Swank 1990; Vet et al. 1988).

Sulfate deposition to the two IFS mountain sites (ST and WF) was strongly influenced by cloud water interception, which contributed 45%–50% of the total SO_4^{2-} loading. As a result, the deposition fluxes at these two sites are substantially higher than would be suggested by their airborne sulfur concentrations, reflecting the influence of site characteristics on different deposition processes. Factors that increase the efficiency of atmosphere to surface exchange of pollutants in mountains, such as cloud immersion, the heterogeneity and high surface area of the mountain conifer forests, and the generally higher windspeed, result in increased loading despite generally lower air concentrations. For example, total SO_4^{2-} deposition at ST exceeded by factors of 2 to 4 the fluxes to the nearby LP and CP low-elevation sites, although these sites exhibited comparable or much higher air concentrations of S. This enhancement is also reflected in throughfall fluxes of SO_4^{2-}, as well as in both deposition and throughfall fluxes of N, H^+, and base cations (discussed in the following chapters). Although wet and dry deposition of SO_4^{2-} were both somewhat higher in the mountains, cloud water interception accounted for most of the enhancement.

Cloud immersion times in the range of 650 to 850 h yr^{-1} were measured at these sites, and annual average cloud water fluxes were estimated to be

16 cm at WF and 41 cm at ST in addition to the 114 and 203 cm of precipitation, respectively (see Chapter 2: for details, see Lindberg and Johnson 1989; Schaefer et al., 1989; Lindberg et al., in manuscript). These results support the idea that mountain forests are exposed to higher atmospheric loading because of climatologic factors and site characteristics, despite their distance from major emission sources (Lovett et al. 1982; Mohnen 1988; Unsworth and Fowler 1988; Saxena and Lin 1989; Lovett and Kinsman, 1990).

Uncertainties in Fluxes

All these deposition estimates are subject to considerable uncertainty, particularly at the high-elevation sites because of the presence of cloud water, which is difficult to quantify (Lovett 1988). The overall uncertainty in total SO_4^{2-} deposition to the ST mountain site was estimated to be of the order of ±50% and that at the nearby LP valley site to be ±20% (Lindberg and Johnson 1989). For the fluxes in throughfall plus stemflow, we estimated overall uncertainties on the order of ±20% for ST and ±10% for LP. These uncertainties were determined from a combination of the measured standard errors in ion concentrations, analytical precision and accuracy, potential bias in certain physical measurements, and published sensitivity analyses of models used in the IFS (Lovett 1984; Matt et al. 1987; Lindberg et al. 1988a; Hicks et al. 1989). For example, using these data and applying physical limitations wherever possible (e.g., hydrologic fluxes based on independent measurements), we established the following bounds to the total SO_4^{2-} deposition at ST: minimum, 1600; best estimate, 2200; maximum, 3900 eq ha^{-1} yr^{-1}. The overall range at this site is asymmetrical about the best estimate, largely because of the range in the volume of cloud water derived from different field methods (Lindberg and Johnson 1989; Schaefer et al. 1990), and suggests that the best estimate is more likely to be an underestimate than an overestimate. By comparison, the interannual variation in estimated total deposition to the ST site was as follows: 2000 eq ha^{-1} yr^{-1} for 1986, 2200 eq ha^{-1} yr^{-1} for 1987, and 2400 eq ha^{-1} yr^{-1} for 1988. Despite this large uncertainty, the sulfur flux to the ST site clearly exceeds that measured at any other IFS site.

Solution Fluxes to the Forest Floor in Throughfall and Stemflow

The throughfall plus stemflow (TF + SF) fluxes of SO_4^{2-} at the IFS sites also cover a wide range, from about 300 to 2500 eq ha^{-1} yr^{-1} (Figure 5.7), and strongly reflect the trends in deposition. For example, the flux of sulfate in TF + SF at the ST site exceeds the next highest flux (WF) by a factor of 2.4 and exceeds that at the geographically nearest site (CP) by a factor of more than 5. The large spatial variability of TF + SF fluxes within forest stands is well known (e.g., Kimmins 1973). The overall uncertainties in TF fluxes at IFS forests from spatial variability are in the range of 10% to 20%

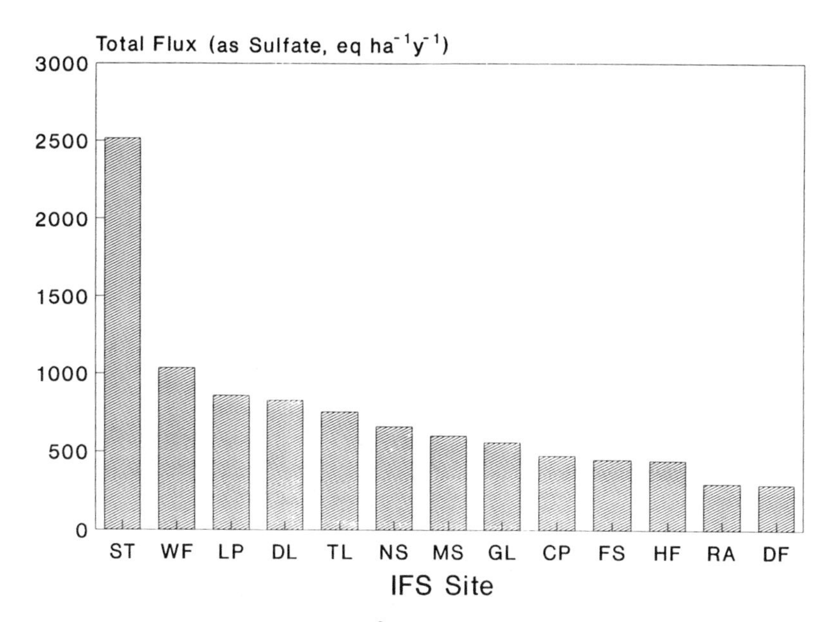

Figure 5.7. Total annual flux of SO_4^{2-} in throughfall plus stemflow below canopies of each IFS site for 1986–1989.

(Lindberg and Johnson 1989). The sites fall into the same three categories discussed earlier for total deposition (highest at the mountain sites and at those with the highest airborne S concentrations, and lowest in Washington). The sources of SO_4^{2-} in TF + SF may be external or internal to the plant and include SO_4^{2-} in precipitation and intercepted cloud water, washoff of dry-deposited aerosols and particles, and leaching of internal plant SO_4^{2-}.

For the nine most complete data sets, there was no significant relationship across the IFS sites between the SO_4^{2-} flux in TF + SF and the soil content of S (r = .41), and only a marginally significant relationship ($p < .10$) between TF + SF and the vegetation content of S (r = .65). However, across all IFS sites, the fluxes in deposition and in TF + SF are strongly related. Figure 5.8 illustrates the relationship between the measured flux in TF + SF and the estimated total annual wet plus dry deposition of SO_4^{2-} to the forest canopy and at each site. At five of the IFS sites the fluxes are essentially the same above and below the canopy, and at the remaining sites the differences are generally small. The correlation between the two is highly significant (r = .99), with a linear regression slope close to 1 [1.1; standard error (SE) of the estimate = 0.1] and an x intercept close to 0 (-150 ± 110 eq ha^{-1} yr^{-1}). This suggests that SO_4^{2-} in TF + SF is largely controlled by atmospheric deposition, as discussed next. In addition to SO_4^{2-}, sulfur in throughfall also includes an organic fraction (Mitchell et al. 1989). However, based on detailed studies at the DF and RA sites and very limited

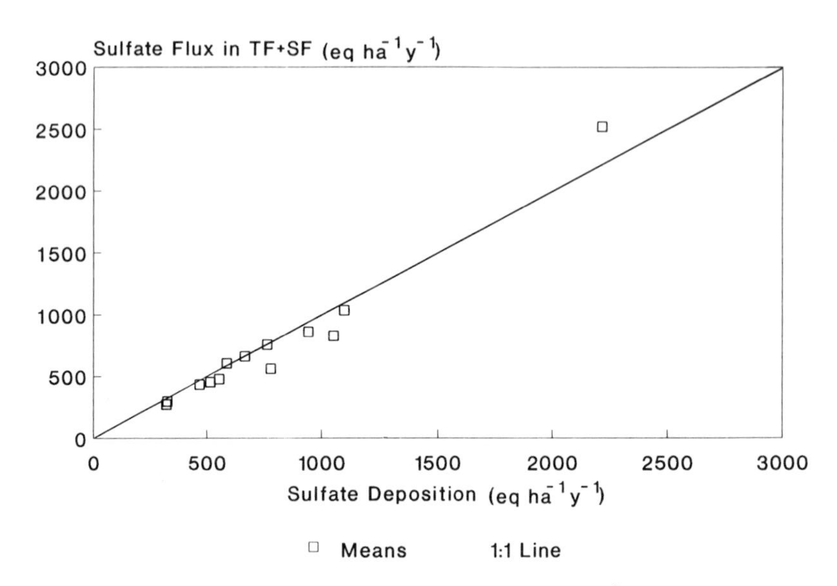

Figure 5.8. Relationship between total annual flux of SO_4^{2-} in throughfall plus stemflow (TF + SF) below canopies of each IFS site and total annual atmospheric deposition of SO_4^{2-} for 1986–1989.

analyses (1–4 samples) at some of the other IFS intensive sites, organic S is a relatively small fraction (~5%–20%) of the total S contained in TF (Homann et al., 1990). Adding estimates of the fluxes of organic S to the fluxes of SO_4^{2-} in TF + SF at these sites does not alter the trends discussed here.

Forest Canopy Interactions of Deposited Sulfur

Comparison of the total deposition of SO_4^{2-} to each forest canopy with the fluxes of SO_4^{2-} below these canopies in TF + SF provides an indication of the interactions and fate of deposited atmospheric sulfur. As illustrated in Figure 5.8, the data indicate that canopy interactions (uptake and leaching) appear to be small, particularly in comparison to the magnitude of the total fluxes. Points above the 1:1 line represent sites at which fluxes in TF + SF exceed total deposition, indicating a net source of SO_4^{2-} in the canopy (foliar leaching); points below the line are sites where total deposition exceeds TF + SF, indicating a net sink for deposited SO_4^{2-} in the canopy (foliar uptake). Points on the line suggest either no interaction or a balance between uptake and leaching.

The distance of each point from the 1:1 line indicates the magnitude of the canopy interaction for sulfur. This can be quantified by the difference between the total atmospheric flux and the flux in TF + SF and is termed the net canopy exchange (NCE):

$$\text{NCE} = (\text{TF} + \text{SF}) - (\text{total deposition}) \quad [5.1]$$

These values are summarized in Figure 5.9. The flux below the canopy exceeds that to the canopy, yielding positive NCE values, only at ST and MS. At all the other IFS forests the estimated atmospheric deposition exceeds the flux in TF + SF, yielding negative NCE values. Given the uncertainty of these fluxes, NCEs of an absolute magnitude less than 100 eq ha^{-1} yr^{-1} are probably not significantly different from 0. This suggests an approximate balance of S in the canopies at the MS, TL, NS, RA, DF, HF, CP, LP, WF, and FS (not shown) sites.

While significant negative NCE values indicate that the canopy is a net sink for deposited S, positive NCE values only indicate that foliar leaching exceeds uptake of SO_4^{2-} in the canopy, but not that uptake is absent. Uptake is most likely in the form of dry-deposited SO_2, which is known to readily enter the substomatal cavity of the leaf or needle (Winner et al. 1985). Although the behavior of the forest canopy as a source or sink is influenced by many factors, the IFS data suggest some interesting generalities. The only

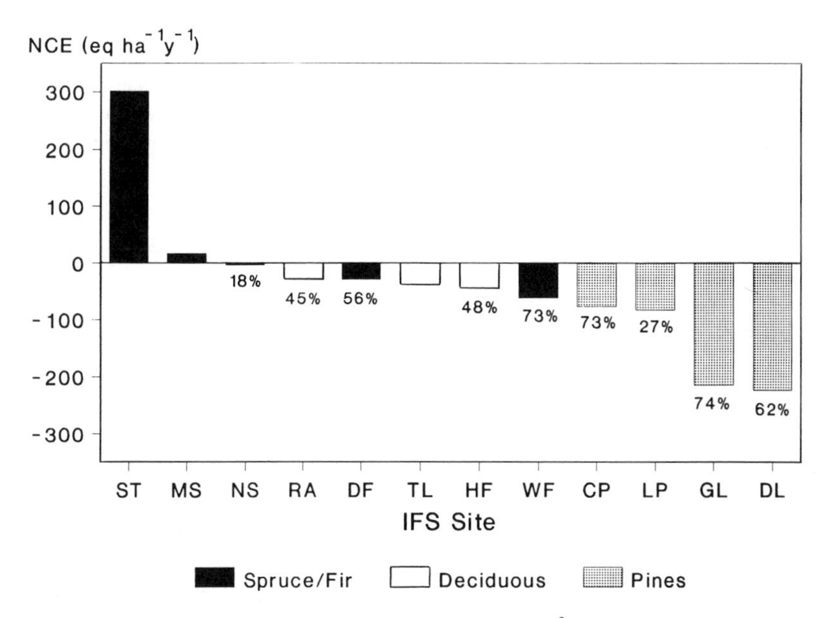

Figure 5.9. Annual net canopy exchange (NCE) for SO_4^{2-} at each IFS site for 1986–1989. NCE = (annual flux in TF + SF) − (total annual deposition to canopy). For sites where NCE is negative, numbers below each bar represent NCE as percent of total SO_2 dry deposition.

significant net "source" canopy is high-elevation spruce (ST), while the most significant "sink" canopies are all southern pine stands (CP, DL, LP, GL, FS). In between, more or less in balance, are a mixture of spruce, fir, and deciduous systems.

If dry deposition of SO_2 through the stomates represents the only mode of airborne S uptake in the canopy, then the relative uptake of SO_2 in the "sink" canopies is given by:

$$(\text{NCE})*100/(SO_2 \text{ dry deposition}) = \text{percent uptake} \quad [5.2]$$

These values are shown below each bar in Figure 5.9. The data suggest that during an annual cycle about 20%–70% of the dry-deposited SO_2 remains fixed in the canopies of these forests or is degassed to the atmosphere following deposition (probably insignificant; Matt et al. 1987). Conversely, these forest canopies release about 30%–80% (average, ~50%) of dry-deposited SO_2 into TF + SF. This compares with the results of Gay and Murphy (1985), who reported, on the basis of field cuvette experiments with $^{35}SO_2$, that roughly half of dry-deposited SO_2 could be washed from needles of loblolly pines.

Although canopy uptake rates estimated from a mass balance are subject to large uncertainty, it is apparent that the southern pine IFS sites exhibited the largest relative uptake of SO_2 and consistently showed the highest absolute uptake of deposited sulfur. The drought conditions at all these sites during the IFS study (Cook et al. 1988) may partially explain this trend. The model used to estimate dry deposition to the IFS forests (Hicks et al. 1987) is thought to overestimate SO_2 deposition velocities during periods of soil moisture stress by approximately 30% (Matt et al. 1987). While this may explain the high uptake rates of sulfur, the conclusion that these forest canopies sequester an important fraction of dry-deposited SO_2 remains valid. However, extreme values for percent uptake of SO_2 indicate significant overestimates of S dry deposition, or underestimates of S fluxes in TF + SF. This is reflected in the NCE for the FS site (60 eq ha^{-1} yr^{-1}), which is substantially higher than of the dry deposition of SO_2 (not included in Figure 5.9; as discussed earlier, the single year's data from this site are preliminary; see the Appendix).

With mass balance calculations of this type, the exact source or sink of any pool of SO_4^{2-} ions cannot accurately be determined because leaching may offset uptake to some extent. Isotopic labeling has been used on mature loblolly pines near the LP site to study sulfur cycling by tracking the actual fate of individual SO_4^{2-} ions internal to the trees (Garten 1990). These data were used to directly quantify foliar leaching and indirectly estimate foliar uptake (Lindberg and Garten 1988). In the study, $Na_2{}^{35}SO_4$ was applied to four pine trees using a stemwell injection method (Garten et al. 1988). Foliar content and speciation measurements confirmed uniform labeling of the proper foliar pool, and measurements of SO_4^{2-} in rain and TF + SF were used with isotope dilution calculations to quantify leaching. Leaching contributed an average of less than 3% of the SO_4^{2-} in TF + SF below the canopies of these trees. Hence, washoff of dry-deposited aerosols and SO_2 was the dominant source of SO_4^{2-} enrichment in TF + SF over rain at this site. These measurements were repeated on a nearby deciduous forest with similar results (leaching <4% of TF + SF; Garten et al. 1988).

Given the IFS mass balance and published isotope data, one might expect the general behavior of deposited sulfur oxides in forest canopies in industrialized areas to lie somewhere between minimal interaction and moderate uptake, depending on local SO_2 dry deposition and tree species. Leaching would be a significant fraction of TF + SF only when dry deposition is minimal. If this is an accurate generalization, the relatively large foliar leaching at the ST mountain site (see Figure 5.9) may simply reflect the inability to accurately quantify the fluxes in highly complex terrain. This apparent leaching could result from an overestimate of the flux in TF + SF or an underestimate of the total S deposition (which is more likely, as suggested by the uncertainty analysis discussed earlier). Assuming that a small amount of foliar leaching does occur in most forests (Johnson 1984; Lindberg and Garten 1988), these data suggest that uptake of dry-deposited SO_2 in many forests is sufficiently large relative to foliar leaching to result in a negative canopy balance for SO_4^{2-}. Thus, leaching of SO_4^{2-} is unimportant at sites with high S loading, where deposition washoff dominates the below canopy flux and cannot be detected in the canopy balance because of uptake of SO_2.

Implications

The IFS data, along with those from published process-level studies cited earlier, can be used to describe the general behavior of sulfur in forest canopies in industrialized regions. Figure 5.10 is a schematic illustrating this behavior. Sulfate is generally the dominant mobile anion in forest canopy throughfall (mean ± SE equivalent ratio of $SO_4^{2-}/NO_3^- = 3.4 \pm 0.6$ at the IFS sites), and is the primary anion accompanying leaching of base cations from foliage (see Chapter 8). The sources of SO_4^{2-} in throughfall are dominated by precipitation and dry deposition washoff, while foliar leaching is minor. Sulfate deposited in rain and cloud droplets appear to pass through the canopy with minor interaction, other than temporary retention during evaporation, which is followed by washoff in the next significant event. Dry-deposited fine aerosols and coarse particles also appear to be quantitatively removed by subsequent rainfall events, at least on seasonal to annual time scales.

Dry-deposited SO_2 does interact with the foliage, with the net uptake probably dominated by that fraction which enters through the stomates. Although the proportion of surface and internal deposition of SO_2 is influenced by plant physiological properties and local climate (Taylor et al. 1983), deposition to surfaces within the stomatal cavity probably dominates the over-

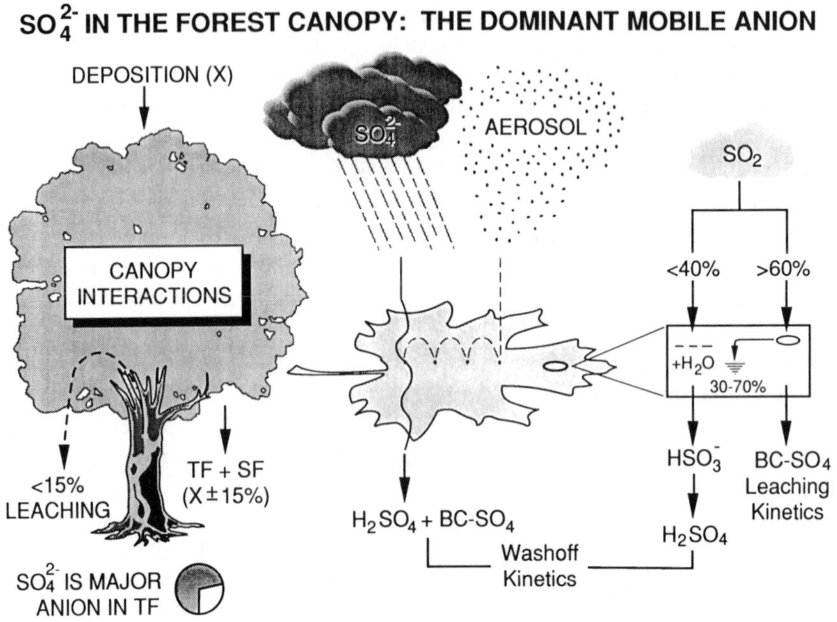

Figure 5.10. Schematic of general behavior of deposited atmospheric sulfur in forest canopy as suggested by data from IFS and previous process-level studies cited in text.

all SO_2 dry deposition, generally accounting for more than 60% of the total. Field measurements and cuvette data suggest that approximately 30%–70% of the total dry-deposited SO_2 is retained in the canopy (Gay and Murphy 1985). Of the SO_4^{2-} that does appear in throughfall, one portion exhibits rapid removal kinetics (washoff), while another exhibits slower removal (leaching) (Lovett and Lindberg 1984; Schaefer and Reiners 1990; cf. discussion of cations in Chapter 8). The chemistry of SO_4^{2-} in throughfall is site specific, but is generally dominated by base cation salts, with H^+ and NH_4^+ able to account for the majority of the SO_4^{2-} only at those IFS sites experiencing the highest SO_2 and cloud water deposition rates (ST, WF, GL, and DL).

The IFS data suggest that canopy interactions are small compared to total deposition. The net result is a remarkable similarity between measured TF + SF fluxes and the estimates of total SO_4^{2-} deposition, as discussed earlier (see Figure 5.8). Figure 5.11 illustrates the general trend in canopy interactions of SO_4^{2-} across the IFS forests, expressing the net canopy exchange (NCE) as a percent of total deposition. In general, NCE is less than 15% of total deposition, and the overall mean for all sites is about 10% of deposition. Only at the GL and DL southern pine sites, which exhibit high loadings of SO_2, is the canopy interaction greater than 15% of deposition

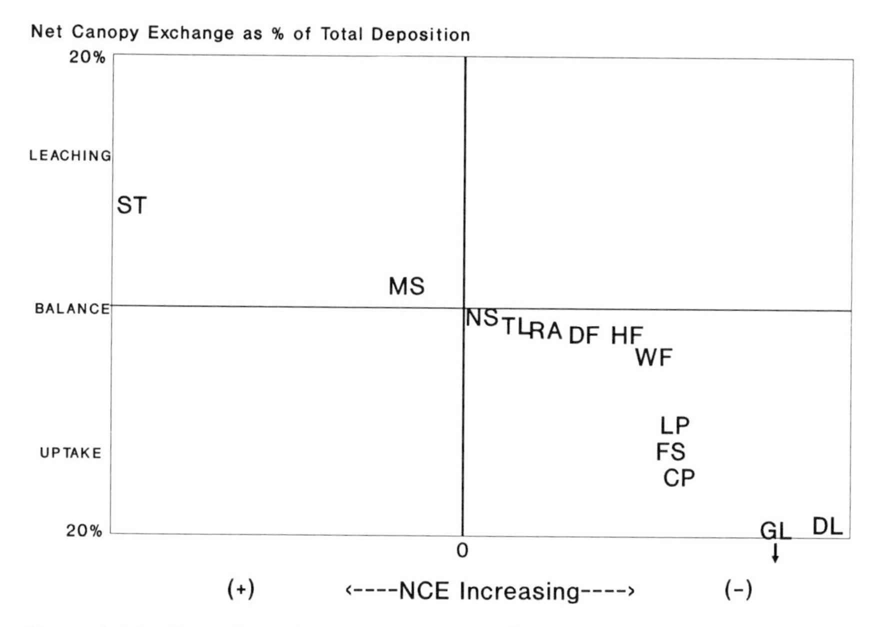

Figure 5.11. General trends in behavior of SO_4^{2-} in canopies at each IFS site for 1986–1989. Magnitude, direction, and relative importance of net canopy exchange (NCE) for SO_4^{2-} was used to place each IFS forest within this behavior schematic (NCE = annual flux in TF + SF − total annual deposition to canopy).

(~20%–30%). The close relationship between TF + SF and deposition may be somewhat fortuitous in that small amounts of foliar leaching appear to offset SO_2 uptake to some extent. However, the quantitative washoff of dry-deposited aerosol sulfur from the canopy by rain has been reported, as has the lack of significant interaction of wet-deposited SO_4^{2-} with foliage (Lindberg and Lovett 1985; McCune and Lauver 1986; Bytnerowicz et al. 1987; Bredemeier 1988). This behavior appears to be generalizable, at least across the range of forest type, climate, and air quality represented by the IFS sites.

The overall lack of a significant canopy interaction of SO_4^{2-} makes TF + SF an excellent indicator of total atmospheric sulfur loading to forests. This hypothesis has been suggested in earlier studies (Mayer and Ulrich 1974; Hultberg 1985; Lindberg et al. 1986) and is strongly supported by recent isotope work (Garten et al. 1988; Lindberg and Garten 1988; Garten 1990), but it has never been demonstrated with a data set as extensive as that of the IFS. The ability of measured annual fluxes of SO_4^{2-} in TF + SF to predict total annual atmospheric deposition of sulfur at IFS forests is illustrated in Figure 5.12 with several years of data. The strength of the relationship is clear, with the variance in TF + SF fluxes able to account for 97% of the variance in estimated total deposition. Although the slope (0.84 ± 0.03) and intercept (170 ± 100) are different from 1 and 0, the

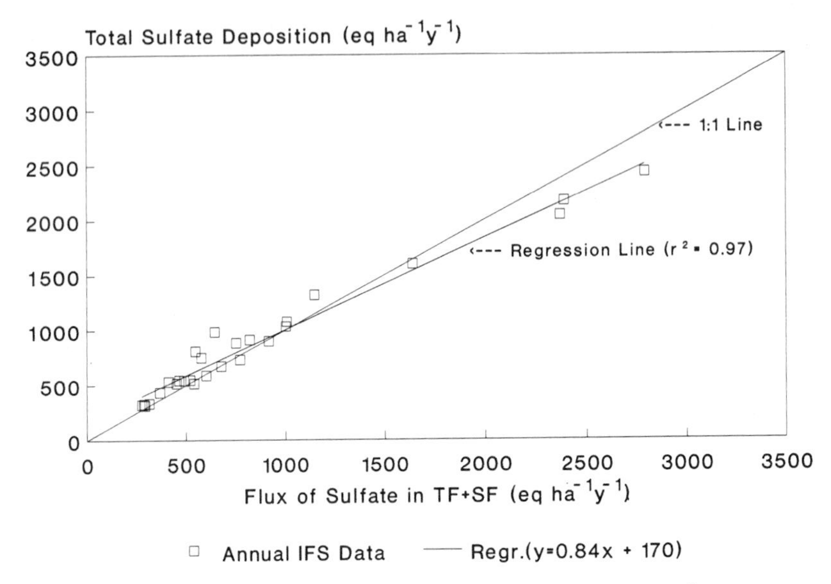

Figure 5.12. Estimated total annual atmospheric deposition of SO_4^{2-} at each IFS site as function of measured flux of SO_4^{2-} in throughfall plus stemflow. Data are plotted for each individual IFS sampling year for 1986–1989; linear regression line is shown (see text).

relative difference between the two fluxes at any site (NCE/total deposition, see Figure 5.11) is well within the typical error of total deposition estimates (Hicks et al. 1986). These results have important implications in many areas. Given the overall uncertainty in all types of deposition estimates, measurements of fluxes of SO_4^{2-} in throughfall beneath forests can provide useful estimates of spatial and temporal trends in the atmospheric deposition of S, particularly in industrialized areas. This application of TF + SF measurements may be especially important in forests in complex terrain where standard micrometeorological methods cannot be routinely used to estimate dry deposition (Hicks et al. 1986). The method also has important applications to deposition model development and evaluation and to prediction of soil and water acidification, and can provide the field data necessary for scaling up from point measurements of atmospheric fluxes to watershed scales. A recent European study has adopted this method to test the results of long-range transport and deposition models (Ivens et al. 1990).

Sulfur Distribution and Cycling in Forest Ecosystems

M.J. Mitchell, R.B. Harrison, J.W. Fitzgerald, D.W. Johnson,
S.E. Lindberg, Y. Zhang, and A. Autry

The distribution and cycling of sulfur in forest ecosystems have been the focus of a number of studies because of the linkage of sulfur with acidic deposition. Previous work has been reviewed by Fitzgerald and Johnson (1982), David et al. (1984), Johnson (1984), and Mitchell et al. (1991). In this section, we focus on the IFS sites but also include information for other selected sites for which detailed information on sulfur is available (Table 5.1). Within this section, information in figures will be presented with sites ranked from high to low atmospheric inputs (Figure 5.13) as indexed by either total deposition (i.e. IFS sites) or bulk throughfall plus stemflow (most other sites) since total deposition is generally not available for most forest ecosystems. This latter measurement, however, gives a close estimate of total sulfur deposition as discussed in the previous section.

Sulfur Content and Constituents

M.J. Mitchell and Y. Zhang

Vegetation

For most sites, vegetation represents a small (<12%) component of the total sulfur content of the ecosystem (Figure 5.14). Similar contributions of vegetation to ecosystem pools are found for other elements including nitrogen (see Chapter 6). In contrast, vegetation in the Coweeta Pine site (CP) and Douglas fir site in Australia (AD) constitutes a larger fraction of the total sulfur mass (17% and 37%, respectively). For the latter site, this high value is attributable to the very small total sulfur pool in the soil and the absence of major anthropogenic sources of sulfur, which results in very low atmospheric sulfur inputs. For the CP site, this higher value also reflects the lower sulfur content in the soil, but this is because its very high coarse fragment (>2 mm) content of the soil mineral horizons (22%–37%) reduces the soil that is biogeochemically active. In contrast, the adjacent hardwood site (CH) has a coarse fragment content that is substantially lower within mineral horizons (9%–12%), resulting in higher sulfur content [see the Appendix of site data for further details]. In examining the nitrogen pools for the IFS sites, it has also been found that the CH site had proportionally more nitrogen in the vegetation (see Chapter 6).

In examining all the sites along the sulfur depositional gradient, including both IFS and non-IFS sites, it is clear that there is no relationship between vegetation content and sulfur inputs ($p > .05$, $r = .10$, $n = 20$). With the possible exception of the low input site in Australia, the sulfur content of the vegetation of these systems shows little variation, especially compared to an element such as nitrogen that more typically may limit forest growth (see Chapter 6). These results concur with previous studies, which also have shown that there is no linear relationship between vegetation content and atmospheric input of sulfur for forest ecosystems (Johnson 1984; Mitchell et al. 1991). The absence of this relationship is a function of sulfur availability, which is in excess of nutritional needs for most forests.

Table 5.1. Selected Non-IFS Forest Sites Used in Comparing Some Aspects of Sulfur Cycling

Name of Site	Abbreviation	Location	Forest Type	References
Ardennes oak	AO	Ardennes, France	*Quercus* spp.	Nys and Ranger (1988)
Ardennes spruce	AS	Ardennes, France	*Picea abies*	Nys and Ranger (1988)
Ispina mixed oak	PO	Ispina, Poland	*Quercus robus*	Karkanis (1976)
Bago, NSW	AD	Australia	*Pseudotsuga menziesii*	Turner and Lambert (1980)
Solling, beech	SF	West Germany	*Fagus silvatica*	Meiwes and Khanna (1981)
Solling, spruce	GS	West Germany	*Picea abies*	Meiwes and Khanna (1981)
Walker Branch (Fullerton soil)	FT	Tennessee, USA	*Quercus prinus* *Carya* spp.	Johnson et al. (1986) Lindberg et al. (1986)
Hubbard Brook, White Mtns.	HB	New Hampshire, USA	*Fagus grandifolia* *Acer* spp.	Likens et al. (1977, 1990) Mitchell et al. (1989)

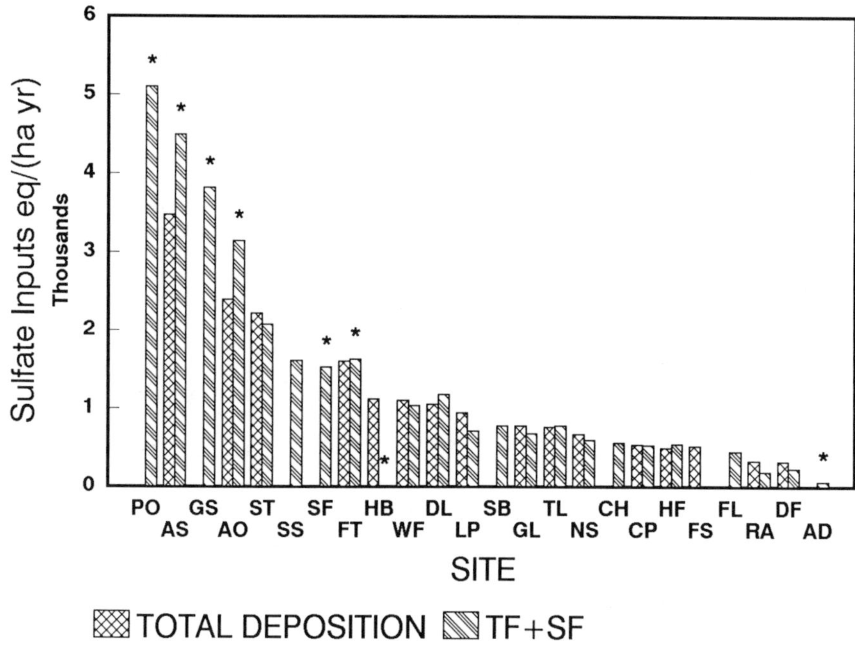

Figure 5.13. Ranking of sites by sulfate inputs using either total deposition or bulk throughfall + stemflow. Codes for IFS sites are explained in text; (*, non-IFS site).

It has been generally assumed that there is a tight linkage between the nitrogen and sulfur concentrations, because the ratios of nitrogen- and sulfur-bearing amino acids are relatively constant and account for almost all the organic nitrogen and sulfur constituents of forest foliage (Turner et al. 1977; Johnson 1984). The ratio between organic nitrogen and organic sulfur is generally presumed to be about 34 on a molar basis or 15 on a mass basis (Kelly and Lambert 1972; Turner and Lambert 1980). This ratio may vary, however, especially from nitrogen fertilization, which may increase the ratio by the formation of compounds such as arginine (Lambert et al. 1976; Homann, personal communication). Lower values would indicate excess sulfur uptake and subsequent storage of SO_4^{2-} in foliage. For additional discussion on elemental ratios in vegetation also see Chapter 7.

An examination of N:S ratios (Figure 5.15) shows that all sites are below 34, except for CH in which the ratio is approximately 35 and the RA site in which it is exceeded (ratio = 40). This latter site has very high nitrogen availability because of autotrophic nitrogen fixation and low sulfur inputs, and thus sulfur may be limiting. In addition, sulfur limitations have been shown for other forests in the northwest region of the United States where the RA site is located (Turner et al. 1977).

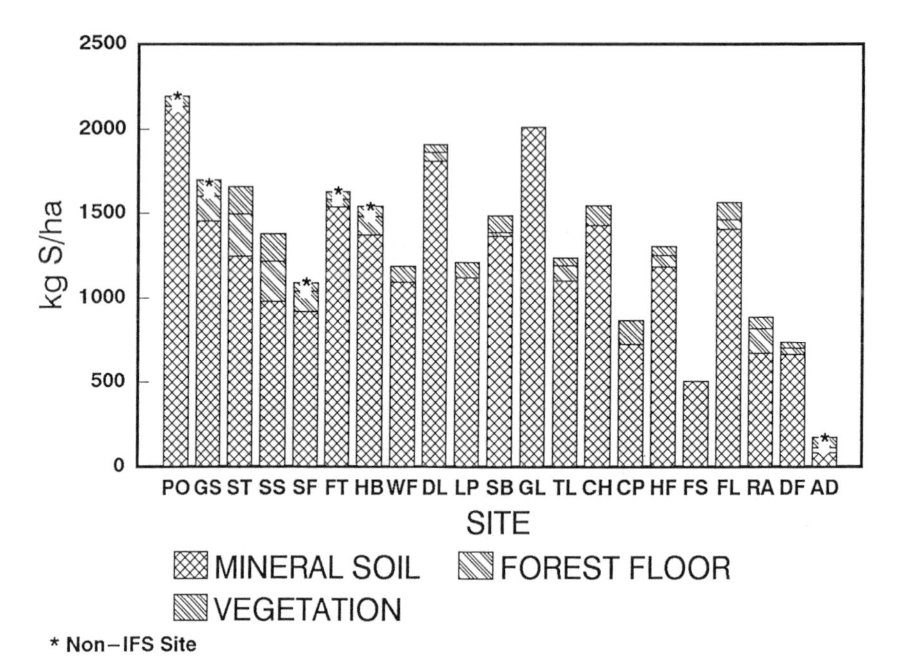

Figure 5.14. Sulfur content in vegetation, forest floor, and mineral soil (*, non-IFS site).

For six selected sites (Table 5.2), additional analysis of foliage was undertaken to obtain more detailed information on total sulfur, inorganic sulfur, total nitrogen, and total carbon. Total carbon was analyzed using a Perkin-Elmer 2400-CHN Analyzer, total sulfur by a Leco SC-132 Sulfur Determinator (David et al. 1989), and inorganic sulfur (assumed to be sulfate) by HI (hydriodic acid) reduction (Landers et al. 1983). Nitrogen concentrations were substantially higher and thus C:N lower in the American beech and maple from HF than in the conifers from the other sites (Figure 5.16). The mean value of sulfate in foliage was approximately 6% of total sulfur ($\sigma = 2.5$, $n = 60$) with a range from 2.2 to 13.2. Of the sites analyzed, loblolly pine had the highest sulfate concentration, 10.4% of total sulfur, and this may be attributed to the very high SO_4^{2-} concentrations found in soil solution from this site (i.e., Bt2 horizon = 312 μeq L^{-1}).

When the inorganic sulfur fraction (HI-S) was subtracted from total S, there was a significant correlation between organic sulfur and total nitrogen for hardwoods ($r = .61$, $p < .01$) and conifers ($r = .46$, $p < .01$), but for both vegetation types there was considerable variation in the mean N:S molar ratios (23.2 hardwoods; 21.9 conifers; on a mass basis, 10.1 and 9.5, respectively). These ratios are lower than typically reported for some forest ecosystems, as was discussed previously. There was no apparent discernible

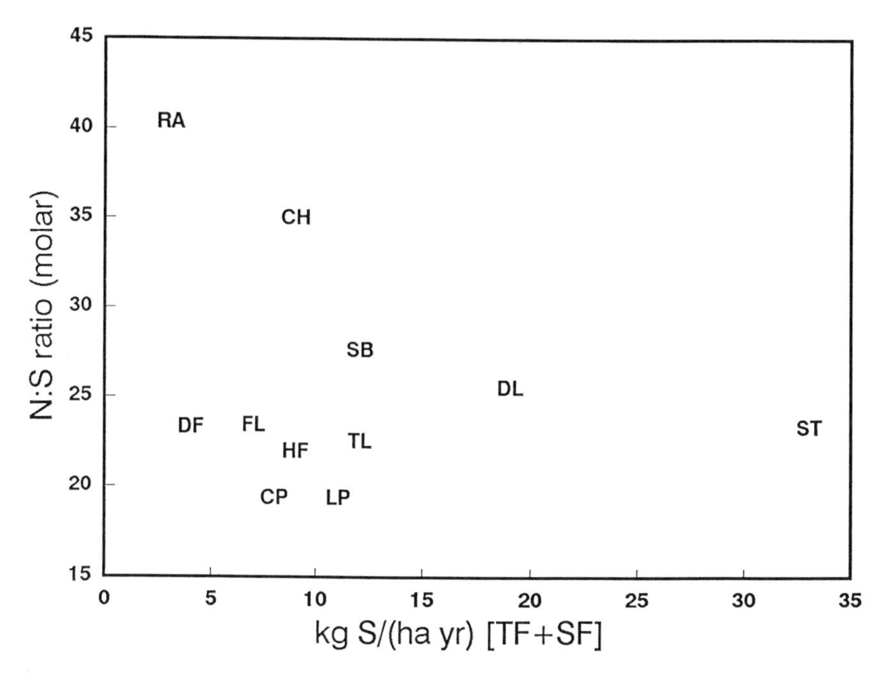

Figure 5.15. N:S ratios in foliage versus inputs in throughfall + stemflow.

pattern in the relationship between organic sulfur and total nitrogen among sites or subsamples on the basis of either solution chemistry or vegetation requirements (Figures 5.17 and 5.18). This lack of pattern may be attributed to the absence of sites in this limited sample showing either low sulfur availability or nitrogen limitations.

Forest Floor and Mineral Soil

The dominant sulfur pool in all sites (Figure 5.19) was the mineral soil (71%–97% of the total sulfur in the ecosystem), with the exception being the Douglas fir site in Australia (AD), which had only 47% of total sulfur in soil. The forest floor contained a similar amount of sulfur as that in the vegetation, and contributed from 2% to 17% of total sulfur in each site. Unlike vegetation, however, there was a significant correlation between total sulfur content in the forest floor ($p < .05$, $r = .57$, $n = 17$) and mineral soil ($p < .05$, $r = .54$, $n = 21$) of these forest ecosystems including both IFS and non-IFS sites and the input of S, but this single factor explained less than 34% and 30%, respectively, of the variation in sulfur content of these ecosystem strata. The correlation of sulfur inputs with the sulfur content of the forest floor and mineral soil is likely not directly linked to sulfur inputs; this correlation is probably more the result of differences in the or-

Table 5.2. Vegetation Samples Analyzed for Total Sulfur, HI Sulfur, Total Nitrogen, and Total Carbon

Site	Vegetation	Replicate Samples	Sample Description
DL	*Pinus taeda*	4	Current flush, four trees
DL	*Pinus taeda*	4	Second flush, four trees
DL	*Pinus taeda*	2	Third flush, four tree
ST	*Pices rubens*	5	New foliage
ST	*Pices rubens*	5	Old foliage
SS	*Pices rubens*	5	New foliage
SS	*Pices rubens*	5	Old foliage
MS	*Pices rubens*	1	New foliage composited from five samples
MS	*Abies balsamea*	1	New foliage composited from two samples
MS	*Pinus strobus*	1	Composited from three samples
NS	*Picea abies*	2	Composites from six trees, pH 6 treatment
NS	*Picea abies*	2	Composites from eight trees, pH 2.5 treatment
NS	*Picea abies*	3	Composites from five trees, A4-R1
NS	*Picea abies*	3	Composites from eight trees, A4-R2
HF	*Fagus grandifolia*	4	Foliage from individual trees in 1986
HF	*Fagus grandifolia*	4	Foliage from individual trees in 1987
HF	*Acer saccharum*	4	Foliage from individual trees in 1986
HF	*Acer saccharum*	4	Foliage from individual trees in 1987

ganic content of soil. For example, the high-elevation sites, which generally have higher sulfur inputs, are characterized by large pools of organic sulfur that correlate closely with carbon content (Mitchell et al. 1991) of these soils (see pp. 119–121) for further details). However, David et al. (1988) also found that forest floor and soil sulfur contents were correlated with atmospheric deposition of sulfur over a broad geographical range in the north-central United States.

As has been established from previous studies, the dominant form of sulfur in the soil is generally organic ($>75\%$ of total S), especially for Durochrepts (DF and RA), Cryandepts (FL), Inceptisols (WF, SS and SB), and Spodosols (HF and TL) (see Figure 5.19). This organic sulfur in soils is composed of two major fractions (Freney 1967; Mitchell et al. 1991): carbon-bonded sulfur and ester sulfates, the composition of which has been determined for selected soil horizons at IFS sites (Autry et al., 1990).

For Hapludults (CH, CP, DL, GL), sulfate is more predominate (see Figure 5.19) and constitutes from 43% to 70% of the total sulfur in the mineral soil. These sites in the southeastern United States are south of the most recent continental glaciation and have soils more highly weathered than those of the northern sites. Thus these Hapludults have higher sulfate adsorption potential and have accumulated sulfate. The importance of weathering as well as sulfate adsorption potential has been previously established for other regional comparisons of soil sulfate retention capacities (Johnson et al. 1980; Rochelle et al. 1987).

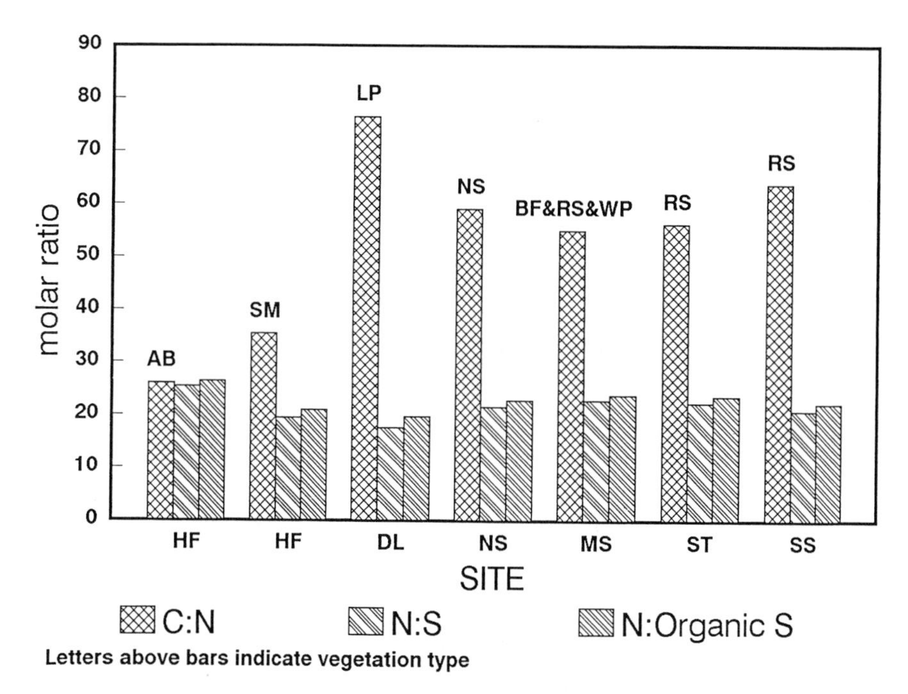

Figure 5.16. Foliage analyses of selected IFS sites (LP, Loblolly pine; RS, red spruce; NS, Norway spruce; WP, white pine; BF, balsam fir; AB, American beech; SM, sugar maple).

Fluxes and Regulating Factors

M.J. Mitchell, D.W. Johnson, and S.E. Lindberg

Throughfall, Stemflow, and Foliar Leaching

The throughfall plus stemflow (TF + SF) fluxes of SO_4^{2-} at the IFS sites cover a wide range, from about 300 to 2500 eq ha^{-1} yr^{-1} (see Figure 5.7). The sites fall into roughly the same three categories discussed earlier in Chapter 5 for total deposition (highest at the mountain sites and at those with the highest airborne sulfur concentrations, and lowest in Washington). The sources of SO_4^{2-} in TF + SF may be external or internal to the plant and include SO_4^{2-} in precipitation and intercepted cloud water, washoff of dry-deposited aerosols and particles, and leaching of internal plant SO_4^{2-}. For the IFS sites with complete data sets on wet-only TF + SF fluxes and sulfur content in vegetation and soils (n = 9), there was no significant relationship between SO_4^{2-} flux in TF + SF and soil content of sulfur (r = .41), and only a marginally significant relationship ($p < 0.10$) between TF + SF and the vegetation content of sulfur (r = .65). However, across all IFS sites,

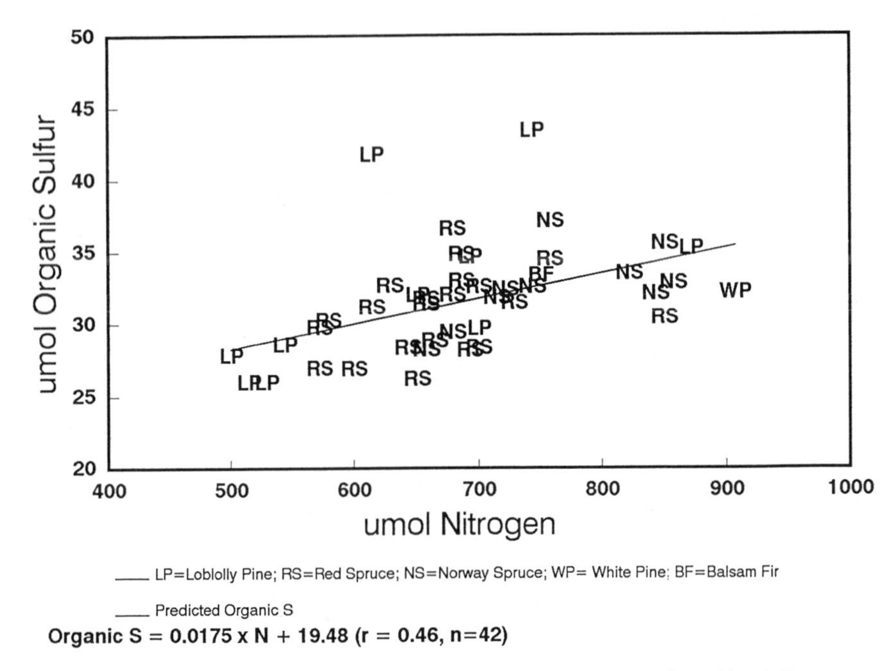

Figure 5.17. Relationship between organic S and total N of conifer foliage.

the fluxes in deposition and TF + SF are strongly correlated (r = .99) (see Figure 5.8). At five IFS sites, the fluxes are essentially the same above and below the canopy, and at the remaining sites the differences are generally small. The IFS and other recent data (Garten et al. 1988; Lindberg and Garten 1988; Garten 1990) suggest that SO_4^{2-} in TF + SF is largely controlled by atmospheric deposition and that foliar leaching is minimal.

Litterfall, Translocation, and Root Turnover

Litterfall inputs of sulfur show little variation among the forest sites used in the present comparison (mean, 3.8 kg ha^{-1} yr^{-1}; σ = 1.8, n = 19). Similar litterfall inputs combined with low variation in sulfur concentration of litter result in the similarities in sulfur litter flux among sites. For those sites with more than 10 kg S ha^{-1} yr^{-1}, litter inputs are less than total atmospheric deposition, which suggests that for these sites the geochemical cycling of sulfur may be more important than its biological cycling. The relative importance of biological cycling among sites can be indexed by calculating whether atmospheric inputs exceed requirements (Figure 5.20). We would hypothesize that for those sites in which requirements are greater than atmospheric inputs, the cycling of sulfur through the vegetation would be es-

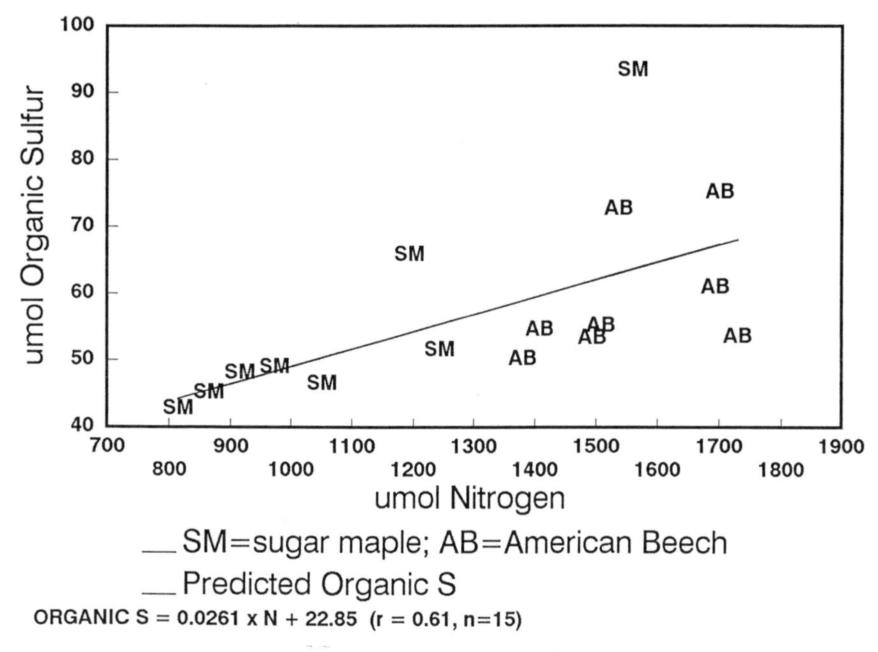

Figure 5.18. Relationship between organic S and total N of hardwood foliage.

pecially important in regulating sulfur flux. For two sites (CP and RA), this criterion is met and thus biological cycling of sulfur through the vegetation would be required to meet nutritional demands. Even for those sites (DF, HF, TL) in which inputs exceed requirements by less than 4 kg S ha^{-1} yr^{-1}, biological cycling of sulfur through the vegetation would also be important. Because inputs are calculated on an annual basis while requirements will be maximal during the growing season, some of the atmospherically derived sulfur may be leached from the soil during dormant periods and thus not be available for plant uptake. Thus, internal recycling of sulfur may be needed to meet plant requirements.

In considering the ways in which sulfur might be retained within the forests of the IFS network, we have made the broad distinction between organic and inorganic processes. Within the category of organic processes, the potential for sulfur incorporation by microbial processes and the inorganic processes are discussed in detail in the third part of this chapter. Neither of these sections addresses a third way in which sulfur could be retained within forest ecosystems, however, which is by tree uptake, return by litterfall, and accumulation within the forest floor. This process cannot be addressed directly because with lack of data on the rate of forest floor accumulation on the IFS sites, but we can perform an approximate analysis of this potential

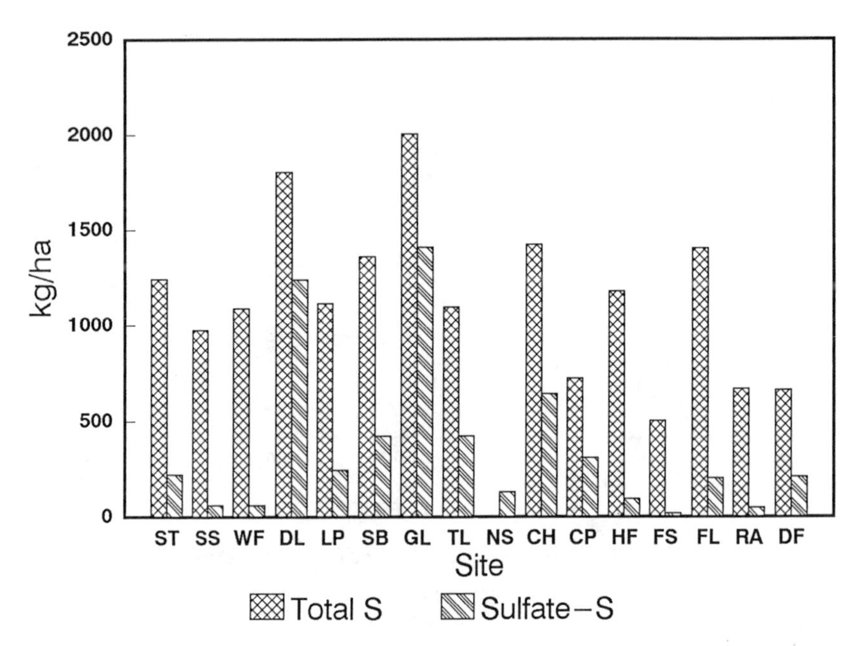

Figure 5.19. Total sulfur and sulfate content of mineral soil.

contribution by comparing litterfall rates with net ecosystem sulfur accumulation rates to see if there is any relationship.

As the data in Figure 5.21 show, litterfall sulfur equals a substantial fraction of net ecosystem sulfur retention in many of the IFS sites. However, only a fraction (if any) of this litterfall sulfur return would be accumulating in the forest floor. For estimating the maximum contribution of this litter input, we can calculate that if all the litterfall accumlated in the forest floor, it could account for half or more of the net ecosystem sulfur accumulation in all sites except those in sites in the Smoky Mountains (SS, ST, SB). If at little as one-tenth of litterfall sulfur accumulated in the forest floor annually, this accumulation would be minor for all sites. Forest floor accumulations are most likely to be occurring at the colder, high-elevation coniferous sites (SS, ST, FL). However, in each of these cases, the potential for sulfur accumulation via this mechanism is not high, given (1) the lack of sulfur retention now occurring or (2) the relatively small amount of litterfall sulfur as compared to ecosystem sulfur accumulation.

Solute Concentrations and Flux

The fluxes of SO_4^{2-} through the various ecosystem strata are given in Figure 5.22. A comparison of total sulfur inputs via atmospheric deposition versus SO_4^{2-} leached from beneath the B horizon is shown in Figure 5.23; in gen-

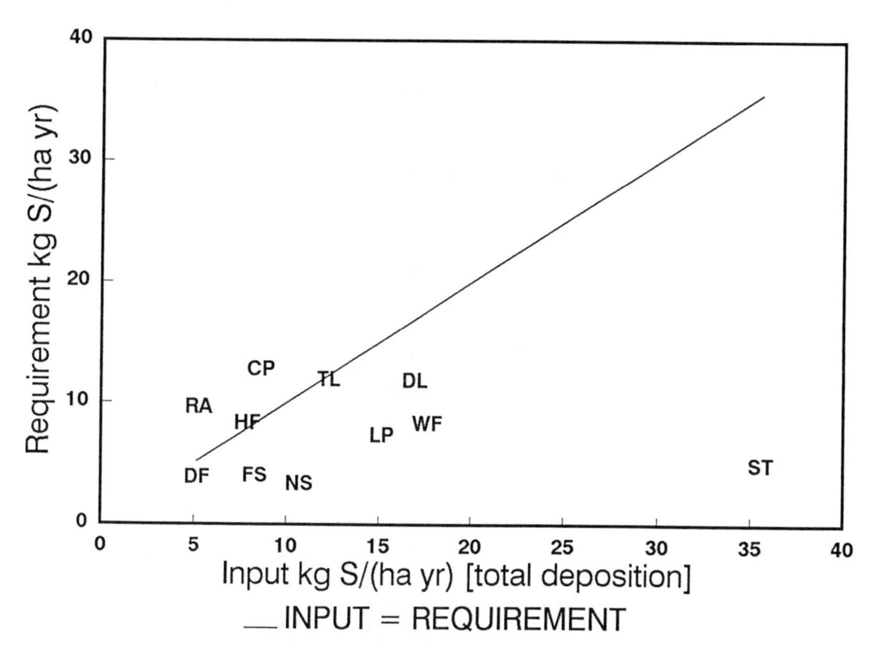

Figure 5.20. Vegetation requirements versus inputs of sulfur.

eral, SO_4^{2-} is behaving as a conservative ion with most of the SO_4^{2-} entering the system being leached from the mineral soil ($p < 0.01$, $r^2 = .94$, $n = 15$, slope = 1.4, constant = 0.37 keq SO_4^{2-} ha^{-1} yr^{-1} leached). Similar results have been reported for other sites, especially where sulfate adsorption potentials are low (Rochelle et al. 1987; Mitchell et al. 1991). Note, however, for some sites there is marked deviation from this relationship (Figure 5.24). The deviations from conservative behavior can result from processes that decrease sulfate retention (e.g., sulfate desorption or sulfur mineralization) not being in steady state with processes which increase sulfate retention (e.g., sulfate adsorption or sulfur immobilization). The relative roles of these processes are evaluated in later sections. In addition, any errors in calculating inputs and outputs may contribute to such differences. The importance of various assumptions in determining sulfur input-output budgets, especially the importance of accurately determining dry deposition, was discussed earlier in Chapter 5 and has been detailed elsewhere for the HF site (Shepard et al. 1989).

For the IFS sites, there is no evidence that either chemical weathering or the formation of secondary sulfate minerals play significant roles (see Chapter 10). These processes, however, cannot be ignored for other forest ecosystems. For example, in forest ecosystems underlain by sulfur-rich sedimentary rocks, weathering inputs can exceed atmospheric inputs (Gibson et

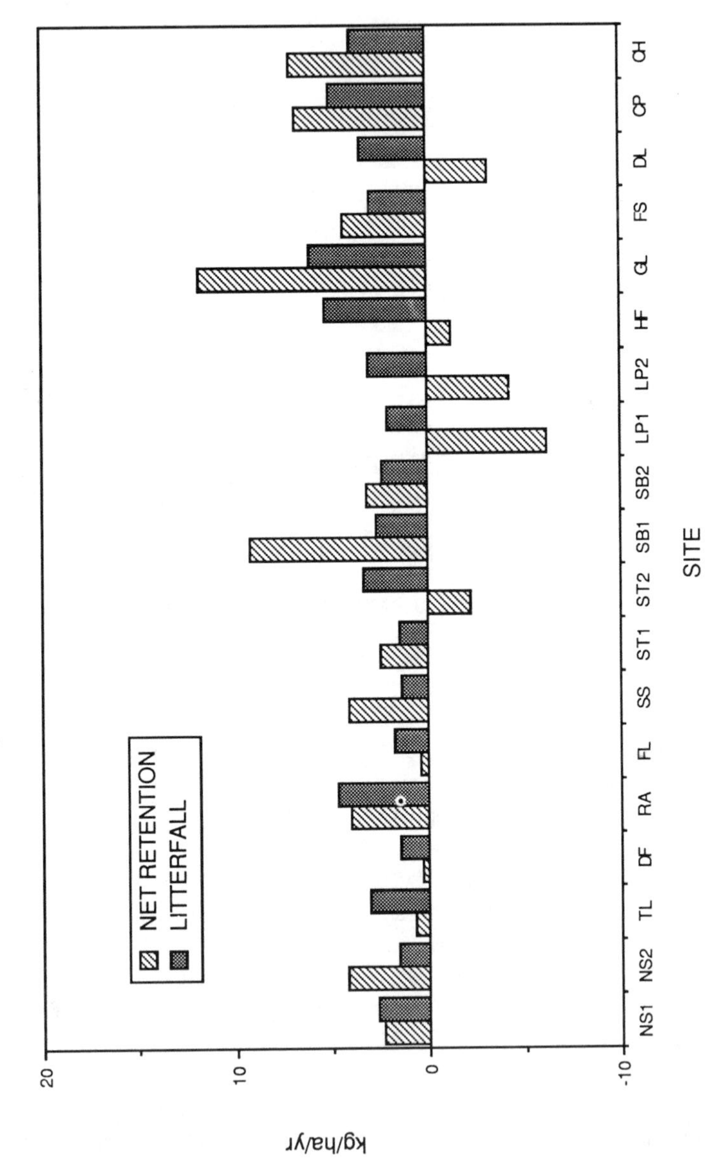

Figure 5.21. Litterfall versus sulfur retention.

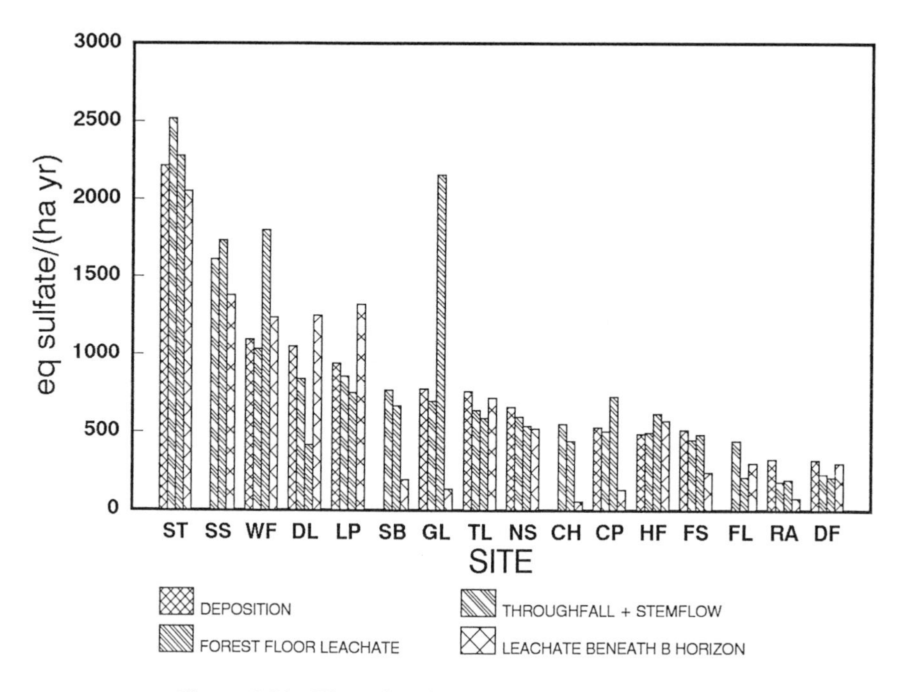

Figure 5.22. Flux of sulfate through ecosystem strata.

al. 1983; Mitchell et al. 1986). In Europe, with substantially higher atmospheric inputs (see Figure 5.13) and concentrations of SO_4^{2-} in the soil solution, the formation of sulfur minerals may also be important (Van Breeman 1973; Adams and Hajek 1978) and may be related to the very high SO_4^{2-} concentrations in certain non-IFS sites (GS and SF).

Within the IFS sites for which organic sulfur in solution was measured (Homann et al. 1990), it contributes to solute flux through the various strata of all ecosystems (Figure 5.25). The importance of organic sulfur as a contributor to sulfur flux through solution decreases with increasing SO_4^{2-} concentration, because the amount of organic sulfur in solution shows a smaller range than that of SO_4^{2-}. The amount of organic sulfur in precipitation appears negligible, but it increases as rain passes through the canopy and the forest floor, the latter of which has the highest organic sulfur concentrations. The concentration of organic sulfur generally shows a marked decrease after passage through the mineral soil. This decrease could be caused by the combined effects of catabolism and chemical precipitation of dissolved organic sulfur in the mineral soil. Although some of the soluble organic sulfur formed by microbial processes may be rapidly catabolized, it has also been suggested that the transport of soluble organic sulfur is an important mechanism for organic sulfur accumulation in soil (Schoenau and Bettany 1987; Mitch-

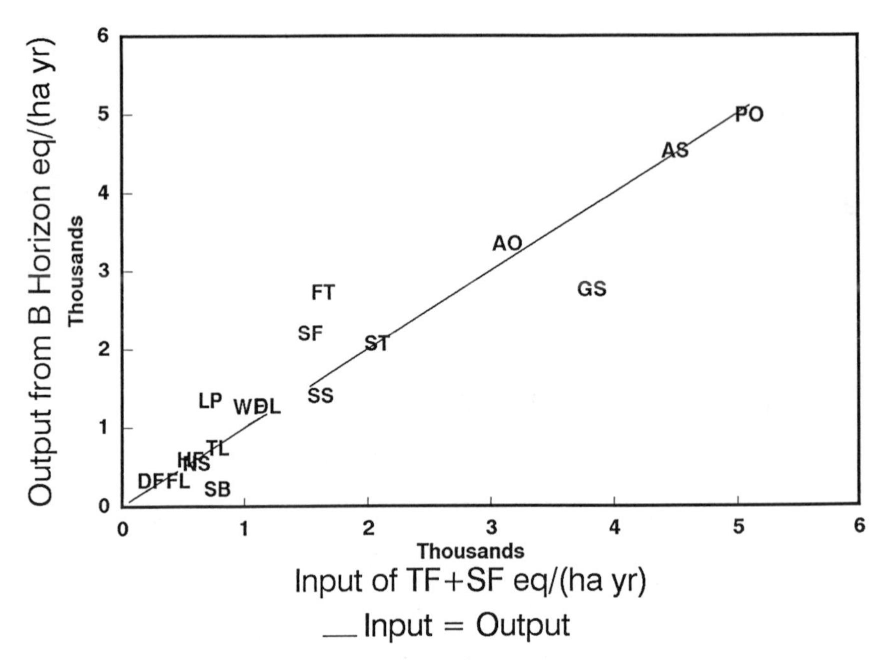

Figure 5.23. Total sulfur deposition versus leaching from B horizon.

ell et al. 1989) as has been shown for organic carbon (McDowell and Likens 1988). Because dissolved organic sulfur that is deposited in the mineral soil is apparently recalcitrant to decomposition, it may contribute to the large organic sulfur pool, most of which shows low biogeochemical activity (McLaren et al. 1985; Schindler and Mitchell 1987). For further details on organic sulfur in solutions for the IFS sites, with emphasis on the RA and DF sites in Washington, Homann et al. (1990) should be consulted.

Inorganic Sulfate Dynamics

R.B. Harrison and D.W. Johnson

Role of Sulfate Adsorption in Regulating Sulfate Flux

The ability of many soils to retain significant quantities of SO_4^{2-} by inorganic adsorption mechanisms has long been recognized (Ensminger 1954; Kamprath et al. 1956; Tikhova 1958; Chao et al. 1962). Because one of the predominant inputs in acidic precipitation is the SO_4^{2-} anion, the role of adsorption in regulating the flux of sulfate and associated cations through soil has been considered to be a factor of great importance (Cole and Johnson 1977; Johnson et al. 1983; Gaston et al. 1986). The mobility of SO_4^{2-} has received considerable attention because net SO_4^{2-} retention by soils can

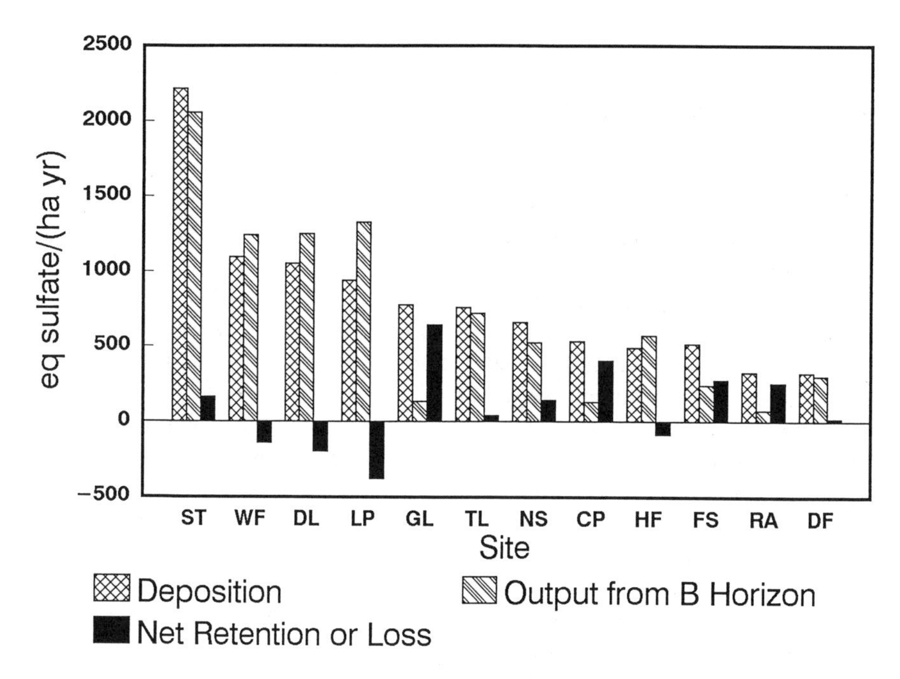

Figure 5.24. Net retention or loss of sulfur from forest ecosystems.

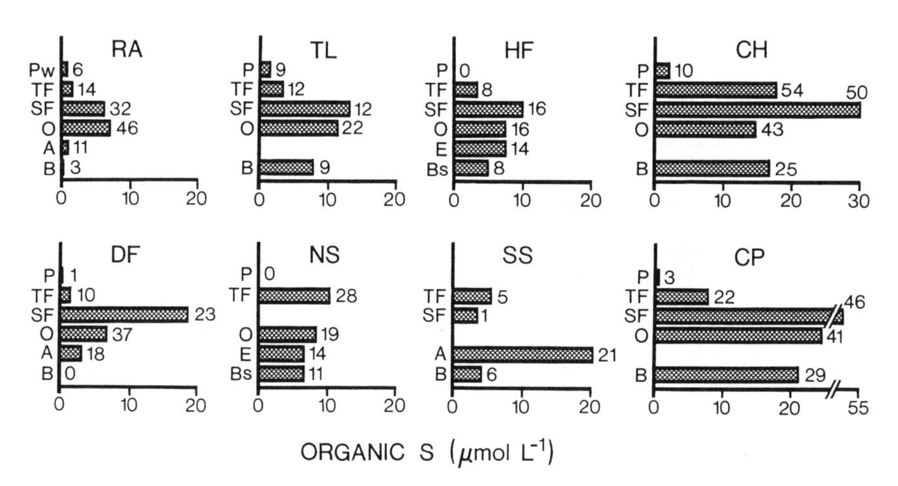

Figure 5.25. Organic S concentrations in various strata of selected IFS sites (P, bulk precipitation; Pw, wet-only precipitation; TF, throughfall; SF, stemflow; others, soil horizon solutions). Values following bars are organic S as percentage of total S. (From Homann et al. 1990).

result in reductions in cation leaching. Where an equivalent displacement of OH^- or other anion does not occur during the SO_4^{2-} retention process, cations are coadsorbed with SO_4^{2-} (Johnson and Cole 1977; Cronan et al. 1978; Singh et al. 1980).

Most research, including that completed as part of the U. S. Environmental Protection Agency Direct/Delayed Response Program (Church et al. 1989), has centered on sulfate adsorption potential as a result of increasing deposition levels only. The effect of SO_4^{2-} on ecosystem processes is considered primarily from the amount of SO_4^{2-} adsorption that will prevent sulfate from leaching through a forest ecosystem as well as the amounts of base cations or aluminum retained within the soil profile and not released to the aquatic environment. However, air pollution reductions in many regions have resulted in a decrease in SO_4^{2-} inputs, and the effect of decreasing SO_4^{2-} deposition levels on soil solution chemistry is equally important. The degree to which SO_4^{2-} adsorption is reversible is an important consideration in projecting recovery of acidified surface waters and acidified soils to preacidified levels following decreases in atmospheric deposition (Galloway et al. 1983; Reuss and Johnson 1986). Reductions in ambient SO_4^{2-} concentrations of solutions in forest ecosystems would be prolonged considerably if SO_4^{2-} (along with an equivalent quantity of acidic and basic cations) desorbs from acidified soils when atmospheric inputs decrease, whereas large decreases in SO_4^{2-} concentrations could be virtually instantaneous if SO_4^{2-} does not desorb or desorbs very slowly. The potential effects of varying degrees of reversibility of SO_4^{2-} adsorption on soil solution chemistry following a reduction in SO_4^{2-} inputs are shown in Figure 5.26. The possibilities for SO_4^{2-} reversibility are (1) more release of SO_4^{2-} than

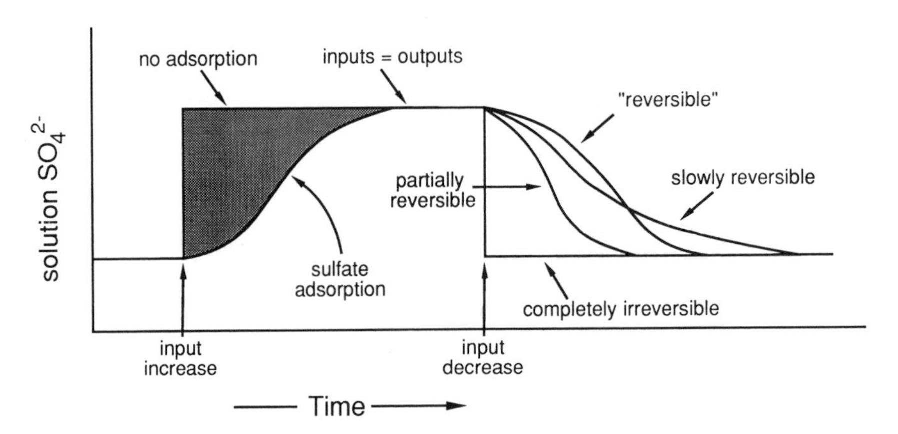

Figure 5.26. Paths of soil solution sulfate concentration with and without adsorption following deposition increases show variety of sulfate reversibility responses following decreases in deposition.

was absorbed (mobilization of other S pools); (2) complete or near-complete reversibility (all adsorbed SO_4^{2-} released); (3) partial reversibility of adsorbed SO_4^{2-}; and (4) little or no reversibility of SO_4^{2-} adsorption.

Researchers have not often reported observing more SO_4^{2-} released by desorption than was retained during previous adsorption studies, which may result from the nature of the soils for which results are published. Research has centered on soils that show a high degree of SO_4^{2-} adsorption. Thus, the soils generally would show more retention than release. However, soils that have been subject to previous high S loading and have a relatively small SO_4^{2-} retention capacity may easily show more release than adsorption. This observation could also be seen in a situation in which an extractant that removed previously insoluble SO_4^{2-} was used to determine SO_4^{2-} adsorption reversibility or because organic S mineralization could release soluble SO_4^{2-}.

For instance, Couto et al. (1979) speculated that a high percentage of adsorption reversibility observed in several samples they studied (more than 100% in the case of an Acrohumox B2 horizon) might be caused by mobilization of a high native amount of SO_4^{2-} in their samples, which would also explain a weak retention of SO_4^{2-} initially. Equally as possible would be the mineralization of organic S to soluble SO_4^{2-}, but no determinations of organic S pools were made in their study.

Chao et al. (1962) found SO_4^{2-} adsorbed by 11 of 15 Oregon soils was readily desorbed by water. Four of the 15 horizons studied demonstrated relatively irreversible SO_4^{2-} adsorption. The previously adsorbed SO_4^{2-} was much more readily extracted with KH_2PO_4 solution, as would be expected by the potential SO_4^{2-} retention mechanisms outlined here, because phosphate could effectively displace inner- and outer-sphere SO_4^{2-} simply by its higher negative valence and associated greater charge density. This supports the inner-sphere concept of SO_4^{2-} adsorption as providing for adsorption irreversibility. Unfortunately, the concentrations of the inputs were not specified in this study, and the evaluation of relative nonadsorbance was based on movement of SO_4^{2-} down into a column of soil, so little can be said of these results in the sense of comparing soil to soil.

In the study mentioned previously (Chao et al. 1962), researchers found that 4 of 15 Oregon soils showed an appreciable ability to hold an unspecified concentration of ^{35}S sulfate ions against leaching. The interpretation of "appreciable" was that little of the SO_4^{2-} added was leached below a 4-cm depth in the soil column with an application of 20 cm of water, and that that soil exhibited a considerable degree of SO_4^{2-} adsorption.

Further studies of two of the sulfate-adsorbing soils (Chao et al. 1962) showed that SO_4^{2-} adsorption was mostly reversible (>70%) after four extractions with deionized water (samples were desorbed with 2-hour extractions successively (soil to solution ratio, 1:5)). Thus, the observation that adsorption reversibility depends on the time period of desorption applies in these studies. In a study of the SO_4^{2-} adsorption reversibility by miscible displacement (Hodges and Johnson 1987), it was estimated that completely desorbing previously adsorbed SO_4^{2-} would require a time period 10 to 20 times greater than the initial adsorption time period, although researchers did not actually follow desorption through its completion.

In a study of SO_4^{2-} adsorption on kaolinite (Aylmore et al. 1967), SO_4^{2-} adsorption was mostly reversible (>50%), but SO_4^{2-} adsorbed onto Fe and Al oxides was essentially irreversibly adsorbed (≤50%). It would be expected that SO_4^{2-} adsorbed onto amorphous Al and Fe oxides might show a higher degree of irreversibility compared to kaolinite, because there would be a higher potential for the SO_4^{2-} ion to enter the inner sphere of the amorphous surface than the crystalline structure of kaolinite. In another study, one of four soil horizons (10–20 cm; sampling down to 30 cm) showed irreversible SO_4^{2-} adsorption (Weaver et al. 1985). The particular horizon showing irreversible adsorption had the lowest pH measured and a lower native SO_4^{2-}-S concentration than soil horizons above it. This would be expected because soil properties that lead to higher initial SO_4^{2-} adsorption appear to decrease adsorption reversibility.

In a study of SO_4^{2-} adsorption reversibility, 10%–30% of the total of previously adsorbed SO_4^{2-} was extracted with dilute HCl (pH 3.5), 20% to 40% with water (pH 5.6), 50% to 60% with 0.01 *M* $Ca(NO_3)_2$, and nearly 100% of adsorbed SO_4^{2-} with 0.01 *M* $Ca(H_2PO_4)_2$ by repeated extractions of several soils (Singh 1984). Soil samples included iron-podzols (highest SO_4^{2-} adsorption) and brown earths (lowest SO_4^{2-} adsorption). In another study, 33% to 76% of adsorbed SO_4^{2-} was desorbed with four KH_2PO_4 (16 m*M* P) extractions from clay and sandy loam soils from the West Indies (Haque and Walmsley 1973).

Other studies (Nodvin et al. 1986; Fuller et al. 1987) noted that the subsurface horizons of some Northeastern spodosols adsorbed up to 0.5 mmol SO_4^{2-} kg^{-1} soil (final solution concentration of 1 m*M* Na_2SO_4). Little of the SO_4^{2-} retained in these horizons was desorbed by water, but desorption was higher with phosphate extraction, ranging from 0.3 to 2.1 mmol kg^{-1}.

Bornhemisza and Llanos (1967) observed leaching of ^{35}S-SO_4^{2-} [applied as $CaSO_4$, K_2SO_4, and $(NH_4)_2SO_4$)] by water in columns of three Costa Rican soils, including the A and B horizons of a regosol, a pumice-originated alluvial soil, and a latosol. They found that ^{35}S did not move to any large degree in the latosol and the alluvial soil, but that the regosol showed little retention. Where orthophosphate was intermixed with SO_4^{2-} applications [$Ca(H_2PO_4)_2$ or triple superphosphate], the added SO_4^{2-} was more mobile because of displacement of SO_4^{2-} by phosphate; however, in the subsoil of the latosol, SO_4^{2-} was retained despite the additions of phosphate.

As mentioned previously, Aylmore et al. (1967) found SO_4^{2-} adsorption on kaolinite to be mostly reversible, but SO_4^{2-} adsorbed onto Fe and Al oxides was essentially irreversible, indicating that soils with high Fe and Al oxide content show high adsorption and low adsorption reversibility. Khanna and Beese (1978) found that leaching of chloride and nitrate salts through a podzolic acid brown soil released little previously adsorbed SO_4^{2-}. Chloride and NO_3^- would not be expected to compete to a high degree with SO_4^{2-} for retention sites because of their lower charge density; however, the high concentrations of these monovalent ions used in the studies of Khanna and Beese should have displaced SO_4^{2-} retained in the outer sphere of the retention surface by mass action. This would be analogous to displacing exchangeable Ca^{2+} by adding high concentrations of KCl to a soil, thus displacing the Ca^{2+} by mass action. These observations tend to support inner-sphere retention of SO_4^{2-} as the primary mechanism for irreversible adsorption.

Adsorption/Desorption/Reversibility Results of IFS Studies

Studies of SO_4^{2-} adsorption potential and the reversibility of SO_4^{2-} adsorption in the IFS soils were undertaken to determine the impact that SO_4^{2-} adsorption properties of soils have on regulating the flux of sulfate and associated cations through the forest ecosystems studies as part of the IFS. Soil samples collected from IFS sites of differing soil and forest type, atmospheric deposition histories, and physiographic locations were analyzed for soil properties, including pH (in water and 0.01 *M* $CaCl_2$), total S, ester SO_4^{2-}, phosphate, and water-extractable SO_4^{2-}, Bray P, total C, extractable Al and Fe (dithionate/citrate, oxalate, and pyrophosphate), cation exchange capacity (CEC), exchangeable Ca, Mg, K, Na, and Al. SO_4^{2-} adsorption capacity was determined by sequential equilibration with a percolating solution (0.25 m*M* $CaSO_4$) and desorption reversibility by leaching with deionized water. Two sequences of adsorption, desorption, and extraction were followed to estimate five sulfate pools as follows (procedure followed and sulfate fraction measured):

1.a. Samples are leached with deionized water until the effluent SO_4^{2-} concentration is less than 0.25 m*M*. This gives a measure of native water-soluble SO_4^{2-}.
1.b. Following procedure 1.a., samples were extracted with 1000 ppm P as NaH_2PO_4 until the SO_4^{2-} concentration dropped below the detectable limit for this matrix. This gives a measure of the native insoluble SO_4.
2.a. Samples are leached with 0.25 m*M* $CaSO_4$ solution until the concentration of the percolate reaches 0.25 m*M*. This gives SO_4^{2-} adsorption from a 0.25 m*M* solution.
2.b. Following procedure 2.a., samples were leached with deionized water as in 1.a. The value "1a + 2a − 2b" gives a measure of reversible SO_4^{2-} adsorption.
2.c. Samples from 2.b. were extracted with phosphate as in 1.b. The results of "2c − 1b" give a measure of irreversibly adsorbed SO_4^{2-} from SO_4^{2-} adsorption sequence in 2a.

The soils vary widely in the quantities of SO_4^{2-} adsorbed, desorbed, and in phosphate-soluble pools. All but one subsurface soil horizon showed net

SO_4^{2-} adsorption (up to 4.5 mmol SO_4^{2-} kg^{-1} soil), indicating many of these soils were not yet saturated with respect to an input concentration of 0.25 m*M* SO_4^{2-}. Most subsurface soil horizons also showed irreversible SO_4^{2-} adsorption as evidenced both by changes in phosphate-extractable SO_4^{2-} pools and by a comparison of desorption of SO_4^{2-} before and after saturation by the 0.25 m*M* SO_4^{2-} solution.

The characteristics all of these pools for each soil are important in determining potential SO_4^{2-} adsorption capacity. For instance, the sequence of adsorption, desorption, and extraction for the Duke Bt1 and Findley Lake B2ir soil horizons are shown in Figure 5.27. Note that the major difference between the Findley Lake and Duke soils is the amount of phosphate-extractable SO_4^{2-} and the solution concentration at which the soil begins to adsorb SO_4^{2-}. The Findley Lake soil represents a relatively young soil in a low-deposition environment capable of retaining a relatively high amount of additional SO_4^{2-} whereas the Duke soil represents a more developed soil with a high SO_4^{2-} adsorption capacity that is nearly filled.

The observations of SO_4^{2-} adsorption, desorption, and extractable pools for several other soils are summarized in Table 5.3. The relative proportion of sizes of the five operational pools varied considerably with site and soil horizon. For instance, the Beech site BC horizon had a large pool of phosphate extractable SO_4^{2-} (about 13 mmol kg^{-1}), and this is probably the primary reason it showed relatively low levels of net SO_4^{2-} retention from a 0.25 m*M* $CaSO_4^{2-}$ solution. Most of the soils showed a relatively close cor-

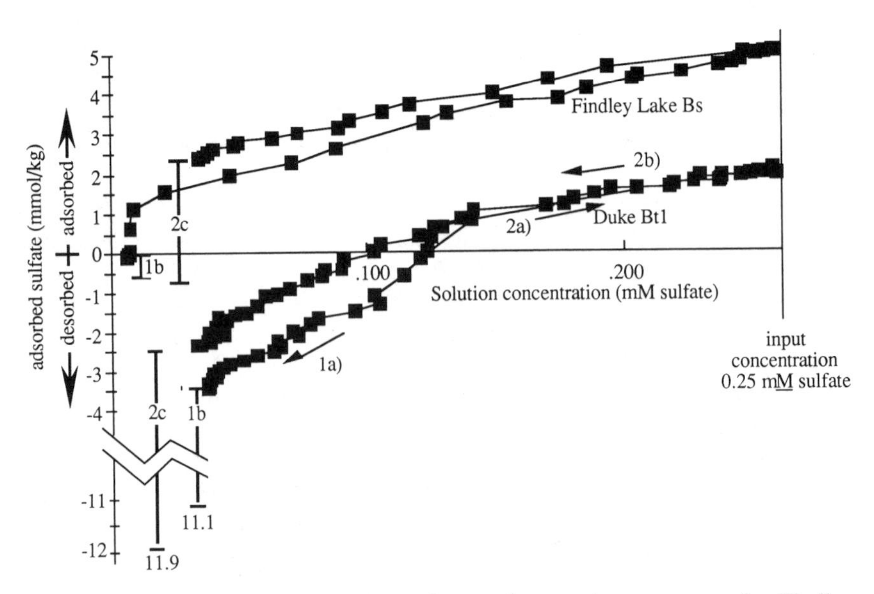

Figure 5.27. Sulfate adsorption, desorption, and extraction sequences for Findley Lake B2ir and Duke Bt1 soils.

relation between SO_4^{2-} irreversibly adsorbed and increases in the phosphate-extractable pools. However, during the time period of the extraction procedure several soils appeared to mineralize organic S to SO_4^{2-}, and one soil sample may have incorporated SO_4^{2-} organically, probably as a result of microbial incorporation.

A measure of the reversibility of SO_4^{2-} adsorption is calculated by making two estimates of irreversibly adsorbed SO_4^{2-}: (1) by difference in phosphate-extractable fractions, and (2) by comparing the desorption of SO_4^{2-} following the adsorption treatment with SO_4^{2-} with desorption from the untreated soil. Comparisons are as follows:

$$\text{Irreversibly adsorbed } SO_4^{2-} = \text{``2c''} - \text{``1b''} \quad [5.3]$$

$$\text{Irreversibly adsorbed } SO_4^{2-} = \text{``2a''} - (\text{``2b''} - \text{``1a''}) \quad [5.4]$$

Evaluation of the reversibility of SO_4^{2-} adsorption is very important because SO_4^{2-} that is irreversibly adsorbed represents a pool removed from inorganic leaching. In the present study there are two measures of SO_4^{2-} adsorption reversibility. One is the increase in phosphate-extractable SO_4^{2-} (Eq. [5.3]) brought about by SO_4^{2-} adsorption from solution during equilibration "2a." It would be expected that irreversibly or slowly reversible adsorbed SO_4^{2-} would increase in an insoluble inorganic pool. The second measure of irreversibly adsorbed SO_4^{2-} is the difference between native water-soluble SO_4^{2-}, SO_4^{2-} adsorbed from $CaSO_4^{2-}$ inputs, and SO_4^{2-} desorbed after $CaSO_4^{2-}$ treatment (Eq. [5.4]). If these two measures of the irreversibility of SO_4^{2-} adsorption are equivalent, there would be a strong indication that organic S incorporation or mineralization processes are not important relative to the inorganic SO_4^{2-} adsorption/desorption processes.

A total of 29 of 36 subsoils showed irreversible adsorption based on increases in phosphate-extractable SO_4^{2-} (Eq. [5.3]) while 28 of 36 subsoils showed irreversible adsorption based on the adsorption/desorption criteria of Eq. [5.4]. Despite air-drying and the potential perturbations associated with laboratory procedures on S incorporation and mineralization processes, most soil samples showed conservation of added sulfate. In other words, SO_4^{2-} adsorbed from solution was recovered quantitatively by water or phosphate extraction, or:

$$\text{``2c''} - \text{``1b''} = \text{``2a''} - (\text{``2b''} - \text{``1b''}) \quad [5.5]$$

A 1:1 graph of these two measures of irreversibly adsorbed SO_4^{2-} is shown in Figure 5.28. Although two soil samples, the Oak Ridge (loblolly pine) B horizon and the Cedar River (Douglas fir) B22 horizon, showed wide deviations in these comparisons, most of the other soil samples showed a strong 1:1 relationship. These observations suggest that inorganic processes are dominant in these air-dried soils, and that inorganic adsorption is likely

Table 5.3. Selected Soil Properties and Sulfate Adsorption Data for Sites

Site	Vegetation	Soil Horizon	Soil pH	Total C	Oxalate Al	Total S
Oak Ridge	Loblolly pine	B	5.2	182	67	3.1
Oak Ridge	Chestnut oak	B1	4.7	13	36	3.1
Oak Ridge	Chestnut oak	B2	4.8	13	37	5.5
Oak Ridge	Yellow poplar	B1	5	23	41	3.9
Oak Ridge	Yellow poplar	B2	5	23	41	3.6
Camp Branch	White oak	B	4.7	—	59	3.2
Camp Branch	White oak	C1	4.9	—	48	3.3
Camp Branch	White oak	C2	5	—	67	4.4
Smoky Mts.	Red spruce 1	B	4.1	3017	170	9.6
Smoky Mts.	Red spruce 1	C	4.7	407	145	2.7
Smoky Mts.	Red spruce 2	B1	4.3	2853	182	9.6
Smoky Mts.	American beech	A	4	5290	133	12.7
Smoky Mts.	American beech	B	4.6	2190	170	13.8
Smoky Mts.	American beech	BC	4.8	1404	245	15.7
Coweeta	Oak-hickory	Bt	5.3	—	37	5.7
Coweeta	White pine	Bt	5.6	375	59	5.3
Coweeta	White pine	BC	5.8	167	41	3.1
Duke Forest	Loblolly pine	Bt1	4.8	214	63	14.3
Duke Forest	Loblolly pine	C	4.8	114	67	14.9
B.F. Grant	Loblolly pine	Bt	5.2	263	47	3.3
B.F. Grant	Loblolly pine	BC	5.2	75	51	3.9
Florida	Slash pine	Bt	4.6	180	51	1.6
Florida	Slash pine	C	4.7	102	32	0.8
Maine	Red spruce	Bs	4.2	0	803	3.1
Cedar River	Red alder	B21	5.1	6524	778	10.9
Cedar River	Red alder	B22	5.2	4033	826	9.4
Cedar River	Douglas fir	B21	5.6	2199	730	6.6
Cedar River	Douglas fir	B22	5.5	1191	693	7.9
Findley Lake	P. silver fir	Bs	4.6	5402	1012	13.0
Whiteface Mt.	Red spruce	Bup	4.8	5436	663	14.4
Huntington Forest	Sugar maple	Bs1	4.9	2787	693	2.2
Turkey Lakes	Sugar maple	BfL1	4.6	4291	296	11.2
Turkey Lakes	Sugar maple	Bfh2	4.9	3281	493	7.4
Turkey Lakes	Sugar maple	Bf	4.9	2245	400	6.3
Norway	Norway spruce	Bs	4.8	795	128	2.2
Norway	Norway spruce	BC	5	78	123	2.4
Norway	Norway spruce	C	5.2	160	70	1.6

to be the most important factor in controlling SO_4^{2-} concentrations in waters equilibrating with these soils. For example, the Bs horizon in the Findley Lake (Pacific Silver fir) site had no native water-soluble SO_4^{2-} and only 0.6 mmol kg^{-1} of native insoluble adsorbed SO_4^{2-} (see Table 5.3). The soil retained 5.0 mmol SO_4^{2-} kg^{-1} on treatment with 0.25 m*M* SO_4^{2-} solution, far more than the native pools combined. A water leaching resulted in the removal of 3.2 mmol kg^{-1} of this retained SO_4^{2-}, and an additional phosphate extraction removed 2.7 mmol SO_4^{2-} kg^{-1}. These results show conservation of added SO_4^{2-}. Applying Eq. [5.5], we see that the net balance is 0.2 mmol SO_4^{2-} kg^{-1}, a relatively small amount.

Ester Sulfate	Extraction Series 2a	2b	2c	1a	1b	Irreversible Adsorbed Sulfate 2c–1b	Irreversible Adsorbed Sulfate 2a–2b–1a
1.6	0.42	2.46	0.35	0.96	0.07	0.28	−1.08
2.4	0.37	1.45	1.09	1.18	0.95	0.14	0.09
3	0.75	1.26	1.82	0.90	1.94	−0.12	0.39
1.8	−0.15	0.40	0.08	0.49	0.11	−0.03	−0.06
1.9	0.03	0.62	0.92	0.92	0.42	0.50	0.33
1.5	0.50	1.16	0.54	0.79	0.30	0.24	0.13
2.2	1.38	1.34	1.82	0.67	1.14	0.69	0.71
3.7	7.22	5.08	5.10	0.10	3.02	2.08	2.23
5.3	0.31	0.88	0.32	1.08	0.20	0.12	0.51
2	0.47	0.97	0.71	0.49	0.15	0.56	−0.01
5.6	0.03	2.24	0.98	2.60	0.89	0.09	0.39
12.3	−0.76	0.86	0.37	1.62	0.27	0.10	−0.01
10.4	0.28	1.21	0.54	1.20	0.44	0.09	0.27
12.7	0.84	3.30	5.95	2.26	5.90	0.05	−0.21
4.4	1.34	1.90	1.92	1.35	1.07	0.85	0.79
4.5	2.67	2.56	2.70	0.56	1.91	0.79	0.67
2.7	3.89	2.33	3.00	0.10	1.46	1.54	1.65
12.6	1.75	3.79	9.29	3.58	7.62	1.67	1.55
14.2	2.78	5.31	11.90	3.62	10.30	1.60	1.10
n.d.	3.09	2.22	3.98	0.00	3.00	0.97	0.87
n.d.	6.39	4.65	5.59	0.00	3.69	1.89	1.74
n.d.	1.00	0.80	0.77	0.37	0.58	0.20	0.56
n.d.	0.93	0.70	0.57	0.00	0.34	0.23	0.23
n.d.	1.05	0.77	2.22	0.00	2.15	0.07	0.28
4.1	2.31	2.19	0.56	0.29	0.14	0.42	0.41
3.4	1.93	1.65	0.67	0.44	0.39	0.28	0.71
4.4	3.18	2.40	3.31	0.87	1.47	1.84	1.65
6.4	2.42	3.53	2.68	0.79	1.79	0.89	−0.32
4.8	4.97	3.17	2.68	0.00	0.64	2.04	1.80
5.8	0.64	1.16	0.49	0.62	0.43	0.07	0.10
3.7	1.54	1.89	1.23	0.63	0.70	0.53	0.28
3.7	1.24	1.58	0.62	0.63	0.29	0.32	0.29
3	0.82	1.34	0.41	0.49	0.33	0.08	−0.03
2.7	0.75	0.93	0.39	0.23	0.37	0.02	0.06
n.d.	−0.41	1.31	0.86	1.93	0.83	0.03	0.22
n.d.	−0.28	0.94	1.78	1.32	1.66	0.11	0.10
n.d.	−0.04	0.48	0.40	0.65	0.28	0.12	0.13

In the sequence of 2a–2b–2c for the Findley Lake Bs, 1.8 mmol SO_4^{2-} kg^{-1} soil or 36% of added SO_4^{2-} was retained irreversibly by the phosphate-extractable SO_4^{2-} pool. In general, soils that adsorbed higher amounts of SO_4^{2-} also showed higher amounts of irreversibly adsorbed SO_4^{2-} (Figure 5.29). On average, approximately 36% of adsorbed SO_4^{2-} was irreversibly adsorbed for all soils, although the data were variable.

The effects of time on the permanence of this retention are questionable because they were incompletely evaluated in this study. Researchers have noted that time has a positive influence on SO_4^{2-} adsorption (Barrow and Shaw 1977), but Singh (1984) saw no correlation between SO_4^{2-} desorption

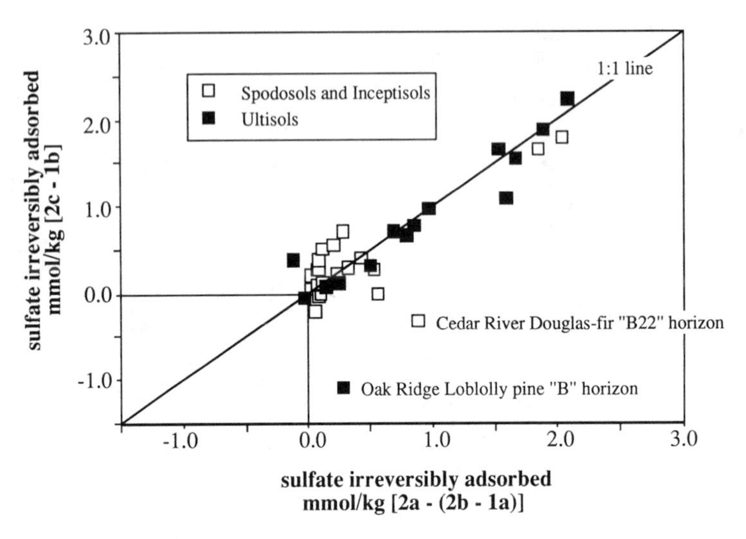

Figure 5.28. Comparisons of two methods for estimation of irreversibly adsorbed SO_4^{2-} for subsoils from IFS sites.

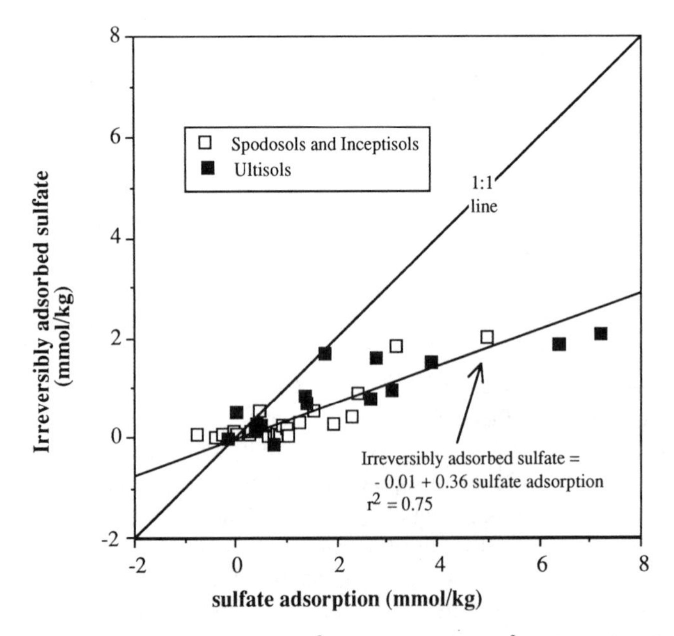

Figure 5.29. Irreversibly adsorbed SO_4^{2-} versus total SO_4^{2-} adsorption from 0.25 m*M* $CaSO_4$ solution for subsoils from IFS sites.

and time. Irreversible SO_4^{2-} adsorption appears to be an important property of many of these soils.

Soil Properties Used in Predicting SO_4^{2-} Adsorption

The exact nature of SO_4^{2-} adsorption in soils is chemically complex. However, SO_4^{2-} adsorption is generally considered to take place by means of two mechanisms, one of which results in an equivalent release of another anion, typically OH^-. The other mechanism results in neutral adsorption of SO_4^{2-} with accompanying cations, which is often called "selective" or "salt" adsorption. The multiple negative charge and the bridging that can potentially occur as a result of the divalent nature of the SO_4^{2-} ion are important for coadsorption of SO_4^{2-} with cations.

Several soil factors have been observed by a variety of researchers to influence SO_4^{2-} adsorption by soils. Adsorption typically increases as (1) the quantity of Al and Fe oxides increases; (2) soil pH decreases; (3) organic matter decreases; and (4) the concentration of similarly adsorbed anions decreases. The relative impact of each of these soil factors varies greatly in its contribution to SO_4^{2-} adsorption, and researchers have sometimes concentrated on one factor in making an evaluation of the relative effect on SO_4^{2-} adsorption. For instance, soils of a finer texture and higher Fe and Al oxide content often have higher levels of organic matter than similar soils of a coarser texture. If one were looking specifically at the effects of organic matter on SO_4^{2-} adsorption, higher organic matter levels might then appear in some studies to enhance SO_4^{2-} adsorption. Procedures that alter a particular soil property experimentally may also change other soil properties, and unless all changes in soil properties are carefully evaluated, their actual effect on SO_4^{2-} adsorption might not be noted.

Effect of Fe and Al Oxides on SO_4^{2-} Adsorption

In general, the relative SO_4^{2-} adsorpion characteristics of soils have been most closely related to Al and Fe hydrous oxide content (Chao et al 1962; Adams and Hajek 1978; Rajan 1978; Couto et al. 1979; Johnson and Henderson 1979; Neary et al. 1987), although the relationship between SO_4^{2-} adsorption and extractable Al and Fe is often quite variable. Such surfaces have a variety of mechanisms for SO_4^{2-} retention, and potential adsorption is altered by pH in the same way as empirical observations occur. Recent studies (e.g., Nodvin et al. 1986) have shown a pH level at which the adsorption reaches a maximum for several soils, and lower pH levels typically result in solubilization of surface coatings of Fe and Al.

Thus, the lower pH levels may destroy the SO_4^{2-} adsorption surfaces and reduce potential SO_4^{2-} retention sites unless formation of new surfaces takes place. One factor typically cannot explain observations fully, because SO_4^{2-} adsorption is considered to be a complex factor of interaction of soil SO_4^{2-}

with organic matter, pH, and extractable Al and Fe (Parfitt and Smart 1978; Johnson and Todd 1983; Dethier et al. 1988).

Effects of pH on SO_4^{2-} Adsorption

Sulfate adsorption typically increases with decreasing soil pH (Kamprath et al. 1956; Elkins and Ensminger 1971; Gebhardt and Coleman 1974; Couto et al. 1979; Singh 1984; Bergseth 1985; Zhang et al. 1987). Thus, if soils acidify by natural soil formation or man-caused processes, they may retain additional SO_4^{2-} and support a reduced SO_4^{2-} concentration in solution for a given level in soil. Some studies in which soils were artificially acidified indicate that there is a pH level below which SO_4^{2-} adsorption begins decreasing (Nodvin et al. 1986). This point at which SO_4^{2-} adsorption began decreasing was associated with the release of soluble aluminum, indicating that the precipitation and surface adsorption of Fe and Al hydroxy-sulfates (Adams and Rawajfih 1977; Rajan 1978; Parfitt and Smart 1978; Singh 1984), which are driven by lower pH, were probably not the primary mechanism for SO_4^{2-} retention in these studies. Rather, retention was associated with surface coatings of Fe and Al oxides, which dissolve at lower pH and release soluble Fe and Al.

As noted, the possible observations of the effects of pH on SO_4^{2-} adsorption are likely to depend on the specific mechanism responsible for SO_4^{2-} adsorption, and these may vary from soil to soil. Most retention mechanisms we can consider, however, would show an increase in adsorption with decreasing pH. Considering the simplest retention mechanism (simple anion adsorption by anion exchange), for instance, the adsorption of SO_4^{2-} would be increased by adding H^+ to the system. Several researchers found SO_4^{2-} adsorption to be correlated to measurements of surface positive charge (Van Raij and Peech 1972; Marsh et al. 1987). Based on the observations of these studies, positive and negative adsorption sites are considered to be separated in space, and the amphoteric nature of the anion exchange capacity gave an increased SO_4^{2-} adsorption with decreasing pH.

As H^+ is forced onto pH-dependent sites, the mineral surface gains positive charge and thus can retain more anions. While supporting a net negative charge at higher pH levels, the nature of the retention sites is changed from negatively charged to neutral to positively charged by protonation. When positively charged, the surface can retain anions such as Cl^-, SO_4^{2-}, and PO_4^{3-}. In the case of multivalent anions such as SO_4^{2-} and PO_4^{3-}, a bridge can form between the anion and the surface, and there is a potential for covalent bonds to form between the anion and the surface by loss of waters (Sposito and Goldberg 1984). Anions that bridge in this way are essentially part of the new surface, are likely to be held tenaciously, and would potentially change the original surface from an anion to a cation exchanger.

Effect of Organic Matter on SO_4^{2-} Adsorption

Organic anions appear to compete directly with SO_4^{2-} for retention sites (Johnson and Todd 1983), although in comparisons of subsoils of highly contrasting physical and chemical properties it is can be observed that SO_4^{2-} adsorption is directly proportional to carbon content (Singh 1984). Observed positive correlations between SO_4^{2-} adsorption and organic matter contents appear to result from the relative stability of organic matter–Al surface complexes in many soil systems. Thus organic matter content may be directly correlated with extractable Al and Fe in subsoils (Harrison et al. 1989).

The mechanism for organic matter competing for SO_4^{2-} adsorption sites can be either that of an anion adsorbing onto and blocking SO_4^{2-} adsorption sites or of solubilizing and removing surface Al and Fe. Solubility of Fe and Al from soil surfaces is typically greatly increased by the addition of organic acids such as citrate and oxalate (Schnitzer 1969; Bartlett and Riego 1972; Hue et al. 1988). Thus, the solubility and removal of SO_4^{2-} adsorption sites may be important in explaining reductions in SO_4^{2-} adsorption with increasing levels of organic matter in soils.

As mentioned previously, the role of organic matter in SO_4^{2-} adsorption may be easily confused with other soil factors, such as surface area, or the presence of Fe and Al oxides, which may increase organic matter retention as well as SO_4^{2-} retention. Studies that have looked at the role of organic matter content in similar soils have typically found a negative relationship between the two. Surface soils, for instance, typically have the highest levels of organic matter and exhibit little SO_4^{2-} adsorption.

Effects of Competing Anions on Sulfate Adsorption

The primary anion competing with SO_4^{2-} for inorganic adsorption in soils is phosphate. A detailed discussion of the interactions of sulfate and phosphate is included in Chapter 7.

Observations from the IFS Study

Several soil properties appeared to be correlated with observed SO_4^{2-} adsorption. The concentration of oxalate extractable Al was most closely correlated with SO_4^{2-} adsorption for Spodosols and Inceptisols. This observation has been seen in several previous studies (Barrow 1969; Haque and Walmsley 1973; Johnson and Todd 1983; Nommik et al. 1984; Singh 1984). For example, Figure 5.30 shows the significant relationship (0.01 level) between SO_4^{2-} adsorption and oxalate-extractable Al for Spodosols and Inceptisols. Very noticeable in this figure is the lack of high levels of oxalate Al in the weathered Ultisols despite a wide range of SO_4^{2-} adsorption. The regression coefficient for Inceptisols and Spodosols of SO_4^{2-} adsorption versus the Al was .63 (r^2).

None of the Al or Fe extractions measured correlated well with observed SO_4^{2-} adsorption for soil horizons collected from Ultisols, although it has been shown in other studies that extractable Fe may be important (Chao et al. 1962; Aylmore et al. 1967; Parfitt and Smart 1978; Johnson et al. 1980; Singh 1984).

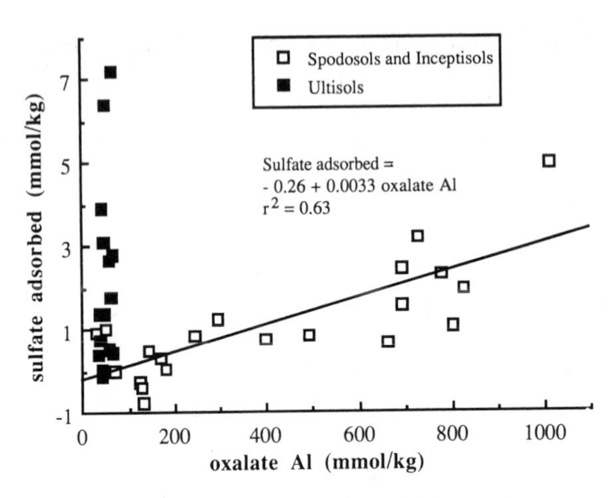

Figure 5.30. Sulfate adsorbed from 0.25 m*M* $CaSO_4$ solution versus oxalate-extractable Al for subsoils from IFS sites; regression line for Spodosols and Inceptisols combined.

It was impossible to assess the effect of organic matter on SO_4^{2-} adsorption in this study. Organic matter levels were directly correlated with extractable Al, so the apparent relationship between SO_4^{2-} adsorption and organic matter was positive and very similar in appearance to Figure 5.30.

A significant positive relationship (0.01 level) between SO_4^{2-} adsorption and pH was seen when all soil samples were plotted simultaneously (Figure 5.31). This direct relationship is in contrast to what has been seen in several other studies of SO_4^{2-} adsorption where soil samples have been acidified by addition of acid solutions (Chao et al. 1964; Harward and Reisenauer 1966; Couto et al. 1979; Huete and McColl 1984). In other studies, sulfate is displaced by the addition of base.

Below about pH 3.5, additional drops in pH have been seen to lower SO_4^{2-} adsorption because of dissolution of Al and possible destruction of adsorption sites (Nodvin et al. 1986; Dongsen and Harrison 1990). Because a range in pH was achieved by natural soil development processes and not by adding acid or base to the same samples of the soil material over a relatively short time period, it is difficult to directly compare the observations of this study with those of the other studies.

Organic Sulfur Dynamics Including Mineralization and Immobilization of Various Organic Fractions

J.W. Fitzgerald and A.R. Autry

Central to assessing the deleterious effects of acidic precipitation on forested ecosystems is a thorough understanding of sulfur accumulation and retention mechanisms in these systems. Sulfate is extremely mobile in the soluble

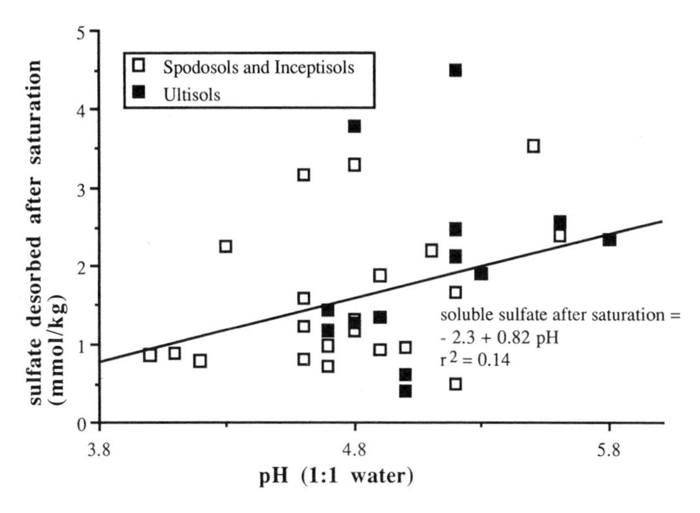

Figure 5.31. Sulfate adsorbed from 0.25 m*M* $CaSO_4$ solution versus soil pH for subsoils from IFS sites.

form (Bormann 1974), and if not immobilized can leach through the soil profile and be lost from the ecosystem (Huete and McColl 1984). A concomitant loss of nutrient cations is associated with this leaching (Johnson 1980), and the combined effects of S loss and cation leaching can have adverse effects on forests. Immobilization and subsequent retention of sulfate can retard cation leaching (Johnson et al. 1982), thus mitigating one of the negative effects of acidic precipitation.

Organic S and Sulfate Pool Sizes of Selected IFS Sites

Sulfate can be immobilized by adsorption onto hydrous oxides and sesquioxides in the soil matrix (Cole and Johnson 1977; Johnson 1980) or by direct incorporation into organic matter, a microbially mediated process termed "organic S formation" (Fitzgerald et al. 1982, 1988). Organic S is the predominant form of S at all depths of the soil profile for various forested sites (David et al. 1983; Bartel-Ortiz and David 1988), and this observation also holds true for sites of the current study (Table 5.4). Methods for fractionation of these forms are provided by Autry and colleagues (1990). Organic S pool sizes greatly exceeded those of adsorbed sulfate for all sites but the GL site, where adsorbed sulfate levels were much higher (see Table 5.4). Soluble sulfate, a biologically available form of S, was present in lower levels than adsorbed sulfate for all sites but LP (Table 5.4). Based on the dominance of organic S levels relative to adsorbed sulfate, organic S formation represents as important an S retention mechanism as does sulfate adsorption. The capacity of a site to retain S as organic S should therefore

Table 5.4. Organic S and Sulfate Pool Sizes of Selected IFS Sites

Site	Sulfur Component (kg S ha^{-1})		
	Organic S	Adsorbed Sulfate	Soluble Sulfate
Norway (NS)	520	150	34
B.F. Grant Forest (GL)	570	1,400	33
Coweeta White Pine (CP)	710	380	39
Loblolly Pine (LP)	850	35	51
Coweeta Hardwood (CH)	1,600	310	17
Becking (SS)	1,600	380	43
Whiteface (WF)	2,300	84	37
Fullerton (FT)	3,500	220	170
Duke (DL)	6,700	2,900	220
Smokies Tower (ST)	18,000	140,000	10,000

be correlated to the capacity of soil samples taken from that site to form organic S.

To predict the response of a given IFS site to increasingly higher anthropogenic inputs of sulfate, we determined the maximum capacity of soil from various depths to form organic S. This maximum capacity is referred to as the "saturation potential" for that site. Moreover, the effect of increasing sulfate concentration on the mineralization rates of organic S must also be assessed to provide a true index of the potential biological accumulation of S. Further, the concentration of added sulfate required to yield saturation of a site with respect to both forming and accumulating organic S must also be determined to ascertain the sulfate deposition rate above which further sulfate deposition will not result in a significant increase in the rate of S accumulation. It should be noted, however, that this technique measures only S accumulation resulting from organic S formation. A similar technique has been used to measure the sulfate adsorption potentials at saturation and has been employed on these sites (A.R. Autry and J.W. Fitzgerald, unpublished data). In all horizons of every site analyzed, the potential for sulfate adsorption at saturation exceeded that for organic S formation. For example, in uppermost horizon soils, 9 of 12 sites analyzed possessed saturation potentials for sulfate adsorption that were 10 fold greater than those for organic S formation. The exceptions to this trend were the CH A, NS E, and HF A horizons. A similar statement can be made for the lowermost (B,C) horizon soils, where 10 of 12 sites examined exhibited potentials for sulfate adsorption at saturation that were 10 fold greater than were those for organic S formation (Autry and Fitzgerald 1991). Exceptions to this trend included the CH C and GL Bt horizons where 4- and 2-fold increases, respectively, for sulfate adsorption relative to organic S formation were noted. In light of the large organic S pool sizes relative to the adsorbed sulfate pool sizes, and bearing in mind that organic S formation exhibits a rate-time dependency whereas sulfate adsorption does not (Autry and Fitzgerald, unpublished data), the data suggest that organic S formation is more important as a long-term S retention mechanism, whereas sulfate adsorption is more important as a short-term S retention mechanism. Further study on the effects of sulfate loading on organic S formation is therefore warranted to better understand long-term S retention and subsequent enhancement of soil fertility.

Kinetic Analysis of Organosulfur Formation

The method employed to determine the saturation potential of a soil with respect to organic S formation is a modification of the "heterotrophic activity method" of Wright and Hobbie (1966), substituting ^{35}S-labeled sulfate for ^{14}C-labeled organic compounds to quantify rates of sulfate incorporation into organic matter by microbial populations in a soil sample. The technique is a kinetic approach utilizing a linearization of the Michaelis–Menten equation for enzyme kinetics yielding the equation, $t/f = A/V_{max} + (Kt + S_n)/V_{max}$, where t is the incubation time, f is the fraction of added substrate taken up or incorporated during time t, V_{max} is the maximal substrate uptake rate, Kt is the half-saturation constant or the substrate concentration yielding 1/2 V_{max}, and S_n is the endogenous substrate concentration. V_{max} is numerically equal to the reciprocal of the slope of the regression line in the Wright–Hobbie plot, while the sum $(Kt + S_n)$ is the x intercept of this line (Wright and Hobbie 1966). As noted by Azam and Hodson (1981), if S_n is known, then Kt can be calculated; if S_n is not known, then the sum $(Kt + S_n)$ provides a useful upper limit on Kt. Doubling this sum can then provide an upper limit on the concentration of added substrate yielding saturation if S_n is not known. Discussions of the application of this technique to organosulfur formation in soil and the validity of this technique have been published (Autry and Fitzgerald 1990).

To assess kinetic parameters for each site, samples (1 g wet weight, not sieved) were incubated 48 hours at 20°C with various amounts of $Na_2{}^{35}SO_4{}^{2-}$ (specific activity 8.33×10^5 Bq $mmol^{-1}$) having added sulfate ranging from 7.5 nmol to 400 $\mu mol\ g^{-1}$ wet weight. Following incubation, samples were extracted to remove adsorbed sulfate, and the ^{35}S incorporated into organic matter was quantified by the method of Fitzgerald and coworkers (1982). Results were subjected to analysis of variance and Duncan's multiple range test ($\alpha = .05$). Linearized data were subjected to linear regression analysis.

As an example of results obtained with samples collected from IFS sites, incorporation data for horizons of the FS site are shown in Figure 5.32. Soil samples from the A horizon were unsaturated with respect to organic S formation even after the addition of 485 $\mu mol\ SO_4{}^{2-}\ g^{-1}$ dry weight (Figure 5.32). A soil is saturated if the mean amounts of organic S formed are not significantly different over 25% of the concentration range examined (ANOVA, Duncan's multiple range test, $\alpha = .05$). If this condition is not satisfied, the horizon is unsaturated. With respect to the E, Bh, and C horizons,

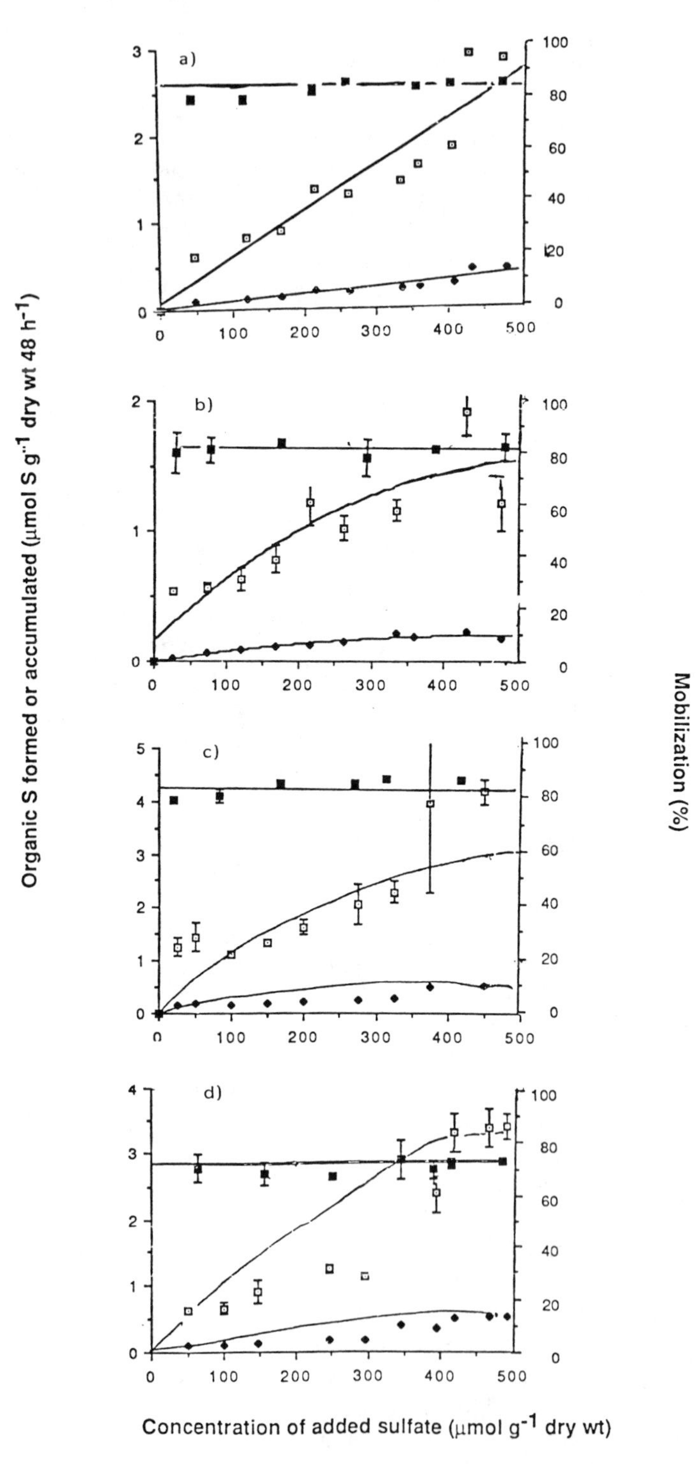

Figure 5.32. Relationship between increasing sulfate concentration and organic S formation (■), organic S mobilization (■), and potential accumulation of organic S (◆) for (a) A, (b) E, (c) Bh, and (d) C horizons of FS site.

saturation of organosulfur formation capacity was observed (see Figure 5.32). Because the A horizon of this site was unsaturated, no linearization of this data was attempted. Data from analysis of the Bh and C horizons exhibited linear Wright-Hobbie plots (r = .96 and .98, respectively) (Figure 5.33b and 5.33c), and kinetic parameters V_{max} and ($Kt + S_n$), calculated from this plot for these horizons are shown in Table 5.5.

For the E horizon, a curvilinear Wright–Hobbie plot was observed (Figure 2a). As first noted by Azam and Hodson (1981), this relationship indicates multiphasic uptake kinetics, which means that multiple microbial populations, each having a different set of kinetic parameters for uptake, are responsible for sulfate incorporation. The analysis of such data is difficult. However, if only the initial, linear portion of the curve, exhibiting first-order kinetics, is considered, a regression line can be fitted and one set of kinetic parameters, the lower limit for the system, can be calculated. The Wright–Hobbie linearization yields a V_{max} for sulfate incorporation per hour. Owing to the linearity of this process with respect to time, this value can be multiplied by 48 to yield the saturation potential of that soil for sulfate incorporation per 48 h.

Complete kinetic data for samples collected from the FS site are summarized in Table 5.5. The observed decrease in saturation potentials with sample depth was not surprising because both microbial biomass and activity usually decrease with increasing depth in the soil profile (Cochran et al. 1989). Similar findings were made with soil samples taken from other IFS sites, in that the bulk of these sites also exhibited general decreases in saturation potentials with increasing depth in the soil profile.

With respect to the sulfate concentration yielding saturation, increasing depth was also associated with general decreases in the magnitude of this parameter. For example, the A horizon required 481 μmol SO_4^{2-} g^{-1} dry weight to yield saturation, while the E, Bh, and C horizons required 297, 114, and 334 μmol added SO_4^{2-} g^{-1} dry weight for saturation to occur. Similar trends for this parameter with increasing depth were noted for samples taken from other IFS sites. A possible explanation for this phenomenon is that the microbial populations in the lower horizons are more efficient utilizers of sulfate (lower collective half-saturation constant for uptake) than are those in the upper horizons.

Mobilization and Potential Accumulation of Organic S at Saturating Sulfate Concentrations

As defined by Strickland and colleagues (1984), mobilization of organic S refers to the depolymerization of large molecular weight organic S constituents, which are then subject to enzymatic desulfation. Unlike organic S formation, which is carried out by living microbial cells, this process is mediated by preformed depolymerases and sulfatases present in the soil matrix (Fitzgerald and Strickland 1987). Increasing concentrations of added

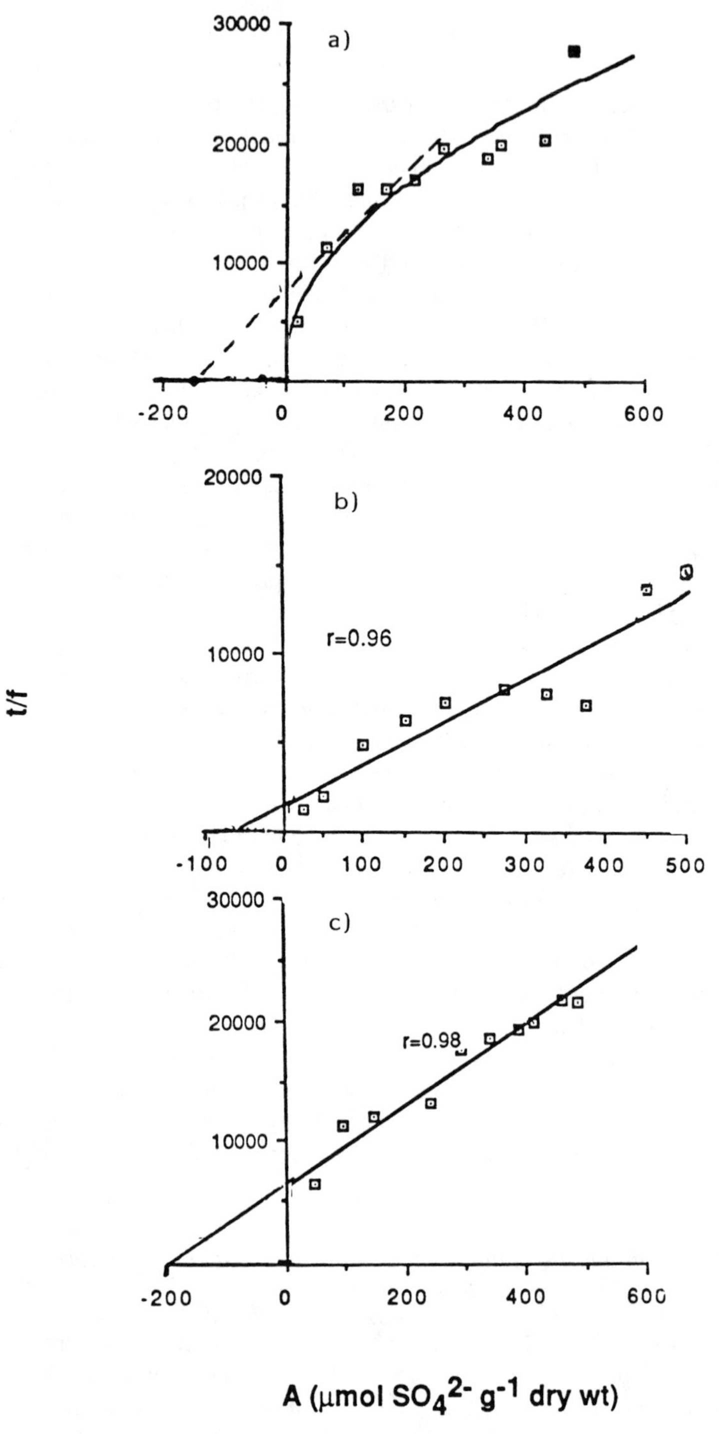

Figure 5.33. Wright–Hobbie plot; t is incubation time, f is fraction of added sulfate incorporated into organic matter, and A is concentration of added sulfate for (a) E, (b) Bh, and (c) C horizons of FS site.

Table 5.5. Summary of Kinetic Parameters for the Florida (FS) Site

	Organic S Formed (μmol g^{-1} dry weight)		
Horizon	Apparent V_{max} ($h^{-1} \times 10^{-2}$)	Potential at Saturation (48 h^{-1})	Sulfate Concentration Yielding Saturation (μmol g^{-1} dry weight[a])
A[b]	4.5	2.2	480
E[c]	2.1	1.0	300
Bh	4.1	2.0	110
C	2.7	1.3	330

[a] Value given is theoretical upper limit and equal to $(Kt + S_n) \times 2$, where Kt is half saturation constant and S_n is endogenous sulfate concentration.
[b] Data given is lower limit; horizon is unsaturated.
[c] Multiphasic system. Data calculated by regression analysis of linear portion of Wright–Hobbie plot, lower limit for this horizon.

sulfate and concomitant increases in organic S did not exert a significant effect ($p < .005$) on mobilization capacities in any horizon examined for the FS site (see Figure 5.32). Mobilization capacities were determined according to the method of Strickland and coworkers (1984). Moreover, this observation was noted for all other IFS sites irregardless of horizon. Mobilization capacities at saturation did decrease slightly with increasing depth, from 87% for the A horizon to 77% for the C horizon of this site (see Figure 5.32). This trend was observed with most of the other IFS sites examined and implies lowered sulfatase levels or activities at these lower depths. Noted decreases in total carbon levels with increasing depth for all study sites examined (data not shown) support this notion. Because mobilization capacities were unaffected by increasing organic S concentration, the potential accumulation of organic S, calculated as the difference between the amount of organic S formed from a given concentration of added sulfate S and the amount of organic S mobilized at that same concentration of added sulfate, also followed saturation kinetics for all horizons of the FS site (see Figure 5.32). Similar statements can be made for the other IFS sites. The potential accumulation of organic S at saturation is numerically equal to the saturation potential less the amount of organic S mobilized at saturation. Unlike the saturation potential for organic S formation and the mobilization capacity for organic S at saturation, the potential accumulation of organic S did not exhibit a significant decrease with increasing depth. Similar trends were noted for the other IFS sites examined.

Site Comparison of Kinetic Parameters

Saturation potentials, mobilization capacities at saturation, and potential accumulation of S as organic S were converted to a kilogram per hectare basis for each horizon and were then multiplied by horizon depth to yield values for each parameter based on a per site, rather than a per horizon, basis.

These values are summarized in Table 5.6. Data expressed in this manner allow for an effective comparison of the ability of a site to form, mobilize, and accumulate organic S. Generally, the LP, SS, and HF sites, having saturation potentials of 59, 259, and 358 kg S ha^{-1} 48 h^{-1}, respectively, were those sites possessing the lowest capacities for organic S formation when exposed to saturating concentrations of sulfate. The GL, DL, and ST sites, having saturation potentials of 1.7×10^4, 1.7×10^4, and 4.1×10^4 kg S ha^{-1} 48 h^{-1}, respectively, were those sites exhibiting the highest capacities for organic S formation under saturating sulfate concentrations (Table 5.6).

The data collectively imply that the resident microbial populations of the former sites have a lower capacity to incorporate S into organic matter (lower collective V_{max}) than do those of the latter sites. Amounts of organic S mobilized at saturation followed a similar trend in that the LP, HF, and SS sites mobilized the least amount of organic S during 48 hours, while the GL, DL, and ST sites mobilized the largest amounts of organic S in a 48-hour period (see Table 5.6). Because the amount of organic S mobilized exceeded 75% of that formed in 48 hours for all sites examined, the organic S pool is dynamic, rather than static, in nature. Further, rapidly mineralizable pools, such as organic S, tend to have high turnover times and accumulate slowly.

Table 5.6. Intersite Comparison of Saturation Potentials, Mobilization Capacities, and Potential Net Accumulation of S as Organic S for Forested Study Sites[a]

Site	Organic S Formation Potential at Saturation[b]	Organic S Mobilized at Saturation	Potential for Accumulation of Organic S at Saturation[c]
Loblolly pine (LP)	58.6	43.1	15.5
Huntington (HF)[d]	358	317.3	40.7
Becking (SS)	239	182.7	56.3
Florida (FS)[d,e]	484	427.5	56.5
Turkey Lakes (TL)[e]	635	530	105
Norway (NS)	451	323	128
Whiteface (WF)[d]	510	345	182
Coweeta hardwood (CH)[e]	1,310	1,128	182
Howland (MS)	794	581	213
Douglas fir (DF)	2,910	1,900	1,010
B.F. Grant Forest (GL)[d]	16,900	14,420	2,480
Duke (DL)	17,100	14,410	2,690
Smokies Tower (ST)	41,200	34,690	6,510

[a] All data are expressed on kg ha^{-1} 48 h^{-1} basis.
[b] Determined by linearization.
[c] Calculated as difference between potential at saturation and amount mobilized at saturation.
[d] Values given are lower limit because samples from one horizon examined were unsaturated.
[e] Multiphasic kinetics present in one horizon; thus, values given are lower limit.

The potential net accumulation of S as organic S at saturation provides the best index of the biological component of the response of a site to increasingly higher concentrations of added sulfate. A comparison of this parameter by site is summarized in Table 5.6. Like saturation potentials and amounts of organic S mobilized at saturation, the LP, HF, and SB sites exhibited the lowest potentials for S accumulation at saturation, while the GL, DL, and ST sites exhibited the highest values for this parameter (see Table 5.6). Moreover, these observations suggest that the former sites would tend to lose more added sulfate at saturation than would the latter sites, because the former sites accumulate less S as organic S from added sulfate than do the latter sites, leaving more sulfate free in solution, potentially able to leach from the soil profile.

Another component of the biological response of a soil to increasing amounts of added sulfate is the concentration of added sulfate yielding saturation. The upper limit for this value was also converted to a kilogram per hectare basis for each horizon and then multiplied by horizon depth, the sum of which provides an estimate of the saturating sulfate concentration for each site.

When expressed in this manner, this estimate defines the degree to which a site is saturated. The higher the concentration of added sulfate to bring a site to saturation, the less-saturated that site is, and, conversely, the lower the concentration of added sulfate required for saturation, the more saturated that site is. This parameter should be a function of the deposition history of a particular site in that less-saturated sites should have lower sulfate deposition rates than the more highly saturated sites. Moreover, sites that are less saturated should continue to form organic S using first-order kinetics at higher concentrations of added sulfate (higher sulfate deposition rates) than should more highly saturated sites. For example, the FS, GL, and ST sites were among the least saturated sites, requiring 1×10^5, 1.6×10^5, and 1.8×10^5 kg SO_4^{2-} ha^{-1}, respectively, to yield saturation, while the LP, SS, and MS sites were among the most saturated, requiring 1.4×10^2, 9.6×10^3, and 1.2×10^4 kg SO_4^{2-} ha^{-1} to yield saturation of the organic S formation capacity (Figure 5.34b).

This observation suggests that the former sites are less susceptible to sulfate leaching and associated cation leaching (Johnson 1980) at higher concentrations of added sulfate (higher sulfate deposition rates). These sites are operating at a sulfate incorporation rate that has not yet reached V_{max}, while the latter sites possess a sulfate incorporation rate which has already reached V_{max} at the concentration of added sulfate of interest. This phenomenon is best explained by the notion that the microbial populations present in the LP, SS, and MS sites are more efficient at forming organic S (lower Kt) than are those populations present in the FS, GL, and MS sites (higher Kt). The disparity in the saturating sulfate values between samples collected from the ST and SS sites, two sites relatively close to each other, is consistent with this hypothesis, owing to the large spatial heterogeneity of microbial

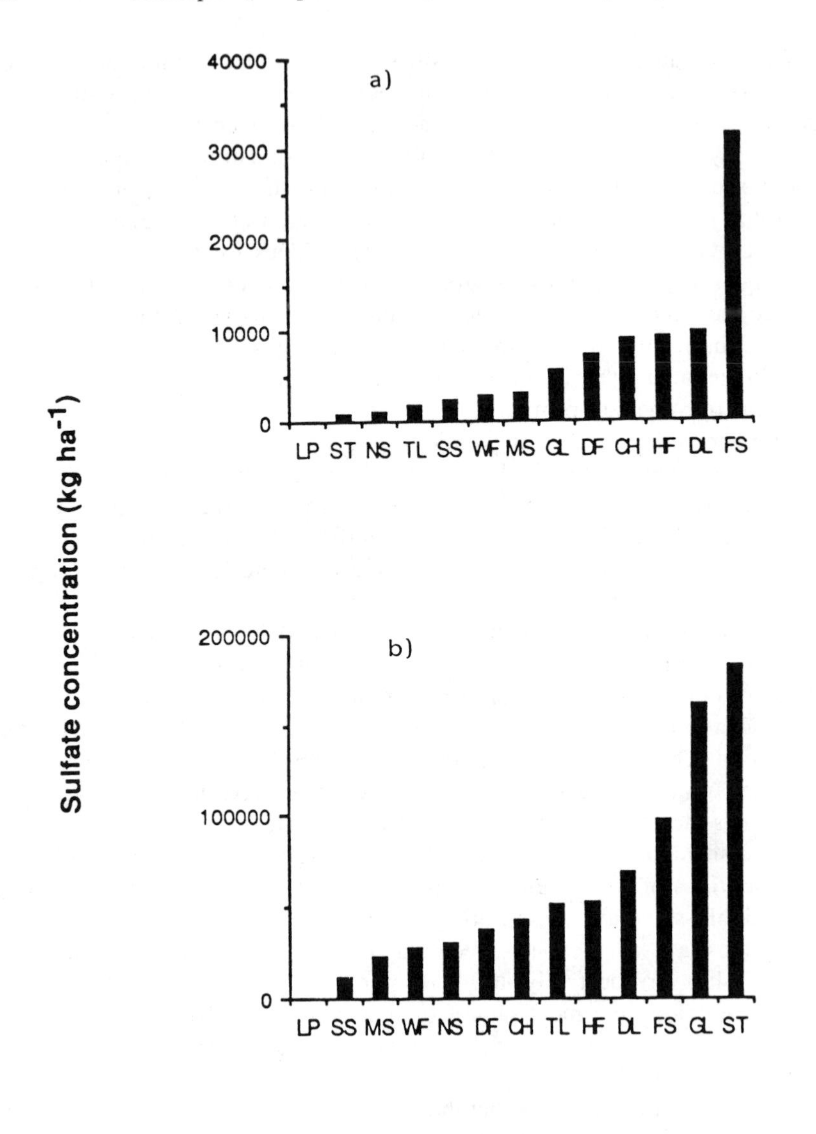

Figure 5.34. Site comparison of concentration of added sulfate yielding saturation. (a) comparison based on uppermost soil horizons only; (b) comparison based on all horizons of soil profile.

populations within a soil ecosystem. This heterogeneity is enhanced by microzones of enrichment and by adsorption of bacterial cells onto soil particles (Balkwill et al. 1988; Beloin et al. 1988).

It can be argued that comparison of this parameter based on all horizons within a site may not provide a true estimate of the saturating sulfate concentration in situ. This is because some of the sulfate may be immobilized in the uppermost mineral horizons and may therefore become unavailable for use by microbial populations in the lower horizons. Figure 5.34 shows that differences in this parameter were prevalent when comparisons were based on the entire profile and upper horizon only. For example, although the LP site was still the most saturated of all the sites, and the FS site was still among the least saturated when comparing only uppermost horizons, the ST site was one of the more saturated sites and the GL site was only sparingly saturated when the uppermost horizons of these sites were compared (Figure 5.34). This contradicts estimates of this parameter based on the entire soil profile in that both the GL and the ST sites were among the least saturated when compared in this manner (Figure 5.34b). It should be noted, however, that rapid organic S mobilization rates in upper horizons may lead to a percolation of sulfate through the soil profile in a concentration only slightly less than that received by the uppermost soil horizons from acidic deposition. Hence, site comparison of the saturating sulfate concentration based on the entire soil profile is still justified.

Conclusions

In summary, the data suggest that the "heterotrophic activity method" of Wright and Hobbie (1966) can be successfully applied to the measurement of organosulfur formation rates in soil ecosystems. The kinetic parameters for sulfate incorporation into organic matter derived from linearization of sulfate uptake data, when extrapolated to a kilogram per hectare basis, provide useful indices of soil responses to increased concentrations of anthropogenically derived sulfate, such as that encountered in acidic precipitation. The concentration of added sulfate yielding saturation for organic S formation provides an index of the relative susceptibility of a site to sulfate leaching, coupled with cation leaching. The magnitude of this leaching is a function of the accumulation of S as organic S, which is in turn dependent on both the saturation potential of the soil and the capacity of soil to mobilize organic S at saturation. Collectively, the data indicate that the kinetic parameters for microbial sulfate uptake and incorporation can provide useful indices of site-specific soil responses to sulfate loading.

Retention or Loss of Sulfur for IFS Sites and Evaluation of Relative Importance of Processes

M.J. Mitchell

The role of biological processes in regulating SO_4^{2-} flux is most important in those sites with low atmospheric inputs of S. There is also evidence that

SO_4^{2-} leaching is linked to sulfate adsorption. For example, in comparing results from laboratory column experiments (R. Harrison, unpublished data) with SO_4^{2-} concentrations obtained from lysimeters in the field (Figure 5.35), it is evident that soils that have the capacity to adsorb more SO_4^{2-} (e.g., RA, CH, DF, FL sites) generally have lower SO_4^{2-} concentrations beneath the B horizon than those soils which have less additional sulfate adsorption capacity (e.g., HF, SS, WF, LP). The major exceptions to this relationship are the DL site, which has unusually high SO_4^{2-} concentrations beneath the B horizon, and the CP and SB sites, which have both low adsorption potential and low SO_4^{2-} concentrations.

The linkage of SO_4^{2-} flux to adsorption is also suggested by the relationship between NO_3^- and SO_4^{2-} in soil solutions (Figure 5.36). For those sites with high NO_3^- export, which would indicate nitrification, SO_4^{2-} concentrations are lower. These lower concentrations can be explained by increased sulfate adsorption caused by pH depression from nitrification. It has been shown for a variety of soils that sulfate adsorption will increase when nitrification is stimulated (Nodvin et al. 1986; Dhamala et al. 1990).

The categorization of sites with respect to the relative importance of abiotic and biotic processes regulating SO_4^{2-} flux is indicated in Figure 5.37. This

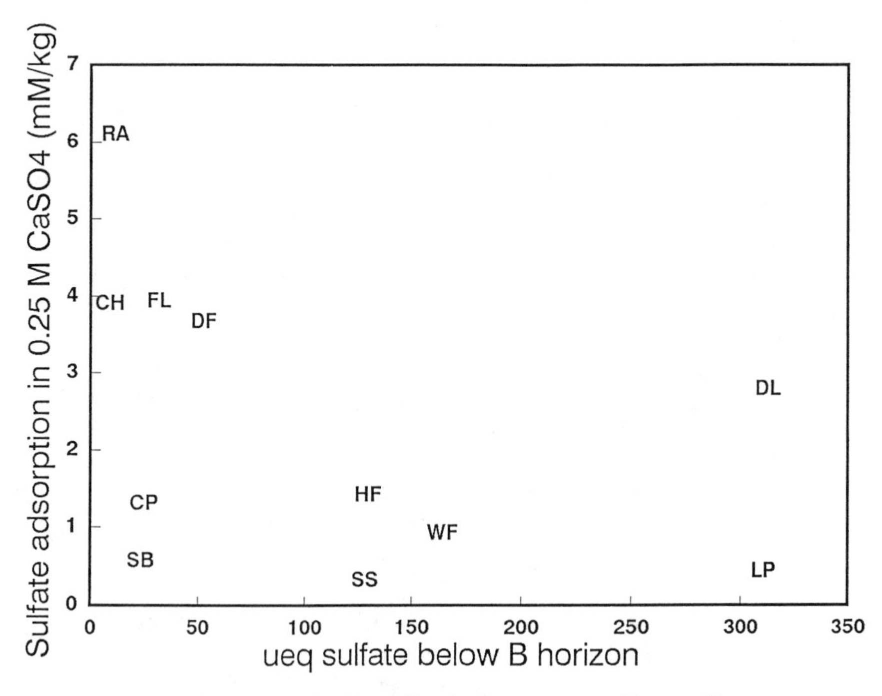

Figure 5.35. Sulfate adsorption in soil columns versus ambient sulfate concentrations in B horizon solutions.

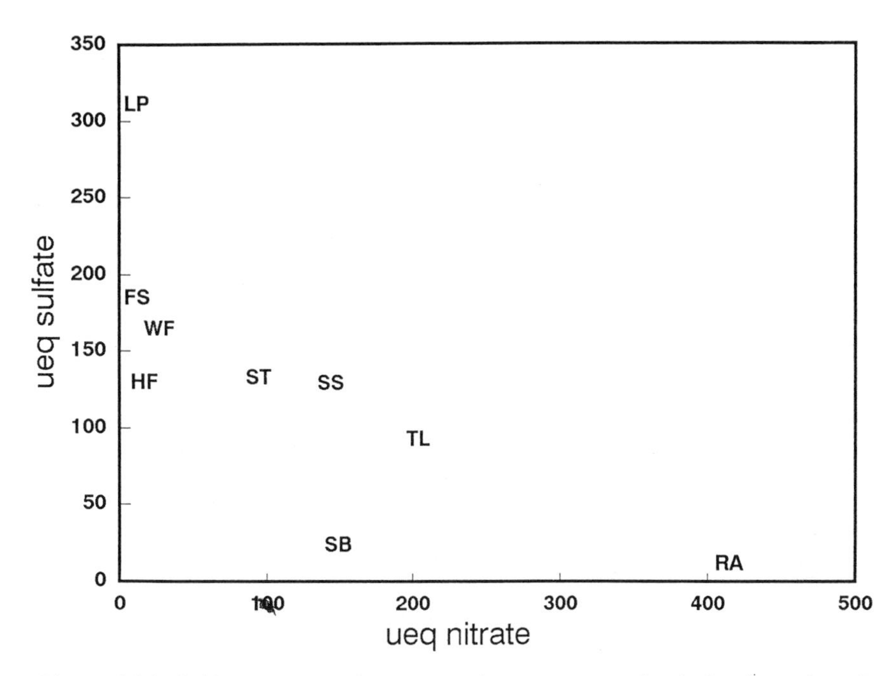

Figure 5.36. Sulfate concentration versus nitrate concentration in leachates from B horizon.

categorization is useful in evaluating the relative importances of biological versus chemical processes in regulating SO_4^{2-} flux. For some sites, inputs and outputs nearly balance and there is little or no net sulfur retention (Category I: ST, TL, NS, and DF).

For those sites with low SO_4^{2-} inputs (Category II: RA and FS), the biotic demand for sulfur including tree uptake and microbial immobilization plays an important role in retaining sulfur within the ecosystem. This biotic demand is most apparent in the RA stand because of high sulfur requirement in the presence of excess nitrogen. However, the retention of sulfur within the RA system may also be attributed to other processes (Van Miegroet and Cole 1984) including sulfate adsorption because high levels of nitrification may enhance this latter process. The net retention of sulfur in some of the sites in the southeastern United States (Category III: GL, CP) may result from enhanced sulfate adsorption because these sites have highly weathered soils with characteristic high sulfate adsorption potential. However, at least for one of the sites (CP), the importance of biological immobilization as a mechanism for sulfur retention cannot be eliminated because other investigations at CP have demonstrated the importance of this sulfur retention mechanism (Swank et al. 1984).

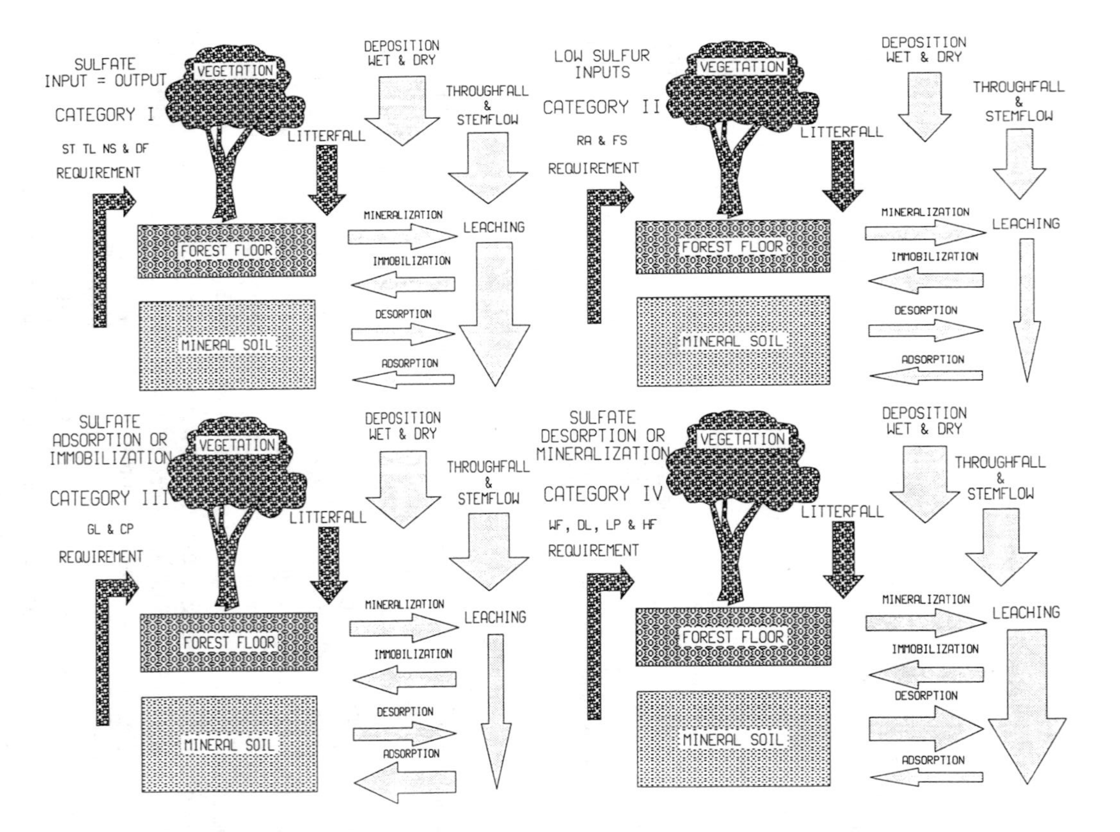

Figure 5.37. Categorization of IFS sites: (I) Sulfate input equals sulfate output; (II) Low sulfate inputs with important role of biological retention; (III) Net sulfate adsorption and immobilization; (IV) Net sulfate desorption or mineralization.

For those sites (Category IV: WF, DL, LP, HF) showing a net loss of sulfur, this may be attributed to either sulfate desorption or sulfur mineralization exceeding sulfur immobilization. Note that for the DL site (see the Appendix for site data) the concentration and hence SO_4^{2-} flux from the Bt2 horizon are markedly greater than for any other strata with a subsequent reduction from 1250 to 320 eq ha^{-1} yr^{-1} after passage through the subsoil to a depth of 2 m. Whether this SO_4^{2-} is being derived from some mineral sulfur source or from SO_4^{2-} desorption is not known. In the northeastern United States, recent decreases in SO_4^{2-} inputs via precipitation are consistent with decreases in SO_2 emissions (Fay et al. 1985). Increased SO_4^{2-} flux in soil solutions under conditions of lower SO_4^{2-} inputs would be expected in these sites if sulfate adsorption is reversible (Mitchell et al. 1991).

Analyses of Selected Sulfur Cycles in Polluted versus Unpolluted Environments

M.J. Mitchell

Although we cannot do a detailed analyses of the nutrient cycles of all the IFS sites, we have selected four sites on the basis of elevation and atmospheric air pollutant inputs. The Smokies ST site (ST) is at a high-elevation (1740 m) and also has large atmospheric inputs of H^+, NO_3^-, and SO_4^{2-} whereas the Findley Lake site (FL) also is at a high-elevation (1100 m) but in a relatively unpolluted region. The low-elevation sites (<220 m) are represented by DL Loblolly site (DL), which has large inputs of mineral acids, and the Douglas fir site (DF), situated in the same region as FL. For these same sites detailed analyses for nitrogen (Chapter 6) and cations (Chapter 8) are also given.

For comparative purposes, a diagram (Figure 5.38a) showing the major sulfur pools and transformations was developed, which is analogous to that utilized in describing cations dynamics of these same sites (see Chapter 8). In general, the pools of sulfur for the two high-altitude sites are similar with respect to the mineral soil and foliage (Figure 5.38b and 5.38c). For the branch and bole pool, the ST site has 2.8 times more sulfur than the FL site. These differences can be attributed to marked differences in bole and branch concentrations of sulfur (e.g., 503 and 39 mg S kg^{-1} dry mass in ST and FL boles, respectively) and not difference in biomass, because the ST site had less (247,650 kg ha^{-1}) biomass of bole and branches than the FL site (540,600 kg ha^{-1}). Similarly, the forest floor at the ST site has lower mass (126,020 kg ha^{-1}) than that at the FL site (548,000 kg ha^{-1}), but the sulfur concentrations were greater in the former (251 versus 88 mg kg^{-1}). In comparing all sites for which sulfur inputs (either total inputs for IFS or TF + SF for non-IFS sites) and content were available, however, a significant correlation was seen between sulfur inputs and sulfur content of the forest floor but not the vegetation (see p. 95).

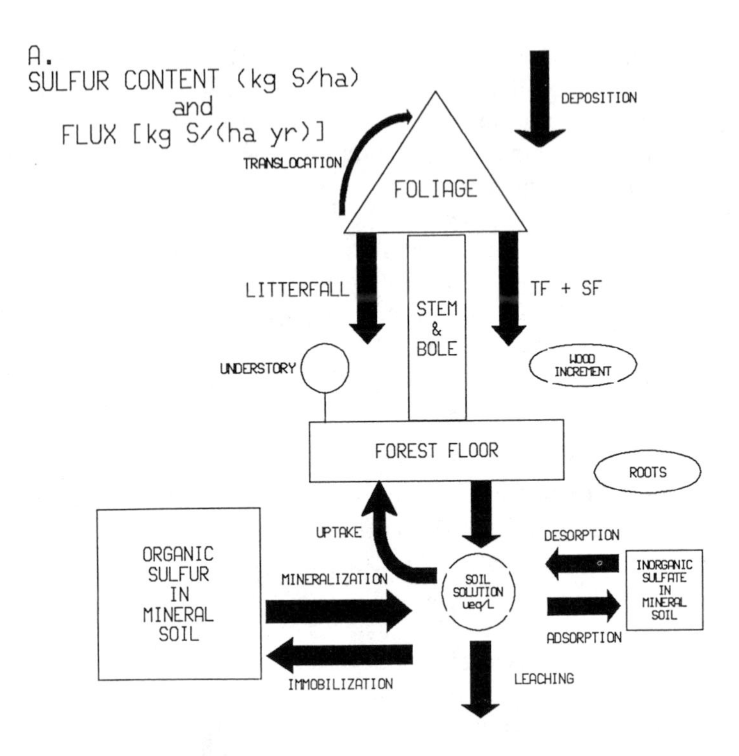

Figure 5.38. (a) Generalized diagram for comparing sulfur cycles in polluted and unpolluted sites; (b) Smokies Tower pools and fluxes; (c) Findley Lake pools and fluxes; (d) Duke loblolly pools and fluxes; (e) Douglas fir pools and fluxes.

In comparing fluxes between these two high-elevation sites, only bulk precipitation values are available for FL, which would certainly be an underestimate of total sulfur input. However, TF + SF fluxes are about six times greater in ST than in FL and, as discussed previously, these values are good estimates of total sulfur inputs. These differences in solute fluxes are also reflected in soil leaching, with mineral soil concentrations and fluxes being four- and sevenfold greater, respectively, at the ST site. Other measured and calculated fluxes including litterfall, uptake, and translocation are not markedly different among the sites because they reflect the biological demand for sulfur that is in excess at both sites. Note, however, for the FL site with lower sulfur atmospheric inputs, the fluxes of sulfur through the biotic components constitute a larger fraction of the total transfer of sulfur through the forest ecosystem than the ST site in which the flux of SO_4^{2-} via mass transfer in solutes dominates. The 2.3-kg difference in TF + SF flux versus solute flux from the B horizon for the FL site may reflect a significant amount of sulfur accumulation when compared to sulfur inputs to this site relative to the situation at the ST site. As a fraction of TF + SF, the B

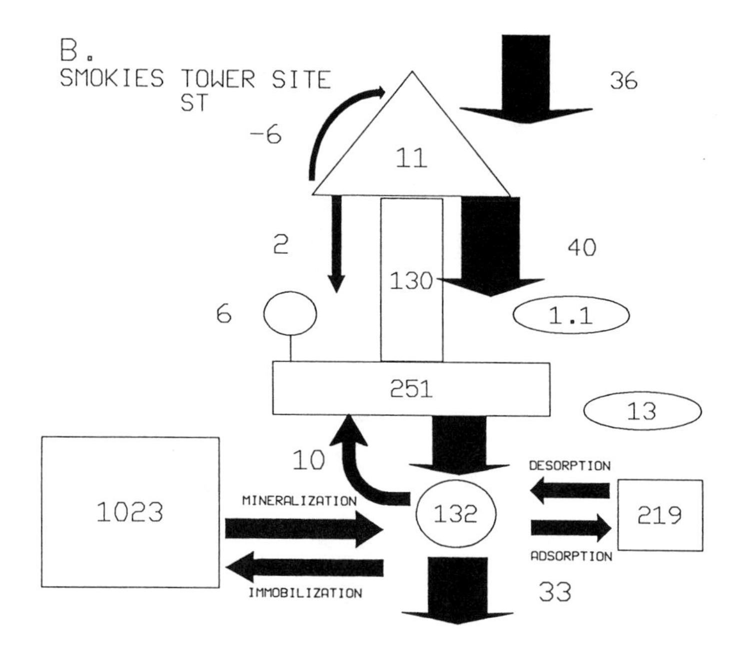

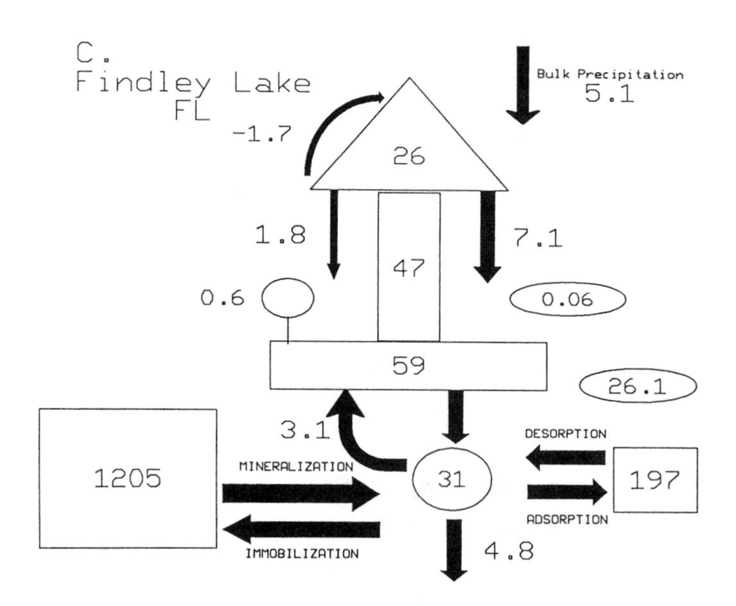

Figure 5.38. *Continued.*

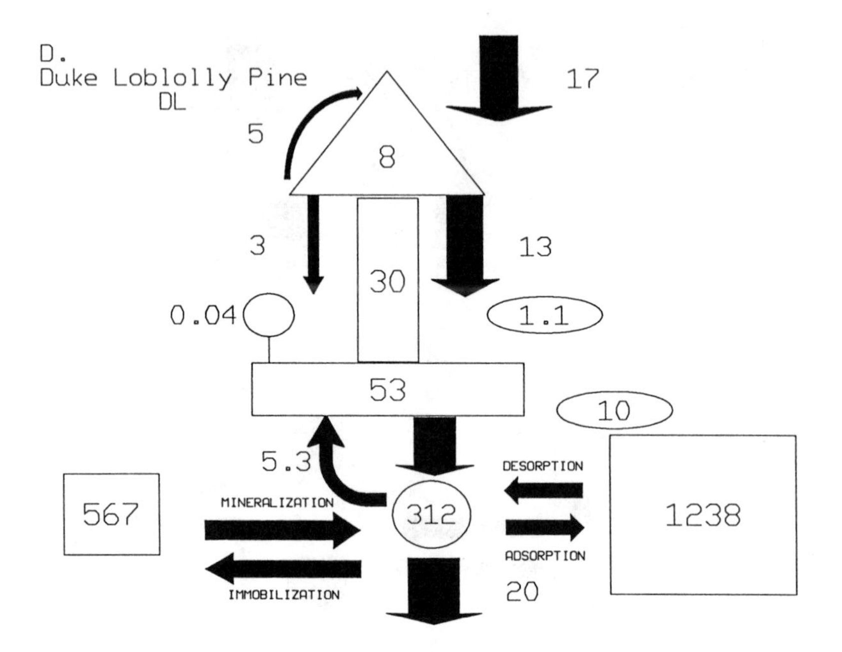

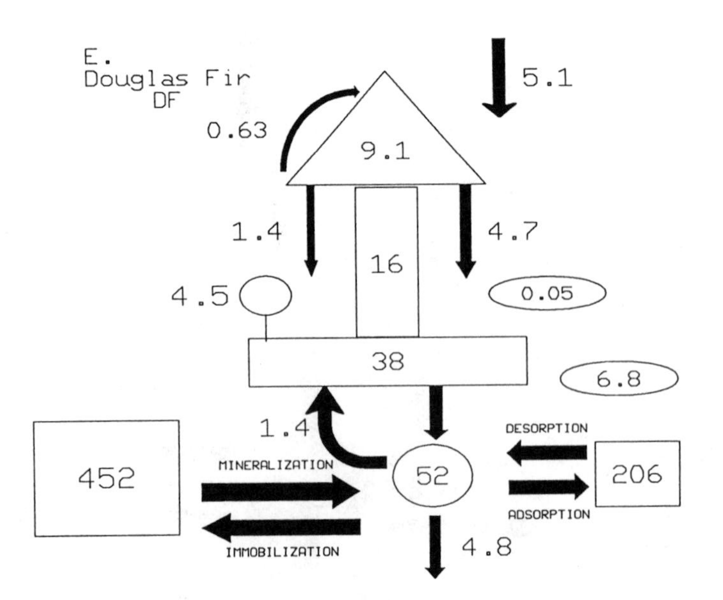

Figure 5.38. *Continued.*

horizon SO_4^{2-} flux constitutes 32% versus 18% of total flux at the FL and ST sites, respectively.

There is little capacity for soils in these sites to retain additional SO_4^{2-} by adsorption at ambient concentrations, but the FL soil may adsorb significantly more SO_4^{2-} if concentrations are increased (Harrison et al. 1989) because it now has very low SO_4^{2-} concentrations. Both sites have soil sulfur pools dominated by organic forms; ST soil is an Umbric Dystrochrept and the FL a Typic Cryohumod (Spodosol) (see pp. 119–120). The greater concentrations and fluxes of SO_4^{2-} through the soil at the ST site results in substantially greater leaching of Ca^{2+}, K^+ and Mg^{2+} (see Chapter 8).

The two low-elevation sites (DL and DF) have substantially lower organic sulfur pools in the mineral soil (see Figure 5.38d and 5.38e). For the DL site this may be partly attributable to the low organic matter content of the mineral soil of this site (34,848 kg C ha^{-1}) compared to the ST (96,339 kg C ha^{-1}) and the FL (279,400 kg C ha^{-1}) sites. However, the DF site has substantial amounts of organic C (109,200 kg C ha^{-1}) but low organic sulfur content. This may be a partial reflection of the historically low sulfur inputs to this latter site. A major difference in the sulfur pools of the DL and DF sites is the large pool of inorganic sulfate in the mineral soil of the DL site. This high level of soil sulfate is characteristic of sites with Hapludult soils that have had substantial inputs of SO_4^{2-}.

The higher deposition rates at DL versus DF are reflected in higher fluxes of SO_4^{2-} in TF + SF and soil leaching in the former site. The higher wood increment, uptake, and translocation of sulfur in DL are attributed to not only higher sulfur concentrations, but also to the more rapid growth of the DL site. For example, the bole and branch increment is about twice as much at DL compared to DF [4825 versus 2340 kg organic matter (ha^{-1} yr^{-1}), respectively]. Despite the greater growth rate of DL, in the DF site a larger fraction of the sulfur is being cycled through the biotic component of the system. For the DF site, however, there appears to be no net accumulation of sulfate in comparing TF + SF versus leaching from the B horizon as these fluxes are almost equal. However, the TF + SF value is less by 0.4 kg ha^{-1} yr^{-1} than the total deposition input and thus there may be direct incorporation of sulfur via SO_2 uptake by the vegetation, which would contribute to the sulfur requirement of the DF site. Litterfall is somewhat greater in the DL versus DF as a fraction of TF + SF (30% and 23%, respectively). The DL site has very high SO_4^{2-} concentrations in the lower B horizon, and there appears to be some source of inorganic sulfur that is contributing to this high value. Similar to the high-elevation polluted site (ST), the greater flux of SO_4^{2-} in DL results in greater leaching (Chapter 8) of base cations (Ca^{2+}, Mg^{2+}, and K^+).

Summary and Conclusions

Atmospheric deposition of sulfur at the IFS forests is highest at the high-elevation and at the southeastern sites, and lowest in the Northwest. Cloud water contributes significantly to input at the mountain sites, dry deposition is comparable to wet at the drier Southeastern sites, and wet deposition dominates input at the northern sites. In all these forests, it appears that deposited sulfur behaves more or less conservatively in the canopy, with SO_2 uptake balanced to some extent by foliar leaching. With few exceptions, estimated fluxes in total deposition are within 15% of the measured fluxes in TF + SF, indicating that useful estimates of total atmospheric deposition of S can be derived from TF networks. These results have important applications to deposition model evaluation, to studies in complex terrain, and to prediction of soil and water acidification.

As shown in previous studies (Mitchell et al. 1991), sulfur plays a major role in elemental cycling by its interactions with other elements in the IFS sites. The movement of SO_4^{2-} in solutions through the strata of forest ecosystems affects the dynamics of other elements, especially basic and acidic cations (see Chapter 8). If H^+ is associated with the SO_4^{2-} anion, this cation has strong affinity for exchange complexes in foliage as well as in soil. Within the canopy there is limited retention or loss of SO_4^{2-}; thus, SO_4^{2-} in TF + SF is generally an accurate predictor of total atmospheric input for most IFS sites (Lindberg and Garten 1988) (see Figure 5.12). At sites with elevated levels of SO_2, dry deposition and canopy assimilation of this gas may be an important source of sulfur to foliage. At those sites with very low inputs, however, canopy leaching of organic and inorganic sulfur in solution may make significant contributions to sulfur fluxes (see pp. 103–105).

In many of the IFS sites that have limited sulfate adsorption capacity and elevated inputs of sulfur, inputs and outputs of this element are generally balanced (see Figures 5.23 and 5.24). As established in previous studies of regions affected by elevated levels of anthropogenic inputs of sulfur (Cogbill and Likens 1974; Ulrich et al. 1980; Glass et al. 1982; Mollitor and Raynal 1983; Foster 1985), SO_4^{2-} is the principal anion for most IFS sites. Exceptions include those in which other mobile anions including NO_3^- (e.g., RA, TL; see Chapter 6), HCO_3^- (FL, DL), and Cl^- (FS) contribute substantially to the soil solution. For the IFS sites, the sulfur pools (see Figure 5.14) and SO_4^{2-} fluxes (see Figure 5.13) are substantially less than at some highly polluted sites in Europe (e.g., AO, AS, SF, GS, and PO sites), but more than at the remote site in Australia (AD).

The cation species leached through the soil are functions of the cation exchange capacity (CEC) and base saturation of the IFS sites (see Chapter 8). An elevated SO_4^{2-} concentration above background levels, between 10 to 30 $\mu eq\ L^{-1}$ (Galloway et al. 1984; Holdren et al. 1984), will result in the increased leaching of cations. All sites except those in Washington (FL, RA, and DF, $\leq 25\ \mu eq\ SO_4^{2-}\ L^{-1}$) have SO_4^{2-} levels in great excess of these background levels (e.g., TL, NS, ST, and WF, ~100–200 $\mu eq\ SO_4^{2-}\ L^{-1}$) in forest floor solutions. Accelerated leaching of cations because of increased atmospheric deposition of mobile anions, especially SO_4^{2-}, is important for many IFS sites (ST, SB, SB, LP, WF, JF, CP, CH). For other sites (DL, FS, GL), the leaching of SO_4^{2-} results in a lesser but still significant role. For the sites in Washington (FL, RA, DF), the contribution is minor. The relative loss of either base cations (Ca^{2+}, Mg^{2+}, and K^+) or acidic cations (H^+ and Al^{3+}) varies among sites with the northern and high-elevation sites (SS, ST, SB, WF, and HF), generally exhibiting the greatest losses of H^+ and Al^{3+} (see Chapter 8). The presence of Al, especially in an inorganic-monomeric form (Al^{3+}) in soil leachates and surface waters, is of major concern because of its toxicity to forest vegetation as well as to aquatic biota (Cronan and Schofield 1979; Driscoll and Schecher 1988).

The loss of base cations may also be deleterious if the depletion rates exceed inputs from weathering and atmospheric inputs (Johnson et al. 1985 1986; Foster et al. 1986; Dewalle et al. 1988; Stottlemyer and Hanson 1989). Within the IFS sites, for which substantially improved estimates of inputs of base cations from both wet and dry deposition are available, as compared to many previous studies, it has been shown that atmospheric inputs of basic cations (Chapter 8) attenuate losses associated with increased leaching rates of these elements. Moreover, differences among sites including the contribution of weathering (see Chapter 10) mean that we cannot generalize that increased SO_4^{2-} inputs are causing deleterious losses of base cations. Chapter 13 gives some predictions of ecosystem fluxes for select IFS sites using the IFS NuCM simulation model under various scenarios of future sulfur deposition loadings assuming complete reversibility of sulfate adsorption.

Accurately predicting sulfur inputs, sulfur retention, and resultant SO_4^{2-} leaching from forested ecosystems is particularly important where acidification of waters is significant. Because atmospheric SO_4^{2-} inputs are changing (Likens et al. 1984; Fay et al. 1985), it is important to ascertain how these changes and resulting changes in solute concentrations (precipitation, throughfall, stemflow, forest floor leachate, and mineral soil leachate) will affect sulfur dynamics. The IFS data suggest that, in general, deposition and TF + SF fluxes of sulfur at any one site are comparable, indicating that reduction in atmospheric deposition from emission controls will be closely reflected by similar reductions in sulfur fluxes to the forest floor (Lindberg and Garten 1988). This is important because the response of ecosystems to SO_4^{2-} inputs will be markedly different if SO_4^{2-} is reversibly rather than irreversibly adsorbed or biologically retained. In the former case, when SO_4^{2-} concentrations decrease, SO_4^{2-} previously stored would be released from the soil, while in the latter case no release would occur. If stored sulfur is released from the soil as SO_4^{2-}, this will accelerate the leaching of either basic or acidic cations. The degree of irreversibility or retention in soil varies among the IFS sites, but 32 of 36 soil horizons analyzed exhibited sulfate adsorption and on average 36% of the adsorbed sulfate was irreversibly retained (see p. 113; Harrison et al. 1989). In addition, studies on these same soils using ^{35}S in laboratory incubations have shown that all IFS soils have the capacity to retain additional sulfur inputs via microbial immobilization (pp. 119–129) the importance of which varies among soils.

The classification of sites based on retention or loss of sulfur as well as the more detailed analyses of sites subjected to different levels of sulfur input shows that the IFS sites exhibit differences in the relative importance of biotic versus abiotic processes in affecting sulfur dynamics. For those sites with elevated loadings of sulfur associated with atmospheric pollutants, the sulfur dynamics during the period of the IFS project can generally be explained by analyses of abiotic sulfate adsorption-desorption processes as has been concluded by previous studies (Johnson et al. 1980; Johnson 1984; Rochelle et al. 1987). However, the role of irreversible sulfate adsorption needs further attention, because our results suggested imbalances of sulfur inputs and outputs for some sites, including net losses of sulfur (Category IV). In addition, for those sites with lower levels of SO_4^{2-} inputs, biotic transformations by uptake, litter production, and storage by forest vegetation as well as microbial transformations (mineralization-immobilization) (Category II) can influence sulfur dynamics because a significant fraction of sulfur flux is associated with passage through biotic constituents (Mitchell et al. 1991). Furthermore, for most IFS sites, the organic sulfur pool in the mineral soil is the dominant sulfur constituent, but the long-term dynamics of this pool including formation rates and net losses are difficult to quantify. Changes in this pool may have important consequences relating to SO_4^{2-} loss and retention in forested ecosystems, and, in conjunction with processes affecting the reversibility of sulfate adsorption, will influence how changes in sulfur loadings from the atmosphere will affect the concentrations and fluxes of SO_4^{2-} in soil and associated surface waters.

References

Adams F., Rawajfih Z. 1977. Basaluminite and alunite: a possible cause of sulfate retention by acid soils. Soil Sci. Soc. Am. J. 41:686–692.

Adams F., Hajek B.F. 1978. Effects of solution sulfate, hydroxide, potassium concentrations on the crystallization of alunite, basaluminite, gibbsite from dilute aluminum solutions. Soil Sci. 126:169–173

Adriano D.C., Havas M. (eds.) 1989. Acidic Precipitation, Vol. 1: Case Studies. Springer-Verlag, New York.

Autry A.R., Fitzgerald J.W., Caldwell P.R. 1990. Sulfur fractions and retention mechanisms in forest soils. Can. J. For. Res. 20:337–341.

Autry A.R., Fitzgerald J.W. 1990. Application of the heterotrophic activity method to organosulfur formation in forest soils. Soil Biol. Biochem. 22:743–748.

Autry A.R., Fitzgerald J.W. In press. Kinetic analysis of organosulfur formation in forest soils:site comparison of kinetic parameters. Soil Biol. Biochem.

Aylmore L.A.G., Karim M., Quirk J.P. 1967. Adsorption and desorption of sulfate ions by soil constituents. Soil Sci. 103:10–15.

Azam F., Hodson R. 1981. Multiphasic kinetics for D-glucose uptake by assemblages of natural marine bacteria. Mar. Ecol. Prog. Ser. 6:213–222.

Balkwill D.L., Leach F.R., Wilson J.T., McNabb J.F., White D.C. 1988. Equivalence of microbial biomass measures based on membrane lipid and cell wall components, adenosine triphosphate, direct counts in subsurface aquifer sediments. Microb. Ecol. 16:73–84.

Barrow N.J. 1969. Effects of adsorption of sulfate by soils on the amount of sulfate present and its availability to plants. Soil Sci. 108:193–201.

Bartel-Ortiz L.M., David M.B. 1988. Sulfur constituents and transformations in upland and floodplain forest soils. Can. J. For. Res. 18:1106–1112.

Bartlett R.J., Riego D.C. 1972. Effect of chelation on the toxicity of aluminum. Plant Soil 37:419–423.

Beloin R.M., Sinclair J.L., Ghiorse W.C. 1988. Distribution and activity of microorganisms in subsurface sediments of a pristine study site in Oklahoma. Microb. Ecol. 16:85–97.

Bergseth H. 1985. Selektierungsvermogen eines eisenoxidehydroxids und einiger tonminerale gegenuber phohphat- und sulfationen. Acta Agric. Scand. 35:375–388.

Bornhemisza E., Llanos R. 1967. Sulfate movement, adsorption, desorption in three Costa Rican soils. Soil Sci. Soc. Am. Proc. 31:356–360.

Brakke D.F., Henrikson A., Norton S.A. 1989. Estimates background concentrations of sulfate in dilute lakes. Water Resour. Res. 35:247–253.

Bredemeier M.J. 1988. Forest canopy transformation of atmospheric deposition. Water Air Soil Pollut. 40:121–138.

Butler T.J. 1988. Composition of particles dry deposited to an inert surface at Ithaca, NY. Atmos. Environ. 22:895–900.

Bytnerowicz A., Miller P.R., Olszyk D.M. 1987. Dry deposition of nitrate, ammonium, sulfate to a *Ceanothus crassifolius* canopy and surrogate surfaces. Atmos. Environ. 21:1805–1814.

Calvert J.G., Lazrus A., Kok G.L., Heikes B.B., Walega J.G., Lind L., Cantrell C.A. 1985. Chemical mechanisms of acid generation in the troposphere. Nature (London) 317:27–35.

Chao T.T., Harward M.E., Fang S.C. 1964. Iron or aluminum coatings in relation to sulfate adsorption characteristics of soils. Soil Sci. Soc. Am. Proc. 28:632–635.

Chao T.T., Harward M.E., Fang S.C. 1962. Movement of S35 tagged sulfate through soil columns. Soil Sci. Soc. Am. Proc. 26:27–32.

Church M.R., Thornton K.W., Shaffer P.W., Stevens D.L., Rochelle B.P., Holdren G.R., Johnson M.G., Lee J.J., Turner R.S., Cassell D.L., Lammers D.A., Campbell W.G., Liff C.I., Brandt C.C., Liegel L.H., Bishop G.D., Mortenson D.C., Pierson S.M., Schmoyer D.D. 1989. Future Effects of Long-Term Sulfur Deposition on Surface Water Chemistry in the Northeast and Southern Blue Ridge Province: Results of the DDRP. EPA/600/3–89/061, U.S. Environmental Protection Agency, Washington, D.C.

Cochran V.L., Elliott L.F., Lewis C.E. 1989. Soil microbial biomass and enzyme activity in subarctic agricultural and forest soils. Biol. Fertil. Soils 7:283–288.

Coe J.M., Lindberg S.E. 1987. The morphology and size distributions of atmospheric particles deposited on foliage and inert surfaces. J. Air Pollut. Control Assoc. 37:237–243.

Cogbill C.V., Likens G.E. 1974. Acid precipitation in the northeastern United States. Water Resour. Res. 10:1133–1137

Cole D.W., Johnson D.W. 1977. Atmospheric sulfate additions and cation leaching in a Douglas-fir ecosystem. Water Resour. Res. 13:313–317.

Cook E.R., Kablack A., Jacoby G.C. 1988. The 1986 drought in the southeastern U.S.: how rare was it? J. Geophys. Res. 93:14257–14260.

Couto W., Lathwell D.J., Bouldin D.R. 1979. Sulfate adsorption by two oxisols and an alfisol of the tropics. Soil Sci. 127:108–116.

Cronan C.S., Reiners W.A., Reynolds R.L., Lang G.E. 1978. Forest floor leaching: Contributions from mineral, organic, and carbonic acids in New Hampshire subalpine forests. Science 200:309–311.

Cronan C.S., Schofield C.L. 1979. Aluminum leaching response to acid precipitation:effects on high-elevation watersheds in the Northeast. Science 204:304–306.

Cronan C.S., Walker W.J., Bloom P.R. 1986. Predicting aqueous aluminum concentrations in natural waters. Nature (London) 324:140–143.

David M.B., Mitchell M.J., Aldcorn D., Harrison R.B. 1989. Analysis of sulfur in soil, plant and sediment materials: sample handling and use of an automated analyzer. Soil Biol. Biochem. 21:119–123.

David M.B., Grigal D.F., Ohmann L.F., Gertner G.Z. 1988. Sulfur, carbon, nitrogen relationships in forest soils across the northern Great Lakes States as affected by atmospheric deposition and vegetation. Can J. For. Res. 18:1386–1391.

David M.B., Schindler S.C., Mitchell M.J., Strick J.E. 1983. Importance of organic and inorganic sulfur to mineralization processes in a forest soil. Soil Biol. Biochem. 15:671–677.

David M.B., Mitchell M.J., Schindler S.C. 1984. Dynamics of organic and inorganic sulfur constituents in hardwood forest soils. In Stone E.L. (ed.) Forest Soils and Treatment Impacts, Proceedings of the Sixth North American Soils Conference, pp. 221–245. University of Tennessee, Knoxville, Tennessee.

Davidson C.I., Chu L. 1981. SEM study of Fe-containing particles on foxtail. Environ. Sci. Technol. 15:198–201.

Davidson C.I., Wu Y.-L. 1990. Dry deposition of particles and vapors. In Lindberg S.E., Page A., Norton S. (eds.) Acidic Precipitation, Vol. 3: Sources, Deposition, Canopy Interactions. Springer-Verlag, New York, pp. 103–216.

Davidson C.I., Miller J.M., Pleskow M.A. 1982. The influence of surface structure on predicted particle dry deposition to natural grass canopies. Water Air Soil Pollut. 18:25–43.

Davidson C.I., Lindberg S.E., Schmidt J., Cartwright L., Landis L. 1985. Dry deposition of sulfate onto surrogate surfaces. J. Geophys. Res. 90:2121–2130.

Dethier D.P., Jones S.B., Feist T.P., Ricker J.E. 1988. Relations among sulfate, aluminum, iron, dissolved organic carbon, pH in upland forest soils of northwestern Massachusetts. Soil Sci. Soc. Am. J. 52:506–512.

Dewalle D.R., Sharpe W.R., Edwards P.J. 1988. Biogeochemistry of two Appalachian deciduous forest sites in relation to episodic stream acidification. Water Air Soil Pollut. 40:143–156.

Dhamala B., Mitchell M.J., Stam A., 1990. Sulfur dynamics of two northern hardwood soils:a column study with ^{35}S. Biogeochemistry 10:143–160.

Dignon J., Hameed S. 1989. Global emissions of nitrogen and sulfur dioxides from 1860 to 1980. J. Air Pollut. Control Assoc. 39:180–186.

Driscoll C.T., Schecher W.D. 1988. Aluminum in the environment. In Sigel H., Sigel A. (eds.) Metal Ions in Biological Systems. Vol. 24. Aluminum and its Role in Biology. Marcel Dekker, New York, pp. 59–122.

Duke S.H., Reisenauer H.M. 1986. Roles and requirements of sulfur in plant nutrition. In Tabatabai M.A. (ed.) Sulfur in Agriculture. Agronomy Monograph 27, American Society of Agronomy, Madison, Wisconsin, pp. 123–168.

Elkins D.M., Ensminger L.E. 1971. Effect of soil pH on the availability of adsorbed sulfate. Soil Sci. Soc. Am. Proc. 35:931–934.

Ensminger L.E. 1954. Some factors affecting the adsorption of sulfate by Alabama soils. Soil Sci. Soc. Am. Proc. 18:259–264.

Fay J.A., Golomb D., Kumar S. 1985. Source apportionment of wet sulfate deposition in eastern North America. Atmos. Environ. 19:1773–1782.

Fitzgerald J.W., Johnson D.W. 1982. Transformations of sulphate in forested and agricultural lands. In Moore A.I. (ed.) Sulphur-82: Proceedings of the International Conference, Vol. II, pp. 411–426. British Sulphur Corporation, London.

Fitzgerald J.W., Strickland T.C. 1987. Mineralization of organic sulphur in the O2 horizon of a hardwood forest:involvement of sulphatase enzymes. Soil Biol. Biochem. 19:779–781.

Fitzgerald J.W., Strickland T.C., Swank W.T. 1982. Metabolic fate of inorganic sulphate in soil samples from undisturbed and managed forest ecosystems. Soil Biol. Biochem. 14:529–536.

Fitzgerald J.W., Swank W.T., Strickland T.C., Ash J.T., Hale D.D., Andrew T.L., Watwood M.E. 1988. Sulfur pools and trans-formations in litter and surface soil of a hardwood forest. In Swank W.T., Crossley D.A. Jr. (eds.) Ecological Studies: Forest Hydrology and Ecology at Coweeta Vol. 66. Springer-Verlag, New York, pp. 246–253.

Fortmann R.C. 1982. Characterization of the Interception and Retention of Individual Atmospheric Particles by Needles of *Taxus*. Ph.D. Thesis, State Universitiy of New York, Syracuse.

Foster N. 1989. Acidic deposition—what is fact, what is speculation, what is needed? Water Air Soil Pollut., 48:299–306.

Foster N.W. 1985. Acid precipitation and soil solution chemistry within a maple-birch forest in Canada. For. Ecol. Manage. 12:215–231.

Foster N.W., Morrison I.K., Nicolson J.A. 1986. Acid deposition and ion leaching from a podzolic soil under hardwood forest. Water Air Soil Pollut. 31:879–889.

Freney J.R. 1967. Sulfur-containing organics. In McLaren A.D., Peterson G.H. (eds.) Soil Biochemistry. Marcel Dekker, New York, pp. 229–252.

Fuller R.D., Driscoll C.T., Lawrence G.B., Nodvin S.C. 1987. Processes regulating sulfate flux after whole-tree harvesting. Nature (London) 325:707–710.

Galloway J.N., Norton S.A., Church M.R. 1983. Freshwater acidification from atmospheric deposition of sulfuric acid: a conceptual model. Environ. Sci. Technol. 17:541a–545a.

Galloway J.N., Likens G.E., Hawley M.E. 1984. Acid precipitation: natural versus anthropogenic components. Science 226:829–831.

Garland J.A. 1978. Dry and wet removal of sulfur from the atmosphere. Atmos. Environ. 12:349–362.

Garsed S.G. 1985. SO_2 uptake and transport. In Winner W.E., Mooney H.A., Goldstein R.A. (eds.) Sulfur Dioxide and Vegetation, Physiology, Ecology, Policy Issues. Stanford University Press, Stanford, California.

Garten C.T. 1990. Foliar leaching, translocation, biogenic emission ^{35}S in radiolabelled loblolly pines. Ecology 71:239–251.

Garten C.T., Bondietti E.A., Lomax R.D. 1988. Contribution of foliar leaching and dry deposition to sulfate in net throughfall below deciduous trees. Atmos. Environ. 22, 1425–1432

Gaston L.A., Mansell R.S., Rhue R.D. 1986. Sulfate mobility in acid soils and implications with respect to cation leaching: a review. Soil Crop Sci. Soc. Fla. Proc. 45:67–72.

Gay D.W., Murphy C.E. Jr. 1985. Final report: The Deposition of SO_2 on Forests. Final Report, E.P.R.I Project R.P.1813–2, Electric Power Research Institute, Palo Alto, California.

Gebhardt H., Coleman N.T. 1974. Anion adsorption by allophanic tropical soils:II. sulfate adsorption. Soil Sci. Soc. Am. Proc. 38:259–262.

Gibson J.H., Galloway J.N., Schofield C., McFee W., Johnson R., McCarley S., Dise N., Herzog D. 1983. Rocky Mountain Acidification Study. FWS/OBS-80/40.17, U.S. Fish and Wildlife Service, Division of Biological Services, Eastern Energy and Land Use Team, Washington, D.C.

Glass N.R., Arnold D.E., Galloway J.N., Hendry G.R., Lee J.J., McFee W.W., Norton S.A. Powers C.F., Rambo D.L., Schofield C.L. 1982. Effects of acid precipitation. Environ. Sci. Technol. 16:162–169.

Goldberg S., Sposito G. 1984. A chemical model of phosphate adsorption by soils: I. Reference oxide minerals. Soil Sci. Soc. Amer. J. 48:772–778

Grennfelt P., Larson S., Leyton P., Olsson B. 1985. Atmospheric deposition in the Lake Gardsjon area, SW Sweden. Ecol. Bull. 37:101–108.

Haque I., Walmsley D. 1973. Adsorption and desorption of sulfate in some soils of the West Indies. Geoderma 9:269–278.

Harrison R.B., Johnson D.W., Todd D.E. 1989. Sulfate adsorption and desorption reversibility in a variety of forest soils. J. Environ. Qual. 18:419–426.

Harward M.E., Reisenauer H.M. 1966. Reactions and movement of inorganic soil sulfur. Soil Sci. 101:326–335.

Henrikson A. 1980. Acidification of freshwater—a large scale titration. p. 68–74. In Drablos D., Tollan A. (eds.) Ecological Impact of Acid Precipitation. Proc. Int. Conf. Sandefjord, Norway, SNSF Project, Oslo, Norway.

Hicks B.B., Wesley M.L., Lindberg S.E., Bromberg S.M. (eds.) 1986. Proceedings of the Dry Deposition Workshop of the National Acid Precipitation Assessment Program, March 25–27, 1986. NOAA/AT.D.D., Oak Ridge, Tennessee.

Hicks B.B., Meyers T.P., Fairall V.A., Mohnen V.A., Dolske D.A. Ratios of dry to wet deposition as derived from preliminary field data. Global Biogeochem. Cycles (in press).

Hicks B.B., Baldocchi D.D., Meyers T.P., Hosker R.P. Jr., Matt D.R. 1987. A preliminary multiple resistance routine for deriving dry deposition velocities from measured quantities. Water Air Soil Pollut. 36:311–330.

Hicks B.B., Matt D.R., McMillen R.T., Womack J.D., Wesely M.L., Hart R.L., Cook D.R., Lindberg S.E., de Pena R.G., Thomson D.W. 1989. A field investigation of sulfate fluxes to a deciduous forest. J. Geophys. Res. 94:13003–13011

Hodges S.C., Johnson G.C. 1987. Kinetics of sulfate adsorption and desorption by Cecil soil using miscible displacement. Soil Sci. Soc. Am. J. 51:323–331.

Holdren G.R., Brunelle T.M., Matisoff G., Wahlen M. 1984. Timing the increase in atmospheric sulphur deposition in the Adirondack Mountains. Nature (London) 311:245–248.

Homann P.S., Mitchell M.J., Van Miegroet H., Cole D.W. 1990. Organic sulfur in throughfall, stemflow, soil solutions from temperate forests. Can. J. For. Res. 20:1535–1539.

Hosker R.P., Lindberg S.E. 1982. Review Article: Atmospheric deposition and plant assimilation of airborne gases and particles. Atmos. Environ. 16:889–910.

Hue N.V., Craddock G.R., Adams F. 1988. Effect of organic acids on aluminum toxicity in subsoils. Soil Sci. Soc. Am. J. 50:28–34.

Huete A.R., McColl J.G. 1984. Soil cation leaching by acid rain with varying nitrate-to-sulfate ratios. J. Environ. Qual. 13:366–371.

Hultberg H. 1985. Budgets of base cations, Cl, N, S in the Lake Gardsjon catchment, SW Sweden. Ecol. Bull. 37:133–157.

Hussar R.B. 1986. Emissions of sulfur dioxide and nitrogen oxides and trends for eastern North America. In Acid Deposition Long Term Trends. National Academy Press, Washington, D.C., pp. 48–92.

Hutchinson T.C., Meema K.M. (eds.) 1987. Effects of Atmospheric Pollutants on Forests, Wetlands and Agricultural Ecosystems. Springer-Verlag, Berlin.

Ivens W., Kauppi P., Alacamo J., Posch M. 1990. Sulfur deposition onto European Forests: throughfall data and model estimates. Tellus 42B:294–303.

Johnson A.H., Siccama T.J. 1983. Acid deposition and forest decline. Environ. Sci. Technol. 17:294–305.

Johnson D.W. 1980. Site susceptibility to leaching by HB_2SO_4 in acid rainfall. In Hutchinson T.C., Harras M. (eds.) Effects of Acid Precipitation on Terrestrial Ecosystems. Plenum, New York, pp. 525–535.

Johnson D.W. 1984. Sulfur cycling in forests. Biogeochemistry 1:29–43.

Johnson D.W., Cole D.W. 1977. Sulfate mobility in an outwash soil in western Washington. Water Air Soil Pollut. 7:489–495.

Johnson D.W., Henderson G.S. 1979. Sulfate adsorption and sulfur fractions in a highly weathered soil under a mixed deciduous forest. Soil Sci. 128:34–40.

Johnson D.W., Todd D.E. 1983. Relationships among iron, aluminum, carbon, sulfate in a variety of forest soils. Soil Sci. Soc. Am. J. 47:792–800.

Johnson D.W., Van Hook R.I. (eds.) 1989. Analysis of Biogeochemical Cycling Processes in Walker Branch Watershed. Springer-Verlag, Berlin.

Johnson D.W., Turner J., Kelly J.M. 1982. The effects of acid rain on forest nutrient status. Water Resour. Res. 18:449–461.

Johnson D.W., Miegroet H.V., Cole D.W., Richter D.D. 1983. Contributions of acid deposition and natural processes to cation leaching from forest soils: a review. J. Air Pollut. Control Assoc. 33:1036–1041.

Johnson D.W., Richter D.D., Lovett G.M., Lindberg S.E. 1985. The effects of atmospheric deposition on potassium, calcium, magnesium cycling in two deciduous forests. Can. J. For. Res. 15:773–782

Johnson D.W., Hornbeck J.W., Kelly J.M., Swank W.T., Todd D.E. 1980. Regional patterns of soil sulfate accumulation:relevance to ecosystem sulfur budgets. In Shriner D.S. et al. (eds.) Atmospheric Sulfur Deposition: Environmental Impact and Health Effects. Ann Arbor Science, Ann Arbor, Michigan, pp. 507–520.

Johnson D.W., Richter D.D., Van Miegroet H., Cole D.W., Kelly J.M. 1986. Sulfur cycling in five forested ecosystems. Water Air Soil Pollut. 30:965–979.

Johnson D.W., Henderson G.S., Huff D.D., Lindberg S.E., Richter D.D., Shriner D.S., Todd D.E., Turner J. 1982. Cycling of organic and inorganic sulphur in a chestnut oak forest. Oecologia 54:141–148.

Joslin J.D., Lindberg S.E., Wolfe M.H., Robarge W.P. 1989. Estimates of deposition to high elevation sites using throughfall measurements. In Olem H. (ed.) T.V.A. Acid Rain Conference Abstracts, Gatlinburg, Tennessee, October 1989. Tennessee Valley Authority, Muscle Shoals, Alabama.

Kamprath E.J., Nelson W.L., Fitts J.W. 1956. The effect of pH, sulfate and phosphate concentrations on the adsorption of sulfate of soils. Soil Sci. Soc. Am. Proc. 26:463–466.

Karkanis M. 1976. The circulation of sulphur in the forest ecosystem *Tilo-Carpinetum* in the northern part of Puszcza Niepolomicka near Ispina. Fragmenta Floristica et Geobotanica 22:351–363.

Kelly J., Lambert M.J. 1972. The relationship between sulphur and nitrogen in the foliage of *Pinus radiata*. Plant Soil 37:395–408.

Khanna P.K., Beese F. 1978. The behavior of sulfate salt input in podzolic brown earth. Soil Sci. 125:16–22.

Kimmins J.P. 1973. Some statistical aspects of sampling throughfall precipitation in nutrient cycling studies in British Columbian coastal forests. Ecology 54:1008–1019.

Klein R.M., Perkins T.D. 1988. Primary and secondary consequences of contemporary forest decline. Bot. Rev. 54:1–43.

Lambert M.J. 1986. Sulphur and nitrogen nutrition and their interactive effects on *Dothistroma* infection in *Pinus radiata*. Can. J. For. Res. 16:1055–1062.

Lambert M.J., Turner J., Edwards D.W. 1976. Effects of sulphur deficiency in forests. In Proceedings of the 16th IUFRO World Congress, Oslo, Norway, November 1976. IUFRO, Oslo.

Landers D.H., David M.B., Mitchell M.J. 1983. Analysis of organic and inorganic sulfur constituents in sediments, soil and water. Int. J. Environ. Anal. Chem. 14:245–256.

Legge A.H., Krupa S.V. (eds.) 1986. Air Pollutants and Their Effects on the Terrestrial Ecosystem. Wiley, New York.

Likens G.E., Bormann F.H., Hedin L.O., Driscoll C.T., Eaton J.S. 1990. Dry deposition of sulfur, a 23 year record for the Hubbard Brook Forest Ecosystem. Tellus, 42B:319–329.

Likens G.E., Bormann F.H., Pierce R.S., Eaton J.S., Munn R.E. 1984. Long-term trends in precipitation chemistry at Hubbard Brook, New Hampshire. Atmos. Environ. 18:2641–2647

Likens G.E., Bormann F.H., Pierce R.S., Eaton J.S., Johnson N.M. 1977. Biogeochemistry of a Forested Ecosystem. Springer-Verlag, New York.

Lindberg S.E., Garten C.T. 1988. Sources of sulfur in forest canopy throughfall. Nature (London) 336:148–151.

Lindberg S.E., Johnson D.W. (eds.) 1989. 1988 Annual Report of the Integrated Forest Study. ORNL/TM 11121, Oak Ridge National Laboratory, Oak Ridge, Tennessee.

Lindberg S.E., Lovett G.M. 1985. Field measurements of particle dry deposition rates to foliage and inert surfaces in a forest canopy. Environ. Sci. Technol. 19:238–244.

Lindberg S.E., Page A., Norton S.A. 1990a. Acidic Precipitation, Vol 3. Sources, Deposition, Canopy Interactions. Springer-Verlag, New York.

Lindberg S.E., Lovett G.M., Richter D.D., Johnson D.W. 1986. Atmospheric deposition and canopy interactions of major ions in a forest. Science 231:141–145.

Lindberg S.E., Lovett G.M., Schaefer D.A., Bredemeier M. 1988a. Coarse aerosol deposition velocities and surface-to-canopy scaling factors from forest canopy throughfall. J. Aerosol Sci. 19:1187–1190.

Lindberg S.E., Silsbee D., Schaefer D.A., Owens J.G., Petty W. 1988b. A comparison of atmospheric exposure conditions at high- and low-elevation forests in the southern Appalachian Mountains. In Unsworth M. (ed.) Processes of Acidic Deposition in Mountainous Terrain. Kluwer, London, pp. 321–344.

Lindberg S.E., Bredemeier M., Schaefer D.A., Qi L. 1990b. Atmospheric concentrations and deposition during the growing season in conifer forests in the United States and West Germany. Atmos. Environ. 24A:2207–2220.

Lovett G.L. Canopy structure and cloud water deposition in subalpine coniferous forests. Tellus (in press).

Lovett G.M. 1984. Rates and mechanisms of cloud water deposition to a subalpine balsam fir forest. Atmos. Environ. 18:361–371.

Lovett G.M. 1988. A comparison of methods for estimating cloud water deposition to a New Hampshire subalpine forest. In Unsworth M., Fowler D. (eds.) Processes of Acidic Deposition in Mountainous Terrain. Kluwer, London, pp. 309–320.

Lovett G.M., Kinsman J.D. 1991. Atmospheric pollutant deposition to high elevation ecosystems. Atmos. Environ. 24A:2767–2786.

Lovett G.M., Lindberg S.E. 1984. Dry deposition and canopy exchange in a mixed oak forest determined from analysis of throughfall. J. Appl. Ecol. 21:1013–1028.

Lovett G.M., Reiners W.A., Olson R.K. 1982. Cloud droplet deposition in subalpine balsam fir forests: Hydrological and chemical inputs. Science 218:1303–1304.

Marsh K.B., Tillman R.W., Syers J.K. 1987. Charge relationships of sulfate sorption by soils. Soil Sci. Soc. Am. J. 51:318–323.

Matt D.R., McMillan R.T., Womack J.D., Hicks B.B. 1987. A comparison of estimated and measured SO_2 deposition velocities. Water Air Soil Pollut. 36:331–347.

Mayer R., Ulrich B. 1974. Conclusions on the filtering action of forests from ecosystem analysis. Oecol. Plant 9:157–168.
McCune D.C., Lauver T.L. 1986. Experimental modeling of the interaction of wet and dry deposition on conifers. In Lee S.D., Schneider T., Grant L.D., Verkerk P.J. (eds.) Aerosols. Lewis, Chelsea, Michigan.
McDowell W.H., Likens G.E. 1988. Origin, composition, flux of dissolved organic carbon in the Hubbard Brook valley. Ecol. Monogr. 58:177–195.
McLaren R.G., Keer J.I., Swift R.S. 1985. Sulphur transformation in soils using sulphur-35 labelling. Soil Biol. Biochem. 17:73–79.
Meiwes K.J., Khanna P.K. 1981. Distribution and cycling of sulphur in the vegetation of two forest ecosystems in an acid rain environment. Plant Soil 60:369–375.
Mitchell M.J., David M.B., Harrison R. 1991. Sulfur dynamics of forest ecosystems. In Howarth R., Stewart J. (eds.) Sulfur Biogeochemistry. SCOPE. Wiley, New York, (in press).
Mitchell M.J., David M.B., Maynard D.G., Telang S.A. 1986. Sulfur constituents in soils and streams of a watershed in the Rocky Mountains of Alberta. Can. J. For. Res. 16:315–320.
Mitchell M.J., Driscoll C.T., Fuller R.D., David M.B., Likens G.E. 1989. Effect of whole-tree harvesting on the sulfur dynamics of a forest soil. Soil Sci. Soc. Am. J. 53:933–940.
Mohnen V.A. 1988. Mountain Cloud Chemistry Project: Wet, Dry, Cloud Deposition. C.R. 813934–01–2., U.S. Environmental Protection Agency, North Carolina Research Triangle Park.
Mollitor A.V., Raynal D.J. 1983. Acid precipitation and ionic movements in Adirondack forest soils. Soil Sci. Soc. Am. J. 46:137–141.
Murphy C.E., Sigmon J.T. 1990. Dry deposition of S and N oxide gases to forest vegetation. In Lindberg S.E., Page A., Norton S. (eds.) Acidic Precipitation, Vol. 3; Sources, Deposition, and Canopy Interactions. Springer-Verlag, New York, pp 217–240.
NADP. 1988. NADP/NTN Annual Data Summary of Precipitation Chemistry in the United States for 1987. U.S. National Atmospheric Deposition Program, Fort Collins, Colorado.
National Research Council (NRC). 1983. Atmospheric Processes in Eastern North America. National Academy of Sciences Press, Washington, D.C.
Neary A.J., Mistry E., Vanderstar L. 1987. Sulfate relationships in some central Ontario forest soils. Can. J. Soil Sci. 67:341–352.
Nodvin S.C., Driscoll C.T., Likens G.E. 1986. The effect of pH on sulfate adsorption by a forest soil. Soil Sci. 142:69–75.
Noll K.E., Pontius A., Frey R., Gould M. 1985. Comparison of atmospheric coarse particles at an urban and non-urban site. Atmos. Environ. 19:1931–1943.
Nommik, H., Larsson K., Lohm U. 1984. Effects of Experimental Acidification and Liming on the Transformation of Carbon, Nitrogen and Sulphur in Forest Soils. SNV-PM-1869,
Nys C., Ranger J. 1988. Influence dune substitution despece sur le fonctionnement biogeochimique de leecosysteme foret Lexemple du cyle du soufre. Ann. Sci. For. 45:169–188.
Parfitt R.L., Smart R.S.C. 1978. The mechanism of sulfate adsorption on iron oxides. Soil Sci. Soc. Am. J. 42:48–50.
Parfitt R.L. 1978. Anion adsorption by soils and soil materials. Adv. Agron. 30:1–50.
Rajan S.S.S. 1978. Sulfate adsorbed on hydrous alumina, ligands displaced, changes in surface charge. Soil Sci. Soc. Am. J. 42:39–44.

Reuss J.O., Johnson D.W. 1986. Acid Deposition and the Acidification of Soils and Waters. Springer-Verlag, New York.

Rochelle B.P., Church M.R., David M.B. 1987. Sulfur retention at intensively studied sites in the U.S., Canada. Water Air Soil Pollut. 33:73–83.

Saxena V.K., Lin N.H. 1989. Cloud chemistry measurements and estimates of acidic deposition on an above cloudbase coniferous forest. Atmos. Environ. 24A:329–352.

Schaefer D.A., Reiners W.A. 1990. Throughfall chemistry and canopy processing mechanisms. In Lindberg S.E., Page A., Norton S. (eds.) Acidic Precipitation, Vol. 3, Sources, Deposition, Canopy Interactions. Springer-Verlag, New York, pp. 241–284.

Schaefer D.A., Lindberg S.E., Lovett G.M. Comparing cloud water flux estimates for a coniferous forest. In Proceedings International Conference on Acidic Deposition. C.E.P. Ltd. (in press).

Schaefer D.A., Lindberg S.E., Hoffman W.A. 1989. Fluxes of undissociated acids to terrestrial ecosystems by atmospheric deposition. *Tellus* 41 B: 207–218.

Schindler S.C., Mitchell M.J. 1987. Dynamics of ^{35}S in horizons and leachates from a hardwood forest spodosol. Soil Biol. Biochem. 19:531–538.

Schindler D.W. 1988. Effects of acid rain on freshwater ecosystems. Science 239:149–157.

Schnitzer M. 1969. Reactions between fulvic acid, a soil humic compound, inorganic soil constituents. Soil Sci. Soc. Am. Proc. 33:75–81.

Schnoor J.L. (ed.) 1984. Modeling of Total Acid Precipitation Impacts. Acid Precipitation Series, Vol. 9, Butterworth, Boston.

Schoenau J.J., Bettany J.R. 1987. Organic matter leaching as a component of carbon, nitrogen, phosphorus, sulfur cycles in a forest, grassland, gleyed soil. Soil Sci. Soc. Am. J. 51:646–651.

Schofield C., Trojnar J. 1980. Aluminum toxicity to fish in acidified waters. In Toribar T., Miller M., Morrow P. (eds.) Polluted Rain. Plenum, New York, pp. 347–366.

Shepard J.P., Mitchell M.J., Scott T.J., Zhang Y.M., Raynal D.J. 1989. Measurements of wet and dry deposition in a northern hardwood forest. Water Air Soil Pollut. 48:225–238.

Shriner D.S., Richmond C.R., Lindberg S.E. (eds.) 1980. Atmospheric Sulfur Deposition, Environmental Impact aid Health Effects Ann Arbor Science, Ann Arbor, Michigan.

Singh B.R. 1984. Sulfate sorption by acid forest soils:2. Sulfate adsorption isotherms with and without organic matter and oxides of aluminum and iron. Soil Sci. 138:294–297.

Singh B.R., Abrahamsen G., Stuanes A. 1980. Effects of simulated acid rain on sulfate movement in acid forest soils. Soil Sci. Soc. Am. J. 44:75–80.

Stottlemyer R., Hanson D.G. Jr. 1989. Atmospheric deposition and ionic concentrations in forest soils of Isle Royale National Park, Michigan. Soil Sci. Soc. Am. J. 53:270–274

Strickland T.C., Fitzgerald J.W., Swank W.T. 1984. Mobilization of recently formed forest soil organic sulfur. Can. J. For. Res. 14:63–67.

Swank W.T., Fitzgerald J.W., Ash J.T. 1984. Microbial transformation of sulfate in forest soils. Science 223:182–184.

Tanner R.L. 1990. Sources of acids, bases, their precursors in the atmosphere. In Lindberg S.E., Page A., Norton S. (eds.) Acidic Precipitation, Vol. 3, Sources, Deposition, Canopy Interactions. Springer-Verlag, New York, pp. 1–19.

Taylor G.E., McLaughlin S.B., Shriner D.S., Selvidge W. 1983. The flux of sulfur gases to vegetation. Atmos. Environ. 17:789–796.

Tikhova, E.P. 1958. Sulfate ion absorption by soil colloids. Trudy Voronezh. Gosudarst. Univ. 45:29–32.
Turner J., Lambert M.J. 1980. Sulfur nutrition of forests. In Shriner D.S., Richmond C.R., Lindberg S.E. (eds.) Atmospheric Sulfur Deposition, Environmental Impact and Health Effects. Ann Arbor Science, Ann Arbor, Michigan. pp. 321–334.
Turner J., Lambert M.J., Gessel S.P. 1977. Use of foliage sulphate concentrations to predict response to urea application by Douglas-Fir. Can. J. For. Res. 7:476–480.
Ulrich B., Pankrath J. (eds.) 1983. Effects of Accumulation of Air Pollutants in Forest Ecosystems. Reidel, Dordrecht.
Ulrich B, Mayer R., Khanna R.K. 1980. Chemical changes due to acid precipitation in a loess-derived soil in central Europe. Soil Sci. 130:193–199
Unsworth M.H., 1985. Pathways of S from the atmosphere to plants and soil. In Winner W.E., Mooney H.A., Goldstein R.A. (eds.) Sulfur Dioxide and Vegetation, Physiology, Ecology, Policy Issues. Stanford University Press, Stanford, California.
Unsworth M.H., Fowler D. (eds.) 1988. Acid Deposition at High Elevation Sites. Kluwer, The Netherlands.
Van Breeman N. 1973. Dissolved aluminum in acid sulfate soils and in acid mine waters. Soil Sci. Soc. Am. Proc. 37:694–697.
Van Miegroet H., Cole D.W. 1984. The impact of nitrification on soil acidification and cation leaching in a red alder forest. J. Environ. Qual. 13:1274–1279.
Van Miegroet H., Cole D.W. 1985. Acidification sources in red alder and Douglas-fir soils: importance of nitrification. Soil Sci. Soc. Am. J. 49:1274–1279.
Van Raij B., Peech M. 1972. Electro-chemical properties of some oxisols and alfisols of the tropics. Soil Sci. Soc. Am. Proc. 36:587–593.
Vet R.J., Sirois A., Jeffries D.S., Semkin R.G., Foster N.W., Hazlett P., Chan C.H., 1988. A comparison of bulk, wet-only and wet-plus-dry deposition measurements at the Turkey Lakes watershed. Can. J. Fish. Aqua. Sci. 45:26–37.
Vose J.M., Swank W.T. 1990. Foliar absorption of 15-N labelled HNO_3 by mature eastern white pine. Can. J. For. Res. 20:857–860.
Weaver G.T., Khanna P.K., Beese F. 1985. Retention and transport of sulfate in a slightly acid forest soil. Soil Sci. Soc. Am. J. 49:746–750.
Winner W.E., Mooney H.A., Goldstein R.A. (eds.) 1985. Sulfur Dioxide and Vegetation, Physiology, Ecology, Policy Issues. Stanford University Press, Stanford, California.
Wolfe G.V. 1988. Atmospheric Deposition to a Forest Ecosystem: Seasonal Variation and Interaction with the Canopy. M.S. Thesis, University of Washington, Seattle.
Wright R.T., Hobbie J.E. 1966. Use of glucose and acetate by bacteria and algae in aquatic ecosystems. Ecology 47:447–464.
Xue D., Harrison R.B. 1991. Sulfate, aluminum, iron, and pH relationships in four Pacific Northwest Forest subsoil horizons. Soil Sci. Soc. Amer. J. 55:837–840.
Zhang G.Y., Zhang X.N., Yu T.R. 1987. Adsorption of sulfate and fluoride by variable charge soils. J. Soil Sci. 38:29–38.

6. Nitrogen Chemistry, Deposition, and Cycling in Forests

Overview

D.W. Cole

At the conception of the Integrated Forest Study (IFS) program, a series of questions were raised specific to the effect of N input on ecosystem nutrient cycling processes including those mechanisms responsible for the retention, transformation, and loss of N through leaching from forest ecosystems. The fact that comparable data were collected at 17 sites (covering a wide array of forest types, elevations, climatic conditions, and atmospheric N input regimes) allowed us to integrate the results from this program and begin to answer some of the questions that could not be addressed from the results generated from any single individual site.

A large portion of this chapter will be centered around the concept of N saturation, its occurrence and causes, and how it is affected by atmospheric N deposition. The term has emerged in the last decade in the context of acid rain effects (e.g., Agren and Bosatta 1988; Dempster and Manning 1988; Aber et al. 1989), although this ecosystem condition can also be achieved without large anthropogenic N inputs, as is discussed later. In simple terms, N saturation can be defined as that forest soil condition whereby the N flux density (i.e., N input from N mineralization plus atmospheric N input) exceeds the system's retention capacity and NO_3^- starts leaching out of the system (e.g., Agren and Bosatta 1988). The choice of measurable NO_3^-

leaching as an indicator of N saturation will be discussed in greater detail later in Chapter 6. The term "N saturation" as defined here is related to, but generally broader than the terms "excess N deposition" [i.e., the point at which the system's N retention capacity is exceeded by atmospheric N inputs (Dempster and Manning 1988)], or "critical N load" [i.e., the maximum allowable N inputs so as to avoid excess NO_3^- leaching (Nilsson and Grennfelt 1988)] in that it also recognizes and accounts for natural N sources and sinks within the ecosystem.

The questions posed at the onset of the IFS program regarding N dynamics were as follows:

- Is N saturation and NO_3^- leaching a common process within our forest ecosystems or is it limited to unique conditions associated with disturbance or special ecosystem types?
- If N saturation and NO_3^- leaching are common processes, do they follow any consistent pattern? Can we understand, or at least predict, where saturation and leaching is most likely to occur by following the cycling processes within the ecosystems under question?
- How important are cloud water interception and dry deposition as mechanisms of N addition to forest ecosystems? Does their importance vary with forest composition, climatic conditions, and other features of the environment or is there consistency between sites? Can these measurements realistically be made?
- Can atmospheric N deposition alone cause N saturation and trigger NO_3^- leaching?
- Does the leaching of nitrate at N-saturated sites significantly change soil solution properties including pH, cation leaching, and Al concentration? Will these changes in turn cause changes to soil properties?
- Are these changes reversible? If so, which ones, how long will it take, and can the process be accelerated?
- Are these changes of sufficient significance to affect ecosystem productivity?

Although it was not feasible within the 3-year monitoring period of this program to adequately answer all these questions, the integrated nature of the study has nevertheless allowed a comprehensive analysis of these questions and has generated answers to many.

To determine if N saturation and nitrification are common in forest ecosystems, whether internal processes or external inputs are of greater significance in triggering nitrification, and to what extent nitrification and NO_3^- leaching significantly alter soil and soil solution properties, it was essential to examine these questions over a broad range of ecosystem conditions and within the context of the mineral cycle. Specifically, at 13 of the sites the various dry and wet N input fluxes were established (see pp. 152–166 Chapter 6). For those sites where only bulk precipitation N inputs were measured, total N inputs were estimated on the basis of ecosystem characteristics, and the relationship between wet and dry N deposition observed at the intensive monitoring sites as discussed later in Chapter 6. In addition, the rates and processes of N retention, transformation, and leaching were documented. The impact of N input and cycling on other ecosystem properties were also followed. Using NO_3^- as an indicator of N saturation, the data sets collected over this broad range of forest types and air pollution regimes allowed us to analyze, through multiple regression analysis, the relative role of various dynamic and static ecosystem properties in explaining variations in NO_3^- leaching losses among the different sites within the IFS network. The results from these efforts are discussed in detail in the following sections.

Atmospheric Deposition and Canopy Interactions of Nitrogen

G.M. Lovett

Processes of N Deposition

The importance of the atmospheric deposition of N to terrestrial ecosystems was firmly established by the pioneering work (and heated controversy) of Liebig, Lawes, Gilbert, and Boussingault in the mid-nineteenth century (Aulie 1970; Paul 1976). Since that time, N has been undoubtedly the most studied element in forest nutrient cycling because of its key role in plant nutrition and the fact that it generally limits forest production in the temperate zone. Nitrogen is also an important element in the atmosphere, where N_2 comprises about 78% by volume, and where reactive N species, although much less abundant, are crucial in atmospheric chemical reactions. These reactive species have natural and anthropogenic sources, the latter including both combustion and agriculture (Singh 1987; Warneck 1988). Therefore, the potential exists for human alteration of the N chemistry of the atmosphere to have a profound effect on the biosphere.

The reactive N species of primary concern from the standpoint of atmosphere–forest interactions include both oxidized forms (N_2O, NO, NO_2, HNO_3, and NO_3^- ion) and reduced forms (NH_3, NH_4^+, and organically bound N). Other N oxides (e.g., N_2O_5, HNO_2, and NO_3 radical) are present in the atmosphere and are thought to play important roles in atmospheric chemistry, but their concentrations are less well quantified (Singh 1987). The atmospheric chemistry of N oxides is complicated and incompletely understood, but it is known that combustion in motor vehicles, power plants, and industrial processes releases primarily NO, which is rather quickly oxidized to NO_2 and then to HNO_3 vapor in the gas phase (Singh 1987). Nitric acid vapor can be dry-deposited, dissolve in rain or cloud droplets, adsorb to atmospheric particles, or react with basic substances (e.g., NH_3) to form fine aerosols (e.g., NH_4NO_3). Natural processes in the soil can also release NO and NO_2, which will undergo the same reactions (Singh 1987).

Most of the NH_3 in the lower atmosphere in North America and Europe appears to be a result of agricultural activities, mainly volatilization from animal wastes and fertilized soils (Soderlund and Svensson 1976; Warneck 1988). Volatilization of NH_3 from unfertilized soils is small unless the soils are alkaline (Warneck 1988), a rare condition in moist temperate forests. Ammonia is the only alkaline trace gas of any significance in the atmosphere, and it reacts quickly with acidic compounds to form NH_4^+, either in dry particles or in droplets (Warneck 1988). Ammonia may also be dry-deposited to vegetation and soil surfaces.

The sources and forms of organic N in the atmosphere are poorly studied, and the concentrations are rarely measured, except in precipitation. Possible sources include particulate material entrained from soils and vegetation (e.g., pollen, soil dust, spores) and reaction products of N oxides with organic compounds (e.g., peroxyacetyl nitrate, PAN).

Thus, both natural and anthropogenic activities result in a wide variety of N compounds in the atmosphere in gaseous, particulate, and dissolved forms. Our primary purpose in this study was to determine the concentrations and deposition rates of the species that contribute the greatest amount of N deposition to forest ecosystems. Therefore, we focused on dry deposition of HNO_3 vapor, particulate NO_3^- and NH_4^+, and dissolved NO_3^- and NH_4^+ in precipitation and cloud water. Concentrations of NH_3 were not measured at most sites, but estimates were made during the growing season at the Oak Ridge loblolly pine (LP) site (Lindberg et al. 1990) and are discussed briefly. Organic N in precipitation was measured at some sites, and NO_2 concentration and deposition were estimated for sites where data were available. Separate experimental studies investigated the factors controlling the deposition of NO_2 and HNO_3 at the leaf and branch level (see pp. 166–177). In addition, we measured N passing out of the canopy as stemflow and throughfall to elucidate the interactions of the various forms of N within the forest canopies.

Methods and Sites

Data collection methods are described in detail in Chapter 2. Briefly, a three-stage filter-pack was used to measure atmospheric concentration of fine aerosols (on a Teflon filter), HNO_3 vapor (on a nylon filter), and SO_2 (on a K_2CO_3-impregnated filter) above each of the forest canopies. Concentrations of NO_2 were measured continuously at or near some sites using chemiluminescence analyzers. Dry deposition of fine particles (approximately <2-μm diameter) and HNO_3 were calculated using a deposition model slightly modified from Hicks et al. (1987), which uses measured canopy structure and meteorological parameters to estimate dry deposition velocities. Dry deposition fluxes were calculated as the product of mean atmospheric concentrations and mean deposition velocities for sampling periods of several days to 1 week. Coarse-particle dry deposition fluxes were measured on plastic petri dishes exposed above the canopies and were extrapolated to full-canopy deposition rates using a scaling factor derived by using Ca^{2+} in throughfall as a "tracer" of coarse-particle deposition. Precipitation and throughfall were measured on an event basis using automated wetfall-only collections. Cloud water was collected for chemical analysis using harplike string collectors above the canopy at the mountain sites and an active collector at the Oak Ridge loblolly pine site. Deposition of cloud water was calculated from analysis of the water balance in the canopy (Lovett 1988).

The study sites are described in Chapter 2. Some sites in the IFS project were not included in this analysis because of the lack of the full complement of deposition data for particulate, gaseous, and dissolved forms of NH_4^+ and NO_3^-. Of the intensive deposition study sites considered here, 2 years of data collection (roughly 1986–1988) are available for all areas except the ST, LP, and CP sites (3 years) and the MS and FS sites (1 year). The data from the FS and MS sites should be interpreted with caution because the record is short. In the bar graphs in this section, the order of sites is as follows: first, the southeastern sites in order of elevation; then, the northeastern sites in order of elevation; then, the western sites; and finally, the site in Norway.

Atmospheric Concentrations and Dry Deposition

The sum of the concentrations of HNO_3 vapor and fine-particle NO_3^- and NH_4^+ ranged between 0.62 and 2.2 μg N m^{-3} (Figure 6.1). The highest values are for the LP and DL sites, probably because these are the IFS sites closest to urban areas. Of these three constituents, fine-particle NH_4^+ consistently dominated, with concentrations of 0.34 to 1.6 μg N m^{-3} (0.43–2.0 μg NH_4^+ m^{-3}). Concentrations of HNO_3 vapor ranged from 0.18 to 0.62 μg N m^{-3} (0.82–2.8 μg HNO_3 m^{-3}), and fine-particle NO_3^- concentrations were generally less than 0.14 μg N m^{-3} (0.62 μg NO_3^- m^{-3}). Note that sampling artifacts in the filter-pack technique may affect the relative proportions of particulate and gaseous NO_3^- (Appel and Tokiwa 1981), but this problem has been shown to be minor in a study near our WF site (Kelly et al. 1984); in any event, the measurement of total atmospheric NO_3^- is not affected. Concentrations of aerosol NO_3^- and NH_4^+ seem to vary in concert between sites, such that the NH_4^+ fraction of the total is always in the range 50% to 70%, and in most cases is between 60% and 70% (Figure 6.2).

Despite the dominance of NH_4^+ in the air, HNO_3 dominates the total dry deposition of N (Figure 6.3) such that the contribution of aerosol NH_4^+ is always less than 50% and usually less than 30% of the total (Figure 6.2). This occurs because HNO_3 is a very reactive gas that adsorbs readily to most surfaces after diffusing through the quasi-laminar boundary layer immediately adjacent to the surface (Huebert and Robert 1985). Particulate NH_4^+, however, is generally in the 0.1- to 1.0-μm size range (Lindberg et al. 1986), a class of particles that cannot be transported easily across boundary layers

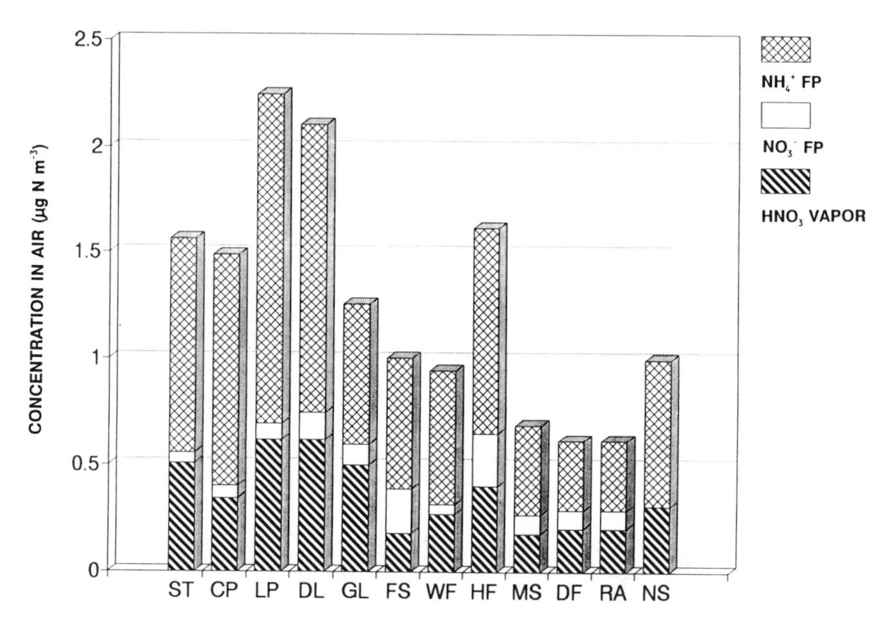

Figure 6.1. Concentrations ($\mu g\ N\ m^{-3}$) of HNO_3 vapor and fine-particle ($<\sim 2\ \mu m$) NH_4^+ and NO_3^- at the intensive study sites (see Chapter 2 for abbreviations).

by impaction, sedimentation, or diffusion. The model-calculated average deposition velocities for HNO_3 vapor range from 1.3 to 6.0 cm s^{-1} (with the highest values at the windy, high-elevation sites), while those for fine particles range from 0.02 to 0.4 cm s^{-1} (see the Appendix).

Concentrations of coarse particles are not included in Figure 6.1 because these larger particles are not sampled efficiently by standard air filtration (Noll et al. 1985). We measured coarse-particle deposition directly on artificial-surface collectors that we previously tested and compared to individual leaves and full canopies (Lindberg and Lovett 1985; Lovett and Lindberg 1986). The coarse-particle deposition fluxes of both NH_4^+ and NO_3^- can add substantially to the total dry deposition of N, contributing up to 60% to 80% at the FS, MS, DF, and RA sites, where HNO_3 vapor concentrations are lowest (Figure 6.3).

In the discussion thus far, we have ignored the dry deposition of NH_3, NO, and NO_2. We did not measure NH_3, but some studies have shown it to be low in both concentration and deposition in the sort of rural, nonagricultural areas that characterize most sites in this study (Tjepkema et al. 1981). A short-term study at the LP site showed NH_3 concentrations of about 0.2 $\mu g\ N\ m^{-3}$ during the growing season, and NH_3 at this concentration was estimated to contribute about 5% of the total growing-season N deposition, assuming an average deposition velocity for NH_3 of 1.0 cm s^{-1} (Lindberg et al. 1990).

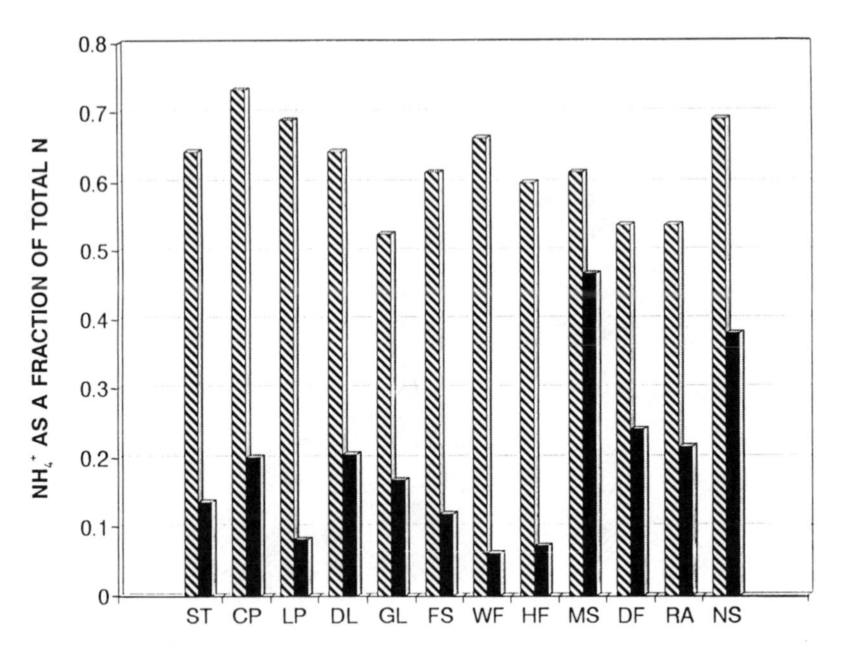

Figure 6.2. Ammonium as fraction of ammonium plus nitrate in atmospheric concentrations and in dry deposition at study sites.

Concentration of NO is generally low outside urban source regions (Warneck 1988), and its deposition velocity was also found to be quite low compared to NO_2 or HNO_3. However, NO_2 can reach quite high concentrations, and occasionally high deposition levels, especially in urban areas. Later in Chapter 6, Hanson et al. report measured NO_2 concentrations of 4.6 μg N m^{-3} at the LP site and 1.2 μg N m^{-3} at the HF and WF sites, and they estimate concentrations of 0.57 to 2.8 μg N m^{-3} at the other sites. These concentrations are equal to or higher than the concentrations of the N species shown in Figure 6.1, especially for the sites closest to urban areas (LP and DL), indicating that NO_2 may be the dominant form of reactive N in the atmosphere. However, because of the low deposition velocity, Hanson et al. (pp. 166–177) estimated the dry deposition of NO_2 to range from only 10 mol ha^{-1} yr^{-1} at the WF site to 82 mol ha^{-1} yr^{-1} at the DL site. This suggests that accounting for NO_2 could increase the total dry deposition by only 2% to 20% over the values shown in Figure 6.3.

Annual, volume-weighted, mean concentrations of NO_3^- and NH_4^+ in precipitation both vary in the range 5–30 μmol L^{-1} among the sites, with NO_3^- concentrations generally slightly higher than NH_4^+ concentrations. Surprisingly, there was no correlation between concentrations of these ions in precipitation and their corresponding concentrations in the air. At the three sites (ST, LP, and WF) that measured cloud or fog water chemistry,

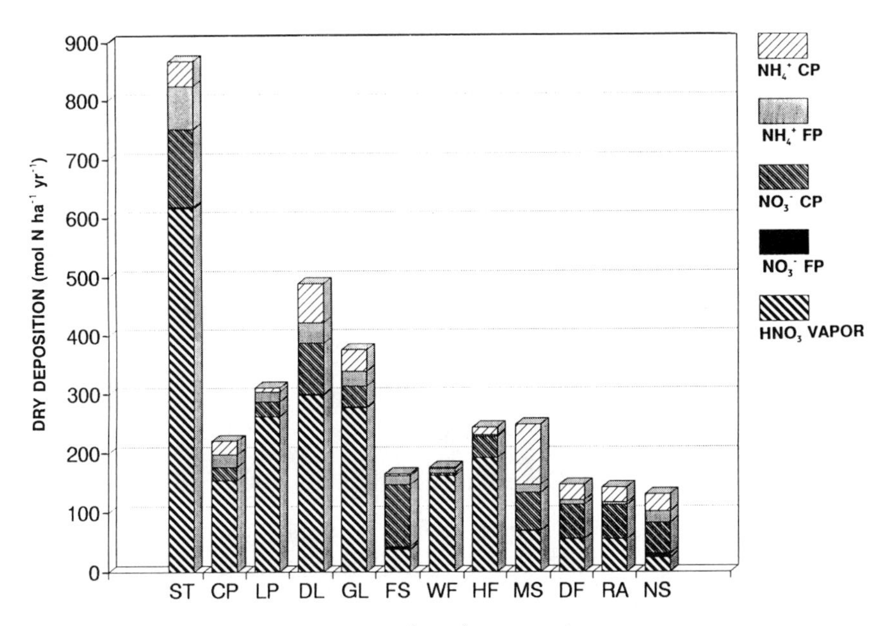

Figure 6.3. Dry deposition (mol ha^{-1} yr^{-1}) of NH_4^+ and NO_3^- via HNO_3 vapor, fine-particle (FP), and coarse-particle (CP) NH_4^+ and NO_3^- to study sites.

concentrations in cloud water were substantially higher than corresponding concentrations in precipitation. For NO_3^-, the cloud to rain ratio for annual mean concentrations was 5.8 at the ST site, 3.8 at LP, and 6.9 at WF. For NH_4^+, the ratios were 8.5, 18, and 11 for the three sites, respectively. While this indicates that cloud and fog water was on average many times more concentrated than rain, and more so for NH_4^+ than for NO_3^-, Weathers et al. (1988) cautioned that these ratios can vary greatly from event to event.

Total Deposition

Total atmospheric deposition of N (the sum of wet, dry, and cloud water deposition) ranged from 340 mol ha^{-1} yr^{-1} (4.7 kg ha^{-1} yr^{-1}) at the RA site to 1900 mol ha^{-1} yr^{-1} (27 kg ha^{-1} yr^{-1}) at the ST site (Figure 6.4). (Note that 50–100 kg N ha^{-1} yr^{-1} is added to the red alder (RA) site by symbiotic N fixation, which could be considered a form of atmospheric deposition.) For the lowland sites in the eastern United States, N deposition varied from 500 to 1000 mol ha^{-1} yr^{-1}, with all except the DL site less than 700 mol ha^{-1} yr^{-1} (roughly 10 kg ha^{-1} yr^{-1}). Although it appears in Figure 6.4 that total N deposition is generally higher in the southeast than in the northeast, this may be an artifact of having chosen sites closer to urban areas in the southeast. Hanson et al. (pp 166–177) estimated that inclusion of NO_2 deposition would add less than 10% to the total deposition at the IFS sites in

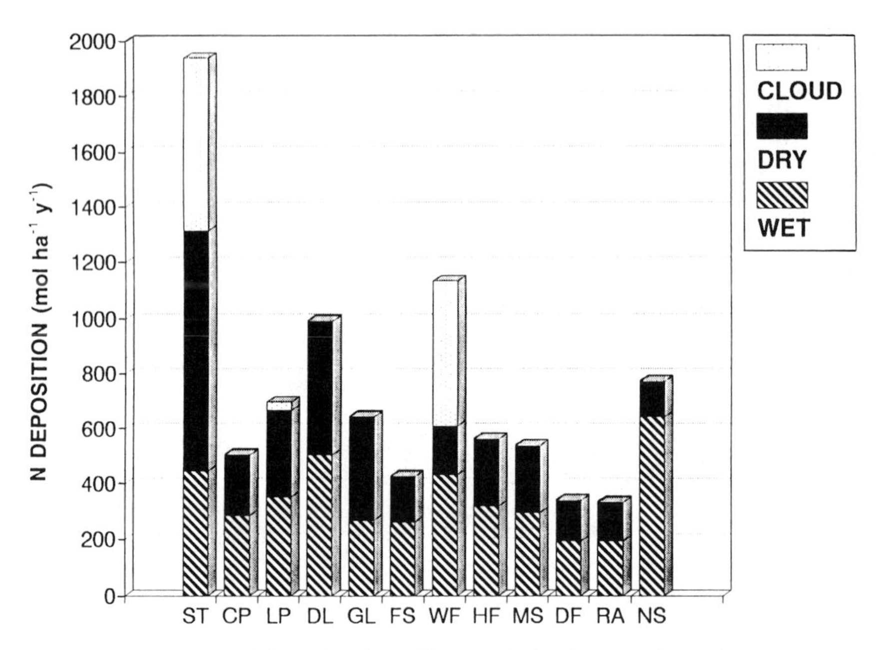

Figure 6.4. Wet deposition, dry deposition, and cloud water deposition of N (mol ha^{-1} yr^{-1}) to study sites.

Figure 6.4. Deposition of NO_3^- (including HNO_3 vapor) exceeded that of NH_4^+ at all sites, usually by a factor of 2 or more.

Dry deposition varied more than wet deposition between sites and made a significant contribution to the total deposition at all sites. Cloud water deposition is a very important source of N for the high-elevation sites and made ST and WF the sites with the highest N deposition in this project. Nonetheless, the cloud water deposition at these sites might be lower than that of many mountain summit sites in the eastern United States (Lovett and Kinsman 1990), because the ST site was estimated to be in cloud only 10% of the time and the WF site 7% of the time. By contrast, cloud immersion frequencies are estimated to be 28%–35% for the summit of Mt. Mitchell, North Carolina, and 40%–45% for the summit of Whiteface Mt., New York (Mohnen 1988).

In an effort to simplify the process for estimating total N deposition, it is valuable to address the question of whether total deposition can be predicted from the more easily measured wet deposition. Figure 6.5 shows the considerable scatter in the relationship between the two variables; regression analysis indicated that wet deposition explains only 35% of the variance in total deposition when all the sites are considered. However, if we restrict the analysis to low-elevation sites in the United States (i.e., exclude ST, WF, and NS), then wet deposition explains 90% of the variance in total deposition (see Figure 6.5). The regression line in Figure 6.5, which is based

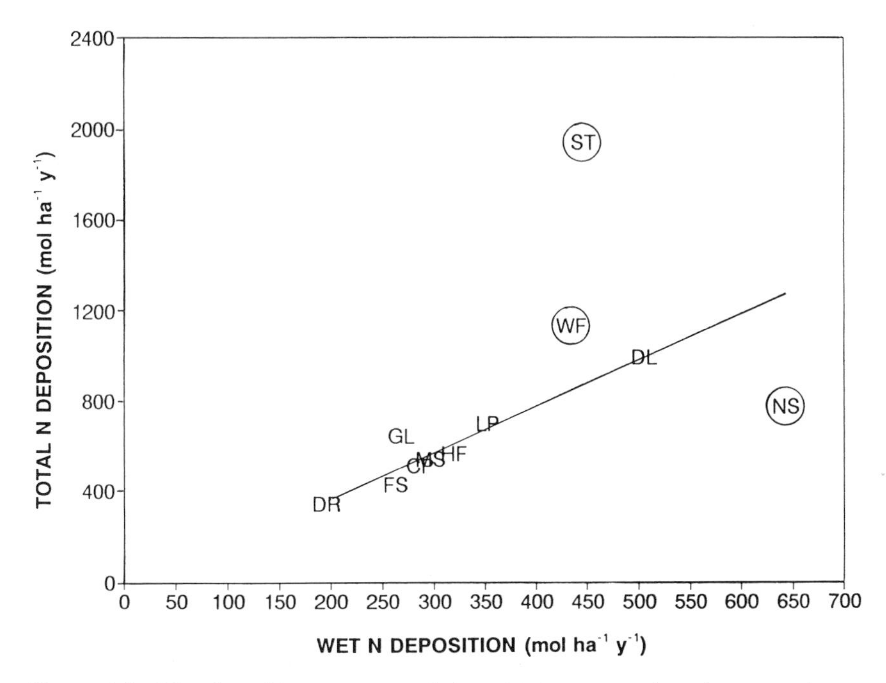

Figure 6.5. Wet deposition versus total deposition (mol ha^{-1} yr^{-1}) of NH_4^+ plus NO_3^- for study sites. Regression line ($y = -49 + 2.1x$, $r^2 = .90$) excludes circled points (see text). DF and RA points have nearly identical coordinates and are plotted together as DR.

on the data excluding the ST, WF, and NS sites, has a slope of 2.1 (±0.3 SE) and an intercept that is not significantly different from zero. Thus, it appears that in general total deposition can be assumed to be roughly twice as great as wet deposition for low-elevation forests in the United States. Considering the sites individually, dry deposition of N contributes a mean of 46% of total deposition, with a range from 39% (FS site) to 59% (GL site); both of these extremes are from sites with only 1 year of data. This indicates that wet and dry deposition are roughly equal for these low-elevation United States sites. For sites subject to high levels of cloud water deposition, total deposition will be further enhanced. Note that at the NS site, wet deposition greatly exceeds dry deposition; this probably results from the location of the NS site far from major NO_x emission sources. These deposition estimates are higher than many N inputs reported in the literature for forested ecosystems, mainly because we specifically included dry and cloud water deposition while many previous estimates were based on bulk deposition alone.

Significant uncertainties are associated with these total deposition estimates because of the combination of measurements and models that are used. We estimate that the overall uncertainty in total N deposition is about ±50%

at the high-elevation sites and ±30% at the low-elevation sites, with the largest potential sources of error associated with the specification of deposition velocities for HNO_3 vapor and the N fluxes with cloud water.

In the past, N input measurements for ecosystem studies were frequently made using continuously open funnels or buckets, called bulk deposition collectors. From our best estimates, total deposition significantly exceeds bulk deposition at most sites where both measurements were made (Table 6.1), especially at those sites where dry and cloud water deposition are important. This suggests that bulk deposition measurements seriously underestimate total N deposition for many forests.

Canopy Exchange

If one considers only precipitation (PPT), throughfall (TF), and stemflow (SF) fluxes, it appears that the canopies at all of these sites release NO_3^- [i.e., PPT < (TF + SF)], and most of them take up NH_4^+ [PPT > (TF + SF)]. Consideration of dry deposition and cloud water deposition changes that conclusion, however. Our best estimate of the true role of the canopy is the net canopy exchange (NCE), which we define as total deposition (wet + dry + cloud water) minus (TF + SF). The NCE for both NO_3^- and NH_4^+ is negative for all sites, indicating that the canopies clearly are sinks for inorganic N (Figure 6.6). The apparent NO_3^- release from the canopies (when only wet deposition is considered) can be completely accounted for by washoff of dry-deposited nitrate. Our data for organic N are less reliable because at some sites organic N was not measured at all and at some sites it was measured only in bulk deposition collectors, which were left in the field for a week or more after a precipitation event, possibly allowing microbial transformation of N in the collectors. However, the available data indicate a release of organic N (i.e., NCE > 0) from all canopies for which it was measured (Figure 6.6).

Table 6.1. Comparison of Total Deposition[a] (wet + dry + cloud/fog) and Bulk Deposition of Inorganic N at IFS Sites Where Both were Measured

	Total Deposition			Bulk Deposition		
Site	NO_3^-	NH_4^+	$NO_3^- + NH_4^+$	NO_3^-	NH_4^+	$NO_3^- + NH_4^+$
ST	1200	700	1900	315	12	327
CP	330	180	510	220	150	370
DL	590	410	1000	270	330	600
GL	470	180	650	190	160	350
HF	420	150	570	260	180	440
DF	240	110	350	140	75	215
RA	240	100	340	140	75	215
NS	420	360	780	310	310	620

[a] Data are annual averages, mol ha^{-1} yr^{-1}.

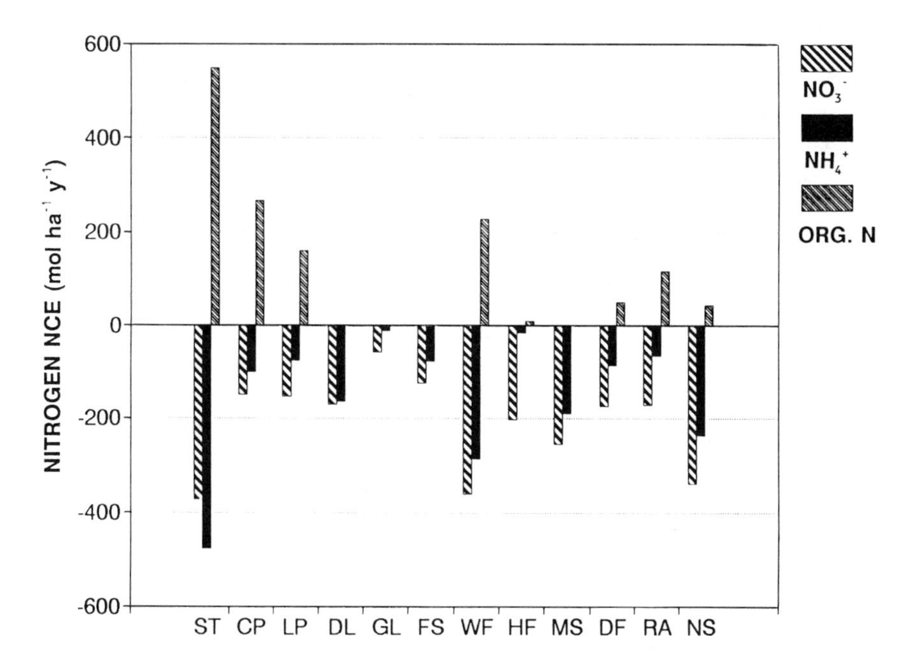

Figure 6.6. Net canopy exchange (NCE, defined as total atmospheric deposition minus throughfall and stemflow deposition) for NO_3^-, NH_4^+, and for some sites, organic N. Negative values indicate sink in canopy; positive values indicate source.

We cannot determine the source of the organic N or the fate of the inorganic N from these data. Both denitrification and nitrification can cause N to be lost via gaseous release (as N_2O, N_2, and NO), but neither is likely to be important in these canopies. Denitrification requires anaerobic sites, which must certainly be rare in forest canopies, except in cases where large amounts of organic detritus accumulate on limbs or in crotches. If nitrification were responsible for the loss of NH_4^+, we would expect a significant release of NO_3^- from the canopies, which does not occur. Chen et al. (1983) considered nitrification to be responsible for the decrease in NH_4^+ and the increase in NO_3^- in precipitation passing through a northern hardwoods canopy. Dry deposition of NO_3^- was inadequately characterized in that study, and the data presented here indicate that dry deposition could probably have accounted for the increase in NO_3^-.

On the other hand, biological uptake of N by canopy surfaces is well documented. Trees have been demonstrated experimentally to be capable of absorbing and incorporating gaseous NO_2 and HNO_3 (see, for example, the next part of Chapter 6 in this volume), as well as NO_3^- and NH_4^+ in solution (Reiners and Olson 1984; Bowden et al. 1989). Epiphytic lichens have also been shown to be active absorbers of NO_3^- and NH_4^+ in solution (Lang et al. 1976, Reiners and Olson 1984). We note that the strongest sinks (most negative NCE values) for inorganic N are seen in the spruce and spruce-fir forests (ST, WF, MS, NS), all of which also have a high biomass of epiphytic lichens.

Leaf-surface bacteria and fungi should also show active uptake of NO_3^- and NH_4^+, although this has not been well researched. Microorganisms are unlikely to be a quantitatively significant sink for deposited N, however, because their populations should quickly reach a steady state on canopy surfaces. Microbes could play an important role in the conversion of inorganic to organic N, if that occurs.

The organic N in TF + SF probably arises from the efflux of N-containing compounds (e.g., amino acids, proteins) from internal pools and surfaces of plants, lichens, and microbes, although microparticulate detritus could also contribute (Schaefer and Olson 1984). The NCEs of organic and inorganic N are not significantly correlated across sites, but it is nonetheless possible that some of the lost inorganic N might be transformed to organic N by biological action in the canopy.

Because of the in-canopy sink for NO_3^- and NH_4^+, the below-canopy flux (TF + SF) is depleted in inorganic N relative to the total deposition flux, and a plot of these two fluxes shows that all points fall below the 1:1 line (Figure 6.7). The slope of the best-fit regression line is 0.59, indicating

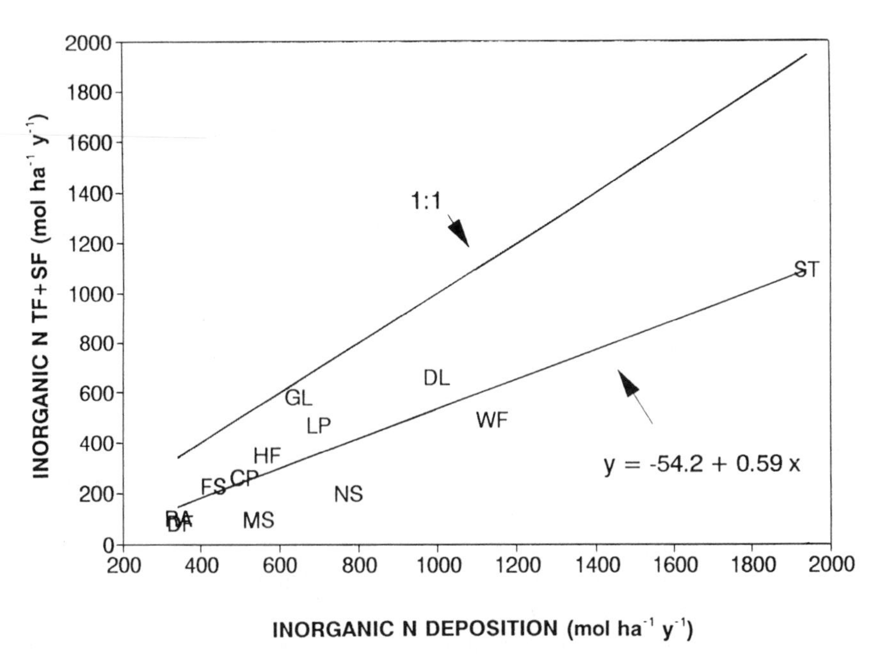

Figure 6.7. Total deposition versus throughfall plus stemflow (TF + SF) for inorganic N (mol ha^{-1} yr^{-1}), including 1:1 line (inputs = outputs for canopy) and regression line (r^2 = .80).

that, on average, about 60% of the inorganic N deposited to these canopies passes through as TF + SF, while the other 40% is taken up by biological surfaces, converted to organic N in TF or SF, or rereleased to the atmosphere. The fact that the slope of this line is less than 1 implies that the sink strength for inorganic N in the canopy is correlated with the amount of deposition, such that increased deposition results in NCEs that are more strongly negative (Figure 6.8). Close examination of Figure 6.7 suggests that the spruce and spruce-fir canopies (MS, NS, WF, ST) fall on one line and that the rest of the sites fall on another, with the latter having a slope close to 1. This is reflected in Figure 6.8 as a strong relationship between N deposition and canopy uptake for the spruce and spruce-fir sites and a very weak relationship for the other sites. One interpretation is that the four "spruce" sites have some characteristic, perhaps the spruce trees themselves or their epiphytic lichens, that causes greater canopy uptake in response to greater N inputs. The other canopies appear to have a moderate level of N uptake (about 200 mol ha^{-1} yr^{-1}), with little response to deposition amount.

For all the sites except CP, the "uptake" of inorganic N exceeds the release of organic N, so the NCE of total N (inorganic + organic) is negative. For the CP site, this number is only slightly positive. The regression line for total deposition versus TF + SF for total N has a slope of 0.84 (Figure 6.9), indicating that only 84% of the change in total N deposition from site

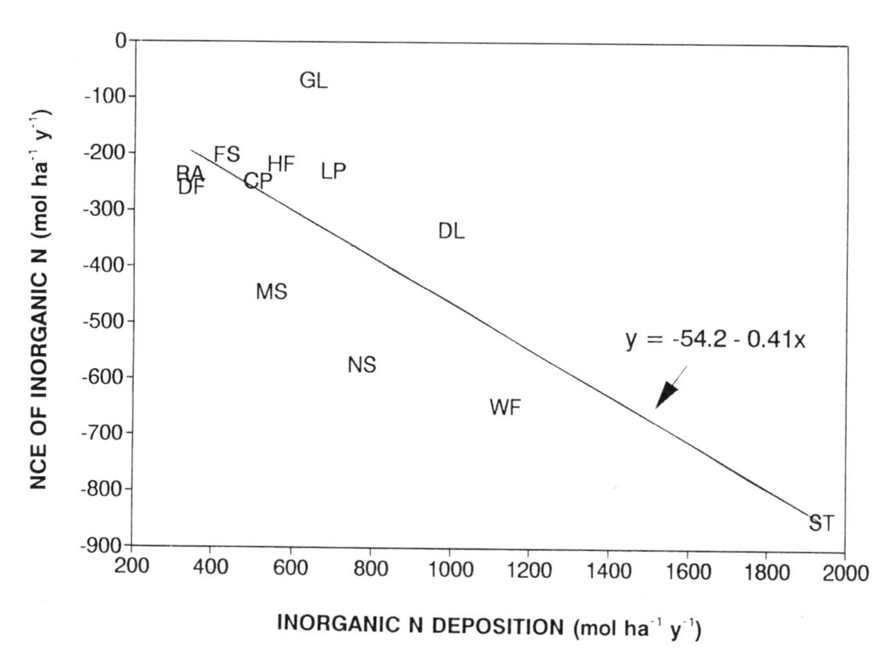

Figure 6.8. Total deposition versus net canopy exchange (NCE) of inorganic N, including regression line (r^2 = .66).

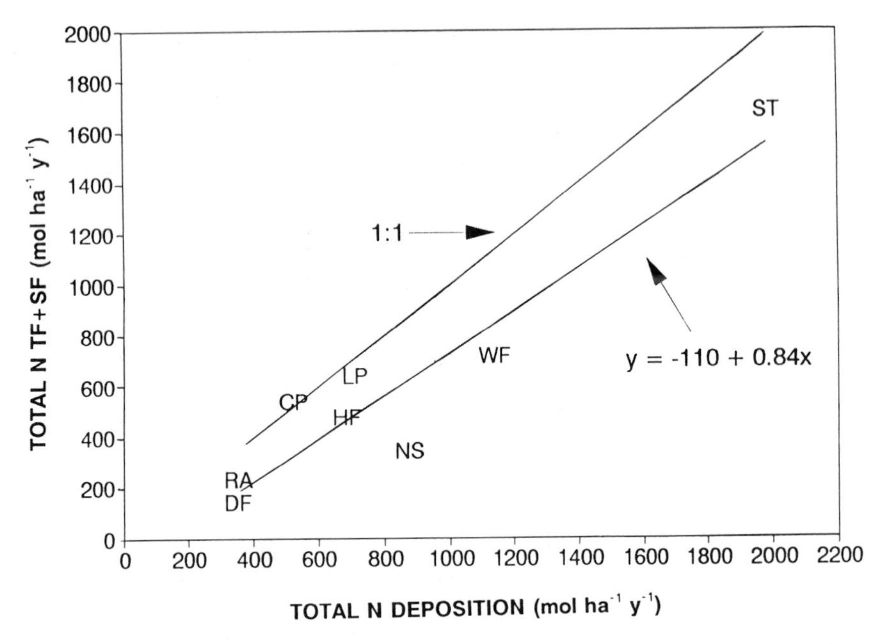

Figure 6.9. Total deposition versus TF + SF for total N (inorganic plus organic), including 1:1 line and regression line (r^2 = .89).

to site is reflected in a corresponding change in the below-canopy flux of N. The remainder (16%) must be taken up by the canopy or rereleased to the atmosphere.

The NCE of total N in these canopies ranges from +21 mol ha^{-1} yr^{-1} (CP site) to −531 mol ha^{-1} yr^{-1} (NS site). Most of the eastern United States sites fall in the range of −200 to −450 mol ha^{-1} yr^{-1} (−2.8 to −6.3 kg ha^{-1} yr^{-1}), with the highest values in the spruce and spruce-fir canopies (MS, WF, ST) and in the DL pine site, which receives high HNO_3 deposition. These NCE values have considerable uncertainty because they are calculated as difference between two numbers, both measured with uncertainty. However, because dry deposition of NO_2, HONO, and NH_3 was ignored in these calculations, we feel the NCE values are probably conservative.

Although most of the NCE for total N probably results from plant uptake, this amount of uptake is generally small compared to the N needs of the forests. Figure 6.10 contrasts the canopy uptake (= −1 × NCE, assuming all NCE is biological uptake) with the annual stand requirement (defined as the sum of the N contained in current growth of foliage, branches, boles, and, for some sites, roots) and foliar increment (N in current foliage only). In most cases, canopy N uptake is only a small fraction of both the requirement and the foliar increment. The obvious exception is the NS site, for

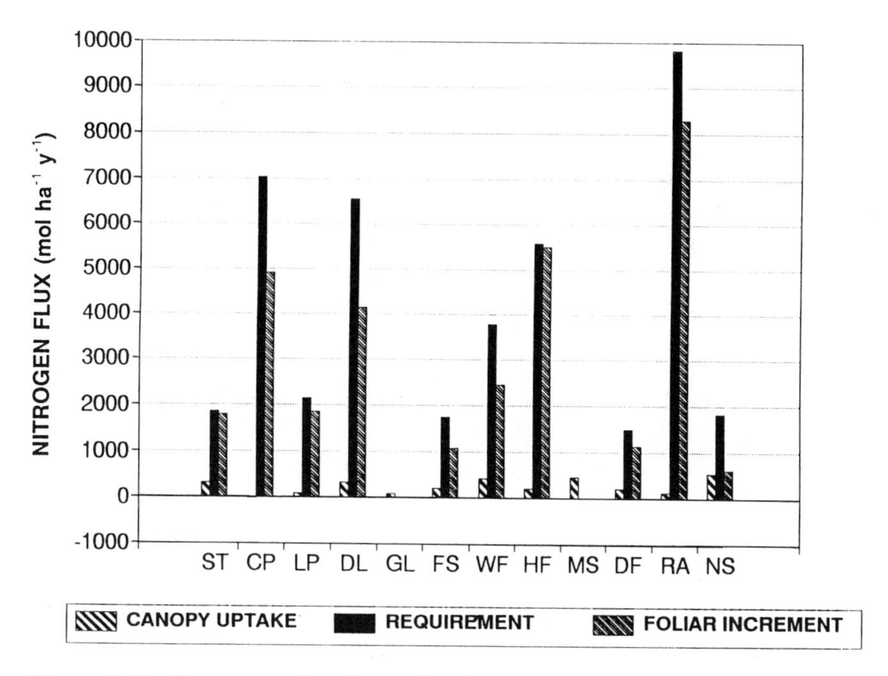

Figure 6.10. Canopy uptake of organic plus inorganic N compared to forest requirement and foliar increment.

which canopy uptake is nearly equal to the foliar increment. Although the canopies are probably absorbing more N from the atmosphere than they were before the advent of air pollution, in general most of their overall N requirement is still being met by root uptake and retranslocation (resorption). Canopy uptake of N is most important nutritionally at the high-elevation sites, which have high uptake and low N requirements.

Summary

Particulate NH_4^+ constitutes 50% to 70% of the total atmospheric N concentrations measured in this study (including HNO_3 vapor and fine-particle NH_4^+ and NO_3^-). Nitric acid vapor is the dominant species contributing to dry deposition, however, and coarse-particle NO_3^- and NH_4^+ can also be quite important. Although NO_2 is probably the reactive N species of highest concentration in the atmosphere at many of the sites, it is estimated to contribute less than 20% of the total dry deposition (and <10% of the total deposition) of N and was not directly measured at most sites. The highest total deposition levels were found at the southeastern sites. Total deposition of N averaged about twice the wet deposition for the low-elevation sites in the United States, indicating that dry deposition was roughly equal to wet deposition. Cloud water deposition added substantially more N at the high-elevation sites ST and WF, which received 1900 and 1100 mol ha^{-1} yr^{-1}, respectively, of total N deposition.

The net canopy exchange of both NO_3^- and NH_4^+ was negative (implying canopy uptake), but the NCE of organic N was always positive. Inorganic N in throughfall and stemflow (TF + SF) was about 60% of the inorganic N deposited to the canopy, suggesting that the canopy retained or transformed the other 40%. Total N (organic + inorganic) in TF + SF was about 84% of total N deposition. Canopy uptake of N ranged up to 531 mol ha^{-1} yr^{-1} but was generally small compared to the N requirements of the forests.

Experimental Laboratory Measurements of Reactive N Gas Deposition to Forest Landscape Surfaces: Biological and Environmental Controls

P.J. Hanson, G.E. Taylor, Jr., and J. Vose

Introduction

As discussed earlier, natural or anthropogenically produced oxides of N occur in the atmosphere in various forms including nitric oxide (NO), N dioxide (NO_2), and nitric acid vapor (HNO_3). Ambient concentrations of these oxides reflect a balance between natural or anthropogenic emissions, atmospheric chemical cycling with ozone, and deposition to landscape surfaces (Russel et al. 1985; Finlayson-Pitts and Pitts 1986). Nitrogen dioxide is typically the most concentrated form of atmospheric N oxides ranging from 1 to 2 nl L^{-1} in rural (pristine) areas, 5 to 10 nl L^{-1} in suburban areas, and as much as 50 nl L^{-1} in highly polluted urban/industrial regions (EPA 1982; Bytnerowicz et al. 1987). However, HNO_3, with concentrations commonly in the range from 0.5 to 1 nl L^{-1}, is the principal chemical sink for removal of NO_x from the atmosphere (Cadle et al. 1982; Galbally and Roy 1983).

Oxides of N can be harmful to plants, depending on a variety of biological and environmental factors. However, the threshold concentrations of NO_2 known to directly impede plant physiological responses are seldom attained under ambient conditions (Hill and Bennett 1970; Furukawa and Totsuka 1979; Kress et al. 1982; Saxe 1986a). Atmospheric sources of N, including NO_2 and HNO_3, have also been hypothesized to contribute to forest decline in several ways: disrupting the normal development of winter hardiness (Nihlgård 1985; Waring 1987), creating nutrient imbalances (Schulze 1989), and increasing shoot:root ratios (McLaughlin 1983). Conversely, as a source of N for plant growth, atmospheric oxides of N have been shown to contribute N for amino acid production within leaves (Rogers et al. 1979a; Kaji et al. 1980), to enhance the nutrition of plants growing under conditions of low fertility (Yoneyama and Sasakawa 1979), and to increase the activity of N-assimilating enzyme systems (Norby et al. 1989).

Deposition of NO_2 and HNO_3 to forest landscape surfaces may represent a potentially significant addition of N to the biogeochemical cycle of forest ecosystems, as discussed earlier in Chapter 6. Unfortunately, quantitative data describing rates and locations of NO_2 and HNO_3 deposition to woody plant tissues are not widely available. Deposition measurements are needed as input data for models of atmospheric chemistry, biogeochemical cycling, and studies of pollutant effects on plants (Hosker and Lindberg 1982). Foliar sites of N deposition (internal versus external) must be determined to predict the fate of dry-deposited gases. In addition, because bark or forest floor surfaces account for between 20% and 40% of the total landscape area available for deposition (100% for broadleaf forests in winter; Halldin 1985), deposition to these surfaces must also be quantified. Laboratory studies were conducted to provide direct deposition data for use in evaluating assumptions of the IFS stand-level deposition model. The objectives of the laboratory studies were to characterize deposition of NO_2 and HNO_3 to a variety of landscape surfaces, to evaluate the significance of the leaf surface and the leaf interior as sites for NO_2 or HNO_3 deposition, and to contrast the deposition characteristics of NO_2 and HNO_3. Because the N budgets summarized previously in Chapter 6 do not include estimates of NO_2 deposition, we assessed the amount of additional N deposition that may have resulted from NO_2 inputs. Laboratory measurements of foliar conductances to NO_2 were used together with available information on ambient NO_2 concentrations to estimate the probable contribution of NO_2 to N deposition in the IFS research sites.

Direct Measurements and Controlling Factors: Laboratory Mass Balance Techniques

Bare-root seedlings of the following species were obtained from commercial nurseries and grown under glasshouse conditions before measurements: red maple (*Acer rubrum* L.), white ash (*Fraxinus americana* L.), tulip poplar (*Liriodendron tulipifera* L.), white oak (*Quercus alba* L.), sycamore (*Platanus occidentalis* L.), red spruce (*Picea rubens* L.), white pine (*Pinus strobus* L.), and loblolly pine (*Pinus taeda* L.). Bark samples were obtained from branches or boles of four mature tree species, including shagbark hickory [*Carya ovata* (Mill.) K. Koch], tulip poplar (*Liriodendron tulipifera*), loblolly pine, and southern red oak (*Quercus falcata* Michx. var. *falcata*). Forest floor samples from loblolly pine or mixed hardwood forest stands near Oak Ridge, Tennessee, were collected as undisturbed cylindrical cores approximately 19 cm in diameter. Hanson et al. (1989) provided complete details of the growing conditions and measurement procedures for these representative forest surfaces.

Mass balance measurements of HNO_3 and NO_2 deposition to elements representative of a forest landscape (e.g., foliage, bark, soil) were conducted in an open gas-exchange system. The system simultaneously monitored the exchange of CO_2, H_2O, and either NO_2 or HNO_3 under controlled conditions of temperature, light, vapor pressure, and soil water availability. Techniques for HNO_3 deposition measurements were similar, but employed a technique based on thermal decomposition (Burkhardt et al. 1988) for measurements of HNO_3 concentration. Deposition rates (nmol m^{-2} s^{-1}) were calculated as the product of flow rate and the inlet-outlet concentration differential normalized for surface area and corrected for losses to chamber walls. Measurements of HNO_3 and NO_2 deposition to foliage shoots were conducted under light and dark conditions to establish patterns of diurnal variability associated with stomatal conductance. Shoot conductance (K_l) to a reactive N gas, a leaf-level measurement analogous to the deposition velocity (V_d), was determined by dividing the rate of deposition by the ambient concentration of the gas being measured.

Field ^{15}N Exposures

To quantify internal deposition of HNO_3, foliage in the upper canopy of mature eastern white pine trees at the Coweeta Hydrologic Lab was exposed to $H^{15}NO_3$ in branch cuvettes (9.0-L volume) constructed from Teflon film. $H^{15}NO_3$ was generated from calibrated permeation tubes (KIN-TEK Laboratories, Texas City, Texas) and mixed with HNO_3-free air created by pulling ambient air through a nylon filter (Nylasorb, Gelman) at a flow rate of 9.0 L min^{-1}. Three treatments were imposed to represent a range of exposure conditions: (1) 10 ppb for 30 h; (2) 50 ppb for 12 h; and (3) 100 ppb for 30 h. After exposures, foliage was immediately rinsed with 1000 ml deionized water to remove surface-deposited $H^{15}NO_3$ (Marshall and Cadle 1989), separated by age-class (current year and 1 year), dried at 60°C for 48 h; and ground to 100 μm using a ball mill. Tissue subsamples (n = 2) were analyzed for total N and atom ^{15}N% using mass spectrometry (Europa Scientific Instruments, ISO-TEC Labs, Miamisburg, Ohio). Excess ^{15}N was determined by subtracting total ^{15}N of exposed foliage from total ^{15}N of unexposed foliage. Additional details of the $H^{15}NO_3$ exposure methodology have been provided by Vose et al. (1989).

Deposition of NO_2 to Plant Foliage

Under daylight conditions and a mean concentration of 33 nl L^{-1}, NO_2 deposition to foliage of forest tree species varied by more than an order of magnitude, ranging from 0.35 (loblolly pine) to 5.75 nmol m^{-2} s^{-1} (sycamore), and the flux to most broadleaf species was greater than deposition to conifers (Table 6.2 and Figure 6.11). For this comparison all surfaces of leaves containing stomata were used as the reference area for calculations (i.e., one side for the broadleaf species and all sides for the conifers). The broadleaf species exhibiting the highest rates of NO_2 deposition had greater shoot conductance to water vapor (Figure 6.11). Interspecific variation in shoot conductance to water vapor reflects variation in stomatal frequencies

Table 6.2. Conductance of Various Terrestrial Surfaces to NO_2 Deposition

Surface	Conductance to NO_2[a] (cm s^{-1})
Distilled water	0.021 ± 0.02
Bark	
Dry	0.047 ± 0.001
Wet	0.093 ± 0.023
Plant shoots[b]	
Deciduous	0.093 ± 0.042
Coniferous	0.049 ± 0.015
Forest floor	
Deciduous	0.47 ± 0.26
Coniferous	0.48 ± 0.12

[a] Data for bark and plant shoots are expressed on total area basis and on planar area for intact forest floor samples.
[b] Data for plant shoots correspond to conditions of maximal stomatal conductance.

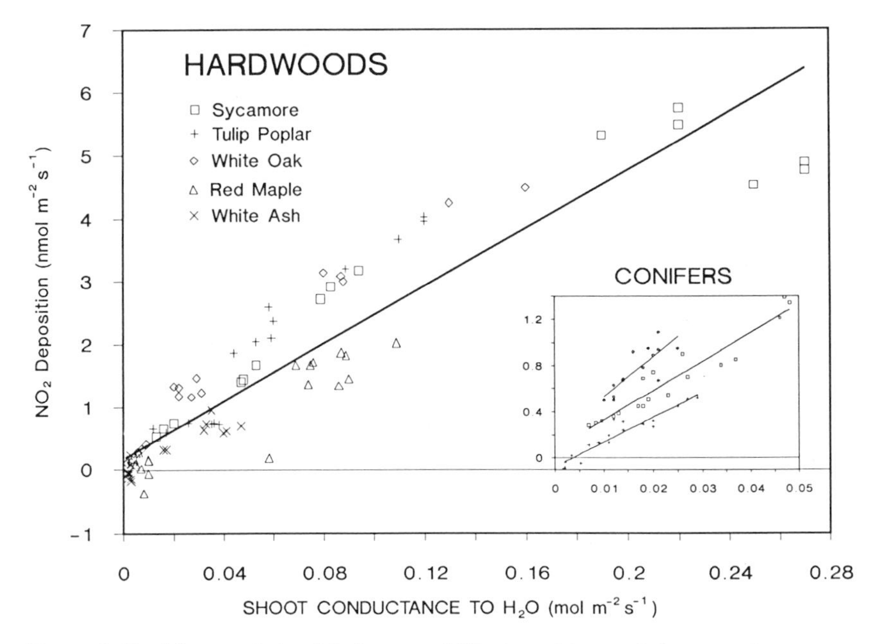

Figure 6.11. Linear relationship between NO_2 deposition and shoot conductance to water vapor. Symbols represent individual measurements for five broadleaf (main graph) and three conifer species (inset). Data for broadleaf species are expressed on projected leaf area basis; those for conifers are on total area basis. Mean NO_2 concentration was 34 nl L^{-1}. (Hanson et al. 1989. Reprinted with permission from Atmospheric Environment, vol. 23, NO_2 Deposition to Elements Representative of a Forest Landscape, Copyright 1989 Pergamon Press PLC.)

and stomatal apertures. Stomatal frequencies of *Quercus* species (540 stomata mm^{-2}) are greater than that of *Acer* and *Fraxinus* species (372 and 210 stomata mm^{-2}; Kramer and Kozlowski 1979). The average rate of NO_2 deposition to these species in the light followed the same trend: *Quercus alba* (2.05 ± 1.44 nmol m^{-2} s^{-1}) > *Acer rubrum* (0.78 ± 0.11) > *Fraxinus americana* (0.57 ± 0.21). Similarly, lower NO_2 deposition to conifers with respect to broadleaf species is consistent with their lower conductance to water vapor (Körner et al. 1979). Okano et al. (1988) reported a positive correlation between NO_2 uptake and stomatal conductance for eight different crop species that followed a trend associated with stomatal densities of the foliage, and Grennfelt et al. (1983) found a similar correlation for *Pinus sylvestris*.

Measurements of NO_2 deposition in the dark provided information to evaluate surface versus internal deposition of NO_2. The contrasting light:dark measurements of NO_2 deposition indicated that the principal foliar site of NO_2 deposition was the leaf interior, constituting typically more than 90% of total deposition to individual leaves (Hanson et al. 1989). These data confirm previous hypotheses of stomatal control over NO_2 and other trace gas deposition (Rogers et al. 1979b; Weseley et al. 1982; Saxe 1986b; Taylor et al. 1988). Accordingly, NO_2 deposition is strongly influenced by stomatal conductance, which is governed, in turn, by the plant's physiological state as well as a host of environmental factors (e.g., light, vapor pressure, water availability).

Because the atmosphere–leaf exchange of NO_2 is strongly controlled by stomatal physiology, conventional modeling approaches of gas exchange operating at the level of individual leaves or within whole canopies (and based on analogy to water vapor) should be appropriate for characterizing NO_2 deposition to vegetation surfaces. Although interspecific variation in NO_2 deposition was high (see Figure 6.11), a linear regression of the combined data set for deposition to broadleaf shoots versus shoot conductance to water vapor accounted for 85% of the variation, indicating that a single relationship might be useful for predicting deposition to broadleaf forests. A linear regression of the combined data on NO_2 deposition to conifer shoots against shoot conductance to water vapor explained 66% of the variation.

Deposition of HNO_3 to Plant Foliage

For similar ambient concentrations and low shoot conductances to water vapor, surface deposition of HNO_3 vapor to plant shoots exceeded that for NO_2 (Figure 6.12). The HNO_3 deposition measurements were necessarily limited to conditions of low humidity leading to plants with low shoot conductance to water vapor. Therefore, comparisons of deposition between HNO_3 and NO_2 were restricted to the NO_2 measurements corresponding to leaves having closed stomata (low water vapor conductances). This limitation means that the data for foliar surfaces in Figure 6.12 should be viewed as a com-

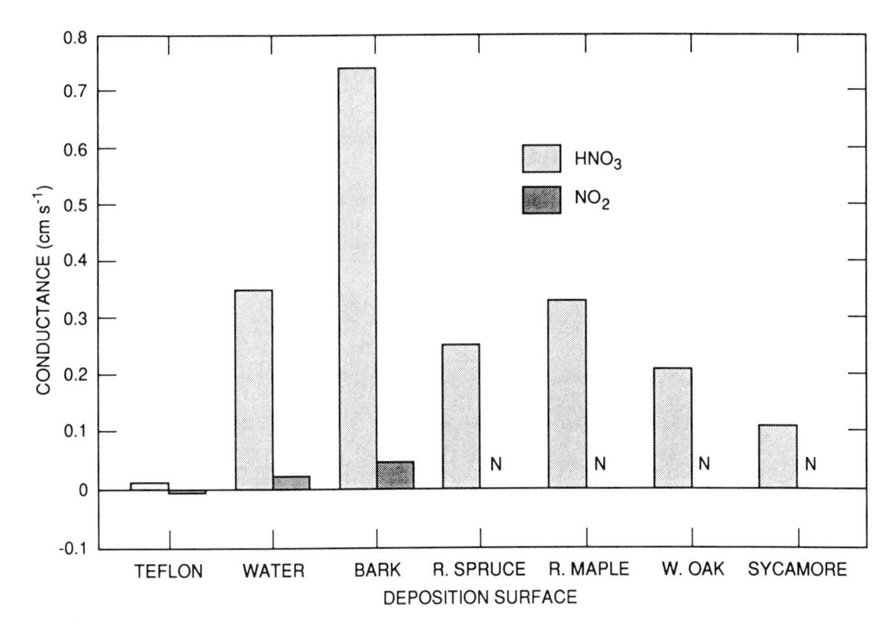

Figure 6.12. Comparison of conductance of NO_2 and HNO_3 to water, bark, and plant cuticular surfaces. Because dry conditions needed to measure HNO_3 exchange shoots resulted in reduced stomatal conductance, data are compared only with NO_2 data from Figure 6.1 that correspond to very low conductances. N, negligible.

parison of the deposition of HNO_3 and NO_2 to external cuticular leaf surfaces. Independent of these constraints on the laboratory data, the HNO_3 and NO_2 data are consistent with field observations that employed micrometeorological techniques (Wesely et al. 1982; Meyers et al. 1989).

Even though large variability and reduced stomatal opening limited rigorous quantification of internal versus external sites of HNO_3 deposition, our data indicated that part of HNO_3 deposition to the leaf interior through leaf stomata is likely to occur coincident with surface deposition (data not shown). More recent experiments employing ^{15}N-labeled HNO_3 have confirmed this dual pathway for HNO_3 deposition (P.J. Hanson, personal communication).

Field exposures of eastern white pine branches to ^{15}N-labeled HNO_3 yielded internal foliar deposition rates ranging from 5 to 53 nmol g^{-1} s^{-1} (Figure 6.13). These deposition rates correspond to a conductance of HNO_3 to white pine foliage of 0.0045 cm s^{-1}, which is far lower than the values obtained using the mass balance approach that measures total deposition. The discrepancy results from efficient removal of surface-deposited HNO_3 during the postexposure rinse of the needles (see methods described previously). Marshall and Cadle (1989) found that 97% and 59% of foliar-deposited HNO_3 was recovered in aqueous leaf washings after 2 and 16 hours, respectively.

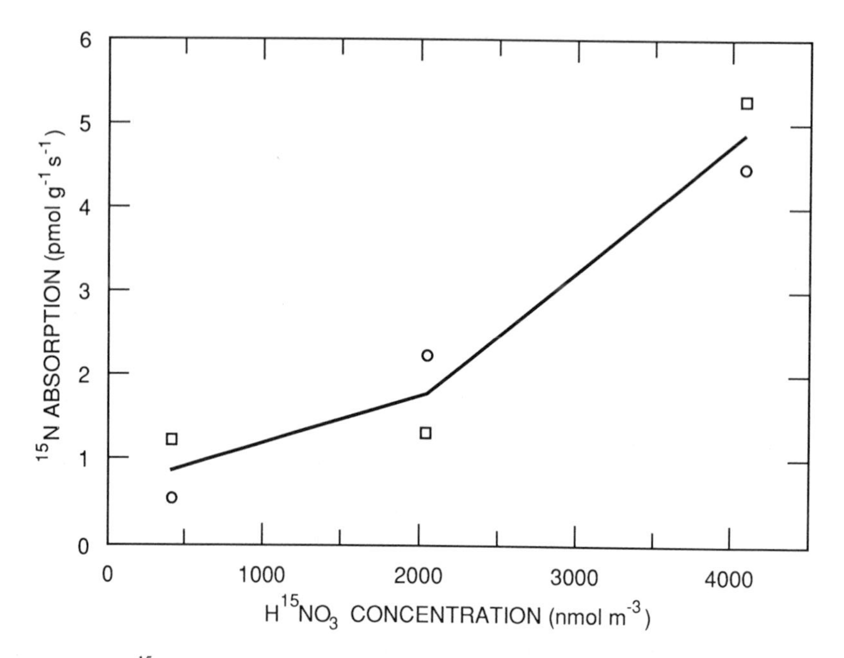

Figure 6.13. ^{15}N absorption rate as function of HNO_3 concentration for white pine foliage of two age-classes. Squares and circles are 1- and 2- year foliage, respectively. Data are normalized to needle dry weight.

Garten and Hanson (1990) have also shown that 70% to 90% of the nitrate ions in solution deposited to foliar surfaces remain available for subsequent removal by aqueous solutions after a 48-hour period. Because HNO_3 is likely to dissociate into H^+ and NO_3^- on contact with foliar surfaces, a similar tendency for removal in an aqueous rinse is plausible, and the ^{15}N-HNO_3 data in Figure 6.13 are probably representative of significant deposition to internal foliar surfaces. Longer residence times for NO_3^- on foliar surfaces between rain events (>48 h) might allow for more internal deposition through the cuticle than was suggested by the work of Garten and Hanson (1990).

Deposition to Nonfoliar Landscape Elements

Deposition of NO_2 varied among forest elements measured. Foliar, bark, and forest floor surfaces typically showed greater conductance to NO_2 than distilled water alone, and forest floor surfaces showed a disproportionately high conductance when compared to bark or foliage (see Table 6.2). The average conductance to NO_2 of the materials measured ranged from -0.0045 to 0.48 cm s^{-1}. The deposition of NO_2 to distilled water was relatively low by comparison to bark and foliage, but the measured conductance of NO_2 from the atmosphere to water (0.021 cm s^{-1}) was similar to the 0.010 cm s^{-1} value reported by van Aalst (1982). Conductance to NO_2 of dry bark was similar to that for conifer shoots (0.047 and 0.049 cm s^{-1}, respectively), and conductance to NO_2 of wet bark was similar to values for broadleaf shoots (0.093 and 0.093 cm s^{-1}, respectively). Conductance to NO_2 was not influenced by species of bark (data not shown); HNO_3 conductance to bark surfaces was 15 times greater than NO_2 conductance to the same surface (see Figure 6.12).

Conductance to NO_2 of the forest floor, based on a ground area basis, was six- to sevenfold greater than conductance to foliar surfaces or dry bark (see Table 6.2). This high conductance to NO_2 is, at least in part, a result of unaccounted-for convolutions in the samples (i.e., high surface area). The true area of the forest floor samples might easily have been two- or threefold times greater than the actual ground area and would have resulted in deposition values closer to those of leaves and bark surfaces. Judeikis and Wren (1978) observed similar or higher conductance of NO_2 to both a sandy loam and an adobe clay soil (average velocity of 0.68 cm s^{-1}), which compares favorably with our conductance values of 0.47 and 0.48 cm s^{-1} for the broadleaf and conifer forest floor samples, respectively. Measured conductance to NO_2 for autoclaved or oven-dried soils (data not shown) were similar to values for unsterilized soil, indicating that soil microorganisms were not responsible for the high conductances. Abeles et al. (1971) and Ghiorse and Alexander (1976) also observed no effect of microorganisms on NO_2 deposition.

Calculated NO_2 and HNO_3 Deposition to Forest Stands

To demonstrate the use of leaf-level data for making estimates of stand-level NO_2 deposition, laboratory observations were extrapolated to 10 of the conifer and hardwood forest canopies in the Integrated Forest Study (see Chapter 2, this volume) and to the forest canopy of the Walker Branch Watershed (WB). The Walker Branch Watershed represents a oak-hickory forest located in a suburban area near Knoxville, Tennessee (Johnson 1989) that has higher levels of gaseous pollutants (Table 6.3). Surface conductances for N oxides (K_l) measured in the laboratory were used to approximate the analogous measure for forest canopies (i.e, the deposition velocity, V_d) using the following equation:

$$V_d = K_l * LAI \qquad [6.1]$$

where LAI is the leaf area index (Jarvis 1971; O'Dell et al. 1977; Hicks et al. 1987). Rates of total growing season N deposition contributed by NO_2 for each forest site were then approximated from the product of the stand V_d, the mean daytime concentration of NO_2 for those stands, and the effective hours of daytime deposition (Table 6.3). A more detailed discussion of this calculation has been provided by Hanson et al. (1989). Calculated growing season rates of HNO_3 deposition were also determined, based on the

Table 6.3. Estimated Growing Season N Deposition from Nitrogen Dioxide (NO_2-N) and Nitric Acid Vapor (HNO_3-N) to 10 IFS Sites, a Suburban Site (WB) and a Hypothetical Urban Deciduous and Coniferous Forest Stand (UD and UC) (modified from Hanson et al. 1989 with permission from Atmospheric Environment, volume 23. Copyright 1989 Pergamon Press PLC.)

Site[a]	LAI (m^2 m^{-2})	Concentration[b] [EDH][c] NO_2 (μg m^{-3})	HNO_3 (μg m^{-3})	Estimated NO_2-N Deposition[d] (mol ha^{-1} yr^{-1})	Estimated HNO_3^- Deposition[d] (mol ha^{-1} yr^{-1})	NO_2-N deposition: N as % of total N deposition[e] (%)
Pine						
FS	5.1	1.9 [3402]	0.8 [6750]	20	44	5
LP	3.6	15.1 [2394]	2.8 [4500]	62	69	9
DL	6.5	9.2 [2859]	2.8 [5375]	82	152	8
CP	16.9	3.8 [2394]	1.8 [4500]	48	213	9
Spruce						
WF	4.6	3.8 [1800]	1.2 [3000]	10	25	<1
MS	5.5	2.8 [2179]	0.8 [3749]	11	26	2
NS	10.5	1.9 [2124]	1.4 [3000]	13	71	2
ST	9.0	3.8 [1995]	2.3 [3750]	21	120	1
Douglas fir						
DF	10.6	1.9 [2940]	0.9 [5040]	23	82	7
Northern hardwoods (maple)						
HF	4.7	3.8 [2179]	1.8 [3749]	54	79	9
Oak-hickory						
WB	5.5	15.1 [2394]	2.8 [4500]	200	109	21
Urban						
UD	5	55.2 [2394]	8.2 [4500]	1258	125	NA[f]
UC	5	55.2 [2394]	8.2 [4500]	331	68	NA

[a] Research site abbreviations are as follows [EPRI sites]: CP, white pine, North Carolina; DF, Douglas fir, Washington; DL, Loblolly pine, North Carolina; FS, slash pine, Florida; HF, N. hardwoods, New York; LP, loblolly pine, Tennessee; MS, red spruce, Maine; NS, Norway spruce, Norway; RA, red alder, Washington; ST, red spruce, North Carolina; WF, red spruce, New York; [Non-EPRI sites]: WB, oak/hickory, Tennessee; UB and UC, hypothetical urban deciduous and coniferous forests.

[b] Ambient NO_2 concentrations for stand LP and WB are from Kelly and Meagher (1986) and from personal communications with investigators of the nine IFS sites (WF, HF); the others are extrapolated from Galbally and Roy (1983) and EPA (1982) for pristine (41 nmol m^{-3}) and rural polluted sites (200 nmol m^{-3}). HNO_3 concentrations are from site data on pp. 154–157.

[c] EDH, Estimated hours of deposition.

[d] Estimates are based on laboratory measurements of conductance of NO_2 and HNO_3 to seedling shoots and assume constant concentration of NO_2 or HNO_3 throughout forest canopy and negligible leaf boundary layer (i.e., well-mixed conditions).

[e] Percent was based on total N deposition values (pp. 157–160) that do not include estimates of NO_2 deposition.

[f] NA, Not available.

assumption that deposition of HNO_3 during the night was to external canopy surfaces and during the day to external surfaces and to internal leaf sites via open stomata. The "growing season" HNO_3 deposition rates included in Table 6.3 and used in Figure 6.14 are provided for direct comparison to the rates estimated for NO_2. The growing season estimates for HNO_3 deposition are not equivalent to the more rigorous estimates obtained from the IFS model, which uses additional meteorological and stand structure characteristics to predict gaseous deposition for an entire year (see pp. 152–166; Lindberg et al. 1989).

Estimates of growing season NO_2-N deposition to the different forest canopy types varied 20 fold from a minimum of 10 for the New York red spruce-fir site (WF) to a maximum of 200 mol NO_2-N ha^{-1} yr^{-1} for the oak-hickory site in Oak Ridge (WB). The mean $\pm$ SE among all forest sites was 49 $\pm$ 17 mol NO_2-N ha^{-1} yr^{-1}. The estimates for the sites with the lowest ambient concentrations of NO_2 (FS, NS, DF) averaged 19 mol NO_2-N ha^{-1} yr^{-1}, whereas the two most polluted sites exhibited NO_2 deposition rates as

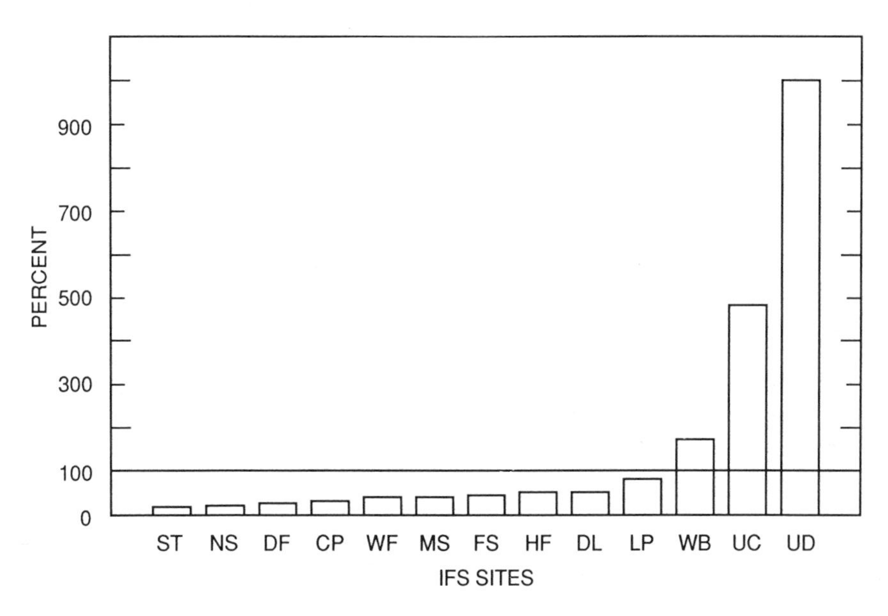

Figure 6.14. Calculated growing season deposition of NO_2-N as percentage of calculated HNO_3-N. Horizontal line at 100% indicates equivalent molar amounts of HNO_3 and NO_2 deposition. Forest site abbreviations are as follows: [EPRI sites] CP, white pine, North Carolina; DF, Douglas fir, Washington; DL, loblolly pine, North Carolina; FS, slash pine, Florida; HF, northern hardwoods, New York; LP, loblolly pine, Tennessee; MS, red spruce, Maine; NS, Norway spruce, Norway; RA, red alder, Washington; ST, red spruce, North Carolina; WF, red spruce, New York; [Non-EPRI sites] WB, oak hickory, Tennessee; UB and UC, hypothetical urban deciduous and coniferous forests.

much as 10 fold higher, of 62 (LP) and 200 mol NO_2-N ha^{-1} yr^{-1} (WB). Higher deposition to the WB and LP site can be accounted for principally by higher NO_2 concentrations and K_l. The high predicted deposition of NO_2 to the CP site is a result of its large LAI. Extrapolated measurements of ^{15}N-HNO_3 deposition from Figure 6.13 to the stand level indicate that of the total HNO_3 deposited perhaps only 10% to 20% is absorbed immediately by the forest canopy. If sufficiently short intervals were to occur between precipitation events, this observation would indicate that most dry-deposited HNO_3 is removed to the forest floor in throughfall. These data corroborate the earlier conclusions in this chapter that NO_3^- losses from the IFS canopies in throughfall were predicted to result from the accumulation of dry-deposited NO_3^- from HNO_3 between rain events.

The dry deposition of NO_2 is not included in the N budgets presented in the preceding portion of Chapter 6. However, growing season NO_2 deposition was less than 10% of annual total N deposition for the IFS sites. Only the non-IFS WB site with mean NO_2 concentrations of 15.1 $\mu g\ m^{-3}$ and high foliar conductances to NO_2 was predicted to have a substantially higher percentage (21%) of N deposition resulting from NO_2 (see Table 6.3). Forest canopies for which estimated NO_2 deposition contributed more than 5% of total N deposition included pristine (DF), intermediate (CP and HF), and polluted sites (DL, LP, and WB).

If we contrast the predicted rates of NO_2 and HNO_3 deposition obtained from the extrapolated laboratory measurements (see Figure 6.14), we find that NO_2 deposition will only exceed HNO_3 deposition to forest canopies under conditions of high ambient NO_2 concentrations [i.e., the WB, urban deciduous (UD), and urban coniferous (UC) forest sites]. In or near urban environments, NO_2 concentrations often reach concentrations as high as 60 nl L^{-1} and are routinely greater than 30 nl L^{-1} throughout the day (Lefohn and Tingey 1984; Bytnerowicz et al. 1987; Laxen and Noordally 1987). To estimate NO_2 deposition to deciduous or coniferous "urban forests" for comparison to the IFS and WB sites, we used calculations similar to those described for the 11 IFS forest canopies (see Table 6.3). Estimated NO_2-N deposition to urban forest canopies ranged from 330 to 1260 mol ha^{-1} yr^{-1} for coniferous and deciduous trees, respectively. These hypothetical urban NO_2-N deposition estimates are an order of magnitude higher than the rates of deposition calculated for "natural" forest canopies, suggesting that dry deposition of NO_2 must be considered in or near urban areas.

The estimates of NO_2 deposition to forest canopies presented in Table 6.3 are first approximations that assume maximum canopy conductance, throughout the growing season. They do not account for the impact of drought conditions on stomatal conductance, which leads to an overestimate of dry deposition. However, they also do not include estimates of deposition to bark and forest floor surfaces, resulting in an underestimate that would partially offset such an overestimate. Independent of the shortcomings of these first-approximation estimates, we believe that the extrapolations from the laboratory measurements indicate that current concentrations of NO_2 in conjunction with other forms of atmospheric N (wet and dry) provide physiologically significant inputs of N to forest systems. The estimated inputs of NO_2-N and HNO_3-N listed in Table 6.3, ranging from 10 to 213 mol N ha^{-1} yr^{-1}, are similar to inputs expected for temperate forest systems from nonsymbiotic N-fixing bacteria (0–214 mol N ha^{-1} yr^{-1}; Waring and Schlesinger 1985).

Summary

Laboratory measurements have shown that comprehensive estimates of atmospheric inputs N to forest stands should consider inputs of NO_2 and HNO_3 along pathways leading to foliage, bark, and forest floor surfaces. For most of the IFS sites, contributions of N from dry-deposited NO_2 would remain small because of low atmospheric NO_2 concentrations. However, where the concentration of NO_2 is significant (near polluted urban areas), the following points need to be considered: NO_2 deposition to foliage of forest tree species varied by more than an order of magnitude, and deposition of NO_2 to most broadleaf species was greater than deposition to conifers. Measurements of NO_2 deposition under light and dark conditions indicated the principal foliar site of NO_2 deposition to be the leaf interior and supported previous observations of stomatal control over NO_2 deposition. Vegetation surfaces typically showed greater conductance to NO_2 uptake than distilled water alone, and forest floor surfaces showed a disproportionately high conductance to NO_2 when compared to bark or foliage surfaces.

For a similar atmospheric molar concentration, the rate of deposition of HNO_3 will always exceed that for NO_2, but the relative contribution from each reactive N gas to dry deposition is largely controlled by existing ambient air concentrations. Unlike NO_2, HNO_3 vapor exhibits significant deposition to leaf cuticular surfaces, but a finite amount of internal deposition also occurs. Models for HNO_3 uptake might be enhanced by recognizing two pathways of uptake: a cuticular path for surface deposition and a stomatal pathway leading to internal surfaces. A portion of surface-deposited HNO_3 is likely to remain available for subsequent removal from foliar surfaces during rain events, in contrast to internal uptake of HNO_3 and NO_2, which are likely assimilated into organic N forms.

Rates of growing season NO_2 and HNO_3 dry deposition to nonurban forests are similar to inputs expected from nonsymbiotic N-fixing bacteria in temperate forest ecosystems (0–214 mol N ha^{-1} yr^{-1}). Dry deposition of NO_2 to the IFS sites was generally less than 10% of the total N deposited from other forms (wet, particle, HNO_3), but NO_2 should not be ignored as a source of dry N deposition when calculating total N loading to forest stands near urban areas.

Nitrogen Distribution and Cycling

H. Van Miegroet, D.W. Cole, and N.W. Foster

Ecosystem N Distribution

The forest ecosystems within the IFS network cover a wide array of climatic conditions (from cold boreal to subtropical), forest type (deciduous versus conifer), stand age, and disturbance history (from young plantations established on recently disturbed sites to overmature climax forests (see Chapter 2). These factors have contributed to the great variation in N distribution and cycling patterns among the 17 IFS locations. The total ecosystem N content varies from less than 120 kmol ha^{-1} in the Florida Slash Pine (FS) to 1200 kmol ha^{-1} in the high-elevation spruce forests at Whiteface Mountain (WF), while several sites contain about or more than 750 kmol N ha^{-1} [i.e., high-elevation spruce and beech forests of the Great Smoky Mountains (ST, SB), hardwood at Turkey Lakes, Ontario (TL), and low-elevation red alder stands in Washington (RA)] (Figure 6.15). The low-elevation southern conifer sites [e.g., loblolly pine at Duke Forest (DL) and Oak Ridge (LP), white pine at Coweeta (CP), and slash pine in Floria (FS)], and the spruce

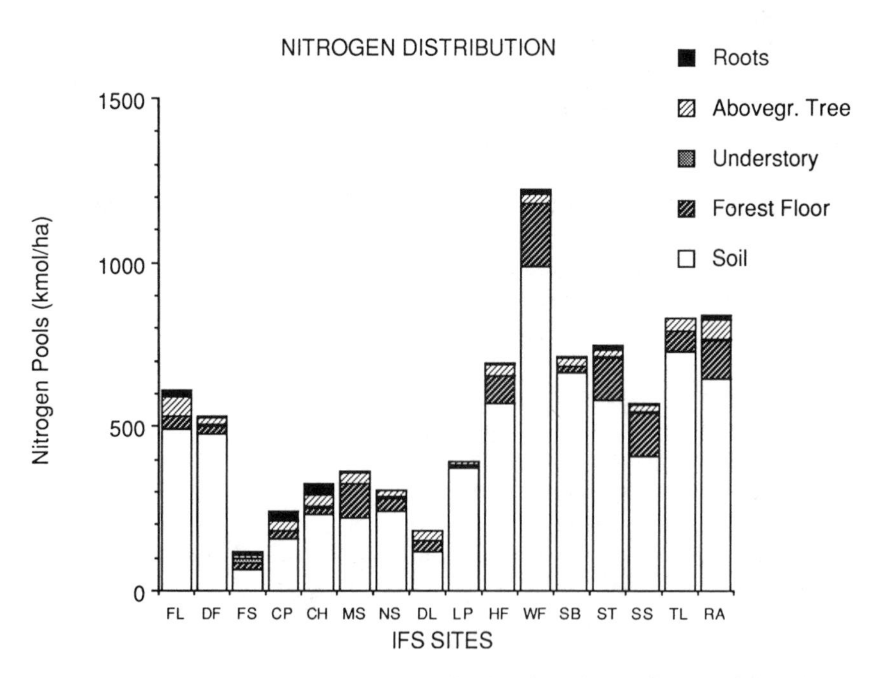

Figure 6.15. Total N content and N distribution in soil (0–60 cm), forest floor, understory vegetation, aboveground tree biomass, and roots (where available) at IFS sites ranked by NO_3^- output.

sites in Maine (MS) and Norway (NS) generally fall at the lower end of the total N content scale.

In all ecosystems, most of the N is sequestered in the soil compartment, which accounts for 55% to 95% or an average of 80% of the total ecosystem N capital. The lowest percentages are found in the relatively young and nutrient-poor sandy soils derived from marine deposits in the FS site, and in the low-elevation Maine spruce site (MS) where a substantially greater portion of the total N pool resides in the forest floor. Soil N values to a rooting zone of approximately 60 cm depth range from less than 64 kmol ha^{-1} (900 kg ha^{-1}) in the FS site to almost 990 kmol ha^{-1} (14,000 kg ha^{-1}) at the WF site. Forest floor N contents and their contribution to total N vary more widely among the IFS sites: the lowest forest floor N content of 8 kmol ha^{-1} (or 110 kg ha^{-1}) is observed in the loblolly pine site at Oak Ridge, Tennessee (LP), compared to a high of 190 kmol ha^{-1} (roughly 2,600 kg ha^{-1}) at WF. The highest forest floor N contents are generally associated with significantly larger organic matter accumulation in the northern and high-elevation sites (ST, SS, WF, MS), probably a result of the colder climate in those areas and its negative effect on the decomposition process (Witkamp 1966; Meentemeyer 1978; Meentemeyer and Berg 1986).

By contrast, the high forest floor N content in the red alder forest (RA) reflects large annual inputs of N-rich litter material by this N-fixing tree species rather than prolonged N accumulation or slow organic matter decomposition (Cole et al. 1978). Understory and tree biomass generally contribute less than 10% of the total ecosystem N content, except in the North Carolina sites (CP, CH, DL) where low soil and forest floor N contents raise the contribution of the aboveground vegetation to approximately 20%. Living biomass N values do not differ greatly among ecosystems compared to total N capitals and range from 14 to 66 kmol ha^{-1} (or 200–900 kg ha^{-1}). Our observations are consistent with results from the International Biological Program (IBP), which similarly indicated a relatively minor role of the living biomass and the major contribution of the soil rooting zone in the overall ecosystem N distribution (Duvigneaud and Denaeyer-De Smet 1970; Cole and Rapp 1981).

Factors Controlling N Retention and Release

The primary goal of this chapter was to evaluate the occurrence and causes of N saturation and excess NO_3^- leaching from forest ecosystems, particularly the role of atmospheric N deposition. A simple ranking of the IFS sites by average NO_3^- leaching output during a 2- to 3-year period (Figure 6.16) demonstrates that some of the sites are clearly leaching large amounts of NO_3^- [e.g., the TL, RA, and Smoky Mountain sites (SS, ST, SB)], whereas there are others that show little or no N leaching loss. Although the Smoky Mountain sites receive the highest N input via dry and wet deposition within the IFS network, and low NO_3^- leaching rates are generally observed in

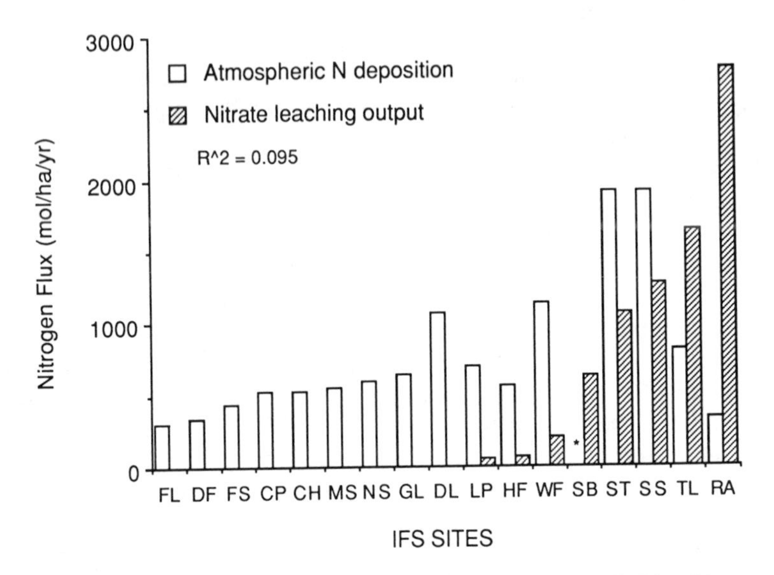

Figure 6.16. Total annual atmospheric N deposition input and NO_3^- leaching output in IFS forest ecosystems (* indicates no deposition data available).

sites that receive low N inputs, there is only a weak positive correlation between annual N input and N output rates ($r^2 \sim .10$) across all sites. In other words, atmospheric N deposition as a single factor is insufficient to explain differences in occurrence and rate of NO_3^- leaching among all sites within the IFS network. However, if only those sites are considered that are currently showing considerable NO_3^- leaching, then a somewhat stronger correlation between atmospheric N deposition and NO_3^- leaching outputs emerges [$r^2 = .26$ excluding N-fixing RA; $r^2 = .71$, including RA with total atmospheric N input to RA = atmospheric N deposition + approximately 6100 mol ha^{-1} yr^{-1} via N fixation (Cole et al. 1978)]. This suggests that once a system reaches N saturation, increasing N inputs will result in concomitant increases in NO_3^- leaching.

General Concepts Involved

Nitrification is a microbial N transformation process mediated by nitrifying organisms whose activity is strongly controlled by N availability (Vitousek et al. 1979; Riha et al. 1986). The NO_3^- formed during this process is highly mobile (Kinjo and Pratt 1971; Johnson and Cole 1980) and will quickly leach through the soil profile and out of the rooting zone unless it is immobilized biologically. Thus, the nitrification potential of a site or the degree of NO_3^- retention following nitrification depends on the relative N source compared to sink strengths within the system. These, in turn, are determined both by natural ecosystem characteristics (e.g., age and type of vegetation, soil age, and organic matter accumulation) and anthropogenic influences on this system (e.g., management practices, pollution). Nitrogen may be added to the belowground compartment from external sources such as N fertilization, N fixation, and direct atmospheric N deposition, or may be released internally via mineralization of soil and forest floor organic matter.

Nitrogen sinks include N uptake by the vegetation, microbial N immobilization, and adsorption to the soil exchange complex (in the case of NH_4^+). Nitrogen may also be lost in gaseous form, that is, through denitrification or NH_3 volatilization, but these output fluxes are expected to play a minor role in the overall N budget of these systems. Volatilization of NH_3 occurs under alkaline conditions and should therefore be insignificant in typically acid forest soils (Warneck 1988). Denitrification requires the presence of NO_3, and the values reported in the literature are generally small, even for systems with seemingly favorable conditions for denitrification [e.g., high-elevation spruce forests in the Southern Appalachians (Wells et al. 1988)].

In that nitrifiers are weak competitors for N against plants and heterotrophic decomposers, nitrification is generally restricted in systems that are N deficient, that is, where potential N sinks exceed or equal N inputs. When the N supply increases or N sinks decline, sufficient N may become available for nitrification after plant and heterotrophic N demands are met, and NO_3^- may leach out of the system. Thus, the production and leaching of NO_3^- suggest N availability in excess of biological retention capacity (i.e., plant and microbial immobilization), making NO_3^- leaching a practical indicator of N saturation.

The N retention capacity and, conversely, the leakiness of the system with respect to NO_3^- or degree of N saturation could also be placed in the context of the long-term ecological history of an area, forest ecosystem maturity or successional stage, and forest stand development and age as represented in Figure 6.17. Ecosystems on more recently developed soils (e.g., following deglaciation, volcanic ash deposition) such as DF and FL in Washington or whose organic C and N pool has been frequently or recently reduced by disturbance (e.g., fire or agricultural cropping) such as the southern conifer plantations (e.g., DL) are still at the stage of active organic matter (C) and N accumulation. Nitrogen retention in such systems should be near or at maximum, irrespective of age or vigor of the forest. In some cases late-rotation N deficiencies develop on sites with low N (e.g., Turner 1981). No NO_3^- leaching occurs, even with small increases in N inputs. The leaching process in such systems is mostly dominated by organic acid anions and bicarbonate resulting from the root respiration and decomposition of soil organic matter or by SO_4^{2-} originating from atmospheric deposition (Johnson and Cole 1980).

At the other extreme of the spectrum one could consider ecosystems that have accumulated large amounts of N and C, either over a long period with-

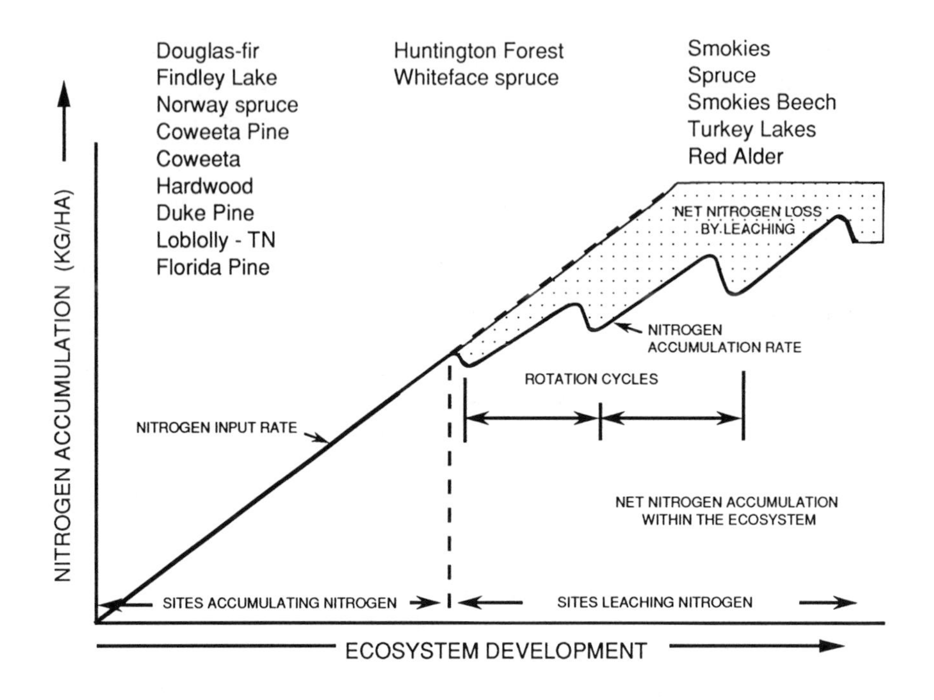

Figure 6.17. Schematic representation of IFS sites with respect to N accumulation and N saturation according to forest ecosystem history and maturity.

out significant reduction by site disturbance (e.g., WF and the Smokies sites), or relatively rapidly with the presence of N fixers (e.g. RA site in Washington). Nitrogen storage capacity of the system has reached its upper limit with respect to its current C content (see section below and Figure 6.18), the level of which, in turn, is regulated by climatic conditions (Post et al. 1982, 1985). Nitrogen inputs in such systems are no longer retained, and NO_3^- leaching occurs. Stand age and vigor play a critical role in regulating the extent of the NO_3^- leaching output (see Figure 6.17): young or highly productive stands are able to curtail NO_3^- losses somewhat through plant N uptake (Vitousek and Reiners 1975). In mature to overmature stands, N uptake rates are generally low, and any disturbance of the stand structure with mortality may actually accelerate N mineralization and nitrification rates. Increases in N inputs to such systems will lead to concomitant increases in NO_3^- leaching (Foster et al. 1989a).

Forests may also be situated in an intermediate position in terms of N accumulation and retention. The combination of total N content, mineralization potential, stand age and vigor, and N input rates, that is, small changes

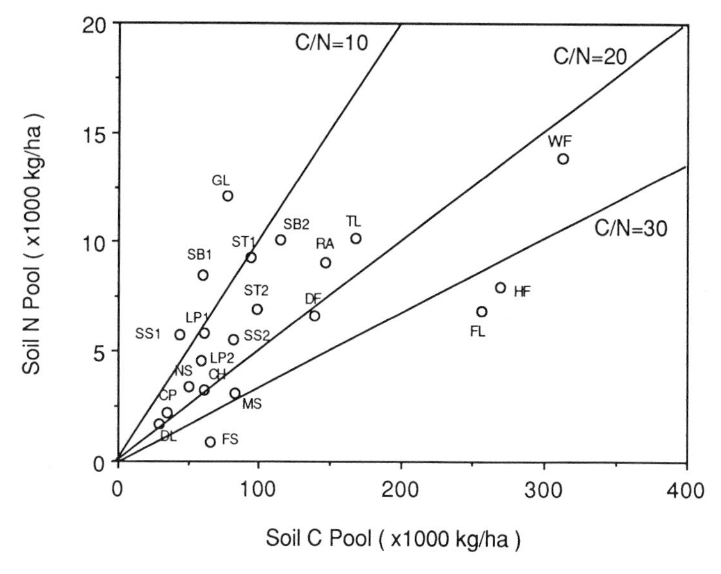

Figure 6.18. Total soil N and C accumulation in IFS sites.

in N source and N sink strengths, will play a critical role in such system in determining the extent of NO_3^- production and leaching.

Nitrogen Release

A crude way to identify sites that are N saturated is by comparing atmospheric N inputs to annual N leaching losses (e.g., Dempster and Manning 1988; Nilsson and Grennfelt 1988). Based on the average annual N input-output budget, only the hardwood site in Ontario (TL) shows a net release from the system and could thus be designated as N saturated on the basis of this approach (see Figure 6.16). Indeed, in the red alder forest, symbiotic N fixation represents an additional N input of the order of 3,500 to 11,000 mol ha^{-1} yr^{-1} (Zavitkovski and Newton 1968; Cole et al. 1978; Binkley 1981) suggesting some N is still being accumulated in the system. However, input-output budgets must be interpreted with some caution because of the uncertainties associated with the quantification of the dry deposition component (pp. 153–166) and of the total water flux from the rooting zone (Chapter 3). Using a simple input-output approach does not account for the role of internal processes and the release through mineralization of N accumulated previously in the ecosystem. Thus, based on these considerations and the foregoing discussions, we will use the occurrence of measurable NO_3^- leaching as an indication that N sources (internal and external) exceed the N sinks in the system; that is, NO_3^- leaching will serve as the operational definition of N saturation.

Averaging solution concentrations over a given time period, however, tends to obscure occasional NO_3^- leaching pulses that may take place either seasonally or at any given year and which indicate that the N retention capacity of the system is at least temporarily exceeded by N inputs. This was the case in the Smokies spruce forests (Johnson et al. 1991), or in the northern hardwoods (Foster et al. 1989b; Shepard et al. 1990). Therefore, some low level of NO_3^- leaching should be deemed "normal" [e.g., 150 mol ha^{-1} yr^{-1} or roughly 2 kg N ha^{-1} yr^{-1} as used in the definition of critical N loads (Nilsson and Grennfelt 1988)] to account for the temporary NO_3^- release during the spring snowmelt period before the onset of the growing season.

The first step in determining the factors that cause N saturation in forest systems is identifying common characteristics among sites that currently show extensive NO_3^- leaching as well as those ecosystem properties that distinctly separate NO_3^- leaching from nonleaching sites. To that effect, simple linear regression and stepwise multiple regression analyses were performed using various N pools and N fluxes as independent variables and NO_3^- leaching as the dependent variable. Table 6.4 summarizes the results from the stepwise multiple regression analysis for those parameters that were found to have the greatest effect on NO_3^- leaching losses. Included in this table are only those sites or replicate plots for which information for all the independent variables was available. The following variables were found most important: N content of the mineral soils (0–60 cm depth) + forest floor; and as N fluxes: atmospheric N input, total tree N uptake (calculated from biomass increment and litterfall data as described in Chapter 2), and relative soil N mineralization rate [annual N mineralization in 0–10 cm of mineral

Table 6.4. Variation in NO_3 Leaching Explained by Different Regulating Factors (16 IFS sites)

Independent Variable	Percent NO_3 Variation Explained
Mineralization[a]	44
Total N[b]	18
N Uptake[c]	15
Input[d]	11
Input, mineralization	64
Input, N uptake	41
Input, total N	25
Input, N uptake, mineralization	67
Input, N uptake, total N	45
Estimated leaching potential[e]	51

[a] Mineralization, N mineralization in 0–10 cm of mineral soil determined through in situ incubation using buried bag technique (Eno 1960).
[b] Total N, total N capital in forest floor and mineral soil (0–60 cm depth).
[c] Uptake, total uptake by overstory.
[d] Input, total input via atmospheric deposition + N fixation (in case of red alder).
[e] For explanation see text.

soil as determined by buried bag technique (Eno 1960) and expressed as percentage of the maximum rate (ST) observed among the IFS sites].

First, all NO_3^- leaching sites have large amounts of N accumulated in the mineral soil and forest floor, in contrast to the sites at the lower end of the N leaching spectrum (see Figure 6.15). Differences in total N content explain approximately 18% of the variation in NO_3^- leaching among the IFS sites (Table 6.4). In the case of the red alder (RA), this N accumulation has occurred over a fairly short time period (50 years), resulting from large N inputs via symbiotic N fixation (Cole et al. 1978; Van Miegroet et al. 1990). For the other N-saturated sites, the N accumulation likely took place over a longer time period. Total soil N content is positively correlated with the amount of organic C accumulated in the soil (see Figure 6.18), a relationship that has also been demonstrated in chronosequence studies on newly developed soils (e.g., Crocker and Major 1955), with forest succession on former agricultural soils (e.g., Hamburg 1984), and for other ecosystems worldwide (Post et al. 1985).

The correlation can be explained by N immobilization in C-rich organic material substrate until an equilibrium is attained between N demands of and C supply to the heterotrophs. When heterotrophic activity is restricted by C supply rather than by N availability, N is no longer retained and a net N release may occur. In some boreal soils where decomposition is restricted by extreme climatic conditions, both C and N availability may be low, causing strong competition for N among the different organism groups and in turn resulting in low NO_3^- leaching rates. The transition point between net N immobilization and net N mineralization is determined by the C:N ratio of the substrate: in general terms, net immobilization occurs at a substrate $C:N > 30$, whereas net N release only occurs at a substrate $C:N < 20$ (Alexander 1977), although Berg and Staaf (1981) have shown that this critical C:N value may vary with climate and organic matter composition.

Soil N and C contents and corresponding C:N ratios are shown in Figure 6.18. Soils that are located below the $C:N = 20$ line on the graph (i.e., soils with relatively low N contents compared to total C) are expected to retain N, whereas sites that are above the $C:N = 20$ line and particularly those above the $C:N = 10$ line (i.e., soils that have a relatively high N content compared to C content) are likely candidates for net N release through mineralization. These projections correspond very well with the results from the in situ mineralization assays. Although the buried bag technique used (Eno 1960) has its limitations in providing an absolute measure of mineralization under actual field conditions, it is nevertheless useful for comparisons between sites and as an indicators of relative mineralization potential.

The IFS sites again cover a wide array of soil mineralization and nitrification rates (Figure 6.19). The highest relative soil N mineralization rates are measured in those soils with high total soil N content (>350 kmol ha^{-1} or 5000 kg ha^{-1}) between 0 and 60 cm soil depth and an overall $C:N < 20$ (e.g., ST, SS, SB, RA, TL). Those are precisely the sites that also showed

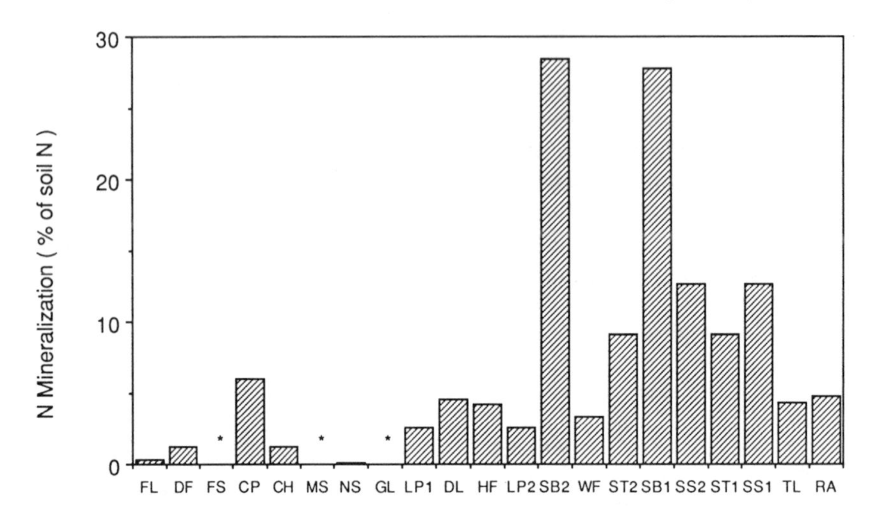

Figure 6.19. Relative soil N mineralization potential (N mineralized as percent of total soil N between 0 and 10 cm) of IFS sites ranked by NO_3^- output (* indicates no mineralization data available).

high NO_3^- leaching rates (i.e., >750 mol ha^{-1} yr^{-1}). On the other hand, the C:N ratio, which is greater than 20 in the soils under spruce (WF) or hardwood cover (HF) in New York and under subalpine fir forest in the Washington Cascades (FL), suggested low N release (mineralization) and high N immobilization potential despite the elevated soil N contents, mostly as a result of the even larger organic C pools (see Figure 6.18). Annual N mineralization at the WF and HF sites was less than half that calculated for the Southern Appalachian ST site. Low temperatures at the high-elevation FL and WF sites should further reduce mineralization rates, and little net N mineralization was measured at FL. As a single factor, differences in soil N mineralization potential (0–10 cm soil depth) are most important in explaining variations in NO_3^- leaching between the IFS sites (see Table 6.4). Nitrogen mineralization and atmospheric N deposition appear to be the most important sources regulating NO_3^- leaching, accounting together for about 64% of the variability in N losses among all IFS sites (see Table 6.4). These results further emphasize that both internal and external N sources play a role in N saturation.

Nitrogen Retention

In contrast to S (see Chapter 5), N is basically retained within in the system through biological rather than physicochemical processes. Microbial N immobilization in soil and forest floor is influenced by past N accumulation and total N content, both in absolute terms and relative to the total organic C pool in soil and forest floor. The forest systems in the lower left corner of Figure 6.18 have lower organic matter and N contents, either because soils are still relatively young or because organic C and N pools have been reduced through prior disturbance(s) or land use. Nitrogen availability in such systems is limited, and because of this low total N content, the forests in that group should have a high capacity to retain and accumulate incoming N. This should in turn reduce the potential for NO_3^- leaching. Field observations confirm that these sites indeed show little or no net N mineralization, but rather periodically exhibit net N immobilization, and that solution NO_3^- concentrations are insignificant.

The high-elevation fir forest in Washington (FL) and the high-elevation spruce (WF) and hardwood forests (HF) in New York are also expected to have substantial N immobilization capacity, not so much because of a low total N content (total soil and forest floor N capital in those systems are 530, 1200, and 650 kmol ha^{-1}, respectively) but because of a high soil C:N ratio. The HF and WF sites currently occupy an intermediate position on the relative N mineralization scale (see Figure 6.19), while inorganic N contents in the FL soil samples either decline (= net immobilization) or show only very small increases during incubation.

In systems where there is measurable net N mineralization, nitrification rates will be further affected by the degree of competition for NH_4 between microbes and vegetation, while NO_3^- leaching can also be reduced by vegetation uptake. Calculated tree N uptake rates differ widely among the IFS sites (Figure 6.20), from a high of 3500 mol ha^{-1} yr^{-1} (roughly 75 kg ha^{-1} yr^{-1}) by alder (RA) to a low of 570 mol ha^{-1} yr^{-1} (roughly 8 kg ha^{-1} yr^{-1}) by subalpine fir (FL) in Washington, a reflection of the differences in forest type, stand age and vigor, and general growth conditions (such as climate and nutrient availability). Uptake rates are generally higher in deciduous than in coniferous forests, because annual replacement of the entire foliage requires more N (Cole and Rapp 1981): the uptake rates are of the order of 2500–3500 mol ha^{-1} yr^{-1} (or 40–50 kg ha^{-1} yr^{-1}) in the hardwood forests (TL, HF, CH), generally exceed 2000 mol ha^{-1} yr^{-1} (30 kg ha^{-1} yr^{-1}) for the different pine species in the southeastern United States (FS, DL, CP), and seldom exceed 1500 mol ha^{-1} yr^{-1} (or 20 kg ha^{-1} yr^{-1}) in the Douglas-fir stand (DF), at GL, and in the spruce-fir forests (FL, NS, MS, SS, ST). The WF and the LP sites with their calculated N uptake of 2900 mol ha^{-1} yr^{-1} and 1300 mol ha^{-1} yr^{-1}, respectively, represent exceptions in this respect. The low N uptake in the Oak Ridge plantation is probably a reflection of the low fertility status of that site.

Over the entire IFS data range one cannot demonstrate a significant correlation between N uptake and NO_3^- leaching (see Table 6.4 and Figure 6.20): high NO_3^- leaching rates occur in association with high (RA), intermediate (TL), or low N uptake rates (SS, ST), whereas low N uptake rates do not necessarily result in high NO_3^- leaching losses (e.g., NS, LP, DF,

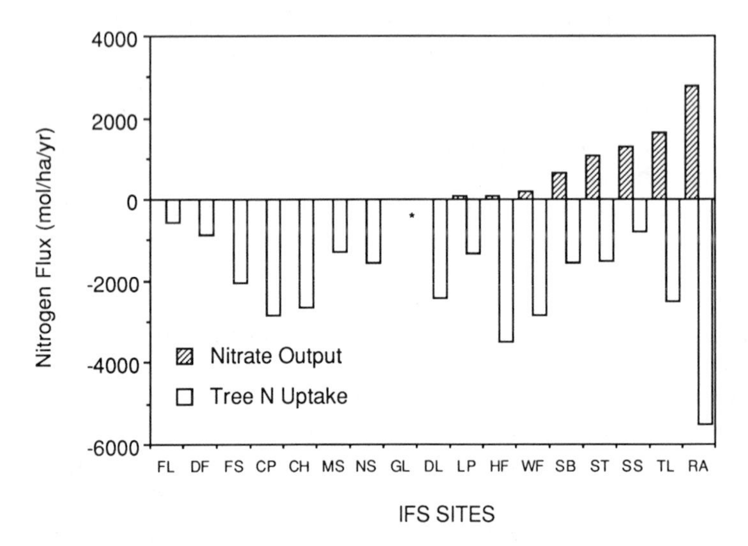

Figure 6.20. Calculated annual tree N uptake as N sink compared to NO_3^- leaching rates in IFS sites ranked by NO_3^- output (* indicates no uptake data available).

FL). Even when only NO_3^- leaching sites are considered (excluding N-fixing RA), the (negative) correlation between N uptake and NO_3^- leaching remains weak, a reflection of the complexity of N source and sink processes that regulate N availability for nitrification. The potential role of the vegetation in N saturation can be illustrated by comparing sites with similar N pool and N flux characteristics, except for N uptake; or by comparing seasonality in the NO_3^- leaching losses with the seasonality of the growth phenomena.

Indeed, in spite of the many similarities that exist between the high-elevation spruce sites in the northern (WF) and in the southern Appalachians (ST, SS), especially in terms of N input regime (see Figure 6.16) and total N content (see Figure 6.15), in which they rank highest among the IFS sites, annual NO_3^- leaching rates are drastically different between them. The higher N demands by the apparently more vigorous spruce-fir forests at Whiteface, which still contain a significant fir component in the overstory, could help to explain the lower nitrification potential and NO_3^- leaching rates compared to the Smokies sites, where fir mortality, by the balsam woolly adelgid (*Adelges piceae*) reduced stand integrity and annual N demands over the last decade. In addition, the canopy disruption created by the tree mortality may have further increased mineralization and nitrification rates through soil warming, as suggested by the high relative soil N turnover rates (see Figure 6.19). Canopy gaps caused by this mortality may also result in higher N deposition to the remaining trees surrounding the gaps.

The two northern hardwood forests (TL, HF) show a distinct seasonality in NO_3^- solution concentrations: peak NO_3^- levels are observed with snowmelt in early spring (Foster et al. 1989b; Shepard et al. 1990) and before significant root activity occurs. At TL they increase as a result of a flush of mineralization and nitrification toward the end of the dormant season. With the onset of the growing season NO_3^- solution levels decline in both forest types, probably as a result of increased root uptake, and at TL the mean growing season NO_3^- concentrations are lower than dormant season NO_3^- levels. At HF, NO_3^- solution concentrations remain fairly low throughout the rest of the year, and total NO_3^- leaching rates are less than 100 mol ha^{-1} yr^{-1} (or 5 kg ha^{-1} yr^{-1}) compared to 1600 mol ha^{-1} yr^{-1} at the TL site. Both hardwood sites are subject to similar N input regimes and are generally comparable in terms of total N content and distribution. One of the features that distinguishes the sites is the slightly higher N uptake at HF; 3500 mol ha^{-1} yr^{-1} (49 kg ha^{-1} yr^{-1}) versus 2500 mol ha^{-1} yr^{-1} (35 kg ha^{-1} yr^{-1}). This difference, in association with a greater potential for N immobilization at the HF site as indicated by the C:N ratio of the soil (see Figure 6.18), may be responsible for the lower total NO_3^- leaching rates that are observed there.

Finally, an attempt was made to integrate the role of the relative N source and sink strengths into one parameter, designated as "leaching potential" (see Table 6.4). It was calculated as the sum of atmospheric N input (including N fixation in case of RA) and relative N mineralization rate minus N removal via tree uptake. Estimated NO_3^- leaching potential and measured NO_3^- leaching values were highly correlated (r = .72+), but the leaching potential overestimated the measured values by a factor of 5:

$$[NO_3^- \text{ (measured)} = 3.4 + 0.19\ NO_3^- \text{ (potential)}]$$

Although our approach proved too crude to accurately predict NO_3^- leaching rates, this parameter was able to clearly distinguish NO_3^- leaching from nonleaching sites (Figure 6.21), and explained more than half (r^2 = .51) of the variability in NO_3^- leaching observed.

Effect of Nitrification and NO_3^- Leaching

Nitrification of N derived from atmospheric deposition or internal mineralization of organic N has a strongly acidifying effect on the soil and soil solution because of the net H^+ release during this oxidation process, but more importantly because of the significant increase in total anion concentration (and ionic strength) of the percolating solution and the mobile nature of NO_3^- (Kinjo and Pratt 1971; Wiklander 1976). To keep the solution electrically neutral, such increase in total anion concentration must be accompanied by a concomitant increase in cations (Nye and Greenland 1960). Whereas the anion production (in this case nitrification) determines the total

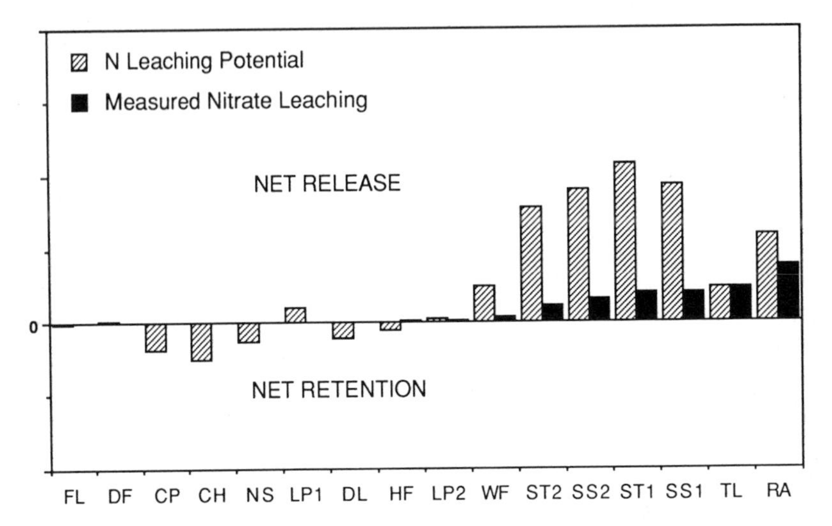

Figure 6.21. Comparison of estimated NO_3^- leaching potential based on atmospheric N inputs, soil N mineralization, and tree N uptake with measured NO_3^- leaching rates at IFS sites (see text for explanation of leaching potential).

increase in positive charges in solution, exchange reactions between the soil water and the cation exchange complex largely determine the relative abundance of the different cations in solution (Reuss and Johnson 1986). The equations most commonly used to describe such exchange reactions are those of Gaines and Thomas (1953) and Gapon (1933). According to these theoretical exchange relationships, which are discussed more extensively in Chapter 8, any increase in ionic strength of the solution will mostly displace multivalent cations such as Ca^{2+} and Mg^{2+} from soils with medium to high base saturation, whereas Al will be preferentially mobilized from more acid soils (i.e., soils with low base saturation). Elevated Al concentrations in solution or accelerated cation loss through NO_3^--mediated leaching may cause nutrient imbalances or deficiencies of essential nutrients, which in turn may negatively impact forest productivity (e.g., Thornton et al. 1987; Shortle and Smith 1988; Kelly et al. 1990; Raynal et al. 1990).

Cation Loss and Base Saturation

The weighted average ion concentrations of B horizon solutions collected at the various IFS sites are given in Figure 6.22, and illustrate the role of natural and anthropogenic anion loading in soil solution chemistry. This section focuses mainly on the contribution of NO_3^- to the overall leaching process.

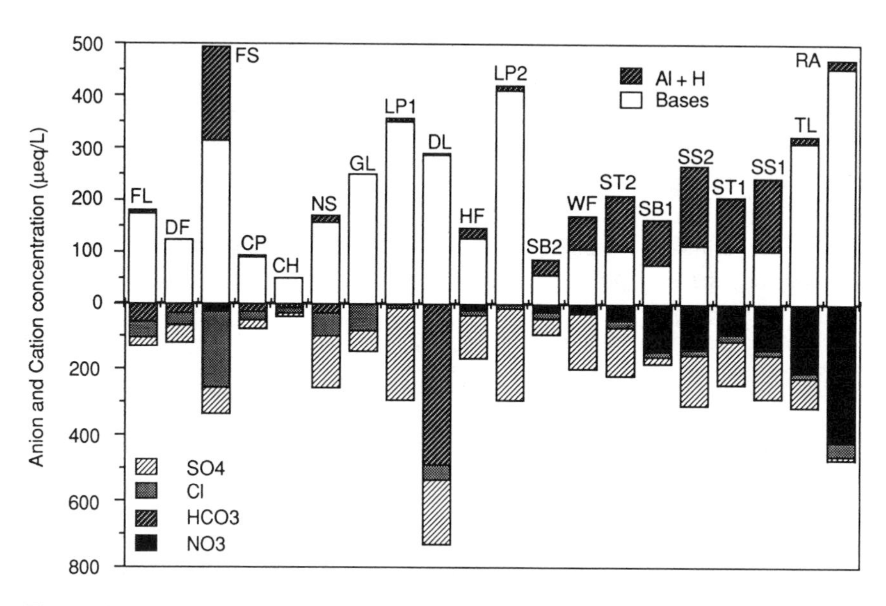

Figure 6.22. Weighted average anion and cation concentrations of B horizon solutions collected in IFS sites.

At the TL and RA sites, soil solution chemistry is largely dominated by NO_3^- leaching. High NO_3^- concentrations in solution cause accelerated cation leaching, predominantly of Ca^{2+} and to a lesser extent of Mg^{2+} (Van Miegroet and Cole 1984; Foster and Nicolson 1988). Calculations for the RA site show that as much as 14% of the exchangeable Ca^{2+}, Mg^{2+}, and K^+ are exported annually below the 40-cm soil depth through NO_3^--mediated leaching. This accelerated leaching has caused a striking redistribution of exchangeable Ca^{2+} from the upper parts of the A horizon to the B horizon, with a similar but less pronounced downward shift for Mg^{2+} and K^+. Such relative cation depletion of the upper soil horizons could at least in part account for the decline in percent base saturation of the A horizon under alder (Van Miegroet and Cole 1984; Van Miegroet et al. 1989). However, despite high annual cation leaching rates, no measurable decline in the total exchangeable cation pool has occurred, suggesting some base replenishment possibly from increased mineral weathering. Indications of differences in weathering rates between the DF and RA sites, which are established on the same soil, are further discussed in Chapter 10.

At TL, base cations are taken up by the hardwood forest from the effective rooting zone between 0- and 60-cm soil depth. Net base retention in this old-growth forest is small relative to the recycling of bases to the soil by litterfall, canopy leaching, and tree mortality. Some Ca^{2+} and Mg^{2+} are removed from decomposing organic matter in the forest floor layers by NO_3^-. Additional Ca^{2+} and Mg^{2+} leached from the organic layers in association with organic anions are exchanged in the the B horizon with H^+ produced by nitrification. Hence, at TL the large annual Ca^{2+} and Mg^{2+} leaching losses from the rooting zone are primarily induced by nitrification in the soil. Strong acids from the atmosphere are not retained by the vegetation or soil, further increasing leaching of bases from the rooting zone. At TL, leaching removes as much as 5% of the exchangeable Ca^{2+} and 10% of the exchangeable Mg^{2+} in the soil annually. Although weathering currently replaces some of the cation loss, the base saturation of the soil is likely declining and the soil acidity increasing.

Al Mobilization

In extremely acid soils [such as those in the spruce forests (ST and SS) of the Smokies] or soil horizons (e.g., A horizon under RA) with an exchange complex largely dominated by exchangeable acidity (i.e., in soils with low base saturation), the NO_3^--induced increase in ionic strength of the solution causes significant Al mobilization (Figure 6.23), which is in agreement with the soil–solution equilibrium reactions discussed earlier (see also Chapter 8). Seasonal NO_3^- peaks are particularly important in displacing Al from the exchange complex, as illustrated for the Smokies spruce site by Johnson et al. (1991) and demonstrated by Reuss (1989) for selected leachates collected at the RA site. As shown in Figure 6.22 for the ST site, Al concentrations in the A horizon leachates mostly range between 30 and 100 μmol L^{-1}, but peak values (associated with NO_3^- or SO_4^{2-} peak levels) occasionally approach threshold toxicity levels reported for red spruce seedlings (Hutchinson et al. 1986; Thornton et al. 1987; Joslin and Wolfe 1988).

It has been hypothesized that increases in Al concentrations induced by atmospheric deposition (SO_4^{2-} as well as NO_3^-) may cause an imbalance in the Ca or Mg nutrition of the trees, which may ultimately result in a reduction in forest productivity (e.g., Shortle and Smith 1988; Schulze 1989). Such causal relationship between soil solution chemistry and productivity decline in high-elevation spruce forests has not been conclusively established, however. It further needs to be pointed out that part of the increase in NO_3^- leaching that has been observed in the spruce forests of the Smoky Mountains is caused by internal N sources (i.e., mineralization) in addition to atmospheric N deposition. The large N pools with a small C:N ratio that currently lead to those large NO_3^- releases via mineralization, are, of course, the reflection of accumulation of atmospherically derived N over a period of perhaps several millennia. Indeed, if these sites have been receiving 25 kg N ha^{-1} yr^{-1} for 30 years (which is an upper-limit assumption), then that accumulated N would represent only 6% of the total N capital of the site. Thus, while high N inputs over the last decades have undoubtedly accelerated the N saturation process, past site history has also significantly contributed to the high NO_3^- leaching losses currently observed.

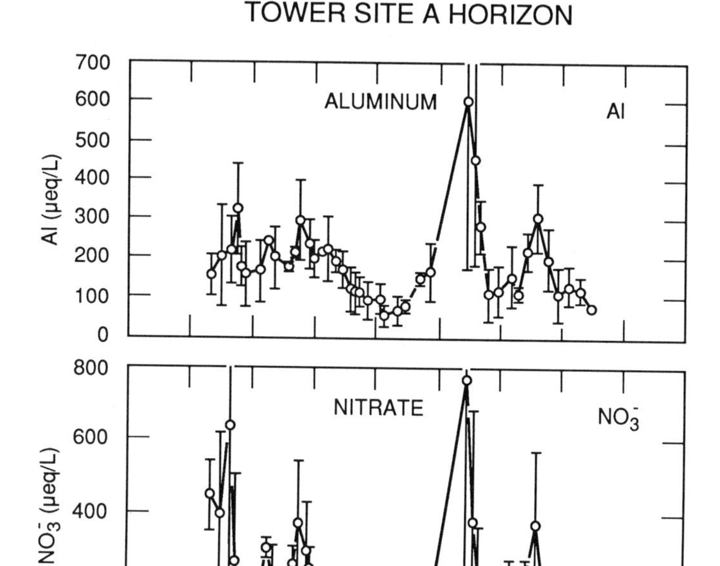

Figure 6.23. Nitrate and aluminum concentrations over time in A horizon solutions at spruce Tower site in Great Smoky Mountains National Park (From Johnson et al., 1991).

Recovery

Nitrification can lead to a series of changes in soil and soil solution properties including a decrease in soil solution pH (caused by proton generation during the nitrification process), an increase in the exchangeable acidity, decrease in soil pH and base saturation, and an increase in the Al concentration in soil solution and on the exchange sites. Consequently it is critical to determine if systems subjected to nitrification can recover and, if so, how long such a process will take. While this subject is discussed further in Chapter 12 (this volume), some aspects of recovery specific to N saturation are discussed here.

Ideally, an experiment on recovery from the effects of nitrification should be conducted on systems that have reached N saturation from atmospheric deposition, but within the timeframe of the IFS program such examination of recovery was not feasible. However, we were able to follow the recovery process in a system where N saturation had been achieved through N inputs

Table 6.5. Average Soil Solution NO_3^- Concentrations (μmol/L) Collected beneath Forest Floor, A and B Horizons, from the 55-Year-Old Red Alder and Douglas Fir Forest Sites at Thompson Research Center

	Forest Type	
Horizon	Red Alder	Douglas Fir
Forest floor (0 cm)	315	1.7
A (0–10 cm)	273	0.5
B (10–40 cm)	229	0.5

via a symbiotic N fixation associated with the red alder. It has been well documented through previous research (Van Miegroet and Cole 1984; Reuss 1989; Van Miegroet et al. 1989; Van Miegroet et al. 1990) that such N additions in the red alder stand result in a number of soil and soil solution properties changes, including increased NO_3^- leaching (Table 6.5), a decline in soil pH (see Figure 6.24), and Al mobilization (Figure 6.25). The "control" or reference used in this study to evaluate these changes was an adjacent stand of Douglas fir established at approximately the same time (55 years ago) on the same soil series.

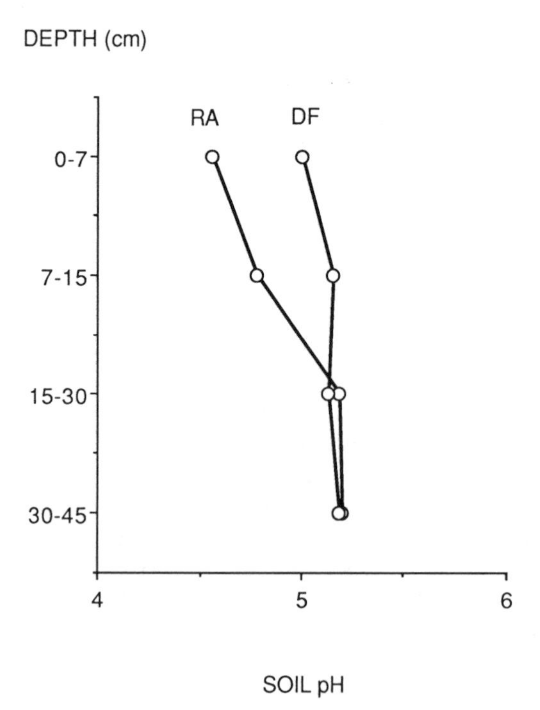

Figure 6.24. Soil pH (water) under mature stand of alder and Douglas fir at Thompson Research Center.

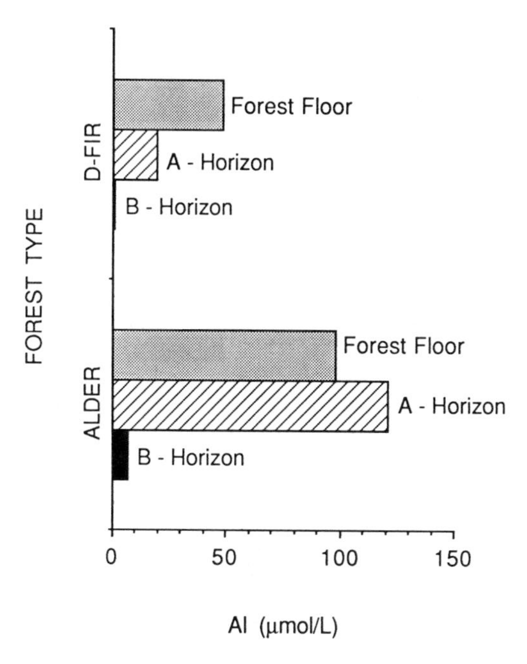

Figure 6.25. Average soil solution Al concentrations (μmol L^{-1}) collected beneath forest floor (FF), A and B horizons from 55-year-old red alder and Douglas fir forest sites at Thompson Research Center.

An experiment was carried out to determine the effect of removing the input of N, in this case through the fixation process. To that end, the alder was harvested in a 1-ha area and 0.5-ha plots of both alder and Douglas fir were subsequently established. The first indication that a site could recover from N saturation was observed approximately 1 year after the removal of the red alder when soil solution NO_3^- concentrations dropped below levels in the uncut control area. Total NO_3^- leaching loss dramatically decreased (Figure 6.26) in both conversion plots. Nitrate solution concentrations reached levels as low as 1 to 2 μmol L^{-1} in the fourth year following harvesting, which were similar to those found under the adjacent Douglas fir stand. This decrease in the NO_3^- leaching was accompanied with an equivalent decrease in the leaching of base cations. We have not, however, seen any corresponding shift in soil solution acidity nor decrease in soil solution Al concentration, nor have we resampled the site to see if there has been any cation replacement on the exchange sites through mineralization and mineral weathering. We believe that recovery of these soil properties will be a long-term process that could not have occurred within this short time period.

While this evidence for recovery from N saturation in terms of soil solution chemistry is only one example, that is, N fixation associated with alder, the processes involved should be similar to those triggered from de-

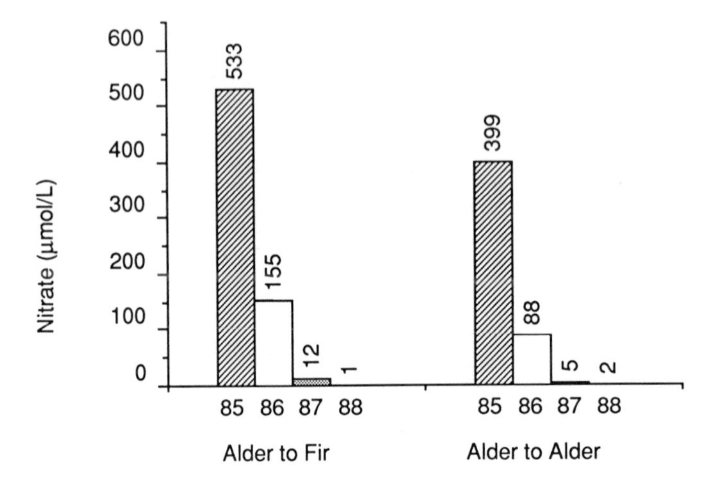

Figure 6.26. Reduction in soil solution NO_3^- concentrations from 1985 to 1988 following harvesting of alder with subsequent establishment of alder and Douglas fir conversion plots. (Expanded from Van Miegroet et al., 1990.)

position. Consequently, we believe that this study of alder provides a preliminary assessment of recovery potential from N saturation caused by atmospheric N deposition.

Retention or Loss of N in IFS Sites and Evaluation of Relative Importance of Processes

D.W. Cole, H. Van Miegroet, and N.W. Foster

Summary of Pathways and Processes

As discussed earlier in Chapter 6, there is a striking difference between the IFS sites in their relative capacity to retain N: some sites are clearly N saturated (ST, SS, TL, RA) as indicated by the consistently high solution NO_3^- level, while there are a few sites (e.g., WF, HF) that appear to be at a transition stage, as suggested by a combination of total soil N content and C:N ratio, N mineralization potential, and measured NO_3^- leaching. The majority of the sites within the IFS network, however, are not yet saturated: they consistently retain incoming atmospheric N inputs and no traces of NO_3^- leaching could be detected.

It also became clear from our analysis that atmospheric N deposition, as a single factor, is insufficient to explain all of these differences in NO_3^- leaching. Other sources and sinks of N within the forest ecosystems must also be considered if we are to explain adequately the differences in N leaching rates between sites. For those sites with measurable NO_3^- leaching there is a stronger positive correlation between atmospheric N input and NO_3^-

leaching rates than across all the sites, indicating that the NO_3^- output from sites that approach or have reached N saturation is more directly influenced by changes in atmospheric N deposition. In the IFS program, a number of sources and sinks were studied and their contribution to the retention or loss of N evaluated. Besides deposition inputs, the factors analyzed included total N content of the soil and forest floor, N mineralization rate in the upper soil layer, C:N ratio of the soil (as an index of N immobilization), and the age, vigor, and species composition of the stand resulting in differences in N uptake. As a single factor, N mineralization potential explains best the variation in NO_3^- leaching between sites (see Table 6.4). Recognizing the difficulty in estimating actual field N mineralization rates accurately, the best practical index of N loss was found to be the C:N ratio of the soil with total soil N. Sites with a narrow C:N ratios (C:N < 20) and with a total soil N content greater than 350 kmol ha^{-1} (5000 kg ha^{-1}) also appear to be the sites most susceptible to nitrification and NO_3^- leaching. Atmospheric N deposition and N mineralization potential together are the two major sources of N that can explain almost two-thirds of the variation in NO_3^- leaching observed, underscoring the equally important role of current and past N deposition history in the degree of N saturation reached by a particular forest ecosystem.

Comparison of Net Losses between Sites

As illustrated in Figure 6.16, there is a major disparity between sites in the extent of NO_3^- leaching taking place. The greatest leaching loss amongst the 17 sites is associated with red alder. Here, additions from N fixation are an order of magnitude larger than those via atmospheric N deposition at the site and exceed by at least a factor of four the total N deposition measured at the other sites of this program. Besides the red alder site (RA), annual N losses in excess of 750 mol ha^{-1} ($\sim$10 kg ha^{-1}) are found only at the sites in the Smokies (SS, ST, SB) and at Turkey Lakes (TL). More than 50% of the sites experience annual leaching losses of less than than 100 mol ha^{-1}, and 4 sites leach between 100 and 200 mol ha^{-1} yr^{-1} of NO_3^-. These results raise the obvious questions: Why do such differences occur? To what extent are they regulated by internal processes and to what extent are they a function of inputs from external sources (e.g., deposition or fixation)? Nitrification results in the introduction of a mobile NO_3^- and H^+ ions in the soil solution, which in turn can cause various changes to take place in the soil and soil solution, including:

- A decrease in the soil solution pH
- An increase in exchangeable acidity, a decrease in base saturation, and a decrease in soil pH
- An increase in Al activity in the soil solution
- Change in the nutrient availability (e.g., P).

It is imperative we understand the processes and environmental conditions that trigger elevated NO_3^- levels in the soil and NO_3^- leaching losses if we are to know when and how extensively these reactions occur in our forest ecosystems. Indeed, the occurrence and rate of NO_3^- leaching may have important implications for site fertility or water quality in these forested systems.

Reasons for N Losses from Ecosystems

It is clear from this analysis that the loss of N from an ecosystem is not simply a function of some current event including atmospheric N deposition or additions through fertilization unless such events are drastic in their magnitude. Rather, the loss of N is regulated through the integration of a series of soil related processes, processes which could well be tempered or accentuated by events predating the current ecosystem occupying the site. For example, the accumulation of N and C at these sites undoubtedly reflects the cumulative history of these sites (temperature, disturbance, stand history, prior N deposition regime) far more than any current event. This is not to say that the deposition input of N is not important and does not lead or contribute to N saturation and NO_3^- leaching. As a matter of fact, deposition inputs can play critical roles in N loss in several ways:

- At sites with a long history of such additions, there undoubtedly will be a cumulative effect over time and N leaching will ultimately take place. With some sites experiencing additions in excess of 1500 mol ha^{-1} (~20 kg ha^{-1}) annually, ultimately the C:N ratio will narrow through N accumulation in excess of organic C accumulation, triggering, in time, nitrification and NO_3^- leaching. The alder site in Washington illustrates this concept. Nitrogen addition through the fixation process has resulted in the accumulation of more than 200 kmol ha^{-1} of N during a 50-year period, increasing the storage of N in this ecosystem from approximately 200 kmol ha^{-1} (or 3000 kg ha^{-1}) to 400 kmol ha^{-1} (5500 kg ha^{-1}). At some point in time predating our research at this site (before 1970 when the stand was only 35 years old), NO_3^- leaching began to take place, that is, N saturation was reached in less than 35 years of high N input levels. There is every reason to believe atmospheric additions will cause the same net effect. If the inputs are less than those caused by fixation (often >7000 mol ha^{-1} yr^{-1} or 100 kg ha^{-1} yr^{-1}), then only the time function will differ. For sites receiving an annual deposition of 1500 mol ha^{-1} (~20 kg ha^{-1}), N saturation and NO_3^- leaching should take about five times longer than that witnessed with alder, assuming of course that all other environmental and ecosystem factors are equal.
- At sites where NO_3^- leaching is already occuring, such as ST, SS, SB, RA, and TL, additional inputs of N, by deposition or any other source will further add to the NO_3^- losses currently taking place. This additional loss could certainly be larger than the N input through stimulation of N

mineralization in systems with already narrow C:N ratios (i.e., low N immobilization potential).

- A few sites, notably the high-elevation spruce forest at Whiteface Mountain and the mixed hardwood site at Huntington Forest in New York, appear to be at a transition stage. They show some, albeit still low, annual NO_3^- leaching loss indicative of periodic nitrification peaks. They can be classified as intermediate in terms of N mineralization potential based on field assays and the soil C:N ratio. Such systems will be particularly sensitive to atmospheric N deposition, especially as the stands further mature and the role of N uptake as a N sink decreases.

Analysis of N Cycles in Polluted versus Unpolluted Environment

H. Van Miegroet, D.W. Johnson, and D.W. Cole

Figure 6.27 summarizes the N distribution and cycling patterns at the Findley Lake and the Smokies Tower sites, which represent high-elevation conifer systems subject to low and high atmospheric N inputs, respectively. Within the IFS network the total atmospheric N input is highest at the Smokies site (28 kg ha^{-1} yr^{-1} or 2000 mol ha^{-1} yr^{-1}) which is more than sixfold higher than the estimated 4 kg ha^{-1} yr^{-1} (300 mol ha^{-1} yr^{-1}) at Findley Lake.

Total ecosystem N content (biomass + soil to a depth of 60 cm) is somewhat higher at the ST sites (12,000 kg ha^{-1} or 900 kmol ha^{-1} in plot 1 and 9,000 kg ha^{-1} or 700 kmol ha^{-1} in plot 2) than at the FL site (8,500 kg ha^{-1} or 600 kmol ha^{-1}), but in view of the observed within-site variability and compared to the range of values for the other IFS sites (see Figure 6.15) this difference can be considered inconsequential. Both systems generally rank at the higher end of total N content among the IFS sites. However, they show distinct differences in terms of N distribution and cycling patterns. The vegetation N content is 60% greater at FL than at the ST site, reflecting its greater aboveground standing biomass. Subordinate trees (such as birch) and understory vegetation account for 15% (plot 2) to 35% (plot 1) of total aboveground N at the ST site, a result of the opening of the stand following adelgid-caused Fraser fir mortality, which is also indicated by a sizable N pool as dead trunks. At the FL site with its dense old-growth subalpine fir forest, the understory component remains small. On the other hand, forest floor N is two- to fourfold higher at the ST sites, accounting for almost one-fifth of the total ecosystem N compared to less than 10% at FL. Soil N is slightly higher at the Smokies sites.

A comparison of the total N input to the ecosystem versus rate and form of N leaching below the rooting zone indicates that the FL site is still accumulating N (N output <1 kg ha^{-1} yr^{-1}). At the Smokies sites substantial NO_3^- leaching outputs are measured, which are similar to or in some years

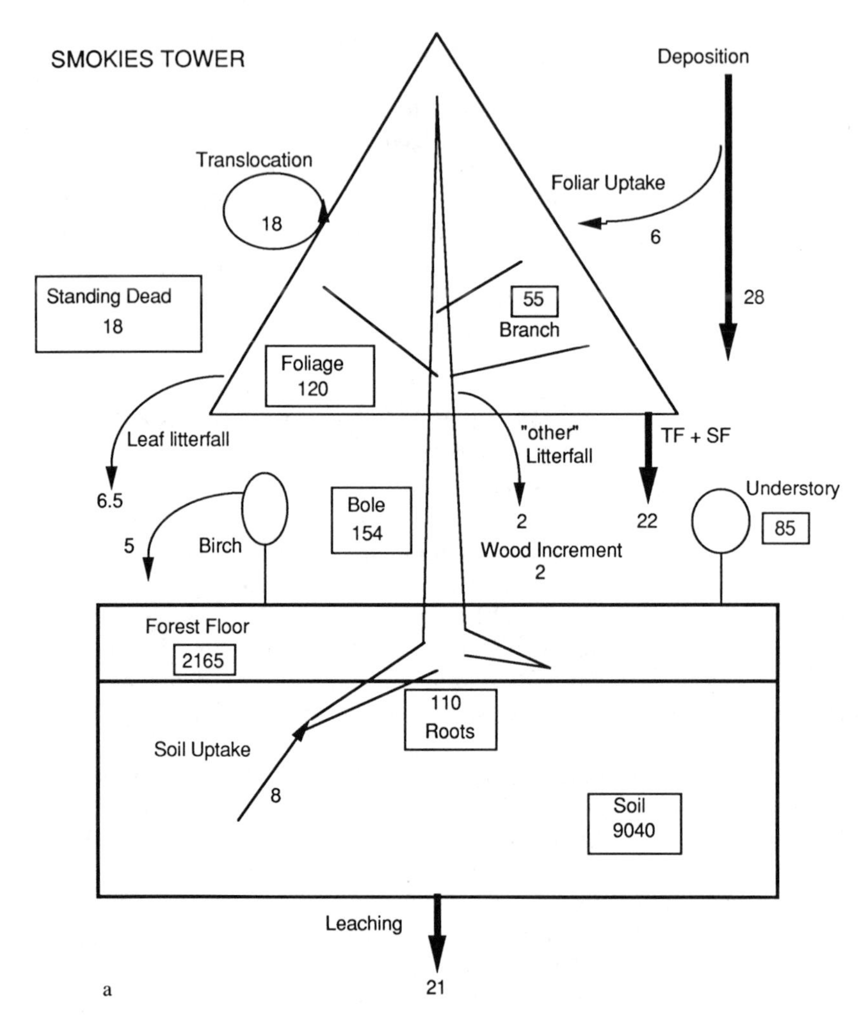

Figure 6.27. Nitrogen distribution and cycling patterns in spruce Tower site at Great Smoky Mountains National Park (a) and at Findley Lake site in Washington (b).

exceed total atmospheric N deposition, suggesting low N retention capacity, as discussed earlier in Chapter 6. This difference in ecosystem N retention cannot be accounted for by concurrent differences in vegetation uptake values. As a matter of fact, the N fluxes to and from the vegetation and internal cycling patterns within the vegetation are quite similar in both ecosystems, except for the somewhat greater role of the subordinate vegetation at the ST site. Wood increments are low (of the order of 2 kg ha^{-1} yr^{-1} or 150 mol ha^{-1} yr^{-1}), and most of the N requirement (12 kg ha^{-1} yr^{-1} at FL and between 21 and 30 kg ha^{-1} yr^{-1} at ST) is associated with new foliage production.

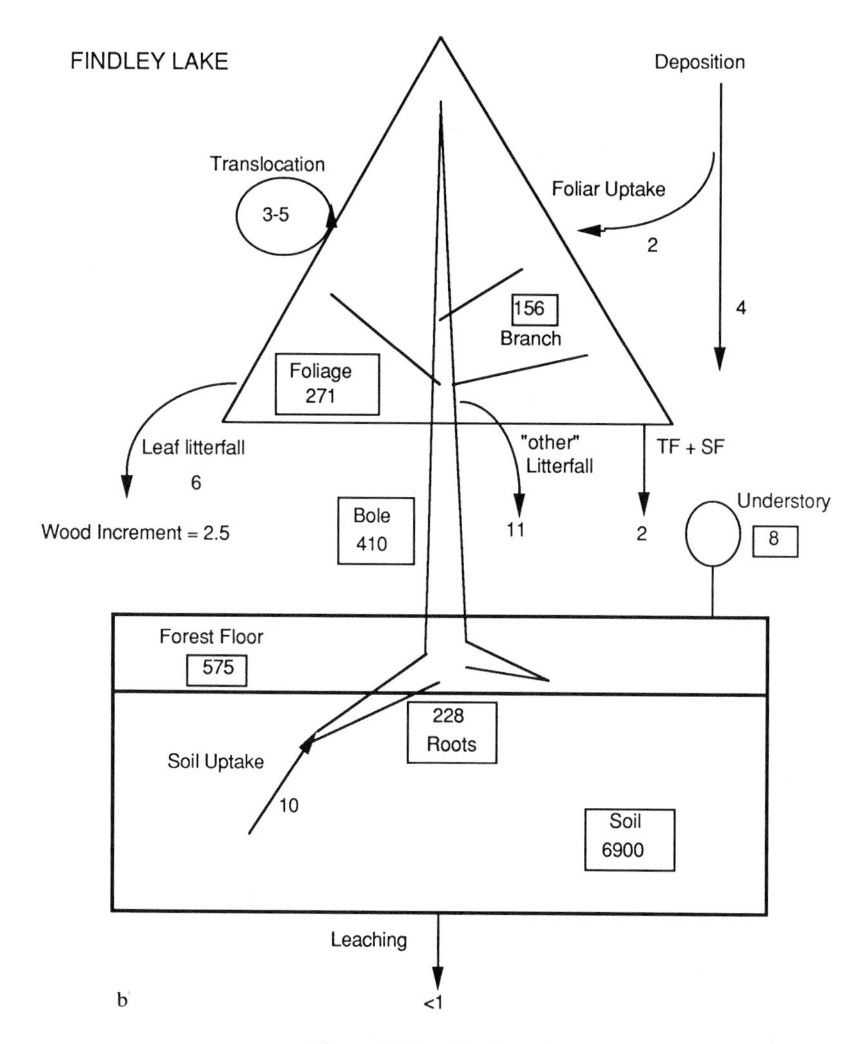

Figure 6.27. *Continued.*

Approximately 30% at FL and 60% at the ST site of annual N requirement is met through retranslocation. Calculated net canopy exchange (throughfall + stemflow − total N deposition) is negative at both sites, suggesting canopy retention of N even at the ST site, with its high levels of inorganic (available) N in the soil (see pp. 178–196). Because throughfall and total N deposition fluxes are based on different methodologies (see Chapter 2), the error value on net canopy retention calculated by difference could be substantial and should be interpreted with caution. The remainder of the N requirement is derived from soil N uptake, which is about 10 kg ha^{-1} yr^{-1} (700 mol ha^{-1} yr^{-1}) in both systems.

Annual N return with needle litterfall is similar in both sites (~7 kg ha^{-1} yr^{-1} or 500 mol ha^{-1} yr^{-1}). This, in association with the significantly larger foliar N pool at the FL site, suggests that needles are retained three to four times longer at that site than in the Smokies, which may be an indication of the less favorable nutrient (N) conditions at FL. As a result of the higher total N deposition at the ST sites, throughfall N input to the forest floor is significantly greater at the Smokies site. At FL, the N entering the belowground compartment via litterfall (needle and other) plus net throughfall is retained completely below ground and no measurable amounts of N leach below a 40-cm soil depth. Root uptake can account for approximately half of this N retention, while the remainder (about 9 kg ha^{-1} yr^{-1} or 700 mol ha^{-1} yr^{-1}) must be accumulation in the soil.

At the Smokies site, by contrast, N input to the soil (via litterfall and net throughfall) minus calculated root uptake of N is about 20 kg ha^{-1} yr^{-1} (or 1500 mol ha^{-1} yr^{-1}), which is close to the average annual NO_3^- leaching value, indicating little or no net N immobilization in the soil. The greater N accumulation potential of the FL soil is further substantiated by the ratio C:N = 37 (see Figure 6.18) and by the insignificant net N mineralization rates measured during in situ incubation (see Figure 6.19 and discussion earlier in Chapter 6). The C:N value of the ST soil (11 for plot 1 and 18 for plot 2) and the results from the field mineralization assays suggest that the Smokies soils, on the other hand, are at a stage of minimum net N accumulation and large net release through nitrification and NO_3^- leaching. The high atmospheric N inputs undoubtedly contribute to the high NO_3^- leaching rates currently observed, but it is clear that past N accumulation and disturbance history (or lack thereof) and not the current N input regime have led to the present soil condition (i.e., N saturation).

Even though both systems have similar (high) total ecosystem N contents, they differ significantly in belowground N dynamics. The FL site exhibits the behavior of a site that is N limited. All incoming N is retained, N mineralization (N availability) is low, canopy N turnover is slow, and no excess N is available for NO_3^- production and leaching. The ST site behaves as a N saturated system: because of past N and C accumulation the soil C:N is such that N in excess of what can be immobilized via vegetation uptake can no longer be retained within the belowground compartment. In addition, fir dieback has opened up the stand, rendering the soil even more susceptible to the direct impact of external influences. This site condition, coupled with the high atmospheric N deposition rates, accounts for the high NO_3^- leaching losses observed at the site.

Summary and Conclusions

H. Van Miegroet, G.M. Lovett, and D.W. Cole

In the course of the IFS project, we have been able to answer some of the research questions that were posed at the onset of the project with respect to the form and amount of atmospheric N input to forested ecosystems, the occurrence and predictability of N saturation, and the role of atmospheric N deposition and internal ecosystem properties in causing or exacerbating N saturation and excess NO_3^- leaching. Nitrogen saturation is defined as that ecosystem condition where N flux density, that is, the N input from N mineralization plus atmospheric N deposition, exceeds the ecosystem's retention capacity and is expressed by measurable NO_3^- leaching out of the tree rooting zone.

Atmospheric N Input and Canopy Interactions of N

In 13 sites of the IFS network, which covered a wide array of forest types, elevations, climatic conditions, and atmospheric deposition regimes, total atmospheric input of NH_4^+ and NO_3^- via wet and dry deposition and through cloud water input was estimated using methods described in Chapter 2.

Despite the dominance of aerosol NH_4^+ in the air (50% to 70% of the atmospheric N concentrations), HNO_3 is the dominant N species contributing to dry deposition because of its more reactive nature and ready absorption to surfaces. Coarse-particle NO_3^- and NH_4^+ can also add substantially to dry deposition fluxes. Deposition of NO_2, while not directly measured at most IFS sites, was investigated in laboratory experiments and found to be potentially important for sites close to urban areas. Particularly high rates of NO_2 deposition were noted for forest floor surfaces, as opposed to leaves and bark, and the deposition to leaves occurred mainly through the stomates. Deposition velocities of HNO_3 are much higher than those of NO_2, and the HNO_3 deposits primarily to the exterior surfaces of the plant although a small amount of deposition does occur through the stomates.

Nitrogen input data suggest that wet and dry deposition contribute roughly equally to total N input to the low-elevation sites in the United States, implying that total atmospheric N input may be approximated by doubling wet deposition measurements at these sites. This generalization does not hold, however, for the high-elevation sites, where cloud water deposition can add substantially to the total N loading, particularly in forests with high cloud immersion frequencies. Total atmospheric deposition of N ranges from 340 to 1900 mol ha^{-1} yr^{-1} (4.7 to 27 kg ha^{-1} yr^{-1}) across the IFS network, with the highest values in the eastern high-elevation sites. Low-elevation sites in the southeastern United States also receive relatively high N loadings, presumably because of the close proximity of the field sites to urban areas. However, the character of the deposition at the low-elevation sites, being predominantly nitrogen oxide gases and precipitation of relatively low N concentration, is quite different from that of the high-elevation sites, which are exposed to highly concentrated cloud water.

A comparison between atmospheric N deposition (including HNO_3) vapor and throughfall plus stemflow N indicates that the forest canopies in all intensive monitoring sites consistently act as sinks for inorganic N. Throughfall N fluxes can therefore not be used as a substitute or simple measure for total atmospheric N deposition as is the case for sulfur. This inorganic N removal partially coincides with a net increase in organic N in the throughfall flux, but the exact nature or cause of this net canopy effect is still unclear. On the average, 40% of the incoming inorganic N is retained in or transformed at passage through the canopy, and in absolute terms the net canopy effect is greatest at those sites receiving the highest atmospheric N inputs. The indication of significant N retention by forest canopies, and the role of foliar N uptake implied by it, has required a modification of the traditional calculation methods for tree N uptake and N requirements (Johnson and Lindberg 1989; Johnson et al. 1991).

Atmospheric N Deposition and N Saturation

Belowground solutions were monitored using fritted-glass lysimeters at 10-kPa tension placed underneath the forest floor and at two depths in the mineral soil. Annual N leaching losses below the rooting zone were calculated from weighted average solution concentrations and water fluxes based on a hydrologic model modified for site-specific climatic conditions (see Chapter 3).

No strong positive correlation was found between estimated total atmospheric N input and measured NO_3^- leaching below the rooting zone across all sites. No NO_3^- leaching below the rooting zone occurs in more than half the sites irrespective of atmospheric N input rates. The relationship between input and output fluxes is somewhat stronger when only those sites are considered that are currently leaching NO_3^- below the rooting zone. These results imply that N saturation is not simply the result of current environmental conditions including the N deposition regime. However, high N inputs tend to accelerate the course of events and exacerbate N saturation once it has been attained. Inherent ecosystem properties such as soil N content and N immobilization or mineralization potential (a reflection of the past C and N accumulation history) and stand composition and vigor (expressed by differences in N uptake) must also be considered.

In this study, atmospheric N deposition and N mineralization potential together explain more than 60% of the variation in NO_3^- leaching rates between the various IFS sites, underscoring the importance of both past and current conditions to the degree of N saturation in a forest ecosystem. As a single factor, differences in soil N mineralization rates accounted for most of the variation (~40%) in NO_3^- leaching between sites. Recognizing the difficulties in establishing field N mineralization or immobilization rates accurately, the soil C:N ratio and soil N content appear to be the best practical indices of internal N release.

Some sites within the IFS network, such as the high-elevation spruce site in the Smoky Mountains and the mixed hardwood forest in Ontario's Turkey Lakes watershed, are clearly N saturated as indicated by consistently high solution NO_3^- levels. They are characterized by a combination of moderate to high atmospheric N deposition, large soil N pools with low C:N ratios, high internal N mineralization rates, and moderate to low tree uptake values. A few sites appear to be at a transition stage, as suggested by the combination of soil N content and C:N ratio, N mineralization rates, and soil solution NO_3^- concentrations that increase in the upper soil profile periodically when tree uptake is low. Such sites are particularly sensitive to increases in atmospheric N inputs; they include both the low-elevation hardwoods and the high-elevation spruce sites in New York. The majority of the IFS sites, however, have not yet reached N saturation: they consistently retain incoming N, have low mineralization and nitrification potentials, and show no traces of NO_3^- leaching.

Nitrogen saturation and excess NO_3^- leaching thus do not appear to be limited to special forest ecosystem types (such as those containing N fixers), but can in principle be reached in any forested ecosystem given sufficient time and the appropriate environmental conditions for N to accumulate until it is incompletely retained by the biological component of the ecosystem (i.e., vegetation and microorganisms). Forest management practices or the occurrence and frequency of site disturbances reducing the N pool or increasing the biological N retention capacity of the system will tend to delay the time to N saturation, whereas an increase in external N sources (e.g., atmospheric N deposition, N fixation, and anthropogenic N additions) tend to speed up the N saturation process.

Effects of N Saturation on Biogeochemical Cycles

The most immediate impact of high NO_3^- production and leaching rates is the accelerated leaching of cations displaced from the exchange complex. In sites with medium base saturation this means increased export of nutrient bases from the rooting zone, potentially resulting in nutrient deficiencies if bases are insufficiently replenished through atmospheric inputs, organic matter decomposition, or mineral weathering. In more acid soils with low base saturation it also entails increased Al mobilization into the soil solution, which in turn may adversely affect root activity and uptake. The IFS network encompasses sites in both categories that illustrate the magnitude of soil intensity effects of N saturation, that is, changes in soil solution chemistry in response to periodic NO_3^- peaks.

Within the short timeframe of this study it would have been unrealistic to expect drastic changes in soil properties (i.e., capacity effects of N saturation and excess NO_3^- leaching) to have taken place. The extent of such changes could nevertheless be illustrated by a low-elevation site in the northwestern United States that is receiving low N inputs from pollution but where N fixation at an average rate in excess of 7000 mol ha^{-1} yr^{-1} (>100 kg N ha^{-1} yr^{-1}) caused N saturation in an alder forest compared to an adjacent conifer stand containing no N fixers. Intensive NO_3^- production and leaching under alder has caused a significant decrease in pH and an increase in Al activity in the shallow soil solutions, a downward redistribution of the exchangeable bases associated with a concomitant increase in exchangeable acidity in the upper soil. These changes have resulted in lower base saturation and lower soil pH in the upper part of the soil profile where most plant roots are located. The first indications of possible changes in nutrient availability (e.g., P limitations) under alder are starting to emerge (Compton and Cole 1989). If those changes result from soil acidification, N addition through pollution could eventually lead to similar end conditions, albeit over a different (longer) timeframe.

Biological flux patterns were also monitored in all IFS field sites, which allowed us to evaluate to what extent the degree of N saturation affects internal N cycling within forests. Estimated fluxes included annual N return from the canopy using throughfall collectors and litter traps; tree requirements and uptake of N calculated from periodic remeasurement of tree growth, mortality, and chemical composition of the tissues; and belowground N mineralization and nitrification rates through field incubation assays. Two high-elevation IFS sites at opposite ends of the N deposition and N saturation spectrum, namely the Findley Lake site in western Washington and the spruce-fir site in the Smoky Mountains, were used to illustrate the influence of N deposition and N saturation on N cycling patterns.

Under the condition of N saturation, N availability to plants and microorganisms is high and the need for N conservation is minimal. This is expressed by a faster turnover of the canopy N, net N release through mineralization in the soil and forest floor, and high nitrification potential. Non-N-saturated systems are more conservative with respect to N, expressed by canopy and belowground N dynamics: Needles are retained longer and foliar N turnover is slow, N input fluxes to the forest floor are immobilized, and there is little net N mineralization. These factors result in low N availability for plant uptake and no measurable nitrification.

At this point, a causal relationship between the soil and solution chemistry changes associated with N saturation and a decline in forest productivity and health (especially at the high-elevation, high-N-input spruce forests) has not been conclusively established. There is still lively debate on this topic. As to the question of a possible recovery of forest soils from changes induced by N saturation and excess NO_3^- leaching, there are some indications of a fairly rapid response of the soil solution chemistry to a decrease in nitrification and NO_3^- leaching. Chemical changes in the soil generally require a longer time to take place, and within the timeframe of this study it was not possible to evaluate the rate or extent of soil recovery in response to a decline in atmospheric N input. Moreover, it is expected that other factors such as site history, forest cover, and management actions will also greatly influence the rate and extent of the recovery process.

There are considerable uncertainties associated with the estimates of N deposition and NO_3^- leaching outputs and of some of the internal N fluxes, particularly those that are derived from other (measured or estimated) N fluxes. Uncertainties in N input–N output budgets arise from the fact that a combination of field measurements and model outputs is needed to estimate both N deposition and NO_3^- leaching losses. Other fluxes, such as net canopy exchange, are calculated as a difference between two or more N fluxes, each estimated with their own level of uncertainty. For example, an error analysis in one of the high-elevation spruce sites in the Smoky Mountains using statistical methods (where possible) and logical constraints (e.g., soil water flux <90% of precipitation) led to the conclusion that only differences between inputs and outputs greater than 50% can be considered statistically significant and meaningful (Johnson et al. 1991). Despite this uncertainty regarding absolute values of N deposition and NO_3^- leaching losses and the magnitude of various internal N fluxes, however, the IFS network has offered a unique opportunity to compare biogeochemical cycling of N in forest ecosystems spanning a wide range of N deposition regimes and at various degrees of N saturation.

References

Abeles F.B., Craker L.E., Forrence L.E., Leather G.R.. 1971. Fate of air pollutants: removal of ethylene, sulfur dioxide, nitrogen dioxide by soil. Science 173:914–916.

Aber J.D., Nadelhoffer K.J., Streudler P., Melillo J.M. 1989. Nitrogen saturation in northern forest ecosystems. Bioscience 39:378–387.

Agren G.I., Bosatta E. 1988. Nitrogen saturation of terrestrial ecosystems. Environ. Pollut. 54:185–197.

Alexander M. 1977. Introduction to Soil Microbiology, 2d Ed. Wiley, New York.

Appel B.R., Tokiwa Y. 1981. Atmosphere particulate nitrate sampling errors due to reactions with particulate and gaseous strong acids. Atmos. Environ. 15:1087–1089.

Aulie R.P. 1970. Boussingault and the nitrogen cycle. Proc. Am. Philos. Soc. 114:439–479.

Berg B., Staaf H. 1981. Leaching, accumulation and release of nitrogen in decomposing forest litter. Ecol. Bull. (Stockholm) 33:163–178.

Binkley D. 1981. Nodule biomass and acetylene reduction rates of red alder and Sitka alder on Vancouver Island, B.C. Can. J. For. Res. 11:281–286.

Bowden R.D., Geballe G.T., Bowden W.B. 1989. Foliar uptake of ^{15}N from cloud water by red spruce (*Picea rubens* Sarg.). Can. J. For. Res. 19:382–386.

Burkhardt M.R., Buhr M.P., Ray J.D., Stedman D.H. 1988. A continuous monitor for nitric acid. Atmos. Environ. 22:1575–1578.

Bytnerowicz A., Miller P.R., Olszyk D.M., Dawson P.J., Fox C.A. 1987. Gaseous and particulate air pollution in the San Gabriel mountains of southern California. Atmos. Environ. 21:1805–1814.

Cadle S.H., Countess R.J., Kelly N.A. 1982. Nitric acid and ammonia in urban and rural locations. Atmos. Environ. 16:2501–2506.

Chen C.W., Hudson R.J.M., Gherini S.A., Dean J.D., Goldstein R.A. 1983. Acid rain model: canopy module. J. Environ. Eng. 109:585–603.

Cole D.W, Rapp M. 1981. Elemental cycling in forest ecosystems. In Reichle D.E. (ed.) Dynamic Properties of Forest Ecosystems. IBP 23, Cambridge University Press, Cambridge, England, pp. 341–409.

Cole D.W., Gessel S.P., Turner J. 1978. Comparative nutrient cycling in red alder and Douglas-fir. In Briggs D.G., DeBell D.S., Atkinson W.A. (eds.) Utilization and Management of Alder. USDA For. Serv. Gen. Tech. Rep. PNW-70, U.S. Government Printing Office, Washington, D.C., pp. 327–336.

Compton J.E, Cole D.W. 1989. Growth and nutrition of second rotation red alder. Agron. Abstr. 1989:300.

Crocker R.L, Major J. 1955. Soil development in relationship to vegetation and surface age at Glacier Bay. J. Ecol. 43:427–448.

Dempster J.P, Manning W.J. (eds.) 1988. Special Issue: Excess Nitrogen Deposition. Environ. Pollut. 54:159–298.

Duvigneaud P., Denaeyer-De Smet S. 1970. Biological cycling of minerals in temperate deciduous forests. In Reichle D.E. (ed.) Analysis of Temperate Forest Ecosystems. Springer-Verlag, New York, pp. 199–225.

Eno C.F. 1960. Nitrate production in the field by incubating the soil in polyethylene bags. Soil Sci. Soc. Am. Proc. 24:277–279.

EPA (U.S. Environmental Protection Agency). 1982. Air Quality Criteria for Oxides of Nitrogen. EPA-600/8–82–026, U.S. Environmental Protection Agency, Research Triangle Park, North Carolina, U.S.A.

Finlayson-Pitts B.J., Pitts J.N. Jr. 1986. Atmospheric Chemistry: Fundamentals and Experimental Techniques. Wiley, New York, pp. 522–586.

Foster N.W., Nicolson J.A. 1988. Acid deposition and nutrient leaching from deciduous vegetation and poolzolic soils at the Turkey Lakes Watershed. Can. J. Fish. Aquat. Sci. 45:96–100.

Foster N.W., Nicolson J.A., Hazlett P.W. 1989b. Temporal variation in nitrate and nutrient cations in drainage waters from a deciduous forest. J. Environ. Qual. 18:238–244.

Foster N.W., Hazlett P.W., Nicolson J.A., Morrison I.K. 1989a. Ion leaching from a sugar maple forest in response to acidic deposition and nitrification. Water Air Soil Pollut. 48:252–261.

Foster N.W, Nicolson J.A. 1988. Acid deposition and nutrient leaching from deciduous vegetation and podzolic soils at the Turkey Lakes Watershed. Can. J. Fish. Aquat. Sci. 45(Suppl. 1):96–100.

Furukawa A., Totsuka T. 1979. Effects of NO_2, SO_2, O_3 alone and in combinations on net photosynthesis in sunflower. Environ. Control Biol. 17:161–166.

Gaines G.L., Thomas H.C. 1953. Adsorption studies in clay minerals: a formulation of the thermodynamics of exchange adsorption. J. Chem. Phys. 21:714–718.

Galbally I.E., Roy C.R. 1983. The fate of nitrogen compounds in the atmosphere. In Freney J.R., Simpson J.R. (eds.) Gaseous Loss of Nitrogen from Plant-Soil Systems. Martinus Nijhoff/Dr. W. Junk Publishers, The Hague, pp. 264–284.

Gapon E.N. 1933. On the theory of exchange adsorption in soils (in russian). J. Gen. Chem. USSR 3:144.

Garten C.T., Hanson P.J. 1990. Foliar retention of ^{15}N-nitrate and ^{15}N-ammonium by red maple (*Acer rubrum*) and white oak (*Quercus alba*) leaves from simulated rain. Environ. Exp. Bot. 30:333–342.

Ghiorse W.C., Alexander M. 1976. Effect of microorganisms on the sorption and fate of sulfur dioxide and nitrogen dioxide in soil. J. Environ. Qual. 5:227–230.

Grennfelt P., Bengtson C., Skärby L. 1983. Dry deposition of nitrogen dioxide to scots pine needles. In Pruppacher H.R., Semonin R.G., Slinn W.G.N. (eds.) Precipitation Scavenging, Dry Deposition, Resuspension. Elsevier, New York, pp. 753–762.

Halldin S. 1985. Leaf and bark area distribution in a pine forest. In Hutchison B.A., Hicks B.B., (eds.) The Forest-Atmosphere Interaction. D. Reidel Publishing Company, London, pp. 39–58

Hamburg S.P. 1984. Effects of forest growth on soil nitrogen and organic matter pools following release from subsistence agriculture. In Stone E.L. (ed.) Forest Soils and Treatment Impacts, Proceedings of 6th North American Forest Soils Conference, Department of Forestry, Wildlife and Fisheries, University of Tennessee, Knoxville, Tennessee, pp. 145–158.

Hanson P.J., Rott K., Taylor G.E. Jr., Gunderson C.A., Lindberg S.E., Ross-Todd M.B. 1989. NO_2 deposition to elements representative of a forest landscape. Atmos. Environ. 23:1783–1794.

Hicks B.B., Baldocchi D.D., Meyers T.P., Hosker R.P. Jr., Matt D.R. 1987. A preliminary multiple resistance routine for deriving deposition velocities from measured quantities. Water Air Soil Pollut. 36:311–330.

Hill A.C, Bennett J.H. 1970. Inhibition of apparent photosynthesis by nitrogen oxides. Atmos. Environ. 4:341–348.

Hosker R.P, Lindberg S.E. 1982. Review: Atmospheric deposition and plant assimilation of gases and particles. Atmos. Environ. 16:889–910.

Huebert B.J, Robert C.H. 1985. The dry deposition of nitric acid to grass. J. Geophys. Res. 90:2085–2090.

Hutchinson T.C., Bozic L., Munoz-Vega G. 1986. Responses of five species of conifer seedlings to aluminum stress. Water Air Soil Pollut. 31:283–294.

Jarvis P.G. 1971. The estimation of resistances to carbon dioxide transfer. In Sestak Z., Catsky J., Jarvis P.G. (eds.) Plant Photosynthetic Production Manual of Methods. Dr. W. Junk, The Hague, pp. 566–631.

Johnson D. 1989. Site description. In Johnson D.W., Van Hook R.I. (eds.) Analysis of Biogeochemical Cycling Processes in Walker Branch Watershed. Springer-Verlag, New York, pp. 6–20.

Johnson D.W, Cole D.W. 1980. Anion mobility: relevance to nutrient transport from forest ecosystems. Environ. Int. 3:79–90.

Johnson D.W, Lindberg S.E. 1989. Implications of recent nitrogen flux measurements in the Integrated Forest Study to interpretation of nitrogen cycling in forests. Suppl. Bull. Ecol. Soc. Am. 70(2):157 (abstr.).

Johnson D.W., Van Miegroet H., Lindberg S.E., Harrison R.B., Todd D.E. 1991. Nutrient cycling in red spruce forests of the Great Smoky Mountains. Can. J. For. Res. 21:769–787.

Joslin J.D., Wolfe M.H. 1988. Response of red spruce seedlings to change in soil aluminum in six amended forest soil horizons. Can J. For. Res. 18:1614–1623.

Judeikis H.S, Wren A.G. 1978. Laboratory measurements of NO and NO_2 depositions onto soil and cement surfaces. Atmos. Environ. 12:2315–2319.

Kaji M., Yoneyama T., Tosuka T., Iwaki H. 1980. Absorption of atmospheric NO_2 by plants and soils VI. Transformation of NO_2 absorbed in the leaves and transfer of the nitrogen through the plants. Res. Rep. Natl. Inst. Environ. Stud. Jpn. 11:51–58.

Kelly J.M, Meagher J.F. 1986. Nitrogen input/output relationships for three forested watersheds in eastern Tennessee. In Correll D.L. (ed.) Watershed Research Perspectives, Smithsonian Institution Press, Washington, D.C., pp. 360–391.

Kelly J.M., Schaedle M., Thornton F.C., Joslin J.D. 1990. Sensitivity of tree seedlings to aluminum: II. Red oak, sugar maple, and european beech. J. Environ. Qual. 19:172–179.

Kelly T.J., Tanner R.L., Newman L., Galvin P.J., Kadlecek J.P. 1984. Trace gas and aerosol measurements at a remote site in the northeastern U.S. Atmos. Environ. 18:2505–2576.

Kinjo T., Pratt P.F. 1971. Nitrate adsorption: II. In competition with chloride, sulfate and phosphate. Soil Sci. Soc. Am. Proc. 35:725–728.

Körner C., Scheel J.A., Bauer H. 1979. Maximum leaf diffusive conductance in vascular plants. Photosynthetica 13:45–82.

Kramer P.J., Kozlowski T.T. 1979. Physiology of Woody Plants. Academic Press, New York.

Kress L.W., Skelly J.M., Hinkelmann K.H. 1982. Growth impact of O_3, NO_2, and/or SO_2 on *Pinus taeda*. Environ. Monitor. Assess. 1:229–239.

Lang G.E., Reiners W.A., Heier R.K. 1976. Potential alteration of precipitation chemistry by epiphytic lichens. Oecologia (Berl.) 25:229–241.

Laxen D.P.H., Noordally E. 1987. Nitrogen dioxide distribution in street canyons. Atmos. Environ. 21:1899–1903.

Lefohn A.S.L., Tingey D.T. 1984. The co-occurrence of potentially phytotoxic concentrations of various gaseous air pollutants. Atmos. Environ. 18:2521–2526.

Lindberg S.E., Johnson D.W. 1989. 1988 Annual Report of the Integrated Forest Study. ORNL/TM-11121, Environmental Sciences Division, Oak Ridge National Laboratory. Oak Ridge, Tennessee.

Lindberg S.E, Lovett G.M. 1985. Field measurements of particle dry deposition rates to foliage and inert surfaces in a forest canopy. Environ. Sci. Technol. 19:238–244.

Lindberg S.E., Bredemeier M., Schaefer D.A., Qi L. 1990. Atmospheric concentrations and deposition during the growing season in conifer forests in the United States and West Germany. Atmos. Environ. 24A:2207–2220.

Lindberg S.E., Lovett G.M., Richter D.D., Johnson D.W. 1986. Atmospheric deposition and canopy interactions of major ions in a forest. Science 231:141–145.

Lindberg S.E., Johnson D.W., Lovett G.M., Van Miegroet H., Taylor G.E. Jr., Owens J.G. 1989. Sampling and Analysis Protocols and Project Description for the Integrated Forest Study. ORNL/TM-11214, Oak Ridge National Laboratory, Oak Ridge, Tennessee.

Lovett G.M. 1988. A comparison of methods for estimating cloud water deposition to a New Hampshire subalpine forest. In Unsworth M., Fowler D. (eds.) Processes of Acidic Deposition in Mountainous Terrain. Kluwer Academic Publishers, London, pp. 309–320.

Lovett G.M., Kinsman J.D. 1990. Atmospheric pollutant deposition to high-elevation ecosystems. Atmos. Environ. 24A:2767–2786.

Lovett G.M., Lindberg S.E. 1986. Dry deposition of nitrate to a deciduous forest. Biogeochemistry 2:137–148.

Marshall J.D., Cadle S.H. 1989. Evidence for trans-cuticular uptake of HNO_3 vapor by foliage of eastern white pine (*Pinus strobus* L.). Environ. Pollut. 60:15–28.

McLaughlin S.B. 1983. Acid rain and tree physiology: an overview of some possible mechanisms of response. In Air Pollution and the Productivity of the Forest, Pennsylvania State University, University Park, pp. 67–75.

Meentemeyer V. 1978. Macroclimate and lignin control of litter decomposition rates. Ecology 59:465–472.

Meentemeyer V., Berg B. 1986. Regional variation in rate of mass loss of *Pinus sylvestris* needle litter in Swedish pine forests as influenced by climate and litter quality. Scand. J. For. Res. 1:167–180.

Meyers T.P., Huebert B.J., Hicks B.B. 1989. HNO_3 deposition to a deciduous forest. Boundary Layer Meteorol. 49:395–410.

Mohnen V.A. 1988. Exposure of Forests to Air Pollutants, Clouds, Precipitation, Climatic Variables. Report of contract #CR 813934-01-2, U.S. Environmental Protection Agency, AREAL, Research Triangle Park, North Carolina.

Nihlgård B. 1985. The ammonium hypothesis—an additional explanation to the forest dieback in Europe. Ambio 14:2–8.

Nilsson J., Grennfelt P. 1988. Critical loads for sulphur and nitrogen—Report from the nordic working group. Nord 1986:11 (Nordic Council of Ministers).

Noll K.E., Pontius A., Frey R., Gould M. 1985. Comparison of atmospheric coarse particles at an urban and non-urban site. Atmos. Environ. 19:1931–1943.

Norby R.J., Weerasuriya Y., Hanson P.J. 1989. Induction of nitrate reductase activity in red spruce needles by NO_2 and HNO_3 vapor. Can. J. For. Res. 19:889–896.

Nye P.H., Greenland D.J. 1960. The Soil under Shifting Cultivation. Commonwealth Bureau of Soil Tech. Comm. 51, Commonwealth Agricultural Bureaux, Farhnam Royal, Bucks, U.K.

O'Dell R.A., Taheri M., Kabel R.L. 1977. A model for uptake of pollutants by vegetation. J. Air Pollut. Control Assoc. 27:1104–1109.

Okano K., Machida T., Totsuka T. 1988. Absorption of atmospheric NO_2 by several herbaceous species: estimation by the ^{15}N dilution method. New Phytol. 109:203–210.

Paul E.A. 1976. Nitrogen cycling in terrestrial ecosystems. In Nriagu J.O. (ed.) Environmental Biogeochemistry, Vol. 1: Carbon Nitrogen, Phosphorus, Sulfur and Selenium Cycles. Ann Arbor Science, Ann Arbor, Michigan, pp. 225–243.

Post W.M., Emanuel W.R., Zinke P.J., Stangenberger A.G. 1982. Soil carbon pools and world life zones. Nature (London) 298:156–159.

Post W.M., Pastor J., Zinke P.J., Stangenberger A.G. 1985. Global patterns of N storage. Nature (London) 317:613–616.

Raynal D.J., Joslin J.D., Thornton F.C., Schaedle M., Henderson G.S. 1990. Sensitivity of tree seedlings to aluminum: III. Red spruce and loblolly pine. J. Environ. Qual. 19:180–187.

Reiners W.A., Olson R.K. 1984. Effects of canopy components on throughfall chemistry: an experimental analysis. Oecologia (Berl.) 63:320–330.

Reuss J.O. 1989. Soil solution equilibria in lysimeter leachates under red alder. In Olson R.K., Lefohn A.S. (eds.) Effects of Air Pollution on Western Forests. APCA Transaction Series No. 16. Air and Waste Management Association, Pittsburgh, pp. 547–559.

Reuss J.O., Johnson D.W. 1986. Acid Deposition and Acidification of Soils and Waters. Ecological Studies 59, Springer-Verlag, New York.

Riha S.J., Campbell G.S., Wolfe J. 1986. A model of competition for ammonium among heterotrophs, nitrifiers, roots. Soil Sci. Soc. Am. J. 50:1463–1466.

Rogers H.H., Campbell J.C., Volk R.J. 1979a. Nitrogen-15 dioxide uptake and incorporation by *Phaseolus vulgaris* (L.). Science 206:333–335.

Rogers H.H., Jeffries H.E., Witherspoon A.M. 1979b. Measuring air pollutant uptake by plants: nitrogen dioxide. J. Environ. Qual. 8:551–557.

Russel A.G., McRae G.J., Cass G.R. 1985. The dynamics of nitric acid production and the fate of nitrogen oxides. Atmos. Environ. 19:893–903.

Saxe H. 1986a Effects of NO, NO_2, and CO_2 on net photosynthesis, dark respiration and transpiration of pot plants. New Phytol. 103:185–197.

Saxe H. 1986b. Stomatal-dependent and stomatal-independent uptake of NO_x. New Phytol. 103:199–205.

Schaefer D.A., Olson R.K. 1984. Forest canopy chemical processing of nitrogen and sulfur during a summer storm. Bull. Ecol. Soc. Am. 65:238.

Schulze E.D. 1989. Air pollution and forest decline in a spruce (*Picea abies*) forest. Science 244:776–783.

Shepard J.P., Mitchell M.J., Scott T.J., Driscoll C.T. 1990. Soil solution chemistry of an Adirondack spodosol: Lysimetry and N dynamics. Can. J. For. Res. 20:818–824.

Shortle W.C., Smith K.T. 1988. Aluminum-induced calcium deficiency syndrome in declining red spruce. Science 220:1017–1018.

Singh H. 1987. Reactive nitrogen in the troposphere. Environ. Sci. Technol. 21:320–327.

Soderlund R., Svensson B.H. 1976. The global nitrogen cycle. In Nitrogen, Phosphorus, Sulfur—Global Cycles. Ecological Bulletin 22, Swedish National Research Council, Stockholm, pp. 23–73.

Taylor G.E. Jr., Hanson P.J., Baldocchi D.D. 1988. Pollutant deposition to individual leaves and plant canopies: sites of regulation and relationship to injury. In Heck W.W., Taylor O.C., Tingey D.T. (eds.) Assessment of Crop Loss from Air Pollutants, Elsevier, New York, pp. 227–257.

Thornton F.C., Schaedle M., Raynal D.J. 1987. Effect of Al on red spruce seedlings in solution culture. Eviron. Exp. Bot. 27:489–498.

Tjepkema J.D., Cartica R.J., Hemond H.F. 1981. Atmospheric concentration of ammonia in Massachusetts and deposition on vegetation. Nature (London) 294:445–446.

Turner J. 1981. Nutrient cycling in an age sequence of western-Washington Douglas-fir stands. Ann. Bot. 48:159–169.

Van Aalst R.M. 1982. Dry deposition of NO_x. In Schneider T., Grant L. (eds.), Air Pollution by Nitrogen Oxides, Elsevier, Amsterdam, pp. 263–270.

Van Miegroet H., Cole D.W. 1984. The impact of nitrification on soil acidification and cation leaching in a red alder forest. J. Environ. Qual. 13:586–590.

Van Miegroet H., Cole D.W., Homann P.S. 1990. The effect of alder forest cover and alder forest conversion on site fertility and productivity. In Gessel S.P., Lacate D.S., Weetman G.F., Powers R.F. (eds.) Sustained Productivity of Forest Soils. Proceedings of 7th North American Forest Soils Conference, University of British Columbia, Faculty of Forestry Publications, Vancouver, B.C., pp. 333–354.

Van Miegroet H., Cole D.W., Binkley D., Sollins P. 1989. The effect of nitrogen accumulation and nitrification on soil chemical properties in alder forests. In Olson R.K., Lefohn A.S. (eds.) Effects of Air Pollution on Western Forests. APCA Transaction Series No. 16. Air and Waste Management Association, Pittsburgh, Pennsylvania, pp. 515–528.

Vitousek P.M., Reiners W.A. 1975. Ecosystem succession and nutrient retention: a hypothesis. BioScience 25:376–381.

Vitousek P.M., Gosz J.R., Grier C.C., Melillo J.M., Reiners W.A., Todd R.L. 1979. Nitrate losses from disturbed ecosystems. Science 204:469–474.

Vose J.M., Swank W.T., Taylor R.W., Dashek W.V., Williams A.L. 1989. Foliar absorption of ^{15}N labeled nitric acid vapor (HNO_3) in mature eastern white pine (*Pinus strobus* L.). In Delleur J.W. (ed.) Atmospheric Deposition, IAHS Publication No. 179, International Association of Hydrological Sciences, Institute of Hydrology, Wallingford, Oxfordshire, United Kingdom, pp. 211–220.

Waring R.H. 1987. Nitrate pollution: a particular danger to boreal and subalpine coniferous forests. In Fujimori T., Kimura M. (eds.) Human Impacts and Management of Mountain Forests. Forestry and Forest Products Research Institute, Ibaraki, Japan, pp. 93–105.

Waring R.H., Schlesinger W.H. 1985. Forest Ecosystems: Concepts and Management. Academic Press, New York.

Warneck P. 1988. Chemistry of the Natural Atmosphere. Academic Press, New York.

Weathers K.C., Likens G.E., Bormann F.H., Bicknell S.H., Bormann B.T., Daube B.C. Jr., Eaton J.S., Galloway J.N., Keene W.C., Kimball K.D., McDowell W.H., Siccama T.G., Smiley D., Tarrant R.A. 1988. Cloudwater chemistry from ten sites in North America. Environ. Sci Technol. 22:1018–1026.

Wells C.G., Jones A., Craig J. 1988. Denitrification in southern Appalachian spruce-fir forests. In Proceedings of the U.S.—FRG Research Symposium: Effects of Atmospheric Pollutants on the Spruce-Fir Forests of the Eastern United States and the Federal Republic of Germany, October 19–23, 1987, Burlington, Vermont. Gen. Tech. Rep. NE-120, U.S.D.A. Forest Service, Broomall, Pennsylvania, pp. 117–122.

Wesely M.L., Eastman J.A., Stedman D.H., Yalvac E.D. 1982. An Eddy-correlation measurement of NO_2 flux to vegetation and comparison to O_3 flux. Atmos. Environ. 16:815–820.

Wiklander L. 1976. The influence of anions on adsorption and leaching of cations. Grundförbättring 26:125–135.

Witkamp M. 1966. Decomposition of leaf litter in relation to environment, microflora, microbial respiration. Ecology 47:194–201.

Yoneyama T., Sasakawa H. 1979. Transformations of atmospheric NO_2 absorbed in spinach leaves. Plant Cell Physiol. 20:263–266.

Zavitkovski J., Newton M. 1968. Effect of organic matter and combined nitrogen on nodulation and nitrogen fixation in red alder. In Trappe J.M., Franklin J.F., Tarrant R.F., Hansen G.M. (eds.) Biology of Alder. USDA PNW Forest Range Exp. Stn., Portland, Oregon, pp. 209–221.

7. Relationships among N, P, and S in Temperate Forest Ecosystems

P.S. Homann and R.B. Harrison

Introduction

Nitrogen (N), phosphorus (P), and sulfur (S) interact in forest ecosystems through two principal mechanisms: as components of organic matter and as inorganic ions. As required plant and microbial nutrients, N, P, and S are taken up by organisms, incorporated into organic matter, and released from the organic matter during decomposition. Therefore, the distribution and cycling of these nutrients can be closely related. However, the quantitative relationships between these elements in vegetation, as expressed by nutrient ratios (N:P and N:S), vary within trees and between biomes (Vitousek et al. 1988). Accumulation of organic matter and associated N and S in soil over long time periods can be strongly influenced by P dynamics (McGill and Christie 1983). Over shorter periods, stabilization and mineralization of organic N and P in soil are controlled by different factors; the control of organic S may be similar to N or P, depending on its chemical form (McGill and Cole 1981).

The major inorganic ions [N as NO_3^- and NH_4^+; P as HPO_4^{2-} or $H_2PO_4^-$ (hereafter called PO_4), and S as SO_4^{2-}] have different chemical properties and different potentials for interaction with the soil. Competition between anions for adsorption sites can result in one anion displacing another anion from the soil solid phase into the soil solution. The movement of the soil solution can transport the displaced anion and redistribute it within the soil.

Further, biological interactions among the nutrients affect the potential production of inorganic ions; conversely, availability of inorganic ions affects the biological relationships (Vitousek et al. 1988).

The purpose of this chapter is to determine the extent to which the distribution and cycling of N, P, and S are interrelated across the broad range of forest and soil types of the Integrated Forest Study (IFS).

Data Analysis

The nutrient masses of vegetation and soil expressed on an area basis [kilograms per hectare ($kg\ ha^{-1}$)] from the following sites were used: Douglas-fir (DF), red alder (RA), Findley Lake (FL), Turkey Lakes (TL), Whiteface Mountain (WF), Huntington Forest (HF), Coweeta pine (CP), Coweeta hardwood (CH), Duke loblolly (DL), loblolly pine (LP1 and LP2), Smokies beech (SB1 and SB2), Smokies tower (ST1 and ST2), Smokies Becking (SS1), and the four Norway spruce (NS) plots. All available data were included except S in SS1 (Smokies Becking, plot 1) soil, because of an anomalous N:S ratio in a lower horizon. Some data were not available for some sites. Three components of aboveground vegetation were considered: woody tissue = overstory bole wood + bole bark + branches; foliage = overstory foliage; understory = total understory. For mineral soil, nutrient masses to soil depth of 45 cm were calculated, except for WF (Whiteface Mountain) where soil depth was limited to 41 cm.

The following abbreviations are used:

N_t = total N, and is assumed to be nearly all organic N.
P_t = total P
S_t = total S
S_o = organic S = S_t − inorganic SO_4^{2-}-S
P_o = organic P = P_t − inorganic PO_4-P

For soils, inorganic SO_4^{2-}-S was extracted with PO_4 and quantified by ion chromatography; inorganic PO_4-P was extracted with HCl/NH_4F. For vegetation, inorganic SO_4^{2-}-S was extracted with HCl or H_2O and quantified by hydriodic acid (HI) reduction, and was analyzed only at select sites. All ratios between nutrients are expressed on a mass (kilogram per kilogram) basis.

Relationships in Living Vegetation

N, P, and S are integral components of organic compounds. They are required nutrients for the growth of plants, animals, and microorganisms and interact physically through their existence in specific organic compounds. For example, nucleotides contain both N and P, while amino acids and proteins contain both N and S. A general balance between a variety of organic compounds is required to maintain life and can lead to general stoichiometric relationships between N, P, and S in plant tissue.

However, the specific stoichiometry can vary within plants, between species, and in response to the environment. Within a plant, various tissues perform different functions that require different organic compounds. The stoichiometries of tissues can differ (Vitousek et al. 1988), depending on the relative quantities of compounds and their specific compositions. However, this is not always the case; for example, Turner et al. (1977) found the N_t:S_o ratio to be the same in foliage, branches, wood, and bark of Douglas fir. Genetic variation may cause differences in stoichiometry between species, and nutrient availability may affect tissue composition. For example, if excess N is present, it can cause accumulation of arginine (Lambert 1986), an amino acid that contains high amounts of N but no S. An additional factor influencing tissue stoichiometry is the presence of inorganic forms of the nutrients, especially SO_4^{2-}.

In the IFS forests, there was a general increase in N_t in aboveground vegetation as P_t increased. The deciduous forests show a strong relationship [N (kg ha^{-1}) = −69 + 19 P (kg ha^{-1}), r^2 = .99, n = 6] while the relationship for the conifers is slightly weaker [N (kg ha^{-1}) = 86 + 6.5 P (kg ha^{-1}), r^2 = .81, n = 14]. Considerably more N exists per unit P in the deciduous forests than the conifers, as indicated by the slopes of regression lines. Strong relationships exist between N_t and P_t for each of the three components of the aboveground vegetation (Figure 7.1). The overstory woody tissue dominates the nutrient masses in the aboveground biomass; therefore, it shows trends similar to the total aboveground vegetation, including the divergence between coniferous and deciduous forests. Divergence between the two forest types is not evident in the overstory foliage and understory, however; in each case, combined data from both forest types yield strong relationships between N_t and P_t (Figure 7.1).

In contrast to N_t and P_t, the relationship between N_t and S_t in aboveground vegetation is poor. Neither coniferous nor deciduous forests show trends between the two elements, because of the poor relationship in the woody tissue (see Figure 7.1). Analytical methodology during the IFS may have contributed to the apparent lack of relationship between N_t and S_t in the woody tissue. For the sites with woody tissue N_t:S_t ratios between 1 and 5, S analyses were performed by high-temperature combustion followed by infrared detection. Although this method has been assessed for its adequacy to measure S in wood (David et al. 1989), for the IFS sites it yielded wood plus bark concentrations in excess of 400 mg S kg^{-1}. These are much higher S concentrations than are typically found in woody tissue, including those of the other IFS sites where S was quantified by other methods. However, in a compilation of N and S data, Vitousek et al. (1988) reported an N_t:S_t ratio of 2.6 for roots from a tropical forest, as well as much higher values for other ecosystems and tissues.

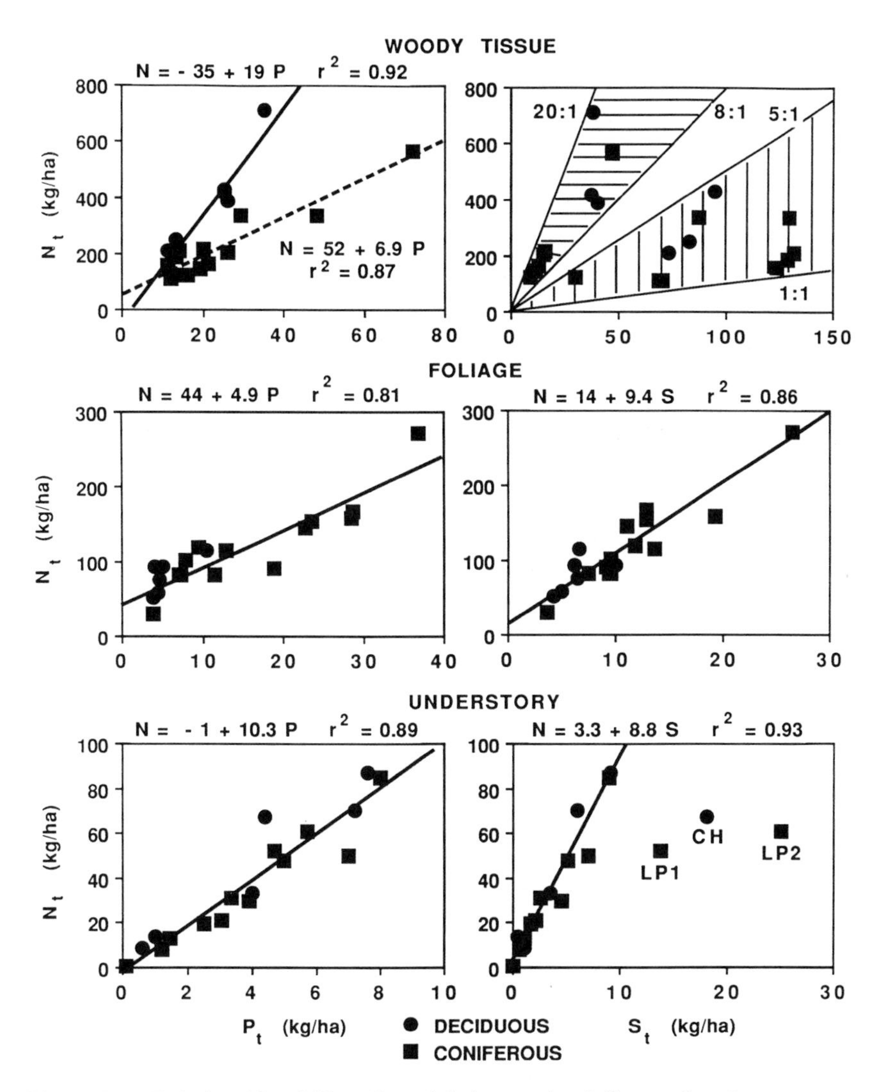

Figure 7.1. Relationship of N_t to P_t and S_t in woody, foliar, and understory components of aboveground vegetation of IFS forests.

While potential analytical discrepancies must be considered, both inherent differences in tissue chemistry among species and influence of environment on tissue chemistry can also contribute to this poor relationship. At Huntington Forest, the N_t:S_t ratio in the wood of the four overstory species varied from 6.4 to 8.8, although the species were growing on the same site (Table 7.1). Similarly, the relationship between N_t and S_t in wood varies among species of the north-central United States (Ohmann and Grigal 1990).

Table 7.1. Ratios (mass basis) of N, P, and S in Wood of Overstory Species at the Huntington Forest (HF) Site

Species	N_t:P_t	N_t:S_t	P_t:S_t
Fagus grandifolia	25.3	8.8	0.35
Acer saccharum	16.6	6.4	0.39
Acer rubrum	17.1	7.0	0.41
Betula alleghaniensis	20.7	7.8	0.38

The influence of environment on wood chemistry is illustrated by the differences that occur in the same species across environmental gradients. Ohmann and Grigal (1990) found that N_t:S_t ratios of wood decreased from northern Minnesota to southern Michigan. This is both a gradient of increasing SO_4^{2-} deposition and a climatic gradient, and therefore the specific cause of the different ratios cannot be determined. However, differences in SO_4^{2-} in the environment have been shown to cause differences in tissue chemistry under controlled conditions (Homann and Cole 1990).

The form of the elements in the tissue is also an important factor in their relationships. S_t can be partitioned into SO_4^{2-} and organic S (S_o), while N_t is almost all organic. A change in the SO_4^{2-} level in the tissue can cause a change in the N_t:S_t ratio, even if N_t and S_o maintain a strong relationship. However, the relationship between N_t and S_o can also vary. For example, fertilization of Douglas fir seedlings with SO_4^{2-} yielded higher foliar S_o and SO_4^{2-}, both of which contributed to decreasing the N_t:S_t ratio (Table 7.2). Although similar measurements of SO_4^{2-} in wood are not known to exist, storage of SO_4^{2-} could dramatically affect N_t:S_t ratios and contribute to the poor relationship between N_t and S_t observed across IFS forests.

A further factor influencing the relationship between N_t and S_t in the woody tissue is distribution of biomass among the different tissue types (Vitousek et al. 1988), which may differ between forests and change during forest growth. Each type of tissue within a tree can have a different element ratio, as shown for Findley Lake forest in Table 7.3. The N_t:S_t ratio of the total woody tissue will depend on the relative contribution of the wood, bark, and branches to the total biomass. Additionally, woody tissues of different ages may have different ratios (Ohmann and Grigal 1990), although

Table 7.2. N_t:S_t and N_t:S_o Ratios (mass basis) in Foliage from SO_4^{2-}-Fertilized and Nonfertilized Douglas Fir Seedlings[a]

	N_t	S_t	S_o	SO_4^{2-}-S	N_t:S_t	N_t:S_o
	(mg kg^{-1})					
Nonfertilized	22,900	1,540	1,040	500	15	22
SO_4^{2-}-fertilized	23,800	3,460	1,560	1,900	7	15

[a] Adapted from Homann and Cole (1990).

Table 7.3. Ratios (mass basis) of N, P, and S in Overstory (*Abies amabilis*) Tissue at Findley Lake (FL) Site

Tissue	N_t:P_t	N_t:S_t	P_t:S_t
Foliage	7.4	10.3	1.40
Branch	6.4	15.6	2.45
Bark	8.6	10.6	1.16
Wood	8.8	12.8	1.46
Roots	3.9	8.7	2.22

it is unclear if this results from formation under different environmental conditions or from natural changes during heartwood development.

In contrast to the poor relationship between N_t and S_t in woody tissue, there is a very good relationship in the overstory foliage (see Figure 7.1) in spite of the potential for environmental factors and variation between species to influence the relationship. The "average" N_t:S_t ratio as indicated by the slope of the regression line is 9.4. In the understory, the relationship between N_t and S_t is also very good for most of the forests, yielding a regression slope of 8.8, similar to the overstory foliage. However, three sites deviated considerably from this relationship, having much higher amounts of S_t than would otherwise be expected. The causes for this deviation may be similar as those for overstory woody tissue: variation in tissue chemistry between species; influence of environment on tissue chemistry; variation among different tissues, especially woody versus foliar tissues; relative distribution of biomass among different tissues; presence of SO_4^{2-} in tissues; and analytical considerations.

Canopy Processes

During leaf senescence, nutrients are mobilized and translocated from the foliage back to the branches or bole. Leaching of nutrients from aging and senescing foliage also occurs, and thus the nutrients in the foliage become partitioned into those that are removed by translocation and leaching and those that remain in the litter. The relationships between N, P, and S that exist in the litter may differ from those in the foliage because only certain compounds may be degraded and their nutrients mobilized.

In the IFS forests, the N_t:P_t ratio of the translocated-plus-leached materials is generally equal to or slightly lower than the foliage, yielding litter that has a ratio equal to or slightly higher than the foliage (Figure 7.2). Thus, in the RA forest, the N_t:P_t ratio of the translocated-and-leached materials is less than the foliage from which it originated, indicating relatively less mobilization of N compared with P. This results in a dramatic increase in the ratio as foliage is converted into litter. In contrast, in the DL forest the opposite pattern occurs, with greater relative mobilization of N compared

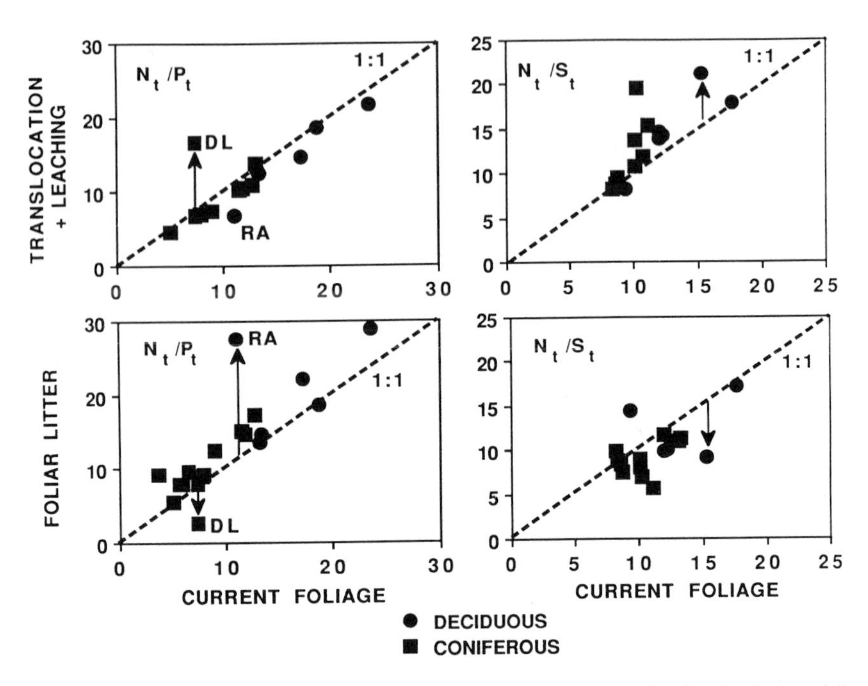

Figure 7.2. N_t:P_t and N_t:S_t ratios (mass basis) in translocated-plus-leached materials and foliar litter compared with current foliage of IFS forests.

with P. These differences may reflect the relative availability of these nutrients and thus the need of the tree to conserve them: the RA is N-fixing red alder, thus N is readily available and does not have to be efficiently conserved by the tree through translocation. The DL is loblolly pine growing on a former agricultural site; it has one of the highest extractable P levels in the soil, an indication of P availability.

The N_t:S_t ratios show a trend opposite those of N_t:P_t ratios. The ratio of the mobilized constituents is equal or greater than that of the foliage, yielding litter with an equivalent or smaller ratio (see Figure 7.2). Thus, N is more readily conserved by trees than S in a number of the IFS forests. This latter pattern has also been observed in *Eucalyptus* (Hurditch and Charley 1982).

Changes during Organic Matter Decomposition

During decomposition, both microbial and chemical processes occur that transform fresh organic matter into decomposed and humified substances. The decomposer organisms may have different nutrient composition than the original substrate. For example, N:P ratios of fungi were less than the N:P

Table 7.4. N_t:S_t and N_t:S_o Ratios (mass basis) in Newly Fallen and Decomposed (1 year) Douglas Fir Needle Litter[a]

	N_t	S_t	SO_4^{2-}-S	S_o	N_t:S_t	N_t:S_o
Newly fallen	5470	1218	586	632	4.5	8.7
Decomposed	5860	589	49	540	9.9	10.9

[a] Adapted from Homann and Cole (1990). Data are mg kg^{-1} original mass.

ratios of the deciduous leaf substrate on which they grew (Frankland et al. 1978), which may lead to decomposed organic matter with nutrient relationships different from those of the original material. The chemical environment of decomposition can also influence the ratios of the substrate. Addition of inorganic N to the decomposition environment increased the organic N:S of decomposing Douglas fir litter, while addition of SO_4^{2-} decreased the organic N:S ratio (Homann and Cole 1990).

Physical processes can also affect the ratios in decomposing organic matter. Inorganic SO_4^{2-} and soluble organic S initially present were readily leached from Douglas fir litter, causing a rapid increase in the N_t:S_t ratio because little leaching of N occurred (Table 7.4). The N_t:S_o ratio also changed, primarily because the leaching of soluble organic S.

The forest floor is the major site of litter decomposition. The N_t:P_t ratios of the forest floors generally differed from the ratios in the total aboveground litter (Figure 7.3); 70% of the sites had higher ratios in the forest floors than the litter and 25% had lower ratios. The CP and CH sites had the largest decreases in ratios between litter and forest floor, suggesting a cause unique to the location of these forests, such as climatic or soil influence. For a number of the sites, the N_t:S_t ratio of the forest floor was nearly equal to the ratio of the total litter (see Figure 7.3). However, the RA and HF sites differed from this pattern. The RA site had a very high ratio in the litter,

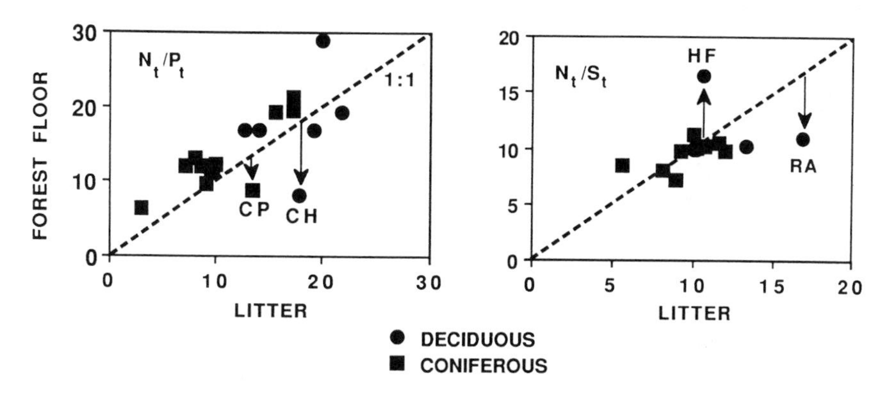

Figure 7.3. N_t:P_t and N_t:S_t ratios (mass basis) in forest floor compared with total litter of IFS forests.

primarily because of the high N concentrations created by N fixation. The ratio in the forest floor was much lower, suggesting large loss of N in contrast to the S during decomposition of the litter; this is supported by nutrient fluxes in percolating solutions. As a result, the ratio for the RA forest floor fell within the range of 7 to 11 found at the other sites (excluding HF). In contrast, litter of the HF site had a ratio within the range of 8 to 12 found at the other sites (excluding RA), while the forest floor had a much higher ratio. The cause of this shift in the ratio is not known. However, an additional consideration is that fine-root turnover might add considerably to the organic matter input into the forest floor in some forests and thus has the potential to influence the nutrient relationships if the fine-root ratios differ from those in the aboveground litter.

Organic Relationships in Mineral Soils

Soil organic matter is composed largely of highly complex macromolecules collectively called humus (Stevenson 1986). Humus is formed through both microbial and chemical pathways (Stevenson 1986). As such, its nutrient composition can be influenced by both biological and chemical processes. Because of its complexity, less of the soil organic matter can be chemically classified in comparison with organic matter in vegetation. However, even basic separation into fulvic acid, humic acid, and humin indicates that these fractions vary considerably with respect to their nutrient stoichiometries (Schoenau and Bettany 1987). Therefore, differences in fulvic acid, humic acid, and humin content may contribute to variation in N, P, and S stoichiometry in the soil.

McGill and Cole (1981) proposed a conceptual model of organic N, P, and S dynamics in soil. Organic N and one form of organic S (C-bonded S) are mineralized as organic C is utilized by microorganisms, while organic P and another form of organic S (ester S) are mineralized in response to microbial enzymes that are synthesized when low amounts of inorganic P and S are available in the soil. This model provides a rationale for the divergent behavior of organic N, P, and S in soil.

The relationships between nutrients in soils are influenced by all the factors that affect soil genesis: parent material, climate, landscape position, topography, organisms, time of genesis, and land use history. For example, in surface soil, Bettany et al. (1973) found N_t:S_t ratios to increase from the semiarid grassland soils of southern Saskatchewan to the forested regions in the north. The cause of this was not difference in inorganic SO_4^{2-}, but lower amounts of organic S, particularly ester S, in the northern soils.

Variation in nutrient ratios occurs within the soil profile (Schoenau and Bettany 1987). At the FL site N_t:P_t and N_t:S_t ratios decreased with depth, while P_t:S_t ratios increased with depth (Table 7.5). Considering only the organic forms of the elements changed the absolute ratios but not the trends

Table 7.5. Ratios (mass basis) of N, P, and S in Mineral Soil (Cryohumod) at Findley Lake (FL) Site

Horizon	N_t:P_t	N_t:S_t	P_t:S_t	N_t:P_o	N_t:S_o	P_o:S_o
E	5.94	9.52	1.60	6.31	10.41	1.65
Bhs	4.76	8.73	1.84	4.83	9.22	1.91
Bs	2.75	5.20	1.89	2.76	5.96	2.16
BC	1.78	4.52	2.54	1.78	5.56	3.11

with depth. These trends have also been found in the alfisol A and B horizons under aspen (Schoenau and Bettany 1987). Similarly, the N_t:P_o ratio decreased with depth in all the IFS profiles except CP. However, the N_t:S_o ratio decreased with depth in the four podsolized profiles (FL, HF, WF, TL), but consistent trends with depth were not evident at the other sites. Schoenau and Bettany (1987) have interpreted these and other trends of solid-phase nutrients as indications of redistribution of organically bound nutrients by leaching from surface horizons and their precipitation at depth in the soil profile. The presence of organic N and organic S in soil solutions (Homann et al. 1990) supports this hypothesis.

In addition to leaching of organic forms of nutrients, the relative rates of mineralization (conversion from organic to inorganic forms) of the nutrients affects their ratio in the remaining material. Sulfur and N mineralization rates in soils may be related or unrelated and may be affected by the mineralization environment (Maynard et al. 1983). The chemical composition of the mineralization environment may affect some microbial processes but not others, as exemplified by the effect of SO_4^{2-} levels on S mineralization but not nitrification (Table 7.6).

As a result of these many factors, a very poor relationship exists between N_t and P_o in the mineral soil of the IFS sites (Figure 7.4). The relationship is not improved by dividing the soils into soil orders. In contrast, N_t in the mineral soil is strongly related to S_o (r^2 = .68; Figure 7.4). The slope of

Table 7.6. Effect of SO_4^{2-} Addition on Apparent Net S Mineralization and Nitrification in Soils from RA and DF Sites[a]

	S Mineralization[b] (mg S kg^{-1} d^{-1})		Nitrification[b] (mg N kg^{-1} d^{-1})	
Horizon	No S Added	S Added[c]	No S Added	S Added[c]
Red alder Oa	0.474[d]	0.549[d]	4.91	5.01
Douglas fir Oa	0.114	0.109	4.50	4.47
Red alder A	0.093[d]	0.039[d]	1.65	1.76
Douglas fir A	0.031[d]	−0.009[d]	1.41	1.42

[a] Data are from Homann (1988).
[b] Based on changes in NH_4Cl-extractable SO_4^{2-} and NO_3^- during a 60-day static incubation.
[c] 38 mg SO_4^{2-}-S added per kg soil at beginning of incubation.
[d] Significant effect ($p < .05$) of SO_4^{2-} addition within horizon.

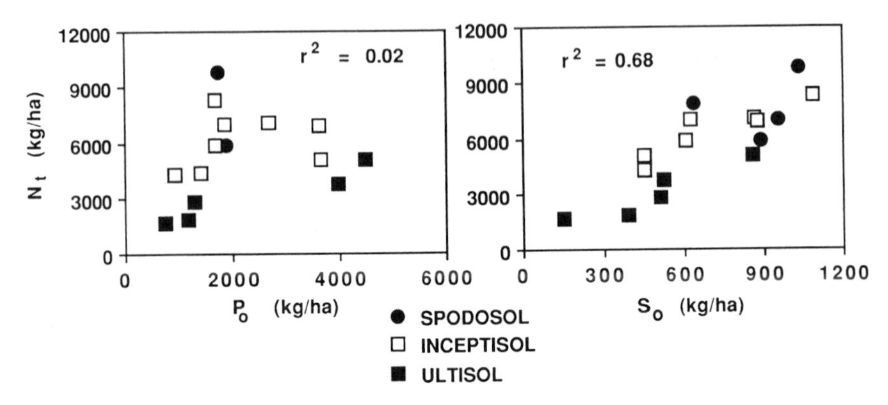

Figure 7.4. Correlation of N_t to P_o and S_o in mineral soil (to 45 cm depth) of IFS forests.

the regression line indicates that on average there is 7.4 fold as much N_t as S_o. Comparison of N_t with S_t yields a weaker correlation ($r^2 = 0.61$), with an average 6.5 fold as much N_t as S_t.

Inorganic Interactions in Mineral Soils

The most important inorganic species of N, P, and S in well-drained soils are NH_4^+, NO_3^-, $H_2PO_4^-$ and HPO_4^{2-} (herein called PO_4), and SO_4^{2-}. The cation NH_4^+ exists in soil solution and is also readily held in a plant-available form on the soil cation exchange complex, although there it is also subject to nitrification to NO_3^-. The anions can exist in solution, adsorbed on the anion exchange complex and, for PO_4 and SO_4^{2-}, as precipitated salts and components of soil minerals. Interaction among the inorganic anions occurs through competition for adsorption sites on the anion exchange complex.

The inorganic anions differ in their potential to be adsorbed (Bohn et al. 1985). NO_3^- is weakly adsorbed through electrostatic interaction with particle surfaces. It is poorly retained except in soils with high anion exchange capacity (Singh and Kanehiro 1969; Kinjo and Pratt 1971). PO_4 is strongly adsorbed through ligand exchange. SO_4^{2-} adsorption occurs through both electrostatic interaction and ligand exchange (Parfitt and Smart 1978; Marsh et al. 1987); SO_4^{2-} is intermediate between NO_3^- and PO_4 in its potential to be adsorbed.

Interaction between PO_4 and SO_4^{2-} is indicated when PO_4 fertilization yields low SO_4^{2-} levels in surface soils (Ensminger 1954), suggesting that SO_4^{2-} has been displaced from anion exchange sites by PO_4 and transported to greater depths. This inverse relationship is evident in the DL site, which was previously under cultivation (Figure 7.5). Similar relationships between

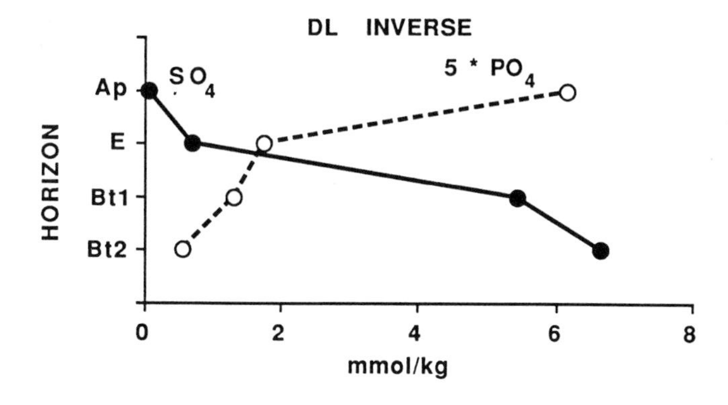

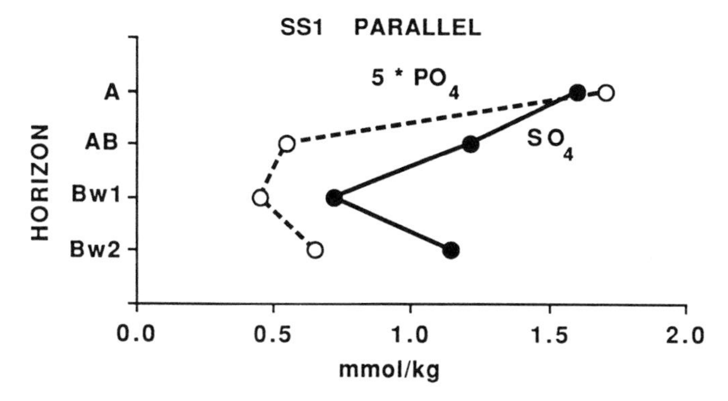

Figure 7.5. Relationship between SO_4^{2-} and PO_4 in DL and SS1 soils.

PO_4 and SO_4^{2-} also occur at several other IFS sites. However, other patterns are also possible, including parallel trends with depth (Figure 7.5).

The influence of NO_3^- on SO_4^{2-} mobility in soils occurs through two contrasting mechanisms. Under some conditions enhanced NO_3^- concentrations can result in displacement of some adsorbed SO_4^{2-} (Singh 1984), thereby increasing SO_4^{2-} concentrations in soil solutions. However, enhanced NO_3^- concentrations in forest soil solutions are generally the result of nitrification, which also produces acidity. The enhanced acidity increases the positive charge of the anion exchange complex, thereby increasing the potential to adsorb more SO_4^{2-} (Nodvin et al. 1988) and decreasing SO_4^{2-} in soil solutions. The relationship between NO_3^- and SO_4^{2-} concentrations in soil solutions will depend on the balance between these opposing mechanisms.

Available N is limited in most forests; however, when N availability exceeds biotic uptake, nitrification can occur and result in high NO_3^- concentrations in soil solutions (Vitousek et al. 1982; Foster et al. 1989). Of the IFS forests, high NO_3^- concentrations were observed in soil solutions from the SB, SS, ST, RA, and TL sites (see Chapter 6). Soil solutions from the SB and RA sites show an inverse relationship between NO_3^- and SO_4^{2-} concentrations (Figure 7.6), indicating that enhancement of adsorbed SO_4^{2-} by acidification of the anion exchange complex is currently more important than NO_3^--induced displacement of adsorbed SO_4^{2-}. This relationship of reduced SO_4^{2-} in solutions from nitrification has also been observed following harvesting of northern hardwood forests (Fuller et al. 1987; Nodvin et al. 1988; Mitchell et al. 1989).

Case Study: Effect of Short-Term N Accumulation

Nitrogen additions have been shown to alter S dynamics in several components of forest ecosystems. The addition of N fertilizer to Douglas fir caused enhanced accumulation of both organic N and organic S in the foliage

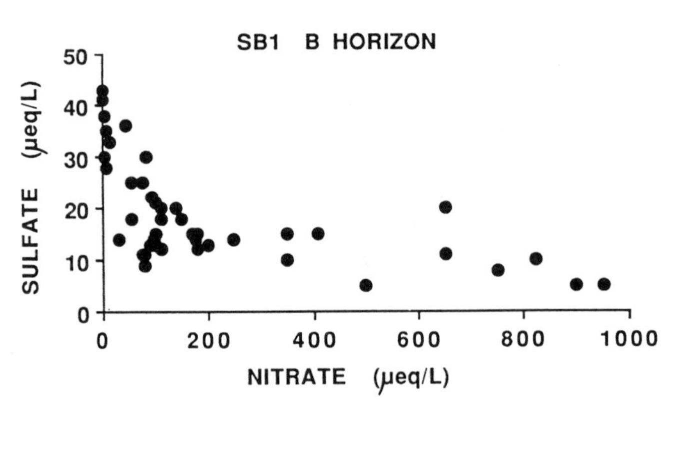

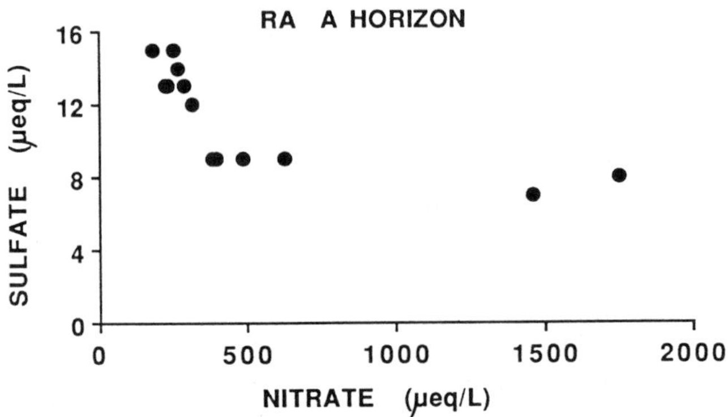

Figure 7.6. Relationship between SO_4^{2-} and NO_3^- in SB1 and RA soil solutions (D.W. Johnson and D.W. Cole, unpublished data).

(Turner et al. 1980). Inorganic N treatments enhanced organic S formation in decomposing litter (Homann and Cole 1990). Addition of inorganic N to soil yielded a reduction in S mineralization (Kowalenko and Lowe 1975). Comparison of red alder (RA) and Douglas fir (DF) stands allows evaluation of the effect of N accumulation on S and P at the ecosystem level.

The RA and DF are adjacent 55-year-old stands that developed on the same soil type following the harvesting of old-growth timber. Red alder is a nitrogen fixer, which has resulted in the rapid accumulation of N in both the vegetation and the soil of RA. This accumulation of N has affected a variety of ecosystem properties, including the S and P accumulation and distribution.

The aboveground vegetation of the RA contains 2.5 fold as much N as the DF (Figure 7.7). This occurs because higher N concentrations in the red alder foliage and wood more than offset the greater biomass of the Douglas fir; additionally, the red alder understory has both higher N concentrations and higher biomass. Related to this accumulation of N is the accumulation of P and S, although they are only 1.25 fold and 1.9 fold as great as in the DF, respectively. This accumulation is dominated by the wood, although the understory also contributes. The red alder foliage has similar P and higher S concentrations compared with Douglas fir but lower P and S masses because of lower biomass.

The forest floor of RA contains 4.3 fold as much N as the DF, because of both higher N concentration and higher mass (Figure 7.8). Paralleling this accumulated N is 2.7 fold as much P and 3.9 fold as much S as the DF. The mineral soil also has more N in RA but considerably less P. Transfer of P to the aboveground vegetation (see Figure 7.7) may partially account for this difference between mineral soil P in the two stands.

The mineral soil contains similar amounts of S in the two stands, but the distribution between S forms differs. The greater S_o in the RA soil parallels

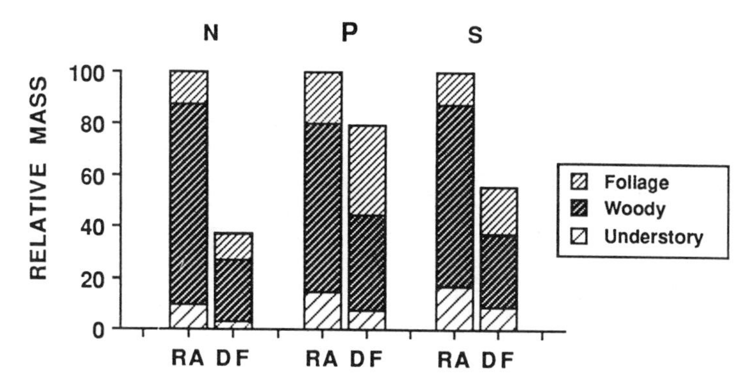

Figure 7.7. Comparison of N_t, P_t, and S_t in aboveground biomass of red alder and Douglas fir forests.

Table 7.7. S_t, S_o, and SO_4^{2-} in B Horizon Soil (15–45 cm) and SO_4^{2-} and NO_3^- in B Horizon Leachates from Red Alder and Douglas Fir Forests[a]

	RA	DF
Soil (mmol S kg^{-1})		
S_t	8.6	7.2
S_o	7.9	4.2
SO_4^{2-}	0.7	3.0
Leachate (μmol(−) L^{-1})		
SO_4^{2-}	11	52
NO_3^-	414	0
Leachate pH	5.1	5.9

[a] Unpublished data of D.W. Cole.

the enhanced N (Figure 7.8) and suggests the conversion of inorganic SO_4^{2-} to an organic form (Table 7.7). Such conversion might occur through uptake and incorporation into plants, followed by production of foliar and root detritus and leaching of organic S from decomposing litter and forest floor (Homann and Cole 1990) or by in situ microbial synthesis (Watwood et al. 1988).

The SO_4^{2-} levels in the RA soil (Table 7.7) might also have been affected by the influence of the nitrification process. Nitrification occurs in the upper profile of the RA soil and produces both high NO_3^- levels and high acidity in soil solution. The acidity can cause enhanced SO_4^{2-} adsorption (Nodvin et al. 1988) and result in inverse relationships between NO_3^- and SO_4^{2-} in soil solution (see Figure 7.6). However, the acidity is largely neutralized in the upper RA profile, resulting in a B horizon soil solution with only slightly lower pH than that of DF, but with very high NO_3^- levels (see Table 7.7). Exposure to high NO_3^- concentrations but similar acidity during the development of the RA stand might have caused displacement of SO_4^{2-} from the

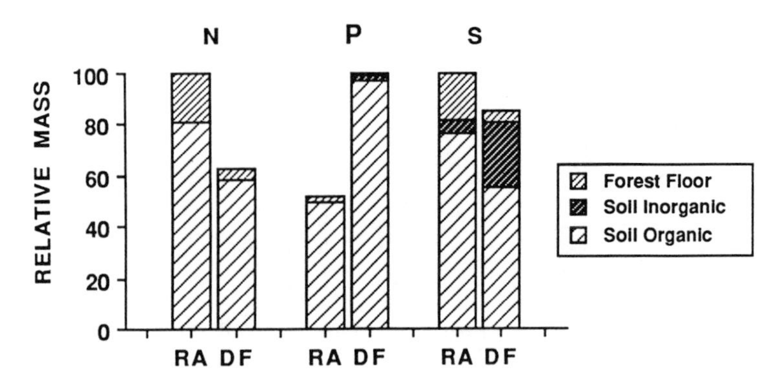

Figure 7.8. Comparison of N_t, P_t, and S_t in forest floor and mineral soil (to 45 cm depth) of red alder and Douglas fir forests.

anion exchange complex into soil solution and subsequent leaching from the soil. Removal of SO_4^{2-} from the RA soil by this mechanism might have contributed to the low levels of SO_4^{2-} now observed in the B horizon (see Table 7.7).

Ecosystem Inputs and Outputs

The interaction of N, P, and S is of potential importance with respect to the retention of these nutrients within the forest. The nutrient that limits tree growth or soil microbial activity will have considerable control over the other nutrients. In a simple scenario, the limiting nutrient prevents further biomass production and possibly ecosystem accumulation, thereby preventing incorporation of other nutrients into biomass.

When accumulation of a limiting nutrient does occur within a forest ecosystem, other nutrients might also be expected to accumulate in a systematic manner. Nitrogen is a limiting nutrient in many forests, and many of the IFS sites show an accumulation of N within the ecosystems, based on estimates of total atmospheric inputs and leaching outputs (see Chapter 6). Because of the general stoichiometric relationships between N and S in ecosystem components, as discussed previously, S accumulation might be expected to parallel N accumulation. Comparison between N and S accumulation, however, shows that they are not well related (Figure 7.9). In only two forests does their ratio fall within a range of 5 to 15, which has been observed in many ecosystem components. In other forests S was lost in spite of N accumulation or accumulated at a much higher rate than would be predicted by the N:S ratio.

The lack of parallel accumulation of N and S may result from the following causes: (1) changes in adsorbed SO_4^{2-} in soil and SO_4^{2-} levels in vegetation, which may be independent of N dynamics; (2) changes in N_t:S_o

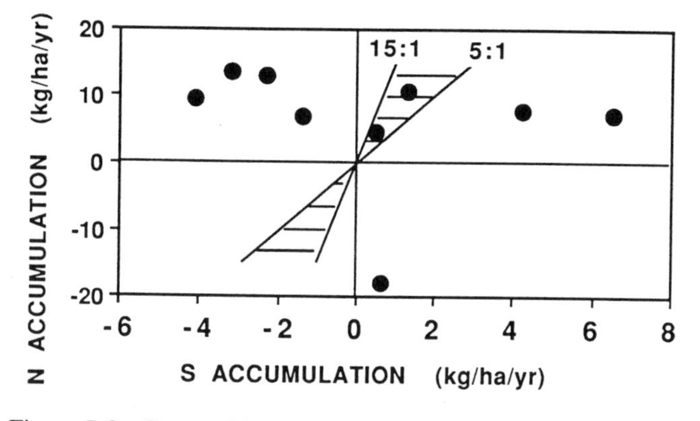

Figure 7.9. Current N_t and S_t accumulation rates in IFS forests.

ratios in vegetation or soil; or (3) incomplete or inaccurate input and output fluxes. Processes that might contribute to fluxes and need further consideration are N fixation, denitrification, and transfers of organic compounds in solutions (Homann et al. 1990). Therefore, stoichiometries of the ecosystem components are not necessarily sufficient to estimate the stoichiometry of current nutrient accumulation.

Summary and Conclusions

The N, P, and S contents of forest vegetation and soils are related because these nutrients are components of organic matter. However, there is considerable variability in the quantitative relationships both within and between ecosystems. This variability for the IFS forests is indicated with the N_t:P_t and N_t:S_t ratios in Table 7.8. There are distinct differences in the N_t:P_t ratios of wood and foliage between deciduous and coniferous species, with deciduous species having higher ratios. Differences do not appear to be as substantial for the understory and forest floor. The N_t:P_t ratios of the soils are much lower than for the vegetation and forest floor, with ultisols having the lowest ratio.

The N_t:S_t ratio of deciduous foliage and forest floor is slightly higher than for coniferous forests. There is considerable variation in N_t:S_t ratios in overstory wood, which may be the result of analytical considerations as well as environmental and species-related differences. The ratios of the soil are much

Table 7.8. N_t:P_t and N_t:S_t Ratios (mass basis) [mean (SE, number of sites)] for Components of IFS Sites by Forest Type or Soil Order

Component	N_t:P_t	N_t:S_t
Woody tissue		
Deciduous	17.7 (0.7, 6)	8.7 (2.6, 6)
Coniferous	9.9 (0.8, 14)	6.4 (1.5, 14)
Foliage		
Deciduous	16.2 (1.9, 6)	13.1 (1.2, 6)
Coniferous	8.4 (0.7, 14)	10.1 (0.5, 14)
Understory		
Deciduous	12.2 (1.1, 6)	11.8 (3.4, 6)
Coniferous	8.5 (0.6, 13)	9.2 (1.0, 11)
Forest floor		
Deciduous	17.8 (2.7, 6)	11.5 (1.2, 5)
Coniferous	13.6 (1.3, 13)	9.3 (0.4, 10)
Mineral soil		
Spodosol[a]	3.9 (1.0, 4)	7.7 (0.8, 4)
Inceptisol	3.2 (0.4, 8)	7.7 (0.6, 7)
Ultisol	1.5 (0.2, 5)	4.5 (0.5, 5)

[a] includes Norway Spruce (NS) soil.

closer to those of the vegetation and forest floor than was the case for N_t:P_t; as with N_t:P_t, the ultisols have the lowest N_t:S_t ratio.

The variation in N_t:P_t and N_t:S_t ratios can be attributed to a large number of differences among the forests, including species, overstory ages and biomass distribution, variation between tissues, availability of nutrients, chemical forms of the nutrients, former land use and management, and pedogenic processes. Stoichiometries of masses of nutrients currently accumulating or being lost from the ecosystems are not necessarily related to the stoichiometries of previously accumulated nutrients. Evaluation of ecosystem-specific controls is required to understand relative rates of current nutrient accumulation.

References

Bettany J.R., Stewart J.W.B., Halstead E.H. 1973. Sulfur fractions and carbon, nitrogen, and sulfur relationships in grassland, forest, and associated transitional soils. Soil Sci. Soc. Am. Proc. 37:915–918.

Bohn H.L., McNeal B.L., O'Connor G.A. 1985. Soil Chemistry, 2d Ed. Wiley, New York.

David M.B., Mitchell M.J., Aldcorn D., Harrison R.B. 1989. Analysis of sulfur in soil, plant and sediment materials: sample handling and use of an automated analyzer. Soil Biol. Biochem. 21:119–125.

Ensminger L.E. 1954. Some factors affecting the adsorption of sulfate by Alabama soils. Soil Sci. Soc. Am. Proc. 18:259–264.

Frankland J.C., Lindley D.K., Swift M.J. 1978. A comparison of two methods for the estimation of mycelial biomass in leaf litter. Soil Biol. Biochem. 10:323–333.

Foster N.W., Nicolson J.A., Hazlett P.W. 1989. Temporal variation in nitrate and nutrient cations in drainage waters from a deciduous forest. J. Environ. Qual. 18:238–244.

Fuller R., Driscoll C.T., Lawrence G., Nodvin S.C. 1987. Processes regulating sulphate flux after whole-tree harvesting. Nature (London) 325:707–710.

Homann P.S. 1988. Sulfur Release and Retention in Forest Soil Organic Matter. Ph.D. Dissertation, University of Washington, Seattle.

Homann P.S., Cole D.W. 1990. Sulfur dynamics in decomposing forest litter: relationship to initial concentrations, ambient sulfate, and nitrogen. Soil Biol. Biochem. 22:621–628.

Homann P.S., Mitchell M.J., Van Miegroet H., Cole D.W. 1990 Organic sulfur in throughfall, stemflow, and soil solutions from temperate forests. Can. J. For. Res. 20:1535–1539.

Hurditch W.J., Charley J.L. 1982. Sulfur and nitrogen relationships in coastal forests. In Freney J.R., Galbally I.E. (eds.) Cycling of Carbon, Nitrogen, Sulfur and Phosphorus in Terrestrial and Aquatic Ecosystems. Springer-Verlag, New York, pp. 25–33.

Kinjo T., Pratt P.F. 1971. Nitrate adsorption: II. In competition with chloride, sulfate and phosphate. Soil Sci. Soc. Am. Proc. 35:725–728.

Kowalenko C.G., Lowe L.E. 1975. Mineralization of sulfur from four soils and its relationship to soil carbon, nitrogen and phosphorus. Can. J. Soil Sci. 88:9–14.

Lambert M.J. 1986. Sulphur and nitrogen nutrition and their interactive effects on *Dothistroma* infection in *Pinus radiata*. Can J. For. Res. 16:1055–1062.

Marsh K.B., Tillman R.W., Syers J.K. 1987. Charge relationships of sulfate sorption by soils. Soil Sci. Soc. Am. J. 51:318–323.

Maynard D.G., Stewart J.W.B., Bettany J.R. 1983. Sulfur and nitrogen mineralization in soils compared using two incubation techniques. Soil Biol. Biochem. 15:251–256.

McGill W.B., Christie E.K. 1983. Biogeochemical aspects of nutrient cycle interactions in soils and organisms. In Bolin B., Cook R.B. (eds.) The Major Biogeochemical Cycles and Their Interactions. SCOPE 21. Wiley, New York pp 271–301.

McGill W.B., Cole C.V. 1981. Comparative aspects of cycling of organic C, N, S and P through soil organic matter. Geoderma 26:267–286.

Mitchell M.J., Driscoll C.T., Fuller R.D., David M.B., Likens G.E. 1989. Effect of whole-tree harvesting on the sulfur dynamics of a forest soil. Soil Sci. Soc. Am. J. 53:933–940.

Nodvin S.C., Driscoll C.T., Likens G.E. 1988. Soil processes and sulfate loss at the Hubbard Brook Experimental Forest. Biogeochemistry 5:185–199.

Ohmann L.F., Grigal D.F. 1990. Spatial and temporal patterns of sulfur and nitrogen in wood of trees across the north central United States. Can. J. For. Res. 20:508–513.

Parfitt R.L., Smart R.S.C. 1978. The mechanism of sulfate adsorption on iron oxides. Soil Sci. Soc. Am. J. 42:48–50.

Schoenau J.J., Bettany J.R. 1987. Organic matter leaching as a component of carbon, nitrogen, phosphorus, and sulfur cycles in a forest, grassland, and gleyed soil. Soil Sci. Soc. Am. J. 51:646–651.

Singh B.R. 1984. Sulfate sorption by acid forest soils: 3. Desorption of sulfate from adsorbed surfaces as a function of time, desorbing ion, pH, and amount of adsorption. Soil Sci. 138:346–353.

Singh B.R., Kanehiro Y. 1969. Adsorption of nitrate in amorphous and kaolinitic Hawaiian soils. Soil Sci. Soc. Am. Proc. 33:681–683.

Stevenson F.J. 1986. Cycles of Soil. Wiley, New York.

Turner J., Johnson D.W., Lambert M.J. 1980. Sulphur cycling in a Douglas-fir forest and its modification by nitrogen applications. Acta Oecologica/Oecologia Plantarum 1:27–35.

Turner J., Lambert M.J., Gessel S.P. 1977. Use of foliage sulphate concentrations to predict response to urea application by Douglas-fir. Can. J. For. Res. 7:476–480.

Vitousek P.M., Gosz J.R., Grier C.C., Melillo J.M., Reiners W.A. 1982. A comparative analysis of potential nitrification and nitrate mobility in forest ecosystems. Ecol. Monogr. 52:155–177.

Vitousek P.M., Fahey T., Johnson D.W., Swift M.J. 1988. Element interactions in forest ecosystems: succession, allometry and input-output budgets. Biogeochemistry 5:7–34.

Watwood M.E., Fitzgerald J.W., Swank W.T., Blood E.R. 1988. Factors involved in potential sulfur accumulation in litter and soil from a coastal pine forest. Biogeochemistry 6:3–19.

8. Base Cations

Introduction

D.W. Johnson

The cycles of base cations differ from those of N, P, and S in several important respects. Inorganically driven processes such as weathering, cation exchange, and soil leaching are often considerably more important in base cation cycles than in the cycles of N, P, and S, whereas organic processes such as heterotrophic immobilization and oxidation are typically less important (Figure 8.1). The fact that Ca, K, and Mg exist primarily as cations in solution whereas N, P, and S exist primarily as anions has major implications for the cycling of these nutrients and the effects of acid deposition upon these cycles. The introduction of H^+, whether by atmospheric deposition or by internal processes, will directly impact the fluxes of Ca, K, and Mg via cation exchange or weathering processes. Thus, soil leaching is often of major importance in cation cycles, and many forest ecosystems show a net loss of base cations (in contrast to net gains of N and P; Cole and Rapp 1981).

The importance of aboveground nutrient cycling processes differs greatly among Ca, K, and Mg. Foliar leaching is of major importance in K cycling, often exceeding litterfall in importance. This is because K stays primarily in ionic form within plant tissue, playing a major role in osmoregulation of stomatal opening and closing. In contrast to K, litterfall is normally the

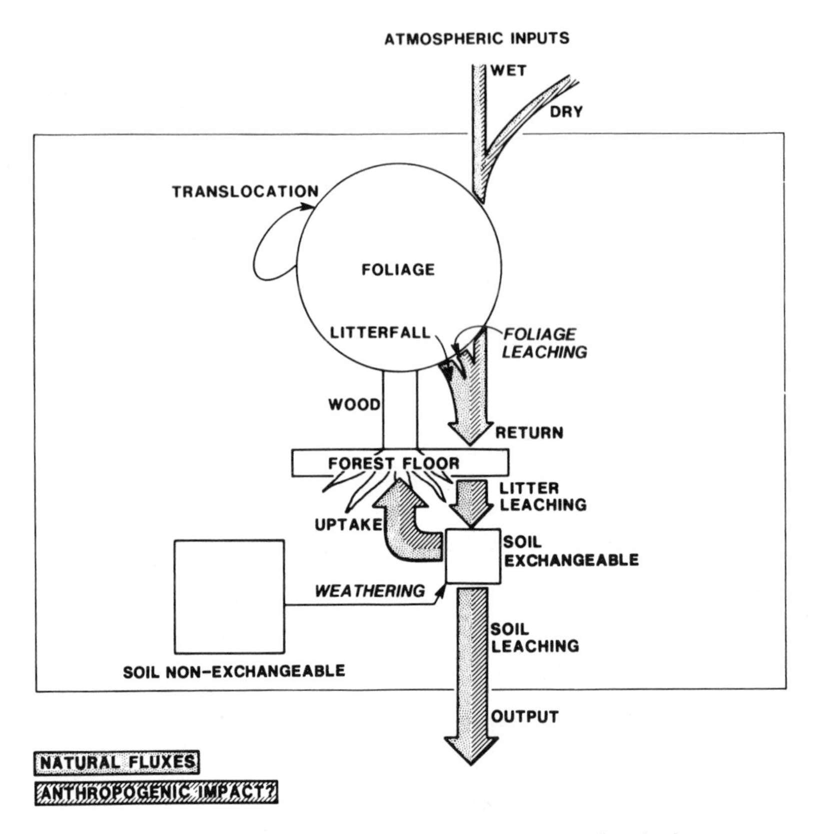

Figure 8.1. Schematic diagram of base cation cycling in forests.

major flux in Ca cycles. Calcium is unique among the macronutrients in that it continues to accumulate in foliage up to the time of senescence, and thus litterfall return of Ca usually exceeds foliar Ca content at maximum leaf biomass (Cole and Rapp 1981). This is because Ca is a major component of permanent plant tissues, such as pectates in cell walls, and has a tendency to precipitate as oxalates. Magnesium is intermediate between K and Ca in terms of its tendency to return from plant foliage compared to litterfall compared to foliar leaching and its role in plant nutrition: it can, like Ca, become part of permanent plant tissue and it also plays a critical role in phototsynthesis within chloroplasts.

Acid deposition can affect base cation cycling in a number of ways. Acid deposition can directly affect the leaching of base cations from both foliage and soil, and a considerable amount of experimental work, both within the Integrated Forest Study (IFS) project and before its inception, has been devoted to estimating the magnitude and importance of acid deposition on both foliar and soil leaching (see reviews by Johnson et al. 1982; Ulrich 1983; Binkley et al. 1989; Johnson and Taylor 1989). The combination of increased foliar and soil leaching may indirectly cause an increase in the uptake of base cations. Specifically, both depletion of foliar base cations and increase in soil solution cation concentrations would tend to cause increased base cation uptake. On the other hand, mobilization of Al in acid soils may impede base cation uptake. If uptake does not keep pace with the increased rate of foliar cation leaching, foliar cation concentrations decrease, perhaps to deficiency levels.

It is now recognized that acid deposition may have two significant effects on soils: (1) an intensity-type effect on soil solution composition and (2) a capacity-type effect on the solid phase of the soil. The most important aspect of the former is the introduction of Al into solution and the most important aspect of the latter is the reduction of base saturation and increase in soil acidity. Anion mobility and cation exchange are the major processes controlling the intensity-type effects (see Chapter 9, this volume), whereas weathering is often the major process controlling the capacity-type effects (see Chapter 10).

In this chapter, we describe the experimental and monitoring activities of the IFS project, which have been devoted to (1) obtaining accurate estimates of base cation inputs, which had not been accomplished previously on a broad regional scale, (2) assessing the effects of acid deposition on foliar leaching, using both field monitoring data and experimental manipulations, and (3) describing the cycles of base cations and assessing the effects of acid deposition on these cycles, with special attention to potentials for Al mobilization and soil acidification.

Atmospheric Deposition and Throughfall Fluxes of Base Cations

H.L. Ragsdale, S.E. Lindberg, G.M. Lovett, and D.A. Schaefer

Introduction

An important problem in estimating atmospheric deposition to plant canopies is the input of base cations. Quantifying deposition of this material, such as calcium and magnesium in soil dust, is necessary to predict the effects of deposition of sulfur and nitrogen oxides on the acidification of forest soils or forested catchments and on forest nutrition. Unlike nitrogen and sulfur, the base cations Ca^{2+}, Mg^{2+}, Na^{+}, and K^{+} have no gaseous phase in the atmosphere and are deposited from the atmosphere by wet deposition, cloud water deposition, and particulate dry deposition, the latter including both coarse particles (>2 μm diameter) and fine aerosols. Coarse particles are deposited on forest surfaces primarily by sedimentation and impaction, while deposition of fine aerosols occurs primarily through turbulent mixing and diffusion through boundary layers of forest surfaces. Wet deposition includes the ions dissolved in rain and snow, and cloud water deposition includes the solutes in cloud and fog droplets that come into direct contact with vegetated surfaces.

Base cations deposited on forest surfaces can move to the forest floor in conjunction with precipitation or by sloughing of tree leaves and branches (litterfall). Precipitation dripping through the canopy is called throughfall (TF), and precipitation collected on tree branch and stem surfaces and flowing down the tree stems is called stemflow (SF). The interaction of precipitation with canopy surfaces may change the base cation chemistry of the water passing through the canopy by washoff or dissolution of previously deposited material and by leaching of cations from the leaf and branch tissues (Parker 1983; Lovett et al. 1985).

Base cation fluxes to the forest floor may be described for specific cations or for the base cations as a group; in this section, total base cation flux is defined as the sum of the Ca^{2+} + Mg^{2+} + K^{+} + Na^{+} fluxes. Individual cations are discussed in terms of both absolute flux (eq ha^{-1} yr^{-1}) and relative contribution (%) to total base cation flux at a site. The fluxes presented here are averaged over 3 years at the LP, CP, and ST sites, for 2 years at the DL, GL, DF, NS, HF, RA, and WF sites, and represent 1-year values at the MS and FS sites (see Chapter 2 for site codes).

Atmospheric Inputs to the Forest Canopy

Dry Deposition

Fluxes of total base cations in coarse particles varied by a factor of 17 across the IFS sites (Table 8.1). The maximum value, 600 eq ha^{-1} yr^{-1}, was measured at the FS site and the minimum value, 36 eq ha^{-1} yr^{-1}, was found at the high-elevation WF site. The median coarse-particle flux at the remaining IFS sites was about 200 eq ha^{-1} yr^{-1}, with a range of 130 to 440 eq ha^{-1} yr^{-1}. These annual fluxes exhibited a weak positive relationship (r = .37, $p < .2$, n = 12) with the total duration of nonrain periods (dry weather) at each site. Specific cations also exhibited considerable variability in coarse-particle flux among sites, ranging over an order of magnitude (Figure 8.2; Table 8.1). While specific cation fluxes were quite variable, Na^{+} and Ca^{2+} were predominant, together constituting approximately 70% or more of the base cation coarse-particle deposition at each site (Figure 8.2, inset). At most sites, Mg^{2+} and K^{+} each accounted for less than 15% of the coarse-particle deposition.

The relative importance of the coarse-particle flux of Na^{+} and Ca^{2+} differed across geographic regions. At sites close to marine influences (DF, RA, FS), 50% or more of the total coarse-particle flux was attributable to Na^{+}. For sites further inland but still having a moderate marine influence (NS, MS, GL, DL), Na^{+} accounted for 20% to 40% of the total coarse-particle flux. Sites isolated from marine influences by the Appalachian mountain range (CP, HF, ST, LP) had relatively low Na^{+} contributions (10%

Table 8.1. Particle Components of Total Dry Deposition of Base Cations[a]

IFS	Coarse Particles					Fine Aerosols				Total Dry Deposition	
Site	K^+	Na^+	Ca^{2+}	Mg^{2+}	Sum of Base Cations	K^+	Na^+	Ca^{2+}	Mg^{2+}	Sum of Base Cations	Sum of Base Cations
LP	9	6.9	180	24	220	0.38	0.67	4.2	0.86	6.1	230
DL	78	94	130	35	340	15	5	8	2	30	370
GL	15	44	51	17	130	5	7	4	1	17	150
CP	26	14	88	17	150	0.50	0.90	1.7	0.4	3.5	150
FS	10	350	160	74	600	3.7	13	—	—	17	620
DF	24	94	44	17	180	2	5	1	1	9	190
NS	40	73	51	19	180	2.7	—	2	1	5.7	190
MS	13	27	79	27	150	0.5	2.6	1.2	0.8	5	160
HF	8.7	8.1	58	9.6	84	0.1	0.2	0.6	0.2	1.1	85
RA	24	94	44	17	180	—	2	1	1	4	180
ST	11	31	350	47	440	2	4	16	3.2	25	470
WF	9.9	14	3.8	8	36	0.20	1	0.70	0.30	2.2	38

[a] Values are fluxes (eq ha^{-1} yr^{-1} averaged over 2 years except LP, CP, and ST, 3-year means; MS and FS, 1-year sums.

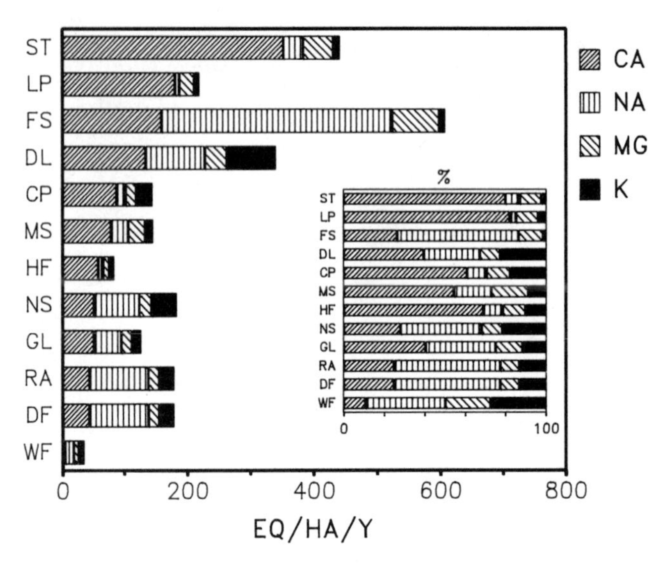

Figure 8.2. Annual mean coarse particle dry deposition rates of cations at IFS sites 1986–1989. Fluxes are expressed in eq ha^{-1} yr^{-1} and presented here as average of two annual periods, except for LP, CP, and ST, which are 3-year averages and except for MS and FS, which are 1-year sums. Sites are ranked from greatest to least calcium flux. Order of remaining cations is based on decreasing magnitude of flux. Inset graph uses same order of sites and shows percent composition by specific cations based on total base cation flux at each site.

or less) to the coarse-particle flux of total base cations. The WF site is somewhat anomalous, in that Na^+ constituted 39% of the coarse-particle flux of base cations although the site is quite distant from the ocean. Two of the southeastern sites, LP and ST, are influenced by wind-blown calcareous soils, and here Ca^{2+} represented about 80% of the coarse-particle flux.

Fine-aerosol dry deposition fluxes of total base cations for the IFS sites were much lower than the corresponding coarse-particle fluxes, but the relative variability in fine-aerosol fluxes was greater (see Table 8.1). Total fine-aerosol deposition of base cations varied by a factor of 30 (from 1 to 30 eq ha^{-1} yr^{-1}) among sites. The highest total fluxes of base cations in fine aerosols (~20–30 eq ha^{-1} yr^{-1}) occurred at the GL, FS, ST, DL sites. Substantially lower total deposition of fine aerosols (~1–10 eq ha^{-1} yr^{-1}) was found at the remaining IFS sites. As for coarse-particles, the relative composition of the fine-aerosol cation flux was similar among sites. Sodium and Ca^{2+} comprised 65% or more of the total fine-aerosol flux for 10 of the 12 IFS sites (see Table 8.1). At both the DL and NS sites, however, Ca^{2+} contributed about 30% and K^+ about 50% of the fine-aerosol flux. Potassium was also important at three other sites, GL, FS, and DF, where it represented 20% to 30% of the fine-aerosol flux.

Generally, fine aerosols contributed 10% or less (median, 4%) of the total dry deposition across the IFS sites (Figure 8.3) for total base cations and for each individual cation; coarse-particles accounted for the remainder. The only significant deviation from this generality for specific cations occurred for K^+ at the three southeastern low-elevation pine forests (DL, GL, FS) (see Table 8.1), where fine-aerosol deposition was 20% to 25% of the total dry deposition of K^+.

Wet Deposition

The flux of base cations in precipitation varied by a factor of 7 among IFS sites (Table 8.2). The highest flux, 630 eq ha^{-1} yr^{-1}, occurred at the FS site. High fluxes, 400 to 550 eq ha^{-1} yr^{-1}, were also measured at the DL, DF, and RA sites. At the remaining sites, the total base cation flux in precipitation ranged between 100 and 200 eq ha^{-1} yr^{-1}. These fluxes were not related to the annual precipitation amounts at each site, indicating that fluxes were primarily influenced by the individual base cation concentrations in precipitation. As seen for dry deposition, Na^+ and Ca^{2+} dominated the base cation flux in precipitation at most sites (Figure 8.4). Together, these two cations accounted for about 70% to 80% of the precipitation base cation flux at the IFS sites (Figure 8.4, inset). The annual base cation flux for specific cations across most sites was quite variable, with K^+ and Ca^{2+} ranging over factors of 6 to 7 and Mg^{2+} and Na^+ ranging from 12 to 15 (Table 8.2).

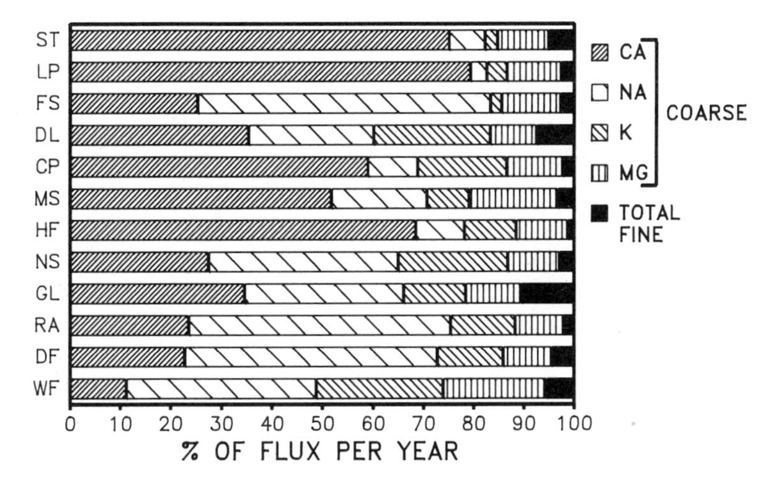

Figure 8.3. Total dry deposition of base cations in particles for IFS sites. Total dry deposition is sum of fine- and coarse-particle deposition for specific cation. Fluxes are presented as percent of total dry deposition base cation flux for each site. Coarse-particle fluxes are shown by specific cation, fine-particle fluxes are shown for sum of base cations.

Table 8.2. Precipitation and Fog/Cloud Components of Total Wet Deposition of Base Cations[a]

IFS Site	Precipitation K^+	Precipitation Na^+	Precipitation Ca^{2+}	Precipitation Mg^{2+}	Sum of Base Cations	Fog/Cloud Deposition K^+	Fog/Cloud Deposition Na^+	Fog/Cloud Deposition Ca^{2+}	Fog/Cloud Deposition Mg^{2+}	Sum of Base Cations
LP	7.3	41	91	26	170	3.5	5.5	9	1.5	20
DL	30	140	200	38	410	—	—	—	—	—
GL	13	73	45	23	150	—	—	—	—	—
CP	7.7	47	53	26	130	—	—	—	—	—
FS	12	240	240	130	620	—	—	—	—	—
DF	38	260	130	50	480	—	—	—	—	—
NS	20	110	48	29	210	—	—	—	—	—
MS	10	66	11	4	91	—	—	—	—	—
HF	32	24	43	12	110	—	—	—	—	—
RA	38	260	130	50	480	—	—	—	—	—
ST	19	48	130	47	240	140	240	300	95	780
WF	9.7	15	93	24	140	4.8	6.4	32	9.7	53

[a] Values are fluxes (eq ha^{-1} yr^{-1} averaged over 2 years; CP, LP, and ST, 3-year means; MS and FS, 1-year sums.

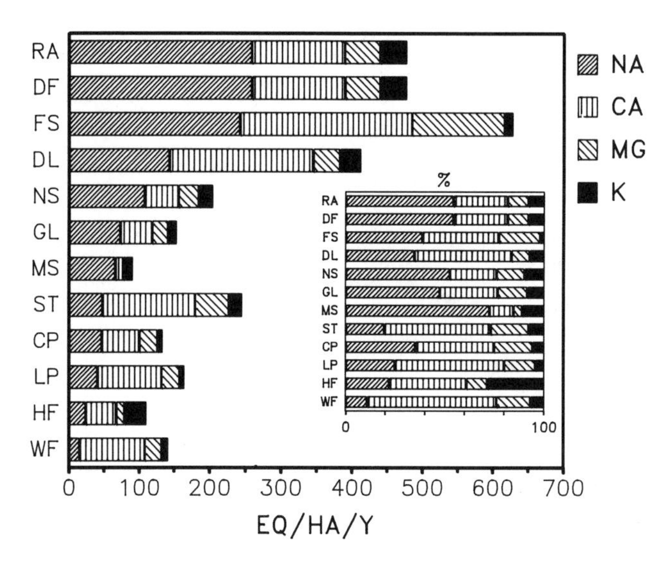

Figure 8.4. Precipitation flux of base cations at IFS sites. Sites are ranked in order of decreasing magnitude of Na^+ flux. Other cations are ranked by decreasing magnitude. Inset graph uses same order of sites and cations but shows percent composition of each cation to total base cation wet deposition.

A geographic pattern similar to that found for dry deposition also existed for the wet deposition flux of Na^+. At sites with a marine influence (NS, MS, DF, RA, GL, FS), Na^+ fluxes represented about 40% to 70% of the total base cation wet deposition flux. The remaining northeastern and southeastern sites exhibited lower relative Na^+ fluxes (25% or less of the total base cation flux in precipitation) and were either inland or isolated by the Appalachian mountain chain or were high-elevation sites.

Droplet Deposition

Droplet deposition from fog/cloud immersion was measurable only at the two high-elevation sites, WF and ST, and the low-elevation, fog-prone LP site. The total base cation flux from fog/cloud interception was relatively low at the LP and WF sites (20 and 53 eq ha^{-1} yr^{-1}, respectively), but was much higher (780 eq ha^{-1} yr^{-1}) at the ST site (Figure 8.5; see Table 8.2). Again, Ca^{2+} and Na^+ were the dominant individual cations in fog/cloud deposition at all three sites (Figure 8.5, inset). Calcium contributed 40% to 60% of the total fog/cloud flux of base cations, and Na^+ accounted for 12% to 31% of the total flux at each site. Together, Na^+ and Ca^{2+} represented about 70% of the fog/cloud flux of base cations. Because cloud droplets are precursors to rainwater, the relative contribution of specific cations to the total base cation flux of each was similar. Cloud water deposition of base cations is clearly an important flux at these high-elevation sites where

Table 8.3. Total Dry and Wet Deposition of Base Cations[a]

IFS Site	Total Dry Deposition K^+	Na^+	Ca^{2+}	Mg^{2+}	Sum of Base Cations	Precipitation K^+	Na^+	Ca^{2+}	Mg^{2+}
LP	9.3	7.5	180	25	220	7.3	41	91	26
DL	93	99	140	37	370	30	140	200	38
GL	20	51	56	18	150	13	73	45	23
CP	27	15	90	17	150	7.7	47	53	26
FS	14	370	160	74	620	12	240	240	130
DF	26	99	45	19	190	38	260	130	50
NS	43	73	53	20	190	20	110	48	29
MS	13	30	81	28	150	10	66	11	4
HF	8.8	8.3	59	8.8	85	32	24	43	12
RA	24	97	44	18	180	38	260	130	50
ST	12	35	370	50	470	19	48	130	47
WF	10	15	4.5	8.3	38	9.7	15	93	24

[a] Values are fluxes (eq ha^{-1} yr^{-1}) averaged over 2 years; LP, CP, and ST, means for 3 years; MS and FS, dry, precipitation and fog/cloud.

it contributed a significant fraction of the total input (Table 8.3). Similar results have been reported at other montane sites (e.g., Lovett et al. 1982; Saxena and Lin 1989).

Total Atmospheric Deposition

The total deposition (wet + dry + cloud water) of base cations ranged from 200 to 1500 eq ha^{-1} yr^{-1} (Table 8.3, Figure 8.6). The high-elevation ST and low-elevation FS sites in the Southeast each received base cation deposition of 1200 eq ha^{-1} yr^{-1} or more. The low-elevation DL, DF, and RA sites received base cation deposition of 660 to 780 eq ha^{-1} yr^{-1}. Deposition at the remaining sites ranged from 200 to 410 eq ha^{-1} yr^{-1} of base cations.

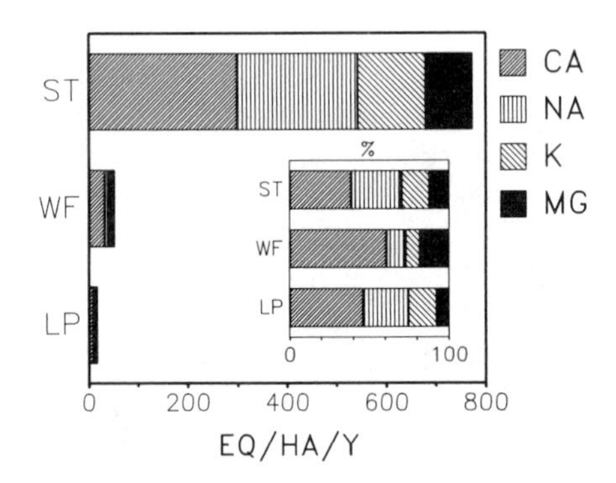

Figure 8.5. Cloud/fog deposition of base cations at IFS sites. Sites are ranked in order of decreasing magnitude of Na^+ flux. Other cations are ranked by decreasing magnitude of flux. Inset graph uses same order of sites and cations but shows percent composition of each cation to base cation total at each site.

Sum of Base Cations	Fog/Cloud Deposition				Sum of Base Cations	Total Deposition (sum of base cations)
	K^+	Na^+	Ca^{2+}	Mg^{2+}		
170	3.5	5.5	9	1.5	20	410
410	—	—	—	—	—	780
150	—	—	—	—	—	300
130	—	—	—	—	—	280
620	—	—	—	—	—	1200
480	—	—	—	—	—	670
210	—	—	—	—	—	390
91	—	—	—	—	—	240
110	—	—	—	—	—	200
480	—	—	—	—	—	660
240	140	240	300	95	780	1500
140	4.8	6.4	32	9.7	53	230

sums for 1 year. Total deposition for specific cation is obtained by summing its fluxes from columns for

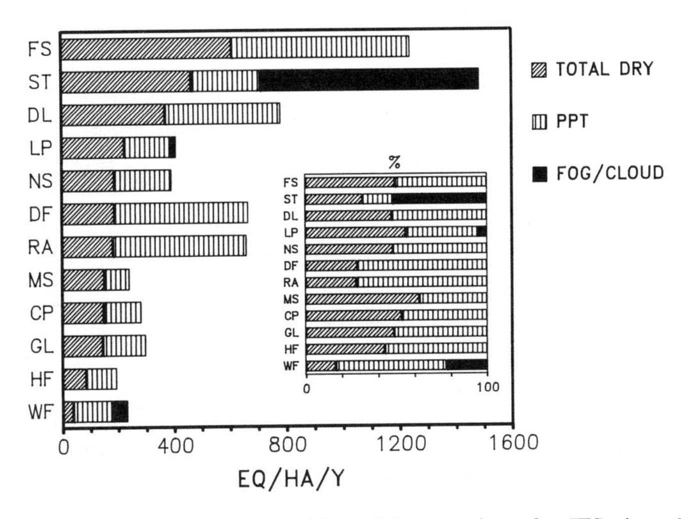

Figure 8.6. Total atmospheric deposition of base cations for IFS sites. Total atmospheric deposition is sum of dry, precipitation, and cloud/fog deposition at site. Each component of total atmospheric deposition is shown as sum of base cations for each site. Inset graph shows percentages of total atmospheric deposition accounted for by dry, precipitation, and cloud/fog fluxes.

The dominant cations in total atmospheric deposition, were Na^+ and Ca^{2+}, together contributed between 70% and 85% of the total base cation atmospheric flux. At 8 of the 11 sites, Na^+ and Ca^{2+} ranked first or second in contribution to the total base cation flux. Potassium ranked third or fourth in relative contribution among all sites and contributed between 5% and 20% of the base cation flux (see Table 8.3).

Wet to Dry Deposition Ratios

The ratio of wet to dry deposition (W:D) reflects the relative importance of these two processes in atmospheric deposition (Figure 8.7). For the total base cation flux, wet and dry deposition processes were generally of equal importance at the low-elevation IFS sites in the eastern United States (W:D ratios ~1, see Figure 8.7). At the northwestern sites (DF and RA), wet deposition consistently exceeded dry with W:D ratios of about 2.5. The two high-elevation sites had differing values: the W:D ratio for ST was 0.53 and that for WF was 3.7. Both montane sites had significant deposition of base cations via cloud water, which was not included in the W:D ratios.

Considering the cations individually, the K^+ W:D ratios at the low-elevation eastern sites were generally less than 1, ranging between 0.29 and 0.86, with the exception of the HF site where the ratio was 3.6. For Ca^{2+}, the ratios were also usually less than 1 at the low-elevation eastern sites, except for DL (1.4) and FS (1.5). In contrast, the W:D ratios for Mg^{2+} for these sites were usually about 1 or greater, except for the MS site (0.14). Sodium W:D ratios were the most variable and showed a strong geographic effect.

Within the southeastern United States, sites remote from a marine influence (LP and CP) showed relatively high W:D ratios of 3.1 to 5.5, while sites with a regional marine influence but geographically removed from the coast (GL and DL) showed only a moderately greater Na^+ deposition from wet processes (W:D = 1.4). For the site with greatest marine influence (FS), a Na^+ W:D ratio less than 1 was found. One interpretation for this trend is that coarse-particle Na^+ generated from sea salt is deposited relatively close to the coast, while fine-aerosol Na^+ can penetrate further inland without dry deposition and is eventually scavenged and deposited in precipitation. The wet climate of the northwestern DF and RA sites led to substantially greater W:D deposition ratios for K^+ (2 W:D), Ca^{2+} and Mg^{2+} (3.5 W:D), and Na^+ (3.6 W:D).

Below-Canopy Fluxes and Sources of Ions in Throughfall and Stemflow

Throughfall and Stemflow

Cation fluxes in throughfall (TF) and stemflow (SF) result from the interception of wet deposition by the forest canopy and boles and the subsequent

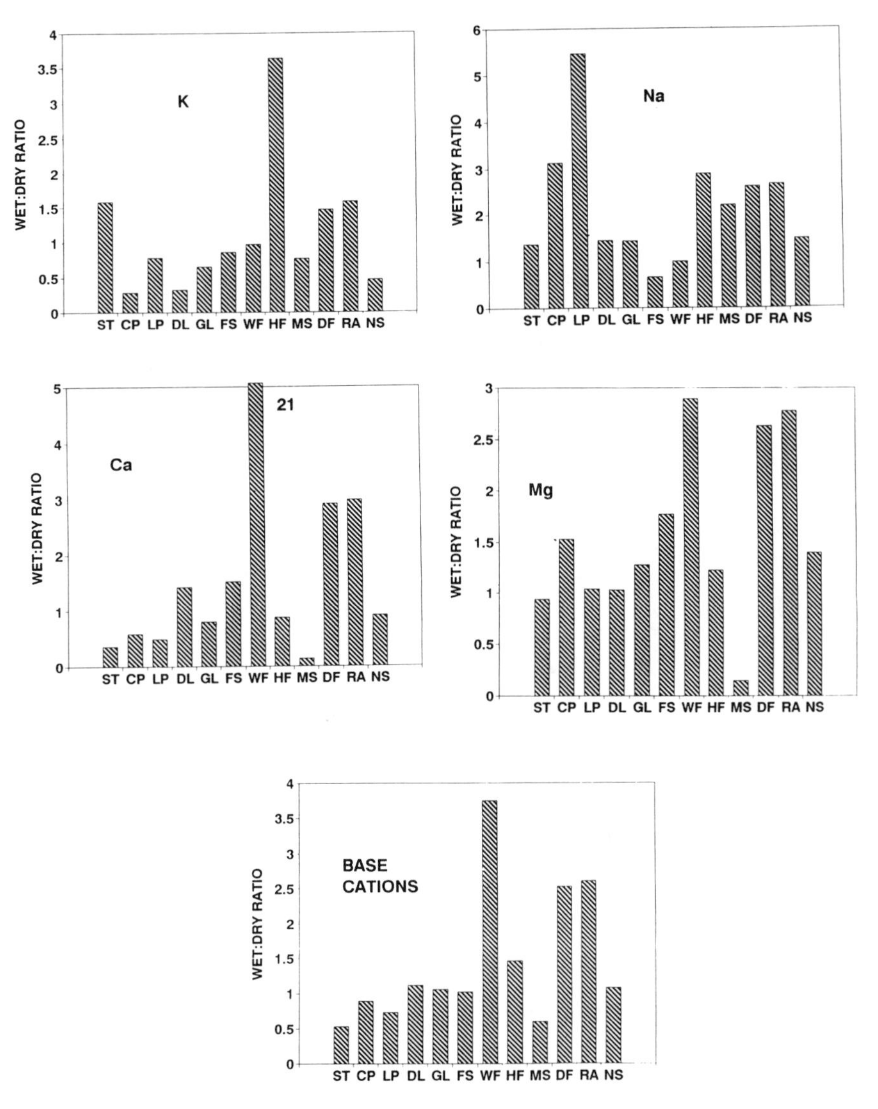

Figure 8.7. Ratio of wet-to-dry deposition fluxes at IFS sites. Wet:dry (W:D) ratios are shown for four cations and sum of base cations.

dissolution of previously deposited material and ions leached from the plant. At most IFS sites, SF fluxes of total base cations were relatively small (<50 eq ha^{-1} yr^{-1}). At two pine sites and one hardwood site (CP, GL, HF), the SF fluxes of base cations were much higher, yielding 120 to 160 eq ha^{-1} yr^{-1} (Table 8.4; Figure 8.8). At these sites the SF hydrologic fluxes ranged from 6% to 10% of the sum of the TF + SF hydrologic fluxes, while at all other sites the SF hydrologic fluxes were less than 2% of TF + SF. Potas-

Table 8.4. Throughfall and Stemflow Components of Total Throughfall Flux of Base Cations[a]

IFS Site	Throughfall Flux K^+	Throughfall Flux Na^+	Throughfall Flux Ca^{2+}	Throughfall Flux Mg^{2+}	Sum of Base Cations	Stemflow Flux K^+	Stemflow Flux Na^+	Stemflow Flux Ca^{2+}	Stemflow Flux Mg^{2+}	Sum of Base Cations	Total Stemflow Plus Throughfall (sum of base cations)
LP	300	61	420	150	930	0.41	0.15	0.90	0.38	1.8	930
DL	200	220	410	120	950	7	4	14	5	30	980
GL	250	93	160	120	620	52	17	38	29	140	760
CP	320	61	290	150	820	85	11	40	25	160	980
FS	57	560	480	250	1400	0.70	5.7	6.8	3.4	17	1400
DF	180	350	220	88	840	18	11	18	5	52	890
NS	320	170	210	110	810	—	—	—	—	—	810
MS	120	110	150	61	440	—	—	—	—	—	440
HF	160	44	170	53	430	95	9.5	13	3.8	120	550
RA	310	400	280	140	1100	7	3	2	1	13	1100
ST	400	290	900	280	1900	12	3.0	2.5	7.1	47	1900
WF	140	46	360	89	640	1.3	0.10	0.90	0.10	2	640

[a] Values are fluxes (eq ha^{-1} yr^{-1}) averaged over 2 years; LP, CP, and ST, 3-year means; MS and FS, 1-year sums.

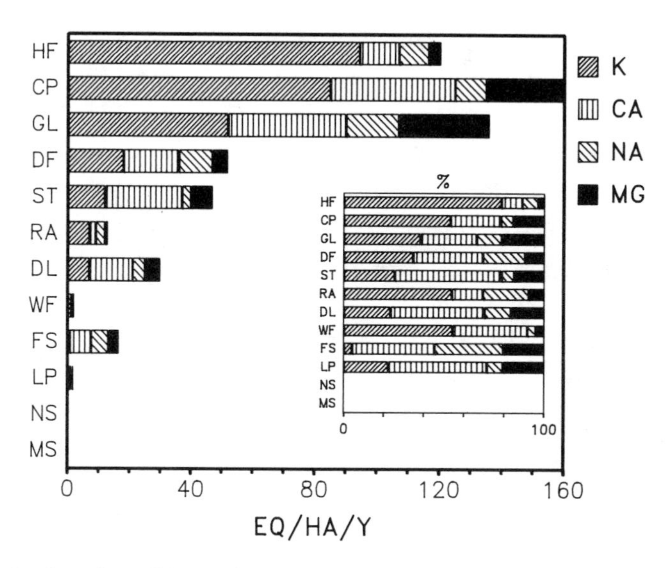

Figure 8.8. Stemflow fluxes of base cations at IFS sites. Inset graph shows four base cations for each site as percentage of sum of base cations.

sium and Ca^{2+} were the dominant base cations in SF, constituting about 70% to 90% of total base cation flux for all sites except the FS site, where Na^{+} and Ca^{2+} were the dominant cations (Figure 8.8, inset).

The variation in magnitude of SF flux was related to the bole density of each forest. The three sites with the highest SF fluxes each have relatively high bole densities compared to the other IFS sites. The HF site is a young, regrowing, mixed deciduous forest dominated by beech and maple, species whose canopy/branch architecture is conducive to stemflow generation. The CP site is a 35-year-old white pine stand with a similar architecture, which has never been thinned. The GL site is a 17-year-old loblolly pine stand planted on 4 × 3 m spacing that has not been thinned and which exhibits the highest stem density of any IFS site (Bondietti 1990). At the remaining IFS forests, the trees are more widely spaced. The DL site is similar to GL in forest type, age, and initial planting on 4 × 3 m spacing, but received a 50% thinning before to the IFS study. Base cation flux in stemflow at the DL site was 30 eq ha^{-1} yr^{-1} as compared to 140 eq ha^{-1} yr^{-1} at the GL site. The relative importance of SF compared to TF also reflects the stem density factor. At most sites, stemflow flux was 3% (DL) or less of the total wet flux of base cations to the forest floor in TF + SF. At the CP, GL, and HF sites, SF contributed from 15% to 20% of the total base cation flux in TF + SF.

Total TF + SF of base cations for the IFS sites varied from 440 eq ha^{-1} yr^{-1} at the MS site to 1900 eq ha^{-1} yr^{-1} at the high-elevation ST site (Table 8.4; Figure 8.9). The three northeastern sites (MS, HF, and WF) exhibited

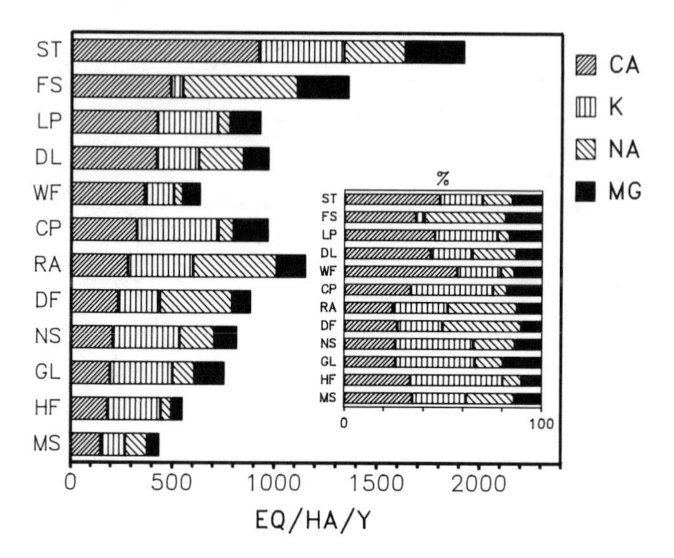

Figure 8.9. Total throughfall flux (throughfall plus stemflow) for IFS sites. Inset graph shows four base cations for each site as percentage of sum of base cations.

TF + SF fluxes less than 650 eq ha^{-1} yr^{-1}, while most sites (LP, GL, DL, CP, DF, NS, RA) had fluxes between about 750 and 1100 eq ha^{-1} yr^{-1}. Base cation fluxes greater than 1400 eq ha^{-1} yr^{-1} were measured at the FS and ST sites.

Potassium and Ca^{2+} were consistently the dominant cations in total TF + SF, together accounting for about 65% to 90% of the total base cation flux (Figure 8.9, inset). This stands in contrast to the dominance of Ca^{2+} and Na^{+} in wet, dry, and cloud water deposition, and the difference reflects the essential and highly mobile nature of K^{+} in plant tissues as opposed to the generally nonessential nature of Na^{+} (Kramer and Kozlowski 1979). The Ca^{2+} flux was exceedingly high (920 eq ha^{-1} yr^{-1}) at the high-elevation ST site when compared to the range of 150 to 490 eq ha^{-1} yr^{-1} for the remaining sites. Sodium fluxes ranged over a factor of 12 (about 45–560 eq ha^{-1} yr^{-1}) among sites. Sodium represented about 7% to 40% of the total base cation TF + SF flux across all sites. The highest proportions of Na^{+} (>20%) were associated with sites having a marine influence (DL, FS, DF, NS, RA, MS). At the inland sites, Na^{+} accounted for less than 15% of the total base cation flux. Magnesium fluxes in TF + SF ranged from about 60 to 290 eq ha^{-1} yr^{-1} and consistently accounted for 10% to 20% of the total base cation flux in TF + SF at each site, tending to be of greater relative importance at the inland sites. In fact, the difference between the relative proportions of Mg^{2+} and Na^{+} in TF + SF could be viewed as indicating the relative importance of the marine influence on ion fluxes to the forest floor at the IFS sites (Figure 8.10).

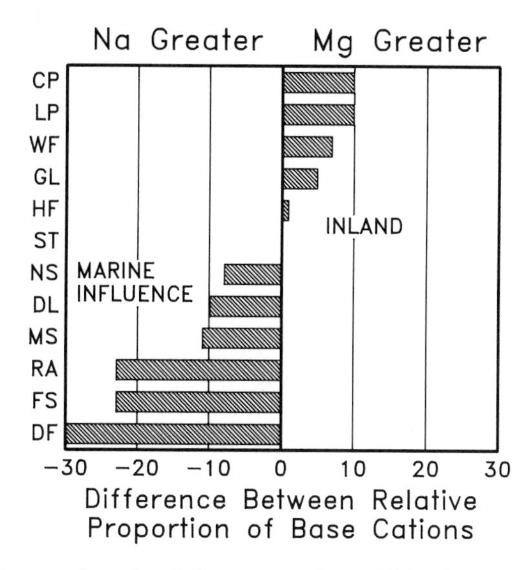

Figure 8.10. Extent of marine influence on throughfall plus stemflow (TF + SF) fluxes at IFS sites. Sites with negative difference (%Mg–%Na) were strongly influenced by proximity to ocean. Sites with positive balance were inland, receiving less marine influence.

Net Canopy Exchange

The canopy of the forest influences the total flux of ions to the forest floor by chemical exchange between surficial deposits (wet and dry) and the internal parts of the plants. This is termed the net canopy exchange (NCE) and is calculated as the difference between TF + SF and total atmospheric deposition (NCE = TF + SF − total atmospheric deposition). If the NCE is positive (e.g., those points above the 1:1 line in Figure 8.11), the canopy has contributed base cations to TF and SF via leaching. Negative values (those points below the 1:1 line) indicate uptake or sorption by the canopy of some portion of the atmospheric deposition. The uncertainty associated with deposition estimates suggests that small positive or negative values of NCE indicate no significant effect of the biotic surfaces on the deposition to the forest floor. In this section we discuss the magnitude of the NCE fluxes for the different sites; a more detailed discussion of the factors regulating the NCE of K^+, Ca^{2+}, and Mg^{2+} is presented in pp. 000–000.

Canopy leaching of base cations ranged from 120 to 690 (eq ha^{-1} yr^{-1}), representing from 14% to 70% of the total base cation wet deposition to the forest floor at the IFS sites in TF + SF, and the NCE of total base cations was positive at all sites (see Figure 8.11), indicating significant foliar leaching sources. Potassium NCE was positive at all sites, indicating leaching, and K^+ was the dominant cation in canopy leaching at most sites, accounting for more than half of the total base cation NCE (Table 8.5; Figure 8.12).

Table 8.5. Net Canopy Exchange (NCE) for Base Cations (NCE = throughfall + stemflow − total deposition)[a]

IFS Site	Net Canopy Exchange				Sum of Base Cations
	K^+	Na^+	Ca^{2+}	Mg^{2+}	
LP	280	6.9	140	99	530
DL	84	−23	78	52	190
GL	270	−14	96	110	460
CP	370	9.7	180	130	690
FS	32	−44	87	47	120
DF	140	−2	60	24	220
NS	260	−6.1	110	63	430
MS	96	14	59	29	200
HF	220	21	83	35	360
RA	260	55	110	77	500
ST	250	−30	120	98	440
WF	120	8.8	230	47	410

[a] Values are fluxes (eq ha^{-1} yr^{-1}) calculated from 2-year averages; LP, CP, ST, 3-year means; and MS and FS, 1-year sums.

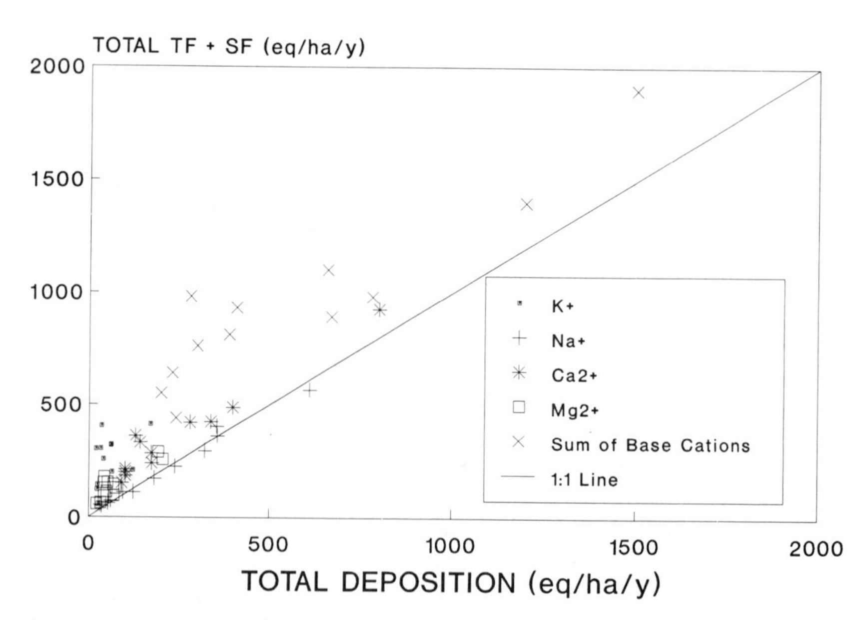

Figure 8.11. Relationship between total annual flux of base cations in throughfall plus stemflow (TF + SF) and total annual atmospheric deposition of base cations at IFS sites. Fluxes of each ion above and below canopy and flux of sum of base cations are plotted.

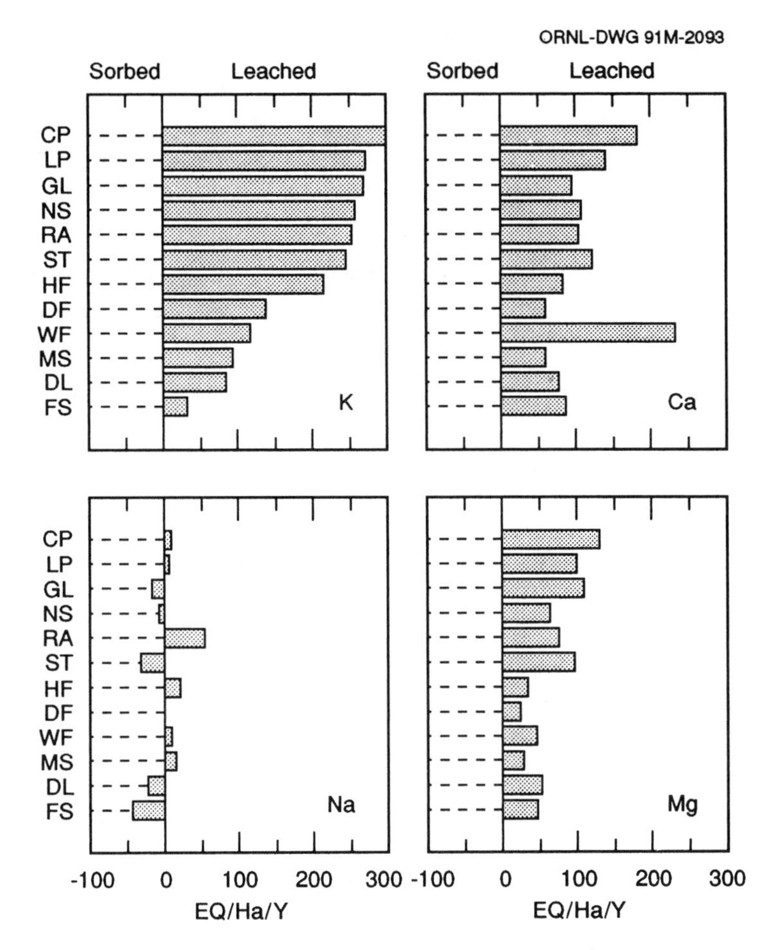

Figure 8.12. Net canopy exchange (NCE = total throughfall flux − total atmospheric deposition) for each IFS site. Negative values indicate sorption of cation in canopy; positive values suggest foliar leaching of cation from canopy.

Potassium leaching was usually more than 70%, and almost always more than 50%, of the total K^+ flux in TF + SF (Figure 8.13).

The positive values of NCE for magnesium and Ca^{2+} also indicated foliar leaching of these cations at all sites, although Mg^{2+} leaching (~30–130 eq ha^{-1} yr^{-1}) was generally less than that of Ca^{2+} (see Figure 8.12). For Mg^{2+}, leaching varied from 20% to 70% and was frequently more than 50% of the total Mg^{2+} TF + SF flux (Figure 8.14). In contrast, the leaching of Ca^{2+} was generally less than 50% of the Ca^{2+} flux in TF + SF (Figure 8.15).

Sodium, unlike the other ions, exhibited both positive and negative values of NCE across the sites, suggesting both leaching and foliar uptake (see Figure 8.12). However, the NCE values for Na^+ were scattered around 0 at

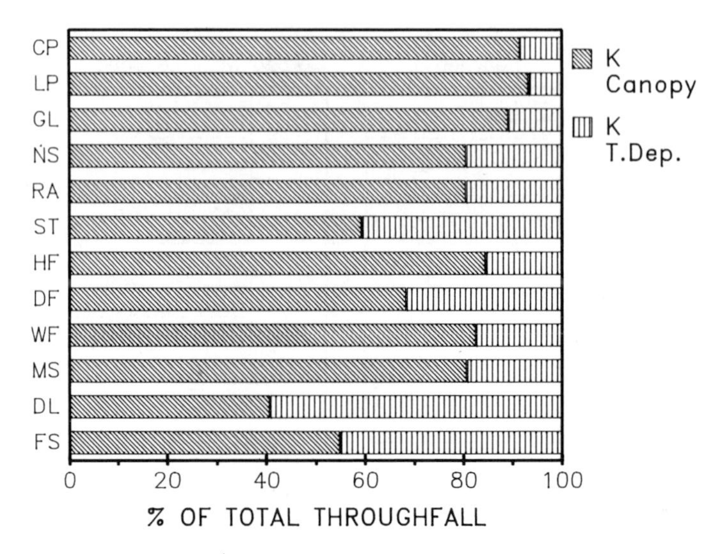

Figure 8.13. Sources of K^+ in throughfall plus stemflow fluxes for IFS sites (estimated as described in text).

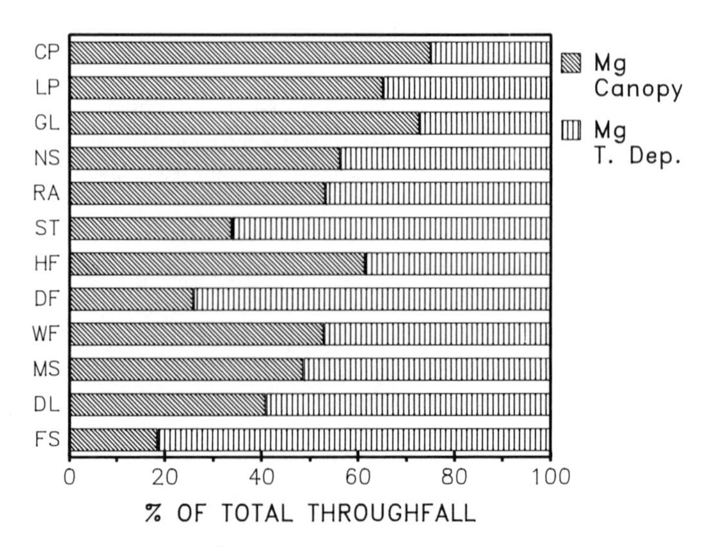

Figure 8.14. Sources of Mg^{2+} in throughfall plus stemflow fluxes for IFS sites (estimated as described in text).

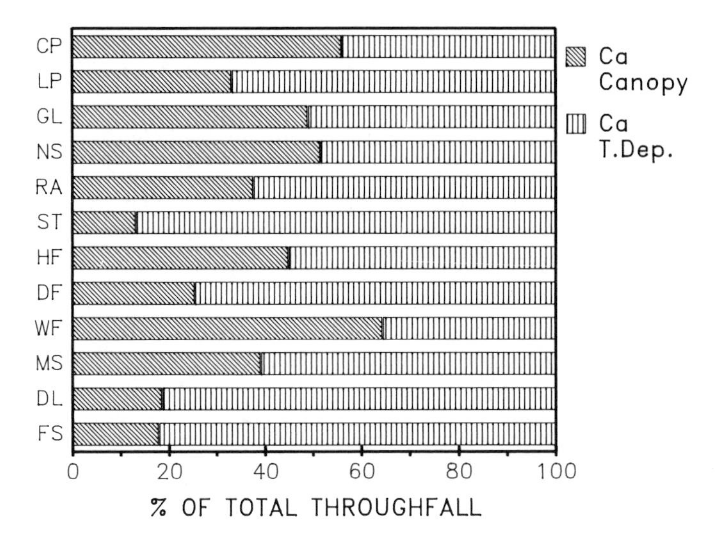

Figure 8.15. Sources of Ca^{2+} in throughfall plus stemflow fluxes for IFS sites (estimated as described in text).

most IFS sites (see Figure 8.11) and were less than 20% of the Na^+ flux in TF + SF (generally <15%; median, 13%), with the exception of the HF site (about 40%). Net canopy exchanges of Na^+ near 0 are expected because concentrations of foliar Na^+ in these canopies are very low (see Appendix) and trees do not require this element (Kramer and Kozlowski 1979). The fact that the NCE values are generally found to be near 0 lends support to the methods used to estimate atmospheric deposition and TF + SF fluxes of base cations in the IFS (Lindberg et al. 1988b; see Chapter 2).

Canopy Interactions of Ca^{2+}, Mg^{2+}, and K^+

G.M. Lovett and D.A. Schaefer

Introduction

Leaching of Ca^{2+}, Mg^{2+}, and K^+ from forest canopies into throughfall and stemflow solutions is an important part of the nutrient cycle of these elements. In this section, we use the term "leaching" to mean the removal by solution of material that is physically or chemically bound to plant surfaces or interiors, thus distinguishing this from the washoff of dry-deposited material that is not bound to the surfaces. Canopy leaching can account for 10 to 80% of the total annual return of these elements from the canopy to the forest floor; the remainder of the return occurs as litterfall (Cole and Rapp 1981; Johnson et al. 1985; Waring and Schlesinger 1985). The annual loss from the canopies resulting from this efflux often represents from 20% to

30% (for Ca^{2+} and Mg^{2+}) and as much as 50% (for K) of the midsummer canopy content of the element (e.g., Eaton et al. 1973; Olson et al. 1981; Johnson et al. 1985).

In recent years, canopy leaching has taken on added significance because of the hypothesis that air pollutants can accelerate the process. Both acid deposition and ozone have been suggested to increase foliar leaching of base cations (K^+, Ca^{2+}, and Mg^{2+} in particular) from trees, but the proposed mechanisms are different. For acid deposition, the mechanism is identified as cation exchange between deposited H^+ and foliar base cations on exchange sites on or in the leaf (Tukey 1980). However, experimental studies have not shown statistically significant effects of acidity on foliar leaching until the pH is decreased to 3.3 or below (Wood and Bormann 1975; Scherbatskoy and Klein 1983; Schier 1987). For ozone, the proposed mechanism is damage to leaf cellular membranes, causing leakage of cell contents that are then leached by subsequent rainfall (Krause et al. 1982). Krause et al. (1982) reported this effect on Norway spruce (*Picea abies*) seedlings, but without statistical analysis of the data. Skeffington and Roberts (1985) could not demonstrate this effect on Scots pine (*Pinus sylvestris*), and Brown and Roberts (1988) attributed at least part of the increase in leaching observed by Krause et al. (1982) to deposition of NO_x formed as an artifact of ozone generation by an electric-discharge ozonator.

Cation exchange and membrane damage are only two of many possible effects of air pollutants on canopy nutrient exchange. Other possible effects include: (1) acute tissue necrosis from severe pollutant exposures, which could cause leakage of cell contents and subsequent leaching from the plant; (2) effects of solution pH on cuticle and cell wall permeability; and (3) indirect effects, such as altering cold tolerance or leaf-surface microbial communities.

In the Integrated Forest Study (IFS), we posed three questions regarding loss of Ca^{2+}, Mg^{2+}, and K^+ from forest canopies:

1. What factors regulate canopy leaching of these elements?
2. Can acid deposition or ozone increase the rate of leaching?
3. Can trees compensate for increased leaching losses by increasing root uptake?

We investigated these questions by a combination of manipulative experiments and observational approaches and summarize that work here.

Cross-Site Comparisons

To provide a context in which to view potential effects of air pollutants, it is important to know what factors, in general, control canopy leaching of Ca^{2+}, Mg^{2+}, and K^+. The IFS sites each measured annual throughfall plus stemflow deposition (TF + SF) and total atmospheric inputs, including wet, dry, and cloud water deposition. We calculated the net canopy exchange

(NCE) of these elements as the difference between TF + SF and total atmospheric deposition. This difference is our best estimate of canopy leaching. For each site, we compiled data on 14 independent variables that might reasonably influence the amount of canopy leaching (Table 8.6) and explored the relationship between these variables and the NCE of Ca^{2+}, Mg^{2+}, and K^{+} using regression analysis. For this analysis, we used primarily 10 sites for which we had dry deposition estimates as well as information on most of the independent variables in Table 8.6. The independent variables included characteristics of acid deposition, canopy biomass, canopy nutrient status, and soil solution nutrient status for Ca^{2+}, Mg^{2+}, and K^{+}.

The results of this analysis were surprising and perplexing. In simple linear regression between the pairs of dependent and independent variables shown in Table 8.6, none of the regressions were statistically significant. The coefficients of determination indicated that in nearly all cases less than 10% of the variation in cation NCE was explained by any one of the independent variables. Apparently, none of the variables we chose exhibits any substantial control over canopy leaching of Ca^{2+}, Mg^{2+}, or K^{+}.

This is a surprising result, given the range of variables examined. We can think of several possible reasons why no significant relationships emerged. First, the data may not be sufficiently accurate to permit us to see underlying relationships. Although dry deposition and ecosystem nutrient content measurements are fraught with uncertainty, all of the data for the IFS project were collected and analyzed by experienced scientists using standardized methods. Although the accuracy may be insufficient, it will be difficult to improve. More likely, perhaps, would be the addition of more sites to the data set, which may permit underlying trends to emerge through the background noise.

Table 8.6. Variables Used in Cross-Site Comparison of Canopy Leaching

Variables
Dependent Variables
Canopy leaching (NCE) of Ca^{2+}, Mg^{2+}, and K^{+} (eq ha^{-1} yr^{-1})
Independent Variables
Precipitation amount (cm yr^{-1})
H^{+} total deposition (eq ha^{-1} yr^{-1})
H^{+} concentration in precipitation (μeq/L)
H^{+} NCE (eq ha^{-1} yr^{-1})
Amount of element in foliage (kg ha^{-1})
Concentration of element in foliage (mg kg^{-1})
Foliar biomass (kg ha^{-1})
Aboveground biomass (kg ha^{-1})
Amount of element in forest floor (kg ha^{-1})
Concentration of element in forest floor (mg kg^{-1})
Soil solution concentration, forest floor (μeq L^{-1})
Soil solution concentration, upper mineral soil (μeq L^{-1})
Exchangeable concentration of element in soil, mean of all horizons (ceq kg^{-1})
Exchangeable amount of element in soil, sum of all horizons (kg ha^{-1})

Second, it is possible that multiple factors may regulate canopy leaching, perhaps with different factors having a dominant role at different sites. With only 10 data points, multiple regression must be used with caution on our data set. We did try, however, to determine whether combinations of two likely factors could explain a significant amount of the variance in NCE of Ca^{2+}, Mg^{2+}, and K^+. We performed multiple regression using all possible pairs of three of the independent variables (precipitation amount, NCE of H^+, and foliar concentration of the element), and found that even so, no statistically significant regressions were generated. In the best case, precipitation amount and NCE of H^+ explained 49% of the variance in Ca^{2+} leaching (i.e., r^2 = .49). The same variables resulted in $r^2 = 0.29$ for Mg^{2+} leaching and $r^2 = 0$ for K^+ leaching.

Third, it is possible that much of the variance is explained by independent variables not considered here; for example, the thickness or chemistry of leaf waxes or the concentrations of peroxides in the air.

Although there were large differences among the sites in the magnitude of the cation NCE fluxes, the relative proportions of Ca^{2+}, Mg^{2+}, and K^+ (i.e., NCE of each ion as a percentage of the sum of all three) were relatively constant among sites: ions contributed on average (with SD in parentheses) CA^{2+}, 29% (11%), Mg^{2+}, 17% (4%), and K^+, 54% (11%). These percentages roughly reflect the average proportions of these ions in foliar tissue at the sites: Ca^{2+}, 38% (9%); Mg^{2+}, 9% (3%); and K^+, 53% (9%). Calcium appears to be slightly depleted and Mg^{2+} slightly enhanced in NCE relative to the foliar chemistry, perhaps indicating differences in the potential for leaching the foliar pools of these two ions.

In conclusion, cross-site analysis revealed how little we know about what regulates foliar leaching of cations. While canopy leaching appears to be understood at the physiological level, at least in general (Tukey 1980), we apparently do not understand or cannot measure what regulates this process quantitatively at the ecosystem level. This clearly points the way for further research on the general regulation of this aspect of nutrient cycling.

Experimental Analysis of the Effects of Ozone and Acid Precipitation

The effects of ozone and acid precipitation on canopy leaching can be determined most effectively by experimental exposure of plants to varying levels of these pollutants. Such experiments are most useful if performed on mature, field-grown trees because of the physiological and morphological differences between leaves of mature trees and seedlings and between field-grown and greenhouse-grown plants. In two series of experiments, we examined pollutant effects on foliar leaching of cations from branches of mature trees growing in natural conditions in the field.

Effects of Ozone and Acid Mist on White Pine and Sugar Maple

We used branch chambers, shown in Figure 8.16 and described in detail by Hubbell and Lovett (1988), to expose upper-canopy branches of mature white

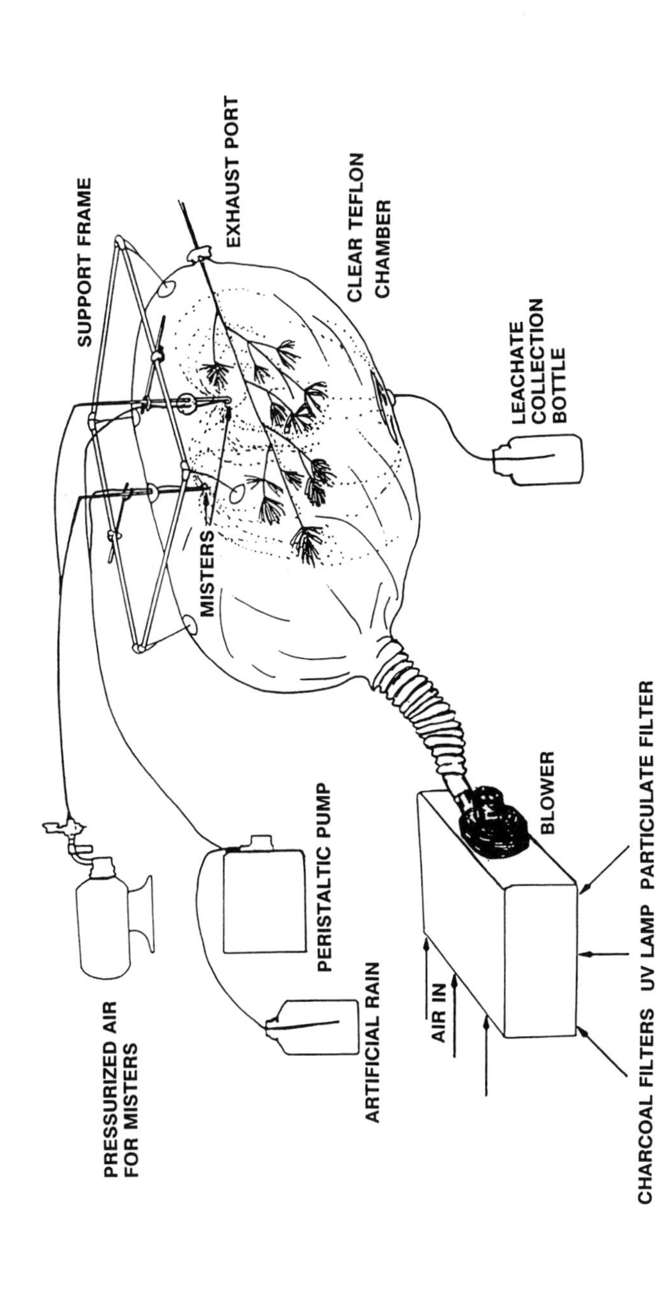

Figure 8.16. Schematic diagram of branch chamber used for ozone/acid mist experiments on sugar maple and white pine. Support frame is suspended from canopy tower; branch remains attached to tree.

pine (*Pinus strobus*) and sugar maple (*Acer saccharum*) to ozone and acid mist and determine the leaching response. The chambers were approximately 0.8 m^3 in volume and enclosed a branch section about 0.5 m long. Ozone was generated using an ultraviolet lamp with a manual voltage control to adjust concentrations. This method of ozone generation does not produce the artifact NO_x described by Brown and Roberts (1988) for electric-discharge ozonators. The air in the chamber was well mixed by the blower, and the air exchange rate (~2 exchanges min^{-1}) was sufficient to maintain the temperature in the chamber within 3°C of ambient. The 3-mil Teflon walls of the chamber transmitted about 75% of incident solar radiation in the photosynthetic wavelengths. The chambers were deployed from the upper part of a 65-ft canopy tower.

The branches were rinsed briefly before each treatment to remove dry-deposited material. Treatments were composed of a 5-h ozone exposure at 25, 70, or 140 ppb, followed by a 1-h leaching with mist acidified to pH 5.0 or 3.8. Thus the treatment sequence mimicked a typical summer day, with elevated ozone concentrations in the afternoon followed by a brief rainshower. Approximately 2500 ml of the mist was applied during the hour, and the solution contained major ions in concentrations approximating the average chemistry of rain in the study area. The experiments were performed at the Cary Arboretum in Millbrook, New York, during the midsummer weeks of 1987 and 1988. The procedures and results have been described in detail by Lovett and Hubbell (1991).

The results of the experiments were very similar for both white pine and sugar maple: In both cases, the 5-h ozone exposures had no effect on leaching of any ions. Increasing mist acidity from pH 5.0 to 3.8 more than doubled the leaching of Ca^{2+} and Mg^{2+} from both species (a highly statistically significant effect) but had no effect on leaching of K^+. For both species, H^+ was released in the pH 5.0 treatment and absorbed in the pH 3.8 treatment, indicating a buffering of the incident precipitation acidity by the foliage. Figures 8.17 and 8.18 show the treatment means for white pine and sugar maple, respectively.

Table 8.7 compares the sugar maple and white pine results for all ions in the two pH treatments. The loss of both cations and anions was higher from sugar maple than from white pine. The anion deficit in Table 8.7 indicates that, in some cases, unmeasured anions were released into the leachate solution. These were probably organic acid anions, which represent part of the pH buffering capacity of these canopies. In all cases except the white pine pH 3.8 treatment, these organic anions provided a major portion of the leached negative charge.

We concluded from these experiments that short-term ozone exposure, in the absence of visible leaf damage, does not affect foliar leaching. However, acid wet deposition, at acidity levels commonly experienced in the eastern United States and Europe, can have a strong effect on canopy exchange of Ca^{2+}, Mg^{2+}, and H^+, but not K^+. Perhaps the most surprising result is that

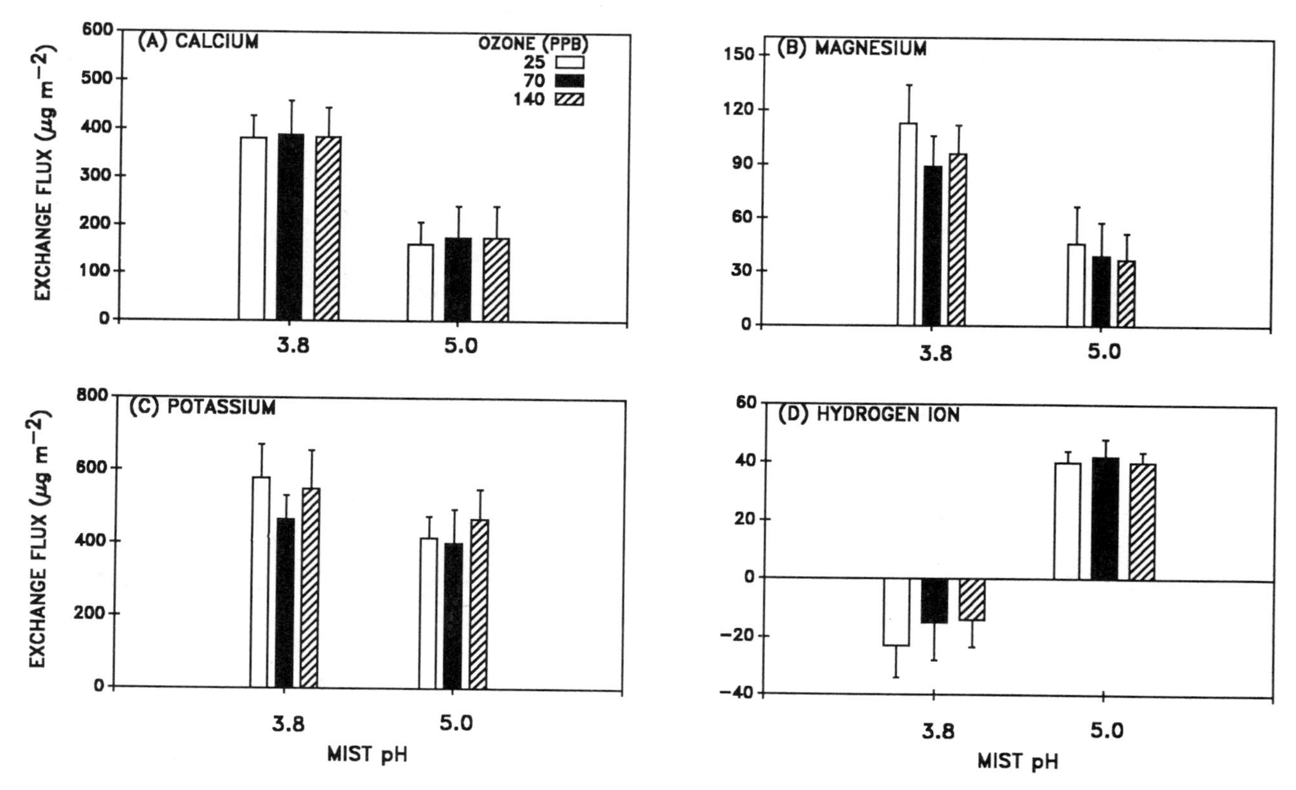

Figure 8.17. Group means and standard error bars for ozone and acid mist treatments on white pine.

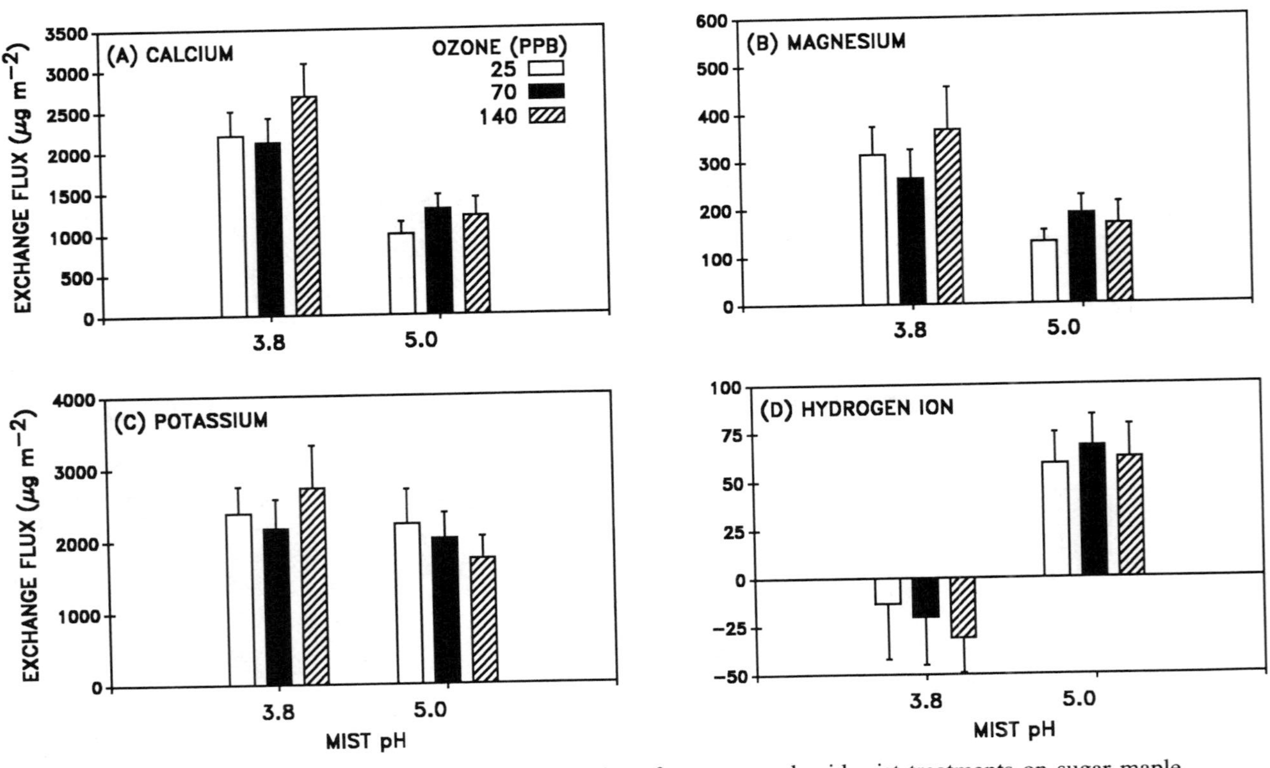

Figure 8.18. Group means and standard error bars for ozone and acid mist treatments on sugar maple.

Table 8.7. Mean Amounts of Canopy Release (positive numbers) or Uptake (negative numbers) from White Pine and Sugar Maple Branches at Two pH Levels[a]

	White Pine		Sugar Maple	
	pH 3.8	pH 5.0	pH 3.8	pH 5.0
Ca^{2+}	19	8	117	60
Mg^{2+}	8	3	26	13
K^+	14	11	62	51
Na^+	8	5	6	7
NH_4^+	8	3	−3	−8
H^+	−17	41	−21	63
Cation Sum	40	71	187	186
SO_4^{2-}	21	21	57	75
NO_3^-	15	11	52	53
Cl^-	7	4	1	3
Anion Sum	43	36	110	131
Anion Deficit	−3	35	77	55

[a] Data expressed as microequivalents per square meter of foliar surface area (one-sided) for 1-h misting treatments.

these two very different species responded similarly to the pollutants, despite having different "baseline" levels of leaching. Both species showed no response to ozone but had Ca^{2+}, Mg^{2+}, and H^+ responses to changes in acidity. In both cases, the Ca^{2+} and Mg^{2+} leaching approximately doubled in response to a pH decrease from 5.0 to 3.8.

Effects of Solution pH on Red Spruce and Loblolly Pine

A series of experiments were conducted in 1986 and 1987 to determine canopy exchange rates in red spruce (*Picea rubens*) and loblolly pine (*Pinus taeda*) and to examine the factors that control exchange. It was anticipated that the incomplete wetting of canopy surfaces during natural or artificial droplet exposure could limit ionic exchange rates. In these experiments, intact foliage was completely immersed in aqueous solutions to determine the maximal exchange rates. Factors controlled in these experiments were foliar immersion time (to compare exchange rates among ions), immersion solution acidity (to determine the acidity effect on exchange rates), and immersion solution concentrations (to determine the concentration gradient effect on exchange rates). Effects of immersion solution acidity on foliar cation release are considered here.

Live, intact branches of loblolly pine and red spruce were placed in polyethylene "Ziploc" bags with 100 ml of immersion solution. After 2 min of gentle agitation, the solution was poured off and another 100 ml was added. Four sequential 2-min immersions were performed on each branch, with a final immersion of as long as 30 min. After the final immersion, branches were severed from the plant and returned to the laboratory for drying and weighing. Foliar surface areas were determined from dry weight/area relationships. In this analysis, only the results from varying the immersion solution pH (values: pH 5.0, 4.3, 4.0, 3.3, and 3.0) are reported.

Immersion solutions were analyzed for pH and major inorganic ions. In this analysis, exchanges during the first 2-min immersion will not be reported, as washoff of dry deposition is expected to dominate during that period. Net foliar fluxes during subsequent immersions are summed and expressed as nanoequivalents per centimeter squared of foliar surface. A final 30-min immersion was not included in all experiments because more than 90% of the cation fluxes took place during the 2-min rinses.

Net foliar fluxes of potassium and calcium from loblolly pine foliage showed a positive response to immersion solution acidity (Figure 8.19). In this (and later) figures, the immersion solution acidity is shown as both the H^+ ion concentration and as pH. This figure shows that the net foliar release of calcium increased from about 5 neq cm^{-2} foliage at pH 4.0 and higher to more than 15 neq cm^{-2} foliage at pH 3.3 and lower. Potassium release from foliage showed a similar pattern (Figure 8.19). For red spruce, there was a two- to fourfold increase of Ca and Mg loss from foliage as immersion solution pH decreased from 4.0 to 3.3, with no additional flux increase from pH 3.3 to 3.0 (Figure 8.20). It could not be determined from these experiments whether the increase occurred gradually over the pH 4.0 to 3.3 range or if there was a "threshold" pH response.

Not all cation fluxes responded to immersion solution acidity. Figure 8.21 shows that red spruce net foliar release of potassium and sodium were in-

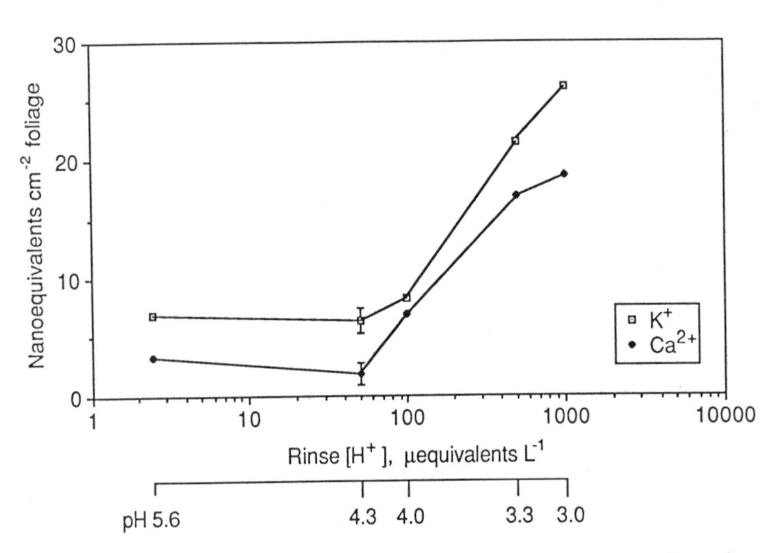

Figure 8.19. Net flux of potassium and calcium from loblolly pine foliage (in nanoequivalents of cations cm^{-2} foliar surface) during immersion in water ranging from pH 5.6 to 3.0.

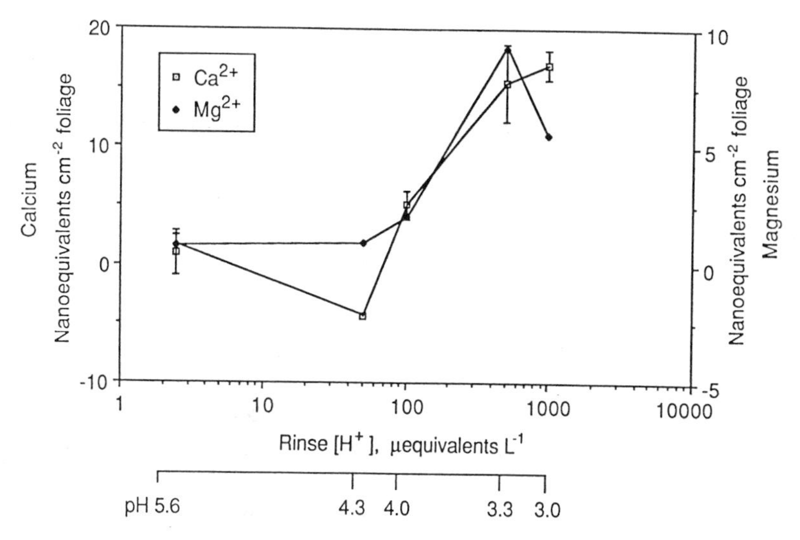

Figure 8.20. Net flux of magnesium and calcium from red spruce foliage (in nanoequivalents of cations cm^{-2} foliar surface) during immersion in water ranging from pH 5.6 to 3.0.

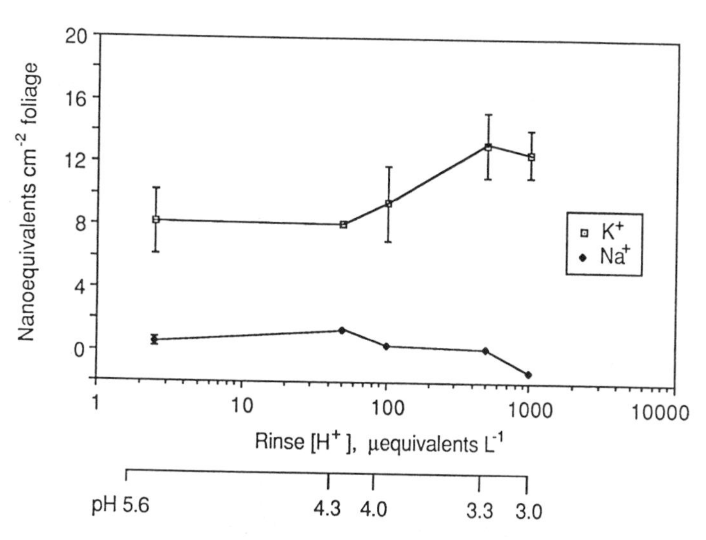

Figure 8.21. Net flux of potassium and sodium from red spruce foliage (in nanoequivalents of cations cm^{-2} foliar surface) during immersion in water ranging from pH 5.6 to 3.0. Negative flux indicates that cations are retained by foliage.

dependent of immersion solution pH over the pH range of 5.6 to 3.0. In general, divalent cation fluxes responded to immersion solution acidity, but monovalent cation fluxes did not.

Figure 8.22 shows that strong acid (measured as pH) retention by both loblolly pine and red spruce foliage increased in those experiments where the immersion solution acidity was the highest and where cation release was the greatest. This figure shows a strong linear relationship between cation release and strong acid retention in loblolly pine and red spruce foliage r^2 = .8 and .65, respectively). The slope of that relationship indicates that for each equivalent of strong acid retained, 0.8 equivalent of cations was released by both loblolly pine and red spruce. This is evidence for foliar cation exchange, but only when the foliage is exposed to water of pH less than 4.0. This pH is below the mean value for rain at all IFS sites, but high-elevation forests in eastern North America are frequently exposed to cloud water with pH substantially less than 4.0. Among the IFS sites, these data indicate that foliar cation exchange is most likely to be observed at the ST and WF sites.

It has been shown that foliar exposure to water of pH 3.3 (or below) can stimulate the release of some 30 neq of cations per centimeter squared foliar surface. Based on approximate leaf area indices of 5 and 10 for loblolly pine and red spruce foliage, respectively, it can be calculated that in one "low-pH" exposure, 150 or 300 eq ha^{-1} of cations, respectively, could be released from foliage. These values are comparable to the total annual "net canopy effect" of divalent cations seen at most IFS sites, exposed to surface

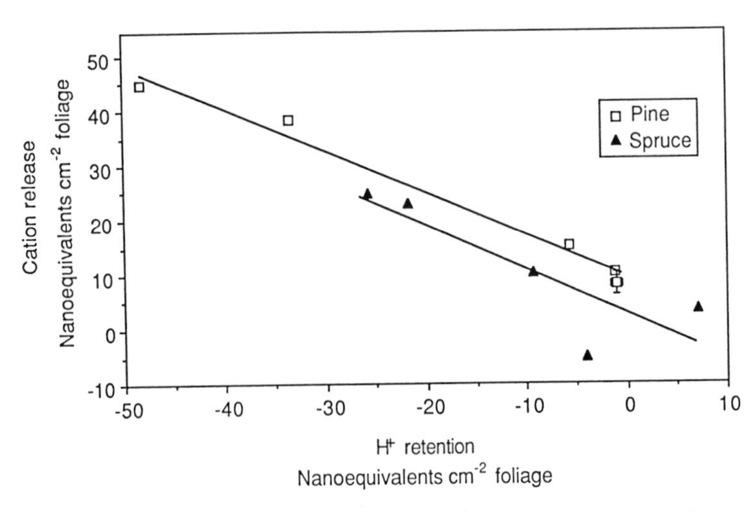

Figure 8.22. Net release of cations from loblolly pine and red spruce foliage versus strong acid retention (in nanoequivalents of cations cm^{-2} foliar surface) from water ranging from pH 5.6 to 3.0. Strong acid retention and cation release showed linear relationship with slope of 0.8, suggesting that cation exchange took place below pH 4.0.

(throughfall) water typically of pH 4.3 to 4.5. If these forests were commonly exposed to water of pH 3.3 or less, it would be reasonable to expect much larger net canopy effects for the divalent cations.

Comparison of Experimental Results

While the two sets of experiments just described used very different methods, both determined the effect of solution pH on cation loss from branches of mature trees. To compare the results of the two experiments quantitatively, we normalized all the leaching data by dividing the leaching flux at each pH level by the flux in the "clean rain" control, pH 5.6 in the red spruce-loblolly pine experiments or pH 5.0 in the sugar maple-white pine experiments. Thus, each determination of cation leaching was expressed as the increase relative to the control in that experiment. Although the controls are at different pH in the two experiments, the difference in acid concentration between the controls (10 compared to 2.6 μeq L^{-1} of H^+) is insignificant compared to the treatments of as much as 500 μeq L^{-1} (pH 3.3). Data from the pH 3.0 treatment in the red spruce-loblolly pine experiments were not used in the analysis, because previous research has shown foliar necrosis in other trees at this acid level (however, tissue damage was not observable in these experiments).

Regression of normalized Ca^{2+} and Mg^{2+} efflux versus solution acidity revealed a significant linear relationship with an r^2 of .69 (Figure 8.23). Calcium and magnesium behave similarly and are plotted together in Figure 8.23, but no significant relationship was seen for K^+. The slope of the

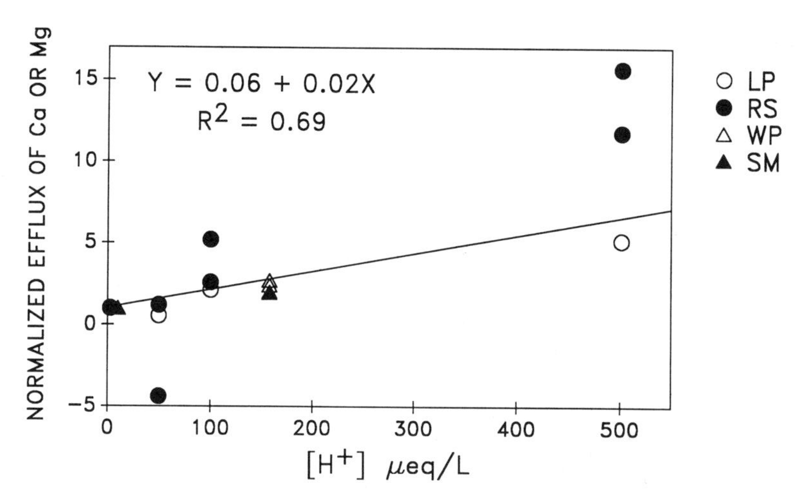

Figure 8.23. Normalized increase in leaching of Ca^{2+} and Mg^{2+} from branches of sugar maple, white pine, loblolly pine, and red spruce as function of acidity of leaching solution. Leaching fluxes are normalized to "clean rain" controls (pH 5.6 or 5.0) for each experiment.

regression line is 0.02, indicating that for all four species taken together, loss of Ca^{2+} or Mg^{2+} increases by 2 normalized units with every 100 μeq L^{-1} increase in the strong acidity of the leaching solution. As a caveat to this analysis, we note the lack of data between 200 and 450 μeq L^{-1} and the fact that the regression is heavily influenced by the points at 500 μeq L^{-1}.

The fact that the results from two series of experiments on two divalent cations and four different tree species can be adequately described with a single line indicates the strength and generality of the effect, in that the trend can be readily seen despite the experimental, chemical, and species-specific "noise." We look forward to confirmation of this trend from other experiments on mature trees, but note that measures should be taken to minimize the influence of dry deposition on the measured cation release.

Foliar Leaching and Compensatory Root Uptake

The strong effects of acid mist on foliar leaching of Ca^{2+} and Mg^{2+} prompted us to ask whether increased root uptake can compensate for losses of nutrients from the foliage. The question is relevant from two points of view. From the perspective of the plant, lack of compensatory uptake could result in a decline in the foliar nutrient pool and thus affect the nutrition of the leaf cells. From the ecosystem perspective, lack of compensatory uptake would indicate an acidification of the foliage, and subsequently the litterfall, in effect creating a delay in the deposition of the acidity to the soil. At the same time, Ca^{2+} and Mg^{2+} return would be increased in throughfall solution and decreased in litterfall. If compensatory uptake did occur, increased hydrogen ion expulsion from the roots would balance the increased cation uptake, in effect resulting in a shunting of the acid deposition from the canopy to the rooting zone and decreasing the potential for neutralization of the acidity in the litter layer of the forest floor (Lovett et al. 1985).

Testing for a decrease in foliar nutrients after exposure to acid deposition would probably not answer this question, because the total foliar pool of cations is large relative to the exchangeable pool, which is located in the foliar apoplasm and provides the nutrients for cell growth. We know of no way to measure this apoplastic pool accurately, so our approach was to examine the relative magnitude of nutrient inputs (uptake) and outputs (foliar leaching) in whole plants under different foliar leaching and cation availability treatments.

In a greenhouse, we grew white pine seedlings in sand culture and calculated their nutrient uptake from measurements of inputs and outputs (in drainage water) from the sand-filled pots. Twenty-four plants were divided into four treatment groups, which received either high or low cation concentrations in nutrient solution and either 3 h wk^{-1} of foliar leaching or no leaching at all. Six sand-filled pots without plants were used as controls. The nitrogen content of the nutrient solution was high and identical for both nutrient solutions. The nitrogen:cation ratios in the high-cation treatment were optimal for pines (Ingestad 1979), whereas in the low-cation treatment, the ratios were approximately one-sixth the optimal ratio. The results were analyzed statistically using an analysis of variance procedure.

Our results included the following:

1. Plants in the high-nutrient treatment took up more Ca^{2+}, Mg^{2+}, and K^+ than those in the low-nutrient treatment, and the cation uptake was linearly proportional to N uptake, indicating incorporation of the cations into tissue.
2. Cation leaching was small relative to cation uptake (e.g., Figure 8.24). Uptake: leaching ratios were generally 10 or greater.
3. The leached plants did not take up significantly more cations than the unleached plants, probably because the leaching rates were small compared to the uptake rates (Figure 8.24).

Our overall impression from these results is that nutrient availability controlled root uptake, and that there was no tight linkage between foliar leaching and root uptake. In additional studies we are investigating this linkage under the conditions of low nutrient availability and high foliar leaching "intensity" characteristic of high-elevation forests.

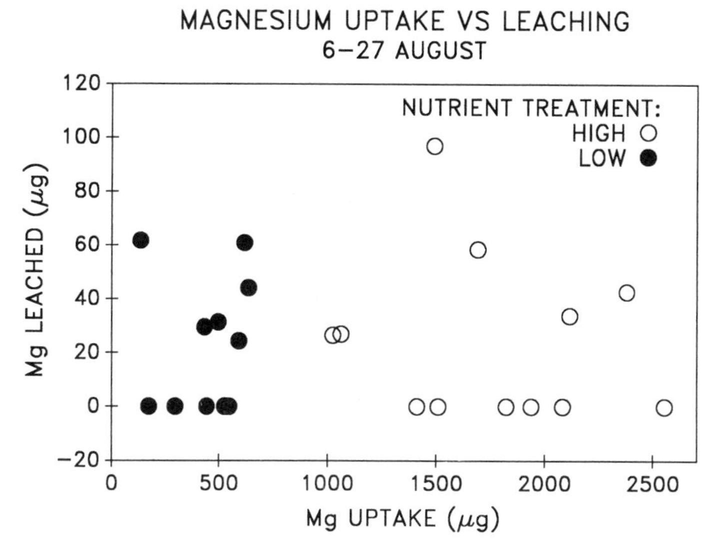

Figure 8.24. Amounts of Mg leached from foliage and taken up by roots of white pine seedlings during 6–27 August 1988. Points showing zero leaching represent plants that were not leached. Qualitatively similar results were obtained for Ca^{2+} and K^+.

Sequential Throughfall Analysis

Plant-surface rinsing experiments indicate that the removal of dry-deposited compounds from foliar surfaces occurs very rapidly (Lindberg and Lovett 1983; Reiners and Olson 1984; Lindberg et al. 1988b; Schaefer et al. 1988a). In contrast, certain compounds present in throughfall (such as potassium ions, sugars, and amino acids) are primarily extracted from the plant tissues themselves, and this extraction can continue as long as the surfaces remain wet (Tukey 1980; Parker 1983; Lovett and Lindberg 1984; Schaefer and Reiners 1990). Sequential ionic fluxes in net throughfall (NTF = throughfall flux − rain flux) can distinguish between rapid dry deposition washoff and the more gradual removal of ions (or retention, for N and H) by the canopy itself (Parker 1983; Johannes et al. 1986; Ivens et al. 1988; Schaefer and Reiners 1990).

Several factors have been identified that can affect net TF fluxes. These include the duration of rain and of the antecedent dry period (Lovett and Lindberg 1984), total rainfall amount (Lindberg et al. 1979), rainfall intensity and surface water/plant concentration gradients (Parker 1983; Reiners and Olson 1984; Reiners et al. 1986; Schaefer et al. 1988a; Schaefer and Reiners 1990), and rainfall acidity (Fairfax and Lepp 1975; Wood and Bormann 1975; Abrahamsen et al. 1977; Reed and Tukey 1978a, 1978b; Abrahamsen et al. 1980; Hindawi et al. 1980; Evans et al. 1981; Cronan and Reiners 1983; Scherbatskoy and Klein 1983; Lovett et al. 1985; Parker et al. 1987; Schaefer et al. 1988a, 1988b; Schaefer and Reiners 1990).

In the experiments reported here, two automatic wet-only samplers with fraction collectors were used in the loblolly pine forest adjacent to the IFS LP site. One was placed in a canopy clearing to collect rain (R) and the other under a particularly dense area of the canopy to collect throughfall (TF). The paired R and TF event subsamples were returned to the laboratory after the rain ended and analyzed for pH, conductivity, and inorganic ions by ion chromatography and atomic absorption spectroscopy (Lovett et al. 1985). The resulting concentration data were expressed as ionic fluxes per unit area, and net TF fluxes (NTF) were determined by differences between TF and R fluxes. All major inorganic ions were examined, but only the results for calcium, magnesium, and potassium ions will be treated here. Eight events of different storm depths, intensities, and antecedent dry deposition "loading" periods were collected and analyzed during 1986 and 1987 (Table 8.8).

Patterns of NTF in eight rainfall events were used to develop ion-specific NTF models. Prestorm conditions were used to scale NTF totals for the conditions before particular rain events. The models were then tested with independent NTF data for all the rain events of an entire growing season from a nearby site in the same loblolly pine forest.

The observed ionic flux patterns are expressed as NTF per rain event subsample (in microequivalents per meter squared) along an axis of cu-

Table 8.8. Characteristics of Sequentially Sampled Rain Events Used in IFS Study[a]

Date	SEQ[a]	IFS[a]	CM[a]	PECM[a]	ADH[a]
10/25/86	10[b]	49	5.3	6.9	253
1/19/87	11	65	7.4	0.7	61
2/17/87	12	70&71	3.0	0.9	285
5/5/87	13	87	2.2	0.2	59
5/18/87	14	88	1.8	2.2	305
6/22/87	16[c]	97	1.9	0.8	19
6/23/87	17	98	4.6	1.9	23
8/13/87	18	106	2.8	1.1	66

[a] SEQ, number of sequential event; IFS, number of same event as sampled in adjacent loblolly pine forest, CM, event rain amount, cm; PECM, previous event rain amount, cm; ADH, dry period before rain event, hr.
[b] Nine earlier events were also collected, but only one sequential sampler was available, and net throughfall fluxes were not determined.
[c] Event 15 was not used due to equipment malfunction.

mulative rainfall depth in millimeters (Figures 8.25–8.27). These figures show that subsample fluxes for these cations generally diminished with rainfall depth, showing that ionic pools accessible to throughfall become depleted during the event. This explains why the average concentrations of these ions in event-sampled throughfall (closely related to NTF fluxes) will generally be lower in events of greater total rainfall depth, as has been noted by Lindberg et al. (1979) and Parker (1983).

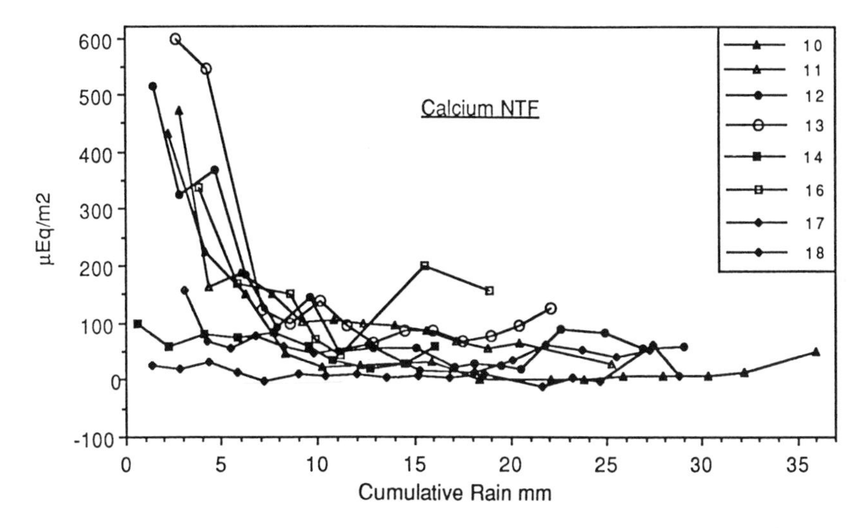

Figure 8.25. Net throughfall fluxes (NTF) of calcium (in microequivalents m^{-2} ground area) in each sequential throughfall subsample versus cumulative millimeters of rainfall in eight rain events in loblolly pine forest in Oak Ridge, Tennessee. Rain event characteristics are described in Table 8.8.

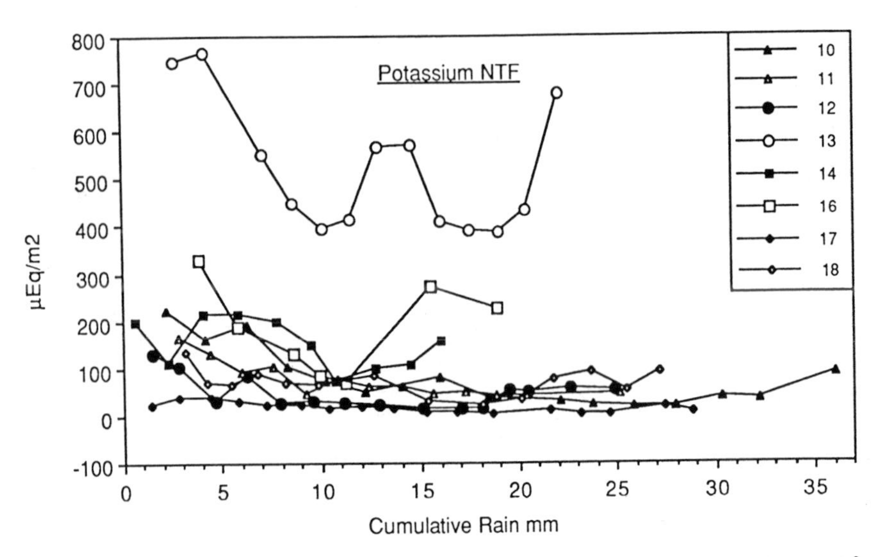

Figure 8.26. Net throughfall fluxes (NTF) of potassium (in microequivalents m^{-2} ground area) in each sequential throughfall subsample versus cumulative millimeter of rainfall in eight rain events in loblolly pine forest in Oak Ridge, Tennessee. Rain event characteristics are described in Table 8.8.

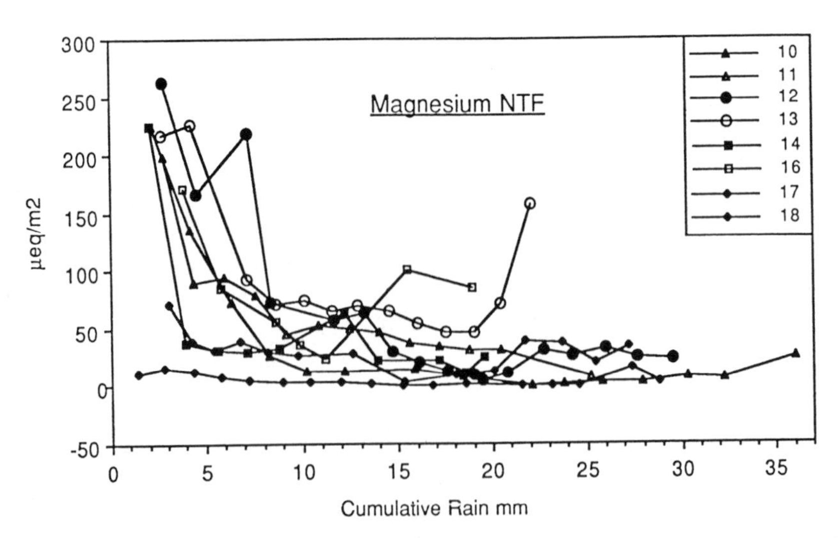

Figure 8.27. Net throughfall fluxes (NTF) of magnesium (in microequivalents m^{-2} ground area) in each sequential throughfall subsample versus cumulative millimeters of rainfall in eight rain events in loblolly pine forest in Oak Ridge, Tennessee. Rain event characteristics are described in Table 8.8.

Calcium NTF occurred almost entirely during the first 5 to 7 mm of rain depth, suggesting that calcium NTF is derived primarily from the washoff of dry deposition or the depletion of readily accessible ionic pools on canopy surfaces (such as ionic exchange sites) (Figure 8.25). Calcium ions are abundant within plant tissue but are regarded as highly immobile there. In only one case (event 16) did calcium fluxes show a substantial increase after the initial washoff period.

In contrast, potassium NTF occurred throughout these storms, suggesting an ionic source not available for immediate "extraction" by throughfall water (Figure 8.26). Other data from the IFS suggest that the dry deposition of potassium is small. In several of the events shown here (for example, 13, 14, and 16), potassium NTF equals or exceeds calcium NTF, a very unlikely outcome if dry deposition washoff dominated NTF for both ions (see Figures 8.25 and 8.26). Potassium NTF also showed increases late in several of these events. Potassium ions are both abundant and mobile within plant tissues, but canopy internal ions must penetrate the superficial layers of plant cuticle and wax before entering throughfall. These resistant layers will delay ionic diffusion, causing a more gradual decline in NTF for an ion derived from the plant interior as compared to an ion derived from surface washoff. Together, these factors point to a generally internal source for potassium in NTF.

Magnesium NTF exhibited more gradual initial declines than did calcium NTF and less NTF increases late in the event than occurred with potassium (see Figure 8.27). This pattern, intermediate between those of calcium and potassium, suggested that magnesium NTF came from a mixture of dry deposition and internal plant sources.

Our approach to the development of ion-specific NTF models was to use functional relationships between in-storm factors and the fraction of available cations that appeared in throughfall. Prestorm conditions were then used to scale ionic pools available for NTF in individual rain events. The in-storm factors that appeared to exert the greatest control over NTF were event rain amount (in centimeters) and rainfall intensity. Potassium NTF in particular increased at low rainfall intensity, implying that a longer contact time between the canopy and the throughfall water on its surfaces facilitates the release of potassium ions from canopy internal pools. Figure 8.28 illustrates this inverse relationship.

The relationship between NTF and rain event (centimeters) was best described by a model of the form used to describe Michaelis–Menten enzyme kinetics (Lehninger 1970). In this model the rainfall depth (in centimeters) is analogous to the "substrate" for the "reaction" describing the appearance of ions in NTF. The rainfall depth at which half the total NTF has appeared (averaged for the eight sequentially sampled events) is the Michaelis–Menten K_m, pertaining to the half-maximal reaction rate, and the NTF total is the V_{max}. The NTF curves for the eight sequentially sampled events were transformed into fractions of event totals, and four were used to estimated

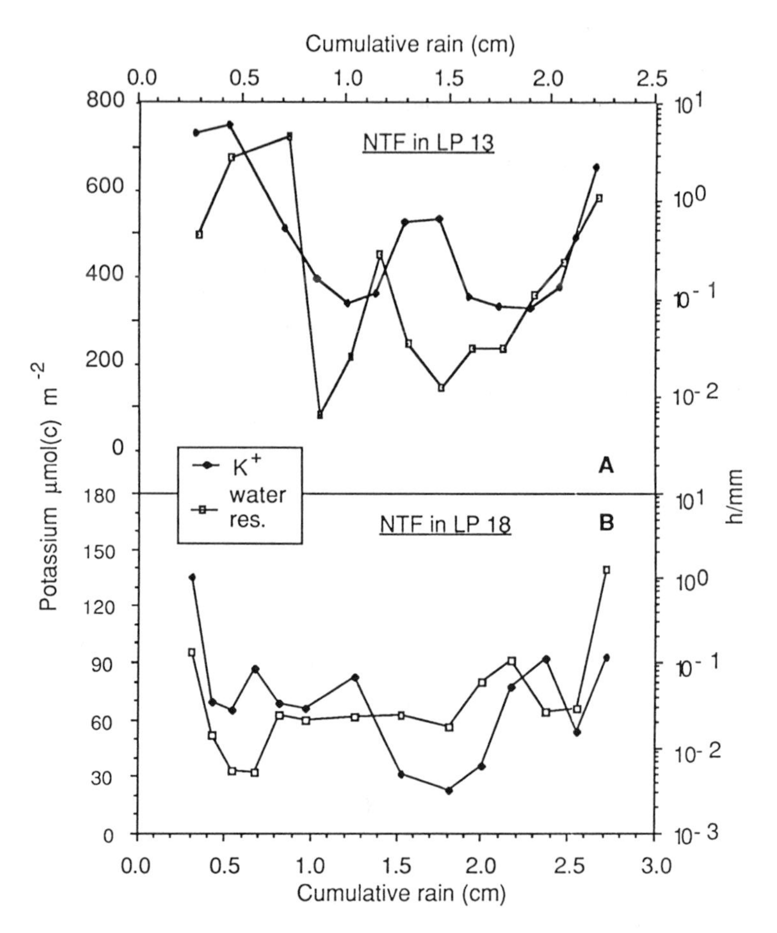

Figure 8.28. Net throughfall fluxes (NTF) of potassium (filled symbols, in microequivalents m^{-2} ground area) in two rain events, plotted with inverse of rainfall intensity (open symbols, canopy water residence time in h $millimeters^{-1}$, log scale) versus cumulative millimeters of rainfall.

K_m and V_{max}. The second step of model development was to use rainfall intensity to evaluate the departure of the fractional NTF from the Michaelis–Menten saturation curve during each of the subevent samples. These relationships were also developed with four of the sequentially sampled events. The fit of these equations to the remaining four sequentially sampled events was tested by multiple linear regression. The equations and significance of the tests are presented in Table 8.9.

In Table 8.9, the ions Ca^{2+}, Mg^{2+} and K^+ are listed in order of increasing contribution of plant internal sources to NTF. Both K_m and V_{max} increase in this order as well. This means that one-half the available Ca^{2+} will appear in NTF in a K_m of about 0.9 cm (1.2 cm for K^+) and that V_{max} increases

Table 8.9. Formulas for Relative Net TF Fluxes for Following Ions (in fraction of available ion pool), Based on Logarithm of Water "Residence Time" in Canopy ($\log_{10}$(hr/cm), and Cumulative Depth of Precipitation (cm) at Particular Sequential Sample (based on Michaelis–Menten formulation as described in text). Formulas are derived from four "parameterization" storms (12,13,17,18). The *r* and F values are derived from those storms, with the r_{test} values based on the four "validation" storms (10,11,14,16)

Ion	Formula	*r*	F	r_{test}
Ca^{2+}	$-.039 + .101 * [(\log_{10}(hr/cm)] + (1.22*cm)/(.912 + cm)$	.85	170.74	
Mg^{2+}	$-.110 + .048 * [(\log_{10}(hr/cm)] + (1.36*cm)/(1.03 + cm)$	.95	540.84	
K^{+}	$-.194 + .023 * [(\log_{10}(hr/cm)] + (1.55*cm)/(1.24 + cm)$	.98	777.94	

from about 1.2 for Ca^{2+} to 1.65 for K^{+}. These correspond to the qualitative observations that Ca^{2+} appears in NTF along a steeply declining curve and that the available pool is removed rapidly compared to the other cations.

For these cations, it is clear that different total pools of NTF ions are available at the beginning of the different events. The maximum NTF that can occur during any rain event should depend on conditions before that event. The antecedent dry hours (ADH) should be related to dry deposition loading of cations, because of both dry deposition and the "recovery" of previously depleted plant ionic pools. Because the entire pool of NTF ions is not necessarily removed in a rain event (especially those <0.5 cm depth), the pool remaining at the end of an event contributes to the ionic pool for the following event. Therefore, the depth of the previous event in centimeters (PECM) should bear some type of inverse relationship with the available NTF pool.

To estimate total available loading, all events from the IFS loblolly pine site of rainfall depth greater than 2.5 cm in 1987 and 1988 were used. For such events, we repeat that essentially all the available cations were extracted (based on the results of the sequentially sampled events), and inter-event differences in NTF would only reflect differences in prestorm history. Those differences include interstorm loading and NTF ions remaining in the canopy after the previous storm event. For each event greater than 2.5 cm, antecedent dry hours, rain depth of the event, and rain depth of the preceding event were regressed against NTF totals by multiple linear regression. The results of these regressions are shown in Table 8.10.

Comparing the results for these cations (Table 8.10) illustrates some important aspects of NTF generation in rain events. First, only about one-half of the variation in NTF among these events is explained by the models. Clearly, factors other than dry hours and rain centimeters affect cation NTF. Air concentrations of these cations (as both aerosol and coarse sedimenting particles) and the respective dry deposition doses (concentration × deposition velocity × exposure hours) were entered by stepwise regression, but did not generate significant relationships. The models for all three ions had significant positive intercepts that were large compared to the ADH and

Table 8.10. Total Available Net Throughfall Flux (NTF_{max}) Scaling Factors Derived from 29 Events of >2.5 cm Rain Depth from Loblolly Pine Canopy During 1987 and 1988. NTF_{max} (millimol(c) m^{-2}) is based on multiple linear regressions of antecedent dry hours (ADH) event rain centimeters (CM) and previous event rain centimeters (PECM) for these events

Formula	r^2	F
$NTF_{max}Ca^{2+}$ = 0.495 + 0.0031*ADH − 0.096*PECM;	r^2 = .52,	F > 14
$NTF_{max}Mg^{2+}$ = 0.164 + 0.00126*ADH − 0.0315*PECM;	r^2 = .51,	F > 13
$NTF_{max}K^{+}$ = 0.203 + 0.00236*ADH − 0.0549*PECM + 0.0661*CM;	r^2 = .48,	F > 7.5

PECM terms; the models therefore indicate that the canopy presents large ion pools available for NTF to any rain event. This point will be addressed again in the model test following. The ADH term was largest for Ca^{2+} and smallest for Mg^{2+}; the values (3.1 and 1.3 μmol(c) m^{-2} h^{-1}) compare favorably to those derived independently by Lindberg et al. (unpublished data) for the same canopy. The NTF of Ca^{2+} was most affected by PECM, and only K^+ showed a significant effect of CM. Even though all these events exceeded 2.5 cm, K^+ NTF still increased with event depth, suggesting that canopy pools were not totally depleted. Other factors that influence NTF remain to be determined.

The complete NTF models can be visualized as "loading" the forest canopy before a rainfall event on the basis of prestorm conditions. When rain occurs, the amount and intensity of the rain depletes a fraction of the NTF pool. The NTF pool is replenished between rain events, and the pattern repeats (Figure 8.29).

The complete model was tested with NTF fluxes for all the wet events from the 1986 growing season (1 April–31 October), with NTF fluxes averaged over five TF event collectors arrayed in the IFS deposition site. A tipping bucket rain gage at the micrometeorological tower logged hourly rainfall and intensity data. Modeled NTF fluxes were based on the available pool of ions, hourly rain intensity, and cumulative rain centimeters during each event. Interstorm canopy "recharge" was based on ADH, CM, and PECM. Modeled NTF fluxes were summed for each hour and compared with the summed net TF event fluxes measured for each ion. Predicted canopy pools and event NTF for Ca^{2+} and K^+ are presented in Figure 8.30, and cumulative modeled and measured NTF for Ca^{2+} and K^+ are presented in Figure 8.31. Results for Mg^{2+} were intermediate between these two extremes and are not shown here.

For Ca^{2+}, predicted NTF tracked the observed data very closely but was always larger by a factor of 2. For K^+, predicted NTF was again larger by a factor of 2 and tended to further overestimate NTF later in the season. Two factors probably contribute to these disparities. First, the model was developed with the sequential TF collector placed under very dense canopy cover, possibly overestimating NTF compared to an area of "average" can-

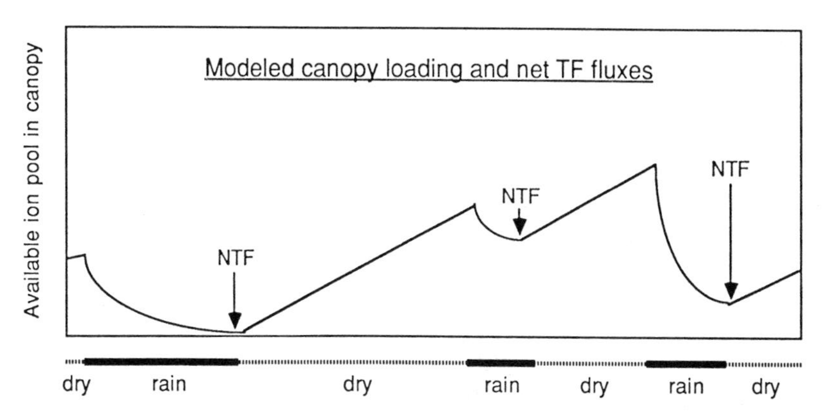

Figure 8.29. Net throughfall flux (NTF) is explicitly modeled as pool of ions in canopy available for NTF that is increased by prestorm canopy loading during a dry period (dotted lines), is depleted by NTF of individual rain events (solid lines), and is then again increased by dry deposition during following dry period. NTF is controlled both by prestorm available pool and intensity and depth of rain event.

opy cover. Second, the interstorm loading section of the model had very large "intercept" terms for each ion. These amounts of ions were added to the canopy pool during every interstorm period, regardless of its length, and were subsequently accessible.

Because K^+ was only gradually removed in NTF, the model showed an accumulation of this ion in the canopy late in the season (see Figure 8.30). With this accumulation, modeled event NTF values exceeded the observed values by increasing margins as the season progressed.

In summary, a detailed TF chemistry model of the factors controlling ionic fluxes in TF was developed from sequentially sampled TF. The major cations in NTF consisted of differing proportions of dry deposition washoff and ions derived from plant tissues. Prestorm conditions such as dry hours and the remaining ionic pool from the previous storm controlled the total amount of ions available for NTF. Intrastorm conditions such as rainfall intensity and total rainfall amount controlled the fraction of the available ions that actually appeared in NTF. The model tracked event-based NTF observations reasonably well but overestimated interstorm loading. In Chapter 14, a much simpler formulation of NTF generation is presented in the IFS nutrient cycling model.

Base Cation Distribution and Cycling

D.W. Johnson

Introduction

Because base cation limitations to forest growth are relatively uncommon (with certain notable exceptions in outwash soils of the northeastern United

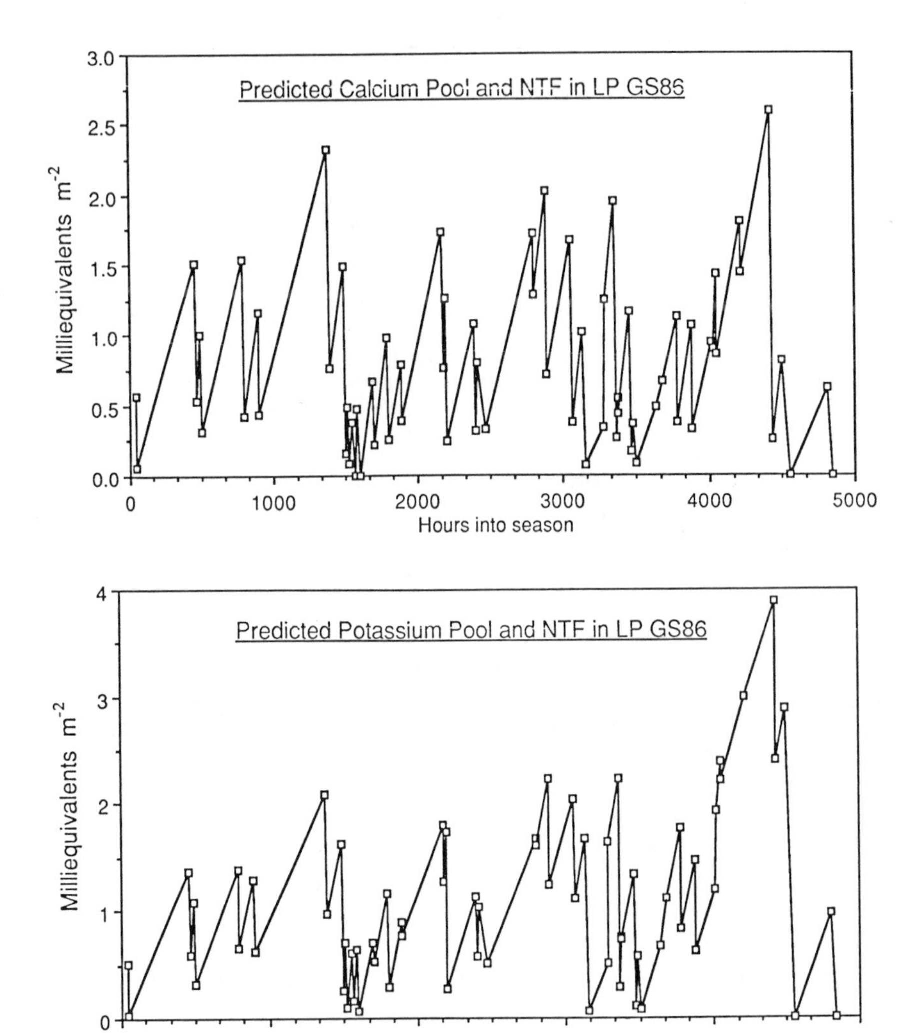

Figure 8.30. Results of net throughfall flux (NTF) model in loblolly pine forest in Oak Ridge, Tennessee. Pool of ions in canopy available for NTF increases during dry periods and decreases (generating NTF) during rain events. Values are plotted against hours of growing season, from April 1–October 31, 1986. Panel a shows calcium; and panel b shows potassium; both in milliequivalents m^{-2} ground area.

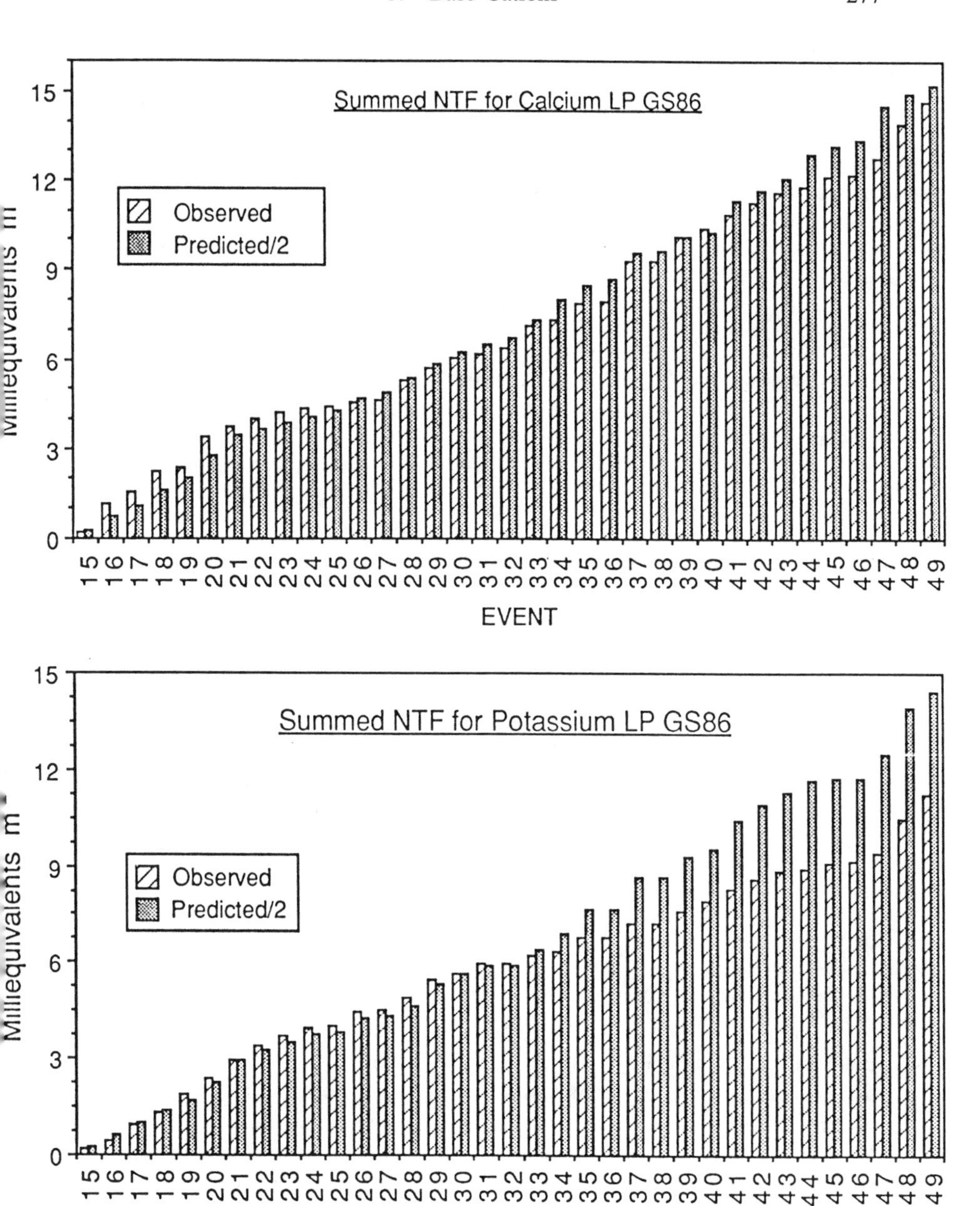

Figure 8.31. Comparison of observed (light bars) and predicted (dark bars) cumulative net throughfall fluxes (NTF) during 1986 growing season (IFS rain event numbers 15–49). Predicted values plotted at one-half their calculated values follow observed values closely. Differences between predicted and observed values are considered in text. Panel a shows calcium; panel b shows potassium; both in milliequivalents m^{-2} ground area.

States; Heiburg and White 1953; Stone and Kszystyniak 1977), considerably less attention has been paid historically to the cycling of base cations than to nitrogen and phosphorus. The cycles of Ca^{2+}, Mg^{2+}, and K^{+} were often characterized along with N and P (e.g., Cole and Rapp 1981), but until recent concerns about the effects of acid deposition and intensive harvesting, base cation cycles have not received the intense scrutiny that N and P cycles have. Interestingly, however, most analyses of nutrient cycling data reveal that Ca rather than N or P is the nutrient most likely to be significantly depleted by leaching and harvesting (Johnson 1981; Johnson et al. 1988b; Federer et al. 1989). Further, there is emerging evidence that the base cation status of soils is changing more rapidly than previously anticipated (Berden et al. 1987; Johnson et al. 1988b; Binkley et al. 1989; Johnson and Taylor 1989).

The primary motivation of the IFS was to evaluate the effects of acid deposition on the nutrient status of forest ecosystems, and the nutrients of most concern were Ca^{2+}, Mg^{2+}, and K^{+}. In this section, an evaluation of the effects of atmospheric deposition on base cation nutrient status is made, with simple, rather crude projections as to potential changes. Much more sophisticated analyses of potential change are presented in Chapter 13 using the IFS Nutrient Cycling model (NuCM).

Neither this analysis nor the NuCM model predictions are meant to be the final word on the subject, either for forest ecosystems in general or for the IFS sites in particular. Rather, these analyses are meant to describe our current state of knowledge and set the stage for future analyses, including both field resampling of soils to test hypotheses as to soil change and further interpretation and modeling of the IFS data set presented in the Appendix.

Base Cation Distribution and Flux

Base saturation is a key variable for the classification of soils and for the categorization of ecosystems with respect to base cation distribution and cycling. Base saturation directly affects not only the distribution but also the flux of base both base and acid (H^{+}, Al) cations in forest ecosystems, and consequently it is useful to order the sites according this parameter. If we stratify according to B horizon base saturation, we find that, with only one exception (the Florida site), the sites can be conveniently divided into two categories: high-elevation and northern sites (low base saturation soils) and low-elevation and southern sites (high base saturation soils) (Figure 8.32). To some extent, the A and E horizons fall into the same categories, but the division is not as clear: some northern and high-elevation A and E horizon soils have rather high base saturations (Whiteface, Huntington Forest).

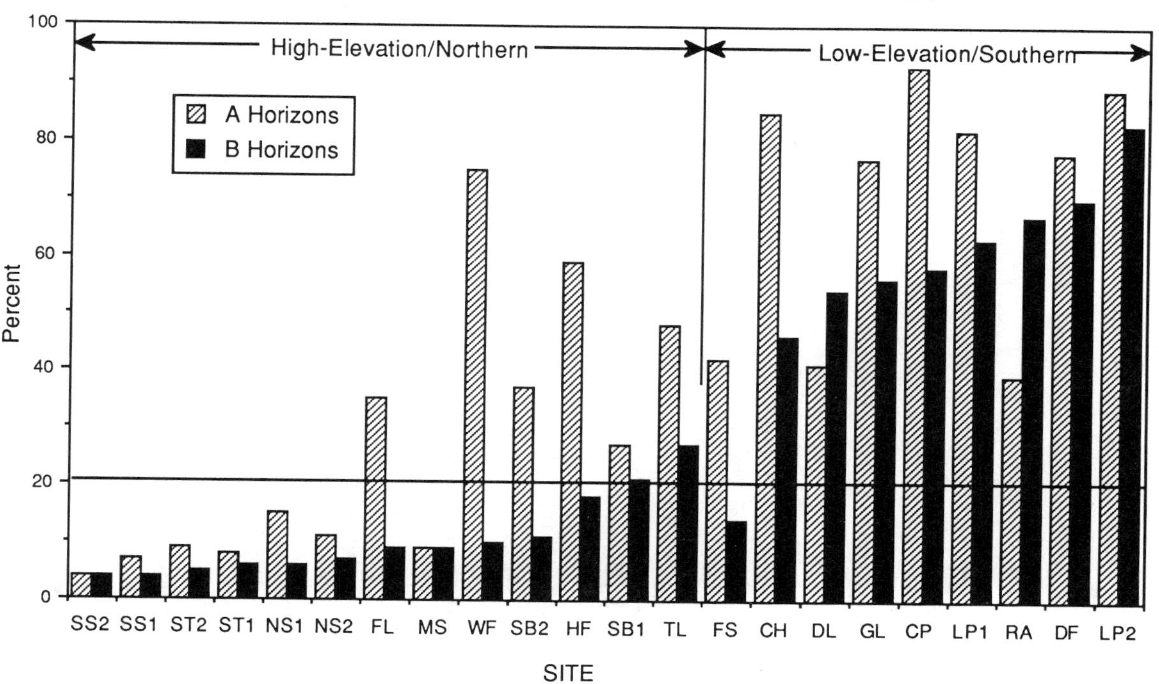

Figure 8.32. Base saturation (NH_4Cl extraction) in B horizon soils from IFS sites. HF, Huntington Forest, NY; TL, Turkey Lakes, Ontario; ST, Smokies Tower, NC; MS, Howland site, Maine; SS, Smokes Becking, North Carolina; SB, Smokies Beech, North Carolina; FL, Findley Lake, Washington; WF, Whiteface Mt., New York; DF, Douglas fir, Thompson site, Washington; NS, Aamli site, Norway; GA, B.F. Grant site, Georgia; CP, White pine site, Coweeta, North Carolina; CH, hardwood site, Coweeta, North Carolina; FS, slash pine site, Florida; DL, Duke loblolly site, North Carolina; RA, red alder site, Thompson site, Washington; LP, loblolly pine site, Oak Ridge, Tennessee. Duplicate abbreviations indicate separate or duplicate plots.

Figures 8.33–8.36 show the distribution of total base cations,[1] Ca^{2+}, Mg^{2+}, and K^+ in the various sites ordered as in Figure 8.32. The upper portion of each figure includes only vegetation, forest floor, and soil exchangeable pools, and the lower figure shows these components along with total soil pools. As is typical of forest ecosystems, the total soil pools are by far the largest pools, constituting more than 90% of total ecosystem base cation, Ca^{2+}, Mg^{2+}, and K^+ content in most cases. Vegetation constitutes less than 5% of total ecosystem base cation, Ca^{2+}, Mg^{2+}, and K^+ content in all sites. The small proportions of base cations in vegetation observed in these sites is typical of most forest ecosystems, but there are exceptions in some hardwood ecosystems in the southeastern United States where a considerable proportion (e.g., 20%–50%) of total ecosystem Ca may lie in vegetation (Johnson et al. 1988a, 1988b).

As is discussed in Chapter 10, the sites from glaciated regions (Whiteface, Turkey Lakes, Huntington Forest, Norway, Findley Lake, Douglas-fir, Red Alder) tend to have greater reserves of primary minerals. Total soil base cation contents do not reflect this pattern very well, however. For instance, the Coweeta hardwood and pine site has quite high soil total base cation, Mg, and K contents, probably as a result of colluvial activity (as noted in Chapter 10). The Red Alder and Douglas-fir sites have rather low total soil Mg and very low total soil K contents (Figures 8.35 and 8.36). The Georgia, Coweeta pine, and Oak Ridge loblolly pine sites all have much higher total soil K contents than most of the glaciated sites (perhaps a result of high content of K-rich phyllosilicates).

To gain a better perspective as to the distribution of soil exchangeable, forest floor, and vegetation pools in these site, these pools are plotted separately in the upper portions of Figures 8.33–8.36. The exchangeable base cation pools follow approximately the same pattern as B horizon base saturation, the deviations resulting from differences in the weight of B horizon soil (which in turn is a function of bulk density, percent gravel, and depth of B horizon). Vegetation and forest floor base cation contents vary from site to site, but exchangeable cation contents vary much more. As a consequence of the large variation in soil exchangeable pools, the ratios of vegetation base cations, Ca, K, and Mg contents to soil exchangeable contents varies from more than 2:1 in the high-elevation and northern sites to less than 0.1:1 in the southern and low-elevation sites (Figures 8.33–8.36).

Deposition and Leaching Fluxes

The deposition and leaching output of total cations, H^+ and Al, base cations, Ca^{2+}, Mg^{2+}, and K^+ in each site are depicted in Figures 8.37–8.39. At first

---

[1]Vegetation and forest floor base cations include Ca, K, and Mg only, whereas soil exchangeable and total base cations include Ca, K, Mg, and Na. In that Na is not a nutrient, its concentrations in plant tissue and forest floor are typically low enough such that it can be ignored as a component of total base cations (Johnson and Henderson 1989).

glance, it would seem that total cation leaching rates are quite closely coupled to atmospheric total cation deposition rates in all but a few sites (Duke, Georgia, Coweeta, Red Alder, Douglas-fir). However, as described later in this chapter and in Chapter 11, there are a variety of internal H^+-generating and -consuming processes that affect total cation leaching rates, and the patterns seen in Figure 8.37 should not be interpreted as evidence that cation leaching is necessarily dominated by atmospheric deposition.

Total cation, H^+, and Al leaching rates are much higher at the Smokies red spruce sites (both Becking and Tower) than at the other sites (see Figure 8.37). H^+ and Al^{3+} are the dominant cations in soil solutions at the Smokies red spruce sites (Figures 8.40 and 8.41), a result that is expected from a combination of high mineral acid anion loading to extremely acid soils. Base cation leaching rates are highest in the Red Alder, Turkey Lakes, and Oak Ridge Loblolly sites (see Figures 8.38 and 8.39). Internal nitrate production is the primary cause of high cation leaching in the Red Alder and Turkey Lakes sites, whereas atmospheric S deposition and sulfate leaching is the major cause of high cation leaching in the Oak Ridge Loblolly site.

Approximately half the sites show a clear net annual loss of base cations Ca^{2+} and K^+ via leaching; the other sites show either a net accumulation or an approximate balance (Figures 8.38 and 8.39). (Leaching data for K and therefore base cations are lacking for the Florida site at the time of this writing.) The majority of sites showed a net annual loss of Mg^{2+}, with only one site showing an accumulation (Florida) and four sites approximately in balance.

Because most of these sites are subject to significant inputs of H^+ from both atmospheric and internal sources (Figure 8.37; see also Chapter 11), the extent to which so many sites retain one or more base cations is somewhat surprising. The net ecosystem retention of base cations is also atypical of the results of previous studies of forest nutrient cycling (e.g., Cole and Rapp 1981). One reason for the difference in these results and those from the previous literature is that much cruder and less efficient methods were used to measure inputs before this study, and it is very likely that base cation deposition was underestimated. In the case of the Smokies red spruce sites, the tendency to conserve base cations and leach acid cations (H^+ and Al^{3+}) is a natural result of extremely acidic soil conditions. In addition to this, however, there are a variety of factors that regulate the retention and losses of individual base cations from forest ecosystems and which need to be evaluated before a full explanation of the patterns observed in Figures 8.37–8.39 can be explained.

A Brief Review of Soil Leaching Processes

The concept of anion mobility has proven to be a useful paradigm for explaining the net retention and loss of cations from soils. This paradigm relies on the simple fact that total cations must balance total anions in soil solution (or any other solution), and, therefore, total cation leaching can be thought

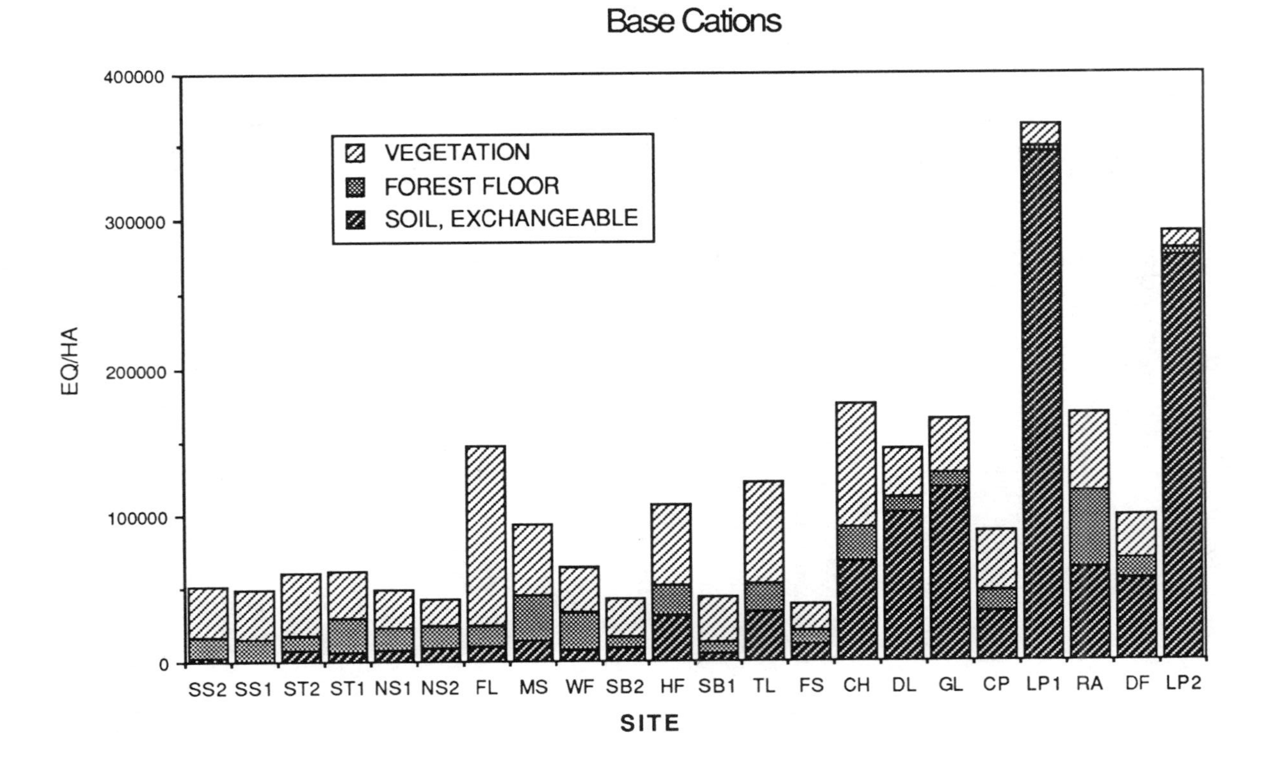
Base Cations
VEGETATION
FOREST FLOOR
SOIL, EXCHANGEABLE
EQ/HA
400000
300000
200000
100000
0
SS2 SS1 ST2 ST1 NS1 NS2 FL MS WF SB2 HF SB1 TL FS CH DL GL CP LP1 RA DF LP2
SITE

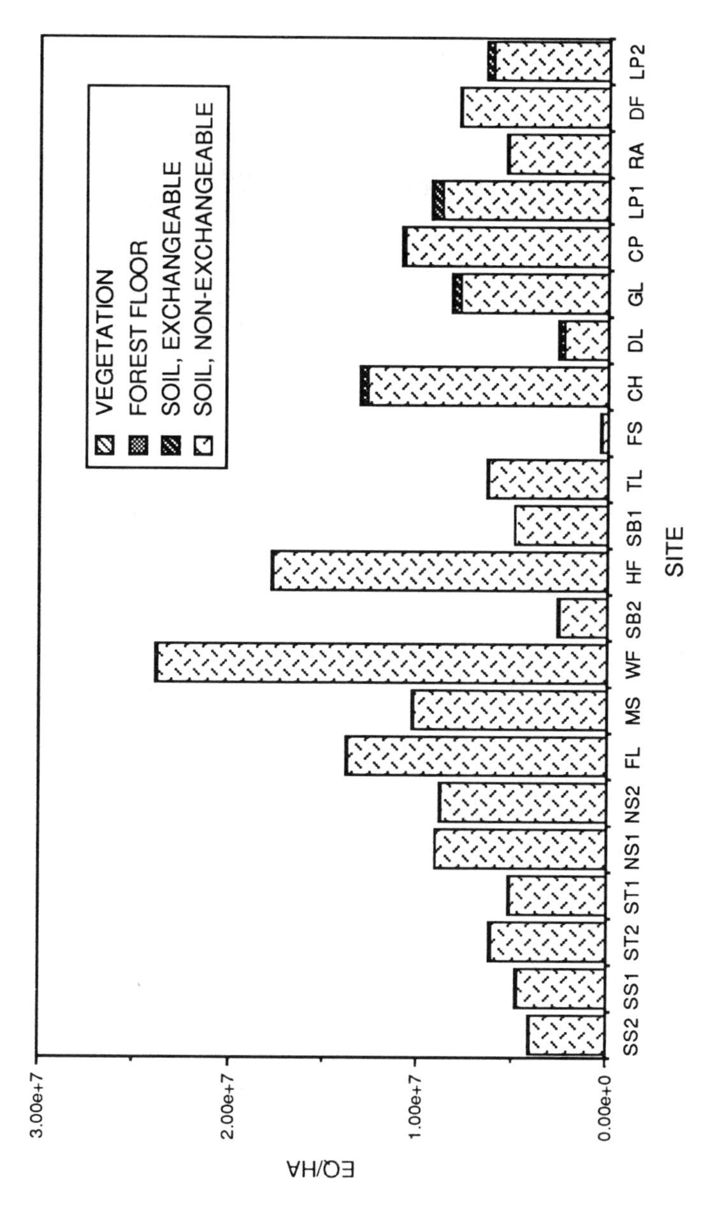

Figure 8.33. Base cation distribution in IFS sites. See Figure 8.32 for legend.

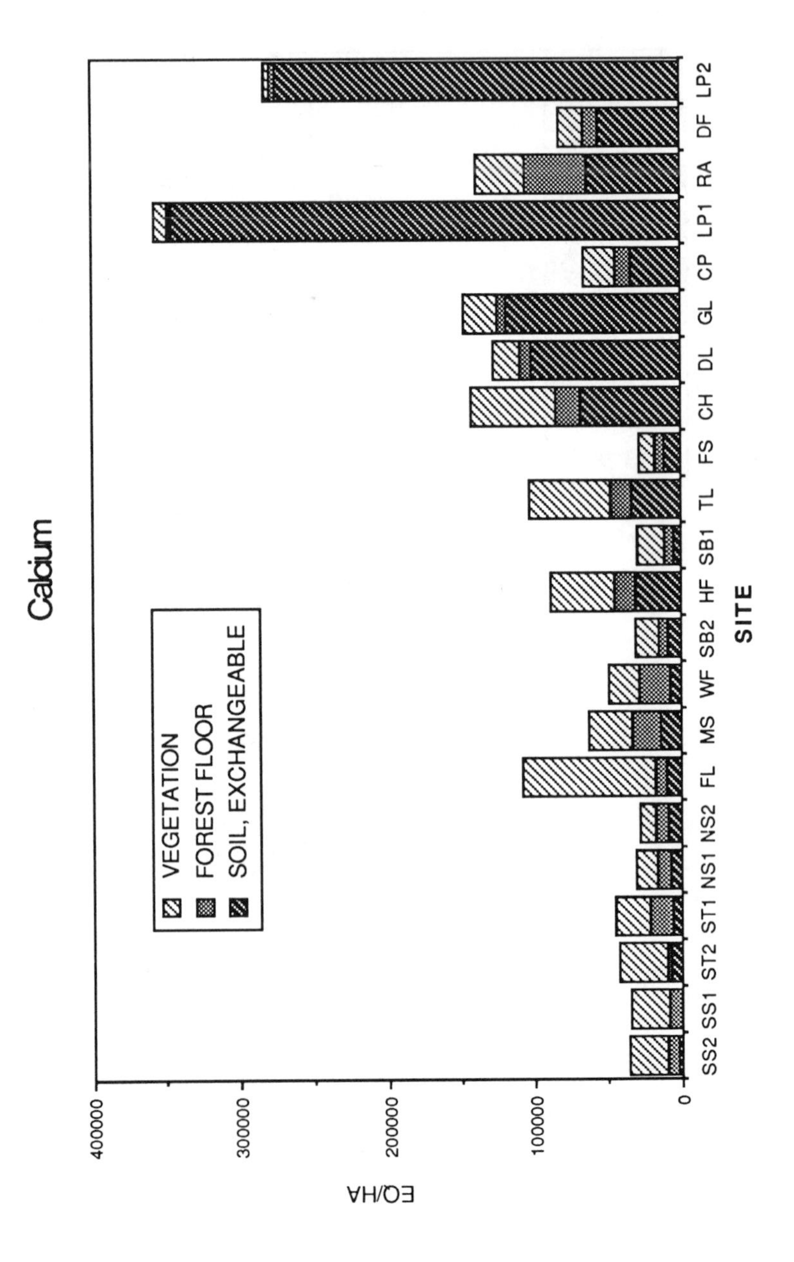
Calcium
VEGETATION
FOREST FLOOR
SOIL, EXCHANGEABLE
EQ/HA
400000
300000
200000
100000
0
SS2 SS1 ST2 ST1 NS1 NS2 FL MS WF SB2 HF SB1 TL FS CH DL GL CP LP1 RA DF LP2
SITE

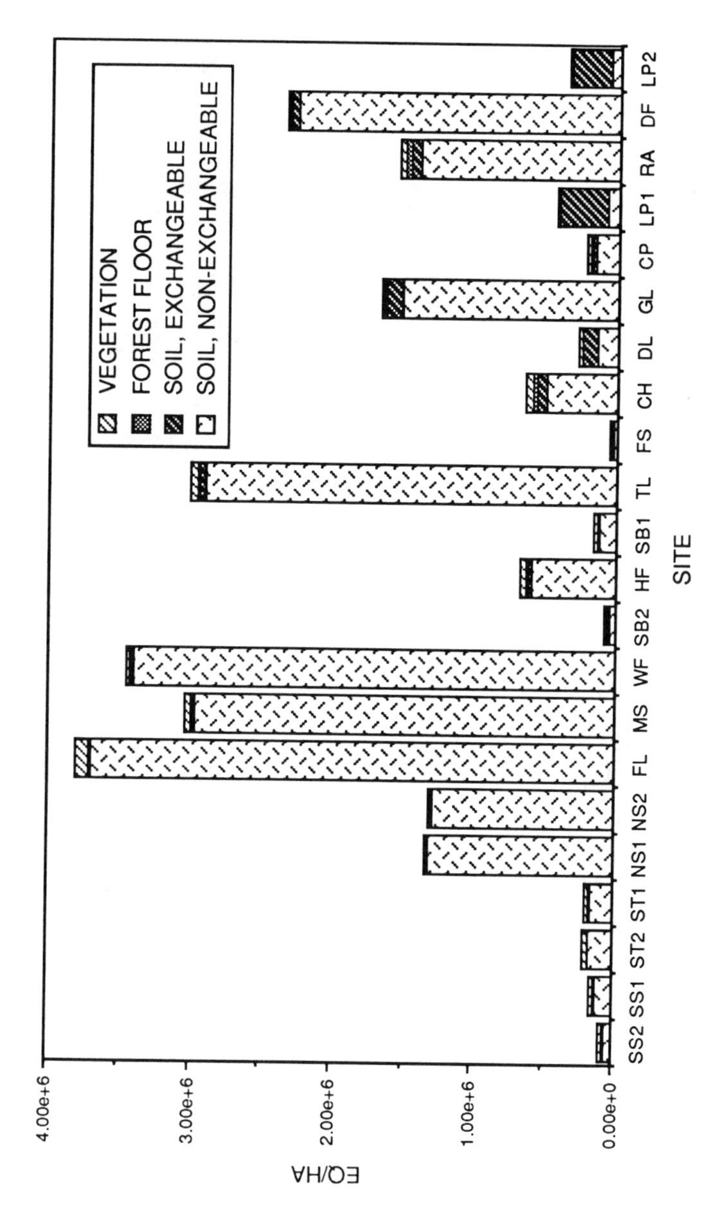

Figure 8.34. Calcium distribution in IFS sites. See Fig 8.32 for legend.

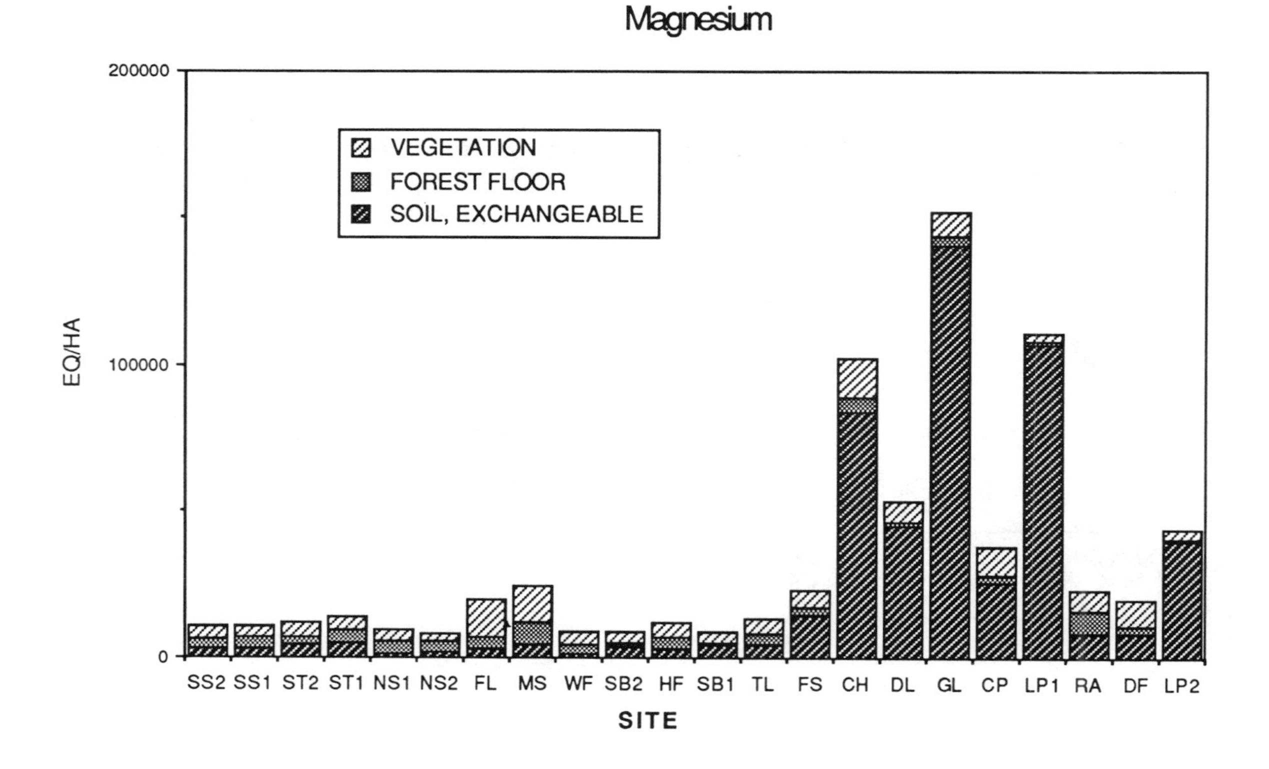
Magnesium
VEGETATION
FOREST FLOOR
SOIL, EXCHANGEABLE
EQ/HA
200000
100000
0
SS2 SS1 ST2 ST1 NS1 NS2 FL MS WF SB2 HF SB1 TL FS CH DL GL CP LP1 RA DF LP2
SITE

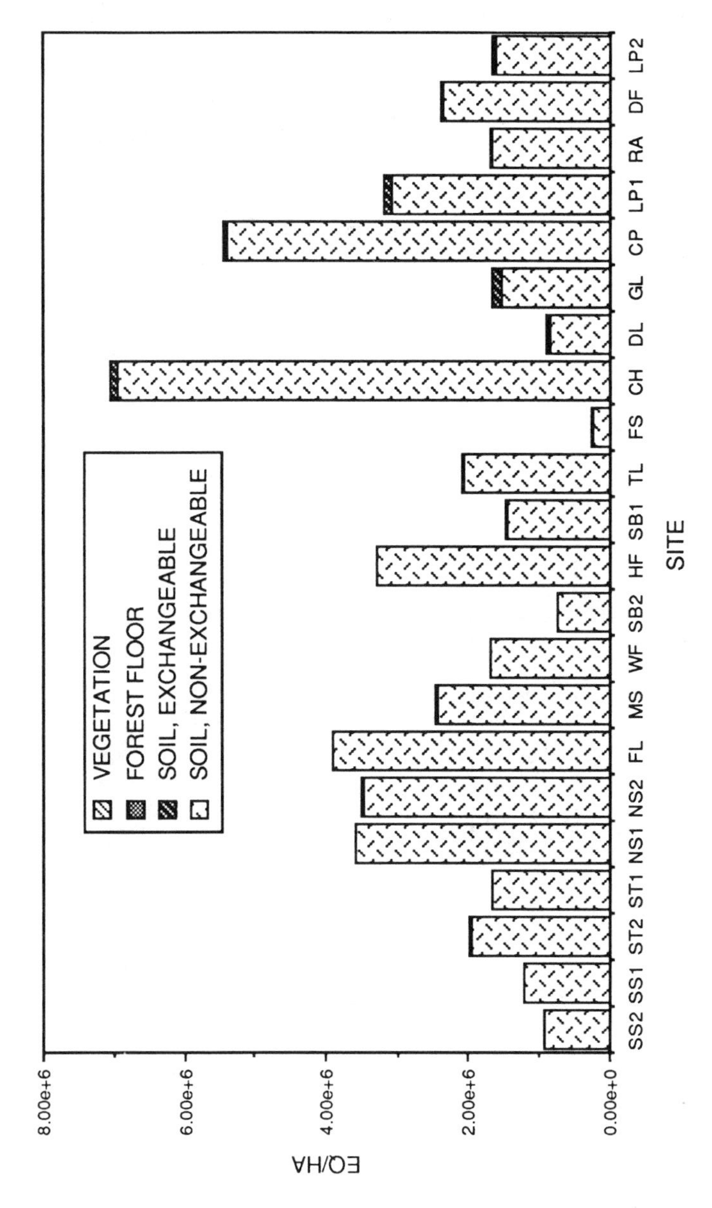

Figure 8.35. Magnesium distribution in IFS sites. See Figure 8.32 for legend.

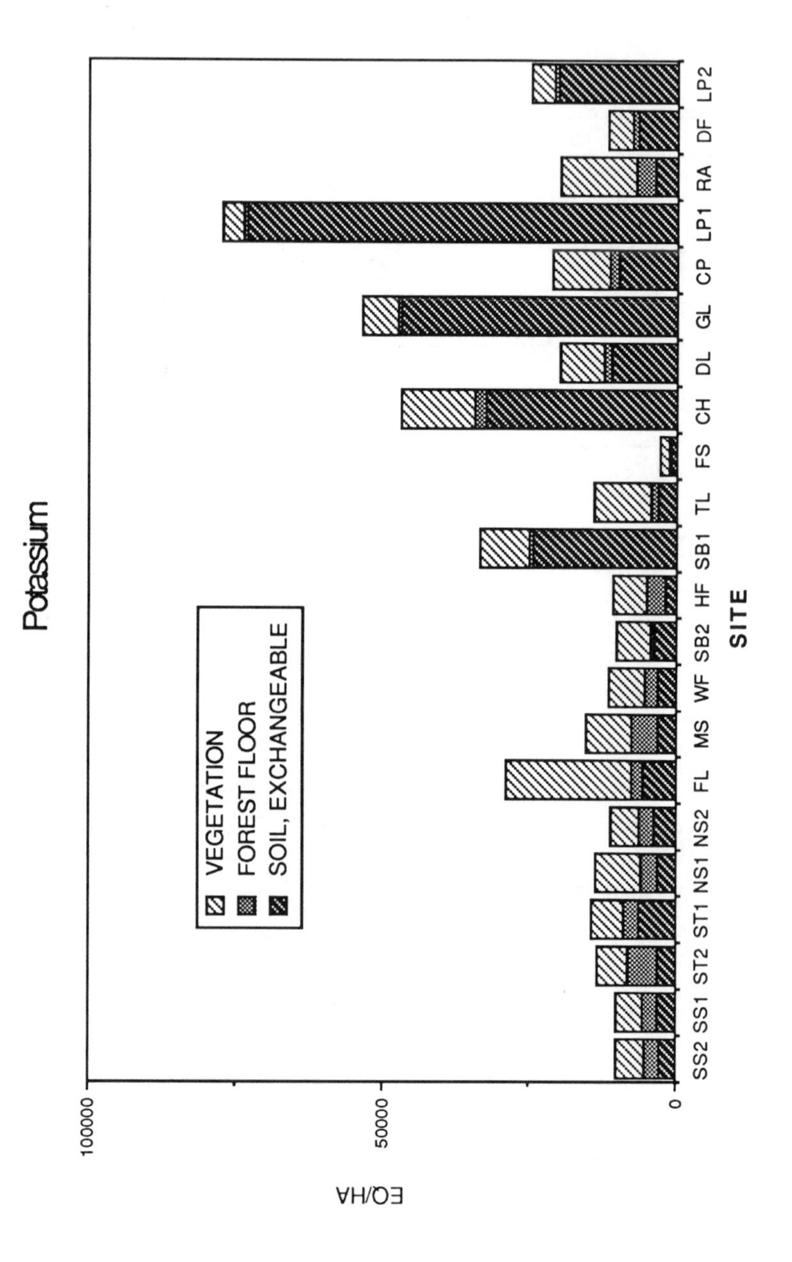
Potassium
VEGETATION
FOREST FLOOR
SOIL, EXCHANGEABLE
EQ/HA
100000
50000
0
SS2
SS1
ST2
ST1
NS1
NS2
FL
MS
WF
SB2
HF
SB1
TL
FS
CH
DL
GL
CP
LP1
RA
DF
LP2
SITE

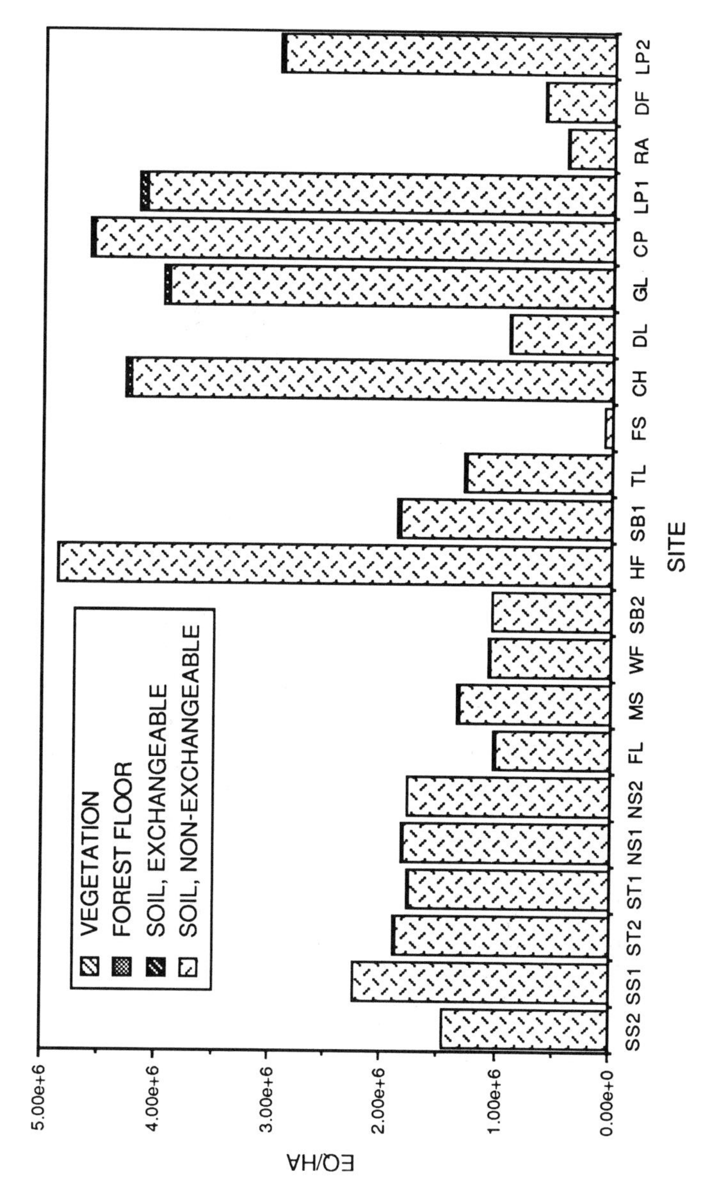

Figure 8.36. Potassium distribution in IFS sites. See Figure 8.32 for legend.

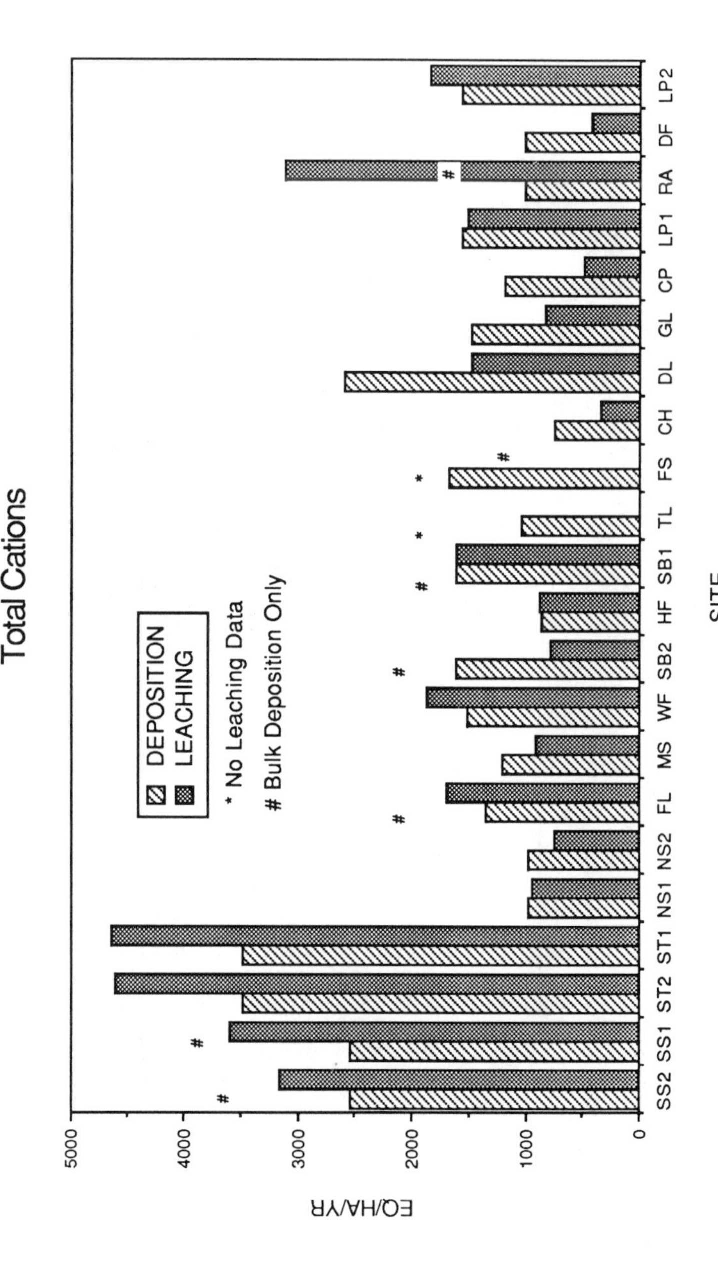
Total Cations
DEPOSITION
LEACHING
* No Leaching Data
# Bulk Deposition Only
EQ/HA/YR
0
1000
2000
3000
4000
5000
SITE
SS2 SS1 ST2 ST1 NS1 NS2 FL MS WF SB2 HF SB1 TL FS CH DL GL CP LP1 RA DF LP2

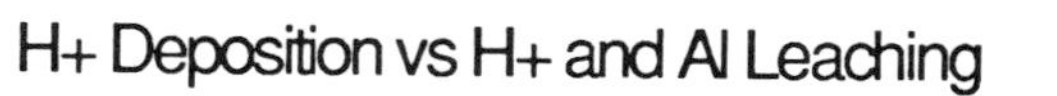

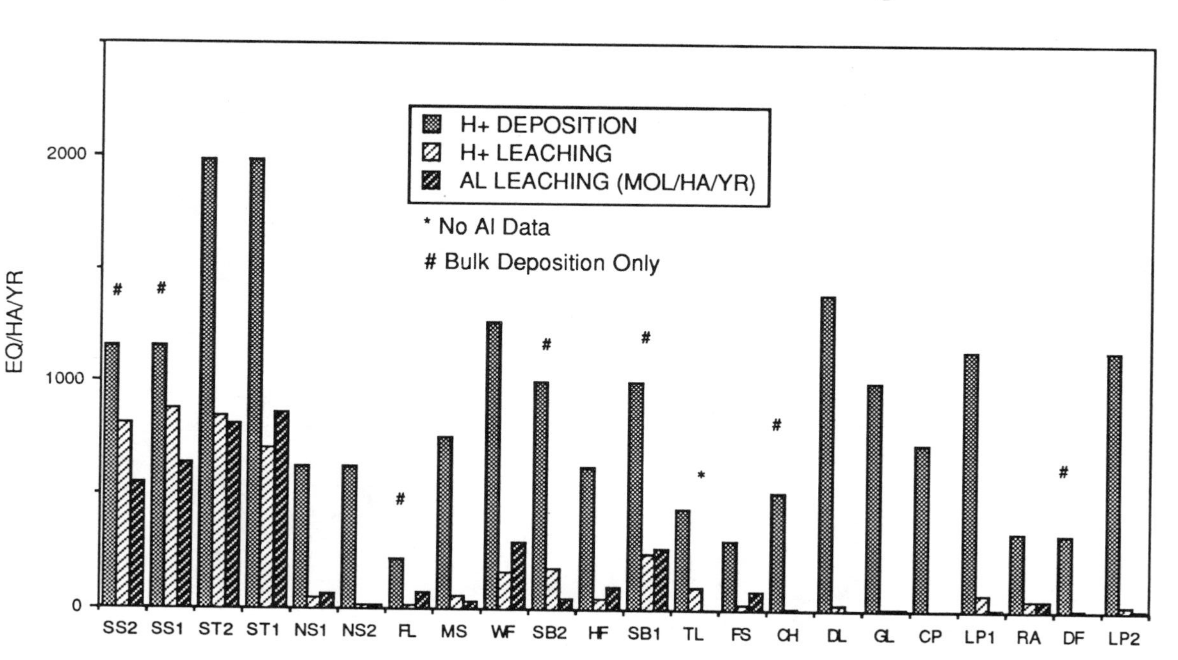

Figure 8.37. Atmospheric deposition and leaching of total cations (top) and H^+ and Al (bottom) in IFS sites. See Figure 8.32 for legend.

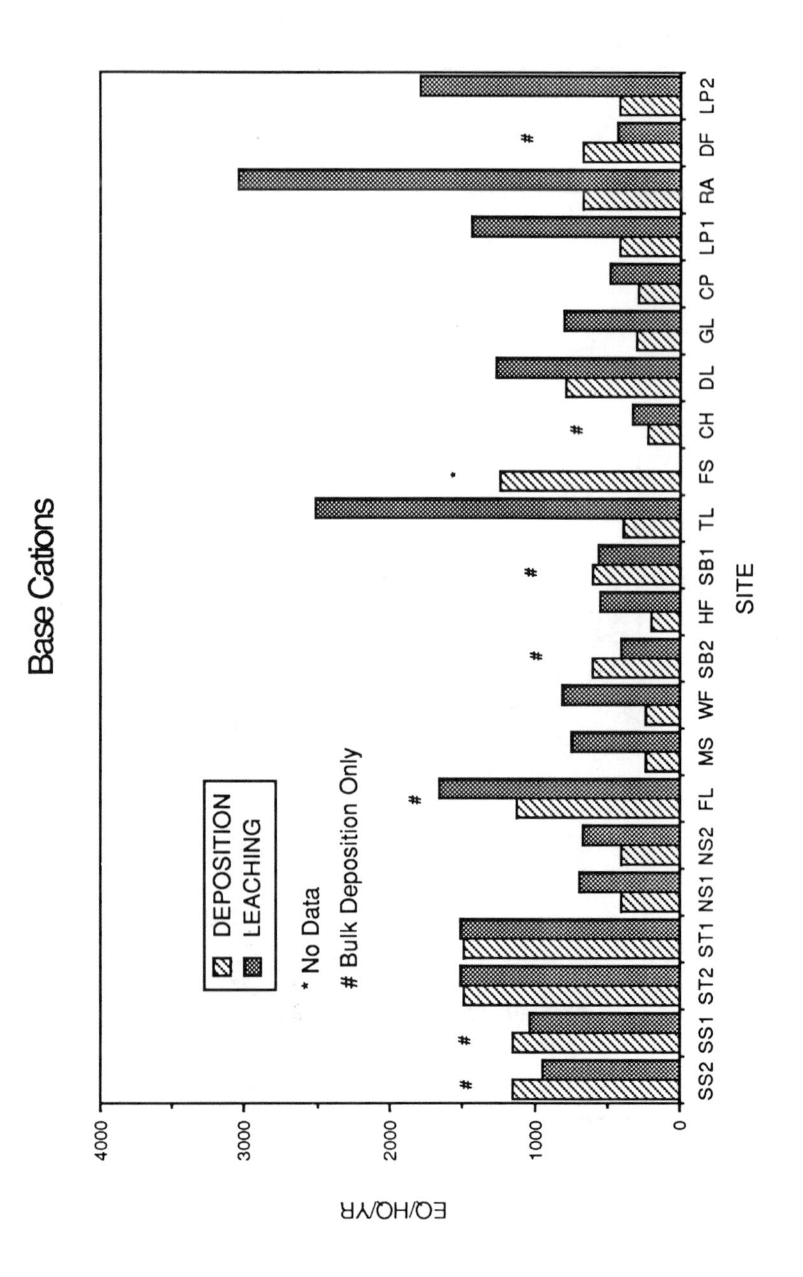
Base Cations
DEPOSITION
LEACHING
* No Data
# Bulk Deposition Only
EQ/HQ/YR
0
1000
2000
3000
4000
SS2 SS1 ST2 ST1 NS1 NS2 FL MS WF SB2 HF SB1 TL FS CH DL GL CP LP1 RA DF LP2
SITE

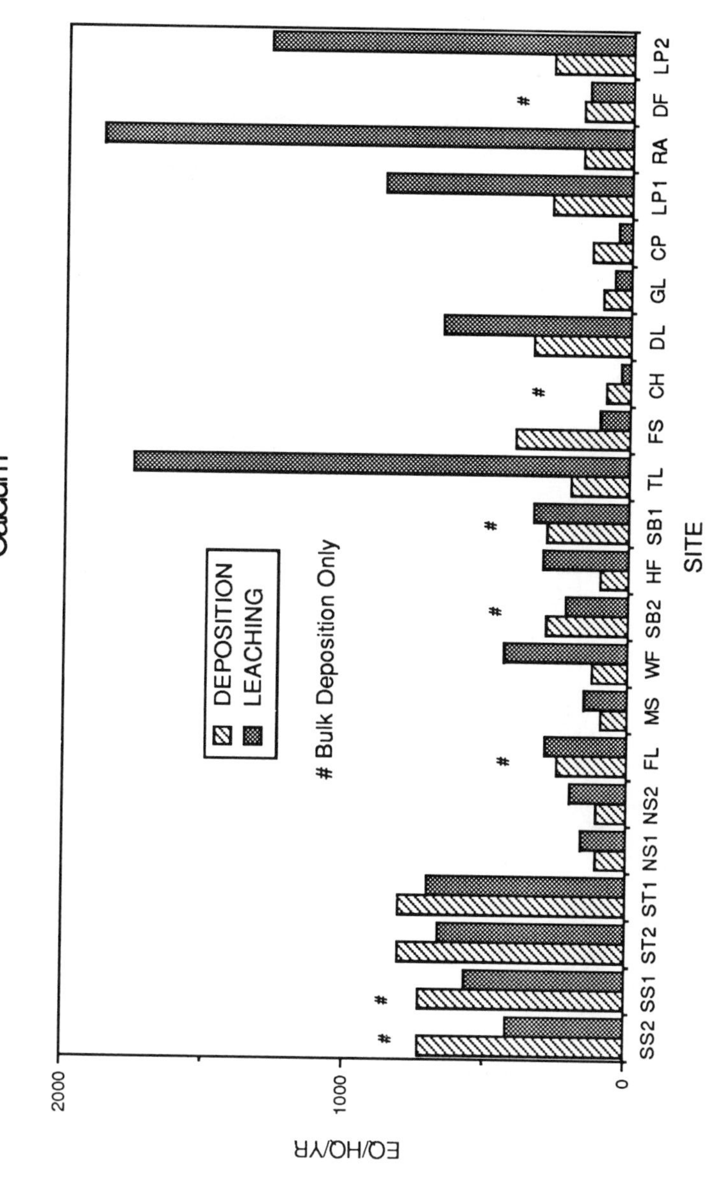

Figure 8.38. Atmospheric deposition and leaching of base cations (top) and calcium (bottom) in IFS sites. See Figure 8.32 for legend.

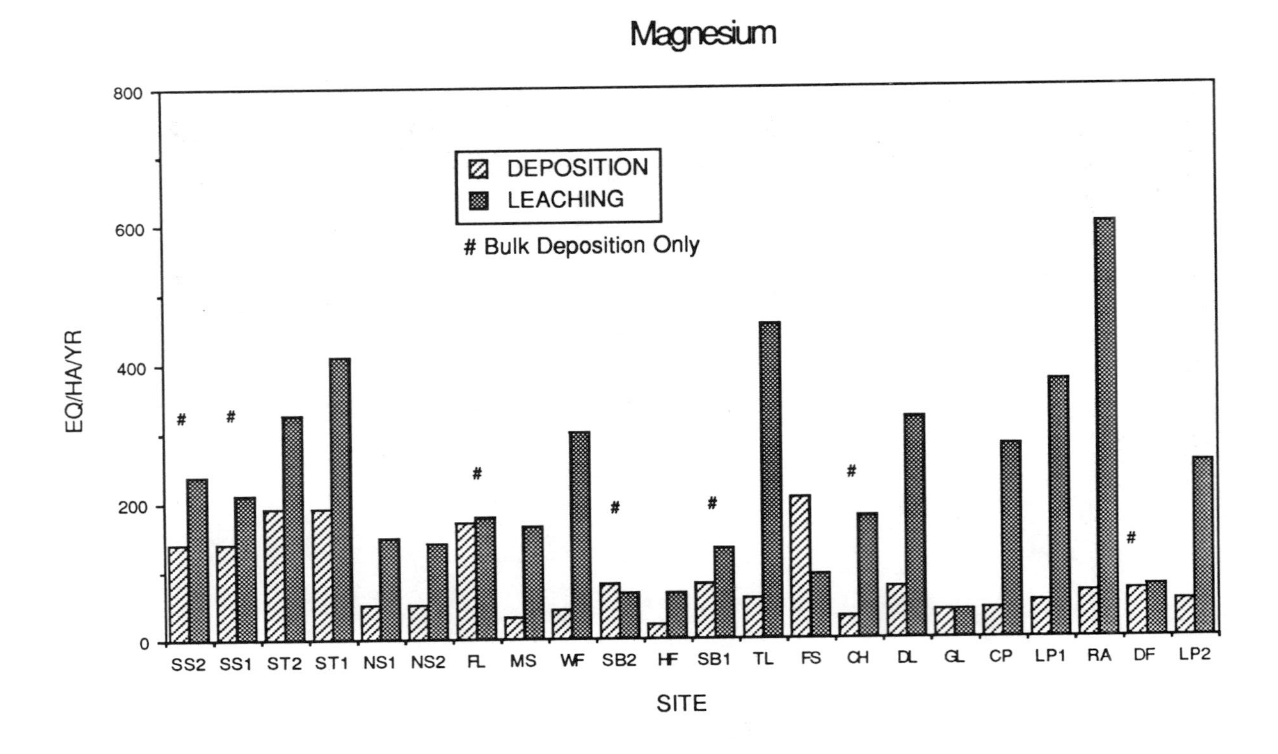
Magnesium
DEPOSITION
LEACHING
Bulk Deposition Only
EQ/HA/YR
800
600
400
200
0
SS2
SS1
ST2
ST1
NS1
NS2
FL
MS
WF
SB2
HF
SB1
TL
FS
CH
DL
GL
CP
LP1
RA
DF
LP2
SITE

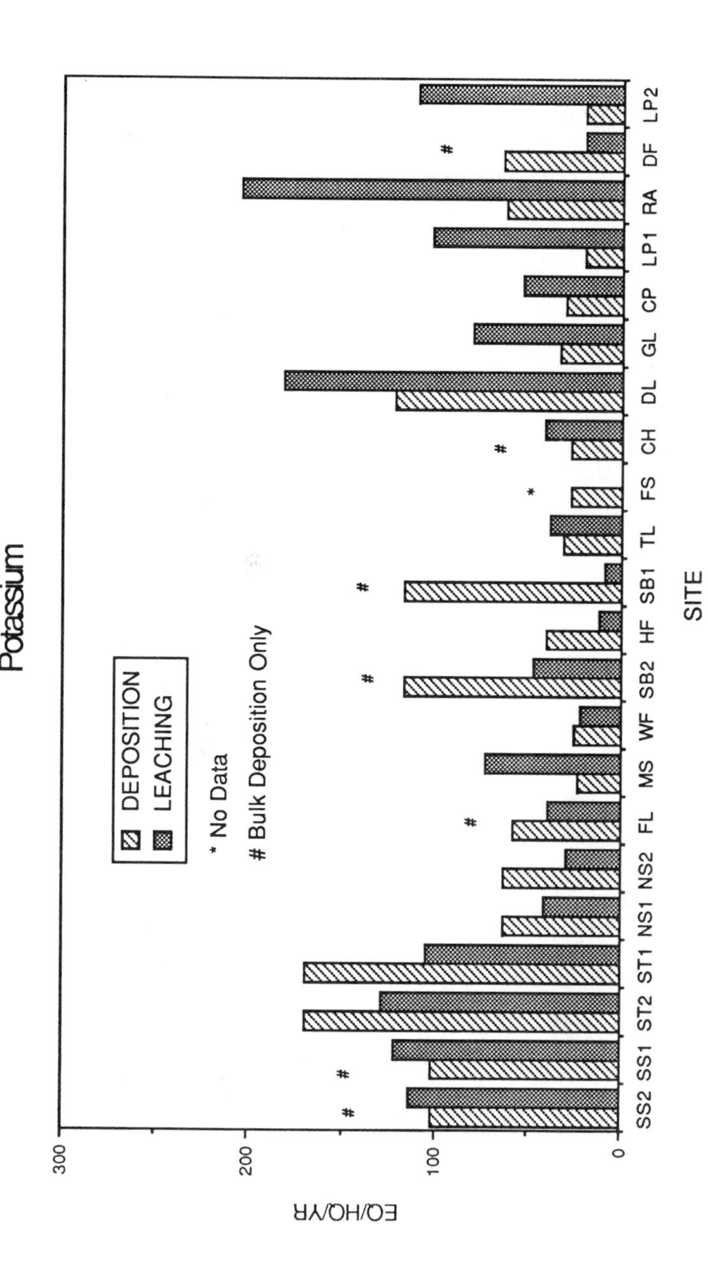

Figure 8.39. Atmospheric deposition and leaching of magnesium (top) and potassium (bottom) in IFS sites. See Figure 8.32 for legend.

A Horizon Solution Anions

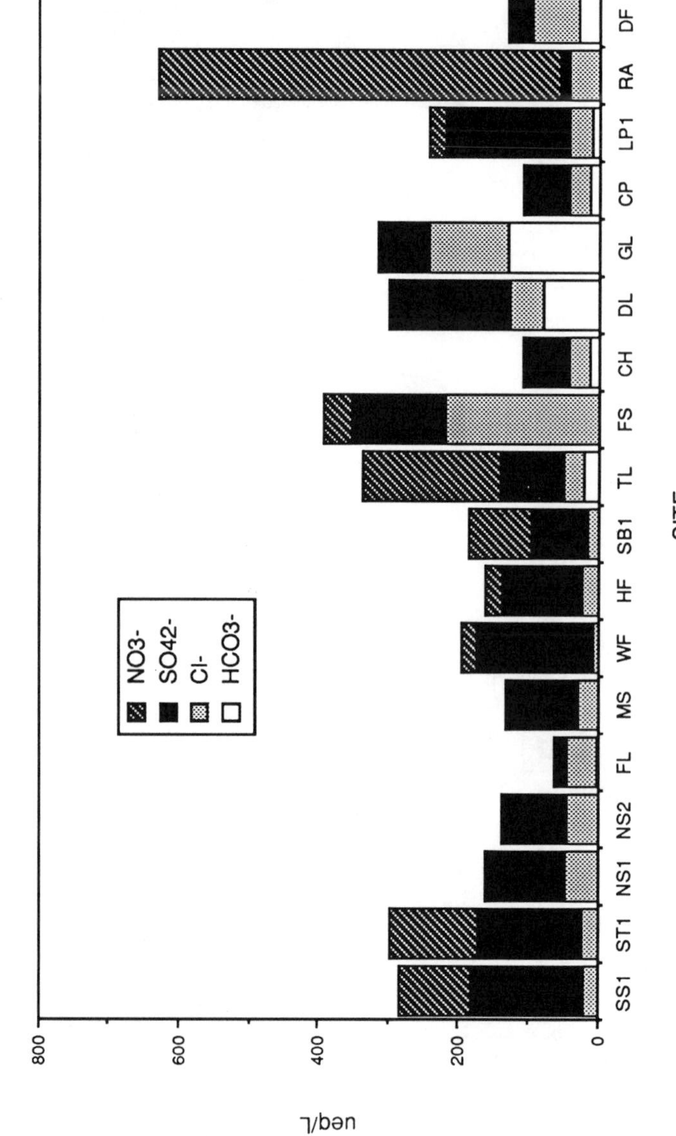
NO3-
SO42-
Cl-
HCO3-
ueq/L
0
200
400
600
800
SS1
ST1
NS1
NS2
FL
MS
WF
HF
SB1
TL
FS
CH
DL
GL
CP
LP1
RA
DF
SITE

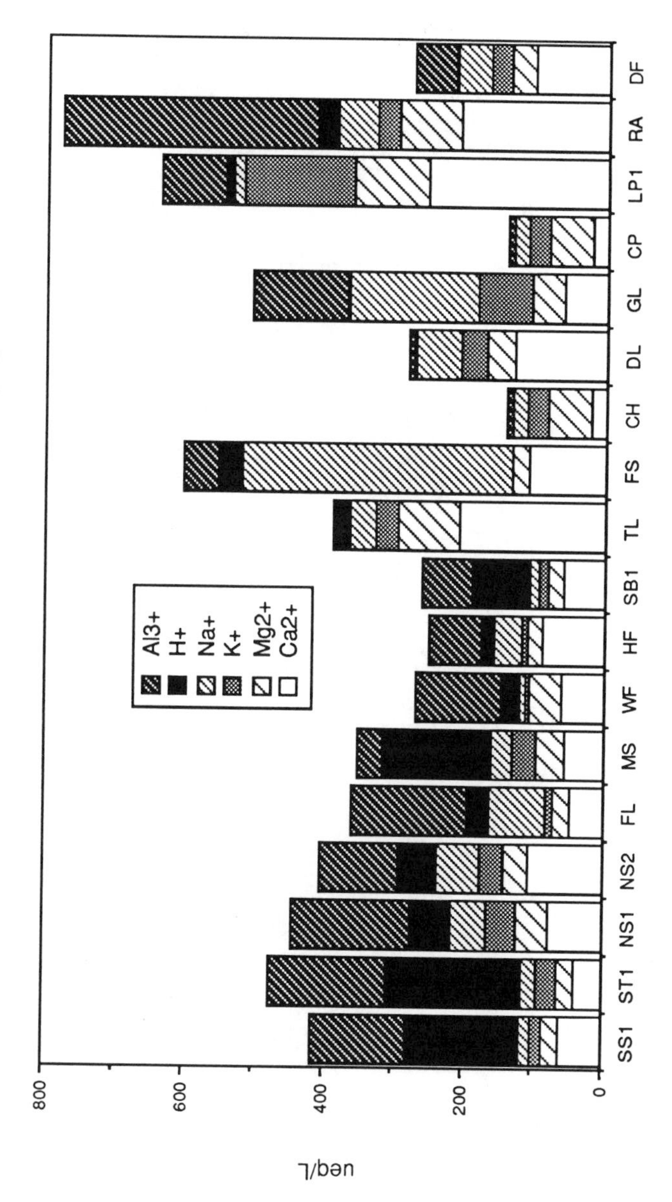

Figure 8.40. A horizon soil solution concentration of anions (top) and cations (bottom) in IFS sites. See Figure 8.32 for legend.

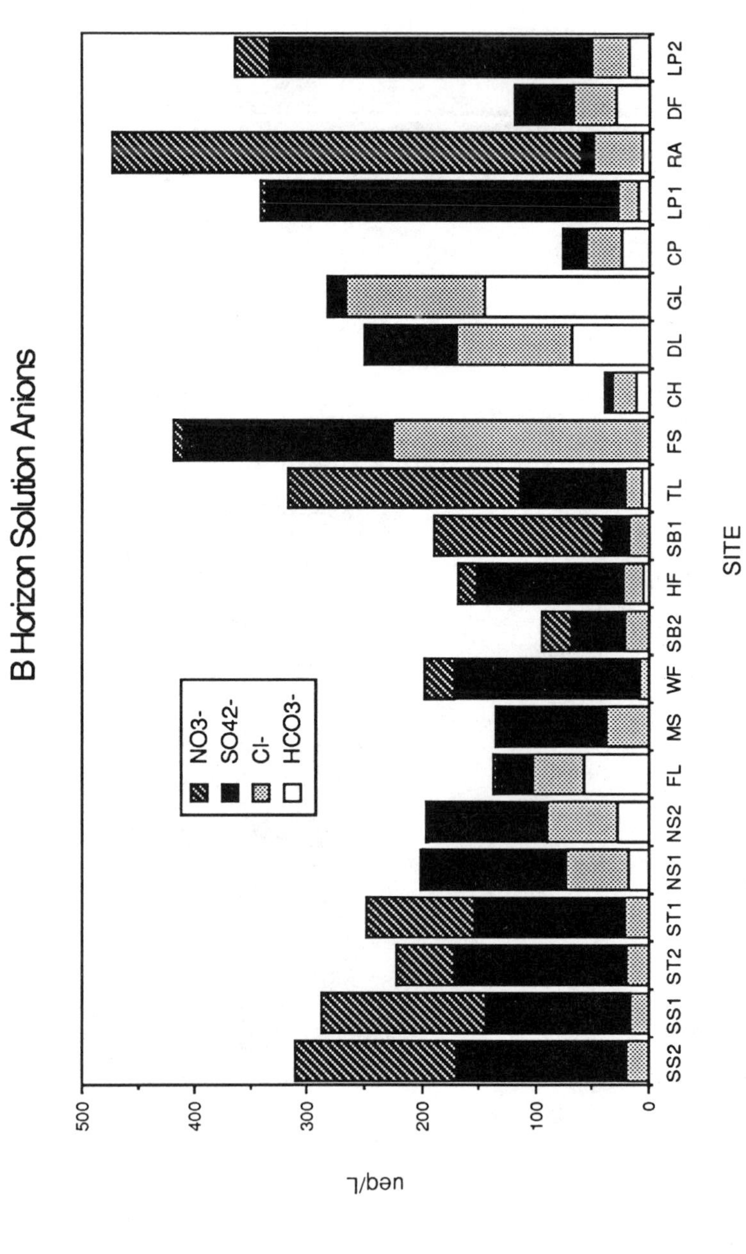
B Horizon Solution Anions
NO3-
SO42-
Cl-
HCO3-
ueq/L
0
100
200
300
400
500
SITE
SS2
SS1
ST2
ST1
NS1
NS2
FL
MS
WF
SB2
HF
SB1
TL
FS
CH
DL
GL
CP
LP1
RA
DF
LP2

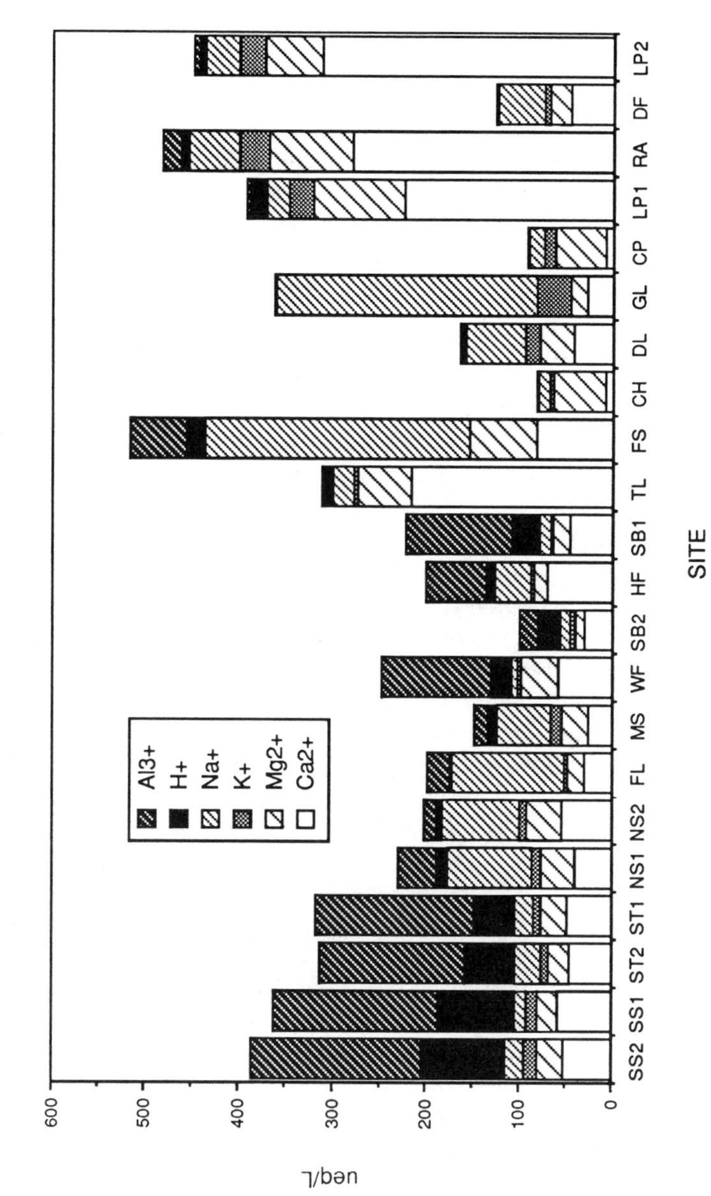

Figure 8.41. B horizon soil solution concentrations of anions (top) and cations (bottom) in IFS sites.

of as a function of total anion leaching (Johnson and Cole 1980). The net production of anions within the soil (e.g., by oxidation or hydrolysis reactions) must result in the net production of cations (normally H^+), whereas the net retention of anions (by either adsorption or biological uptake) must result in the net retention of cations.

The major anions in most soil solutions are Cl^-, SO_4^{2-}, NO_3^-, HCO_3^-, and organic anions (Johnson and Cole 1980). (Phosphate, borate, fluoride, and other anions may be present in trace amounts but very seldom constitute major components of total anion charge in soil solution.) Chloride may dominate soil solutions in coastal areas (e.g., Johnson 1981; Wright et al. 1988), but in that it usually enters the ecosystem as a salt it does not usually cause a net loss of base cations from the ecosystem. (For reasons discussed next, however, base cations entering with Cl^- may be retained by very acid soils.)

The other four major anions mentioned often undergo oxidation-reduction and hydrolysis reactions that cause the net production or consumption of H^+, which in turn strongly affects the net retention or release of base cations from the ecosystem. Carbonic acid, which is formed by the dissolution and hydrolysis of CO_2 in water, is the major natural leaching agent in many temperate and tropical ecosystems (McColl and Cole 1968; Johnson et al. 1977). Carbonic acid concentrations in soil solution are many times greater than in precipitation or throughfall because of the high levels of CO_2 in the soil atmosphere. On dissociation, carbonic acid forms H^+ and HCO_3^-, H^+ exchanges for a base cation (or causes the dissolution of a mineral), and a bicarbonate salt leaches from the system. Because carbonic acid is a very weak acid, HCO_3^- becomes associated with H^+ to form carbonic acid at low soil solution pH. Thus, the carbonic acid leaching mechanism is self-limiting, and eventually becomes inoperable during the soil acidification process (Reuss and Johnson 1986).

Organic acids are often the major cation leaching agents in extremely acid soils that occupy both boreal and tropical-subtropical regions (Kononova 1966; Johnson et al. 1977; Ugolini et al. 1977). These acids are stronger than carbonic acid and can produce low solution pH while providing organic counter-anions for cation leaching. Organic acids are also responsible for the chelation and transport of Fe and Al from surface (E or albic) to subsurface (Bs or spodic) horizons during the podzolization process (Kononova 1966). In theory, organic acid leaching, like carbonic acid leaching, is self-limiting because organic acids are typically weak acids. However, organic acids can produce very low soil and soil solution pHs and still remain active in leaching processes (Johnson et al. 1977; Ugolini et al. 1977).

Sulfate and nitrate are the anions of strong acids, and therefore leaching by sulfuric and nitric acids is not self-limiting because of solution pH alone. The deposition or other inputs of either S or N in excess of biological demands for these nutrients will ultimately cause an increase in SO_4^{2-} or NO_3^- availability within the soil. In the case of SO_4^{2-}, inorganic adsorption processes may prevent increased leaching, but this is seldom true in the case of NO_3^-. In that soil SO_4^{2-} adsorption is strongly pH dependent (increases with decreasing pH); there can be a negative feedback involved in sulfuric acid leaching that is at least partially self-limiting. The processes regulating the net retention and release of S and N and the implications of these and several other processes on ecosystem H^+ retention or loss are discussed in detail in Chapters 5, 6, 7, and 11.

Differences in the chemical and biological properties among soil horizons often cause great differences in the rates of anion production, consumption, and therefore leaching at different depths of a particular soil profile. Nitrate production, for example, is almost invariably greater in litter and A horizons than in B horizons, whereas nitrate consumption by uptake may occur throughout the profile. Sulfate adsorption almost invariably occurs in B horizons (Johnson and Todd 1983), often causing markedly lower total anion and cation leaching rates from system boundaries defined in subsurface horizons than within the system in surface horizons (e.g., Johnson et al. 1985). Organic acids are produced primarily in litter layers and often play a dominant role in surface horizon leaching, but they usually adsorb in Bh or Bhs horizons, resulting in an increase in pH and bicarbonate concentrations in systems not dominated by nitrate or sulfate leaching (Johnson et al. 1977; Ugolini et al. 1977). Thus, the depth in the soil at which the system boundaries are drawn has a considerable influence upon calculated exports of nutrients.

The anion mobility paradigm is a very precise and simple means of accounting for total cation leaching from soils, but it does not explain or account for the net retention or loss of individual cations. If soil exchange sites rather than primary mineral dissolution is the major factor controlling soil solution cation composition, then the composition of the cations in soil solution can be described fairly accurately by cation exchange equations developed more than 50 years ago (see review by Reuss 1983). The Gapon equation is one such formulation that we have found particularly useful in interpreting the results of the IFS data:

$$\frac{EM^{a+}\,(M^{b+})^{1/b}}{EM^{b+}\,(M^{a+})^{1/a}} = Q \qquad [1]$$

where X = exchange phase equivalent fraction, [ ] = soil solution activity ($molL^{-1}$), M^{a+} = cation of valence a, M^{b+} = cation of valence b, and Q = selectivity coefficient (constant).

In essence, Eq. [1] predicts that the concentration of a given cation in soil solution is governed by the proportion of this cation on the soil exchange complex and the total ionic concentration in soil solution. Reuss (1983) pointed out some very interesting and pertinent aspects of these equations with respect to the effects of increases in total ionic concentration in soil solution: as total ionic concentration increases, the ratios of tri- to di- to monovalent cations increase. In other words, as ionic concentration increases, cation concentrations increase as follows: $Al^{3+} > Ca^{2+}, Mg^{2+} > K^+, Na^+, H^+$. To illustrate this, consider $M^{2+} - Al^{3+}$ exchange (where $M^{2+} = Ca^{2+} + Mg^{2+}$) and solve Eq. [1] for Al^{3+} concentration:

$$Al^{3+} = \frac{(EAl^{3+})^3 (M^{2+})^{3/2}}{(EM^{2+})^3 Q^3} \qquad [2]$$

If we assume that the exchangeable fractions are constant (a reasonable assumption for the short term), we have:

$$(Al^{3+}) = K' \cdot (M^{2+})^{3/2}$$

$$\text{Where } K' = \frac{(XAl^{3+})^3}{(XM^{2+})^3 Q^3} = \text{a constant.}$$

Equation [3] indicates that soil solution Al^{3+} concentration increases to the $\frac{3}{2}$ power of soil solution M^{2+} concentration if exchange processes are controlling soil solution composition. By the same principles, soil solution Al^{3+} concentrations increases to the third power and soil solution M^+ ($K^+ + Na^+ + H^+$) concentration, and M^{2+} concentration increases to the second power of soil solution M^+ concentration.

Thus, Al^{3+} concentrations in soil solution increase not only as the soil acidifies (i.e., as the proportion of Al^{3+} on the exchange complex increases) but also as the total ionic concentration of soil solution increases. Equation [1] also implies that, of the major cation nutrients, K^+ leaching will be the least affected by increases in total anion or cation leaching rate.

Equation [1] also shows that the concentration of H^+ increases to some extent (although not as much as Al^{3+}, Ca^{2+}, or Mg^{2+}) with increases in total ionic concentration. This increase in H^+ (decrease in pH) will cause HCO_3^- and organic anions to protonate (become H_2CO_3 and uncharged organic acids), leading to a decrease in the concentrations of these natural anions. This replacement of natural, weak acid anions with atmospherically introduced, strong acid anions (SO_4^{2-} and NO_3^-) is referred to as an "anion shift," and it leads to less increase in the total leaching rate than would occur with the simple addition of SO_4^{2-} and NO_3^- to the natural leaching of HCO_3^- and organic anions. Krug and Frink (1983) argued that this anion shift can result in no net increase in total cation leaching, even though the anion composition of soil solution changes dramatically. This is an extremely unlikely, if not impossible, scenario in that it would require that H^+ concentration increase on a one-for-one basis with SO_4^{2-} and NO_3^-. However, the mitigative effect of the anion shift can be significant in acid soils, where H^+ is a significant component of soil solution cation composition.

The composition of cations on the exchange sites exerts a strong influence on the net retention or release of base cations from soils. It is intuitively apparent that cations in shortest supply on the exchange sites will be the least subject to leaching for a given a total anion and cation leaching rate. Equation [1] supports this intuition.

If we consider $H^+ - Ca^{2+}$ exchange and solve for H^+, we have:

$$(H^+) = \frac{(EH^+)\,(Ca^{2+})^{1/2}}{(ECa^{2+})\,Q} \qquad [4]$$

If we assume that the exchangeable fractions are constant (a reasonable assumption for the short-term), we have:

$$(H^+) = K' \cdot (Ca^{2+})^{1/2} \qquad [5]$$

as in Equation [3]. Taking the negative log of Equation [5], we have the lime potential used by soil scientists to evaluate the lime requirement of soils (see Reuss 1983):

$$K_L = pH - \tfrac{1}{2}pCa \qquad [6]$$

where $K_L = -[\log(K')]$.

As noted by Reuss (1983), precipitation will cause the soil to lose Ca (acidify) if its lime potential is less than that of the soil, and precipitation will cause the soil to gain Ca (basify) if its lime potential is greater than that of the soil. By the same principles, precipitation may cause the soil to gain or lose any other cation depending on the ratios of cations in precipitation and the composition of the exchange complex. It is noteworthy that the potential for precipitation to acidify or basify the soil is not solely a function of pH. It is possible that precipitation of low pH and high Ca concentration will be less acidifying to a given soil than precipitation of higher pH and lower Ca concentration.

There obviously is an inverse relationship between the acidification of soil and the acidification of soil solution. That is, if H^+ leaves the soil solution to exchange for a base cation, the soil is acidified and the soil solution is basified (e.g., when the lime potential of precipitation is less than that of the soil). Conversely, if Ca^{2+} or another base cation leaves the soil solution to displace H^+ or Al^{3+}, the soil is basified and the soil solution is acidified. It is important to note that both extremely acid soil *and* the presence of strong organic or mineral acid anions are necessary for the mobilization of Al^{3+} from exchange sites into solution; extremely acid soils are a necessary but not sufficient condition for Al^{3+} mobilization.

It is not possible to generalize as to whether a given class of ecosystems will always gain or lose base cations. However, certain general patterns should hold true if cation exchange rather than weathering is the primary regulator of cation leaching. For a given total anion (or cation) leaching rate, the cation exchange equations predict that the more acid northern, high-elevation sites will have lower rates of base cation leaching and higher rates of Al and H^+ leaching than the less acid southern, low-elevation soils. Total cation leaching rates are not constant, yet the pattern of greater Al and H^+ leaching in northern and high-elevation sites still emerges (see Figure 8.37). Notable exceptions to this pattern are the Norway and Findley Lake, Washington sites. Although subsoils from these sites are extremely acid (B horizon base saturation less than 10%, sum of cations method), B horizon soil solution Al and H^+ are very low and HCO_3^- is present in significant amounts (see Figures 8.40 and 8.41). The presence of HCO_3^- combined with the fact that Na^+ is the dominant cation in B horizon solutions from these sites suggests that weathering of a sodium-bearing mineral rather than cation exchange is the major factor regulating base cation leaching. For the Findley Lake site, this hypothesis is supported by the fact that N-rich plagioclase feldspars are present and that the Findley Lake soil is second highest in average weight of NaO_2 content among all the IFS sites tested (see Chapter 10, Figures 10.2 and 10.3). Mineral acid anion concentrations are very low in the Findley Lake soil solutions (the site is virtually unpolluted; see Figures 8.40 and 8.41), and thus there is no counter ion for the mobilization of Al^{3+}. A horizon soil solutions at the Findley Lake site are quite acid ($pH < 4.5$), high in total Al, and devoid of HCO_3^-, however, because of the organic acids (as was noted in previous studies by Johnson et al. 1977 and Ugolini et al. 1977).

An Evaluation of the Effects of Atmospheric Deposition on Soil Leaching in the IFS Sites

There are many philosophies concerning how the effects of acid deposition on forest soil cation leaching should be evaluated, most of which revolve around an analyses of H^+ budgets in one form or another. This subject is discussed in detail in Chapter 11. Here, we will take only a brief, simple overview of the probable role of acid deposition in causing the observed leaching patterns based simply on gross cation budgets and soil solution anion composition.

As a first approximation of the role of acid deposition in soil leaching, we can examine the soil solution anion composition and make certain simplifying assumptions as to reasonable background (unpolluted) levels of SO_4^{2-} and NO_3^- in soil solution (Cole and Johnson 1977; Mollitor and Raynal 1982). A reasonable background level for SO_4^{2-} is between 20 and 30 μeq L^{-1}; SO_4^{2-} does not exceed these levels even in coastal areas of unpolluted southeast Alaska (Johnson 1981). In A horizon solutions, only the sites in the less polluted northwestern United States (Findley Lake, red alder, and Douglas fir) have SO_4^{2-} concentrations approaching these low levels. All the sites from the eastern United States as well as Turkey Lakes, Ontario, and Norway have A horizon solution SO_4^{2-} levels that very likely reflect pollutant inputs of S. In many of the latter sites, SO_4^{2-} is in fact the dominant anion in A horizon solutions (Norway, Whiteface, both Coweeta sites, Duke, and Oak Ridge loblolly; see Figure 8.40).

It is somewhat more difficult to assign a reasonable background level of NO_3^- in soil solution, because there are larger reserves of N within even natural systems that can be mobilized under certain conditions (e.g., insect or disease attack, or simply with natural stand senescence; Vitousek et al. 1979; Swank et al. 1982). It has also been shown that the excessive fixation of N by N fixing (e.g., red alder) can cause high levels of NO_3^- in soil solution (Van Miegroet and Cole 1984). The many factors entering into the retention and release of NO_3^- and other forms of N are discussed at length in Chapter 6.

Within the IFS network, we have five sites where NO_3^- is either one of the major anions or the dominant anion in soil solution: Smokies, Becking; Smokies, Tower; Smokies, Beech; Turkey Lakes; and Red Alder (see Figures 8.40 and 8.41). As discussed in Chapter 6, atmospheric N depositon is a major factor in the high NO_3^- levels in the Smokies sites, but atmospheric deposition cannot be ascribed as the major cause of high NO_3^- at the Turkey Lakes and Red Alder sites. Both sites are naturally N rich, and in the case of red alder, this enrichment results from excessive N fixation (see Chapter 6).

Comparisons of Leaching, Vegetation Increment, and Uptake

As noted, cation leaching can be described as a function of purely physical and chemical processes (even though these processes are affected by other biological processes). In contrast, uptake is driven by biological requirements, which vary considerably among nutrients and species and with stand age and growing conditions, but are not controlled by soil exchangeable reserves. Soil exchangeable reserves may have some effect on uptake but by no means regulate it.

When evaluating the potential for soil change, it is imperative to consider the effects of vegetation uptake and increment as well as leaching. Vegetation increment, or the accumulation of nutrients in permanent woody tissues, represents a net export from the soil that can be of considerable significance, especially with regard to Ca (e.g., Johnson and Todd 1987). That part of vegetation uptake that is recycled annually will not result in a net export of nutrient from the litter/soil system, but can result in considerable redistribution (for example, from subsoil to surface soil or litter; Alban 1982; Johnson et al. 1989).

The effects of uptake on soil change and the interactions between uptake and leaching deserve special consideration, especially in managed forests (Duke, Florida, Georgia, Oak Ridge loblolly, Douglas fir, Coweeta pine and hardwood, Norway). Even if there is little or no harvesting, however, tree uptake and recycling can have a major effect on the distribution of cation nutrients in the soil and consequently on leaching rates. Alban (1982) showed that species which take up and recycle large amounts of Ca tend to acidify subsoils and basify surface soils whereas species which take up and recycle low amounts of Ca tend to acidify surface soils and leave subsoils at higher base status. On a whole-soil basis, this may or may not result in a large change in exchangeable Ca^{2+} reserves, but the changes in the distribution of exchangeable Ca among horizons may change leaching export of Ca^{2+} considerably. For instance, Johnson and Todd (1987) found that high rates of Ca^{2+} uptake and recycling in a mixed oak forest had depleted subsoil soil exchangeable Ca^{2+} and reduced base saturation relative to a loblolly pine site in eastern Tennessee. This in turn caused lower Ca^{2+} and total cation leaching but greater Mg^{2+} leaching in the mixed oak site than in the loblolly pine site.

Thus, tree uptake can result in the conservation of cations in short supply by depleting soil exchangeable reserves. Cations in most demand by the forest can be effectively conserved by depletion of soil exchangeable pools. This depletion by uptake could progress well beyond the levels possible through leaching alone, because plants selectively take up nutrients on an individual basis rather than on the basis of their availability on exchange sites, as is the case with leaching. Indeed, uptake can deplete soil pools to the extent that the system begins to accumulate one or more base cations, even with elevated inputs of H^+ from the atmosphere (Ulrich et al. 1980; Johnson and Richter 1984). This conservation is not always without negative side effects, however. In cases where all base cations are being conserved, such as at the Solling site (Ulrich et al. 1980), soil solution aluminum levels must rise (to balance continuing anion loading) with all the potential problems with toxicity to trees and aquatic systems.

The increment (net annual accumulation in woody tissue), uptake, and leaching of base cations, Ca, K, and Mg are shown in Figures 8.42 and 8.43. The increments of base cations, Ca, K, and Mg vary considerably from site to site but exhibit no particular pattern relative to the northern high-elevation versus southern high-elevation ordering. This is not especially surprising, in that increment is a function not only of primary productivity (which would be expected to be greater in the southern low-elevation sites) but also of stand age and species composition. There is as much as a 10-fold variation in the cation concentration in woody tissues among species (Marion 1979; Johnson et al. 1988b). This variation occurs within both angiosperms and gymnosperms and seems to be primarily a function of genus. For instance, the angiosperms *Quercus* and *Carya* and the gymnosperms *Thuja* and *Juniperus* have unusually high wood and bark Ca concentrations (Marion 1979; Alban 1982; Johnson et al. 1982, 1988a).

The interactions of productivity and wood cation concentration in determining increment can be seen by comparing Ca increments in the Duke loblolly pine, the Smokies beech (plot 1), and the Coweeta hardwood sites. These sites have the greatest increments of Ca (see Figure 8.42), but for very different reasons: The Duke loblolly site is very young and has a very high rate of net biomass accumulation (see Table DL-2, Appendix) and thus a high rate of Ca accumulation despite the fact that loblolly pine has very low wood Ca concentrations. The net biomass accumulations in the Smokies beech and Coweeta hardwood sites are much lower than in the Duke loblolly site (see Tables SB-2 and CH-2, Appendix), but the wood Ca concentrations are much higher.

There is little north-south or low- to high-elevation pattern in increment, but base cation uptake is considerably greater in the southern and low-elevation sites than in the northern and high-elevation sites, a pattern that naturally emerges from systematic geographical patterns in primary productivity (Cole and Rapp 1981). Thus, the potential for net depletion of base cations from the litter/soil system via increment biomass and nutrient increment does not relate to geographical location, but the potential for base cation redistribution within the litter/soil system (for instance, from lower to upper horizons, or from soil to litter) via nutrient uptake and recycling is greater in the southern, low-elevation sites than in the northern, high-elevation sites.

Within the IFS network, there are several sites where base cation leaching rivals or even exceeds uptake (see Figure 8.42). Some of these sites are in the northern high-elevation category, where uptake is lower because of lower primary productivity (Smokies Becking and Tower, Findley Lake, Turkey Lakes), but others are in the southern low-elevation category where leaching rates are high (Oak Ridge loblolly, Red alder). In these cases, leaching greatly exceeds vegetation increment, and leaching will likely be a major factor in any observed soil change. In three of the sites where leaching equals or exceeds uptake, leaching is dominated by atmospheric deposition (Smokies Becking, Smokies Tower, Oak Ridge Loblolly). Vegetation base cation increment is a major factor in several other sites, where it equals or exceeds leaching (Norway, Whiteface, Smokies Beech, Coweeta Hardwood, Duke, Georgia, and Coweeta Pine) (Figure 8.44). Base cation uptake in these sites greatly exceeds leaching and will likely be a major factor in any observed soil change.

The patterns for Ca and Mg leaching, increment, and uptake are similar to those of base cations in many respects, but there are some notable exceptions (see Figures 8.42 and 8.43). Ca uptake clearly exceeds leaching in the Findley Lake site, whereas the reverse was true for total base cations (see Figure 8.41). Ca increment greatly exceeds leaching in the Coweeta Hardwood and Georgia sites. Leaching takes on a much more important role for Mg in the Whiteface site, exceeding increment and equaling uptake. In contrast, increment plays a lesser role for Mg in the Coweeta Hardwood and Pine sites.

The patterns for potassium differ from those of the other cations in that uptake exceeds leaching in all sites, and increment exceeds leaching in several sites (Norway, Findley Lake, Whiteface, Smokies Beech, Coweeta Hardwood, Georgia, and Coweeta Pine) (see Figure 8.43).

To summarize, it is clear that the uptake and recycling of all base cations is much greater in the southern low-elevation sites than in the northern high-elevation sites (see Figures 8.42 and 8.43), and thus we can expect that the

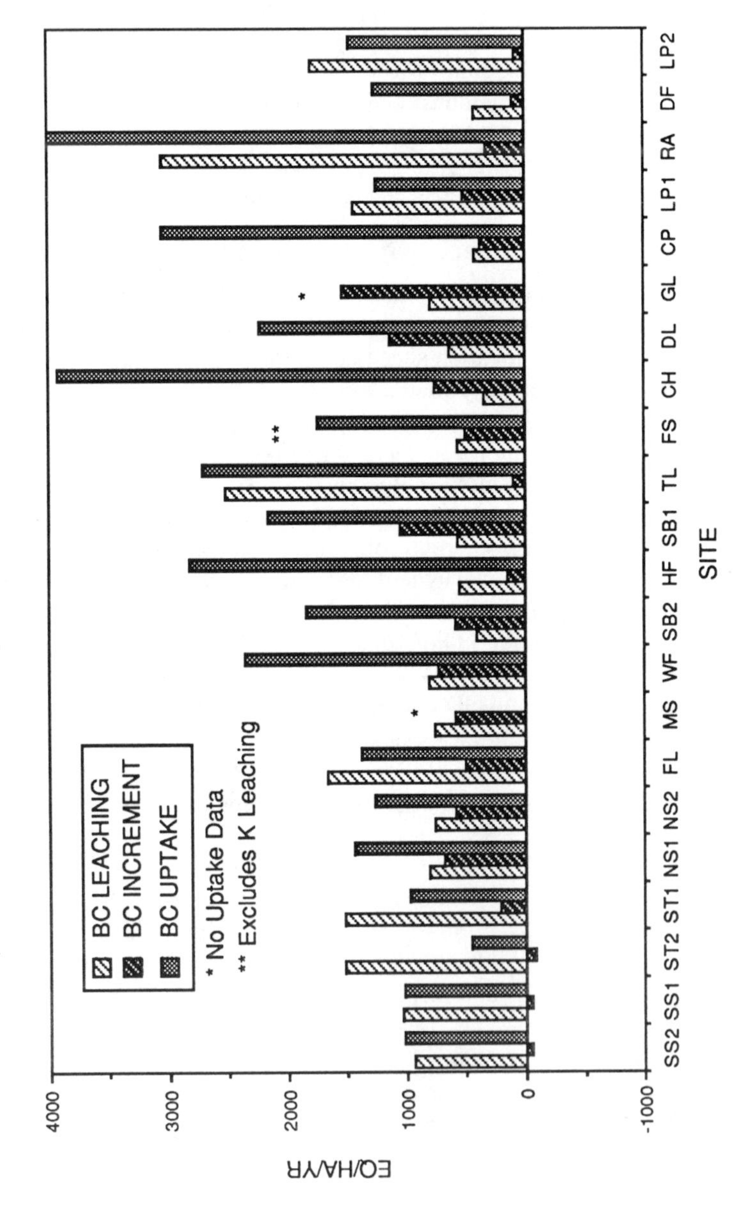
Base Cation Leaching, Vegetation Increment, and Uptake
BC LEACHING
BC INCREMENT
BC UPTAKE
* No Uptake Data
** Excludes K Leaching
EQ/HA/YR
4000
3000
2000
1000
0
-1000
SITE
SS2 SS1 ST2 ST1 NS1 NS2 FL MS WF SB2 HF SB1 TL FS CH DL GL CP LP1 RA DF LP2

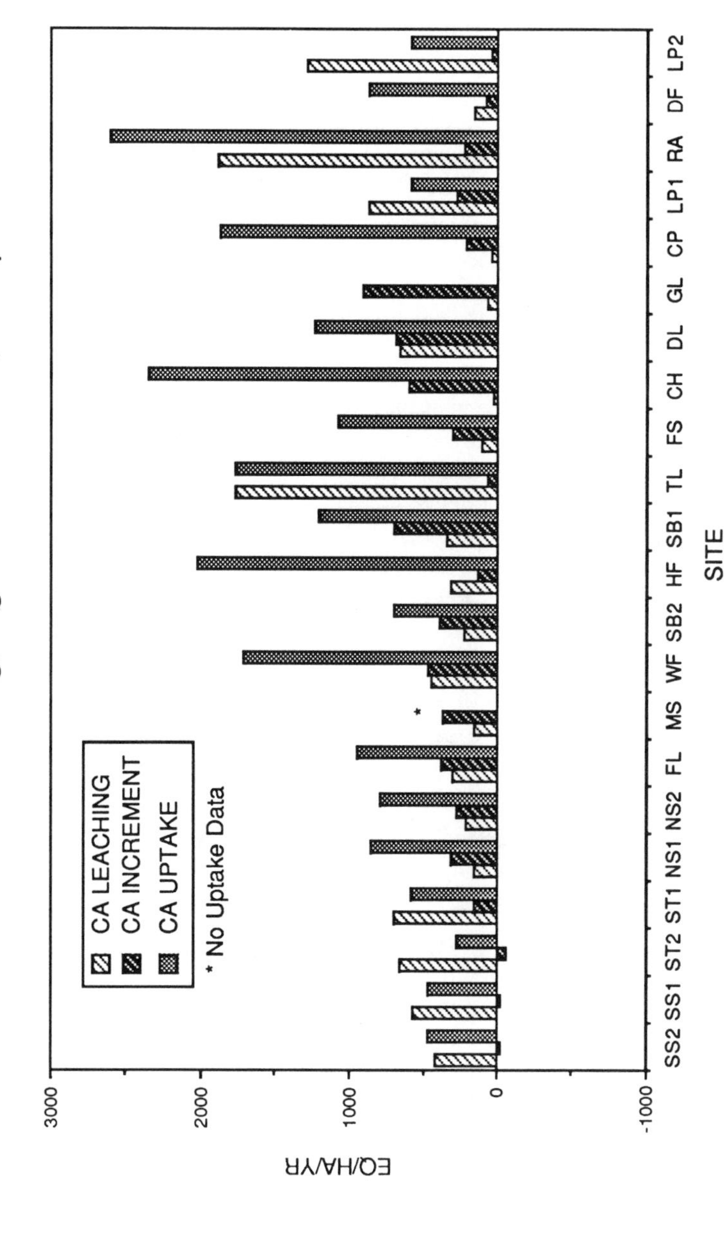

Figure 8.42. Base cation (top) and calcium (bottom) fluxes via leaching, vegetation increment, and uptake in IFS sites. See Figure 8.32 for legend.

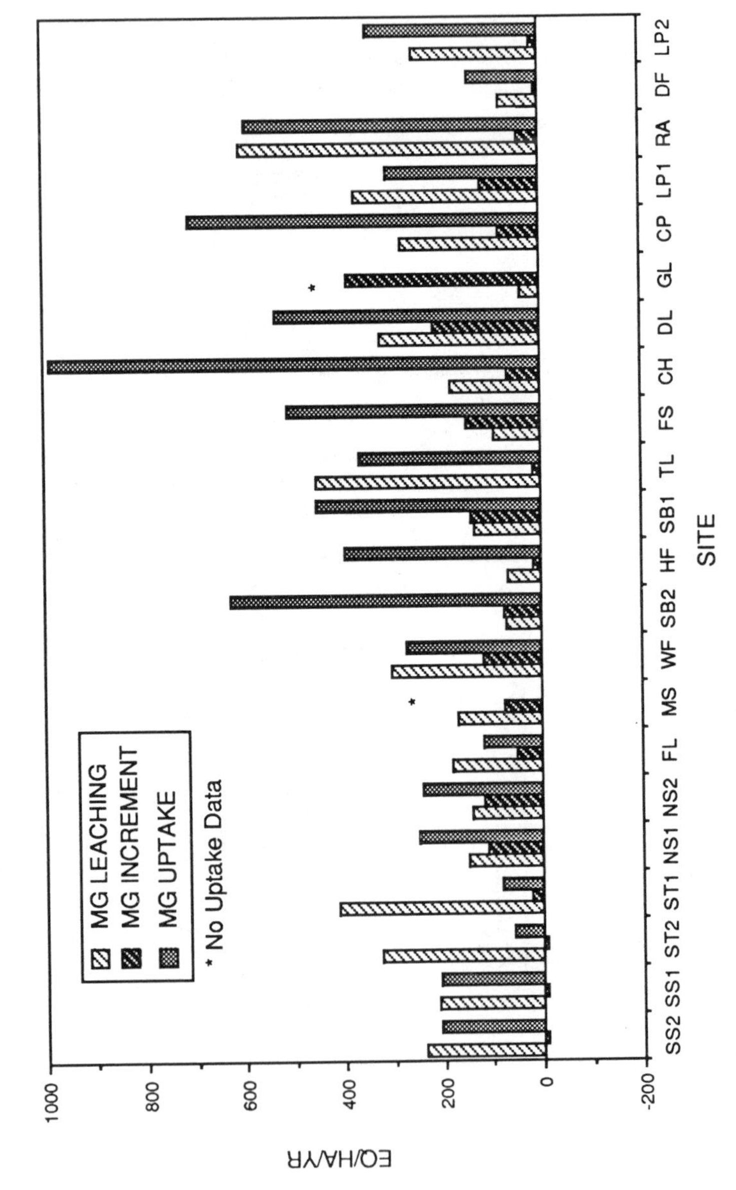
Magnesium Leaching, Vegetation Increment, and Uptake
MG LEACHING
MG INCREMENT
MG UPTAKE
* No Uptake Data
EQ/HA/YR
1000
800
600
400
200
0
-200
SS2 SS1 ST2 ST1 NS1 NS2 FL MS WF SB2 HF SB1 TL FS CH DL GL CP LP1 RA DF LP2
SITE

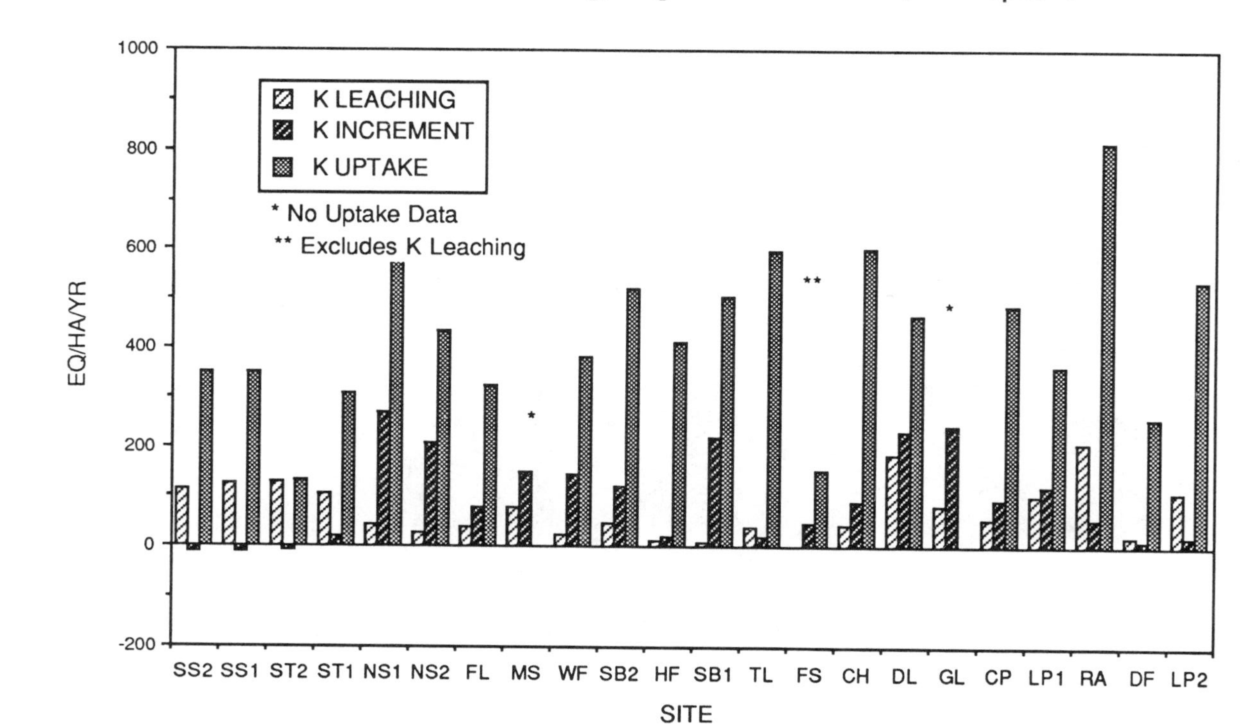

Figure 8.43. Magnesium (top) and potassium (bottom) fluxes via leaching, vegetation increment, and uptake in IFS sites. See Figure 8.32 for legend.

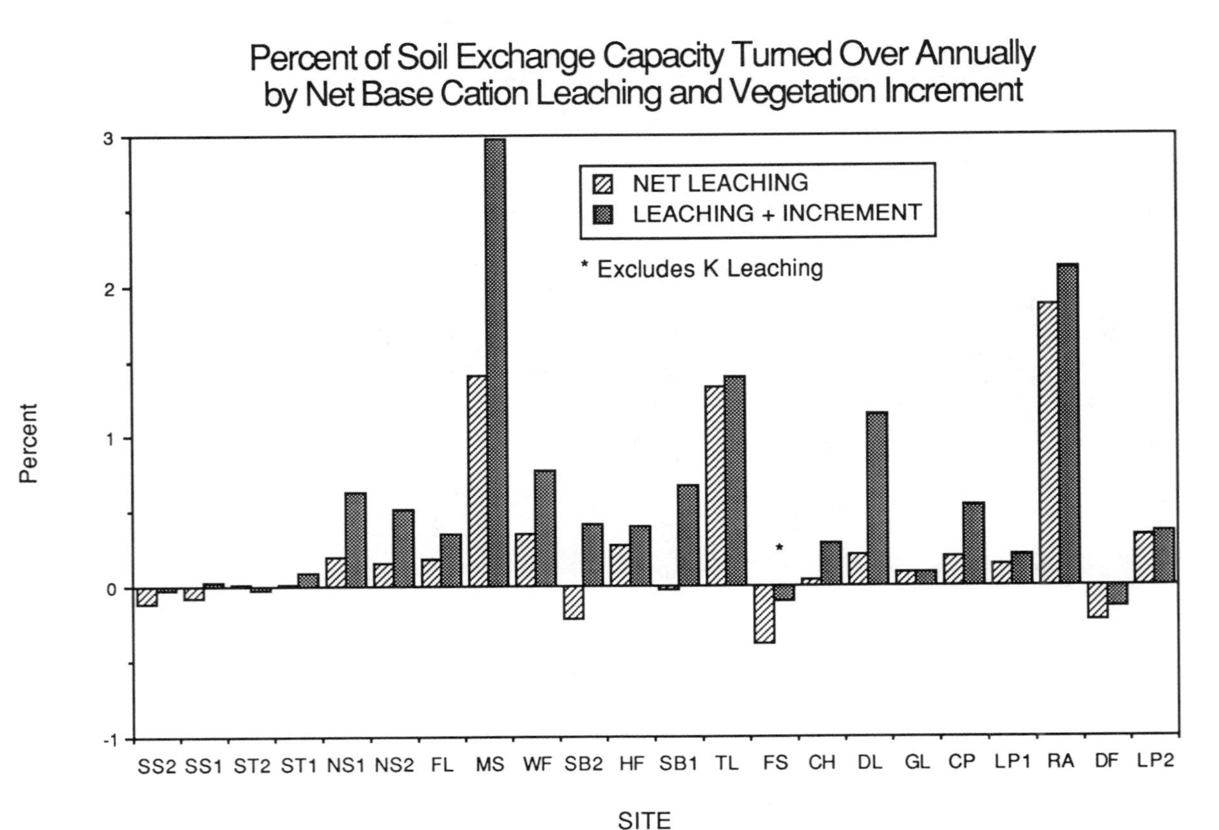

Figure 8.44. Base cation export via leaching and vegetation increment as percentage of soil cation exchange capacity (sum of cations in IFS sites. See Figure 8.32 for legend.

conservation of potentially limiting base cation nutrients via cycling will tend to be greater in the former sites than in the latter sites. As to the actual extent to which such conservation occurs, however, we can only speculate from the data at hand. With the use of the NuCM model (see Chapter 13), we can simulate the interactions of uptake and recycling and leaching in any given site and gain some appreciation as to the potential for uptake to act as a conservation mechanism for potentially limiting base cations. The situation in many northern high-elevation sites where leaching equals or exceeds uptake is quite interesting in that it has not been previously observed in other summarizations of base cation cycles (Cole and Rapp 1981). It is clear that leaching will play a major role in observed soil change (if any) in these sites.

Potential for Soil Change in the IFS Sites

It is very difficult to evaluate the potential for soil base cation change, even with an excellent data set on leaching, increment, and uptake such as that available in the IFS study, because of great uncertainties in the weathering rate. Chapter 10 gives us a detailed evaluation of the weathering at the IFS sites, but quantitative values for field weathering rates in most sites are still lacking. In some southern low-elevation sites (e.g., Duke, Oak Ridge Loblolly), the primary mineral pools of Ca are so low that a weathering rate of zero can be assumed (see Chapter 10), but in most cases the primary mineral pools are considerably larger than the exchangeable pools and weathering remains an unquantified but potentially major factor in projecting soil change.

One way to put sideboards on the problem is to compare the net fluxes of cations from the soil to exchangeable pools for a maximum estimate of potential change and to total soil pools for a minimum estimate of potential change. Such comparisons are made in Figures 8.44–8.48. In each figure, the net cation budgets with leaching only (leaching − deposition) and leaching plus biomass increment (leaching − deposition − vegetation increment) are expressed as a percent of exchangeable and total base cation pools.

To assess the potential for a change in soil acidity, the net base cation exports via leaching and increment for each site are shown as a percent of soil cation exchange capacity (by sum of cations) in Figure 8.44. The percent export of exchangeable base cation capital (shown in Figure 8.45 and discussed later) gives an index of the maximum potential for exchangeable base cation depletion but is not necessarily a good index of potential change in soil acidity. Specifically, in very acid soils, a large percentage change in base saturation (say, from 10% to 5%, for a 50% change) will be accompanied by relatively small change in exchangeable acidity (from 90% to 95%, for a 5.5% change). Thus, expressing potential changes in soil acidity by comparing exchangeable base cation pools alone would greatly exaggerate this potential for the more acid northern high-elevation sites.

The Red Alder, Turkey Lakes, and Maine sites show the greatest percent turnover of soil cation exchange capacity by leaching, suggesting the great-

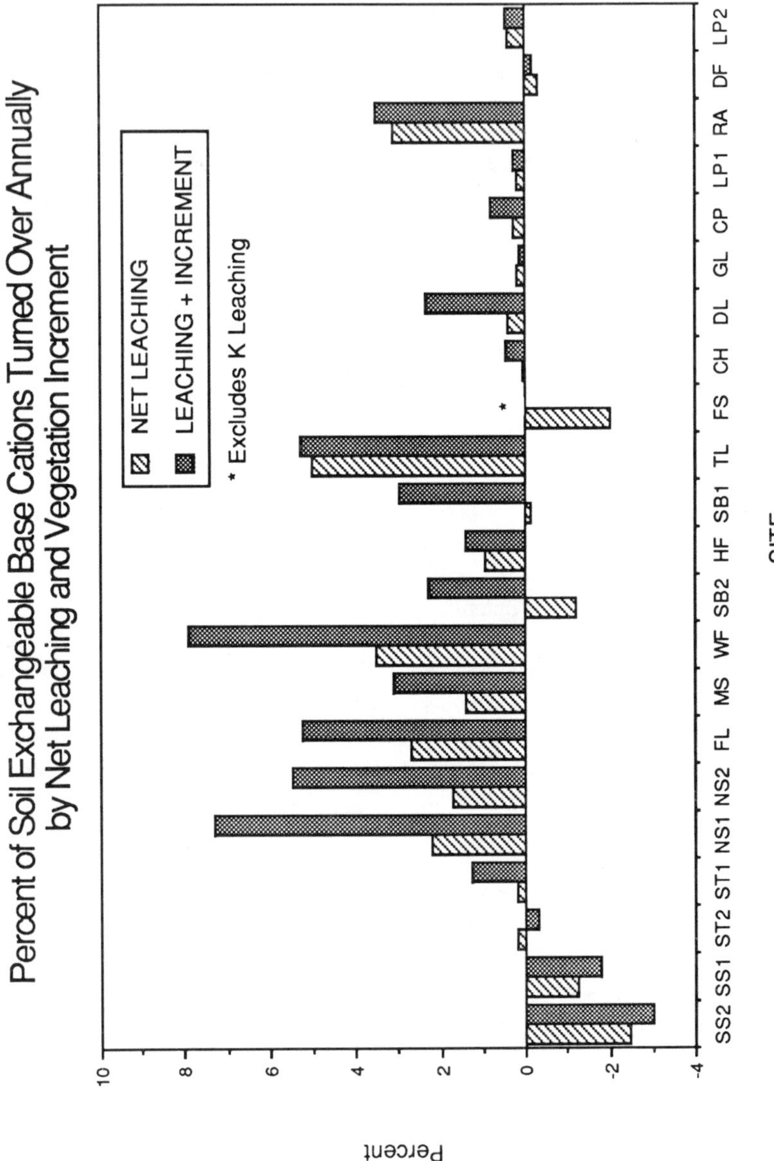
Percent of Soil Exchangeable Base Cations Turned Over Annually
by Net Leaching and Vegetation Increment
NET LEACHING
LEACHING + INCREMENT
* Excludes K Leaching
*
Percent
10
8
6
4
2
0
-2
-4
SS2 SS1 ST2 ST1 NS1 NS2 FL MS WF SB2 HF SB1 TL FS CH DL GL CP LP1 RA DF LP2
SITE

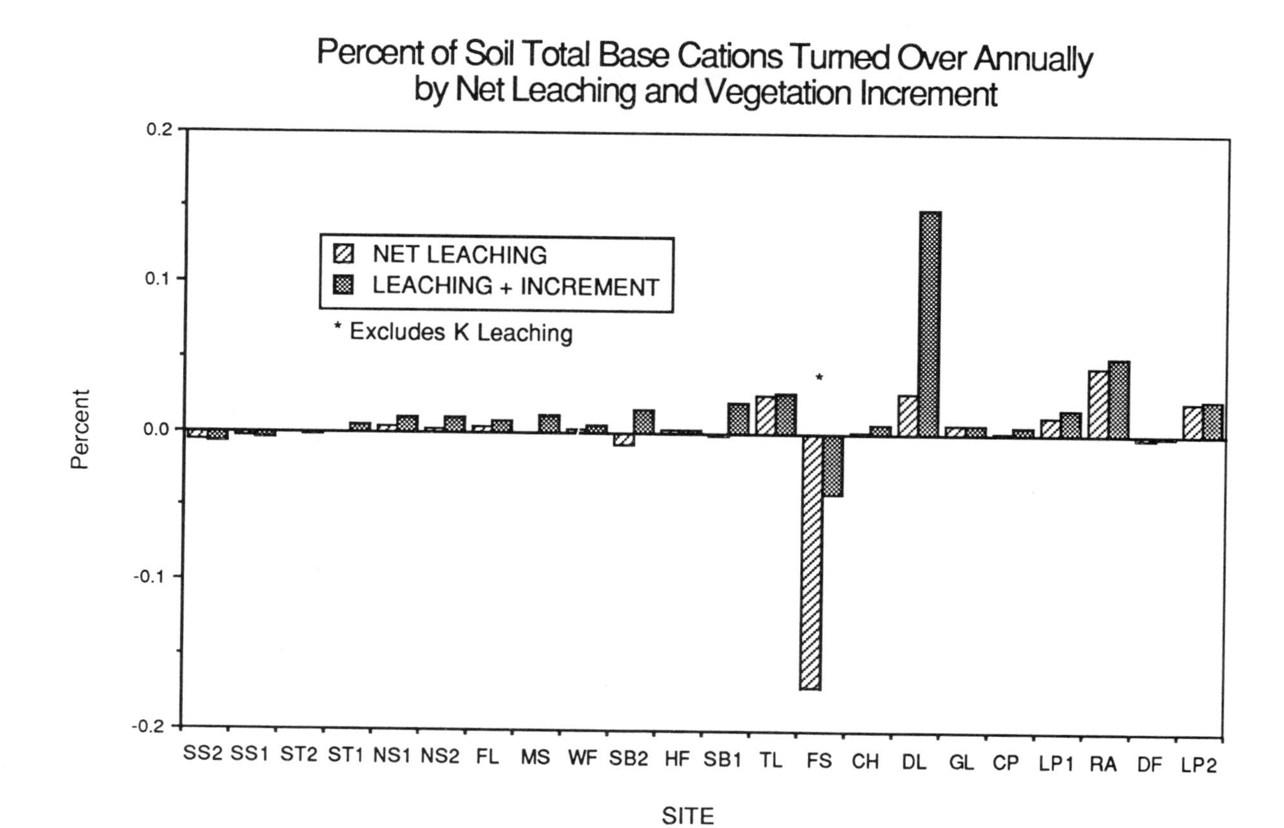

Figure 8.45. Base cation export via leaching and vegetation increment as percentage of exchangeable base cations (top) and total soil base cations (bottom) in IFS sites. See Figure 8.32 for legend.

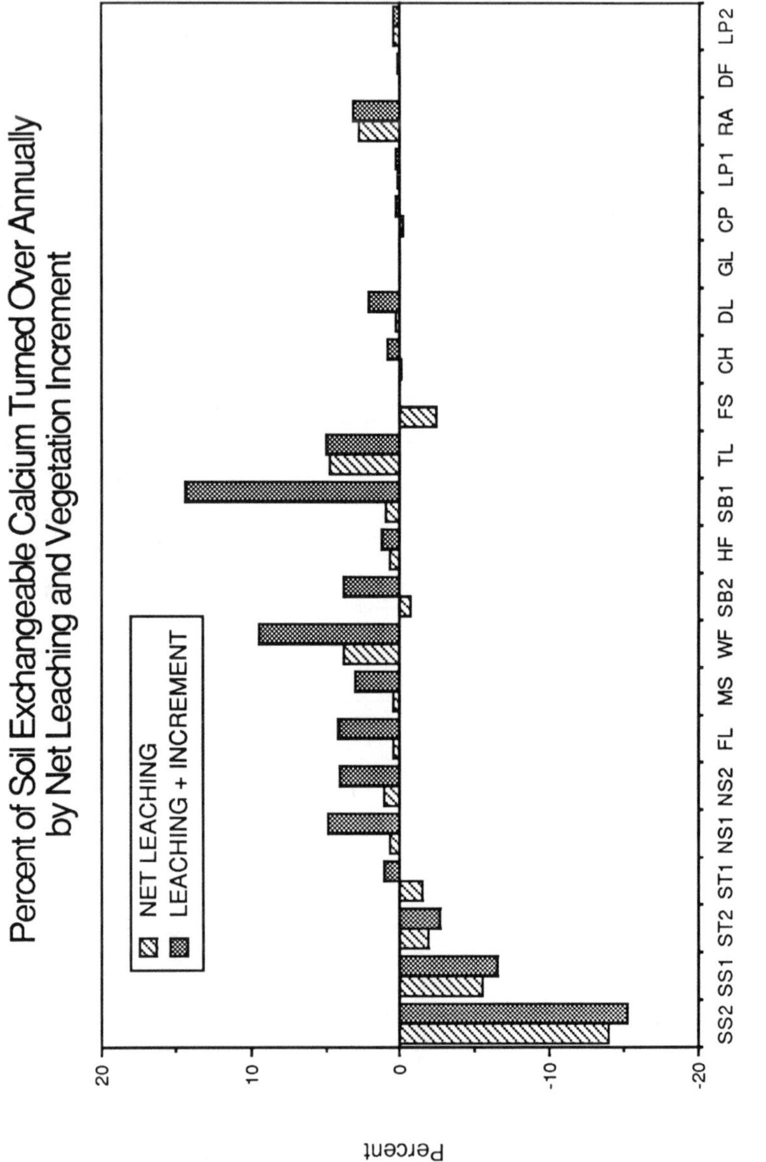
Percent of Soil Exchangeable Calcium Turned Over Annually
by Net Leaching and Vegetation Increment
NET LEACHING
LEACHING + INCREMENT
Percent
20
10
0
-10
-20
SS2 SS1 ST2 ST1 NS1 NS2 FL MS WF SB2 HF SB1 TL FS CH DL GL CP LP1 RA DF LP2
SITE

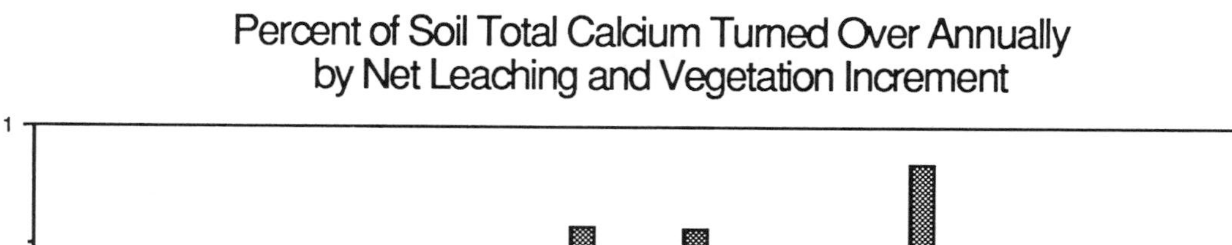
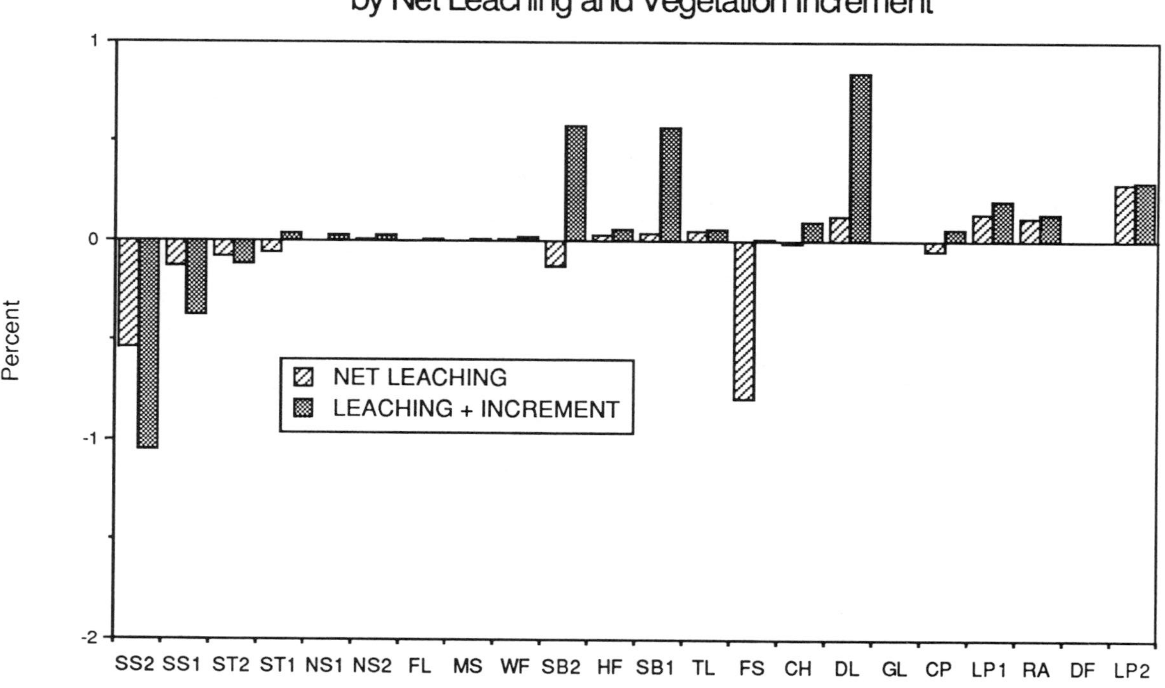

Figure 8.46. Calcium export via leaching and vegetation increment as percentage of exchangeable calcium (top) and total soil calcium (bottom) in IFS sites. See Figure 8.32 for legend.

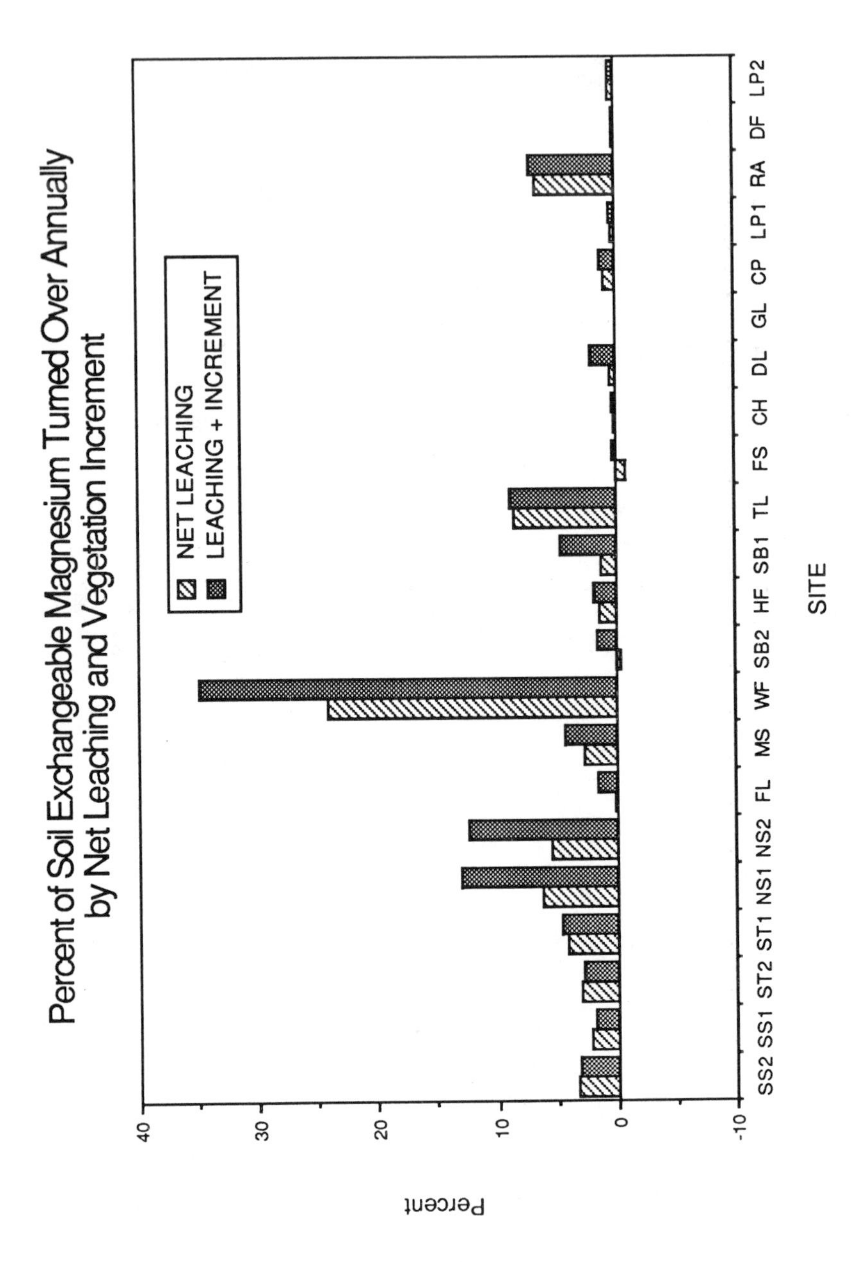
Percent of Soil Exchangeable Magnesium Turned Over Annually
by Net Leaching and Vegetation Increment
NET LEACHING
LEACHING + INCREMENT
Percent
40
30
20
10
0
-10
SS2 SS1 ST2 ST1 NS1 NS2 FL MS WF SB2 HF SB1 TL FS CH DL GL CP LP1 RA DF LP2
SITE

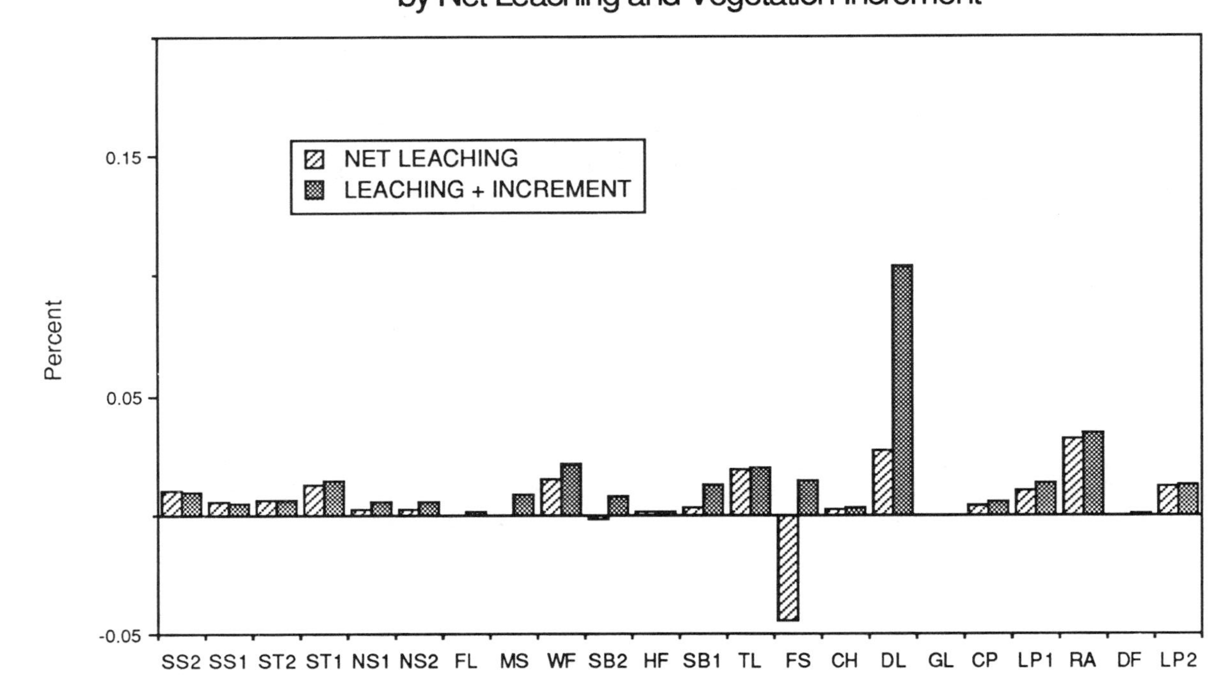

Figure 8.47. Magnesium export via leaching and vegetation increment as percentage of exchangeable magnesium (top) and total soil magnesium (bottom) in IFS sites. See Figure 8.32 for legend.

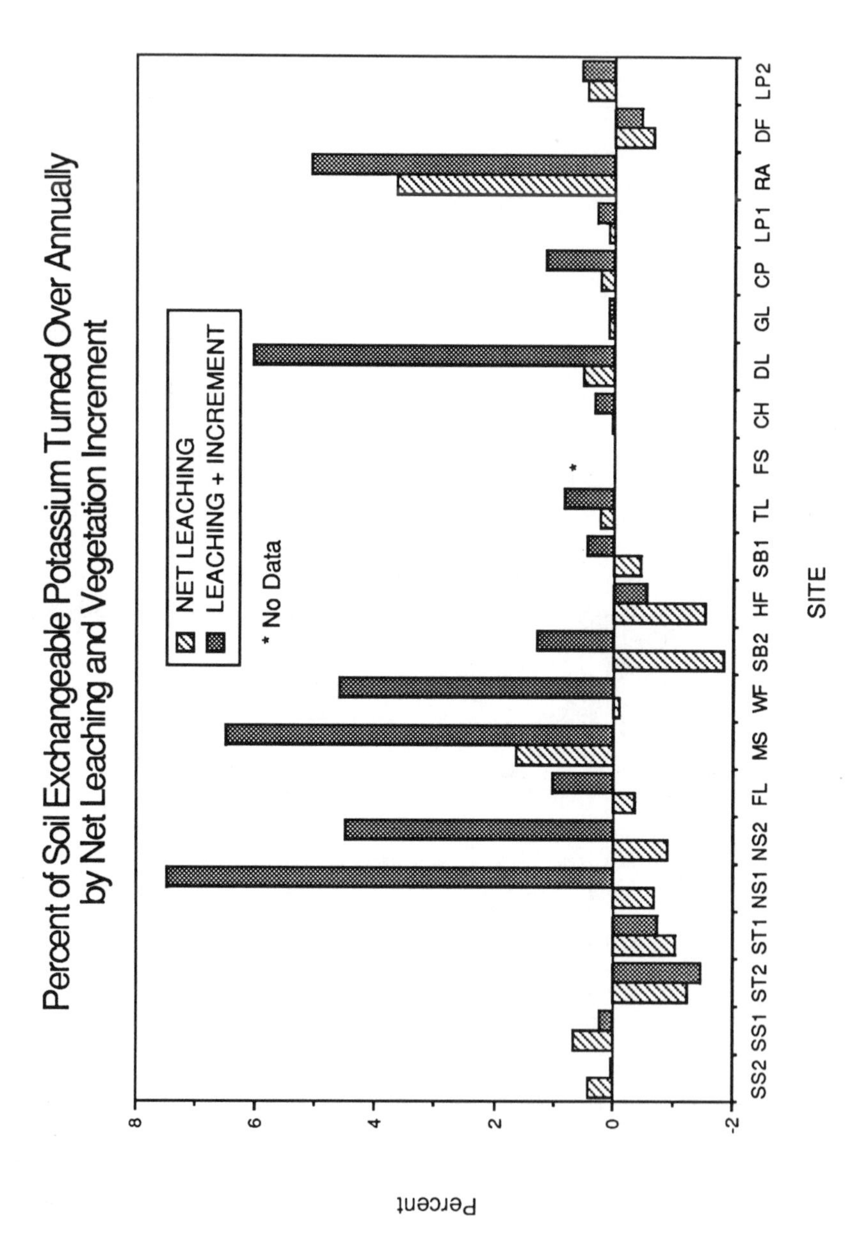
Percent of Soil Exchangeable Potassium Turned Over Annually
by Net Leaching and Vegetation Increment
NET LEACHING
LEACHING + INCREMENT
* No Data
*
Percent
8
6
4
2
0
-2
SS2 SS1 ST2 ST1 NS1 NS2 FL MS WF SB2 HF SB1 TL FS CH DL GL CP LP1 RA DF LP2
SITE

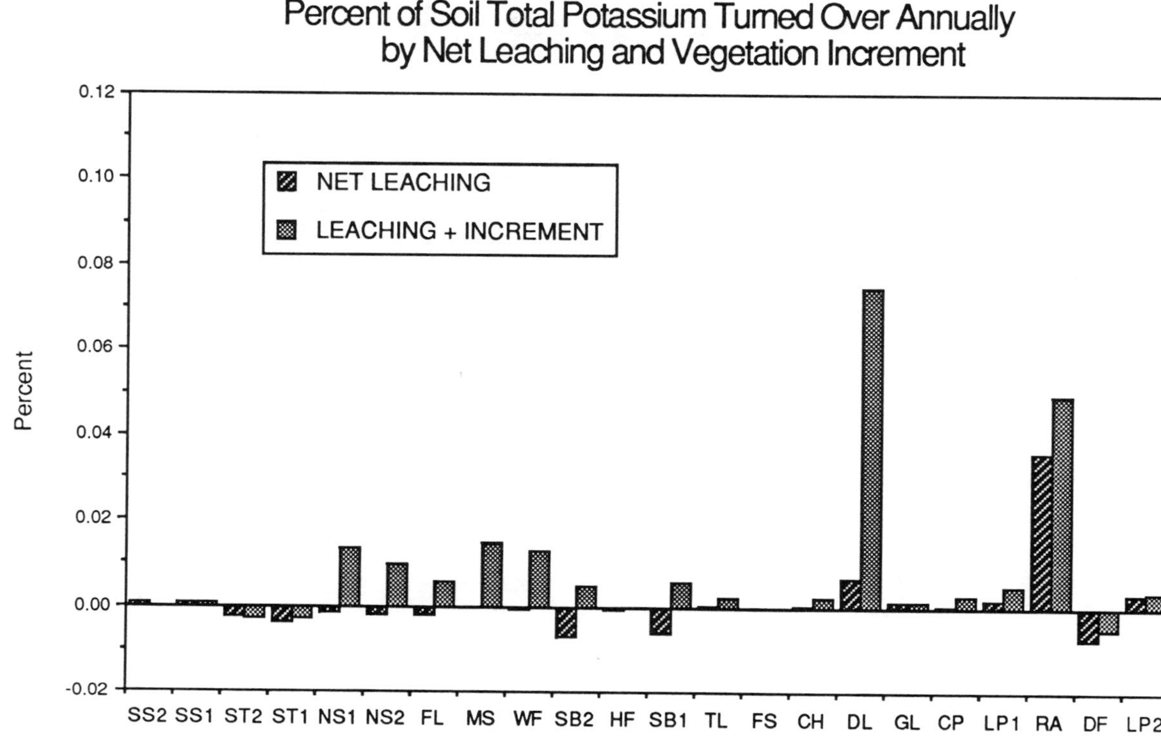

Figure 8.48. Potassium export via leaching and vegetation increment as percentage of exchangeable potassium (top) and total soil potassium (bottom) in IFS sites. See Figure 8.32 for legend.

est potential for soil acidification. Approximately 1% to 2% of the exchangeable cation pools could be leached from these sites annually, meaning that it would take about 25 to 50 years for a 50% change in base saturation in the absence of weathering (see Figure 8.44). When increment is added to leaching, the Maine site shows the greatest turnover of cation exchange capacity (~3%). The Duke site has a relatively low percentage for net leaching (0.2%), but a relatively high percentage for leaching + increment, indicating that biomass increment is the major factor in soil base cation depletion at this site. Binkley et al. (1989) found substantial reductions in exchangeable base cations during a 25-year period in soils near the Duke site and attributed the changes primarily to uptake and retention by forest vegetation. At the other end of the spectrum, the Smokies Becking and Tower sites have very low percentages of cation exchange capacity exported by leaching and increment, indicating very little possibility for soil change. This is as expected, in that (1) these soils are already extremely acid, (2) H^+ and Al^{3+} are the dominant cations in soil solution, and (3) the sites are now either accumulating or near balance with respect to base cation inputs and outputs.

As noted in Chapter 4, both simulation modeling (Reuss 1983) and empirical evidence indicate that fairly minor changes in base saturation within the 10% to 20% range can cause quite large increases in soil solution Al^{3+} concentration (see Figure 8.1; Chapter 10). Given this guideline, it would seem that several sites within the IFS network are sensitive to Al^{3+} mobilization (i.e., have base saturations less than 20%) (see Figures 8.1 and 10.2). Conversely, a reduction in Al^{3+} can be expected in those sites with currently high Al^{3+} (e.g., Smokies spruce sites) given a reduction in the mineral acid anion level. However, the sensitivity to Al mobilization will also depend on the availability of readily weatherable minerals. As noted previously, cation leaching in those sites with acid soils and low Al concentrations (Findley Lake, Norway) seems to be dominated by weathering of Na-bearing minerals rather than exchange processes.

Annual percentages of exchangeable base cation pools removed by net leaching and increment are highest in the Turkey Lakes, Whiteface, Norway, Findley Lake, Red Alder, Norway, and Maine sites (see Figure 8.45). Approximately 2% to 5% of the exchangeable base cations in these sites are removed from the soil annually by leaching, meaning that it would take approximately 10 to 25 years to turn over 50% of the exchangeable base cation pools. Vegetation increment increases this percentage in all but the Turkey Lakes and Red Alder sites. (Net leaching plus increment equal 4%–8% of exchangeable base cations, meaning it would take 6–12 years to turn over 50% of exchangeable base cation pools.) Sites with the lowest percentages of exchangeable base cation pools removed are the Smokies Becking, Smokies Tower, Coweeta Hardwood, Georgia Loblolly, and Oak Ridge Loblolly sites (less than 0.5% of exchangeable base cations are removed from the soil annually, or 100 years to turn over 50% of the exchangeable base cation pools). Sites showing negative percentages (Smokies Becking, Smokies Tower, Douglas fir) are accumulating base cations.

The Duke site shows by far the greatest annual percent of total base cation pool export (0.15%, or 300 years to deplete total soil cations by 50%), most of which is in vegetation increment. This reflects the very low reserve of primary minerals in this soil (Chapter 10) and the high rate of net biomass accumulation (see Figure 8.45). The Red Alder, Turkey Lakes, Oak Ridge Loblolly, and Smokies Beech sites have the next largest percentages of total base cation pools exported, but all are less than or equal to 0.05% (1000 years for a 50% depletion).

The patterns for individual cations differs somewhat from that of total base cations. Because of net ecosystem accumulation of Ca from atmospheric depositon, several sites have negative percentages with leaching alone (Smokies Becking, Tower, and Beech plot 2; Florida; and Coweeta Pine), and some have negative percentages even when increment is added (Smokies Becking, Tower, plot 2) (Figure 8.46). The Whiteface, Turkey Lakes, and Red Alder sites show the largest net annual losses from the exchangeable Ca^{2+} pools from leaching (3%–4% per year, or 13–17 years for a 50% depletion). Increment adds significantly to this in all but the Turkey Lakes and Red Alder sites, raising the percent turnover to 3% to 15%, or 3 to 17 years for a 50% turnover. The other sites show percent losses of exchangeable Ca^{2+} of less than 1% (50 years for a 50% turnover.)

The pattern of percent total soil Ca export differs somewhat from that of exchangeable Ca^{2+}: the Duke, Smokies Beech, Oak Ridge Loblolly, and Red Alder sites show the highest percentages (0.2%–0.8%, or 60–250 years for a 50% reduction; see Figure 8.46). Increment is the major factor in the Smokies Beech and Duke sites, but leaching is the major factor in the Red Alder and Oak Ridge Loblolly sites. For the Oak Ridge, Duke, and Smokies Beech sites, the high percentages are primarily a result of very low total soil Ca (see Tables LP-2 and DL-2, Appendix) and for the Red Alder site, the high percentages are a consequence of high leaching rates.

The percentages of exchangeable Mg^{2+} removed annually are almost all positive (Figure 8.47), indicating a net loss of Mg from almost all sites (as shown in Figure 8.39). The Smokies Beech, plot 2, and Florida sites show a slight negative percent loss with leaching only, but a negative percent with leaching + increment. The Whiteface site has by far the greatest percentage loss of exchangeable Mg^{2+}: 35%, or about 2 years for a 50% turnover, most of which is from leaching. The Norway, Turkey Lakes, and Red Alder sites also show quite high percent annual losses from exchangeable Mg^{2+}: 7% to 13%, or about 4 to 8 years for a 50% turnover. Increment is less important than leaching in all but the Norway site. As noted in Chapter 4, however, these sites have rather large reserves of weatherable Mg and the percentages of Mg export to total Mg reserves in the sites mentioned indicate a net annual loss of only 0.01% to 0.03%, or 3,000 to 10,000 years for a 50% reduction. The Duke site has the highest percentage of total soil Mg export (0.1%, or

500 years for a 50% reduction, with vegetation increment as the major factor), but the value for Duke and all other sites falls well below those for percent of total soil Ca export.

The percentages of exchangeable K exported are negative in several sites, but not in the same sites in which Ca percentages were negative. In the case of K, the Smokies Tower (both plots), Huntington Forest, Norway (both plots) Findley Lake, Smokies, Beech (both plots), and Douglas fir plots show negative percentages with leaching only, and the Smokies Tower (both plots), Huntington Forest, and Douglas fir sites show a negative increment with leaching plus increment as well (Figure 8.48). Of those sites that show a positive percentages, biomass increment is the dominant factor in most cases (exceptions being Smokies Becking, Oak Ridge Loblolly, and Red Alder sites). The sites with the largest negative percentages—Norway, Maine, Whiteface, Duke, and Red Alder—show a losses of approximately 4% to 7.5% annually from the exchangeable K^+ pools or (7–13 years for a 50% turnover). The Duke and Red Alder sites showed the highest percentages of soil total K export: 0.06% to 0.075% (670–830 years for a 50% reduction).

To summarize, there is considerable potential for rapid turnover of exchangeable base cations in many sites, especially with respect to Mg^{2+} and Ca^{2+}. Whether this turnover results in depletion or not depends on the as-yet-unquantified rates of weathering at these sites. The greatest potential for significant depletion of total soil base cations exists at the Duke site, where primary mineral reserves are quite low. Vegetation increment is the major factor in soil base cation depletion at this site. Other sites with potential for significant losses of total soil base cations are the Red Alder, Smokies Beech, and Oak Ridge Loblolly sites.

Analysis of Selected Cation Cycles in Polluted versus Unpolluted Environments

A full analysis of the nutrient cycles and effects of acid depositon upon them for each site is well beyond the scope of this chapter. The reader is referred to specific journal articles for such a detailed analysis. It is useful to compare selected sites, however, that represent a range of environmental and pollutant input conditions. We have selected the Findley Lake and Smokies Tower sites from the northern high-elevation category as relative pristine and polluted sites, respectively, and the Douglas fir and Duke loblolly sites from the southern low-elevation category as their counterparts (relatively unpolluted and polluted, respectively). The Ca, K, and Mg cycles of the Smokies Tower and Findley Lake sites are shown in Figures 8.49–8.52, and the Ca, K, and Mg cycles of the Duke and Douglas fir sites are shown in Figures 8.53–8.55. The Smokies Tower site is subject not only to higher inputs of S, N, and H^+ than the Findley Lake site, but also to greater inputs of Ca, K, and Mg (see Figures 8.37–8.39). Thus, while leaching rates of all three base cations are much greater in the Smokies than in the Findley

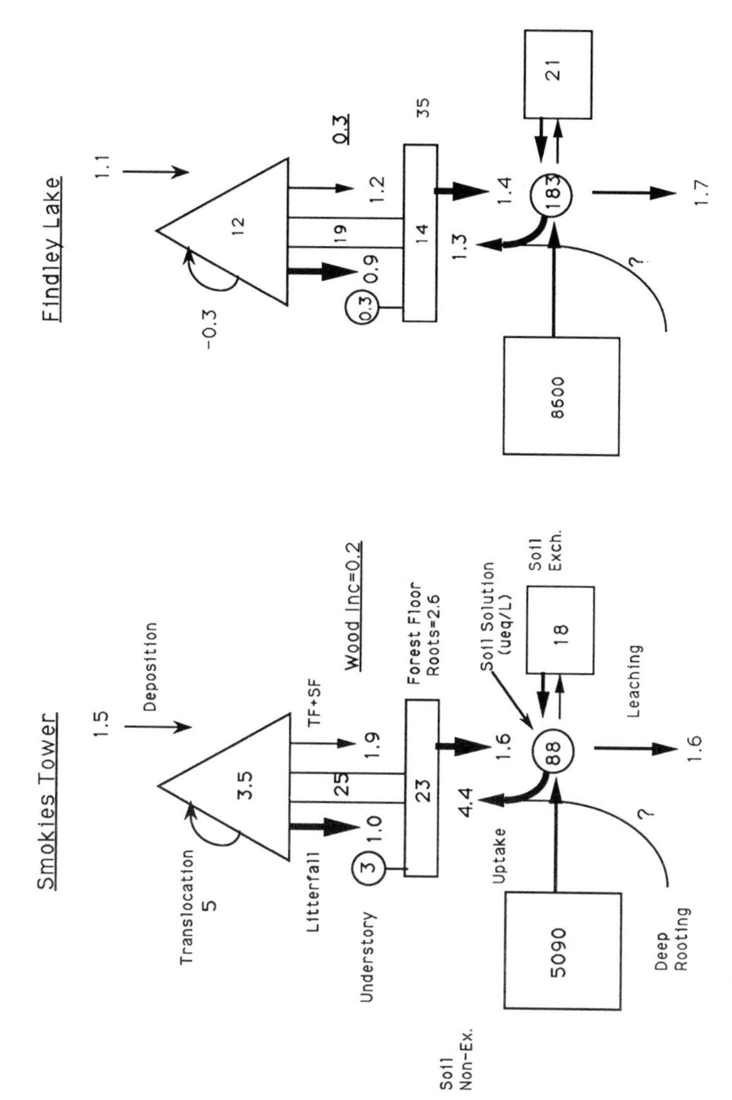

Figure 8.49. Cycling of base cations in Smokies Tower and Findley Lake sites.

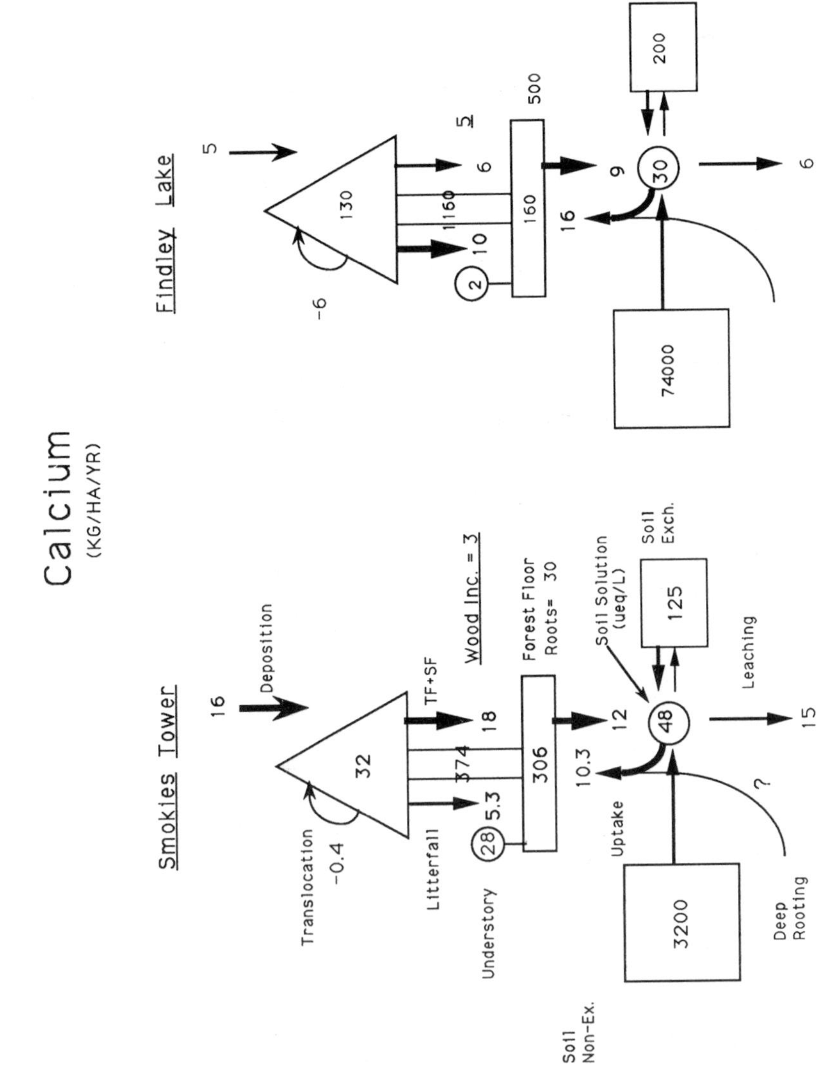

Figure 8.50. Cycling of calcium in Smokies Tower and Findley Lake sites.

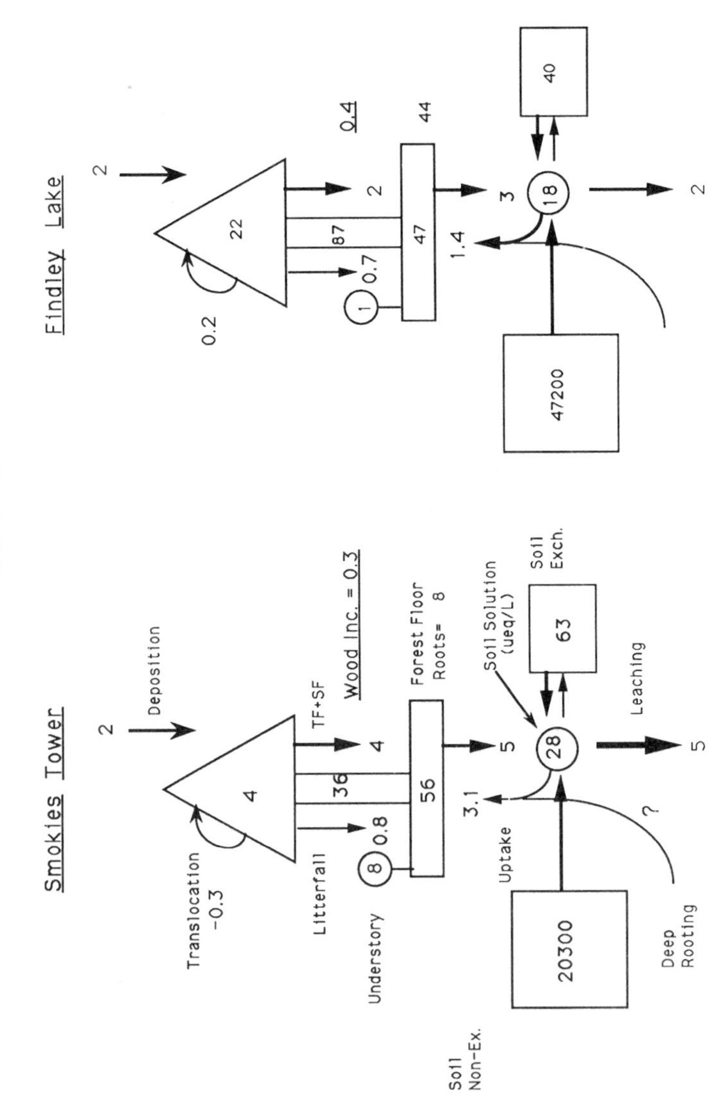

Figure 8.51. Cycling of magnesium in Smokies Tower and Findley Lake sites.

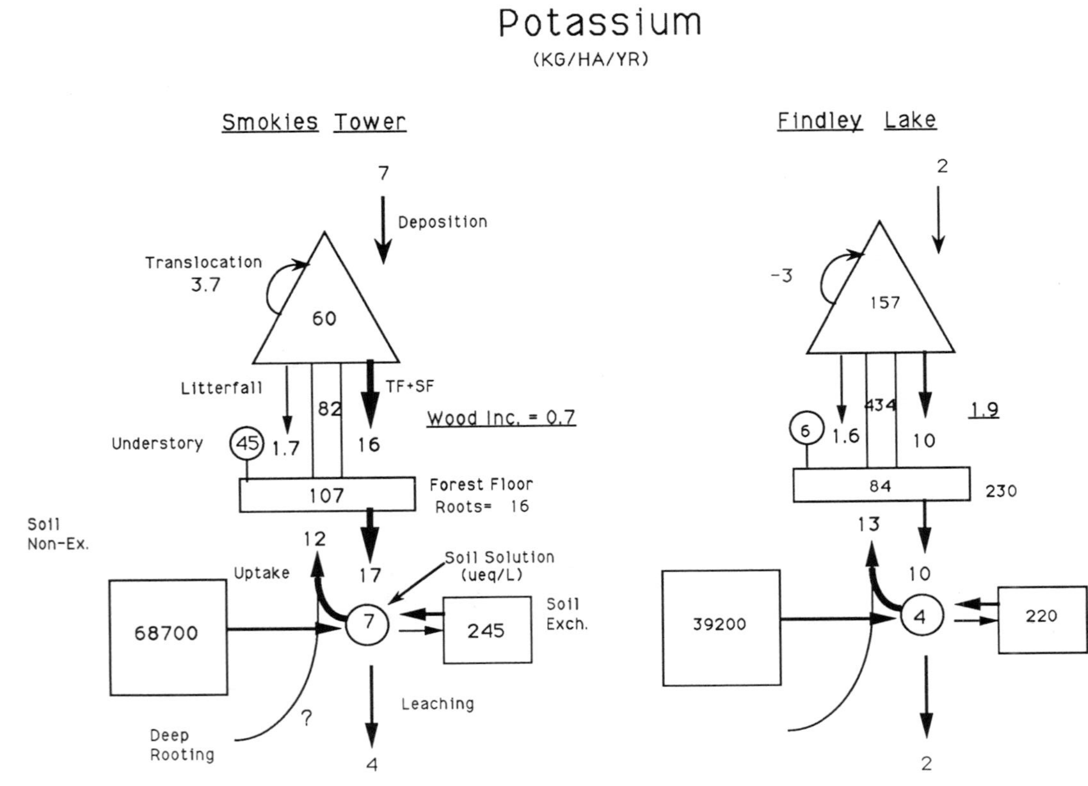

Figure 8.52. Cycling of potassium in Smokies Tower and Findley Lake sites.

Lake site, the Smokies site shows net annual accumulations of Ca and K, whereas the Findley Lake site shows slight net losses of these elements. Only in the case of Mg is the Findley Lake site slightly more conservative.

The distribution and internal cycling of Ca, K, and Mg differ considerably between these two sites. The Findley Lake site has over an order of magnitude more total soil Ca, nearly twice as much exchangeable Ca, and three times more vegetation Ca than the Smokies site. Only in the forest floor compartment does the Smokies site contain more Ca. Calcium cycles twice as fast at the Findley Lake site as at the Smokies site, yet the rates of K and Mg cycling are similar. In the case of K, the Smokies site has nearly twice as much as the Findley Lake site in the soil, nonexchangeable pool, and the size of the exchangeable pools is similar. The Findley Lake site has much greater vegetation K content, reflecting its greater biomass.The Findley Lake has over twice as much soil, nonexchangeable Mg as the Smokies site, yet the Smokies site has 50% more exchangeable Mg.

In terms of net export from the exchangeable pool, the Findley Lake site exceeds the Smokies site for Ca (6 kg ha^{-1} yr^{-1} versus a net import of 3 kg ha^{-1} yr^{-1} at the Smokies site), and K (1.9 kg ha^{-1} yr^{-1} versus a net import of 2.3 kg ha^{-1} yr^{-1} at the Smokies site), but the Smokies site exceeds the Findley Lake site with respect to Mg (1.3 kg ha^{-1} yr^{-1} versus 0.4 kg ha^{-1} yr^{-1} at the Findley Lake site) (see Figures 8.49–8.52).

Thus, it seems that in most respects the Findley Lake site is more susceptible to soil change than the Smokies site, despite the fact that acid deposition is much greater at the Smokies site. This comparison provides an excellent illustration of the fallacy of assuming that acid deposition will necessarily cause ecosystem acidification and that pristine ecosystems are at steady state. This is not to say that acid deposition has no effect on the Smokies site, however; it has clearly caused a major increase in soil solution Al. Whereas soil solution Ca, K, and Mg concentration are within approximately 50% of one another at these two sites, soil solution total Al concentrations in the Smokies red spruce site are nearly an order of magnitude greater than in the Findley Lake site (see Figures 8.49–8.52).[2] Soil solution total Al concentrations in both Smokies red spruce sites (Becking and Tower) have approached toxicity thresholds (Johnson et al. 1989, 1991), and for this reason, follow-up studies on the potential for Al toxicity at these sites are under way. However, the Smokies red spruce sites are not in a particularly severe or unusual state of decline as yet.

The Duke loblolly–Douglas fir comparison provides an interesting contrast to the Smokies Tower–Findley Lake comparison. In comparing Duke

[2]The somewhat high soil solution Al in the Findley Lake A horizon solution is presumably organically complexed, in that the level of mineral acid anions are too low to balance it in the monomeric form. Spot checks in the Smokies sites have shown that 80% to 90% of soil solution Al in all horizons is monomeric (Johnson et al., 1991).

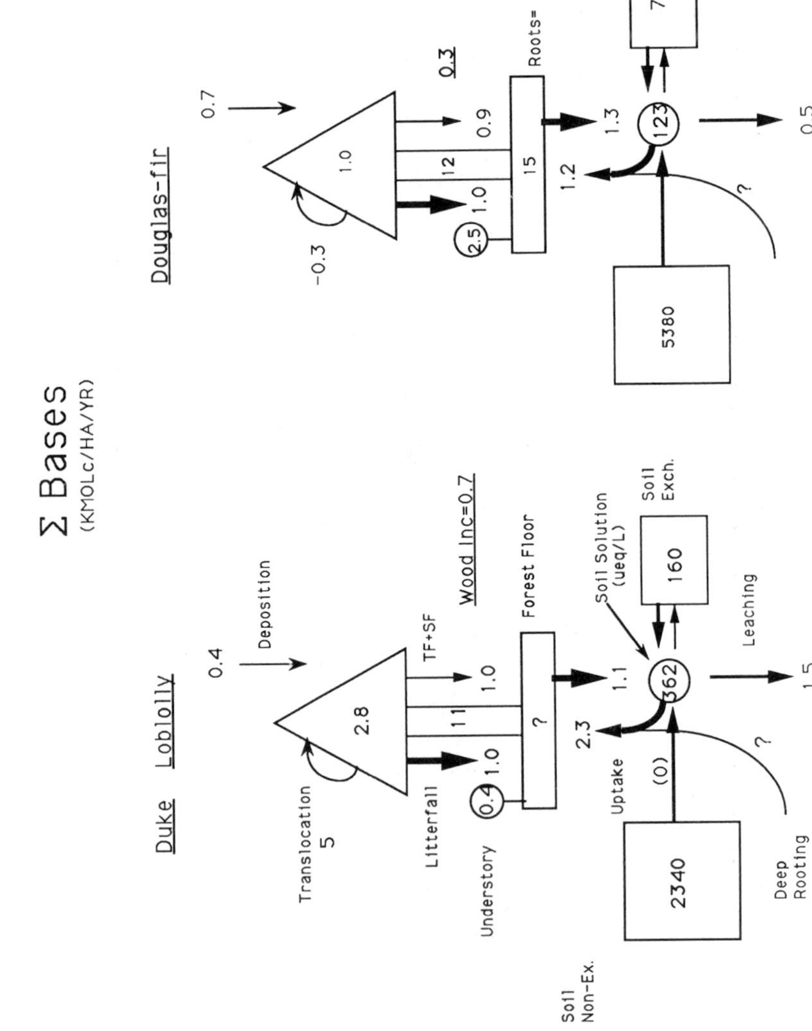

Figure 8.53. Cycling of base cations in Duke and Douglas fir sites.

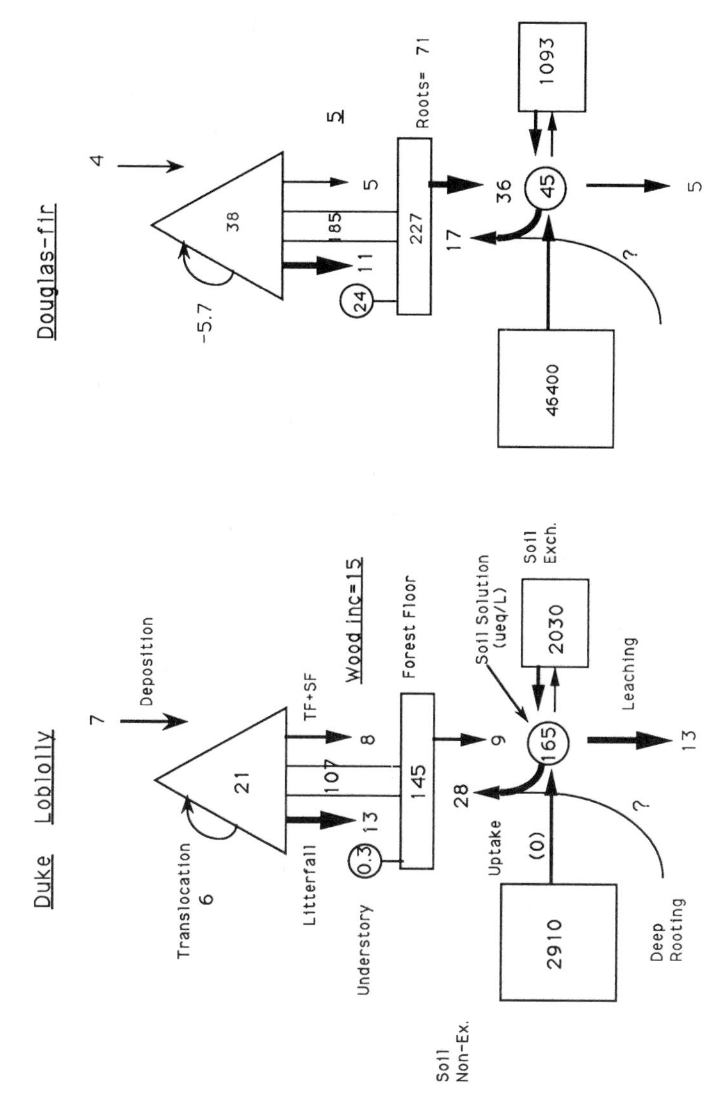

Figure 8.54. Cycling of calcium in Duke and Douglas fir sites.

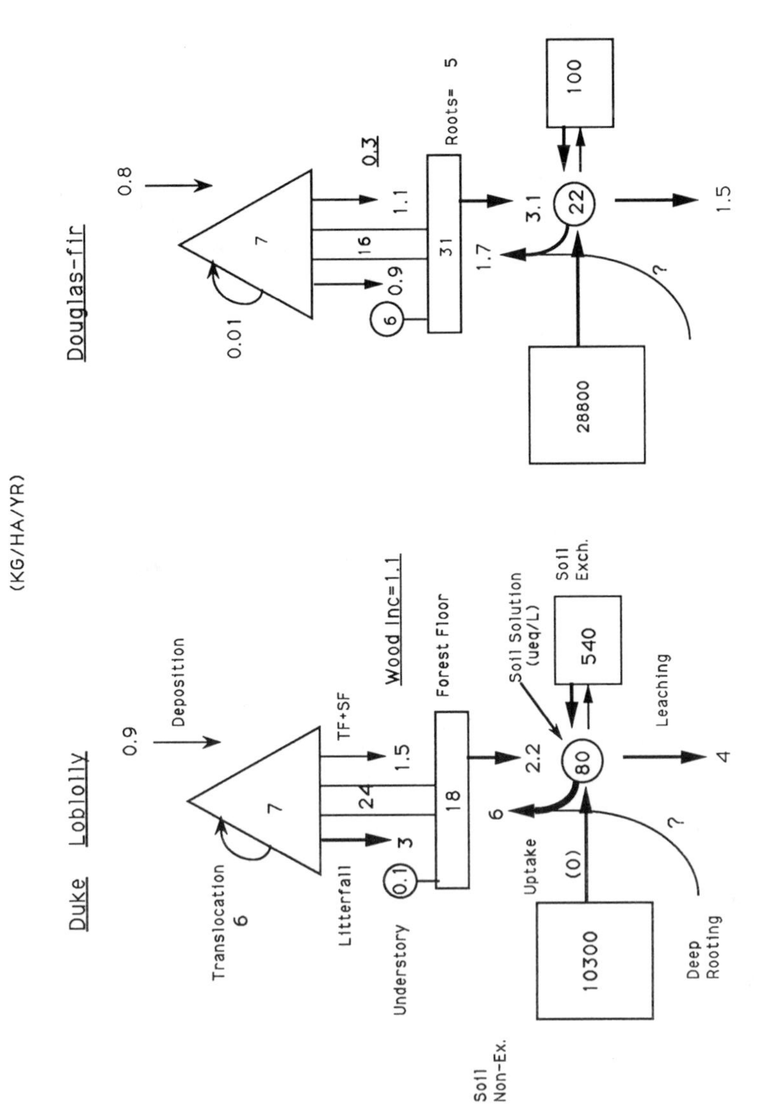

Figure 8.55. Cycling of magnesium in Duke and Douglas fir sites.

with Douglas fir, the more polluted site (Duke) clearly has greater rates of Ca, K, and Mg loss via leaching (see Figures 8.53–8.56). Soil solution base cation concentrations are three- to sevenfold greater at the Duke site than at the Douglas fir site, and it is tempting to conclude that atmospheric deposition is the major reason for the difference. A closer examination of the soil solution anion composition reveals that most of the difference in total anion and cation concentration in B horizons is caused by greater HCO_3^- and Cl^- at the Duke site, however (see Figure 8.41).

The source of the Cl^- is unclear at this time, but the source of HCO_3^- is certainly internal carbonic acid production. The differences in atmospheric deposition are clearly manifested in the anion composition in A horizon soil solutions, where SO_4^{2-} concentrations at the Duke site are much greater than at the Douglas fir site (see Figure 8.42). The Duke soil is strongly sulfate adsorbing, however, a factor that strongly mitigates the effects of sulfur deposition on cation leaching (see Chapter 5). Even with the greater surface soil solution SO_4^{2-} concentration at the Duke site, the rate of Ca leaching from the Douglas fir site forest floor is fourfold greater than at the Duke site and the rates of K and Mg leaching are similar (Figures 8.53–8.56). Thus, once again, natural acid production is of considerable importance in cation flux and soil acidification pressure.

The total soil contents of Ca and, to a lesser extent, Mg are greater at the Douglas fir than at the Duke site, presumably because of the greater reserves of weatherable minerals at the Douglas fir site (Figures 8.53–8.56). Interestingly, however, the soil exchangeable contents of Ca^{2+}, Mg^{2+}, and K^+ are considerably (two- to fivefold) greater at the Duke site. The Duke site has somewhat greater soil total K, perhaps a result of K trapped in interlayers of 2:1 clay minerals.

The Duke site has considerably greater rates of Ca, K, and Mg uptake than the Douglas fir site, mostly because of a very high rate of increment in woody biomass. The net result is that there is a greater net export of Ca (21 kg ha^{-1} yr^{-1} versus 6 kg ha^{-1} yr^{-1} at the Douglas fir site), K (11 kg ha^{-1} yr^{-1} versus a net gain of 1 kg ha^{-1} yr^{-1} at the Douglas fir site), and Mg (4.2 kg ha^{-1} yr^{-1} versus 0.9 kg ha^{-1} yr^{-1} at the Douglas fir site) (see Figures 8.53–8.56). Relative to either exchangeable or total soil pool sizes, the Duke site is apparently more susceptible to change than the Douglas fir site, as noted. It must be remembered, however, that the differences in net base cation export via leaching are not caused by atmospheric depositon and that biomass increment plays a major role at the Duke site (equaling or exceeding Ca and K export via leaching).

Summary and Conclusions

The total atmospheric deposition of the base cations Ca^{2+}, Na^+, Mg^{2+}, and K^+ to the forest canopy and forest floor is strongly influenced by elevation,

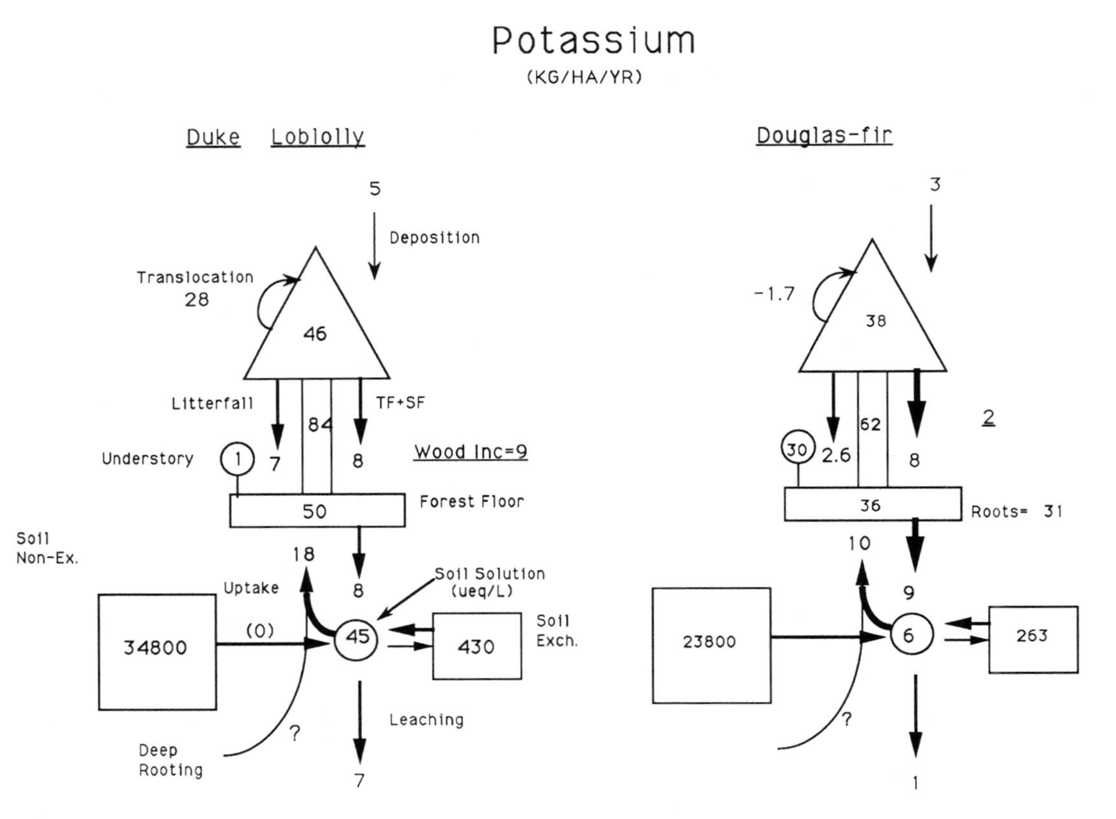

Figure 8.56. Cycling of potassium in Duke and Douglas fir sites.

geographic location, and the forest canopy. Coarse-particles accounted for 95% or more of total dry deposition of the base cations. Sodium and Ca^{2+} dominated all forms of atmospheric deposition, accounting for 70% or more of total flux of base cations in dry deposition and cloud water and 80% or more of the flux in precipitation. Cloud water deposition was important at high-elevation sites, where it accounted for about one-fourth to one-half of total atmospheric flux. The relative composition of base cations in these fog/cloud fluxes was quite similar to that in precipitation. Fluxes of base cations in coarse-particles are accompanied by HCO_3^- or CO_3^{2-} and can thus act to neutralize acidic inputs. In contrast, cloud water fluxes of base cations are exceeded by SO_4^{2-} and NO_3^- and have no neutralizing effect.

Wet and dry deposition of total base cations were approximately equal in importance at most low-elevation sites. However, wet deposition was greater than dry at low-elevation sites in the Pacific Northwest and at all the high-elevation sites. The relative importance of wet and dry deposition processes differed among the cations. At most sites, wet deposition of Mg^{2+} equaled, or generally exceeded, dry deposition. Wet deposition of Na^+ exceeded dry deposition for virtually all sites, and in the southeastern United States the wet to dry deposition ratio increased with increasing distance from the sea. Calcium dry deposition slightly exceeded Ca^{2+} wet deposition at most low-elevation sites, while the atmospheric deposition of K^+ was generally dominated by dry deposition at most low-elevation sites.

Three geographic influences were evident in the fluxes that contribute to the total atmospheric deposition of these cations. Near-coastal sites experienced a marine influence that was evident in both coarse-particle and precipitation fluxes. Sodium dominated these fluxes at the coastal sites, generally decreasing in importance with distance from the coast. A second geographic effect was the Ca^{2+} enrichment in the coarse-particle flux at sites located in calcareous areas. The third geographic effect was the influence of a humid coastal climate on increasing the relative importance of wet over dry deposition processes for low-elevation sites.

Leaching from canopy surfaces adds substantial amounts of K^+, Ca^{2+}, and Mg^{2+} to the atmospherically deposited base cations, so that the base cation flux in TF + SF exceeds that in total atmospheric deposition. The contribution of foliar leaching to TF + SF varies from considerably more than half of the K^+, one-half to two-thirds of Mg^{2+}, and one-third to one-half of the Ca^{2+}, to essentially none for Na^+ (see Figure 8.11). Thus, because of canopy exchange processes, the relative composition of the base cation pool is altered from Ca^{2+}, $Na^+ \gg Mg^{2+}$, K^+ in total atmospheric deposition to K^+, $Ca^{2+} > Na^+ > Mg^{2+}$ in TF + SF.

A number of experimental and comparative studies designed to elucidate the factors that control the efflux of Ca^{2+}, Mg^{2+}, and K^+ from forest canopies were conducted. This is clearly a complicated biogeochemical process, but one which is important for understanding both basic nutrient cycling and air pollution effects in forests. The results of these studies suggested several generalizations and conclusions: differences between sites in the annual average amount of net canopy exchange (NCE) of K^+, Ca^{2+}, and Mg^{2+} could not be explained on the basis of canopy and soil characteristics that we measured at the sites. However, the relative proportions of K^+, Ca^{2+}, and Mg^{2+} in NCE reflect roughly their relative proportions in foliar biomass. A more complicated model, involving atmospheric and rain event characteristics at a single site (LP), qualitatively tracked the pattern of net throughfall flux (NTF) from event to event for Ca^{2+} and K^+ but quantitatively overestimated observed NTF by a factor of 2. Short-term ozone exposures did not affect cation leaching from mature white pine or sugar maple in field experiments. Increasing acidity of the leaching solution increased efflux of Ca^{2+} and Mg^{2+} from white pine, sugar maple, loblolly pine, and red spruce. The responses to acidity were quantitatively similar between the two ions and among the four species. The increase in cation efflux is associated with a decrease of free acidity of the solution, and appears to be a result of cation exchange reactions. For white pine seedlings under moderate leaching conditions, foliar leaching rates are small compared to root uptake rates of Ca^{2+}, Mg^{2+}, and K^+.

Base saturation is a key variable for the classification of the IFS sites in terms of base cation distribution, cycling, and Al mobilization. When stratified by B horizon base saturation, the IFS sites can be fairly neatly divided into northern, high-elevation and southern, low-elevation categories. The northern high-elevation sites are more acid, are more susceptible to Al mobilization into soil solution, and cycle base cations at a slower rate than the southern low-elevation sites. Uptake, and to a lesser extent vegetation increment, can be a major factor in potential soil cation depletion in the southern low-elevation sites, whereas leaching plays a much more dominant role in the northern high-elevations sites.

With some reasonable assumptions as to background levels, it can be concluded from soil solution anion concentrations that soil cation leaching (H^+, Al^{3+}, and base cations) is either dominated by or strongly affected by atmospheric deposition of sulfate nitrate anions in the Smokies (Becking, Tower, and Beech), Whiteface, Huntington Forest, and Oak Ridge Loblolly sites. Atmospheric deposition plays a lesser but still significant role (approximately 25% or more of total soil solution cation leaching) in the Turkey Lakes, Coweeta Pine and Hardwood, Duke, Florida, and Georgia sites. Atmospheric deposition plays a very minor role in soil solution cation leaching in the Washington (Findley Lake, Douglas fir and Red Alder) sites. However, the actual magnitude of base cation leaching does not correspond well to the degree to which sites are impacted by atmospheric deposition. Specifically, Red Alder, Turkey Lakes, Oak Ridge Loblolly, and Findley Lake sites have the highest rates of base cation leaching, yet atmospheric deposition dominates leaching only in the Oak Ridge Loblolly site, and internal NO_3^- production dominates leaching at the Turkey Lakes and Red Alder sites.

There is considerable potential for rapid turnover of exchangeable base cations in many sites, but the extent to which this results in observable depletion of exchangeable base cations or base saturation depends on weathering rates. The greatest potential for significant depletion of total soil base cations exists at the Duke site, where vegetation increment rather than leaching is the major factor in soil base cation depletion. Other sites that have potential for significant losses of total soil base cations include the Red Alder, Smokies Beech, and Oak Ridge Loblolly sites. Only in the latter case would atmospheric deposition play a major role in soil change.

References

Alban D.H. 1982. Effects of nutrient accumulation by aspen, spruce, pine on soil properties. Soil Sci. Soc. Am. J. 46:853–861.

Abrahamsen G., Hornvedt R., Tveite B. 1977. Impacts of acid precipitation on forest ecosystems. Water Air Soil Pollut. 8:57–73.

Abrahamsen G., Bjor K., Hornvedt R., Tveite B. 1980. Effects of acid precipitation on coniferous forest. In Drablos D., Tollan A., (eds.) Proceedings of the International Conference on the Ecological Impact of Acid Precipitation, pp. 58–63. SNSF Project, Oslo, Norway.

Berden M., Nilsson S.I., Rosen K., Tyler G. 1987. Soil acidification: Extent, Causes, Consequences. Report 3292, National Swedish Environmental Protection Board, Solna, Sweden.

Binkley D., Valentine D., Wells C., Valentine U. 1989. An empirical analysis of factors contributing to 20-year decline in soil pH in an old-field plantation of loblolly pine. Biogeochemistry 8:39–54.

Bondietti E.A. 1990. Descriptions of Research Sites in the Integrated Forest Study. ORNL/TM 11149, Oak Ridge National Laboratory, Oak Ridge, Tennessee.

Brown K.A., Roberts T.M. 1988. Effects of ozone on foliar leaching in Norway spruce (*Picea abies* L. Karst): confounding factors due to NO_x production during ozone generation. Environ. Pollut. 55:55–73.

Cole D.W., Johnson D.W. 1977. Atmospheric sulphate additions and cation leaching in a Douglas-fir ecosystem. Water Resour. Res. 13:313–317.

Cole D.W., Rapp M. 1981. Elemental cycling in forest ecosystems. In Reichle D.E. (ed.) Dynamic Properties of Forest Ecosystems. Cambridge University Press, London, pp. 341–409.

Cronan C.S., Reiners W.A. 1983. Canopy processing of acidic precipitation by coniferous and deciduous forests in New England. Oecologia 59:216–223.

Eaton J.S., Likens G.E., Bormann F.H. 1973. Throughfall and stemflow chemistry in a northern hardwoods forest. J. Ecol. 61:495–508.

Evans L.S., Curry T.M., Lewin K.F. 1981. Responses of leaves of *Phaesolus vulgaris* L. to simulated acid rain. New Phytol. 88:403–420.

Fairfax J.A.W., Lepp N.W. 1975. Effects of simulated "acid rain" on cation loss from leaves. Nature (London) 255:324–325.

Federer C.A., Hornbeck J.W., Tritton L.M., Martin C.W., Pierce R.S., Smith C.T. 1989. Long-term depletion of calcium and other nutrients in eastern U.S. forests. Environ. Manage. 13:593–601.

Heiberg S.O., White D.P. 1953. Potassium deficiency of reforested pine and spruce stands in northern New York. Soil Sci. Soc. Am. Proc. 15:369–376.

Hindawi I.J., Rea J.A., Griffiths W.L. 1980. Response of bush bean exposed to acid mist. Am. J. Bot. 67:168–172.

Hubbell J.G., Lovett G.M. 1988. A chamber for ozone exposure and misting of branches on mature trees. In Proceeding of Workshop on the Role of Branch Studies, pp. 120–129. U.S. Environmental Protection Agency, Corvallis, Oregon.

Ingestad, T. 1979. Mineral nutrient requirements of *Pinus sylvestris* and *Picea abies* seedlings. Physiol. Plant. 45:373–380.

Ivens W.P.M.F., Draaijers G.P.J., Bos M.M., Bleuten W. 1988. Dutch Forests as Air Pollutant Sinks in Agricultural Areas. Report 37–09, Dutch Priority Programme on Acidification, State University of Utrecht, Holland.

Johannes A.J., Chen Y.L., Dackson K., Suleski T. 1986. Modeling throughfall chemistry and indirect measurement of dry deposition. Water Air Soil Pollut. 30:211–216.

Johnson D.W. 1981. The natural acidity of some unpolluted waters in southeastern Alaska and potential impacts of acid rain. Water Air Soil Pollut. 16:243–252.

Johnson D.W., Cole D.W. 1980. Anion mobility in soils: relevance to nutrient transport from terrestrial ecosystems. Environ. Int. 3:79–90.

Johnson D.W., Richter D.D. 1984. The combined effects of atmospheric deposition, internal acid production, harvesting on nutrient gains and losses from forest ecosystems. In TAPPI Research and Development conference, Appleton, Wisconsin, Sept. 30–Oct. 3, 1984, pp. 149–156. TAPPI Press, Atlanta, Georgia.

Johnson D.W., Taylor G.E. 1989. Role of air pollution in forest decline in eastern North America. Water Air Soil Pollut. 48:21–43.

Johnson D.W., Todd D.E. 1983. Some relationships among aluminum, carbon, sulfate in a variety of forest soils. Soil Sci. Soc. Am. J. 47:792–800.

Johnson D.W., Todd D.E. 1987. Nutrient export by leaching and whole-tree harvesting in a loblolly pine and mixed oak forest. Plant Soil 102:99–109.

Johnson D.W., Henderson G.S., Todd D.E. 1988a. Changes in nutrient distribution in forests and soils of Walker Branch Watershed, Tennessee, over an eleven-year period. Biogeochemistry 5:275–293.

Johnson D.W., Richter D.D., Lovett G.M., Lindberg S.E. 1985. The effects of atmospheric deposition on potassium, calcium, magnesium cycling in two deciduous forests. Can. J. For. Res. 15:773–782.

Johnson D.W., West D.C., Todd D.E., Mann L.K. 1982. Effects of sawlog vs. whole-tree harvesting on the nitrogen, phosphorus, potassium, calcium budgets of an upland mixed oak forest. Soil Sci. Soc. Am. J. 46:1304–1309.

Johnson D.W., Cole D.W., Gessel S.P., Singer M.J., Minden R.V. 1977. Carbonic acid leaching in a tropical, temperate, subalpine, northern forest soil. Arc. Alp. Res. 9:329–343.

Johnson D.W., Kelly J.M., Swank W.T., Cole D.W., Van Miegroet H., Hornbeck J.W., Pierce R.S., Van Lear D. 1988b. The effects of leaching and whole-tree harvesting on cation budgets of several forests. J. Environ. Qual. 17:418–424

Johnson D.W., Friedland A.J., Van Miegroet H., Harrison R.B., Miller E., Lindberg S.E., Cole D.W., Schaefer D.A., Todd D.E. 1989. Nutrient status of some contrasting high-elevation forests in the eastern and western United States. In Proceedings of the U.S.-German Research Symposium, Burlington, VT, Oct. 18–23, 1987.

Johnson D.W., Van Miegroet H., Lindberg S.E., Todd D.E., Harrison R.B. 1991. Nutrient cycling in red spruce forests in the Great Smoky Mountains. Can. J. For. Res. 21:769–787.

Kononova M. 1966. Soil Organic Matter: Its Nature, Its Role in Soil Formation and Soil Fertility. Pergamon Press, New York.

Kramer P.J., Kozlowski T.T. 1979. Physiology of Woody Plants. Academic Press, New York.

Krause G.H.M., Prinz B., Jung K.D. 1982. Forest effects in West Germany. In Air Pollution and the Productivity of the Forest. Izaak Walton League of America, Washington, DC, pp. 297–332.

Krug E.C., Frink C.R. 1983. Acid rain on acid soil: a new perspective. Science 221:520–525.

Lehninger R.A. 1970. Biochemistry. Worth, New York.

Lindberg S.E., Lovett G.M. 1983. Application of surrogate surface and leaf extraction methods to estimation of dry deposition to plant canopies. In Pruppacher H.R., Semonin R.G., Slinn W.G.N., (eds.) Precipitation Scavenging, Dry Deposition, Resuspension. Elsevier, New York, pp. 837–848.

Lindberg S.E., Silsbee D., Schaefer D.A., Owens J.G., Petty W. 1988a. A comparison of atmospheric exposure conditions at high- and low-elevation forests in the southern Appalachian Mountains. In Unsworth M., Fowler D. (eds.) Processes of Acidic Deposition in Mountainous Terrain. Kluwer Academic Publishers, London, pp. 321–344.

Lindberg S.E., Harriss R.C., Turner R.R., Shriner D.S., Huff D.D. 1979. Mechanisms and Rates of Atmospheric Seposition of Selected Trace Elements and Sulfate to a Deciduous Forest Watershed. ORNL/TM-6674, Oak Ridge National Laboratory, Oak Ridge, Tennessee.

Lindberg S.E., Lovett G.M., Schaefer D.A., Mitchell M., Cole D., Swank W., Foster N., Knoerr K. 1988b. In Lindberg S.E., Johnson D. (eds.) Third Annual Progress Report on the Integrated Forest Study. Electric Power Research Institute, Palo Alto, California, pp. 9–14.

Lovett G.M., Hubbell J.G. Effects of ozone and acid mist on foliar leaching from sugar maple and white pine. J. For. Res. (in press).

Lovett G.M., Lindberg S.E. 1984. Dry deposition and canopy exchange in a mixed oak forest as determined by analysis of throughfall. J. Appl. Ecol. 21:1013–1027.

Lovett G.M., Reiners W.A., Olson R.K. 1982. Cloud droplet deposition in subalpine balsam fir forests: hydrological and chemical inputs. Science 218:1303–1304.

Lovett G.M., Lindberg S.E., Richter D.D., Johnson D.D. 1985. The effects of acidic deposition on cation leaching from a deciduous forest canopy. Can. J. For. Res. 15:1055–1060.

Marion G.K. 1979. Biomass and nutrient removal in long-rotation stands. In Leaf A.L. (ed.) Impact of Intensive Harvesting on Forest Nutrient Cycling. State University of New York, Syracuse, pp. 98–110.

McColl J.G., Cole D.W. 1968. A mechanism of cation transport in a forest soil. Northwest Sci. 42:132–140.

Mollitor A.V., Raynal D.J. 1982. Acid precipitation and ionic movements in Adirondack forest soils Soil Sci. Soc. Am. J. 46:133–137.

Olson R.K., Reiners W.A., Cronan C.S., Lang G.E. 1981. The chemistry and flux of throughfall and stemflow in subalpine balsam fir forests. Holarct. Ecol. 4:291–300.

Parker, G.G. 1983. Throughfall and stemflow in the forest nutrient cycle. Adv. Ecol. Res. 13:57–133.

Parker, G.G., Lovett G.M., Mickler R. 1987. Effects of low ozone dosage and acidified rain on foliar leaching in ozone-sensitive and -insensitive clones of *Populus tremuloides*. Bull. Ecol. Soc. Am. 68:383 (abst.).

Reed D.W., Tukey H.B., Jr. 1978b. Effect of pH on absorption of rubidium compounds by chrysanthemums. J. Am. Soc. Hortic. Sci. 103:815–817.

Reed W.D., Tukey H.B., Jr. 1978a. Effect of pH on foliar absorption of phosphorous compounds by chrysanthemums. J. Am. Soc. Hortic. Sci. 103:336–340.

Reiners W.A., Lovett G.M., Olson R.K. 1986. Chemical interactions of a forest canopy with the atmosphere. In Proceedings of the Forest/Atmosphere Interaction Workshop. pp. 111–146. U.S. Department of Energy, Office of Energy Research, Washington, D.C.

Reiners W.A., Olson R.K. 1984. Effects of canopy components on throughfall chemistry: an experimental analysis. Oecologia 63:320–330.

Reuss J.O. 1983. Implications of the Ca-Al exchange system for the effect of acid precipitation on soils J. Environ. Qual. 12:591–595.

Reuss J.O., Johnson D.W. 1986. Acid Deposition and the Acidification of Soil and Water. Ecological Studies No. 59, Springer-Verlag, New York.

Saxena V.K., Lin N.-H. 1989. Cloud Chemistry Measurements and Estimates of Acidic Deposition on an Above Cloudbase Coniferous Forest. Atmos. Environ. 23:001–024.

Schaefer D.A., Reiners W.A. 1990. Throughfall chemistry and canopy processing mechanisms. In Lindberg W.E., Page A.L., Norton S.A., (eds.) Advances in Environmental Science, Volume 3, Acid Precipitation: Sources, Deposition, Canopy Interactions. Springer-Verlag, New York, pp. 241–284.

Schaefer D.A., Lindberg S.E., Nicolai D.R. 1988a. Ionic exchange rates as determined by foliar surface rinsing. Bull. Ecol. Soc. Am. 69:285–286 (abstr.).

Schaefer D.A., Reiners W.A., Olson R.K. 1988b. Factors controlling the chemical alteration of throughfall in a subalpine balsam fir canopy. Environ. Exp. Bot. 28:175–189.

Scherbatskoy T., Klein R.M. 1983. Response of spruce and birch foliage to leaching by acidic mists. J. Environ. Qual. 12:189–193.

Schier G.A. 1987. Throughfall chemistry in a red maple provenance plantation sprayed with “acid rain.” Can. J. For. Res. 17:660–665.

Skeffington R.A., Roberts T.M. 1985. The effects of ozone and acid mist on Scots pine saplings. Oecologia (Berlin) 65:201–206.

Stone E.L., Kszystyniak R. 1977. Conservation of potassium in the *Pinus resinosa* ecosystem. Science 198:192–194.

Swank W.T., Caskey W.H. 1982. Nitrate depletion in a second-order mountain stream. J. Environ. Qual. 11:581–584.

Tukey H.B. Jr. 1980. Some effects of rain and mist on plants, with implications for acid precipitation. In Hutchinson T.C., Havas M. (eds.) Effects of Acid Precipitation on Terrestrial Ecosystems. Plenum, New York, pp. 141–149.

Ugolini F.C., Minden R., Dawson H., Zachara J. 1977. An example of soil processes in the *Abies amabilis* zone of Central Cascades, Washington. Soil Sci. 124:291–302.

Ulrich B. 1983. Soil acidity and its relation to acid deposition. In Ulrich B., Pankrath J. (eds.) Effects of Accumulation of Air Pollutants in Ecosystems. Reidel, Dordrecht, pp. 127–146.

Ulrich B., Mayer R., Khanna P.K. 1980. Chemical changes due to acid precipitation in a loess-derived soil in central Europe. Soil Sci. 130:193–199.

Van Miegroet H., Cole D.W. 1984. The impact of nitrification on soil acidification and cation leaching in red alderecosystem. J. Environ. Qual. 13:586–590.

Vitousek P.M., Gosz J.R., Grier C.C., Melillo J.M., Reiners W.A., Todd R.L. 1979. Nitrate losses from disturbed ecosystems. Science 204:469–474.

Waring R.H., Schlesinger W.H. 1985. Forest Ecosystems: Concepts and Management. Academic Press, New York.

Wood T., Bormann F.H. 1975. Increases in foliar leaching caused by acidification of an artificial mist. Ambio 4:169–171.

Wright R.F., Norton S.A., Brakke D.F., Frogner T. 1988. Experimental verification of episodic acidification of freshwaters by sea salts. Nature (London) 334:422–424.

9. Cation Exchange and Al Mobilization in Soils

Cation Exchange Reactions in Acid Forested Soils: Effects of Atmospheric Pollutant Deposition

D.D. Richter, D.W. Johnson, and K.H. Dai

Introduction

Predicting the equilibria of cations distributed between soil solutions and exchange sites is a classical soil chemistry problem. Interest in such cation distribution problems has been motivated by a need to improve fertilization to maintain well-balanced concentrations of soil exchangeable Ca, Mg, and K for optimum crop growth; to control potentially toxic Al with Ca and Mg additions in acidic agronomic soils; and to control excess Na with Ca amendments in irrigated soils. The effects of atmospheric deposition of pollutant sulfate on soils is a similar problem and should alter cation distributions between solutions and exchange sites in predictable ways. The consequences of sulfate deposition are potentially serious: displacement of greater concentrations of exchangeable Al, Ca, and Mg (polyvalent cations) into soil solutions, and long-term depletions of exchangeable nutrient cations, for example, Ca and Mg, in highly weathered soils with limited mineral weathering rates.

There are several reasons to suspect that electrolyte chemistry of soil solution systems in natural forest ecosystems contrasts markedly with soil systems that are most intensely studied, that is, agricultural soils that are fertilized, limed, or irrigated. In contrast to many forest soil systems, these better studied systems typically have exchangers that are Ca- or Na saturated, either naturally or as a consequence of liming or fertilization; they are often dominated by constant surface charge, and they have soil solutions that are relatively high in ionic strength because of fertilization or high salt content. Natural forest soils typically have conditions in which exchange sites are Al saturated; they have surfaces that are a mixture of constant and variable charge, and in addition they have relatively dilute soil solutions. Despite their biological significance, cation exchange reactions in forested soils have not yet been well examined.

Acidic cations dominate cation exchange sites of forested soils that encompass a wide range of soil taxa: Ultisols, Spodosols, Entisols, Inceptisols, Histosols, Alfisols, Oxisols, and Andisols. Exchangeable Al commonly counters more than 75% of effective cation exchange capacity (CECe), and a variety of monomeric and polymeric Al–OH cations may counter more than 90% of total CEC (CECt). Such soils may have exchangeable and total Al acidity in the first meter below the surface that total hundreds to thousands of kilomoles per hectare (Richter 1986; Richter et al., 1989), whereas exchangeable nutrient cations in such acidic soils may total only hundreds to tens of kilomoles per hectare. Moreover, the soil solutions of forested soils that are not fertilized are very low in ionic strength. Electrolyte concentration of gravitational soil water in unfertilized Udults, Inceptisols, and Orthods ranges between about 0.1 and 1.0 mmol(c) L^{-1} (Johnson et al. 1977; Richter et al. 1983; van Miegroet and Cole 1985; Driscoll et al. 1985).

Exchange and leaching of cation species in forest ecosystems appear highly responsive to small changes in solution ionic strength (Richter et al. 1988). Ionic strength of soil solutions varies as a function of many ecological processes in forest environments, changes that should affect the activities and distributions of cations in the soil solution. Such processes include atmospheric precipitation, evapotranspiration, leaching of ions from deposited litterfall, microbially mediated mineralization of soil organic matter, chemical weathering reactions of minerals, nitrogen fixation that is followed by N mineralization and nitrification, and atmospheric deposition of natural elements and air pollutants. Quantifying how each of these processes determines ionic strength of soil solutions remains to be accomplished.

Regional sulfur air pollution has increased electrolytes of forest soil solutions of the order of 0.1 to 0.5 mmol(c) L^{-1} in eastern North America and central and northern Europe (Ulrich 1980; van Breeman et al. 1982; Reuss 1983; Richter et al. 1983; Driscoll et al. 1985; Johnson et al. 1985). Electrolytes in soil solutions in some central Europe appear to be elevated more than 0.5 mmol(c) L^{-1} by high sulfur deposition and high evapotranspiration to precipitation ratios. Several models of soil solution chemistry suggest that Ca, Mg, and Al are exchanging and leaching from acid forest soils as a result of such increases in ionic strength (Cosby et al. 1985; Reuss and Johnson 1986). Experimental demonstration of these exchange reactions are notably absent, however. Experiments are especially needed to examine how cations in solutions of very low ionic strength exchange with those that are exchangeable in soils that have mixtures of variable and constant charge.

The overall purpose of this analysis of cation exchange was twofold: first, to examine exchange reactions in acid forest soils of the Integrated Forest Study (see Tables 9.1 and 9.2), especially exchange reactions that are induced by relatively small increases in ionic strength; and second, to evaluate commonly used exchange models (Gapon and Gaines–Thomas equations, Schofield activity ratio) for predicting exchange between cations of different charges. Specific objectives were to use a variety of acid soils with dilute soil solutions to determine: (1) how Al saturation of CECe was quantitatively related to Al displacement into dilute soil solutions; (2) how exchange of heterovalent cations can be quantified; and (3) the replacing abilities of individual cations. Throughout these studies, emphasis was placed on soil solution systems of very low ionic strength.

Methods

Laboratory experiments were conducted with 31 soils collected from most forest stands within the IFS project (see Table 9.1). Samples were air-dried, sieved to pass a 2-mm screen, and analyzed for basic chemical properties (D.W. Johnson, R.B. Harrison, and D.E. Todd, unpublished data).

Exchangeable base cations and Al were determined by displacing soil cations with 1 *M* NH_4Cl, and analyzing filtrates with an atomic absorption spectrophotometer. This method for NH_4Cl-exchangeable Al compared well with that of 1 *M* KCl extraction of soil acidity followed by titration (K.H. Dai and D.D. Richter, unpublished data), a method described by Thomas (1982). Total acidity was determined with a $BaCl_2$-Triethanolamine (TEA) method buffered at pH 8.2 (Thomas 1982). The CEC was thus determined in two ways, by summing NH_4Cl-exchangeable cations (Al, Ca, Mg, K, and Na) for effective CEC (CECe), or by summing $BaCl_2$-TEA acidity with exchangeable Ca, Mg, K, and Na for total CEC (CECt). Soil pH was determined in deionized water and 0.01 *M* $CaCl_2$, (pHw and pHs, respectively). Carbon was determined by dry combustion with a Leco furnace, or by loss on ignition (LOI) for O horizons and assuming C was 50% of LOI.

Al Exchange Compared to Al Saturation

Experiments were designed to determine how Al saturation of CECe was related to Al displacement into solution in response to small increases in low ionic strength of the soil solution. Because of a wide variation in water-soluble electrolytes in these soils, soils were initially contacted with deionized water to reduce water-soluble electrolytes to comparatively low levels in all soils. A water extraction (25 ml water per 2.5 g soil) was repeated twice, and the second extract had a low specific conductance, less than 10

Table 9.1. Summary of 31 Soils Taken from 15 Soil Profiles of Cation Exchange Studies

Soil Code	Soil Series	Horizon	Soil Suborder	Dominant Tree Species	Site Code	Site Location
1	Alderwood	A1	Ochrept	Red alder	RA	Thompson Forest, WA
2		A12				
3	Alderwood	A12	Ochrept	Douglas fir	DF	Thompson Forest, WA
4		B21				
5		B22				
6	Burton	A11	Umbrept	Red spruce	ST	Tower Site, Great Smoky Mtn. Nat. Park, NC (GSMNP)
7		A12				
8		B1				
9		B				
10	Burton	B	Umbrept	Red spruce	SS	Becking Site (GSMNP)
11	Chikamin	E	Andept (or Humod)	Fir and hemlock	FL	Findlay Lake, WA
12		B2ir				
13		B3				
14	Turkey	BfL1	Orthod	Sugar maple	TL	Turkey Lake, Ontario
15	Burton	OA	Umbrept	Red spruce	ST	Tower Site (GSMNP)
16	Becket	Bs3	Orthod	Sugar maple	HF	Huntington Forest, NY
17		C				
18	Becket	B	Humod	Red spruce	WF	Whiteface Mt., NY
19		C				
20	Tarklin	Ap	Udult	Yellow poplar	NC	Oak Ridge, TN
21		B1				
22		B2				
23	Fullerton	E	Udult	Chestnut oak		Oak Ridge, TN
24		B1				
25	Lily	B2	Udult	White oak	CB	Fall Creek Falls, TN
26	Appling	B1	Udult	Loblolly pine	DL	Mebane, NC
27		B2				
28	Fannin	Bt	Udult	White pine	CP	Coweeta Hydro Lab, NC (Watershed 1)
29		BC				
30	Fannin	BA	Udult	Oak-hickory	CH	Coweeta Hydro Lab, NC (Watershed 2)
31		Bt				

Table 9.2. Summary of Soil Chemistry of the 31 Soils

					Exchangeable						
					CECe	CECt	Al	Ca	Mg	K	Na
Soil Code[a]	C (g kg^{-1})	pHw	pHs	BSe (%)	(μMc g^{-1})						
1	111	4.4	4.2	43	129	836	73.2	45.7	6.2	3.4	0.6
2	48	5.2	4.7	67	57	435	18.6	31.1	3.4	2.8	0.6
3	52	5.3	4.6	69	61	430	19.0	34.9	3.7	2.9	0.6
4	26	5.6	5.1	74	11	308	2.8	5.5	0.9	1.1	0.5
5	14	5.5	5.3	58	6	244	2.6	1.9	0.6	0.8	0.3
6	64	3.8	3.4	6	101	476	95.4	2.1	2.0	1.4	0.2
7	57	3.9	3.5	5	87	478	82.1	1.2	1.6	1.3	0.4
8	34	4.3	3.8	6	52	361	48.8	0.7	1.0	1.0	0.3
9	4.2	5.0	4.7	11	6	89	5.2	0.0	0.1	0.3	0.2
10	17	4.1	3.6	5	75	389	71.0	1.0	1.4	1.3	0.3
11	26	4.0	3.4	15	26	93	22.1	2.1	0.9	0.6	0.4
12	51	4.2	3.8	10	92	560	82.1	6.3	1.4	1.1	0.7
13	26	4.8	4.6	5	37	367	35.4	0.9	0.2	0.6	0.2
14	52	4.6	4.0	23	66	502	51.0	12.0	1.9	1.0	0.2
15	442	3.2	2.7	43	148	1585	84.3	53.0	9.4	0.0	0.9
16	16	5.1	4.6	9	13	244	11.7	1.0	0.1	0.1	0.0
17	3.4	5.5	4.9	27	4	54	3.0	0.6	0.1	0.2	0.2
18	103	4.4	3.9	20	61	693	48.8	2.7	0.8	5.4	3.4
19	19	5.3	4.6	20	17	240	13.2	1.1	0.0	1.1	1.0
20	15	5.1	4.7	59	20	125	8.1	8.2	1.9	1.4	0.0
21	7.8	5.0	4.4	48	20	98	10.1	6.4	1.9	1.1	0.0
22	5.3	5.0	4.5	53	27	104	12.8	10.5	2.8	1.1	0.1
23	6.7	4.9	4.5	25	11	98	8.1	1.5	0.6	0.7	0.0
24	2.9	4.7	4.2	21	24	97	18.8	2.2	1.7	0.9	0.1
25	1.5	5.0	4.1	13	58	146	51.0	0.6	5.3	1.4	0.0
26	2.5	4.8	4.2	55	74	255	33.2	29.5	8.5	2.5	0.3
27	1.4	4.8	4.1	44	86	181	48.8	21.5	13.7	2.1	0.2
28	7.6	5.5	4.7	52	15	113	7.2	2.2	3.8	1.6	0.1
29	1.9	5.9	4.8	63	14	96	5.2	0.8	6.4	1.5	0.2
30	12	5.2	4.4	27	23	163	17.2	0.8	2.4	3.0	0.1
31	4.5	5.6	4.3	52	25	143	12.1	0.7	9.2	3.0	0.0

[a] See Table 9.1 for key to soil code.

dS m^{-1}, in all soils. Soils were immediately equilibrated with 25 ml 0.5 mmol(c) L^{-1} $BaCl_2$ by shaking for 15 min. Cation exchange reactions induced by $BaCl_2$ were thus conducted at low but relatively constant ionic strength in all soils. The $BaCl_2$ suspensions were centrifuged for 15 min at 3000 rpm and decanted before analysis.

Extract solutions were analyzed in duplicate for Ca, Mg, K, and Na by a Perkin-Elmer 403 atomic absorption spectrophotometer with $LaCl_3$ additions (Isaac and Kerber 1975) and for monomeric Al by pyrocatechol violet (PCV) colorimetry (Dougan and Wilson 1974). The complexing ability of PCV for monomeric Al (with 15-min color development) is somewhat stronger than that of 8-hydroxyquinoline (HQ-Al), a common method for determining monomeric Al in soil extracts and natural waters (Bloom et al. 1978). In 0.01 *M* $CaCl_2$ extracts of 20 soils used in this study, PCV-Al was linearly related to HQ-Al. The linear regression had a slope significantly greater than 1.0, but an intercept not different from zero ($y = 0.123 + 1.218x$; $r^2 = .95$, $n = 20$), where x was HQ-Al and y was PCV-Al, both in μmoles per gram. Sullivan et al. (1986) found that PCV recovered slightly more Al than HQ in natural surface waters in Scandinavia, most especially in samples containing relatively high concentrations of dissolved organic carbon.

To evaluate Al exchange reactions under field conditions, soil solutions from the field sites were collected by tension lysimeters (set at 10–30 kPa tension). Solutions were collected at monthly intervals, or more often, and analyzed for major cations and anions, pH, conductivity, and total Al. Volumes collected by each lysimeter were recorded at each collection to volume-weight means for each constituent.

Cation Exchange Compared to Solution Ionic Strength

To determine how cation exchange was affected by small changes in low ionic strength, 20 soils were extracted in the laboratory with deionized water or $CaCl_2$ at 0.46, 1.81, 7.2, and 17.0 mmol(c) L^{-1}. Twenty-five milliliters of water or $CaCl_2$ was added to centrifuge tubes that contained 2.5-g soil samples. At least two extractions were conducted of each soil sample and chemically analyzed as in the $BaCl_2$ experiment described previously. Activities of cations in $CaCl_2$ solutions were estimated with extended Debye–Huckel equations (Lindsay 1979).

Schofield's activity ratios (Russell 1973) and selection coefficients for Gapon and Gaines–Thomas equations were calculated with the experimental data obtained from the $CaCl_2$ suspensions. Selection coefficients were estimated for Gapon and Gaines–Thomas equations, respectively, using the Al–Mg exchange as an example:

$$0.33\ AlCl_3 + 0.5\ Mg^{2+} - x \leftrightarrow 0.5\ MgCl_2 + 0.33\ Al^{3+} - x \quad [1]$$

$$Kg = \frac{[\text{Exch Al}]}{[\text{Exch Mg}]} * \frac{(Mg^{2+})^{0.5}}{(Al^{3+})^{0.33}} \quad [2]$$

$$Kgt = \frac{[\text{Ex Al}]^2}{[\text{Ex Mg}]^3} * \frac{(Mg^{2+})^3}{(Al^{3+})^2} \quad [3]$$

where x represents one cation exchange site, Kg or Kgt are selection coefficients for Gapon or Gaines–Thomas equations, respectively, brackets signify a cation's equivalent fraction of CECe, and parentheses signify molar activities.

Exchange of Al as Controlled by Displacing Cation

A third laboratory experiment was designed to evaluate the replacing power of individual cations under conditions of extremely low ionic strength. Replicate suspensions were prepared with 25 ml of 0.3 mmol(c) L^{-1} NaCl, HCl, $CaCl_2$, or $BaCl_2$ were added to 2.5 g of soil. Suspensions were centrifuged and analyzed for Al, Ca, Mg, and K according to procedures outlined.

Results and Discussion

Al Exchange versus Al Saturation

Exchange of soil Al by 0.5 mmol(c) L^{-1} $BaCl_2$ was controlled strongly by the fraction of CECe occupied by exchangeable Al (Figure 9.1). This low concentration of Ba displaced little Al from soils that had less than 70% Al saturation of CECe. In 11 of the 17 soils with Al saturation greater than about 70%, Al represented more than 20% of the total cationic charge in

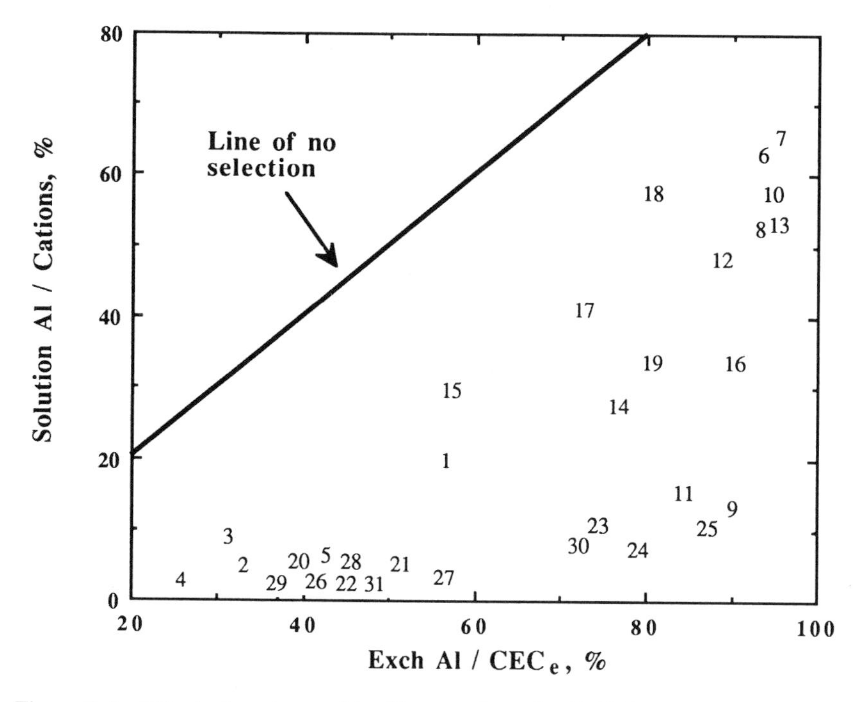

Figure 9.1. Effect of exchangeable Al saturation of soil CECe on Al displaced by 0.5 mmol(c) L^{-1} $BaCl_2$.

solution (Figure 9.1). In 6 of the 8 soils with CECe greater than 90% Al saturation, Al represented more than 40% of the total cationic charge in solution. Despite variation among extremely acid soils (Al saturation >70%) in their release of Al into solution, Figure 9.1 represents experimental verification of the important isothermic relationship emphasized by Reuss (1983) to exist between exchange-phase and solution-phase Al under low ionic strength conditions. Soils with more than 70% Al saturation of effective exchange sites (CECe) that released high concentrations of Al into solution included several Umbrepts from the Great Smoky Mountains in North Carolina (soils #6, 7, 8, and 10); a Humod from Whiteface Mountain in New York (soil # 18); two Orthods, one from Huntington Forest in New York (soil # 16) and the other from Turkey Lake in Ontario (soil # 14); and several horizons of a spodic Andisol from Findley Lake in the Cascade Mountains of Washington state (soils # 11 to 13). These soils are apt to release Al to solution in response to small increases in their otherwise low ionic strength soil solutions.

Several soils that belonged to the group of soils with more than 70% Al saturation of CECe held Al tightly against exchange. These soils included surface and subsoil horizons of Udults of the Fullerton, Lily, and Fannin series (in Figure 9.1, soils # 23 to 25, and 30, respectively), and an 89% Al-saturated Umbrept B horizon (soil #9) from the Great Smoky Mountains. Hypothetically, if these soils were to acidify (such that they increased in Al saturation), increases in ionic strength might displace additional Al into solution. This possibility is investigated in the $CaCl_2$ experiment, the results of which are described next.

Alderwood A1, and Burton Oa (in Figure 9.1, soils # 1, 15, respectively) averaged about 60% in exchangeable Al saturation of CECe, but released some monomeric to solution (see Figure 9.1). These three soils were highest in organic C of the 31 in the study (see Table 9.2), and we attribute to pyrocatechol violet (PCV) a higher recovery of Al, especially from natural organic complexes (that were high in these organic-rich surface soils). Sullivan et al. (1986) also indicated that the PCV-Al procedure recovered at least a fraction of monomeric organic-complexed Al in natural surface waters that had high concentrations of dissolved organic matter.

Factors controlling the concentration of Al in soil solution have been discussed at length by Reuss (1983) and Reuss and Johnson (1986). With the use of a simulation model, Reuss (1983) showed that fairly minor changes in base saturation within the 10% to 20% range (Al saturation within the 80% to 90% range) can cause quite large increases in soil solution Al concentrations. In one of the Ca/Al systems, Al^{3+} increases from about 10% to about 50% of total soil solution cations with a change of only 15% (from 20% to 5%) in exchangeable base cations, equivalent to increasing exchangeable Al (from 80% to 95% of CECe) saturation (Figure 9.2, top). This rapid shift toward Al dominance of soil solutions is referred to by Ulrich (1983) as the aluminum buffering. As a comparison to this model pre-

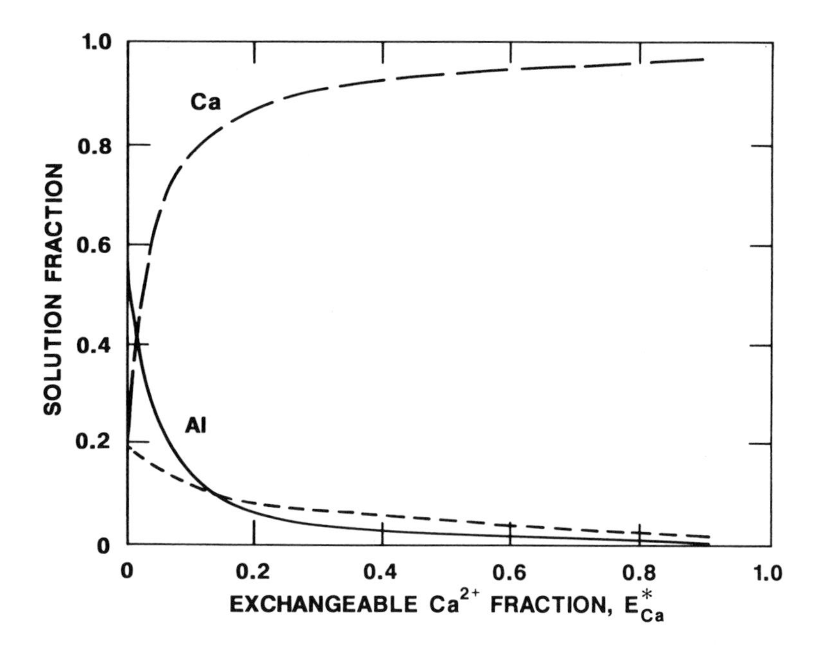

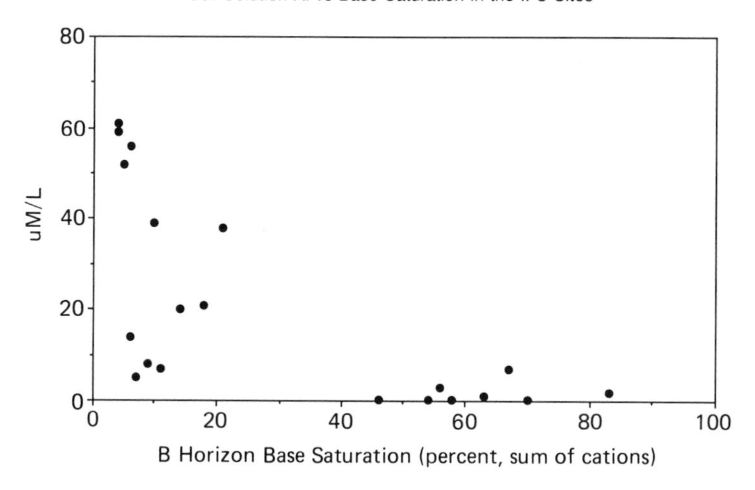

Figure 9.2. Top: Theoretical plot of soil solution fractions of Al and Ca versus exchangeable Ca fraction (after Reuss 1983). Bottom: Soil solution Al versus base saturation in B horizons in IFS sites. HF, Huntington Forest site, New York; TL, Turkey Lakes site, Ontario; ST, Smokies Tower site, North Carolina; MS, Howland site, Maine; SS, Smokies Becking site, North Carolina; SB, Smokies Beech site, North Carolina; FL, Findley Lake site, Washington; WF, Whiteface Mt. site, New York; DF, Douglas fir site, Thompson site, Washington; NS, Norway spruce site, Nordman, Norway; GA, B.F. Grant site, Georgia; CP, white pine site, Coweeta, North Carolina; CB, Camp Branch site, Tennessee; FS, slash pine site, Florida; FT, Fullerton site, Walker Branch, Tennessee; RA, red alder site, Thompson site, Washington; LP, loblolly pine site, Oak Ridge, Tennessee. Duplicate abbreviations indicate separate or duplicate plots.

diction, soil solution total Al concentrations are plotted as a function of soil base saturation in the bottom half of Figure 9.2. While this comparison involves the use of total soluble Al rather than Al^{3+} or monomeric Al, it indicates a sharp increase in Al between 10% and 20% base saturation as predicted by the Reuss (1983) model and experimental laboratory data (see Figure 9.1).

There are several data points in the lower left-hand corner of Figure 9.2, indicating very low soil solution Al associated with very low B horizon base saturation (high exchangeable Al saturation). As noted, extremely acid soils are a necessary but not sufficient condition for the mobilization of Al; either mineral acid or oranic anions must also be present. A third axis is plotted on Figure 9.3 to show mineral acid anion concentrations. Here it is clear that several sites appear extremely sensitive to Al^{3+} mobilization (i.e., with base saturations less than about 30% (Al saturation, >70% of CECe). With the addition of mineral acid anions, Al^{3+} mobilization can be expected at any of these sites with currently low base saturation (e.g., Findley Lake). Conversely, a reduction in Al^{3+} mobilization would likely occur in sites with currently high Al^{3+} (e.g., Smokies spruce sites) given a significant enough reduction in mineral acid concentration.

Al Exchange versus Solution Ionic Strength

To quantify and evaluate exchange reactions of cations as a function of small changes in solution ionic strength, 20 soils of the 31 used in the $BaCl_2$ experiment were equilibrated with deionized water or with one of four concentrations of $CaCl_2$ (0.46, 1.81, 7.2, and 17.0 mmol(c) L^{-1}). Increasing ionic strength shifted solutions from those dominated by monovalent cations to those that were dominated by polyvalents (Figure 9.4). The specific polyvalent cations displaced to solution depended strongly on the proportions of

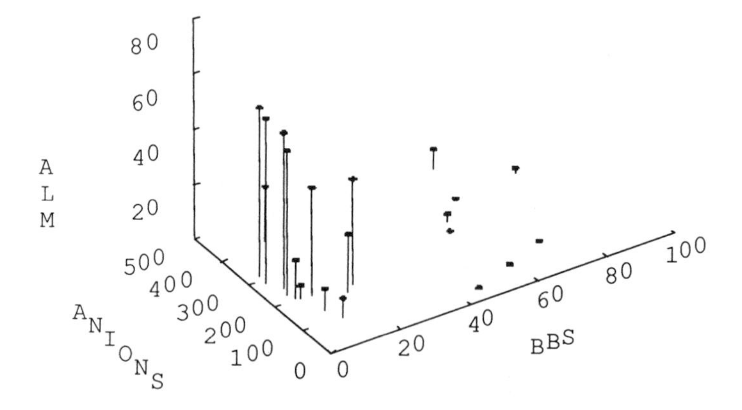

Figure 9.3. Soil solution Al (ALM) versus base saturation (BBS) and soil solution total anion concentrations in B horizons of IFS sites. See Figure 9.2 for legend.

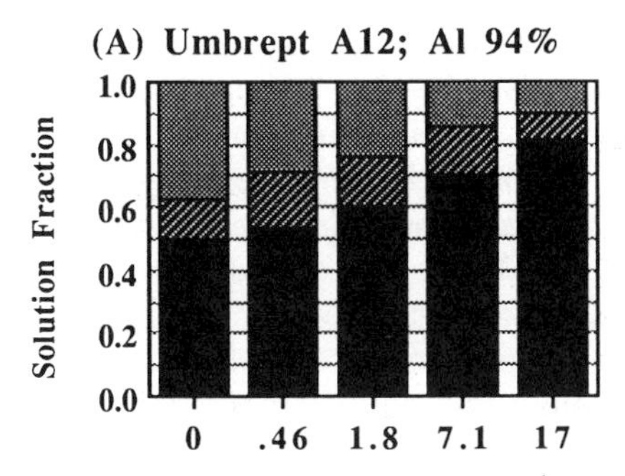

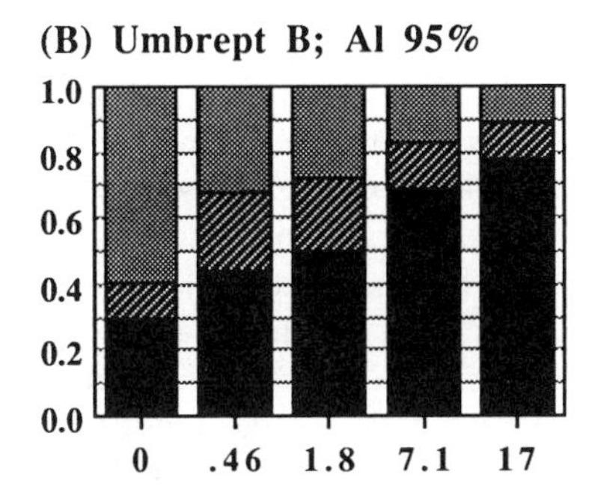

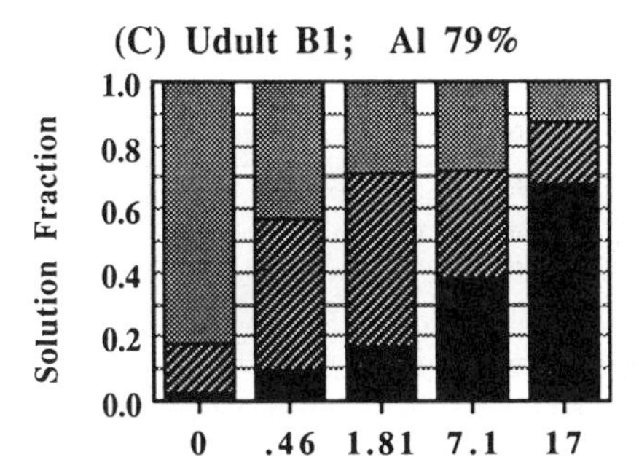

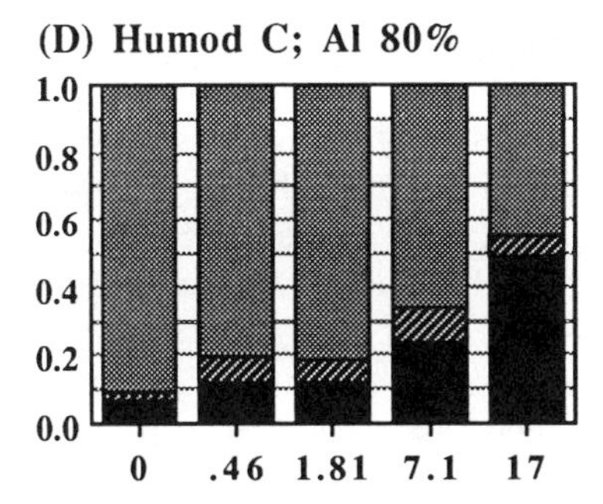

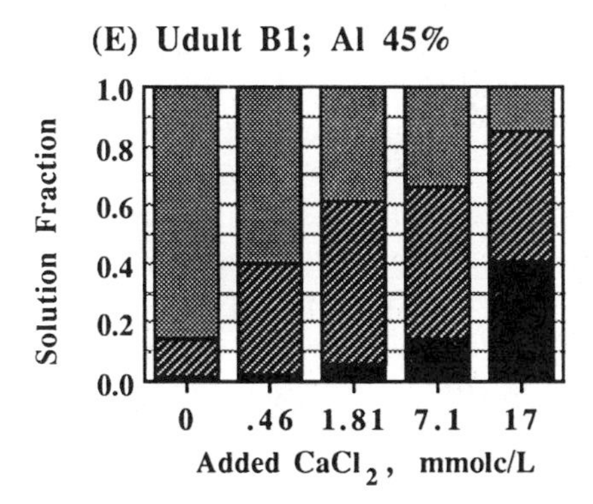

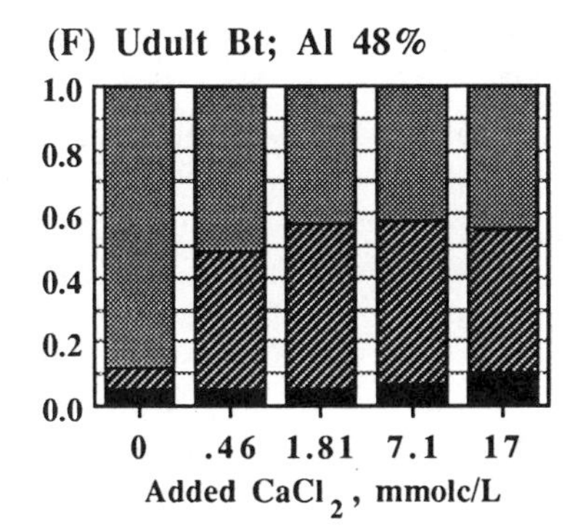

Figure 9.4. Effect of ionic strength on displacement of Al^{3+} (dark), Mg^{2+} (striped), and K^+ plus Na^+ (light) in six soils that range from 95% to 45% Al saturation of CECe. Figures 9.4 (a) to (f) represent soils #7, 10, 24, 19, 26, and 31, respectively.

cations adsorbed to cation exchange sites. In moderately acid soils, such as Fullerton B1, Appling B1, and Fannin Bt (soils # 24, 26, and 31, respectively), Mg rather than Al entered solution in response to small increases in additions of 0.46 mmol(c) L^{-1} $CaCl_2$ (Figure 9.4). In the extremely acid soils such as the Burton A12 and B (soils #7 and 10, respectively), both with greater than 90% Al saturation of CECe, Al represented 30% to 50% of the total cationic charge even in the deionized water extract. In such soils, solution Al concentrations increased in response to increased ionic strength (Figure 9.4). In the moderately acid soils (e.g., Fullerton B1, Becket C, and Appling B1; soils # 24, 19, and 26, respectively), Al was displaced only into solutions with relatively high ionic strength.

Activity Ratios for Al, Mg,and K Exchange

Although Schofield's (1947) activity ratio has rarely been tested with soil solutions at less than 10 mmol(c) L^{-1}, in this experiment such activity ratios successfully predicted soil exchange of Al, Mg, and K in solutions of very low ionic strength. Figure 9.5 illustrates that activity ratios of $Al^{0.33}/Mg^{0.5}$ and $Al^{0.33}/K$ were remarkably constant for individual soils despite variations in $CaCl_2$ concentrations from 0.46 to 7.2 mmol(c) L^{-1}. Activity ratios for $Al^{0.33}/Mg^{0.5}$ and $Al^{0.33}/K$ in 0.46 mmol(c) L^{-1} $CaCl_2$ were well correlated with activity ratios in 7.2 mmol(c) L^{-1} solutions (r^2 = 0.94 and 0.99, respectively), and thus were relatively constant for a given soil over about a 15-fold range in electrolyte concentration.

That the slopes in Figure 9.5 approach 1.0 is important from another perspective, that of whether surface charge affects the activity ratio at low ionic strengths. The literature is a bit unclear about this point, although soils with constant charge have been suggested to follow the "Activity Ratio Law" better than those with variable charge (Sumner and Marques 1968; Russell 1973). Many of the soils in Figure 9.5 have considerable pH-dependent, variable charge (see Table 9.2), and Figure 9.5 suggests that soils with mixtures of variable and constant charge may have relatively constant activity ratios over the range of ionic strengths found under field conditions. This is a highly appropriate area for further research, given the regional significance of soils with variable charge (Richter and Babbar, in press).

Selection Coefficients versus Ionic Strength

The Gapon equation had relatively constant selection coefficients (Kg) for Al–Mg exchange in solutions of widely different ionic strengths (Figure 9.6a). Even Kg values for the 20 soils in contact with 0.46 and 17 mmol(c) L^{-1} were significantly correlated (r^2 = .71). Similar regressions for Gaines–Thomas coefficients (Kgt) calculated from the same Al and Mg data were much less significant or were not significant at all (Figure 9.6b).

In the Gapon and Gaines–Thomas equations (Eq. [2] and [3]), exponents are roots and powers, respectively, a difference that may have large nu-

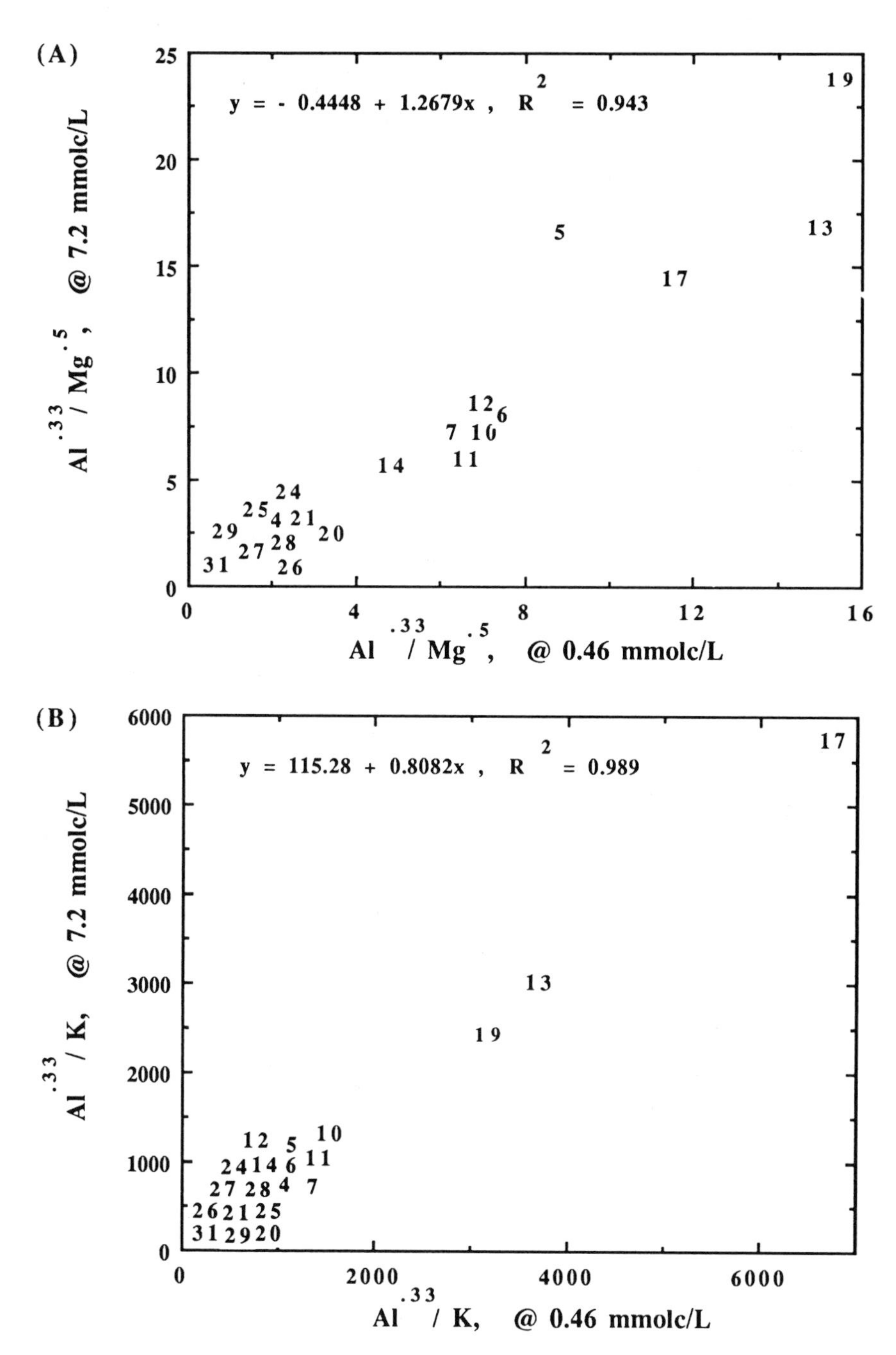

Figure 9.5. Schofield's activity ratios of Al-Mg (a) and Al-K (b) exchange at 0.46 and 7.2 mmol(c) L^{-1} $CaCl_2$, respectively.

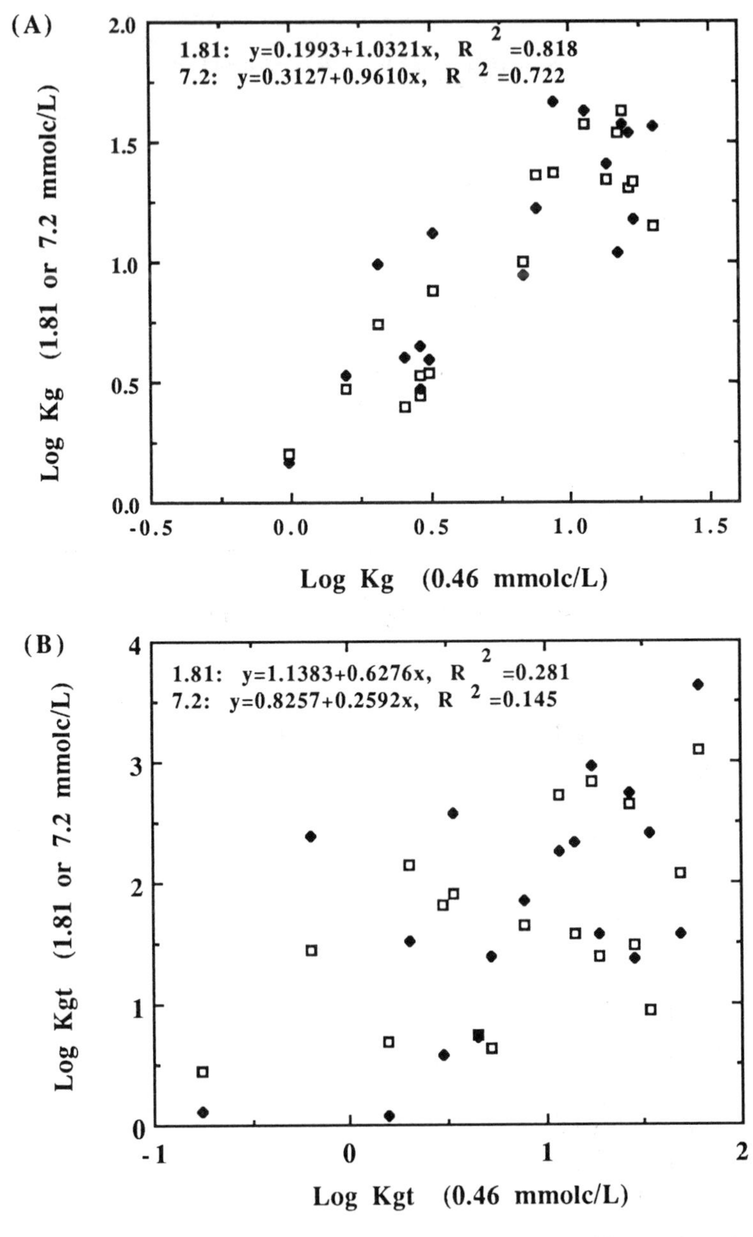

Figure 9.6. Gapon (a) and Gaines–Thomas (b) selection coefficients for Al-Mg exchange at 0.46 as compared with 1.81 or 7.2 mmol(c) L^{-1} $CaCl_2$ (open and closed symbols, respectively).

merical effects on selection coefficients. Such differences in equations produce selection coefficients that respond differently to common data. The Gapon equation appears relatively robust in describing exchange of cations with different charges, in part because of exponents that are roots of very small numbers (i.e., $\ll 1.0$ mol L^{-1}) rather than squares or cubes of small numbers in the Gaines–Thomas equation. Such differences in selection equations also tends to increase the range of selection coefficients of the Gaines–Thomas equation compared to those of the Gapon. The Gapon and Gaines–Thomas coefficients ranged over about 1.5 and nearly 4 orders of magnitude, respectively (Figure 9.6). Somewhat similar patterns were observed with the Al–K exchange data, although ionic strength appeared to have much less effect on selection coefficients compared with Al–Mg exchange (Figure 9.7).

The effect of ionic strength on both selectivity coefficients and activity ratios is critical to modeling changes in soil and solution chemistry. Several models use constant selection coefficients or activity ratios to predict chemistry of soil solutions and assume such constancy over a wide range of exchange conditions. Experimental data that support such an approach are not abundant (McBride and Bloom 1977; Bohn et al. 1985; Richter et al. 1988). Results of these experiments indicate that at the very low ionic strength conditions typical of field conditions, ionic strength of soil solution has relatively little effect on Schofield activity ratios, but that selection coefficients may vary widely depending on the exchange equation. Bohn et al. (1985) made a similar point, but only by analyzing the limited data of Schachtschabel (1940). Soil chemistry models should benefit greatly from having a relatively flexible approach to estimating cation exchange reactions, if only so that effects of different exchange formulations on predicted soil solution chemistry can be more precisely evaluated.

Cation Exchange by Low Concentrations of H, Ba, Ca, and Na

Displacement of exchangeable cations from 9 soils by H, Ba, Ca, and Na was determined with 0.3 mmol(c) L^{-1} chloride solutions, to evaluate if differences in exchange between species of added cations could be detected at such very low concentrations. Even with such small additions of electrolytes, however, differences in cation exchange were detectably different among the four cations added to soils (see Table 9.3). Chlorides of H^+, Ba^{2+}, and Ca^{2+} exchanged and elevated concentrations of polyvalent cations Al^{3+}, Mg^{2}, and Ca^{2+} with much greater efficiency than added Na. Cation exchange depended on the cation being added to the soil solution system and on the cation being displaced from exchange sites. Table 9.3 indicates the relative effectiveness for displacing soil K^+ was $Na^+ = Ca^{2+} = Ba^{2+}$ less than H^+; for displacing soil Ca^{2+} or Mg, the series was Na^+ less than $Ca^{2+} = Ba^{2+} = H^+$; and for displacing soil Al^{3+}, the series was Na^+ less than $Ca^{2+} = Ba^{2+}$ less than H^+. Results indicated that added H^+ displaced exchangeable K^+

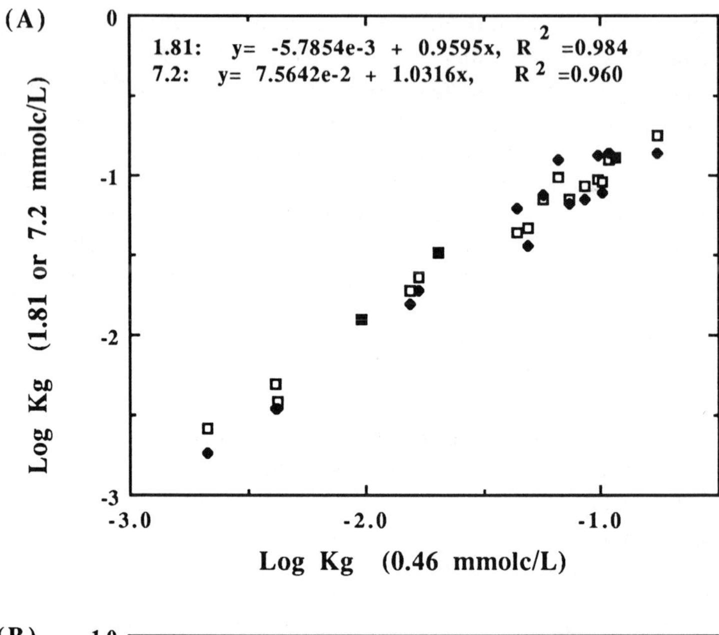

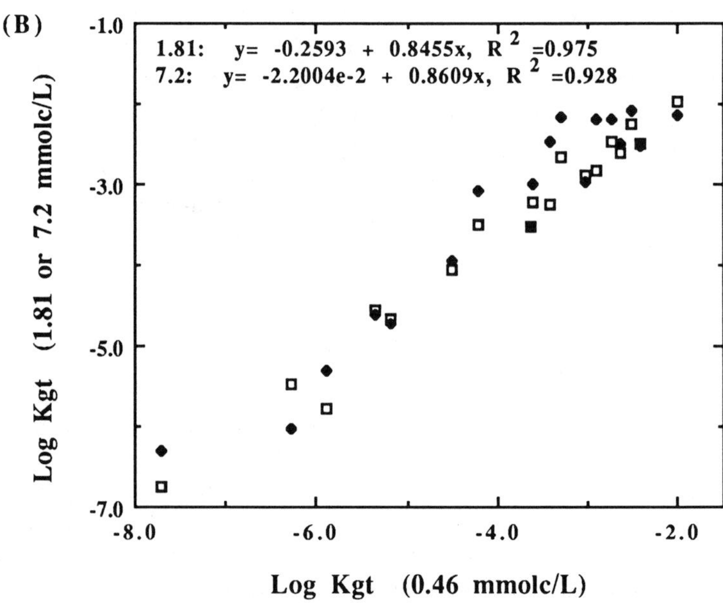

Figure 9.7. Gapon (a) and Gaines-Thomas (b) selection coefficients for Al-K exchange at 0.46 as compared with 1.81 and 7.2 mmol(c) L^{-1} $CaCl_2$ (open and closed symbols, respectively).

Table 9.3. Exchange of Soil Cations on Addition of 0.3 mmol(c) L^{-1} of Na^+, Ca^{2+}, Ba^{2+}, or H^+ as Chloride Solutions to Soils #1, 3, 6, 8, 11, 14, 18, 20, and 21[a]

Na	Ca	Ba	H	WD	
K	0.749	0.813	0.834	1.222	0.2030[b]
Ca	0.895	na[c]	1.641	1.486	0.1915
Mg	0.503	0.830	0.836	0.831	0.0905
Al	1.662	1.890	1.848	2.056	0.0974

[a] In suspensions, displacing cations were added at equivalent of 3 μmol(c) g^{-1} of soil. Units, μmol(c) g^{-1}.
[b] Means within a row that are not significantly different are connected by an underline.
[c] Not applicable.

and Al^{3+} with greater efficiency than added Ba^{2+} and Ca^{2+} or that nonexchangeable K^+ and Al^{3+} were slightly susceptible to acid dissolution (see Table 9.3).

Wiklander and Andersson (1972) suggested that the efficiency with which H ions displace exchangeable Ca^{2+}, Mg^{2+}, and K^+ is a significant soil property that controls rates of soil acidification and cation leaching. Wiklander and Andersson (1972) quantified H ion replacing efficiency by dM/H, where dM was nutrient cations Ca^{2+}, Mg^{2+}, and K^+ displaced per unit H ions added. The approach represented a simplified exchange equation involving exchange between H and the sum of the cations Ca^{2+}, Mg^{2+}, and K^+ as a function of soil base saturation (i.e., the lower the base saturation, the lower the efficiency of H ion exchange for such cations).

Results of these studies (see Figures 9.1 and 9.4, Table 9.3) indicate that the concept of H ion replacing efficiency can be broadened to predict that introductions into the soil solution of any cation with a strong adsorption affinity for exchange sites (e.g., H^+, Al^{3+}, Ca^{2+}, Mg^{2+}, or Ba^{2+}, not just H^+) will displace exchangeable polyvalent cations into soil solution and increase polyvalent cation leaching. In coastal watersheds in western Norway, for example, precipitation laden with sea salts elevates Al^{3+} in runoff from extremely acid soils (Skartveit 1980; Richter 1984; Wright et al. 1988). Such changes may result primarily from sea salt-derived Mg rather than Na^+ exchanging for Al^{3+}, because Mg^{2+} has a much higher adsorption affinity than Na^+ for exchange sites, even at very low concentrations. Acid deposition thus accelerates leaching of exchangeable cations by altering the interaction of short- and long-term phenomenon. In the short term, acid deposition increases the ionic strength of the soil solution and displaces otherwise strongly adsorbed cations into the soil solution. Over the long term, the distributions of soil exchangeable cations may well be altered by leaching losses from acid deposition inputs, thereby altering not only soil exchangeable populations but also the exchange reaction products themselves.

Conclusions

Cation exchange reactions especially those involving polyvalent cations are predictable using soil samples contacted with variable but low ionic strength solutions. In nearly all soils tested, increasing solution ionic strength by even small amounts increased exchange of the polyvalent cations Al^{3+}, Ca^{2+} and Mg^{2+} whereas exchange of monovalent ions K^+ and Na^+ was much less dependent on ionic strength. Schofield activity ratios were relatively constant for Al^{3+}, Mg^{2+}, and K^+ exchange over a wide range of low ionic strength, even for soils that have a mixture of constant and variable charge and a mixture of different exchangeable cations. Similarly, Gapon selection coefficients (which make use of the Schofield activity ratio as part of their formulation) were relatively constant over a wide range of ionic strength, whereas Gaines–Thomas coefficients were not. Some but not all exchange formulations appear to predict cation exchange reaction products.

Such results help explain the effects of acid deposition on the soil leaching process. Over the long term, accelerated soil leaching from increased ionic strength (by acid deposition) is much more likely to increase leaching and deplete soil exchangeable polyvalent cations such as Ca^{2+} and Mg^{2+} rather than K^+ in highly weathered soils. In acid soils, exchangeable Al^{3+} is potentially a major leachable cation that responds to increased ionic strength, provided that Al^{3+} saturates more than about 75% of the soil's effective cation exchange sites. Acid soils are a necessary but not sufficient condition for drainage water to be acidic and to contain Al^{3+}. Elevated ionic strength in combination with extreme soil acidity can elevate Al^{3+} concentrations, resulting in enhanced leaching of Al^{3+} within the soil profile and potentially to drainage waters.

Evidence of Historical Influences of Acidic Deposition on Wood and Soil Chemistry

E.A. Bondietti and S.B. McLaughlin

Introduction

The study of current forest nutrient cycles in relationship to atmospheric deposition must be viewed from the perspective of past, current, and future influences of natural and anthropogenic processes on those cycles. A principal focus of the IFS has been to compare and contrast the influence of these processes on nutrient fluxes through diverse forest types. In viewing the health of the subject forests and the extent to which they may have been impacted by atmospheric deposition, it is also important to evaluate evidence of their historical nutrient status. Although the causes and consequences of acidic deposition have been studied intensively in the United States during only about the past 10 years, much of the eastern United States has experienced relatively high emissions of SO_X during the past 60 years (Husar 1986). Thus the question of whether current nutrient budgets adequately depict expected responses of preindustrial nutrient cycles is a valid issue in evaluating past and present effects of acid deposition on these cycles.

This chapter examines evidence of historical changes in forest nutrient availability using two approaches to characterize past nutrient status: (1) evaluation of historical patterns of tree ring chemistry, and (2) historical and gradient analysis of soil nutrient status. In the first approach we use historical changes in the chemical composition of wood as an indicator of changes in the soil solutions from which nutrients were derived over time. The latter approach examines historical and recent analyses of soils for evidence of changes in nutrient content as well as analyzes foliar nutrient status along gradients of increasing acidic deposition.

In approaching this task, it is important to first consider the theoretical basis for anticipating qualitative and quantitative changes in soil and plant tissue chemistry with increasing atmospheric deposition. Ulrich et al. (1980) interpreted the increased availability of Fe and Al^{3+} of soils under historically heavy loading of atmospheric emissions as a consequence of mobilization of these elements by acidic deposition. Significant losses of base cations were predicted for forest soils in Central Europe based on chronic leaching losses from acidic deposition (Ulrich 1983). The theoretical basis for these reactions is focused around influences of inputs of hydrogen ions and the strong anions, SO_4^{2-} and NO_x, on cation exchange processes (Reuss and Johnson 1986) and has been addressed in Chapters 8 and 9 of this volume.

While the direction of response of forest soils to continued heavy atmospheric loading may be very predictable from the perspective of soil chemistry, the timing of such changes and the relationship they bear to natural acidification processes, nutrient uptake, and cycling have been critical to the aims of the IFS. In addition to the fluxes of nutrients through the system, the availability of those nutrients in the primary rooting zones of the upper soil horizons is of extreme importance to what the trees experience physiologically. Aluminum released by acidification, for example, may reduce uptake of nutrients either directly through toxicity to fine roots or through competitive inhibition of uptake of calcium, magnesium, or phosphorous (see review by Sucoff et al. 1990). The relationship that changes in nutrient cycles and nutrient availability bear to the growth and stability of forests is critical to the relevance of IFS to the broader national effort to assess the influence of acidic deposition on the future health of our nation's forests.

The input of anthropogenically derived sulfur and nitrogen to forests has been viewed from a forest nutritional perspective as likely to produce a wide spectrum of effects depending on the length of the exposure and the nutrient status of the forests at the time at which inputs first significantly intensified. Abrahamsen (1980), for example, noted that the initial response of many forests would likely be positive as nutrient cations were mobilized and nitrogen was added to the system. Negative impacts on growth were seen as the longer term consequence of continued nutrient mobilization and loss from the system. In the United States, studies in both the Northeast (Federer et al. 1989) and the Southeast (Johnson et al. 1985) have indicated that current acid deposition levels have substantially accelerated losses of base cations above levels of loss by natural processes. Federer et al. (1989) estimated that leaching losses of calcium currently exceeded natural calcium resupply processes by more than twofold for eastern forests.

The question that we address in this chapter is whether additional leaching of base cations by acid deposition has caused measurable changes in plant-available levels of these cations, particularly calcium, and whether these changes have left a chemical signature on the soils and associated trees that is mechanistically predictable and historically verifiable. In this context it is perhaps useful to extend, and slightly modify with respect to acidic deposition, the hypothetical analogy applied by Smith (1974) in classifying general stages of forest responses to air pollutants:

Class I. No effects: Preindustrial emissions added insignificant additional nitrogen and did not add measurably to natural soil acidity or natural leaching losses of base cations from forest ecosystems.

Class II. Positive effects (a component of Smith's original Class 1): With increasing inputs of acidity, increased rates of weathering and increased exchange of cations from cation exchange complexes would increase cation availability for plant uptake and leaching losses from the system. On sites that were initially marginal in base status or nitrogen availability, increased uptake of nutrients and increased tree growth would be anticipated.

Class III. Negative effects: Accelerated mineralization and leaching would eventually lead to depletion of soil reserves of base cations and increased relative availability of more abundant and more readily mobilized aluminum and iron. The increased availability of these metals further amplifies the effects of low base status on trees by competitive inhibition of uptake of Ca^{2+}, Mg, and P. At this stage, tree physiological function will ultimately be impacted and growth be limited by reduced nutrient supply or metal toxicity.

Class IV and Beyond: Continued strong anion loading to systems would reduce nutrient pools to very low levels at which plant controlled components, including uptake, litterfall, and death of more sensitive individuals, dominate the small residual plant-available pools of some nutrients such as calcium and magnesium. Additional atmospheric nitrogen only amplifies the situation by further leaching soils and shifting carbon from roots to shoots through foliar fertilization. Plants become increasingly sensitive to other biotic and abiotic stresses.

In summary, if significant base cation mobilization and depletion of base cations from our eastern forest soils had occurred, we would expect to see a chronosequence of changes in tree uptake patterns and perhaps in tree growth as well. Initial increases in calcium availability on some sites would have increased growth only if these systems were initially Ca^{2+} deficient. Continued debasification of the system would be indicated by increasing uptake of aluminum and iron, by decreasing uptake of calcium, and ultimately by decreased growth. Patterns of tree ring chemistry for high-elevation red spruce in the eastern United States suggest that this chronosequence has already occurred (Bondietti et al. 1989, 1990). To test this hypothesis, however, we must establish the reliability of tree ring chemistry as an indicator of soil solution chemistry.

Tree Ring Chemistry as an Indicator of Soil Chemistry

Detection of a close correspondence between elemental patterns of shortleaf pine stemwood and historical SO_2 emissions from the Copper Hill Smelter in East Tennessee (Baes and McLaughlin 1984; Figure 9.8) provided early evidence that chemical changes in tree ring chemistry reflects changing inputs of regional pollutants to forests. Increasing levels of iron (Fe) were found in those tree cores during the approximately 50 years of open pit smelting operations (1860–1910) at Copper Hill, Tennessee, 50 miles upwind from the study site in Cades Cove, in the Great Smoky Mountains National Park (GSMNP). After emissions were reduced to nearly preindustrial levels in 1910, levels of iron were significantly lower for 40 years before increasing significantly again during the past 30 years. The recent trends paralleled significant regional increases in emissions in SO_2. It is interesting to note that increased mobilization of Al was not observed during early smelter operations at the Cades Cove site; however, increases in Al were apparent in connection with increases in Fe concentrations during the past 30 years. Further, as the studies were extended to other species such as red spruce along elevational gradients in the GSMNP (Baes and McLaughlin 1986), it was found that the Al mobilization patterns were more strongly developed on upper elevation sites, accompanied by strong increases in Fe levels, and that initial increases in both Al and Fe preceded a transitory increase in Ca content of wood (Figure 9.9).

Historical markers of changes in Al and Fe availability in the GSMNP (Baes and McLaughlin 1986) have indicated that these elements were valid indicators of temporal change, but sharply decreasing levels of Ca in recent wood observed originally for shortleaf pine and the transitory "calcium bulge" observed for red spruce were more difficult to interpret without knowing the extent to which Ca was translocated. Recent analysis of the tree ring chemistry of red spruce and eastern hemlock used the historical deposition patterns of ^{90}Sr from world thermonuclear bomb testing to evaluate the behavior of this surrogate for Ca. Using this approach, Sr mobility was found to be limited to an approximate 30-year window with maximum concentrations occurring in wood found within 5 years of the period of maximum input of ^{90}Sr to soils (Bondietti et al. 1989). Thus, recent decreases in Ca levels in

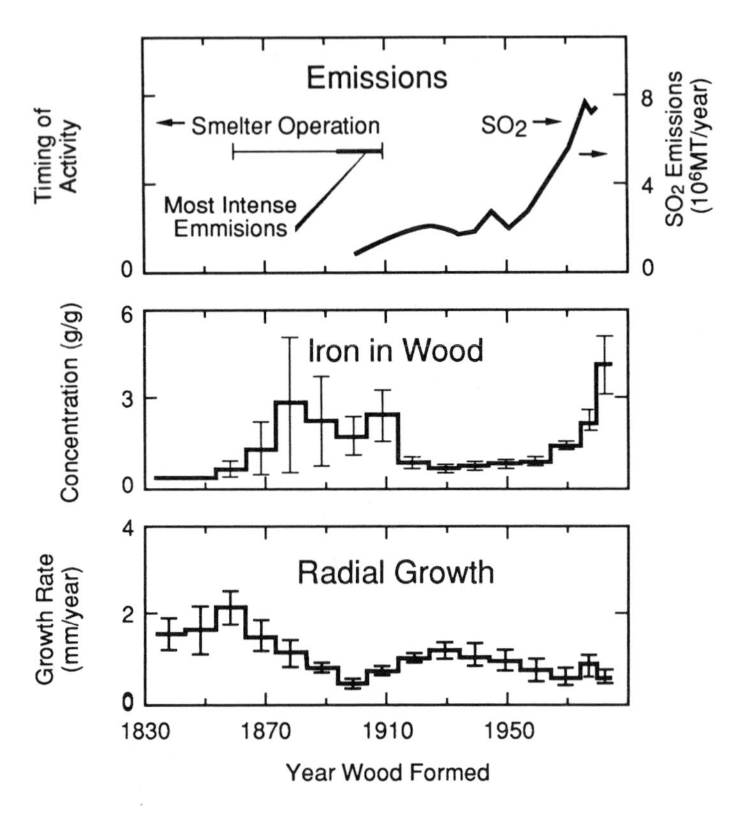

Figure 9.8. Mean growth rate and wood Fe concentrations of shortleaf pine at Cades Cove, Great Smoky Mountains National Park, and estimated timing of changes in smelter emissions from Copper Hill, Tennessee (80 km upwind of Cades Cove). Levels of more recent SO_2 emissions are based on average emission density of areas from within 900 km southeast-southwest of east Tennessee. Error bars represent standard error of mean of eight trees (after Baes and McLaughlin 1984).

wood were seen as indicative of reduced availability of Ca to these trees and not a function of strong inward mobility.

Increasing ratios of Al to Ca over the past 30 to 40 years ago were found to have occurred in wood at many sites in the Smoky Mountains of Tennessee (as shown by an example core in Figure 9.9) as well at a northern test site at Camels Hump in VT. The regional, elevational, and temporal aspects of these patterns were viewed as evidence of the effects of strong anion-induced mobilization processes under acid soil conditions (Bondietti et al. 1989). The implications of Sr distribution patterns to most recent changes in Ca distribution patterns are developed further next.

The extent to which Ca availability in wood reflects the changing availability of Ca in soil solutions is pivotal to its use as an indicator, and this led Momoshima and Bondietti (1990) to explore the physiochemical basis

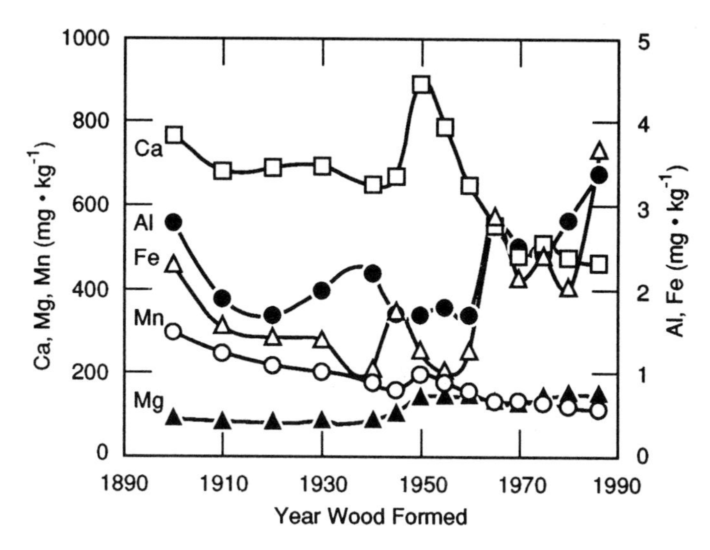

Figure 9.9. Radial trends in cations in red spruce stem section from Noland Divide, near IFS tower site, in GSMNP.

of changing Ca chemistry in wood. They found that ion exchange characteristics of wood, particularly the divalent pectate binding sites in tracheids, control the Ca exchange capacity of wood and lead to its functioning as an ion exchange column. Further, cation binding capacity (CBC) is reduced in a very predictable way by changes in sap solution pH and by the innate composition of wood formed at increased distances from the pith. The expected pattern of Ca binding in wood is a steady decrease in CBC and hence Ca content with age. However, wood, once formed, has the capacity to act as an ion exchange column that is both sensitive to availability of Ca in the sap solution and has an absorption capacity that can be reduced by increases in sap acidity (Momoshima and Bondietti, 1990). This process is of course restricted to sapwood actively involved in transport of sap.

Changes in Patterns of Calcium Accumulation

Analysis of changes in wood chemistry from samples across several sites in the eastern United States (Bondietti et al. 1990) indicated that during the past 30 to 40 years there have been some substantial departures from the expected linear decreases (Momoshima and Bondietti, in press) in calcium accumulation patterns in wood. In Figure 9.10a–9.10c, expected patterns of Ca accumulation in wood based on the analyses of Momoshima and Bondietti (1990) are contrasted with observed patterns for example cores from Crawford Notch (Mt. Washington, New Hampshire), Big Moose (Lake), New York, and Mount Abraham, Vermont. A consistent trend noted in all three of these chemical chronologies is the midcentury increase in Ca above

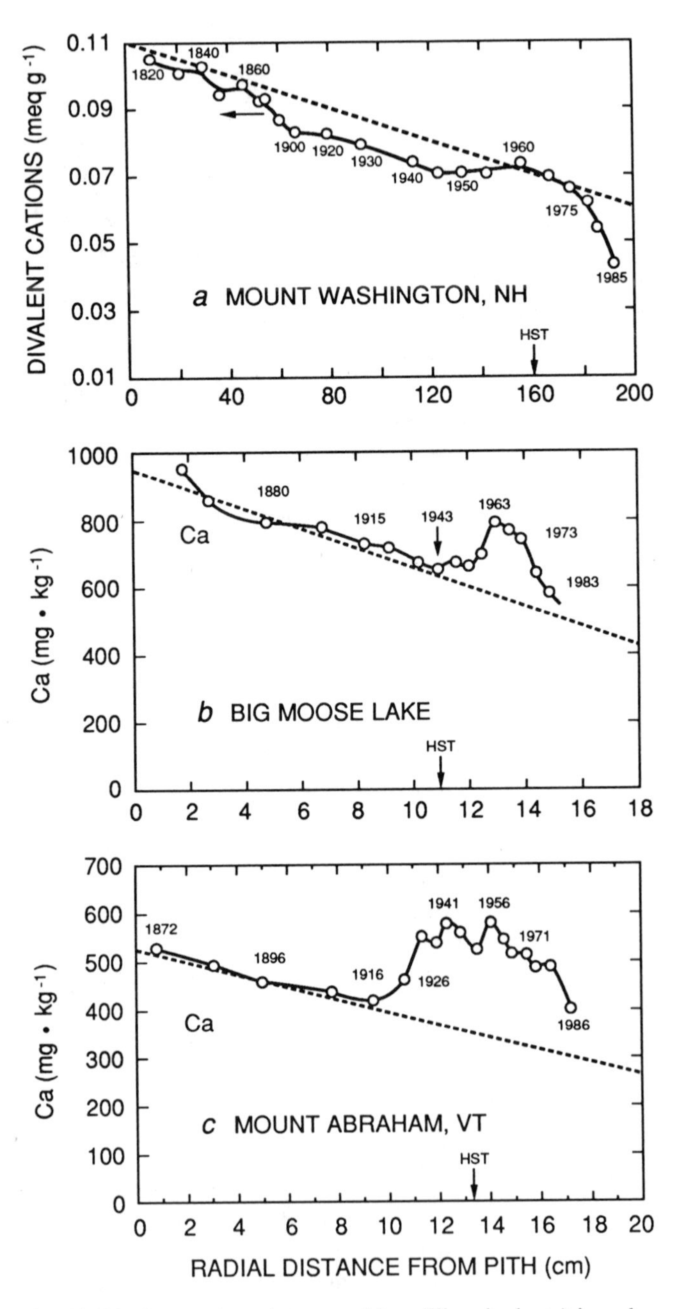

Figure 9.10. (a) Divalent cations (expressed in milliequivalents) in red spruce bolewood from Crawford Notch (RSCN) New Hampshire, Maine, (B) Big Moose, New York, and (c) Mount Abraham, Vermont. Measurements are plotted as a function of radial distance from pith. Dashed lines are predicted trends in wood divalent cation storage under constant sap conditions.

expected levels, followed by a subsequent decrease to levels below (Figure 9.10a) or near (Figure 9.10b, 9.10c) expected levels. The unexpected increase in Ca levels, which parallels the pattern shown previously for the IFS Tower site in GSMNP (see Figure 9.9), suggests that an initial increase in Ca supply followed by decreasing supply coincided with the regional increase and subsequent decline in radial growth of red spruce reported by McLaughlin et al. (1987). It is important to note that, although the trends presented in Figures 9.9 and 9.10 represent more than 50% of the more than 80 red spruce cores we have measured from Tennessee to New Brunswick, not all cores within a stand always show the same pattern. This is not surprising however, considering the influence that heterogeneity of site quality, stand composition, and local disturbance history can have on tree and stand nutrient dynamics. It should also be noted that while these same variables might affect the timing of a response to nutrient disturbance, they occur in heterogenous patterns on a regional basis and would not be expected to produce temporal consistency across the region.

To interpret the patterns of increase and decrease in wood Ca, it is useful to reexamine the analysis of ^{90}Sr patterns discussed earlier. In Figure 9.11, coincident patterns of Sr, ^{90}Sr, and Ca^{2+} are examined for a representative tree from a high-elevation site in the GSMNP. The superimposition of historical ^{90}Sr deposition patterns on observed trends in wood chemistry highlights the temporal shift in peak ^{90}Sr in wood relative to the 1965 peak in

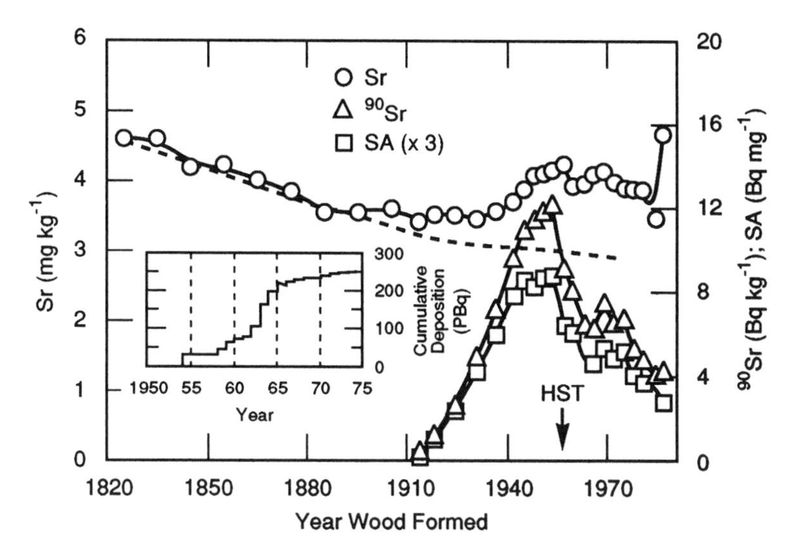

Figure 9.11. Concentration trends in fallout ^{90}Sr, stable strontium, and ^{90}Sr specific activity (SA) in red spruce from Forks Ridge trail, Great Smoky Mountains National Park. Lower left inset illustrates cumulative deposition trend of ^{90}Sr in Northern Hemisphere. All ^{90}Sr data were corrected for radioactive decay.

cumulative ^{90}Sr fallout. Native Sr in wood increases around the turn of the century with a sharp increase apparent around 1950. This later increase coincides with the increase and peak in ^{90}Sr levels. Although Sr levels remained high during the past three decades, both the ^{90}Sr and the Sr:^{90}Sr ratio decreased sharply during the past three decades. This decrease is the expected consequence of continued and significant losses from a limited pool of soil ^{90}Sr with no weatherable source of resupply.

The timing of ^{90}Sr deposition and Sr accumulation in wood raises the question, "Why does peak activity of ^{90}Sr in wood predate the period of maximum total deposition of ^{90}Sr, and how can ^{90}Sr levels found in wood from as far back as 1920 be reconciled with the initiation of bomb testing in the 1950s?" The answer lies in the realization that approximately 30 years of wood were involved in xylem transport of solutes from the soil solution, as noted by the heartwood–sapwood transition indicated in Figure 9.11. Thus at the time of the initial releases of activity in 1954, the sapwood would have extended back to approximately 1924, and small amounts of ^{90}Sr taken up in 1954 were exchanged with wood formed 30 years earlier.

Low levels of ^{90}Sr incorporation for wood formed in the earlier part of this century reflect marginal involvement of older wood in current xylem transport and cation exchange. By the same token, the increases in ^{90}Sr accumulation from around 1920 until the peak in 1955 reflect a weighted averaging process indicating increased involvement of more recently formed wood in the transport and cation exchange processes. Similarly, leaching of ^{90}Sr from the soil after testing stopped in 1965 would have "washed out" the true peak in activity that likely existed around 1965 as much lower concentrations of ^{90}Sr in the ascending xylem sap were equilibrated with ^{90}Sr deposited in wood formed in earlier years. Thus the wood content of cations represents a time-weighted average of sap solution chemistry during as much as approximately 30 years, with the influence on that average being inversely related to the time since the wood was formed.

The 1955 "breakpoint" for ^{90}Sr observed in Figure 9.11 shows that wood formed within 10 years of a substantial decrease in cation input to the system reflected that change. The pattern of decreases in the ^{90}Sr:Sr ratio after 1960 reflects largely the much more rapid loss of ^{90}Sr from the system because the specific activity curve rather closely parallels the ^{90}Sr curve. Figure 9.11 suggests that Sr has been mobilized and incorporated to unexpectedly high levels during the past 40 to 50 years. The biggest increases are projected to have occurred within the 1965–1975 interval, approximately 30 years after the initiation of a sharp increase in wood Sr around 1940.

The decreasing trends in wood Ca seen in the past 15 years (see Figure 9.10a) to 40 years (see Figure 9.9) from four sites suggest that Ca mobilization has been followed by reduced accumulation rates in wood presumably associated with decreasing Ca availability in soil. These decreasing Ca trends observed recently can also be expected to have diminished the magnitude of the preceding Ca bulge through the previously discussed dilution effect.

The pattern of Ca increase in wood followed by decreasing values in more recent years is dominant among approximately 25 red spruce trees from New England and 18 trees from GSMNP examined by Bondietti et al. (1990). The point of transition from increasing to decreasing wood Ca occurred around 1960 in New England and in the mid-1950s in the GSMNP (Figure 9.12).

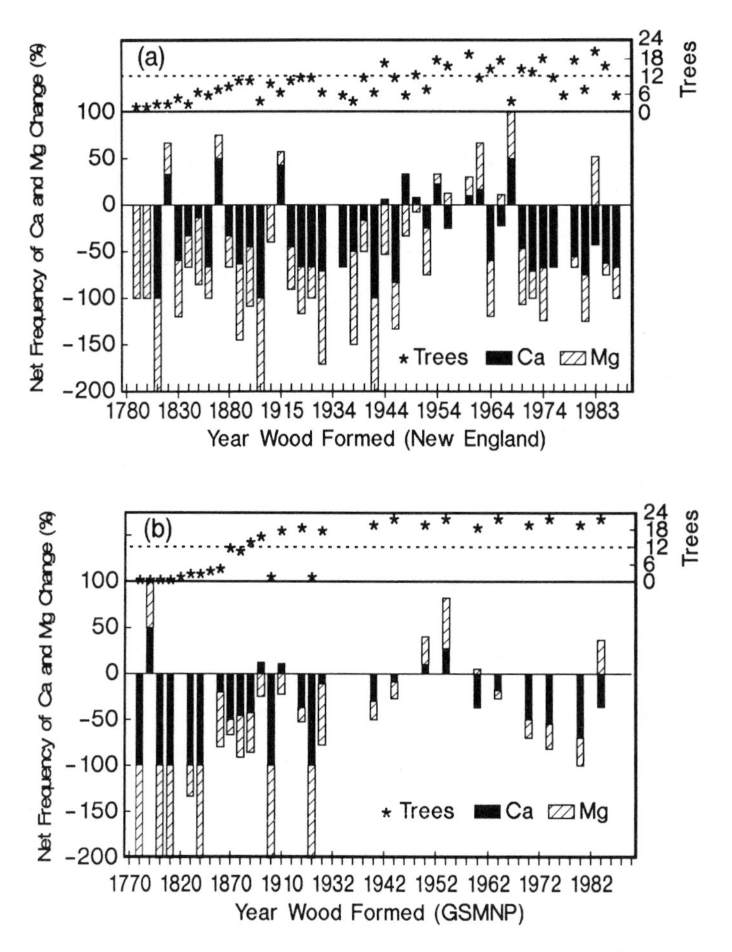

Figure 9.12. Summary of relative concentration trends of Ca and Mg in red spruce obtained from (a) New England and (b) Great Smoky Mountains National Park. Bars indicate net frequency of trees for which calcium either increased (+) or decreased (-) in concentration for each increment relative to previous sampling increment. Asterisks in upper portion of each graph indicate number of trees sampled for each observation. Note that dates are not on linear scale before 1915.

The earlier initiation of increasing wood Ca in the southern trees was suggested as a consequence of a proportionately larger sapwood area and hence longer "bleedback" effect in the milder southern climate (Bondietti et al. 1990). Alternatively, if the currently much higher deposition levels in the southern high-elevation sites (see Chapter 6) are representative of historical levels, the GSMNP Ca^{2+} trends may actually reflect longer term influences of long-distance regional emissions.

Evaluating the Evidence of Calcium Depletion in Forest Ecosystems

The foregoing discussion suggests that there have been significant changes in the availability of Ca to red spruce during the past 50 years. If this has occurred, it is important to evaluate the significance of these changes to both the forest soils from which the Ca was removed and to the trees supported by those soils.

Soil-Forming Processes in Coniferous Forests

The surficial stratification of exchangeable bases in the forest floor of mature forests, particularly under species such as spruce or hemlock that form their own rooting environment, tends to isolate the nutrient cycle of individual trees from the mineral soil. The podzolization process that occurs under low Ca litter is a well-known consequence of this phenomenon because of the almost complete extraction of mobile calcium by the surficial fine-root system in the mor layer. In these situations the replenishment of base cation losses from the mineral cycle is highly ombrotrophic (e.g., Mattson and Karlsson 1944; Sjors 1961), making the base status of these mor soils highly sensitive to impoverishment by acidic deposition. Thus, forest mor displays the same inherent sensitivity to acidic deposition as does bog peat. The consequences are potentially more significant for the forest mor because of the more complex vegetation cover with its higher Ca requirement and the generally better drained condition of the raw humus layer of upland forests.

Historical Survey of Nutritional Status of Soils Supporting Growth of Spruce

To evaluate possible historical changes in base status of forest soils, an extensive literature survey was made (Shortle and Bondietti, in press). The survey focused on historical studies around the world in which the base status of the humus soil layer was analyzed for old-growth spruce, fir, or hemlock forests. The humus layer was the focus of this study because of its role as a primary rooting zone and one from which base depletion by historical increases in strong anion inputs in acidic deposition would be expected to be most likely to occur. More than 36 reports of soil nutrient status made earlier than 1960 were examined. To ensure compatibility of data across time, a set of preconditions was established in comparing historical data with those from more recent studies. These conditions were as follows:

1. Study sites were older spruce, fir, or hemlock forests (>40 yr old) with mor type soils
2. Raw humus had to be at least 5 cm in depth and at a pH of below 4.5 (usually <pH 4.1) and have at least a 70% weight loss on ignition.
3. Total and exchangeable Ca^{2+} and Mg were convertible to standard units of milliequivalents of Ca^{2+} + Mg per 100 g of ash-free soil.

The restriction of comparisons to mor type soils was designed to create a more homogeneous data set both spatially (humus varies less that mineral soil from site to site) and over time (because mor decomposes very slowly). The humus layer is also quantitatively an extremely important rooting zone for spruce and fir (Kelly and Mays 1989) and can be considered the zone of most active nutrient uptake under most conditions. In the southern spruce-fir system examined by Kelly and Mays (1989), roots were heavily concentrated in the humus and very little rooting was noted in the mineral soil.

In summary, comparisons across a wide variety of sites in eastern North America indicated that the base status of humus under spruce consistently had a combined exchangeable Ca + Mg status of 150 to 300 meq kg^{-1} soil before 1960. The range of pre-1960 values conform to Ponomareva's (1969) generalization that humus Ca^{2+} + Mg concentrations under taiga vegetation in the U.S.S.R. reached equilibrium with low Ca litter at 250 mmol(+)*kg^{-1}. The similarity in Ca + Mg values between mor and peat reflects the fact that around pH 4.0 the base exchange capacity of mor (Shortle and Bondietti, in press) and peat (Belkevich and Chistova 1963) is about 250 mmol(+)*kg^{-1}. After 1960, exchangeable levels of Ca^{2+} and Mg in humus were found to be typically less than 150 mmol(+) * kg^{-1} with values less than 10 mmol(+) * kg^{-1} not uncommon (Shortle and Bondietti, in press).

Such changes in base content over time could result from such causes as a substantial depletion of Ca and Mg from the humus layer through uptake by vegetation or leaching by natural processes augmented by acid deposition. A systematic sampling bias in either time interval, or differences in methods of analysis between the time intervals, could also contribute to differences found between time periods. Expanding the database to consider a wider variety of sites, including data from bogs and sites outside North America, produced similar results for data obtained before 1960. This consistency across sites provided reinforcement that techniques of analysis and site factors did not introduce major sources of variability. Thus we assume that such values represent a reasonable baseline for levels of exchangeable

Ca + Mg levels in areas that have not experienced significant acidic deposition (Shortle and Bondietti, in press).[1]

An additional consideration in observed changes in humus Ca levels is that of any changes in the composition of the humus quality mediated by temporal changes in the quality of litter input. Data on turnover times for Ca in soil organic matter in boreal conifer forests indicate that turnover rates are very slow to moderate. Time intervals from around 149 years for northern boreal forests in Alaska to 5.9 years (average of 13 conifer stands for temperate deciduous forests) have been noted (Cole and Rapp 1981). However, whether from reduced uptake of Ca and associated reduced return in litter or from increased leaching of Ca from the humus of the forest floor, substantial changes in Ca in humus reflect a basis of concern for sustained availability of adequate levels of Ca to trees growing on these sites.

Repetitive Analysis of Soil Nutrient Status Over Time

In addition to the historical survey approach described previously, resampling of the same soil over time has been used to estimate relative impacts of acidic deposition and natural processes on base cation status of Swedish forest soils (Tamm and Hallbachen 1988). An increase in the rate of acidification found in the southwestern sites during the 55-year resampling interval was not explainable on the basis of forest history, climate, or soil mineralogy. The unexplained decrease of 0.5–0.7 pH units of deeper soils (70 cm) was suggested as being most likely related to differences in acidic deposition between the southwestern and northeastern sites.

In North America, a careful resampling of the Heimburger (1934) soil plots located on mixed deciduous-coniferous forests in the Adirondack region near Newcomb, New York, was conducted by Anderson (1988). Comparison of changes in the Ca budgets over the 54-year interval (1930–1984) indicated that significant Ca losses had occurred from the OE (F), OA(H), and E (A) horizons (67, 121, and 191 kg Ca ha^{-1}, respectively). The total loss of Ca from these soil horizons (379 kg ha^{-1}) was comparable to calculated 54-year leaching losses estimated from current weathering rates in the northeast (346 kg Ca ha^{-1}). Changes in the humus layer represented an approximate 15% loss from original soil values; in the A horizon, approximately 88% of the original Ca was lost. The greater losses from deeper horizons on these plots may have reflected the influence of nutrient uptake

---

[1]The untimely death of Dr. Bondietti in February 1990 has prevented a more timely and definitive analysis of the historical database on soil nutrient changes. Those analyses are now being completed by Dr. Walter Shortle, USFS, and a manuscript is in preparation. The intriguing trends observed in the initial analyses and presented at the NAPAP review earlier in February 1990 by Dr. Bondietti warrant mention here, but should be considered preliminary until Dr. Shortle has completed these analyses.

and recycling by the deeper rooting hardwood component of these mixed forests.

Anderson (1988) constructed budgets of Ca changes in the total system that indicated that Ca uptake by vegetation (526 kg Ca ha^{-1}) was far larger than Ca lost from the upper three soil horizons. Major limitations with this budget, however, appear to be efforts to retroactively estimate original Ca on the site from increment core chemistry that assumed a uniform distribution of Ca across the core. This would lead to an underestimate of original Ca and an overestimate of recent Ca based on the decreasing Ca content with diameter shown by Momoshima and Bondietti (1990). The failure to consider Ca added back in to the system in slash and roots from harvesting operations that occurred after the original soil sampling also would have resulted in an underestimation of original Ca pools for the site. Nevertheless, this research provides some interesting insights to the range and distribution of soil chemical conditions across an array of sites before the recent increases in emissions of SO_2 and NO_2 across the region as well as a documentation of soil changes from those original conditions.

An additional study of historical changes in soil properties, conducted by Bondietti and Shortle (unpublished data), involved revisitation of a virgin red spruce site near Waterville, New York, and careful relocation of the specific site sampled approximately 60 years earlier by Lunt (1932). In 1929, exchangeable Ca + Mg of the humus was more than 200 mmol (+)*kg^{-1}; recent analysis (1989) indicated that current levels are about 50 mmol(+)*kg^{-1}.

Analysis of root distribution patterns at this site indicated that the depth of the primary rooting zone had decreased substantially from the earlier period and that roots were now confined principally to the upper humus layer. Titration of the recently collected soil with $Ba(OH)_2$ indicated that base exchange capacity (BEC) decreased from around 250 mmol(+) at pH 4.0 to about 100 mmol(+)*kg^{-1} at pH 3.5. Thus, experimentally increasing the soil solution H concentration by a factor of 3.2 (0.5 pH units) decreased the BEC of this soil by 60%. The sensitivity of humus to base mobilization and replacement from exchange complexes by H^+ is the reason why mor and peat are good suppliers of Ca to plant roots despite their naturally high acidity: it also suggests that increased inputs of H^+ from atmospheric deposition could substantially influence humus BEC over time.

Sampling along Deposition Gradients

Recent surveys of foliar nutrient analysis along elevational gradients by Friedland et al. (1988) in the Northeast and Robarge et al. (1989) in the Southeast provided evidence that foliar base status is lowest in the high-elevation sites. These sites also receive the highest inputs of acidic deposition because of the high inputs of cloud moisture. Unfortunately, we have no historical record of foliar chemistry to determine the extent to which current low nutrient levels at high elevations may represent a change from previous conditions. It is known that higher deposition of water can, of itself, increase leaching rates of base cations from humus leading to nutrient deficiency (Jenny 1980). Thus high leaching rates should be expected at high-elevation sites where water deposition is high. The current high levels of acidity and strong anion content deposited at high-elevation sites (see Chapters 6, 7, and 11) however can be expected to have accelerated leaching losses substantially under these conditions of high hydrologic input. Limits on these leaching losses of base cations may be controlled by Al mobilization and dominance as a major cation (Reuss and Johnson 1986). Under these conditions, base cation availability to plants will be reduced by competitive inhibition of Ca uptake by Al (or Fe) at the root surface (Sucoff et al. 1990).

It is in just this context that the tree-ring chemistry record is most relevant. Both initial increases and subsequent decreases in Ca uptake and incorporation in stemwood have occurred in red spruce trees across a large area (Bondietti et al. 1989, 1990). This record provides a valuable additional perspective on the extent to which current conditions represent a change from those of a few decades ago.

Base Cation Availability and Tree Growth

At present we do not have adequate data to evaluate whether current levels of Ca are sufficient for good growth of high-elevation spruce forests. Regional analysis of spruce growth patterns indicates that an unexplained regional decline in radial growth occurred at high elevations around 1960 in the Northeast and typically 5 years later in the Southeast (McLaughlin et al. 1987). These analyses suggested that regional-scale atmospheric pollution contributed to the observed growth patterns but were not intended to test mechanisms of action.

Analysis of foliar nutrient levels by regional surveys (Friedland et al. 1988; Robarge et al. 1989) has been suggested that Ca levels are not yet in the deficiency range. Magnesium levels, on the other hand, were in the initial deficiency zone. However, the basis for these valuations was data from 6-month-old red spruce seedlings (gram size range) under controlled greenhouse conditions (Swan 1971); such data are of questionable relevance to large trees under natural conditions. Nutritional data for mature Norway spruce, a species that in the seedling stage exhibited sensitivity to Ca and Mg levels similar to that of red spruce (Swan 1972), indicated that growth of mature trees is reduced at foliar Ca levels approximately twice those of seedlings (Huettl 1986). Such values are well within the range of currently reported foliar values for red spruce in the Northeast. Recent analysis of carbon economy, specifically the photosynthesis: dark respiration ratio, of sapling red spruce trees in the Smoky Mountains (GSMNP) (McLaughlin et al. 1991) indicated that physiology and growth of these trees may be limited by low foliar Ca and associated high foliar Al levels.

Shortle and Smith (1988) have proposed that current low Ca:Al ratios in fine roots of declining red spruce in the northeastern United States indicate Ca deficiency induced by mobilization of Al by atmospheric deposition. Support for this hypothesis has been provided both by foliar nutrient analysis and by growth correlations for red spruce in the field. Joslin et al. (1988) found that Ca and Mg in spruce foliage and fine roots from five field research sites in the United States and Europe were inversely related to lysimeter levels of monomeric Al. Their data also indicated that annual sulfur deposition, which varied more than 10 fold among their sites, was positively and linearly correlated with Al:Ca ratios in current foliage ($r^2 = .91$) and in O horizon fine roots ($r^2 = .82$).

Analysis of the relationship of radial growth of red spruce, Fraser fir, and hemlock to Al levels in wood (Bondietti et al. 1989) has shown a shift in the GSMNP: 10% of the approximately 33 trees examined showed growth rates that were inversely related ($p < .05$) to Al levels before 1938, and more than 50% showed this relationship since that time. Collectively, these cross-correlations suggest that there have been changes in the chemistry of wood of red spruce that reflect alterations of cation availability and associated changes in stem growth over time.

Summary and Conclusions

We have examined temporal trends in tree ring chemistry as an indicator of historical changes in the chemical environment of red spruce. These analyses, covering principally high-elevation sites in the eastern United States, lead us to conclude that significant changes in soil chemistry have occurred at many of these sites during recent decades. These changes are spatially and temporally consistent with changes in emissions of SO_2 and NO_2 across the region, suggesting that increased acidification of forest soils has occurred. Observed changes include increases in the levels of Al and Fe that typically occur as base cations are removed from soils. Ca, on the other hand, shows a regionwide increase above expected levels followed by a decrease that suggests that increased mobilization began perhaps 30 to 40 years ago. The Ca mobilization period coincides with a regionwide increase in growth rate of red spruce, while the period of decreasing wood Ca in the past 20 to 30 years corresponds temporally with patterns of decreasing radial growth at high-elevation sites throughout the region. The wood Ca patterns across the region suggest that Ca loss may have been accelerated to the point at which base saturation of soils has been reduced.

Preliminary examination of historical reports of exchangeable Ca + Mg for mor type soils developed under spruce, fir, and hemlock suggests that exchangeable Ca + Mg levels reported in eastern North America before 1960 or more recently in areas remote from high acidic deposition are now substantially reduced (>40%). Limited resampling of soils at the same approximate sites to examine changes in base status over time suggests that significant reductions in base status in humus and deeper soils have occurred during the past 50 years. Broad spatial consistencies in pre-1960 base status of humus on mor soils suggest that a baseline condition has been altered. Attempts to budget losses on a whole-stand basis suggest that both leaching losses and uptake of base cations by vegetation could be important contributors to these changes. Both historical survey approaches and efforts to budget total stand uptake require additional analyses to evaluate the methological techniques used in historical reconstruction.

These analyses collectively provide evidence that forest soils and forest trees in the United States may have been responding to anthropogenic inputs of acidity for three to five decades, and well before we began to look for evidence of current impacts. If significant depletion of base cations has already occurred at many of these sites, it is important that we evaluate current nutrient budgets and the impact of current and additional chemical inputs to those systems in the light of those changes. Current physiological research in the southern Appalachians indicates that red spruce saplings may now be experiencing reduced growth as a consequence of limiting Ca availability and potential interferences from corresponding high levels of Al (McLaughlin et al. 1990). Under these conditions nutrient budgets must be viewed in the perspective of nutrient availability to plants as well as with respect to net fluxes through the system.

Acknowledgments. We wish to thank Drs. J.M. Kelly, J.O.Reuss, M.R. Walbridge, and S.B. Weed for their review of earlier versions of this manuscript. This research was jointly sponsored by the Electric Power Research Institute and by the USDA, National Acid Deposition Assessment Program Southern Commercial Pine Cooperative and Spruce-Fir Research Cooperative under Interagency Agreement 40-1647-45 with the U.S. Department of Energy under contract DE-ACO5-840R21400 with Martin Marietta Energy Systems, Inc.

References

Abrahamsen G. 1980. Acid precipitation, plant nutrients, forest growth. In Drablos D., Tollan A. (eds.) SNSF Proceedings, Ecological Impact of Acid Precipitation, pp. 58–63, March 11–14, 1980, Sandefjord, Norway. Johs. Grefslie Trykkeri, Mysen, Norway.

Anderson S.B. 1988. Long-term Changes (1930–32 to 1984) in the Acid-Base Status of Forest Soils in the Adirondacks of New York. Ph. D. Dissertation, Department of Geology, University of Pennsylvania, Philadelphia.

Baes C.F. III, McLaughlin S.B. 1984. Trace elements in tree rings: evidence of recent and historical air pollution. Science 224:494–497.

Baes C.F. III, McLaughlin S.B. 1986. Multielemental Analysis of Tree Rings: A Survey of Coniferous Trees in the Great Smoky Mountins National Park. ORNL-6155, Oak Ridge National Laboratory, Oak Ridge, Tennesee.

Belkevitch P.J., Chistova L.R. 1963. Exchange capacity of peat with respect to alkali and alkaline earths. In Robertson R.A. (ed.) Proceedings of the 3rd International Peat Congress, Leningrad. pp. 904–918.

Bloom P.R., Weaver R.M., McBride M.B. 1978. The spectrophotometric and fluorometric determination of aluminum with 8-hydroxyquinoline and butyl acetate extraction. Soil Sci. Soc. Am. J. 42:713–716.

Bohn H.L., McNeal B.L., O'Connor G.A. 1985. Soil Chemistry. Wiley, New York.

Bondietti E.A., Baes C.F. III, McLaughlin S.B. 1989. Radial trends in cation ratios in tree rings as indicators of the impact of atmospheric deposition on forests. Can. J. For. Res 19:586–594.

Bondietti E.A., Momoshima N., Shortle W.C., Smith K.T. 1990. A historical perspective on changes in divalent cation availability to red spruce in relationship to acidic deposition. Can. J. For. Res. 20:1850–1858.

Cole D.W., Rapp M. 1981. Elemental cycling in forest ecosystems. In Reichle D.E. (ed.) Dynamic Properties of Forest Ecosystems. IBP-23, Cambridge University Press, Cambridge, pp. 342–409.

Cosby B.J., Hornberger G.M., Galloway J.N., Wright R.F. 1985. Modeling the effects of acid deposition: assessment of a lumped-parameter model of soil water and streamwater chemistry. Water Resour. Res. 21:51–63.

Dougan W.K., Wilson A.L. 1974. The absorptiometric determination of aluminum in water: a comparison of some chromogenic reagents and the development of an improved method. Analyst 99:413–430.

Driscoll C.T., van Breeman N., Mulder J. 1985. Aluminum chemistry in a forested Spodosol. Soil Sci. Soc. Am. J. 49:1584–1589.

Federer C.A., Hornbeck J.W., Tritton L.M., Martin C.W., Pierce R.S., Smith C.T. 1989. Environ. Manage. 13: 593–601.

Friedland A.J., Hawley G.J., Gregory R.A. 1988. Red spruce (*Picea rubens* Sarg.) foliar chemistry in Northern Vermont and New York, USA. Plant Soil 105:189–193.

Heimburger C.C. 1934. Forest-type studies in the Adirondack region. Cornell (Ithaca) Univ. Agr. Exp. Sta. Mem. 165.

Huettl Z.F 1986. Forest fertilization results from Germany, France, the Nordic Countries. In The Fertilization Society Proc No. 250, pp. 1–40. Greenhill House, London.

Husar R.B. 1986. Emissions of sulfur dioxide and nitrogen dioxides and trends for eastern North America. In Acid Deposition Long-Term Trends. National Research Council, National Academy Press, Washington, D.C., Chap. 2, pp. 49–92.

Isaac R.A., Kerber J.D. 1975. Atomic absorption and flame photometry: techniques and uses in soils, plant, water analysis. In Walsh L.M. (ed.) Instrumental Methods for Analysis of Soil and Plant Tissue. Soil Science Society America, Madison, Wisconsin, pp. 17–37.

Jenny H. 1980. The Soil Resource: Origin and Behavior. Springer-Verlag, New York, p. 377.

Johnson D.W., Richter D.D., Lovett G.M., Lindberg S.E. 1985. The effects of atmospheric deposition on potassium, calcium, magnesium cycling in two deciduous forests. Can. J. For. Res. 15:773–782.

Johnson D.W., Cole D.W., Gessel S.P., Singer M.J., Minden R.V. 1977. Carbonic acid leaching in a tropical, temperate, subalpine, northern forest soil. Alpine Res. 9:329–343.

Joslin J.D., Kelly J.M., Wolfe M.H., Rustad L.E. 1988. Elemental patterns in roots and foliage of mature spruce across a gradient of soil aluminum. Water Air Soil Pollut. 40:375–390.

Kelly M.J., Mays P.A. 1989. Root zone physical and chemical characteristics in southeastern spruce fir stands. Soil Sci. Soc. Am. J.53:1248–1255.

Lindsay W.L. 1979. Chemical Equilibria in Soils. Wiley, New York.

Lunt H.A. 1932. Profile characteristics of New England forest soils. Conn. Agr. Exp. Sta. Bull. 342.

Mattson S., Karlsson N. 1944. The pedography of hydrologic soil series: VI. The composition and base status of the vegetation in relationship to the soil. Ann. Agr. Coll. Sweden 12:186–203.

McBride M.B., Bloom P.R. 1977. Adsorption of aluminum by a smectite: II. An Al^{3+} Ca^{2+} Exchange Model. Soil Sci. Soc. Am. J. 41: 1073–1077.

McLaughlin S.B., Andersen C.P., Edwards R.W.K., Layton P.A. 1990. Seasonal patterns of photosynthsis and respiration of red spruce saplings from two elevations in declining southern Appalachian stands. Can. J. For. Res. 20:485–495.

McLaughlin S.B., Anderson C.P., Hanson P.J., Tjoelker M.J., Roy W.K. 1991. Increases in dark respiration of red spruce associated with apparent calcium deficiency at high elevation southern appalachian mountain sites. Manuscript submitted to Can J. For Res. (in press).

McLaughlin S.B., Downing D.J., Blasing T.J., Cook E.R., Adams H.S. 1987. An analysis of climate and competition as contributors to decline of red spruce in high elevation Appalachian forests of the eastern United States. Oecologia 72:487–501.

Momoshima N., Bondietti E.A. 1990. Cation binding in wood: applications to understanding historical changes in divalent cation availability to red spruce. Can. J. For. Res. 20:1840–1849.

Ponomareva V.V. 1969. Theory of Podzolization. Israel Program for Scientific Translations. TT 68-50442, National Technical Information Service, Springfield, Virginia.

Reuss J.O. 1983. Implications of the calcium-aluminum exchange system for the effect of acid precipitation on soils. J. Environ. Qual. 12:591–595.

Reuss J.O., Johnson D.W. 1986. Acid Deposition, Soil and Waters. Ecological Studies Series No. 50, Springer-Verlag, New York.

Richter D.D. 1984. Comment on comment on "Acid precipitation in historical perspective" and "Effects of acid precipitation." Environ. Sci. Technol. 18:632–634.

Richter D.D. 1986. Sources of acidity in some forested Udults. Soil Sci. Soc. Am. J. 50:1584–1589.

Richter, D.D., D.W. Johnson, D.E. Todd 1983. Atmospheric sulfur deposition, neutralization, ion leaching in two deciduous forest ecosystems. J. Environ. Qual. 12:263–270.

Richter D.D., King K.S., Witter J.A. 1989. Moisture and nutrient status of extremely acid Umbrepts in the Black Mountains of North Carolina. Soil Sci. Soc. Am. J. 53:1222–1228.

Richter D.D., Comer P.J., King K.S., Sawin H.S., Wright D.W. 1988. Effects of low ionic strength solutions on pH of acid forested soils. Soil Sci. Soc. Am. J. 52:261–264.

Robarge W.P., Pye J.M., Bruck R.J. 1989. Foliar elemental composition of spruce-fir in the southern blue ridge province. Plant Soil 114:19–34.

Russell E.W. 1973. Soil Conditions and Plant Growth, 10th Ed. Longman, London.

Schachtschabel P. 1940. Untersuchungen uber die Sorption de Tonmineralien und organischen Boden-Kolloide, und die Bestimmung des Anteils dieser Kolloide an der Sorption im Boden. Kolloid-Beshefte 51:199–276.

Schofield R.K. 1947. A ratio law governing the equilibrium of cations in the soil solution. In 11th International Congress of Pure and Applied Chemistry, Pergamon, Oxford. pp. 257–261.

Shortle W.C., Bondietti E.A. (in press) Timing, magnitude, and impact of acid deposition on sensitive forest sites. Water Air Soil Pollut.

Shortle W.C., Smith K.T. 1988. Aluminum-induced calcium deficiency syndrome in declining red spruce. Science 240:239–240.

Sjors H. 1961. Some chemical properties of the humus layer in Swedish natural soils. Bull. R. Sch. Forest, Stockholm, Sweden, No. 37.

Skartveit A. 1980. Observed relationships between ionic composition of precipitation and runoff. In Drablos D., Tollan A. (eds.) Ecological Impact of Acid Precipitation, Proceedings of the International Conference on Ecological Impacts of Acid Precipitation, pp. 242–244. SNSF Project, Oslo, Norway.

Smith W.H. 1974. Air pollution—Effects on the structure and function of the temperate forest ecosystem. Environ. Pollut. 6:111–129.

Sucoff E., Thorton F.C., Joslin J.D. 1990. Sensitivity of tree seedlings to aluminum: in honey locust. J. Environ. Qual 19:163–187.

Sullivan T., Seip H.M., Muniz I.P. 1986. A comparison of frequently used methods for the determination of aqueous aluminum. Int. J. Environ. Chem. 26:61–75.

Sumner M.E., Marques J.M. 1968. Applicability of Schofield's ratio law to a ferrallitic clay. Agrochimica 22:191–195.

Swan H.S.D. 1971. Relationship between Nutrient Supply, Growth, Nutrient Concentrations in Foliage of White and Red Spruce. Woodlands Paper No. 29, Pulp and Paper Research Institute of Canada, Ottowa.

Swan H.S.D. 1972. Foliar Nutrient Concentrations in Norway Spruce as Indicators of Tree Nutrient Status and Fertilizer Requirement. Woodlands Report WR/40, Pulp and Paper Research Institute of Canada, Ottowa

Tamm C.O., Hallbacken L. Changes in soil acidity in two forest areas with different acid deposition: 1920s to 1980s. Ambio 17:56–61.

Thomas G. 1982. Exchangeable cations. In Page A.L. (ed.) Methods of Soil Analysis, Part 2, 2d Ed. Agronomy 9:191–195.

Ulrich B. 1980. Production and consumption of hydrogen ions in the ecosphere. In Hutchinson T.C., Havas M. (eds.) Effects of Acid Precipitation on Terrestrial Ecosystems. Plenum, New York, pp. 255–282.

Ulrich B. 1983. Soil acidity and its relations to acid deposition. In Ulrich B., Pankrath J. (eds.) Effects of Accumulation of Air Pollutants in Forest Ecosystems. D. Reidel, Boston, pp. 127–146.

Ulrich B., Mayer R., Khanana P.K. 1980. Chemical changes due to acid precipitation in a loess-derived soil in central Europe. Soil Sci. 130:193–199.

van Miegroet H., Cole D.W. 1985. Acidification sources in red alder and Douglas-fir soils—importance of nitrification. Soil Sci. Soc. Am. J. 49:1274–1279.

van Breeman H., Burrough P.A., Velthorst E.J., Van Dobben H.F., deWit T., Ridder T.B., Reijnders H.F.R. 1982. Soil acidification from atmospheric ammonium sulphate in forest canopy throughfall. Nature, (London) 299:548–550.

Wiklander L., Andersson A. 1972. The replacing efficiency of hydrogen ion in relation to base saturation and pH. Geoderma 7:159–165.

Wright R.F., Norton S.A., Brakke D.F., Frogner T. 1988. Experimental verification of episodic acidification of freshwaters by sea salts. Nature (London) 334:422–424.

10. Mineralogy and Mineral Weathering

R. April and R. Newton

Introduction

Elements such as Ca, Mg, and K, which are required for plant growth, are important components of the nutrient cycle in forested ecosystems, and by far the largest store of these elements in North American and European forests is within the minerals constituting the forest soil. Although external inputs from the atmosphere in both the dissolved and particulate load can provide a portion of these elements to a growing forest, the ultimate source of most inorganic elemental nutrients is provided through cation exchange and mineral weathering reactions that take place in the soil profile. Mineral inventories and determinations of the physical characteristics, mineralogy and chemistry of soil components, and mineral weathering reactions that occur in soils must be an integral part of any study that attempts to document the nutrient status of a forested ecosystem.

Minerals are inherently unstable at conditions encountered at the surface of the Earth. Although physical weathering can reduce the grain size of most earth materials through mechanical stresses brought on by temperature fluctuations, freezing and thawing, or simply abrasion of one grain against another, chemical weathering can transform primary minerals into secondary products (incongruent dissolution), can completely dissolve minerals into solution (congruent dissolution), and can lead to the precipitation of new minerals (neoformation). Minerals present in soils are particularly susceptible to weathering because in this narrow zone, which represents the transitional boundary between Earth and its atmosphere, chemical weathering is driven by the confluence of atmospheric, hydrologic, biological, and geologic processes. The stability of different minerals in soils is generally well known (minerals formed at high temperatures and pressures and containing mafic elements such as Mg, Fe, and Ca are more susceptible to weathering than those formed at low temperatures and pressures and containing K, Na, Si, and Al), but factors such as climate, annual precipitation, soil temperature, soil water content and chemistry, soil pH, vegetation, mineral grain size, and others can influence the rate and extent of mineral weathering in a given region.

The purpose of this chapter is to report on and summarize the geologic data collected during the IFS project, to describe the important reactions involving mineral weathering in the soils, and to compare the "weathering potential" and the factors controlling base cation release from weathering at the Integrated Forest Study (IFS) sites. Because of the large amount of information gathered and the number of sites involved in the IFS project, it would be impractical to provide a complete and detailed analysis of the mineralogy, soil characteristics, and weathering reactions determined for each IFS site. Instead, many summary figures and tables, comparative regional analyses (i.e., northern sites versus southern sites), and detailed descriptions of some selected sites that seem representative of an entire region are given. Information regarding the bedrock geology and soil types present at each of the IFS sites is given in Chapter 2.

Site Characteristics

Grain-Size Distributions

Grain size is a basic physical property of soil that greatly influences many other soil characteristics such as hydraulic conductivity and porosity. Determinations of grain size were accomplished using standard wet sieve, dry sieve, and hydrometer techniques (Folk 1968). Particle distribution was divided into the following three size classes on the basis of measurements of the equivalent spherical diameter (e.s.d.): (1) sand, 2000 to 62.5 μm e.s.d.; silt, 62.5 to 3.9 μm e.s.d.; and clay, less than 3.9 μm e.s.d.. In general, the silt- and clay-size material contained high concentrations of clay minerals (phyllosilicates), whereas primary minerals (e.g., quartz, feldspars, and heavy minerals) constituted the bulk of the sand-size fraction.

Soils from the IFS sites display a wide range of grain-size distributions (Figure 10.1), but the most pronounced regional difference among the sites is in the amount of clay present. Soils from the northern, glaciated sites contain little clay-size material (<6% by weight) whereas those from the southern, unglaciated areas (with the exception of the high-elevation Smoky

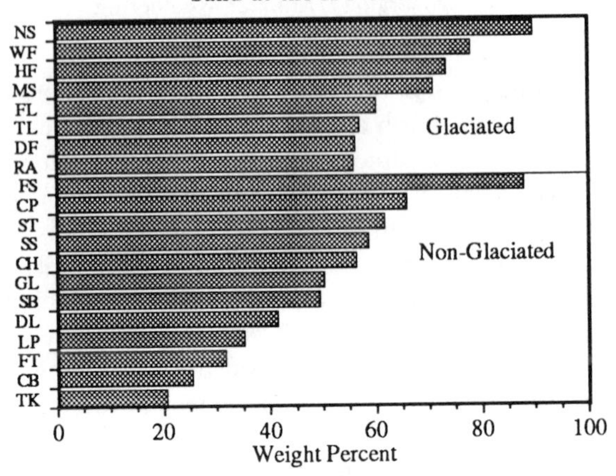

Silt at the IFS Sites

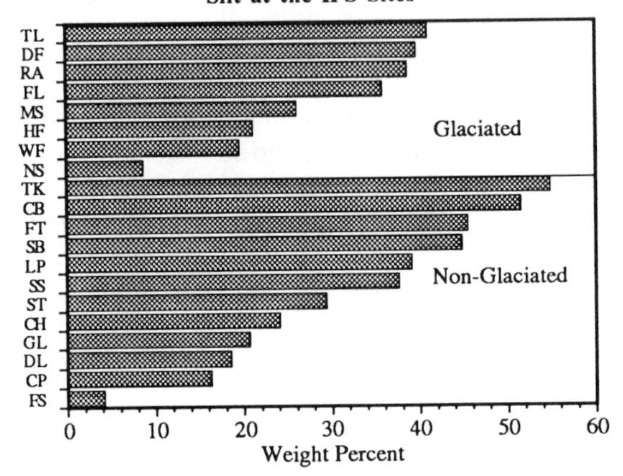

Clay at the IFS Sites

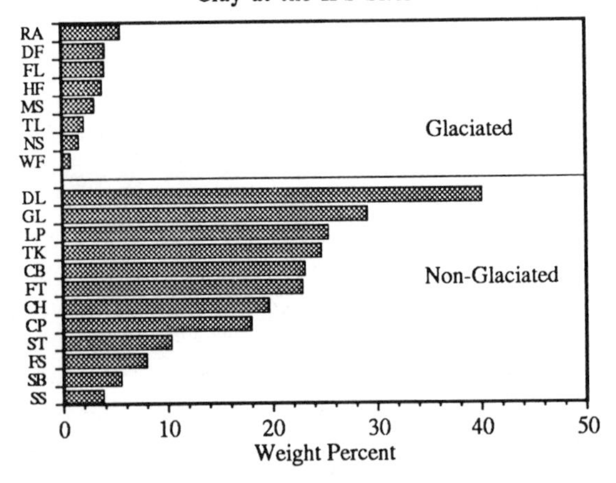

Figure 10.1. Distribution of sand, silt, and clay at IFS sites.

Mountain sites) contain abundant clay (as much as 40%). Because clay-size material is composed predominantly of the products of weathering, the concentrations of clay in these soils can be explained, in part, by the length of time over which pedogenic processes have operated. Glaciation essentially reset the clock for soil development in the northern sites approximately 15,000 years ago. In contrast, soils in the unglaciated, southern sites have been developing for the past tens to hundreds of thousands of years. Of equal importance is the intensity of the weathering process itself as it is related to climatic factors such as temperature and rainfall. In general, soils in the northern IFS sites developed under a cool, temperate climate whereas soils in the southern IFS sites formed under relatively warmer and, in some instances, wetter conditions.

The textural development of soils is also strongly influenced by physical processes that can remove clay-size material from soil horizons, for example through washing and eluviation, and replace it with coarser grained, fresh minerals derived from bedrock. On steep slopes in mountain areas erosion rates are relatively high so that soils are likely to contain less clay and more coarse-grained, fresh material transported downslope by mass wasting; soils mantling areas with low relief and gentle slopes may contain more clay-size material as it simply accumulates over time.

The vertical distribution of clay-size material in the IFS soils varies and is distinctly different between the glaciated and nonglaciated sites. Glaciated sites tend to have the highest concentration of clay in the upper soil horizons, but in nonglaciated sites the clay content is highest deeper within the soil profiles. The Duke Forest (DL) site is typical of southern, unglaciated soils in that the soil-clay content is less than 10% in near-surface samples, increasing to more than 60% at depths greater than 35 cm. In contrast, Spodosols in Huntington Forest (HF) have as much as 9% clay in near surface horizons and 2% or less in the C horizon. This difference results from both the degree of eluviation of clay and the position of the zone of maximum chemical weathering in the soil profile. In the northern, glaciated soils containing abundant surface organic matter and a fresh supply of weatherable minerals, the zone of maximum chemical weathering is in the uppermost mineral soil horizons. In most southern soils, however, maximum weathering occurs at some greater depth in the profile as most of the unstable minerals near the surface have already been depleted.

Difference in grain-size distributions between northern and southern soils also affects the flow of water through the profile. Hydraulic conductivities estimated from particle-size distributions indicate that water can move fairly easily as intergranular flow through the soil profiles of the sandy, northern soils whereas the high clay content in the B horizon of many southern soils would inhibit vertical movement of water, except through macropores.

Soil Mineralogy

The mineralogy and chemistry of forest soils is influenced by many factors, not least of which is the composition of the underlying bedrock from which all saprolitic and residual soils are derived. Even in glaciated areas, where soils have formed on till, outwash sands, and other transported surficial deposits, the mineralogy of the parent material and the soil is strongly influenced by the composition of the regional bedrock. Bulk soil mineralogy tells us much about the nature of the original material from which a soil is derived, but clay mineralogy can provide significant information regarding the extent and type of weathering processes that occur in a given soil pedon.

For ease of reporting, mineral species in the IFS soils have been grouped into three basic classes: (1) light minerals, having a specific gravity less than 2.95; (2) heavy minerals, having a specific gravity greater than 2.95; and (3) clay minerals, having a particle size less than 2 μm in diameter.

Light Minerals in the Soils

The mineralogy of the light mineral fraction was determined by energy dispersive x-ray spectroscopy (EDS) and scanning electron microscopy (SEM). An aliquot of mineral grains from the sand-size fraction of each sample was mounted in epoxy on a glass slide that was subsequently ground to yield a highly polished, flat surface. After inspection of the sample under a petrographic microscope, a Kevex Analyst 1000 EDS/Image Analyzing System was used to automatically scan and chemically analyze a large number of mineral grains under the SEM. A computer routine determined whether the analyzed grain was quartz, plagioclase feldspar, or potassium feldspar, based on its chemical composition.

Results of the light mineral analysis are shown in Figure 10.2. In general, the light mineral group is dominated by quartz and the feldspars. Quartz is relatively nonreactive in soil weathering environments whereas the feldspars, which include both the plagioclase series and the alkali, K-bearing feldspars such as orthoclase and microcline ($KAlSi_3O_8$), break down more readily. The plagioclase feldspars represent a solid solution series ranging in composition from the albite ($NaAlSi_3O_8$) end member to the anorthite ($CaAl_2Si_2O_8$) end member (Figure 10.3), with the latter generally regarded as more susceptible to weathering. Some minerals such as biotite and muscovite, which have specific gravities very close to 2.95 (that of the heavy liquid tetrabromoethane used to separate light from heavy minerals) and have platy habits, may be found in both the light and heavy mineral fractions. Muscovite is a K-rich phyllosilicate that is quite resistant to chemical alteration. Biotite is a Mg- and Fe-rich mica that usually alters rapidly in soil environments to other phyllosilicates such as vermiculite, kaolinite, and mixed-layered clays.

Results from the light mineral analyses show that the feldspars are an abundant component of the mineral fraction of soils in the northern, glaciated sites and that soils in the southern, nonglaciated sites are composed primarily of quartz. This difference in mineralogy is the direct result of the maturity of the landscapes and the extent of weathering that has taken place

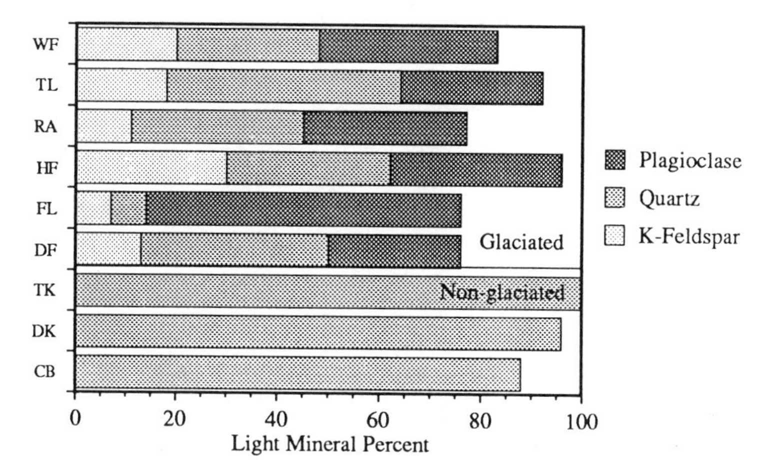

Figure 10.2. Percent plagioclase feldspar, K feldspar, and quartz at selected IFS sites.

over time. The younger northern, glaciated sites contain a large proportion of feldspar minerals, in contrast to the older southern sites in which most of these reactive minerals have been removed by weathering.

Heavy Minerals in the Soils

The heavy mineral fraction of IFS soils contains a wide range of minerals (Table 10.1), including the amphiboles and the pyroxenes, zircon, tourmaline, rutile, and the oxide group of minerals that includes many Fe-bear-

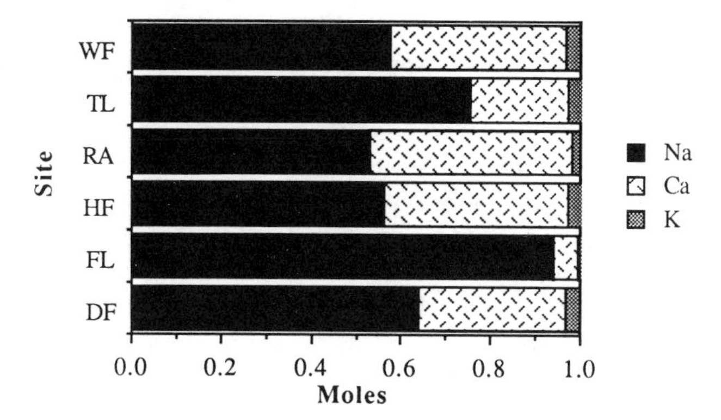

Figure 10.3. Base cation composition of plagioclase feldspars from selected northern IFS sites.

Table 10.1. Common Primary Minerals at the IFS Sites

Light Minerals	
Quartz	SiO_2
Plagioclase[a]	$(Na,Ca)(Si,Al)_4O_8$
K-feldspar[a]	KSi_3AlO_8
Muscovite	$KAl_2(AlSi_3O_{10})(OH)_2$
Biotite[a]	$K(Mg,Fe)_3(AlSi_3O_{10})(OH)_2$
Chlorite[a]	$(Mg,Fe,Al)_6(AlSi)_4O_{10}(OH)_8$
Heavy Minerals	
Staurolite[a]	$(Fe^{+2},Mg)_2(Al,Fe^{+3})_9O_6(SiO_4)_4(O,OH)_2$
Kyanite	Al_2SiO_5
Sillimanite	Al_2SiO_5
Zircon	$Zr(SiO_4)$
Epidote[a]	$Ca_2Fe^{+3}Al_2O \cdot OH \cdot Si_2O_7SiO_4$
Zoisite[a]	$Ca_2Al \cdot Al_2O \cdot OH \cdot Si_2O_7 \cdot SiO_4$
Garnet[a]	$(Mg,Ca,Fe^{+2})_3Al_2Si_3O_{12}$
Tourmaline	$Na(Mg,Fe^{+2},Mn,Li,Al)_3Al_6(Si_6O_{18})(BO_3)_3(OH,F)_4$
Hypersthene[a]	$(Mg,Fe^{+2})SiO_3$
Diopside[a]	$Ca(Mg,Fe^{+2})Si_2O_6$
Augite[a]	$(Ca,Na,Mg,Fe^{+2},Mn,Fe^{+3}Al,Ti)_2(Si,Al)_2O_6$
Hornblende[a]	$(Na,K)_{0-1}Ca_2(Mg,Fe^{+2},Fe^{+3}Al)_5(Si_{6-7}Al_{2-1})O_{22}(OH,F)_2$
Magnetite	Fe_3O_4
Hematite	Fe_2O_3
Ilmenite	$FeTiO_3$
Apatite	$Ca_5(PO_4)_3(OH,F,Cl)$

[a] denotes reactive minerals.

ing minerals such as ilmenite, magnetite, and hematite. In general, we found that soils with high percentages of heavy minerals contained many more of the reactive minerals listed in Table 10.1; soils with low concentrations of heavy minerals contained more of the unreactive heavy minerals. Although the heavy mineral fraction rarely made up more than 20% of any soil we analyzed, it is important to note that dissolution of small amounts of reactive heavy minerals can release disproportionately large amounts of base cations to the soil solution.

The IFS sites fall into two categories with respect to heavy mineral abundance (Figure 10.4). In general, soils in the northern, glaciated sites contain greater than 5% heavy minerals (mean, 11.02%) whereas soils in the southern, unglaciated sites have less than 5% (mean, 2.44%). Not only do the southern sites have a lower heavy mineral content overall, but they also contain fewer of the reactive heavy minerals (Figure 10.5) such as the amphiboles, pyroxenes, and epidote. The chemically resistant heavy minerals that constitute much of the heavy mineral fraction in the southern soils include the opaque minerals ilmenite, magnetite, and hematite, as well as zircon, tourmaline, and rutile.

The glaciated soils developed from sediments deposited at the end of the Pleistocene epoch approximately 10,000 to 15,000 years before present. The glaciers deposited fresh mineral sediment derived from the erosion of the underlying bedrock, and there has been insufficient time for weathering pro-

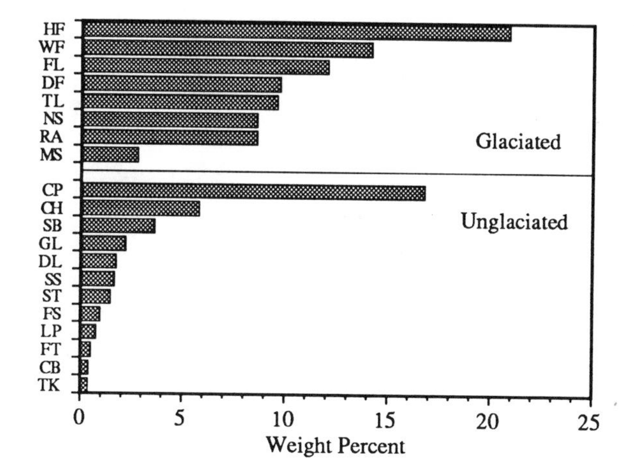

Figure 10.4. Diagram shows amount of heavy minerals at each IFS site: (1) northern, glaciated sites with high heavy mineral contents; and (2) southern, nonglaciated sites with low heavy mineral contents.

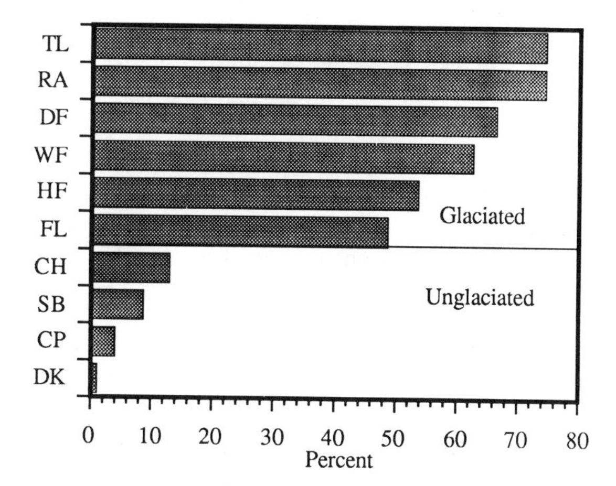

Figure 10.5. Diagram of percentage of reactive heavy minerals in selected IFS sites. Northern, glaciated sites contain abundant reactive heavy minerals compared with southern, nonglaciated sites, which contain mostly unreactive heavy mineral assemblages.

cesses to remove the unstable heavy minerals from these soils. Soils in the unglaciated areas are generally much older and were derived from the chemical and physical weathering of the underlying bedrock. Most of the unstable heavy minerals originally present have been destroyed by chemical weathering during the formation of the soils.

As Figure 10.5 shows, not all the unglaciated sites have a paucity of reactive heavy minerals. The two Coweeta sites (CH and CP) contain 6% to 17% heavy minerals, of which a large proportion is of the reactive type. This can be explained by the very steep slopes in this area, which cause unroofing of fresh bedrock at high elevations, mass movements, and a continuous supply of heavy minerals downslope that are incorporated into the soil.

Clay Minerals in the Soils

Clay minerals in soils are inherited from the parent material, neoformed as precipitates, or formed from the weathering and transformation of primary minerals. This group of minerals includes chlorite, illite, smectite, vermiculite, kaolinite, gibbsite, and a variety of mixed-layered clays. By the very nature of their existence and formation, most clay minerals tend to be stable, or at least metastable, in soil environments. Some clays, for example the smectites, may contain significant amounts of base cations in their structure, exchange, and interlayer sites, whereas others, like kaolinite and gibbsite, contain none. Because of their small particle size and ability to undergo atomic substitution and cation exchange, many clay minerals react dynamically to the soil chemical environment, making them good indicators of both past and present soil conditions.

The clay mineralogy of the IFS soils is most conveniently presented by region, because the most dramatic differences are noted when comparing northern and southern soils. Therefore, a general description of the clay mineralogy of each region is given, followed by detailed descriptions of the clay mineral assemblages from some representative sites.

Southern Sites

The clay-sized fraction of the southern soils is dominated by vermiculite and kaolinite with gibbsite appearing as a minor constituent at most sites (Table 10.2). Mica and mixed-layered mica/vermiculite are present in considerable amounts in the Smoky Mountains and in Coweeta. Figures 10.6 through 10.10 show representative series of x-ray diffractograms illustrating the clay mineralogy and clay mineral trends found within selected soil profiles.

Coweeta. The clay mineralogy of the Coweeta white pine plantation is characterized by vermiculite, kaolinite, and gibbsite, with relatively minor amounts of mica and a hydrous aluminum phase that is distinct from gibbsite (Figure 10.6). Muscovite is the dominant mica species present, as indicated by a

Table 10.2. Semiquantitative Ranking of Clay Mineral Species Present in <2-μm Fraction

Location	Site Code	Vermiculite[a]	Mica[b]	Kaolinite	Gibbsite	Chlorite	Smectite[c]	Mixed-Layer M/V[d]
Northern								
Thompson (DF)	DF	70	5	10	0	15	tr	0
Thompson (RA)	RA	65	5	20	0	5	5	0
Findley Lake	FL	65	0	15	0	10	10	(imogolite)
Huntington	HF	90	0	10	0	0	tr	0
Maine	MS	70	10	10	0	0	10	0
Norway	NS	35	20	10	0	20	tr	15
Turkey Lake	TL	60	5	20	0	15	tr	0
Whiteface	WF	80	5	15	0	0	0	0
Southern								
Camp Branch	CB	55	5	30	5	5	0	0
Coweeta WS2	CH	40	20	20	10	0	0	10
Coweeta WS1	CP	40	15	25	15	0	0	5
Duke	DL	30	0	65	5	0	0	0
Florida	FS	15	0	80	5	0	0	0
Georgia	GL	15	0	85	tr	0	0	0
Oak Ridge	LP	30	20	40	10	0	0	tr
Smoky Mtn (SB)	SB	35	20	10	10	0	0	25
Smoky Mtn (SS)	SS	25	10	20	20	0	tr	25
Smoky Mtn (ST)	ST	30	15	20	15	0	0	20
Walker Branch (FT)	FT	65	0	25	5	0	5	0
Walker Branch (TK)	TK	70	0	25	5	0	tr	0

[a] hydroxy-interlayered vermiculite present in many samples.
[b] includes both muscovite and biotite.
[c] smectite in northern soils often present in E horizon.
[d] M/V = mica/vermiculite.

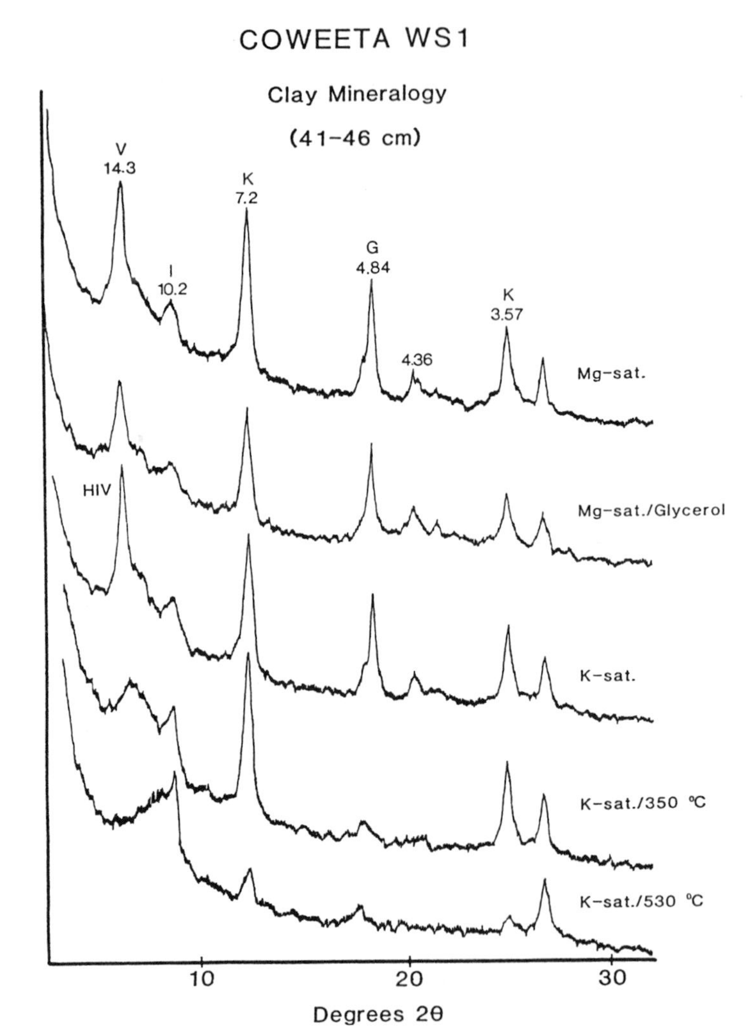

Figure 10.6. X-ray diffraction traces for <2-μm fraction of sample CP-16. Patterns illustrate clay mineral suite of Coweeta watershed #1, characterized by abundant hydroxy-aluminum interlayered vermiculite (HIV & V), kaolinite (K), and gibbsite (G), and minor amounts of illite (I). Small doublet reflections at 4.84 and 4.36 Å suggest presence of an aluminum trihydroxide polymorph.

d(060) spacing at 1.49 Å in random clay mount diffractograms and the occurrence of defined 10 Å, 5.0 Å,and 3.3 Å basal peaks for oriented specimens. Doublet peaks are apparent at 4.73–4.84 Å and 4.31–4.38 Å, indicating the presence of gibbsite and a second discrete hydrous aluminum phase. Basal spacings in these regions are common to the aluminum hydroxide polymorphs, nordstrandite and bayerite (Brown 1980). Traces of a mixed-layered clay are suggested by a small diffraction shoulder at 24–25

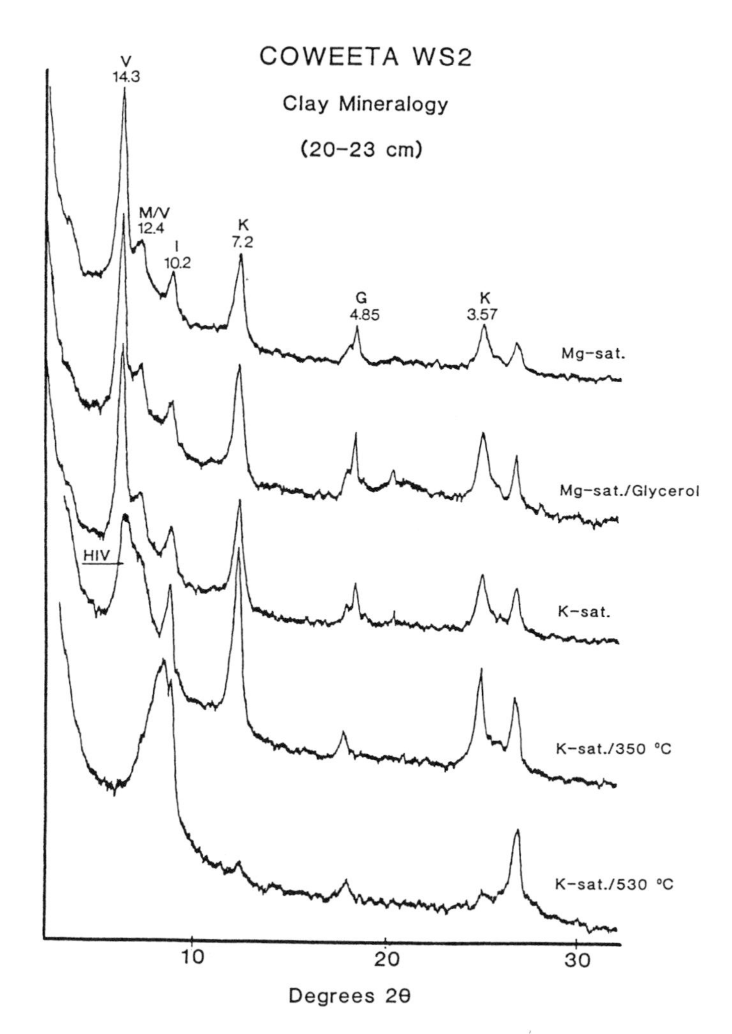

Figure 10.7. X-ray diffraction traces for <2-μm fraction of sample CH-14. Patterns display characteristic mineral suite of Coweeta watershed #2; however, relative amounts of minerals vary somewhat according to depth. Hydroxy-aluminum interlayered vermiculite (HIV & V) and kaolinite (K) are abundantly present. Gibbsite (G) and illite (I) constitute minor phases.

Å. Smectite and chlorite were also detected in trace amounts in some of the diffractograms. In every horizon, the 14-Å vermiculite peak never collapsed completely to 10 Å, even after potassium saturation and heat treatments to of 530°C, indicating the presence of hydroxy-aluminum interlayers throughout the soil profile (Douglas 1977; April et al. 1986).

Vermiculite and kaolinite also comprise the major clay phases at the Coweeta hardwood (CH) site. Minor amounts of gibbsite, mica and a mixed-layered clay are also present (Figure 10.7). The occurrence of a hydrous

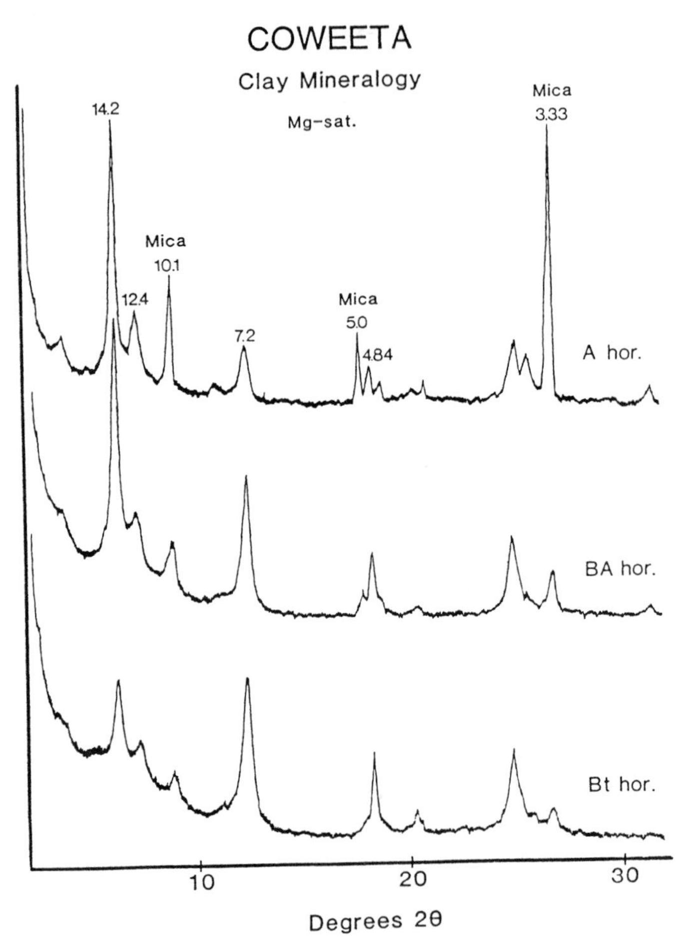

Figure 10.8. X-ray diffraction traces for <2-μm fraction of soils from different depths within Coweeta watershed #2 profile. Diffractograms show apparent increase in mica abundance toward surface.

aluminum phase other than gibbsite is again indicated by the appearance of doublet peaks at 4.74–4.85 Å and 4.31–4.36 Å. Muscovite is the dominate mica and apparently increases in abundance toward the top of the soil profile (Figure 10.8). The presence of an ordered, mixed-layered phase is shown by the appearance of a slight shoulder at 24–25 Å. As in the Coweeta white pine site, hydroxy-aluminum interlayering of vermiculite throughout the soil profile is indicated by the incomplete collapse of the 14-Å peak when potassium saturated and heated to 530°C.

Smoky Mountains. XRD traces of the clay fraction of the Smoky Mountain beech site show these soils to contain abundant vermiculite and mica, with

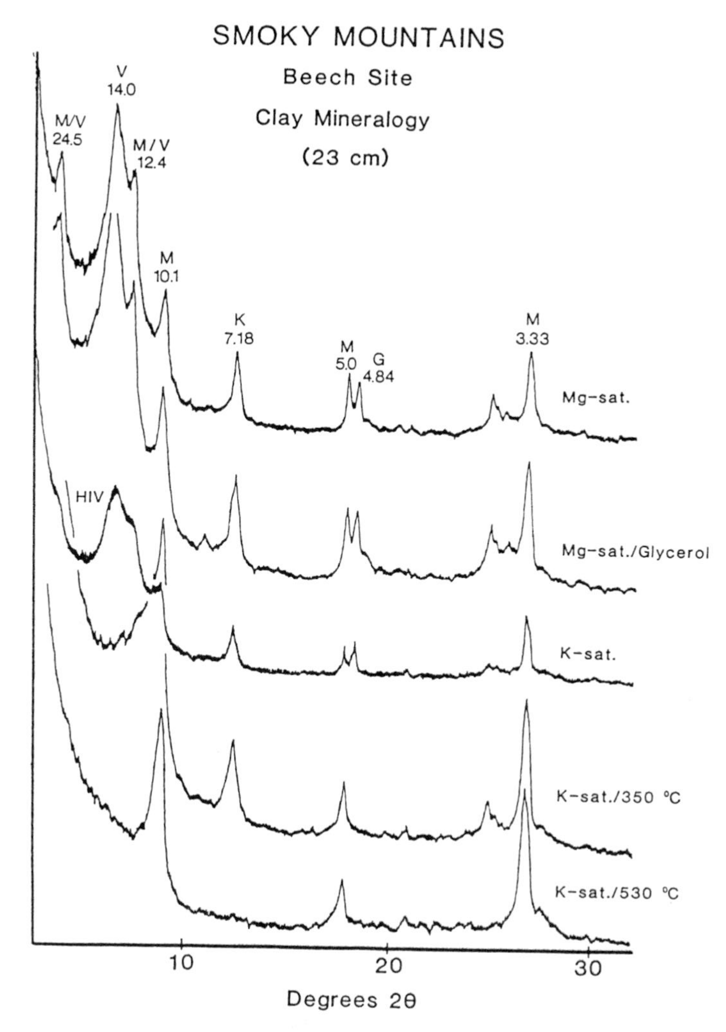

Figure 10.9. X-ray diffraction traces of <2-μm fraction of sample SB-11 illustrate clay mineral suite characteristic of soils from Smoky Mountain Beech site. Minerals present include: hydroxy-aluminum interlayered vermiculite (HIV & V), mica (M), kaolinite (K), and gibbsite (G). Sharp peaks at 24.5 Å and 12.5 Å indicate presence of hydrobiotite, a regularly-ordered, interstratified biotite/vermiculite (M/V).

minor amounts of gibbsite, kaolinite, and a highly ordered mixed-layered clay (Figure 10.9). The occurrence of a sharp superlattice peak at 24.5 Å is characteristic of the ordered, mixed-layered mineral hydrobiotite, which is composed of regularly interstratified biotite and vermiculite layers (Brindley et al. 1983). Muscovite is the dominant mica phase present, and vermiculite displays hydroxyaluminum interlayering throughout the profile.

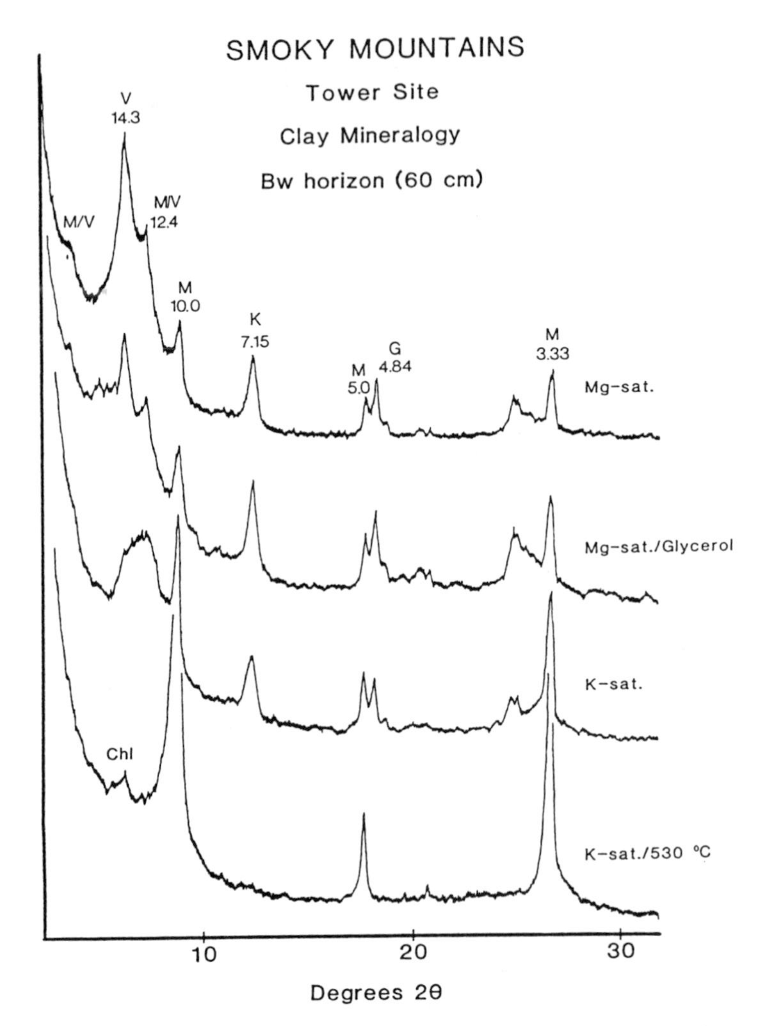

Figure 10.10. X-ray diffraction traces of <2-μm fraction of sample ST-15 illustrate characteristic clay mineralogy of Smoky Mountain Tower site. Hydroxy-aluminum interlayered vermiculite (V), kaolinite (K), mica (M), and gibbsite (G) constitute major clay phases. Minor mixed-layer mica/vermiculite (M/V) and traces of chlorite (Chl) are also present.

Boiling the sample for 2 h in 1 N hydrochloric acid failed to entirely eliminate the 14 Å basal reflection of vermiculite, indicating that it may be a dioctahedral variety (Brown and Brindley 1980). Traces of chlorite and smectite were also identified in some samples.

The clay mineralogy of the Smoky Mountain tower site includes abundant vermiculite and kaolinite with minor gibbsite, mica (both muscovite and biotite), and a less ordered, mixed-layered mineral (Figure 10.10). The mixed-layered clay is significantly different from the ordered hydrobiotite found at the beech site; only a poorly resolved superlattice peak occurs at 24 Å, and the structure collapses readily to 10 Å on potassium-saturation. Vermiculite at the tower site displays hydroxy-aluminum interlayering in all horizons, except in a sample taken around fine roots in the A horizon. Hot acid treatments failed to destroy the 14-Å vermiculite peak, suggesting that the vermiculite is similar to the one found at the beech site. Chlorite and smectite are present in trace amounts in these samples.

Northern Sites

Vermiculite predominates in the clay fraction of the northern soils with micas, mixed-layered mica/vermiculite, and minor amounts of kaolinite also present (see Table 10.2). A few samples contain smectite, chlorite, or talc, likely derived from the local bedrock or, in the case of smectite, formed as a weathering product. Figures 10.11 through 10.14 show series of x-ray diffractograms illustrating the clay mineralogy and clay mineral transformations in soil profiles of selected northern sites.

Huntington Forest. Huntington Forest soils contain abundant vermiculite and smectite and only trace to minor amounts of mica, kaolinite, mixed-layered minerals, and chlorite (Figure 10.11). Smectite is the dominant clay in the upper soil horizons, as evidenced by an intense 18-Å diffraction peak in the XRD traces of the magnesium-saturated, glycerol-solvated aliquots of the O/A and E horizons (Figure 10.12). Beginning in the Bh horizon and continuing downward in the profile, vermiculite replaces smectite as the dominant clay species, as indicated by a 14-Å (001) reflection that remains stable on magnesium saturation and glycerol solvation. Within the Bh horizon, the presence of both low- and high-charge vermiculite is indicated by the expansion of a portion of the (001) reflection to 17 Å on ethylene glycol solvation (Figure 10.13) (Walker 1975). No expansion is evident in the 14-Å peak of either the ethylene glycol or glycerol solvation XRD traces for the lower B and C horizons, which is consistent with the behavior of a high-charge vermiculite. Hydroxy-aluminum interlayers are present in the Huntington vermiculites, as indicated by the resistance of the structure to collapse with potassium saturation and mild heating (see Figure 10.11).

Whiteface Mountain. XRD traces of the clay fraction of Whiteface soils generally show poor resolution because of interferences from organic matter and the low clay content of these soils. Vermiculite is predominant in the clay fraction, as indicated by a prominent 14-Å (001) reflection that remains stable with magnesium saturation and glycerol solvation (Figure 10.14). Kaolinite, muscovite, and biotite are present in minor amounts and a small shoulder at 23 Å in some of the diffractograms indicates the presence of small amounts of a mixed-layered clay. Other minerals that occur in trace amounts include chlorite, talc, and smectite.

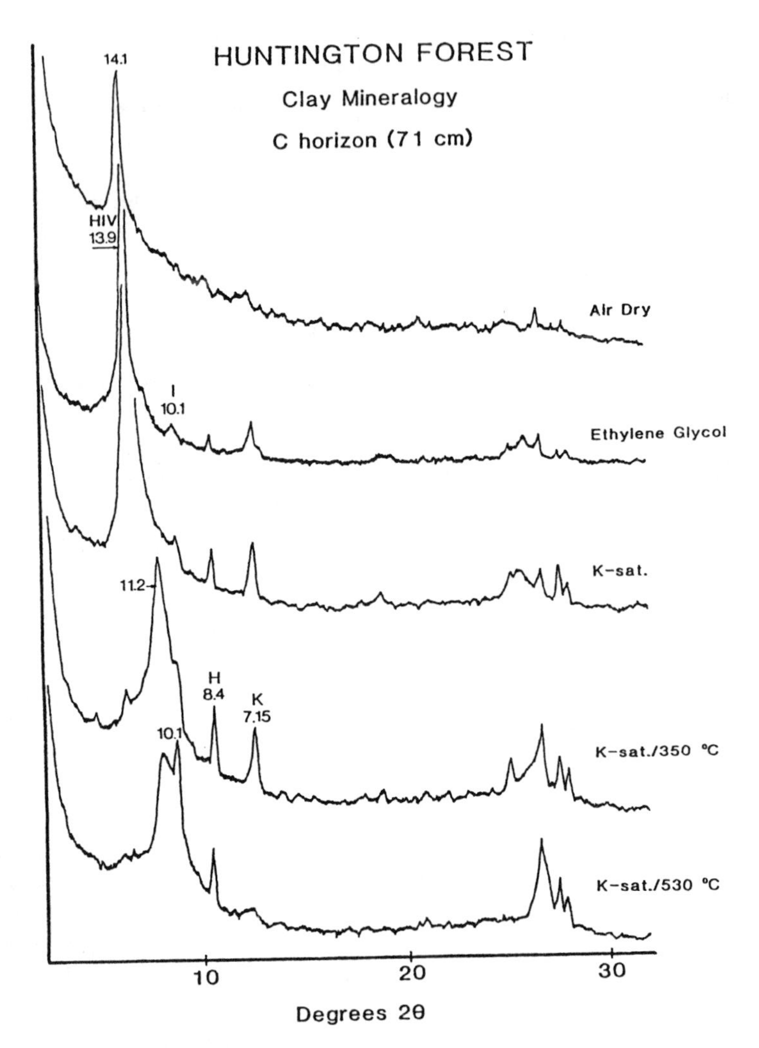

Figure 10.11. X-ray diffraction traces for <2-μm fraction of sample HF-1-1. Patterns represent lower horizons of Huntington profile. Hydroxy-aluminum interlayered vermiculite (HIV) is dominant; kaolinite (K) is minor constituent; traces of illite (I). Hornblende (H) essentially disappears upward in profile.

As described earilier, the most obvious differences in clay mineralogy among the IFS sites are noted when comparing northern and southern soils. Grain-size analyses show that clays are a far greater percentage of the mineral matter in southern soils than in northern soils (20%–40% and 3%–5%, respectively). In addition, the type and abundance of clay minerals present in the clay fraction vary distinctively. For example, northern soils contain no detectable traces of gibbsite and only minor to trace amounts of kaolinite;

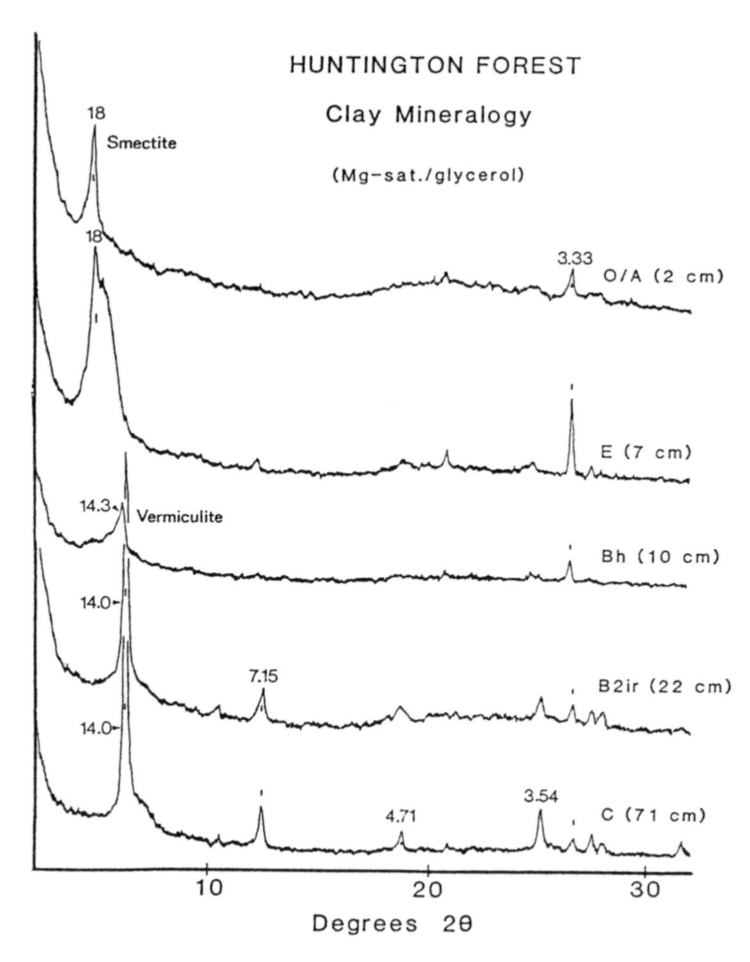

Figure 10.12. X-ray diffraction traces for magnesium-saturated/glycerol-solvated aliquots of <2-μm fraction of soils taken at various depths within Huntington Forest profile. The 14-Å reflection in diffractograms of C and B horizons indicates the vermiculite; 18-Å peak in E and OA horizons indicates expandable smectite in these upper soil horizons.

in sharp contrast, gibbsite and kaolinite are commonly abundant components of the clay fraction of southern soils and are likely derived from the weathering of primary silicate minerals. The abundance of gibbsite and kaolinite, as well as the great abundance of clays in general, suggests that the southern soils have experienced a relatively higher degree of weathering than have the northern soils. As used here, the term "degree of weathering" may be considered as a rate or intensity function influenced by several environmental factors including, but not limited to, time, climate, soil solution chemistry, and mineralogy.

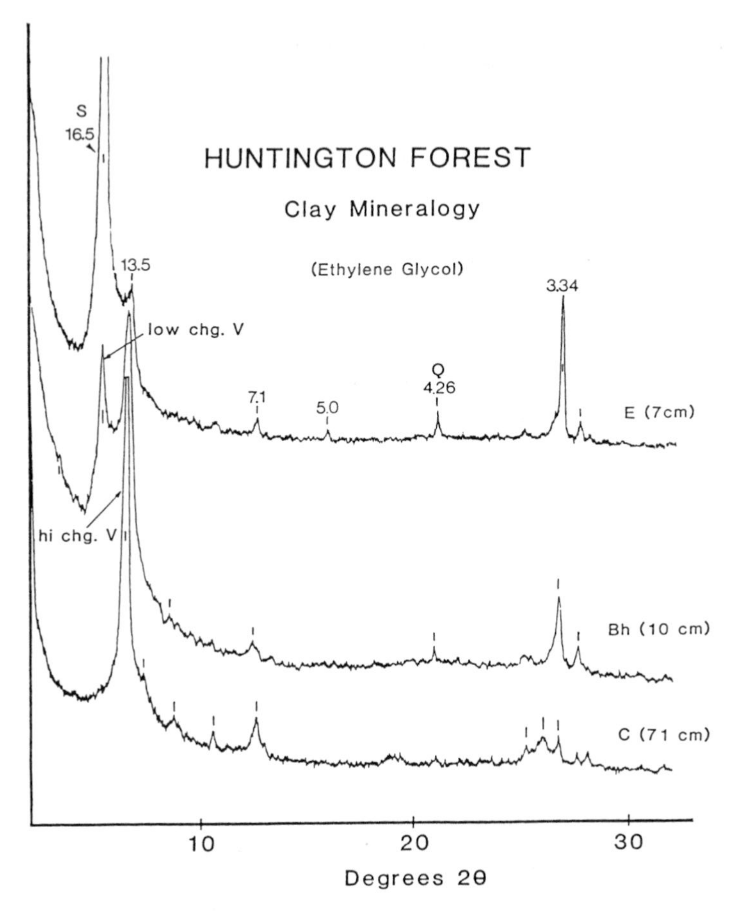

Figure 10.13. X-ray diffraction traces for ethylene glycol-solvated aliquots of <2-μm fraction of soils taken at various depths within Huntington Forest profile. Peaks at 13.5 and 16.5 Å in Bh horizon diffractogram indicate both high- and low-charge vermiculite ("hi chg. V" and "low chg. V," respectively).

With respect to the IFS sites, the factors thought to be most influential in accounting for the regional distinctions in clay mineralogy are (1) the glaciation of the northern sites, which provided fresh mineral surfaces and reset the clock on weathering approximately 15,000 years ago; (2) the mineralogical make-up of the parent material; and (3) the climate characterizing each region. All exert a significant influence on the extent and rate of mineral weathering and on the assemblages of clay minerals present in the soils.

Soil Chemistry

Results of major element determinations by XRF on bulk samples of forest soil from all IFS sites are summarized in Table 10.3. Although elemental

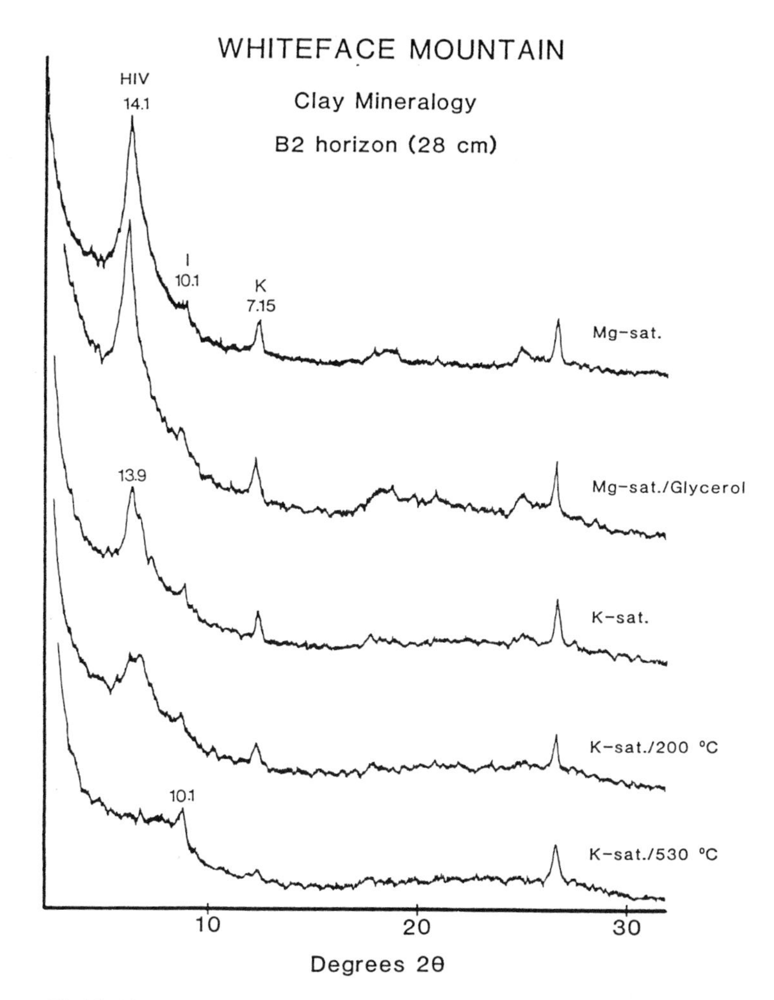

Figure 10.14. X-ray diffraction traces for <2-μm fraction of sample WF-1–3. Patterns represent clay mineral suite present in Whiteface soils, consisting predominantly of hydroxy-aluminum interlayered vermiculite (HIV) and minor to trace amounts of kaolinite (K) and illite (I). Poor peak definition is result of low clay content and high organic matter content that characterize these soils.

abundances vary from site to site, and can even differ substantially among soil pits sampled at any single site, in general the data show that the southern sites are depleted in base cations Ca, Mg, and Na but are enriched in the more conservative elements Al, Fe, and Si. Exceptions can be explained by looking at the mineralogical data for a given locality. For example, at the Florida (FS) site the soil is essentially a pure quartz sand and therefore all oxide weight percents are very low, except for SiO_2, which accounts for 97.90% (by weight) of the soil. Quartz is also the predominant mineral in

Table 10.3. Average Major Element Chemistry of IFS Soils[a]

Site Code	Forest	Fe_2O_3	MnO	TiO_2	CaO	K_2O	P_2O_5	SiO_2	Al_2O_3	MgO	Na_2O	Total
Northern												
FL	Findley	6.43	0.09	0.93	3.10	1.11	0.12	63.93	18.97	2.76	3.11	100.54
HF	Huntington	7.24	0.08	1.46	3.56	4.02	0.17	64.75	14.69	1.71	2.33	100.00
MS	Maine	5.43	0.03	0.86	0.71	1.81	0.07	77.52	11.22	0.53	1.50	99.64
NS	Norway	3.40	0.05	0.54	1.24	1.99	0.32	79.68	9.93	1.12	1.58	99.86
DF	Thompson DF	6.08	0.16	0.93	2.97	1.12	0.23	68.01	16.44	2.16	2.75	100.86
RA	Thompson RA	6.54	0.20	0.97	3.00	1.12	0.24	66.73	17.42	2.42	2.51	101.14
TL	Turkey Lake	5.66	0.07	0.66	3.29	1.83	0.09	69.53	14.75	2.02	3.56	101.45
WF	Whiteface	5.27	0.07	1.11	4.99	2.26	0.30	67.59	14.78	2.19	1.99	100.55
Southern												
GL	B.F. Grant	5.56	0.12	1.00	0.07	1.07	0.05	77.69	14.36	0.08	0.21	100.22
CB	Camp Branch	3.54	0.02	1.18	0.10	0.71	0.04	85.92	8.46	0.01	0.03	99.90
CH	Coweeta	12.32	0.09	2.01	0.10	2.21	0.07	62.24	20.04	1.43	0.30	100.69
CP	Coweeta	11.33	0.08	1.79	0.10	3.71	0.09	59.22	21.99	1.66	0.24	100.10
DL	Duke	6.21	0.02	0.85	0.10	0.49	0.04	76.71	15.78	0.02	0.01	100.11
FS	Florida	0.15	0.01	0.39	0.03	0.01	0.03	97.90	1.82	0.01	0.13	100.47
LP	Oak Ridge	6.55	0.14	1.05	0.09	2.33	0.16	75.55	13.38	0.46	0.16	99.84
SS	SM Becking	6.08	0.07	1.16	0.10	2.93	0.17	73.19	15.10	0.66	1.19	100.56
SB	SM Beech	8.54	0.34	1.16	0.10	2.94	0.34	68.30	16.58	0.63	0.65	99.41
ST	SM Tower	5.66	0.07	1.10	0.10	2.43	0.12	74.73	14.11	0.59	1.13	99.95
TK	Walker Branch	2.31	0.05	0.99	0.09	0.42	0.08	90.73	5.23	0.01	0.10	99.92

[a] Values represent average wt% oxide for all soil horizons.

soils at Camp Branch (CB) and Walker Branch (TK). As Figure 10.15 shows, K is the only base cation that is generally more abundant in southern soils than in northern soils, a situation that results from the abundance and stability of muscovite in many of these soils. If the sum of the averages of the oxides of the four base cations—Ca, Mg, K, and Na—are plotted for each site, a clear distinction can be seen between the base cation-poor southern sites and the base cation-rich northern sites (Figure 10.16). Notable exceptions are the Smoky Mountain and Coweeta sites, where the base cation concentrations seem to be nearly as high as in the northern sites. As discussed previously, these sites are located on steep slopes and are influenced by mass wasting, which provides a continual supply of fresh mineral grains to the upper soil horizons.

Mineral Weathering at the IFS Sites

Mineral Weathering Reactions

Base cations in soils are ultimately derived from either primary silicate weathering or atmospheric deposition. Primary silicate weathering is most important in soils where there is an abundance of fresh, reactive mineral grains such as in the northern, glaciated IFS sites. Atmospheric deposition can be an important source of base cations in the intensively weathered southern soils that contain low concentrations of reactive minerals. The cation exchange pool represents a reservoir of base cations derived from and sustained by sources such as mineral weathering, litter decay, and atmospheric inputs. This exchangeable pool may be depleted over time if hydrogen ions replace base cations faster than they can be resupplied by the processes just mentioned. This is likely to occur in Ultisols impacted by acidic deposition, for example, because these soils contain few weatherable primary minerals in near-surface horizons.

The amount of base cations released from mineral weathering is determined by three factors: (1) the abundance of weatherable minerals in the soil; (2) the rate at which these minerals weather; and (3) the stoichiometry of mineral weathering reactions. In addition, base cations released by mineral weathering may be involved in subsequent reactions that cause precipitation of new mineral species in the soil environment. Although the stoichiometry of a weathering reaction can be determined if both the chemistry of the mineral of interest (i.e., the average chemical composition of the mineral) and the nature of its weathering products are known, it is often difficult to link unequivocally the products of weathering in soils with specific mineral precursors. With detailed mineralogical and chemical information, however, it is possible to write chemical reactions that closely approach those likely occurring in the soil. For example, chemical analyses of the clay mineral products of weathering in many of the IFS soils revealed

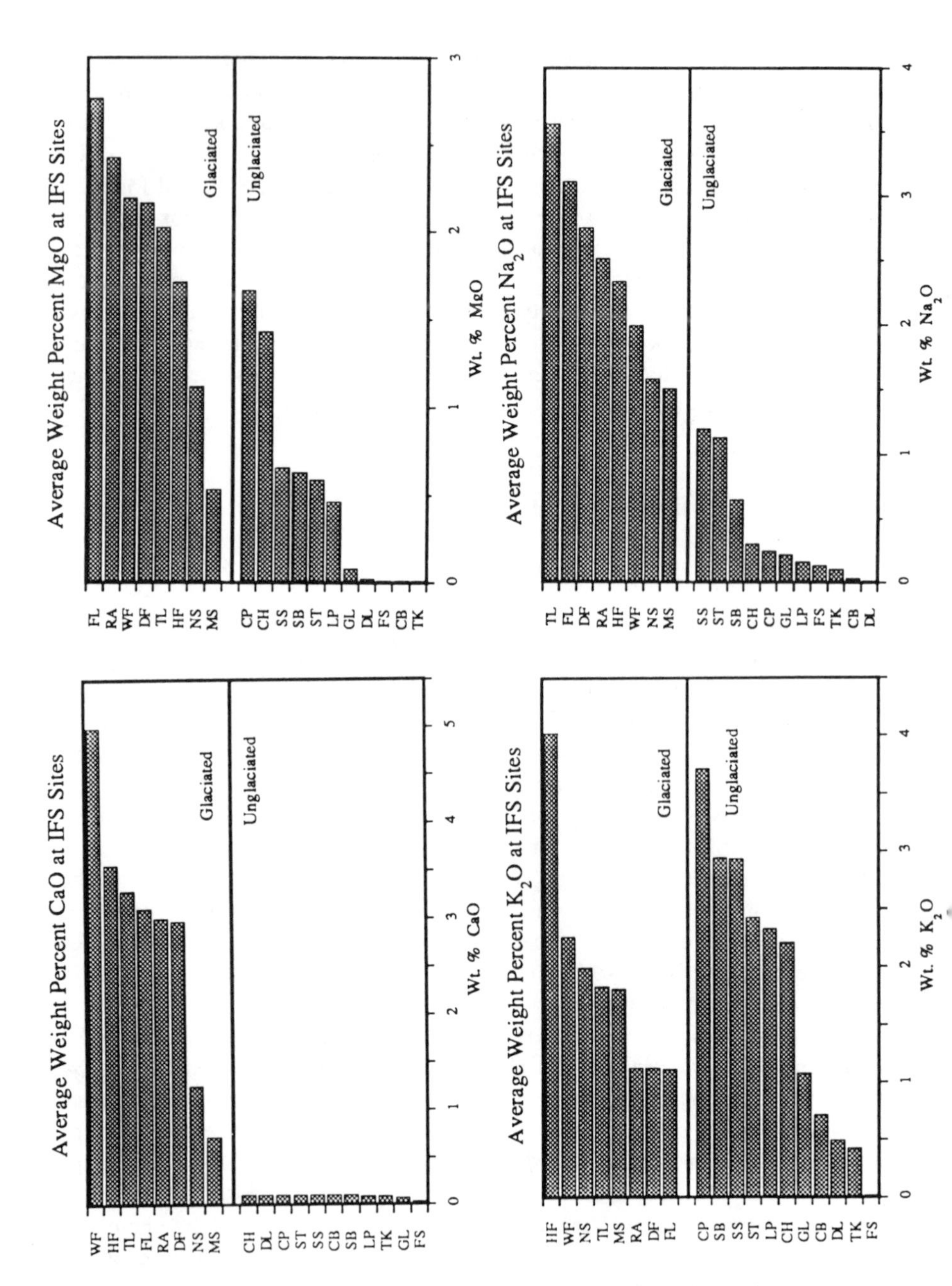

Figure 10.15. Average weight percent (wt%) CaO, MgO, K_2O, and Na_2O in soils from IFS sites.

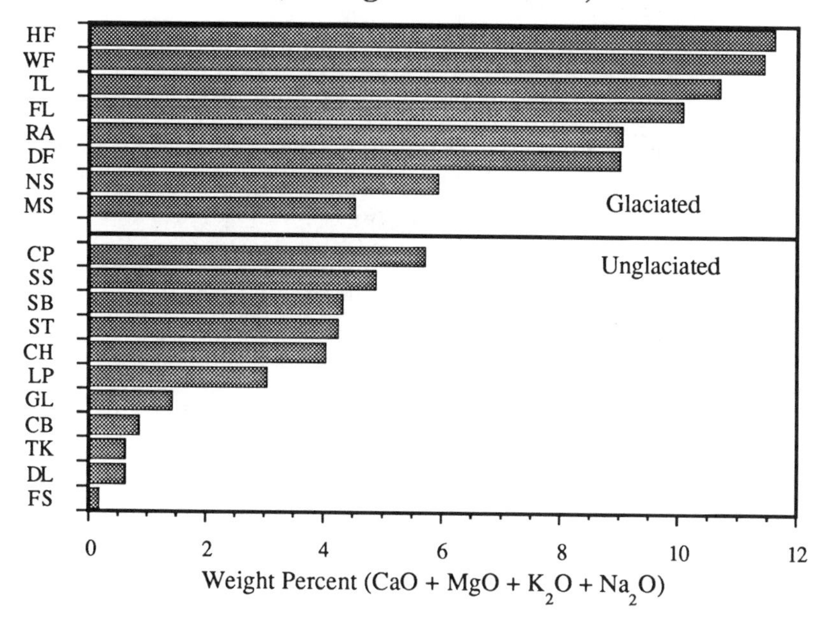

Figure 10.16. Average weight percent (wt%) CaO + MgO + K_2O + Na_2O in soils from IFS sites.

that they are essentially devoid of base cations. Therefore, it can be concluded that the stoichiometry of the primary weathering reactions involving these clay minerals is essentially congruent with respect to the base cations. Because many of these secondary clay minerals contain both silica and aluminum, it can also be assumed that the dissolution reactions are incongruent with respect to these two elements.

A general series of weathering reactions is shown in Table 10.4 for all the volumetrically important weatherable minerals found at the IFS sites. These reactions are listed approximately in order of relative ease of weathering. In general, the relative weathering rate of a mineral can be determined from its position on Bowens' reaction series (Goldich 1938; Pettijohn 1941). In this set of reactions, complete congruent dissolution of all mineral species is assumed. Because some of the silicon and aluminum released by aluminosilicate weathering is subsequently incorporated into secondary clay minerals, these reactions likely overestimate the amount of silicic acid and aluminum hydroxide generated. However, SEM analysis of most weathered primary mineral grains did not show significant amounts of secondary mineral development on the surfaces of the mineral grains. This is consistent with the observations of many other workers who have investigated silicate

Table 10.4. General Weathering Reactions for Minerals Commonly Present in Soils at the IFS Sites, in Approximate Rank Order for Ease of Weathering

Augite (x = mole fraction of iron)
$Ca_2MgAl_{(1-x)}Fe_xSi_3AlO_{12(s)} + 6H^+ + 6H_2O \rightarrow$
$2Ca^{2+} + Mg^{2+} + 3H_4SiO_4 + xFe(OH)_3 + (2 - x)Al(OH)_3$
Hypersthene (x = mole fraction of iron)
$Mg_{(1-x)}Fe_xSiO_{3(s)} + (2 - 2x)H^+ + (1 + 2x)H_2O \rightarrow (1 - x)Mg^{2+} + H_4SiO_4 + xFe(OH)_2$
Diopside (x = mole fraction of iron)
$CaMg_{(1-x)}Fe_xSi_2O_{2(s)} + (4 - 2x)H^+ + (2 + 2x)H_2O \rightarrow Ca^{2+} + (1 - x)Mg^{2+} + 2H_4SiO_4 +$
$xFe(OH)_2$
Hornblende (x = mole fraction of iron)
$NaCa_2Mg_{(4-x)}Fe_xAlSi_6Al_2O_{22}(OH)_{2(s)} + (13 - 2x)H^+ + (9 + 2x)H_2O \rightarrow$
$Na^+ + 2Ca^{2+} (4 - x)Mg^{2+} + 6H_4SiO_4 + xFe(OH)_2 + 3Al(OH)_3$
Zoisite
$Ca_2Al_3Si_3O_{12}(OH)_{(s)} + 4H^+ + 8H_2O \rightarrow 2Ca^{2+} + 3H_4SiO_4 + 3Al(OH)_3$
Epidote (x = mole fraction of iron)
$Ca_2Al_{(3-x)}Fe_xSi_3O_{12}(OH)_{(s)} + 4H^+ + 8H_2O \rightarrow$
$2Ca^{2+} + 3H_4SiO_4 + (3 - x)Al(OH)_3 + xFe(OH)_2$
Garnet
$CaMgFeAl_2Si_3O_{12(s)} + 4H^+ + 8H_2O \rightarrow Fe(OH)_2 + 2Al(OH)_3 + 3H_4SiO_4 + Ca^{2+} + Mg^{2+}$
Plagioclase Feldspar (x = mole fraction of anorthite)
$Na_{(1-x)}Ca_xSi_{(3-x)}Al_{1+x)}O_{8(s)} + (1 + x)H^+ + (7 - x)H_2O \rightarrow$
$(1 - x)Na^+ + xCa^{2+} + (1 + x)Al(OH)_3 + (3 - x)H_4SiO_4$
Chlorite (x = mole fraction of iron)
$Mg_4Al_{(2-x)}Fe_xSi_2Al_2O_{10}(OH)_{8(s)} + 8H^+ + 2H_2O \rightarrow$
$4Mg^{2+} + 2H_4SiO_4 + (4 - x)Al(OH)_3 + xFe(OH)_3$
Biotite (x = mole fraction of magnesium)
$KMg_xFe_{(3-x)}AlSi_3O_{10}(OH)_{2(s)} + (1 + 2x)H^+ + (9 - 2x)H_2O \rightarrow$
$K^+ + xMg^{2+} + 3H_4SiO_4 + Al(OH)_3 + Fe(OH)_2$
Staurolite (x = mole fraction of iron)
$Mg_{(2-x)}Fe_xAl_9O_6Si_4O_{16}O(OH) + (4 - 2x)H^+ + (19 + x)H_2O \rightarrow$
$xMg^{2+} + 4H_4SiO_4 + xFe(OH)_2 + 9Al(OH)_3$
Orthoclase (K feldspar)
$KSi_3AlO_{8(s)} + H^+ + 7H_2O \rightarrow K^+ + Al(OH)_3 + 3H_4SiO_4$
Apatite
$Ca_5(PO_4)_3(OH)_{(s)} \rightarrow 5Ca^{2+} + 3(PO_4^{3-}) + OH^-$

mineral weathering (e.g., Berner and Holdren 1979; Berner and Schott 1982; Velbel 1984).

Site-specific weathering reactions can be written for each mineral if the chemistry of both the primary and secondary minerals is known. Whether or not the reaction is written congruently or incongruently often depends on inferences made from secondary mineralogical and chemical evidence. For example, at the Huntington Forest site SEM analyses suggested that feldspar, hornblende, and diopside are weathering congruently because, although the grains are pitted and etched, they show no sign of altering or transforming into secondary weathering products (Figure 10.17). Acquiring the average chemical analyses of these minerals allows us to formulate the following congruent weathering reactions for the Huntington Forest site:

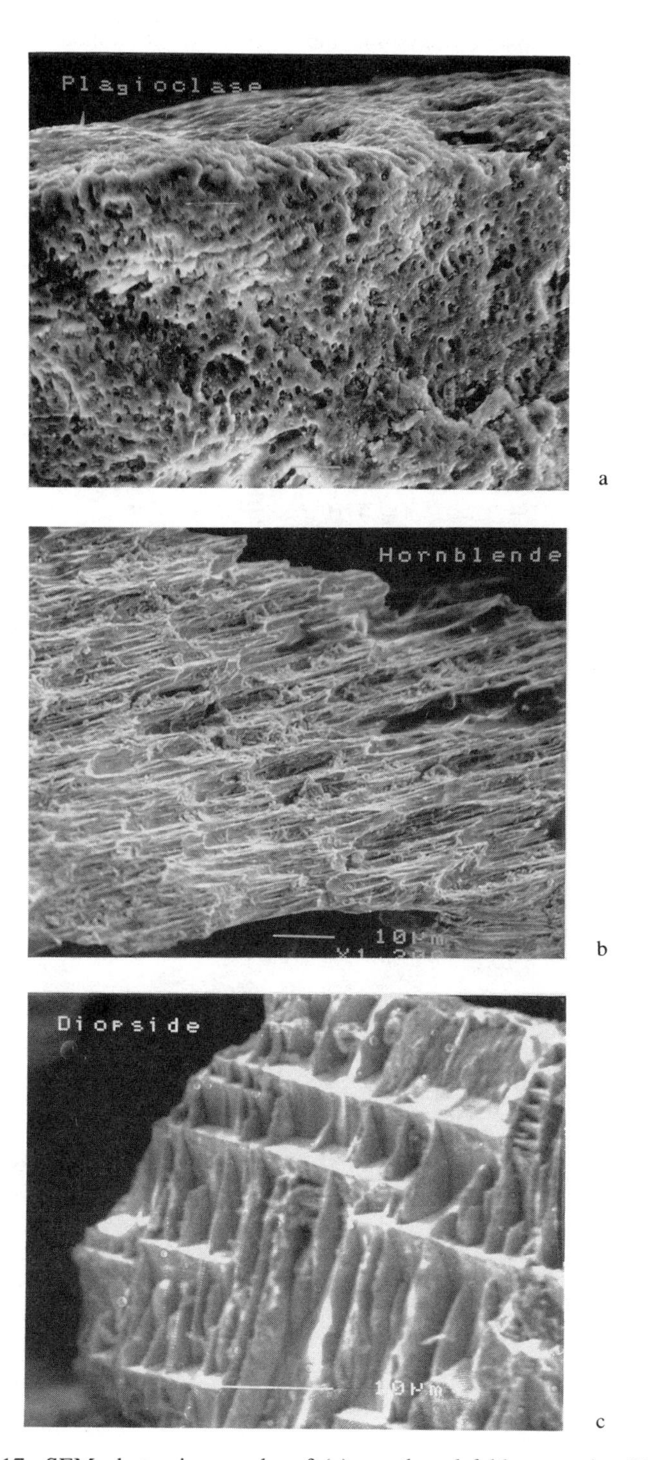

Figure 10.17. SEM photomicrographs of (a) weathered feldspar grain; (b) weathered hornblende grain; and (c) weathered diopside grain show well-developed etch pits and no indications of secondary mineral precipitation or surface coatings.

Plagioclase Feldspar

$$Na_{0.5}Ca_{0.47}K_{0.02}Si_{2.64}Al_{1.33}O_{8(s)} + 1.46H^+ + 6.55H_2O \rightarrow$$
$$\rightarrow 0.5Na^+ + 0.47Ca^{2+} + 0.02K^+ + 1.33Al(OH)_3 + 2.64H_4SiO_4$$

Hornblende

$$K_{0.31}Ca_{1.92}Mg_{1.74}Fe_{1.96}Al_{1.3}Si_{6.96}Al_{0.9}O_{22}(OH)_{2(s)} + 7.63H^+ + 14.37H_2O \rightarrow$$
$$\rightarrow 0.31K^+ + 1.96Ca^{2+} + 1.74Mg^{2+} + 6.96H_4SiO_4 + 1.96Fe(OH)_2 + 2.2Al(OH)_3$$

Diopside

$$CaMg_{0.63}Fe_{0.37}Si_2O_{6(s)} + 3.26H^+ + 2.47H_2O \rightarrow$$
$$\rightarrow Ca^{2+} + 0.63Mg^{2+} + 2H_4SiO_4 + 0.37Fe(OH)_2$$

From these reactions it can be seen that weathering of plagioclase feldspar, hornblende, and diopside will release considerable amounts of Ca and Mg but only minor amounts of Na and K.

A second example of determining the stoichiometry of weathering reactions, this time involving incongruent dissolution, involves biotites in soils at both the Huntington Forest and Coweeta sites. The alteration product (kaolinite and vermiculite) of biotite weathering at each site was determined by both XRD and SEM/EDS analyses. The chemistry of both the biotites and their alteration products was determined by taking numerous spot analyses on fresh mineral grains as well as on altered grains and rims and averaging the results. Stoichiometrically correct weathering reactions for the biotite at each site were then formulated:

Huntington Forest (biotite to vermiculite)

$$\underset{\text{Biotite}}{K_{0.81}(Mg_{0.66}Fe_{1.82}Ti_{0.16}Al_{0.24})(Al_{1.13}Si_{2.87})O_{10}(OH)_2} + 0.94H^+ + 1.44H_2O \rightarrow$$
$$\rightarrow 0.74\underset{\text{Vermiculite}}{[(K_{0.12}Mg \cdot H_2O_{0.25})(Mg_{0.46}Fe_{1.53}Ti_{0.07}Al_{0.74})(Al_{1.12}Si_{2.88})O_{10}(OH)_2]}$$
$$+ 0.72K^+ + 0.13Mg^{2+} + 0.69Fe(OH)_2 + 0.11TiO_2 + 0.74H_4SiO_4$$

Coweeta (biotite to kaolinite)

$$\underset{\text{Biotite}}{K_{0.85}(Mg_{1.22}Fe_{0.95}Ti_{0.11}Al_{0.53})(Al_{1.21}Si_{2.79})O_{10}(OH)_2} + 3.28H^+ + 2.15H_2O \rightarrow$$
$$\rightarrow 0.87\underset{\text{Kaolinite}}{Al_2Si_2O_5(OH)_4} + 0.95Fe(OH)_2 + TiO_2 + 0.85K^+ + 1.22Mg^{2+}$$
$$+ 1.05H_4SiO_4$$

The results show that biotite weathering proceeds further in Coweeta soils than in soils from Huntington Forest. Biotite grains in Coweeta soils are typically altered to kaolinite at their rim, whereas biotite grains in the soil at Huntington Forest weather only so far as to display vermiculitic edges (Keller 1988).

Weathering Rates at the IFS Sites

Base cations in soil solutions can be derived from a variety of sources and processes including cation exchange, decomposition of organic matter, atmospheric inputs, and mineral weathering. Because minerals are the ultimate source of most cations in soils, understanding the mechanisms and rates of mineral weathering is essential when attempting to characterize nutrient cycles in forest soils. Unfortunately, this is no easy task considering that primary mineral weathering in soils depends on a complex set of variables. These include mineralogical factors such as mineral composition and abundance; climatic factors such as mean annual temperature and rainfall; hydrologic factors such as hydraulic conductivity and permeability of the soil; physiographic factors such as slope, aspect, relief, and elevation; and chemical factors such as precipitation and throughfall chemistry, soil pH, and the nature and quantity of organic compounds generated by the decomposition of organic matter. The interactions among these variables ultimately determine the weathering rate within a defined setting.

Because of the large number of sites involved in this project and the fact that soil change data were not available for most, a mass balance input-output approach to compare cation denudation rates among the sites was not possible. Instead, a simple and straightforward experimental approach was used to help determine relative weathering rates among the study sites. In these experiments, a vacuum extractor was used to apply a series of solutions to small volumes of soil sample in a manner similar to the method used by Dahlgren et al. (1990) (Figure 10.18). The vacuum extractor has the advantage over more conventional gravity-fed leaching columns in that leaching of all samples occurs at an identical rate regardless of the difference in grain size and hydraulic conductivity of the sample. A blank and a control sample containing Ottawa Sand (quartz) were included in all the treatments to quantify background ion concentrations. An application rate of 2.5 cm h^{-1} was used throughout these experiments. The samples were first leached for 13 days with deionized water, then by a series of treatments with progressively more acidic solutions according to the following scheme:

10 days using pH 4 sulfuric acid, followed by
10 days using pH 3.5 sulfuric acid, followed by
8 days using pH 3.0 sulfuric acid, followed by
8 days using pH 2.5 sulfuric acid.

Column leachate was analyzed for Ca^{2+}, Mg^{2+}, Na^{+}, and K^{+} by atomic adsorption; for pH by electrode; and for SiO_2 by colorimetry (APHA 1981).

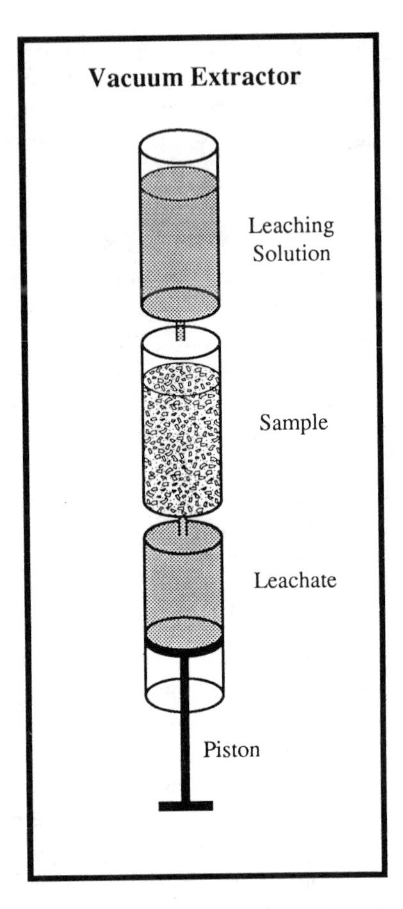

Figure 10.18. Vacuum extractor pulls leachate solution down through sample as vacuum is created by movement of piston assembly. Column leachate is collected in syringe located above piston. Multiple samples can be treated identically because all pistons are moved by single plate.

Exchangeable bases and exchangeable acidity were measured both before the first treatment and after application of the final solution.

It was hypothesized that the progressive acid leaching would stimulate mineral weathering and allow quantification of the hydrogen ion dependency of total mineral weathering for each soil sample (Cronan 1985). However, most of the soils in this study were so well buffered that there was no appreciable change in soil leachate pH until application of the final pH 2 treatment (Figure 10.19). The pH of the leachate following application of the initial deionized water solution varied widely among sites with no significant differences observed between the glacial and nonglacial soils. In addition, there were no significant differences in base cation and silica leaching rates during application of the pH 4.0, pH 3.5, and pH 3.0 acid treatments. The

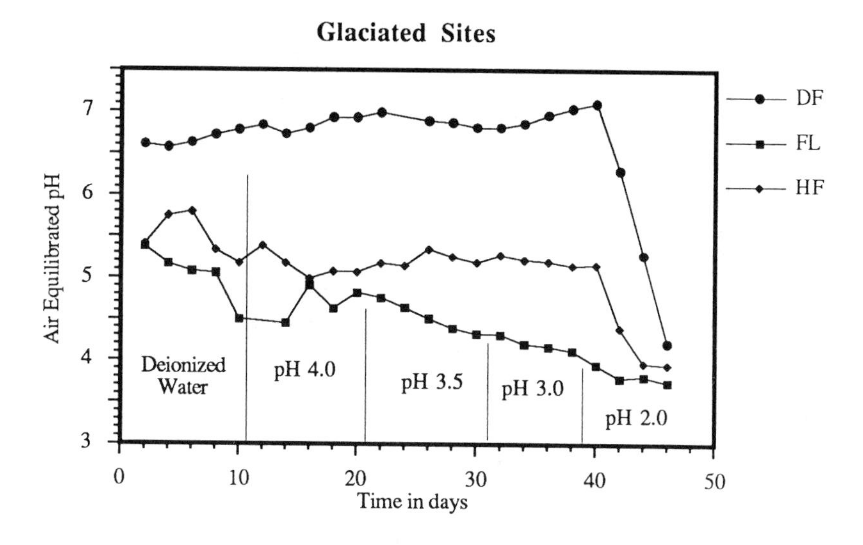

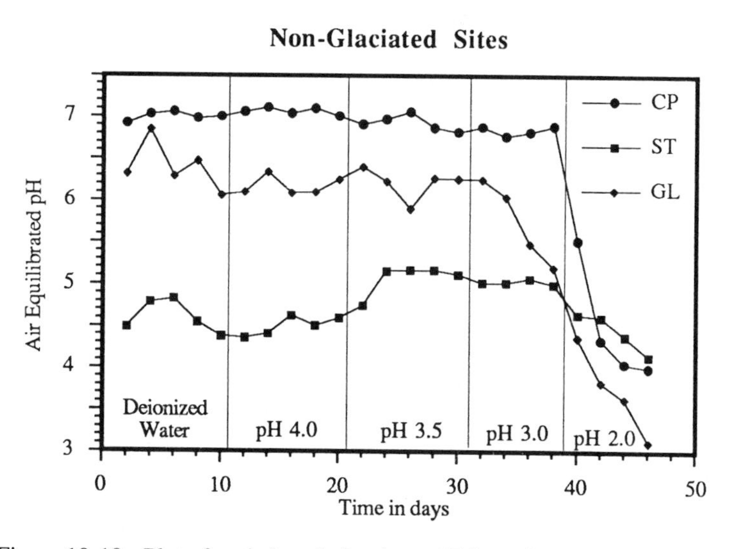

Figure 10.19. Plot of variations in leachate pH for soils from selected glaciated (top) and nonglaciated (bottom) sites. Column leachate showed little change in pH until after final pH 2 treatment.

pH 2 treatment did significantly increase cation release rates but, for reasons that are discussed later, these rates appeared to be associated with the breakdown of secondary clay minerals rather than with the weathering of primary minerals.

In general, application of the the deionized water at the beginning of the experiment and the pH 2 solution at the end of the experiment resulted in the release of most of the base cations and silica as shown, for example, in Figure 10.20 for the Thompson Forest (DF) soil. The leaching experiment resulted in the release of 70% of the Na^+ during the deionized water treatment and an additional 14% during the pH 2 portion of the experiment. Similarly, most of the K^+ and Mg^{2+} was released during the initial deionized water treatment. Ca^{2+} and SiO_2 behaved differently in that maximum release of these species was associated with the pH 2 treatment.

One approach to evaluating the source of the base cations in these experiments is to determine whether correlations exist between base cation and the SiO_2 fluxes. If silicate mineral weathering is responsible for the observed fluxes, in whole or in part, good correlations should be apparent, with the ratios of base cations to SiO_2 a function of the stoichiometry of the major weathering reactions. On the other hand, lack of good correlations would suggest other primary sources of base cations, from cation exchange or from the weathering of nonsilicates, for example.

To illustrate, Figure 10.21 graphically shows the relationship between Ca^{2+} and SiO_2 concentrations from the Thompson Forest Douglas Fir (DF) soil column experiment. The other base cations showed similar relationships. Although the good correlations seem to indicate that Ca^{2+} was derived from silicate mineral weathering, other observations bring this conclusion into question. For example, the highest base cation and SiO_2 fluxes occurred with the deionized water treatment, decreased with application of the pH 4.0, pH 3.5, and pH 3.0 solutions, then increased again with application of the pH 2 solution (see Figure 10.20). This behavior seems somewhat inconsistent with results from laboratory weathering experiments using pure silicate minerals, because progressively more acidic solutions usually stim-

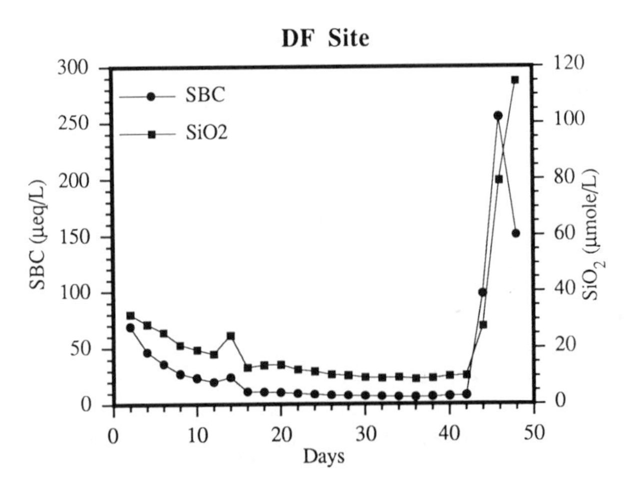

Figure 10.20. Plot of sum of base cation and SiO_2 fluxes versus time for leachate recovered from Thompson Forest (DF) soil sample.

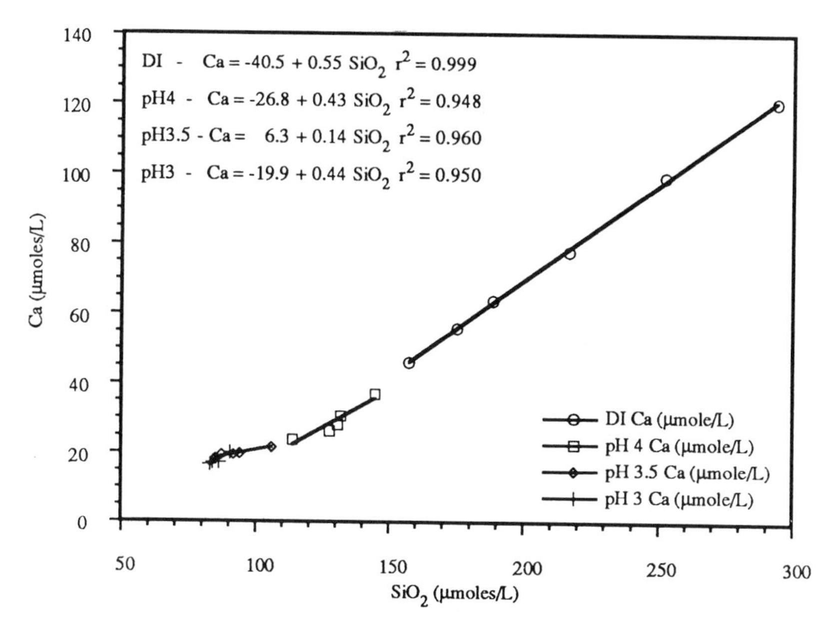

Figure 10.21. Plot of relationship between Ca^{2+} and SiO_2 flux for deionized water (DI), pH 4, pH 3.5, and pH 3 treatments of Thompson Forest (DF) soil sample.

ulate more weathering and higher base cation and silica fluxes. In soils, however, processes other than mineral weathering, such as sulfate adsorption, cation exchange, or the dissolution of Al sesquioxides, can consume hydrogen ions and influence the pH and chemistry of the leachate solution. (Unfortunately, neither SO_4^{2-} nor Al^{3+} concentrations were monitored during the experiments.) The high correlations between base cation and silica fluxes might also be explained by dilution effects as the initial pore water was slowly washed from the column during the experiment.

Perhaps a better way to determine the amount of base cations released by weathering during the leaching experiment is to identify those cations released by exchange processes and then subtract them from the total flux using the following equation:

$$X_{\text{Weathering}} = X_{\text{Total Flux}} - \delta X_{\text{Exchangeable}}$$

where $X_{\text{Total Flux}}$ is the total flux of base cation X, $\delta X_{\text{Exchangeable}}$ is the amount of exchangeable base cation X before leaching minus the amount of exchangeable base cation X after leaching, and $X_{\text{Weathering}}$ is then the base cation flux from weathering. The total flux for each base cation and silica is given in Figure 10.22. Subtracting the total exchangeable base pool measured before treatment from the total exchangeable base pool measured after all treat-

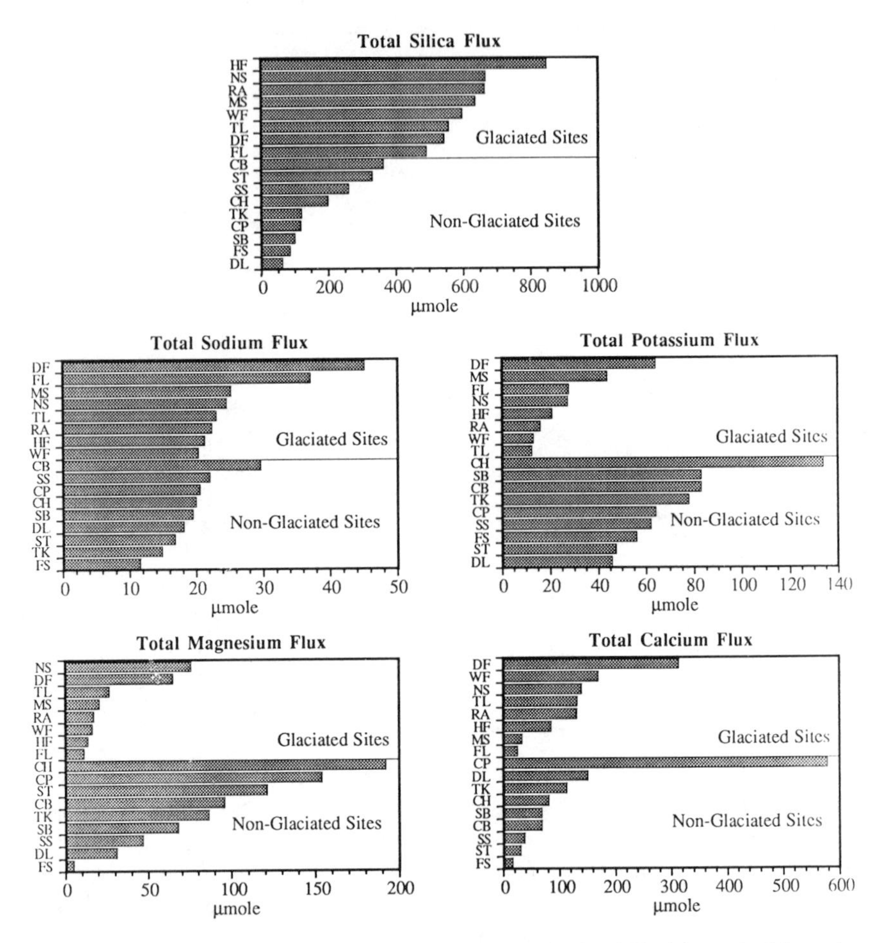

Figure 10.22. Total flux of SiO_2, Na, K, Mg, and Ca for IFS sites ranked in order of decreasing flux for glaciated and nonglaciated sites.

ments yields the amount of base cations, $\delta X_{\text{Exchangeable}}$, contributing to the total cation flux (Figure 10.23). A gain represents more exchangeable bases present after treatment, whereas a loss indicates that fewer exchangeable bases were present after treatment. As the figures show, Ca is by far the most important exchangeable base cation, contributing up to 600 μmol to the total column flux (CP). Soils with initially high pH (e.g., DF and CP) had the highest Ca values, whereas some of the more acid soils (e.g., WF and SS) had the lowest Ca values. The Whiteface (WF) and Nordmoen (NS) samples both showed a net gain in exchangeable Ca at the end of the experiment.

Magnesium was the next most important exchangeable base, contributing up to 150 μmol to the total flux. In general, the nonglaciated sites as a group

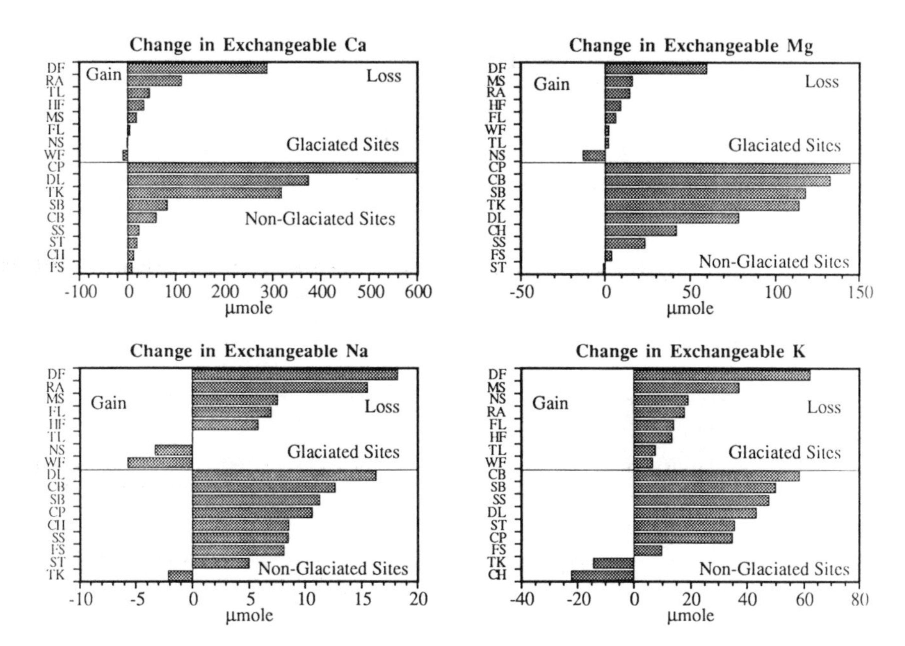

Figure 10.23. Changes in exchangeable pool of Ca, Mg, Na, and K ranked in decreasing order for glaciated and nonglaciated sites. Negative result indicates increase in exchangeable pool of base cation after end of column experiment.

contributed more exchangeable Mg than the glaciated sites. However, both the Nordmoen (NS) and the Smoky Mountain (ST) samples showed a small net gain. Exchangeable K and Na were much less important, accounting for a maximum of 60 and 20 μmol, respectively, to the total base cation flux.

Subtracting the change in exchangeable cations ($\delta X_{Exchangeable}$) from the total flux ($X_{Total\ Flux}$) yields the quantity of base cations derived from weathering. As Figure 10.24 illustrates, the weathering component contribution to the total flux of each base cation varies considerably among the sites. Overall, significantly more Ca, Mg, and K than Na are released by weathering. The glaciated sites in general release more Ca than the nonglaciated sites. Some sites show a large net negative weathering component for certain cations, meaning that more of these cations were released from exchange sites than were measured leaving the columns in leachate solutions. In these cases the cations must be retained within the soil in nonexchangeable sites, perhaps as interlayer cations in clay minerals. For example, although Ca is released by weathering in all the glaciated soils, some of the nonglaciated soils (e.g., DL and TK) appear to be significant sinks for Ca. In turn, these are the soils that contain the highest content of silt and clay (see Figure 10.1).

The distribution of Mg derived from weathering is similar to that of Ca in that all the glaciated sites released at least some Mg whereas a number

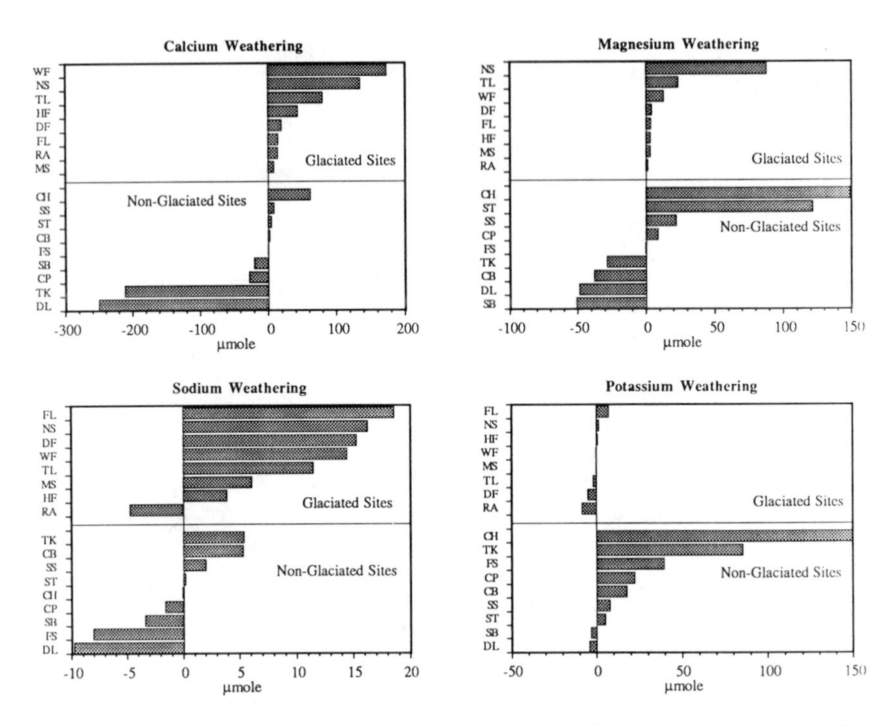

Figure 10.24. Base cations released by weathering, ranked in decreasing order for glaciated and nonglaciated sites. Negative result indicates that cation is retained in soil, perhaps by clay minerals.

of the nonglaciated sites acted as a Mg sink. High amounts of Mg and K derived from weathering in some of the nonglaciated soils (CH and ST for Mg; CH and TK for K) indicated the presence of weatherable Mg- and K-rich silicate minerals. The amount of Na released by weathering is comparatively small relative to the other cations, but in general it appears that more is released from the glaciated sites than from the nonglaciated sites.

Overall, the results from the column experiments strongly suggest that the potential for silicate mineral weathering is greater in the northern, glaciated sites than in the southern, nonglaciated sites. Perhaps this is most apparent from the silica fluxes shown in Figure 10.22. Because silicon is not involved in exchange reactions, these values provide at least a rough measure of the relative amount of silicate mineral weathering in these soils. The northern soils leach significantly more silica than the southern soils, and this is not surprising given the differences in the quantity and chemistry of the light and heavy mineral assemblages previously discussed. The individual total cation fluxes from the glacial and nonglacial soils (see Figure 10.22), however, suggest that K- and Mg-bearing silicates are important weatherable minerals in a number of the southern sites whereas Ca- and Na-bearing minerals are most important in many of the northern sites. This is supported by the mineralogical data, which show biotite to be an important component of many southern soils (especially in Coweeta and the Smoky Mountain sites) and plagioclase feldspar to be abundant in most northern soils. The extremely low cation and silica fluxes from some of the sites (e.g., DL and FS) suggest little or no weathering is occurring. Again, this is consistent with the mineralogy and bulk soil chemical analyses, which show that these sites have few weatherable minerals and less than 1 wt% total base oxides.

Because the ratios of base cations (released by weathering) to silica are related to the sum of individual silicate mineral weathering reactions, the flux data also can be used, together with the mineralogical information, to predict which weathering reactions are important at each site. [Ratios of base cations to silica were calculated from the column flux data for the deionized water treatment only to minimize the effects of cation exchange reactions on the totals (Table 10.5)]. For example, the light mineral analyses for the Huntington Forest (HF) site showed that the B soil horizon is composed of 34% plagioclase feldspar with an average composition of about An_{50}. Using the stoichiometry of the weathering reaction for plagioclase feldspar given in Table 10.4, a Na^+:SiO_2 ratio of 0.19 should result from the congruent dissolution of a plagioclase with a composition of An_{50}, which is close to the observed value of 0.25 that was derived from the column flux data (Table 10.5). With the mineralogical and chemical data showing no other major potential mineral sources of Na in the soil, it seems certain that plagioclase weathering is the most important contributor to the Na flux in Huntington Forest soil solutions. The same reasoning can be applied to the remaining base cations, and the results suggest that plagioclase weathering along with the weathering of two heavy minerals, diopside and hornblende, account for most of the flux of Ca at the Huntington Forest site. Small quantities of K

Table 10.5. Cation to Silica Ratios for Deionized Water Treatment at Selected IFS Sites

Site	Na^+/SiO_2	K^+/SiO_2	Ca^{2+}/SiO_2	Mg^{2+}/SiO_2
HF	0.25	0.17	0.71	0.11
WF	0.25	0.14	0.34	0.04
TL	0.05	0.03	0.09	0.03
MS	0.12	0.21	0.05	0.12
FL	0.34	0.27	0.08	0.11
DF	0.19	0.19	0.35	0.20
RA	0.06	0.02	0.24	0.06
ST	0.16	0.50	0.18	0.35
SS	0.17	0.43	0.19	0.26
SB	0.54	2.61	1.69	1.50
CP	0.94	1.00	3.31	1.90

Table 10.6. Weathering Potential Ranking of the IFS Sites

High potential	HF[a]
	WF[a]
	TL[a]
	NS[a],RA[a],FL[a]
	DF[a]
	CP
	MS[a]
	SS
	CH
	ST
	SB
	CB
	TK,DL
Low potential	FS

[a] Denotes glaciated site.

and Mg are derived mainly from the weathering of biotite, a mineral that is present in small amounts at this site.

Discrepancies between the calculated ratios of base cations to silica from the flux data and the ratios predicted by the stoichiometry of mineral weathering reactions are common in this sort of exercise and can result from such factors as the incongruent weathering of aluminosilicates, the presence of trace quantities of highly weatherable minerals (e.g., carbonates and salts) that remained undetected in mineralogical analyses, precipitation or neoformation of mineral phases (e.g., secondary clay minerals and amorphous to poorly crystalline aluminosilicates) in the soil profile, and errors associated with the mineralogical and chemical characterization of soil materials.

Finally, the results of the experimental work, taken together with the physical, chemical, and mineralogical data, provide sufficient information to estimate the relative weathering potentials of the IFS sites (Table 10.6). This ranking was devised by combining the base cation and silica flux data (see Figure 10.22) with the rankings of percent heavy minerals (see Figure 10.5) and sum of the base oxides (see Figure 10.16). Sites for which a complete data set was not available were not ranked. Sites at the very bottom of the scale, such as the Florida (FS) site, which is almost completely devoid of weatherable minerals, virtually have no base cations supplied to the soil solution from primary mineral weathering. Those sites at the top of the scale (e.g., HF and WF) have a very high weathering potential and over time are least likely to have soil solutions become depleted in base cations.

Observations from the Rhizosphere

The rhizosphere is defined as the narrow zone of soil subject to the influence of living roots. Although the bulk of forest soil lies outside of the rhizosphere, soil in the root zone interacts most directly with forest plants and is, therefore, a significant subcomponent of the forest nutrient cycle. Because few studies have addressed mineral–root dynamics in natural soils (e.g., Mortland et al. 1956; Barshad 1964; Spyridakis et al. 1967) and because the deficiency of essential cations such as Ca and Mg, the increased mobility of competing ions such as Al, and the presence of heavy metals in soils have all been linked to the forest decline phenomenon, we decided to examine the reciprocal effects resulting from the interaction of plant roots and minerals in the rhizosphere from selected IFS sites. The intent of the investigation (Keller 1988; April and Keller 1990) was twofold: first, to examine and compare the mineralogical differences that exist between rhizosphere soils and bulk forest soils (as discussed in previous sections) and, second, to observe the effects (both mechanical and chemical) of the interactions between roots and mineral grains in the rhizosphere.

Sampling locations were selected on the basis of soil type, climate, elevation, forest type, and geology and included the two Adirondack sites in New York State (Huntington Forest and Whiteface Mountain) and four of the forest stands in the southern Appalachian Mountains of North Carolina (the two sites at the Coweeta Hydrologic Laboratory and two sites in the Great Smoky Mountains National Park). These six sites were chosen to provide examples of high-elevation/low-elevation, deciduous/coniferous, and glaciated/nonglaciated forested ecosystems.

Three types of samples were collected at each site: bulk soil samples, cores and root specimens. Bulk samples were collected at vertical intervals within hand-dug sampling pits (~1 m deep) according to soil horizon or depth (depending on how well soil horizons could be distinguished). Where roots were abundant, attempts were made to collect both "rhizosphere" soil taken within 1 to 2 cm of root surfaces and "bulk forest" soil situated away from roots, so that the rhizosphere soil might be compared to soil from outside the root zone. Core samples consisted of roots with their surrounding soil intact around them. At each site, cores were taken at increasing distances from one or two "target trees" that represented the dominant tree species of the area. Core samples were kept upright and refrigerated until processing which began within 48 hours of core extraction. Two sets of root specimens were collected: fine (<2 mm diameter) to medium (2 mm–20 mm diameter) roots cut from the root systems of adult trees, and entire root masses taken from several sizes of representative saplings. Care was taken to preserve, intact, as much of the soil surrounding these roots as possible. Although these root samples were disturbed, many retained either soil masses or soil peds intact around the root bodies. As with the core samples, root specimens were refrigerated immediately after extraction.

Methods used to characterize the mineralogy and chemistry of forest soils, rhizosphere soils (taken in pits within 2 cm of root bodies and shaken from root masses), and rhizoplane soils (soil particles directly contacting the root surface) included x-ray diffraction (XRD) and x-ray fluorescence (XRF) analysis and scanning electron microscopy (SEM) with energy-dispersive spectrometry (EDS).

Results showed that rhizosphere soils displayed several qualities that were distinct from those of bulk forest soils. Interaction between roots and rhizosphere minerals was suggested by several physical, chemical, and mineralogical properties that characterized these root zone soils.

Physical Properties

Physical attributes of rhizosphere minerals that suggested mechanical alteration by adjacent roots included the realignment and bending of phyllosilicates, along with the fracturing of mineral grains. Reorientation of phyllosilicate grains by the mechanical forces exerted by growing roots was suggested by the alignment of the long axis of these grains tangential to the root surface. This tangential arrangement was frequently observed in SEM analyses of both longitudinal and transverse cross sections of root bodies and the surrounding soil (Figure 10.25a,b). Phyllosilicates, apparently bent by adjacent roots, were observed less often (Figure 10.25c). Finally, SEM analyses revealed in numerous cases that the edges of mineral grains oriented toward root surfaces were fragmented to a greater degree than the sides that faced away from the root (Figure 10.25d). Previous studies (Nye and Tinker 1977; Curl and Truelove 1986) noted compaction of rhizosphere minerals around the root but mentioned little regarding physical effects. Cockcroft et al. (1969) and Lund (1965) found that mineral grains in the rhizosphere were only slightly oriented by root growth.

Breakage and alignment of mineral grains are important because they effectively increase the amount of mineral surface area exposed to the weathering regime surrounding the root. Fracturing increases the surface to volume ratio, thereby subjecting more fresh mineral material to the rhizosphere soil conditions. In addition, the tangential alignment of phyllosilicates orients more of the grain surface toward the root where root-induced chemical gradients may accelerate mineral degradation.

Chemical Interactions

Two types of chemical interactions between roots and minerals in the rhizosphere were noted in SEM analyses. Mineral grains displaying an edge that replicated the form of a contiguous root body were interpreted as evidence of preferential mineral dissolution by the root (Figure 10.25e). Such grains were uncommon (in fact, few grains large enough to permit observation of dissolution were found at root surfaces); therefore, although dissolution processes were suggested by this feature, they were not well documented. In addition to the preferential dissolution of mineral grains, precipitation of mineral material within root cells was commonly observed.

Aluminum, silicon, and calcium were identified by SEM/EDS analyses as constituents of the three compounds commonly observed within root cells

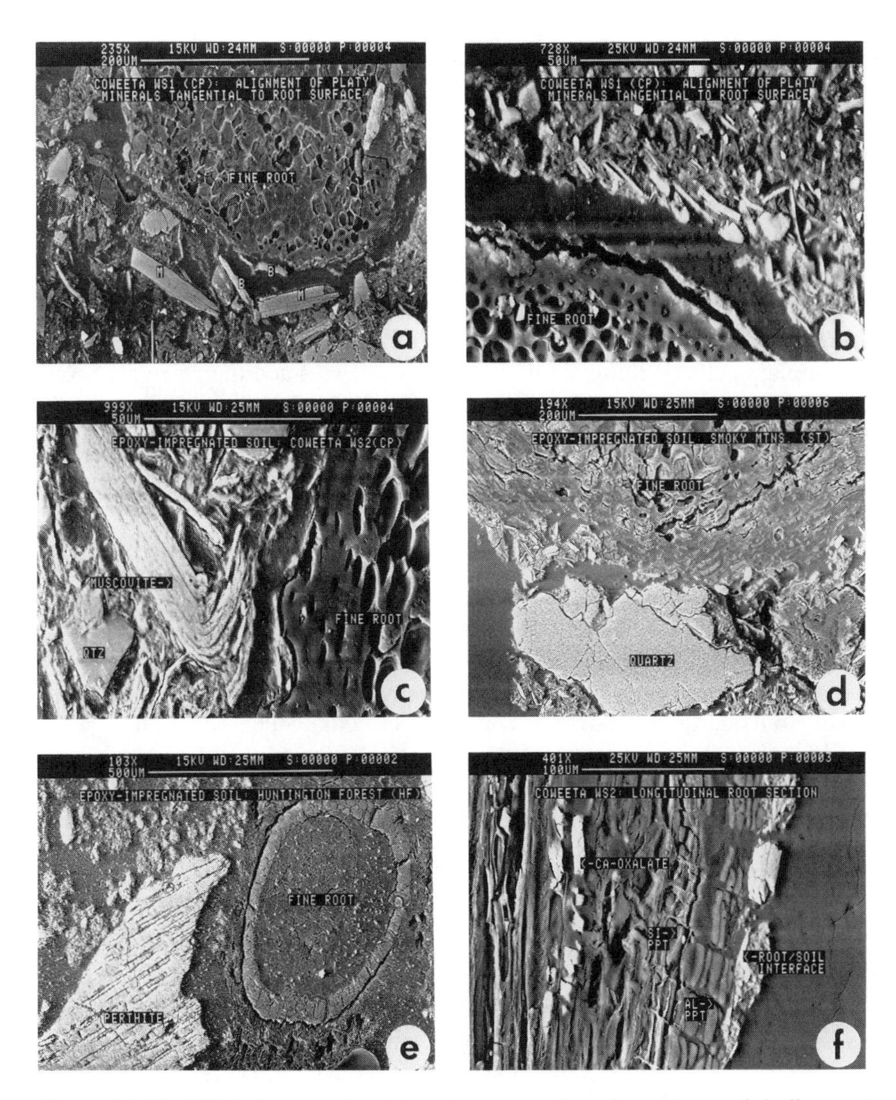

Figure 10.25. SEM photographs: (a) root cross section shows tangential alignment of several silt- and sand-sized muscovite (M) and biotite (B) grains to root surface; (b) finer scale image of root cross section with tangentially aligned phyllosilicates; (c) cross section of fine root next to which silt-sized muscovite grain appears to have been bent by growing root; (d) cross section of fine root with adjacent quartz grain shows better developed fractures at edge abutting root body; (e) shape of adjacent edge of perthite grain replicates root form, suggesting possible preferential dissolution of grain boundary nearest root; (f) longitudinal cross section of fine root contains three cell-filling materials—aluminum precipitate, silica precipitate, and calcium oxalate—within discrete regions of differentiated root cells.

(Figure 10.25f). Each compound was distinguished by the presence of either Al, Si, or Ca. Other elements in each compound were not resolved by SEM/EDS techniques, suggesting that they were lighter than neon, because the EDS system does not detect elements with an atomic number less than 10. Although previous workers have noted the precipitation of aluminum in plant cells, they did not characterize the nature of the mineral species (Huttermann and Ulrich 1984; Cronan et al. 1987; Tepper and Schaedle 1987; Shortle and Smith 1988). The aluminum compounds found within these root cells had no discernible crystal structure and are likely amorphous oxides and hydroxides. Silica gel and opaline silica have been reported as precipitates within plant cells (Wilding et al. 1977; Arnott 1982) and therefore are probably the silica phases present in these root cells. Calcium is precipitated most frequently as calcium oxalate on or in plant material (Graustein et al. 1977; Arnott 1982; Cronan et al. 1987). The morphology of the calcium compound found in the roots of this study corresponds well with documented Ca oxalate habits (Graustein et al. 1977).

These Al, Si, and Ca compounds reside in discrete regions within the cells of mature, differentiated roots (see Figure 10.25f). Aluminous compounds were found in the outermost portion, within the peridermal root cells. This occurrence is consistent with the findings of Cronan et al. (1987) and Tepper and Schaedle (1987), who found aluminum to be restricted largely to cells outside the endodermal zone of root cells. Silica compounds were found precipitated within the phellodermal cells, just inward of the aluminous material. The innermost precipitates noted in these roots were calcium oxalate crystals in cortical parenchymal cells.

Mineralogical Effects

In general, rhizosphere and bulk forest soils contained similar mineral suites; however, two apparent distinctions were noted. For several samples from the Smoky Mountains and Coweeta, the thermal stability of kaolin minerals from the rhizosphere and bulk soils was different (Figure 10.26). In other samples from the Smoky Mountains (tower site; ST), XRD results indicated that rhizoplane soils contained abundant, well-crystallized muscovite and vermiculite, whereas muscovite and degraded biotite were present in rhizosphere soils from the same root masses (Figure 10.27). These differences in clay mineralogy may be related to variations in the intensity of weathering, ionic activities, and the abundance of organic ligands in the rhizoplane, rhizosphere, and bulk forest soil environments.

Summary of Rhizosphere Studies

Preliminary studies of the mineralogy and chemistry of bulk forest soils and rhizosphere soils from six IFS sites in the eastern United States have led to the following observations:

1. Root–soil interactions can change the character of the mineralogy (especially clay mineralogy) in the rhizosphere, suggesting that the pedogenic processes in the root zone differ from those operating in the bulk forest soil.
2. Mechanical effects of the root on adjacent mineral grains thought to expose greater amounts of fresh, mineral material to the weathering regime of the rhizosphere include (a) fracturing in areas of the grain that abutted root bodies, and (b) tangential alignment and bending of phyllosilicate minerals.
3. Preferential dissolution of a few mineral grains was suggested by replication of the shape of the root surface on adjacent grain edges.
4. Precipitation of amorphous aluminum oxides/hydroxides, opaline and amorphous silica, and calcium oxalate was common in the cells of mature root bodies.
5. Greater thermal stability of rhizosphere kaolinite compared to bulk soil kaolin indicates that the former was better crystallized. Mechanisms possible for this enhanced crystallinity include (a) recrystallization caused by a more intensive weathering environment within the rhizosphere, and (b) interference by organic ligands, causing more aluminum to remain in a form that is available for kaolinite formation.
6. Muscovite is abundant within rhizoplane soils sampled in the Smoky Mountains, whereas degraded micas and mixed-layered clays predominate within rhizosphere soils further from root surfaces. Interferences by polyvalent cations, especially aluminum, may cause K enrichment in rhizoplane soil solutions, resulting in less mica weathering at the root surface. Alternatively, biotite may weather preferentially in the root zone microenvironment, leaving muscovite as the predominant mica mineral in rhizoplane soils. The presence of vermiculite in the rhizoplane seems to support the biotite depletion hypothesis.

Summary

In this investigation of the mineralogy and geochemistry of the IFS soils, we determined and characterized the minerals present at each of the sites and identified those regarded most important in influencing the chemical composition of soil water. Stoichiometrically "correct" weathering reactions were formulated for each site using average mineral chemistries of both reactants and products identified in the forest soil. The relative "weathering potential" of the IFS sites was estimated by devising a semiquantitative ranking system using such factors as the percent "reactive" heavy minerals in the soil, the weight percent base oxides, and the base cation and silica flux values obtained from the laboratory column leaching experiment.

Results from this exercise showed that many of the southern sites containing highly weathered soils with small percentages of weatherable min-

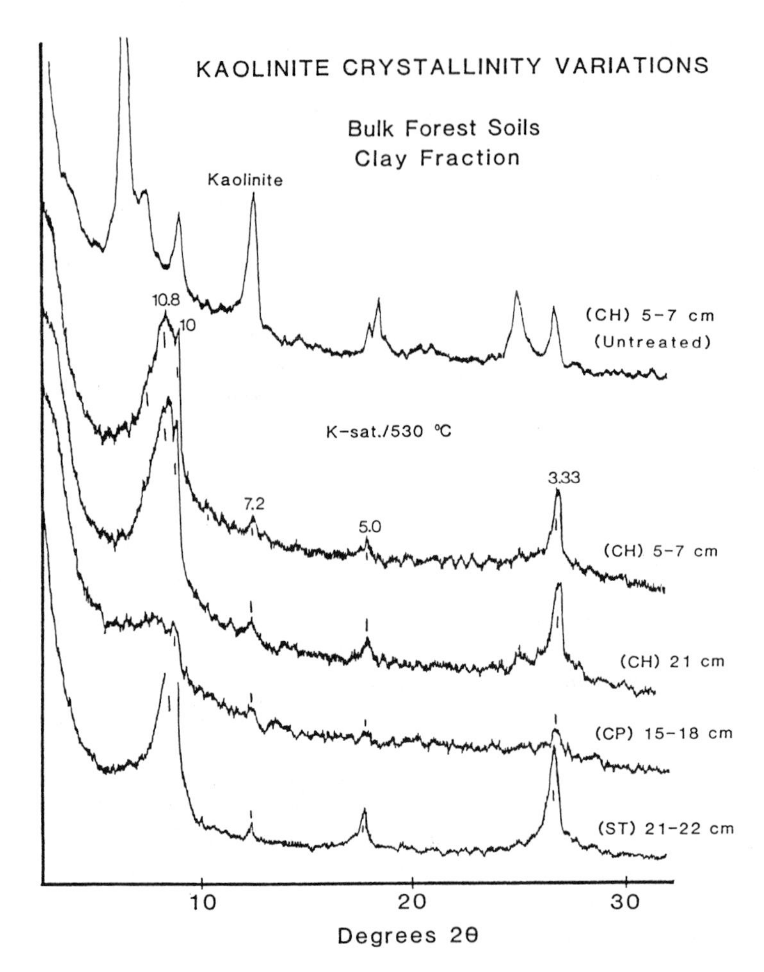

Figure 10.26. (a) X-ray diffractograms of <2-mm fraction of bulk forest soils from different depths at Coweeta and Smoky Mountains (tower site). Comparison of diffractograms for untreated, air-dried samples (top pattern) against samples heated to 530°C (lower patterns) shows marked decrease in (001) 7-Å kaolin peak in latter, suggesting destruction of this clay mineral by low thermal stability. (b) X-ray diffractograms of <2-mm fraction of rhizosphere soils from same depths as seen in (a). Comparison of diffractograms shows little diminishment of (001) 7-Å kaolin peaks, suggesting greater thermal stability for this clay.

erals and large proportions of clay rank low on the scale; the weathering potential of most northern sites mantled by younger, clay-poor but base cation-rich soils that developed on glacial drift is high. (At this writing, we are using the beta version of the NuCM model to simulate mineral weathering reactions at selected IFS sites and to quantify mineral weathering and cation exchange rates and the "weathering potential" rankings given in Table

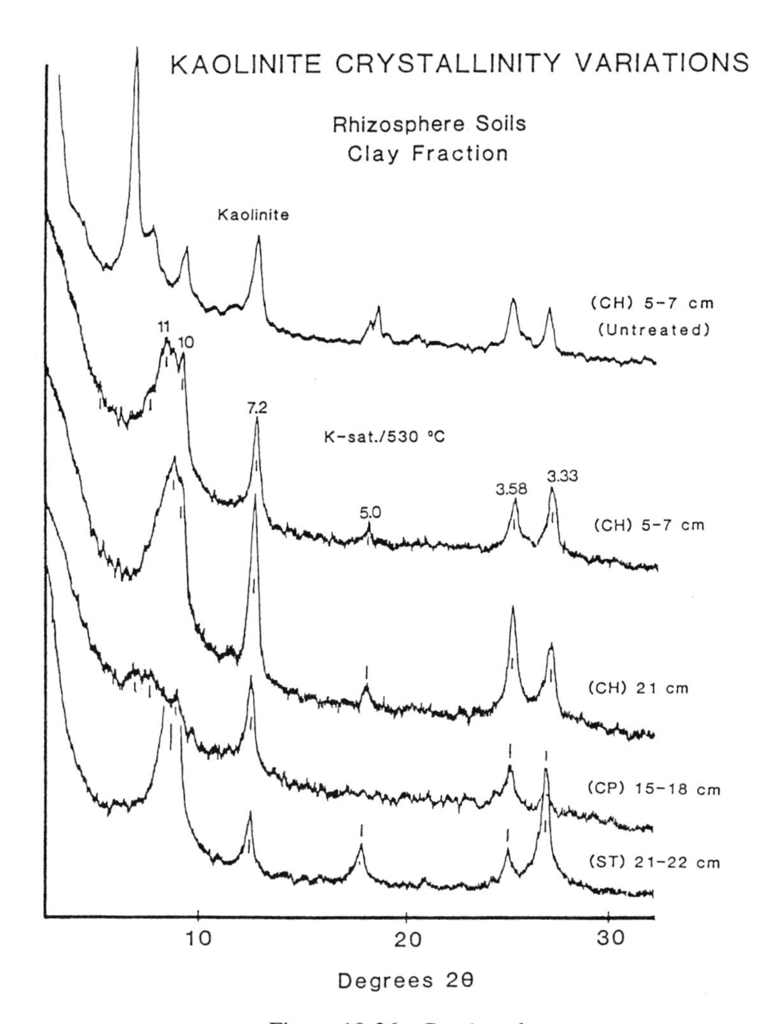

Figure 10.26. *Continued.*

10.6). Results of preliminary mineralogical and geochemical studies of the rhizosphere suggest that pedogenic processes in this zone differ significantly from those operating in the bulk forest soil and deserve more detailed investigations in the future.

Geological materials are the ultimate source of most base cations required for plant growth in natural ecosystems, but the processes and rates by which chemical weathering proceeds in soil environments are still far from being completely understood. In studies of nutrient cycling in forested systems where hydrologic boundaries are often neither well defined nor coincident with the extent of the forested landscape, limited availability of data on hydrology, water fluxes, and input-output budgets may make it difficult to

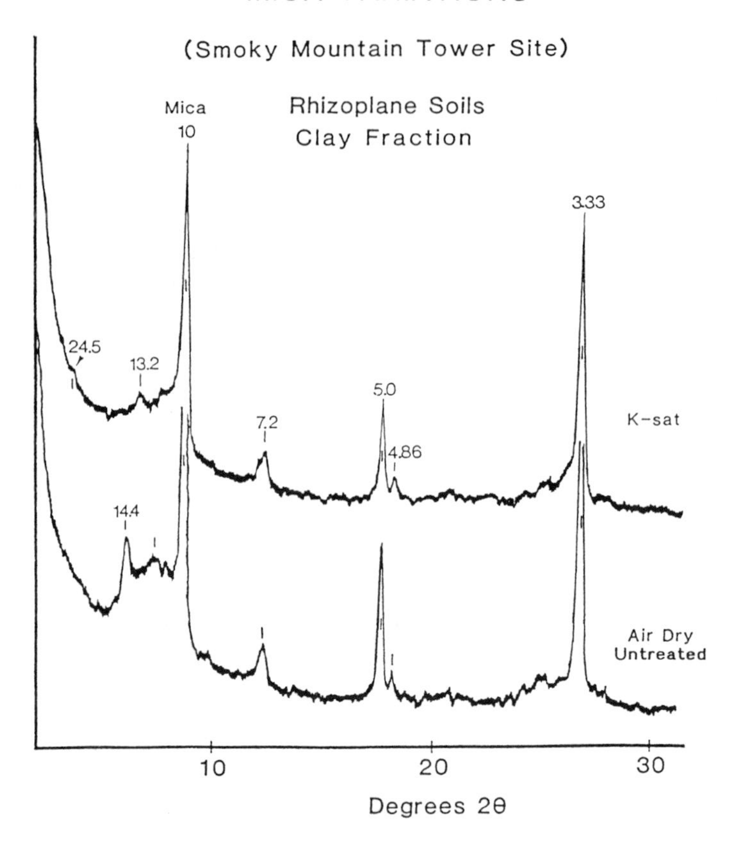

Figure 10.27. (a) X-ray diffractograms of <2-mm fraction of rhizoplane soils (washed from root surfaces) from Smoky Mountains tower site. (001)-10 Å mica peak in diffractogram for untreated clay (lower pattern) is similar to mica peak in artificially potassium-saturated aliquot of same sample (upper pattern). Most vermiculite has collapsed from 14 Å to 10 Å on K saturation. (b) X-ray diffractograms of <2-mm fraction of rhizosphere soils shaken off same root sample as in (a). Diffractogram for untreated clay (lower pattern) shows diffuse band of peaks between 10 and 14-Å, in contrast to defined 10-Å peak in artificially potassium-saturated aliquot of same sample (upper pattern). Clay assemblage includes degraded mica (muscovite and biotite), randomly interstratified mica/vermiculite, and gibbsite.

calculate, with any accuracy and using a mass balance approach, rates of chemical weathering or cation denudation on a catchment or watershed scale.

Indeed, as illustrated in this chapter, other means of estimating weathering and its potential to supply nutrient cations to the soil solution must be used. In so doing, however, results may provide only semiquantitative, relative, or comparative rankings of chemical weathering in soils for the sys-

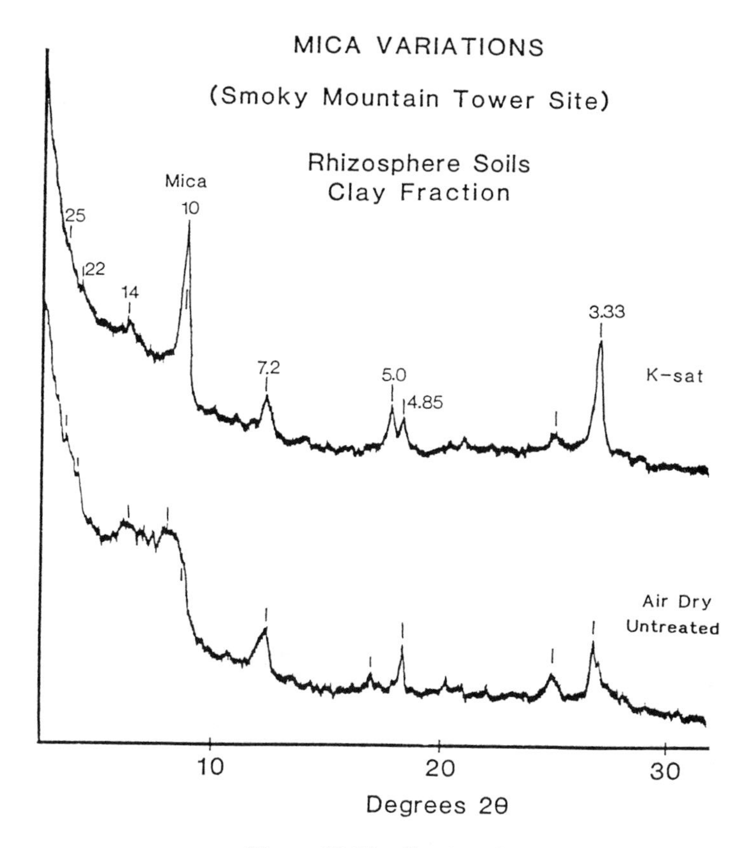

Figure 10.27. *Continued.*

tems under study. Field work followed by laboratory characterization of the physical, mineralogical, and geochemical properties of a soil can provide the necessary database to identify mineral reactants and products of weathering, primary and secondary mineral sources of nutrient base cations, the stoichiometry of mineral weathering reactions, and, with good hydrologic controls on the system, estimates of rates of chemical weathering. If (long-term) hydrologic data are unavailable, laboratory methods such as the column leaching experiment described in this chapter can provide additional useful information for identifying sources and sinks of base cations in soils and the relative importance of cation exchange and mineral weathering in influencing soil water chemistry.

Acknowledgements. Thanks go to Judith Tarplee who assisted in the field and in the laboratory and to Nicole Ruderman, Lisa Hu, Beth Van Schaack, John Figurelli, Veronica Blette and Jill Newton for their assistance with sample preparation, analytical procedures, and preparation of the manuscript. Special gratitude is extended to Dianne Keller whose Master's thesis research at Colgate University contributed greatly to the completion and overall goals of this project, and for her valuable comments and assistance in preparing this manuscript.

References

APHA (American Public Health Association). 1981. Standard Methods for Examinations of Water and Wastewater, 15th Ed. American Public Health Association, Washington, D.C.

April R.H., Hluchy M.M., Newton R.M. 1986. The nature of vermiculite in Adirondack soils and till. Clays Clay Minerals 34(5):549–556.

April R., Keller D. 1990. Mineralogy of the rhizosphere in forest soils of the eastern United States. Biogeochemistry 9:1–18.

Arnott H.J. 1982. Three systems of biomineralization in plants with comments on the associated organic matrix. In Nancollas G.H. (ed.) Biological Mineralization and Demineralization, Dahlem Konferenzen, Berlin, Heidelberg. New York, Springer-Verlag, pp. 199–218.

Barshad I. 1964. Chemistry of soil development. In Bear F.E. (ed.) Chemistry of the Soil. Reinhold, New York, pp. 1–70.

Berner R.A., Holdren G.R. Jr. 1979. Mechanism of feldspar weathering—II. Observations of feldspar from soils. Geochim. Cosmochim. Acta 43: pp. 1173–1186.

Berner R.A., Schott J. 1982. Mechanism of pyroxene and amphibole weathering—II. Observations of soil grains. Am. J. Sci. 282:1214–1231.

Brindley G.W., Zalba P.E., Bethke C.M. 1983. Hydrobiotite, a regular 1:1 interstratification of biotite and vermiculite layers. Am. Miner. 68:420–425.

Brown G. 1980. Associated minerals. In Brindley G.W., Brown G. (eds.) Crystal Structures of Clay Minerals and Their X-Ray Identification. Mineralogical Society, London, pp. 361–411.

Brown G., Brindley G.W. 1980. X-ray diffraction procedures for clay mineral identification: in Brindley G.W., Brown G. (eds.) Crystal Structures of Clay Minerals and their X-ray Identification, Mineralogical Society, London, pp. 305–361.

Cockcroft B., Barley K.P., Greacen E.L. 1969. The penetration of clays by fine probes and root tips. Aust. J. Soil Res. 7:333–348.

Cronan C.S. 1985. Chemical weathering and solution chemistry in acid forest soils: differential influence of soil type, biotic processes, and H^+ deposition. In Drever J.I. (ed.) The Chemistry of Weathering. D. Reidel, Dordrecht, Holland, pp. 175–195.

Cronan C.S., Kelly J.M., Schofield C.L., Goldstein R.A. 1987. Aluminum geochemistry and tree toxicity in forests exposed to acidic deposition. In Proceedings of the International Conference on Acid Rain, Sept. 1–3, 1987, Lisbon, Portugal.

Curl E.A., Truelove B. 1986. The Rhizosphere. Springer-Verlag, Berlin-Heidelberg.

Dahlgren R.A., McAvoy D.C., Driscoll C.T. 1990. Acidification and recovery of a spodosol Bs horizon from acidic deposition. Environ. Sci. Technol. 24, pp. 531–537.

Douglas L.A. 1977. Vermiculites. In Dixon J.B., Weed S.B. (eds.) Minerals in Soil Environments. Soil Science Society of America, Madison, Wisconsin, pp. 259–292.

Folk R.L. 1968. Petrology of Sedimentary Rocks. Hemphills, Austin.

Goldich S.S. 1938. A study in rock weathering, J. Geol. 46:17–58.

Graustein W.C., Cromack K., Sollins P. 1977. Calcium oxalate: occurrence in soils and effect on nutrient and geochemical cycles. Science 198:1252–1254.

Huttermann A., Ulrich B. 1984. Solid phase-solution-root interactions in soils subjected to acid deposition. Philos. Trans. R. Soc. London Ser. B 305:353–368.

Keller D.M. 1988. The Chemistry and Mineralogy of Forest and Rhizosphere Soils in the Eastern United States. M.A. Thesis, Colgate University, Hamilton, New York.

Lund Z.F. 1965. A technique for making thin sections of soil with roots in place. Proc. Soil Sci. Soc. Am. 29:633–635.

Mortland M.M., Lawton K., Uehara G. 1956. Alteration of biotite to vermiculite by plant growth. Soil Sci. 82:477–481.

Nye P.H., Tinker P.B. 1977. Solute Movement in the Soil-Root System. University of California Press, Berkley and Los Angeles.

Pettijohn F.J. 1941. Persistence of heavy minerals and geologic age. J. Geol. 49:610–625.

Shortle W.C., Smith K.T. 1988. Aluminum-induced calcium deficiency syndrome in declining red spruce, Science 240:1017–1018.

Spyridakis D.E., Chesters G., Wilde S.A. 1967. Kaolinization of biotite as a result of coniferous and deciduous seedling growth. Proc. Soil Sci. Soc. Am. 31:203–210.

Tepper H.B., Schaedle M. 1987. Patterns and processes of aluminum uptake and transport. In Proceedings of the International Conference on Acid Rain, Sept. 1–3, 1987, Lisbon, Portugal.

Velbel M.A. 1984. Rate controls during the natural weathering of almandine garnet. Geology 12:631–634.

Walker G.F. 1975. Vermiculites. In Soil Components, Vol. 2: Inorganic Components. Springer-Verlag, New York: 155–189.

Wilding L.P., Smeck N.E., Drees L.R. 1977. Silica in soils: quartz, cristobalite, tridymite, and opal. In Dixon J.B., Weed S.B. (eds.) Minerals in Soil Environments, Soil Sci. Soc. Am., Madison, Wisconsin, pp. 471–553.

11. Processing of Acidic Deposition

Introduction

Deposition of atmospheric acidity has been implicated in negative effects both to vegetation surfaces (e.g., Fairfax and Lepp 1975; Evans et al. 1981; Lovett et al. 1985) and to soil chemistry (e.g., Ulrich 1983; Binkley et al. 1989a). An important goal of the Integrated Forest Studies (IFS) was to monitor important pathways of acid deposition to a wide range of forest types, with consistent protocols over the same period of years. With this information we could examine patterns in nutrient cycling or ecosystem nutrient release across sites in relationship to acid deposition.

The processing of acid deposition within forest ecosystems involves a wide range of potential processes. The free H^+ deposited onto leaf surfaces may exchange with nutrient cations, essentially protonating weak acid anions in the leaf. The weak acids may be either bound within the leaf or wash out into throughfall. In either case, the H^+ is neutralized, but may be released at a later time. If the H^+ passes through the canopy, a variety of possible reactions occur in the soil, including:

- exchange with nutrient cations on the exchange complex
- exchange with aluminum on the exchange complex
- mineral weathering
- consumption in reduction reactions mediated by microbes and plants

- consumption in the oxidation of organic matter containing nutrient cations
- consumption by OH^- released from adsorption of sulfate
- leaching from the soil into aquatic systems (only small quantities follow this path).

Many of these potential reactions are strongly influenced the processing of anions deposited from the atmosphere. For example, deposition of nitrate may lead to nitrate assimilation, which is a reduction reaction that consumes H^+ (Reuss and Johnson 1986; Binkley and Richter 1987).

In addition to H^+ deposited from the atmosphere, natural processes within forest ecosystems also generate H^+:

- accumulation of cation nutrients in biomass
- desorption of adsorbed sulfate
- oxidation reactions mediated by microbes
- decomposition and respiration releases of CO_2 leading to carbonic acid formation.

The net influence of acidic deposition can only be assessed through an accounting of all the major net sources and sinks for H^+ (Binkley and Richter 1987), or by accounting for the net changes in chemicals species (such as bases cations) that are stoichiometrically related to H^+ fluxes (Reuss and Johnson 1986).

The three major sections of this chapter summarize the deposition of acidity at the IFS sites, canopy interactions with acid deposition, and the overall H^+ budgets of the sites.

Atmospheric Deposition of Acids

D.A. Schaefer, P. Conklin, and K. Knoerr

Numerous earlier studies of acid deposition have measured wet deposition alone (Likens et al. 1977; Cole and Rapp 1981; Richter and Lindberg 1988), estimated dry deposition as well (Mayer and Ulrich 1982; Lindberg et al. 1986), or compared deposition estimates with estimates of acidity generated within the ecosystem (van Breeman et al. 1984; Van Miegroet and Cole 1985; Binkley et al. 1989b). To our knowledge, no previous study has attempted to measure both wet and dry deposition of acidity simultaneously to a variety of sites across such a large region.

Using a variety of methods described in related IFS publications (e.g., Lindberg et al. 1989) and briefly described earlier in Chapter 2, wet, dry, and cloud deposition of "strong" (dissociated) and "weak" (undissociated) acidity were estimated to the 12 IFS sites in North America and Norway that collected sufficient meteorological data to model dry deposition. Independent estimates of wet and dry deposition to TL were also available, so that site is treated here as well.

Deposition of Strong Acids

Atmospheric Deposition Pathways of Strong Acidity to the Canopy

H^+ ions are deposited to a forest canopy in several ways. Dissociated acids occur in rain and snow by the condensation of water on acidic nuclei and subsequent scavenging of acid vapors and particles during precipitation (rain and snowfall). Nonprecipitating cloud droplets may deposit directly on canopy surfaces during immersion. Particles and vapors will also accumulate directly on the canopy during precipitation-free periods by diffusion into stomata, surface adsorption, impaction, or sedimentation, depending on the nature of the depositing species.

Vapor-phase deposition of HNO_3 and SO_2 were estimated by the IFS. For both of these species, it was assumed that equivalent fluxes of strong acidity were also deposited on the canopy. For HNO_3 this was simply complete dissociation of the strong acid, while for SO_2, hydration to H_2SO_4 was assumed to precede the analogous dissociation. Results from the TL site are included here, using independent estimates of dry deposition and bulk precipitation (Vet et al., 1988). To the extent that deposition-related conclusions require comparable collection methods, TL would therefore be excluded.

Geographic Patterns of H^+ Deposition

The major H^+ deposition processes to the 13 IFS sites are quantified in Figure 11.1, arranged in decreasing order of total H^+ deposition. Total H^+ deposition varied by a factor of 5 across IFS sites, from slightly more than 300 to nearly 2000 eq ha^{-1} yr^{-1}. Other studies have reported total H^+ deposition values of 300 (H.J. Andrews Forest, Oregon; Sollins et al. 1980), 1600 (Walker Branch, Tennessee; Richter and Lindberg 1988), and as high as 3700 eq ha^{-1} yr^{-1} (Solling Fir Forest, Germany; Matzner et al. 1982).

At IFS sites, precipitation inputs of strong acidity ranged from 27% (ST) to 92% (NS) of total atmospheric deposition, emphasizing that precipitation inputs alone are generally inadequate to characterize atmospheric deposition. For precipitation, there was little evidence of geographic or elevational patterns, other than that the lowest deposition was received by the coastal sites RA, DF, and FS where the air masses affecting these sites came primarily from marine source regions (Figure 11.1). The NS site did not fall in this group, even though it could be regarded as coastal. Air mass trajectory and precipitation studies suggest that southern Norway receives precipitation along with polluted air masses from continental Europe (e.g., Hauhs et al. 1989).

Although the average annual precipitation at the IFS sites (excluding ST) ranged from 79 to 142 cm, the correlation between deposition and the amount of precipitation was not strong (Figure 11.2). However, there was a strong correlation between the H^+ concentration of the precipitation and the amount of deposition (Figure 11.3). Thus the air quality conditions in the source

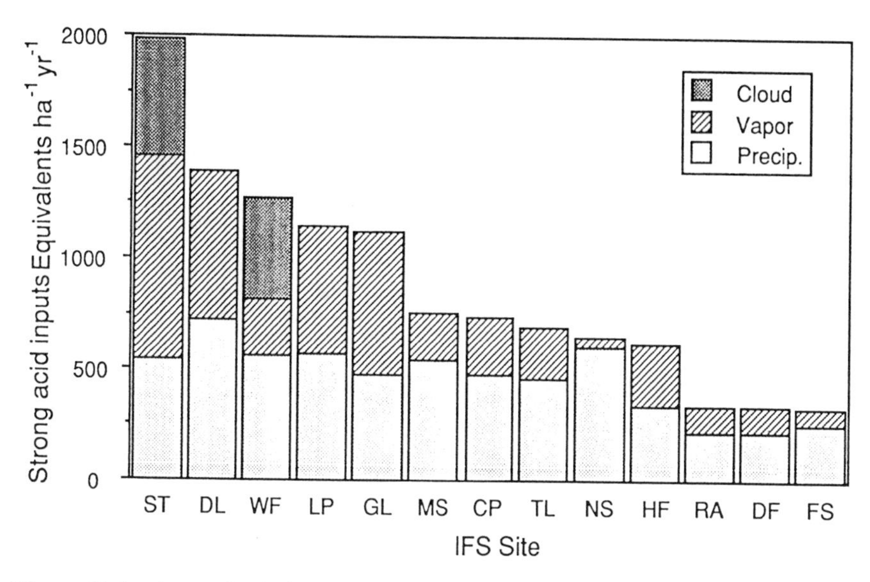

Figure 11.1. Atmospheric input fluxes of strong acidity to 13 IFS sites in equivalents per hectare per year (eq ha^{-1} yr^{-1}) based on multiyear annual averages at most sites and single-year fluxes at FS and MS. Vapor and aerosol fluxes are included in "Vapor;" "Cloud" was measured only at ST, WF, and LP. Sites are ordered according to total atmospheric deposition of strong acidity.

regions for air masses influencing precipitation events has more influence on annual wet deposition than does precipitation amount. Average annual (volume-weighted) pH values for precipitation and throughfall (TF) are summarized in Table 11.1.

For vapor-phase acid deposition (derived from HNO_3 and SO_2), the greatest fluxes were to the southeastern sites (see Figure 11.1). As with wet deposition, a dominant factor is the character of the source regions for the air masses affecting the dry events. It has also been reported that the source regions for dry deposition are generally nearer to the site than for wet deposition (Hicks 1989; Summers and Fricke 1989).

The southeastern United Stated has a high frequency of air masses from high pollution source regions of the Tennessee and Ohio River Valleys and nearby urban regions (King and Vukovich 1981; Vukovich and Fishman 1986; Schwartz 1989). The southeast also exhibits frequent air mass stagnations, increasing the effects of local pollution sources (Albritton et al. 1988).

In the northeastern United States, both air (Galvin et al. 1978; Husain and Samson 1979; Husain et al. 1982; Kelley et al. 1984) and rain concentrations (Raynor and Hayes 1981, 1982a, 1982b) are greater in air masses arriving from the west and southwest. Under other meteorological conditions, acid inputs would be smaller when influenced by "clean" air masses

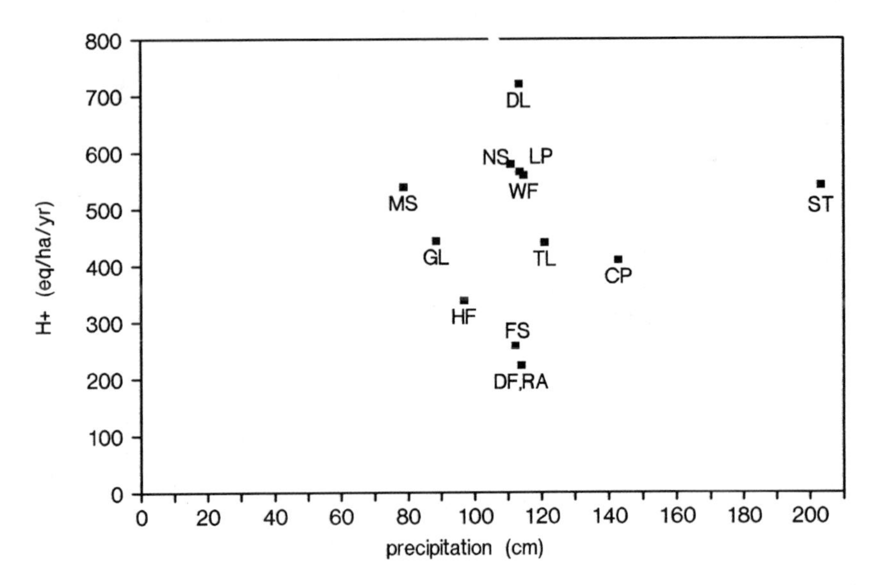

Figure 11.2. Relationship between precipitation fluxes of strong acidity (eq ha^{-1} yr^{-1}) and mean precipitation depth (cm yr^{-1}). Correlation coefficient is 0.1.

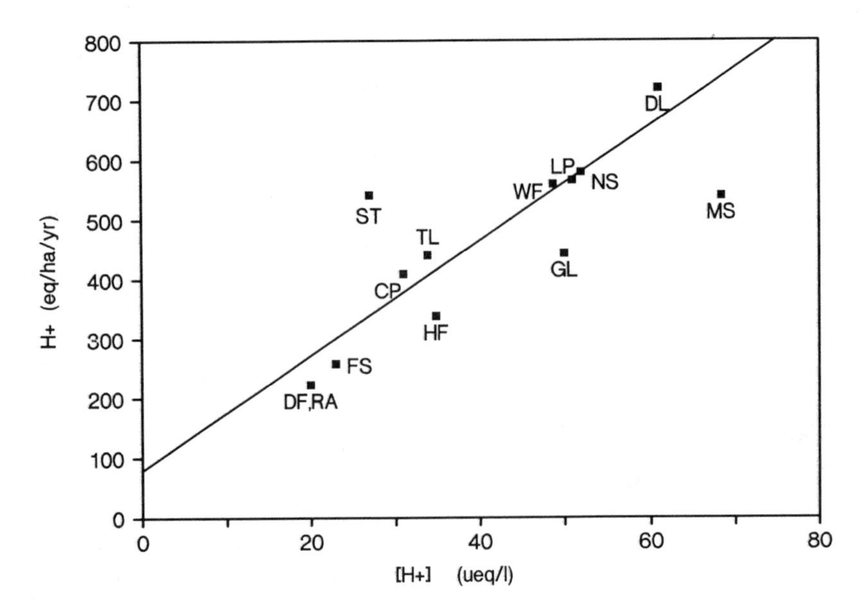

Figure 11.3. Relationship between precipitation fluxes of strong acidity in (eq ha-1 yr^{-1}) and strong acidity (μeq L^{-1}). Correlation coefficient of line is 0.9.

Table 11.1. Cross-Site Comparison of Long-Term Volume-Weighted Precipitation (PPT) and Throughfall (TF) pH values at IFS Sites[a]

Site	PPT pH	TF pH	[HNO_3]	[SO_2]
CP	4.54	4.78	1.8	3.3
DF	4.71	4.64	0.94	1.7
DL	4.20	4.00	2.8	8.9
FS	5.64	5.54	0.82	0.97
GL	4.27	4.44	2.5	5.8
HF	4.46	4.65	1.8	5.8
LP	4.30	4.24	2.9	12
MS	5.16	5.12	0.82	4.8
NS	4.28	4.49	1.4	1.3
RA	4.71	4.86	0.94	1.7
TL	4.44	4.89	nd[b]	nd[b]
ST	4.58	3.89	2.4	3.9
WF	4.31	4.14	1.2	3.2

[a] Long-term time-weighted concentrations of HNO_3 and SO_2 are presented as micrograms (N or S) m^{-3}. Data represent monitoring durations of 1 (FS, MS, TL), 2 (DF, DL, GL, HF, NS, RA, WF), or 3 (CP, LP, ST) years at these sites. Complete concentration and flux data are compiled in the Appendix.

[b] Vapor-phase concentrations were not determined at the Turkey Lakes site.

from Canada. For the coastal sites (Pacific Northwest, Florida, and Norway), the air masses during dry periods may come predominantly from marine sources.

For cloud deposition, only the high-elevation sites (ST and WF) had substantial inputs (27 and 36% of total deposition of strong acidity, respectively; see Figure 11.1). Detailed fog deposition measurements were necessary at only one low-elevation site (LP) as fog was infrequent at all other sites. Nocturnal radiation fog frequently occurred at LP, but it was an insignificant contribution to the total acid (and other chemical) deposition there (Figure 11.1). The Appendix summarizes annual average fluxes of strong acidity in wet, dry, and cloud deposition.

Chemical Patterns of H^+ Deposition

As shown in Figure 11.4, for some of the sites the precipitation equivalence ratio of H^+ to (SO_4^{2-} + NO_3^-) was almost 1, a near balance between the strong acidity and the sulfate plus nitrate ions. However for other sites (e.g., FS, TL), this ratio was considerably less than 1. In such cases the precipitation acidity was partially neutralized by basic aerosols, or some SO_4^{2-} and NO_3^- came from nonacidic sources. As described in Chapter 5, sea salt was a substantial contributor to sulfate at the coastal sites. The SO_4^{2-}:NO_3^- equivalence ratio ranged from 1.5 to 2.9, with the largest ratios in the southeastern region, because of the higher SO_2 concentrations in that region (described in previous chapters).

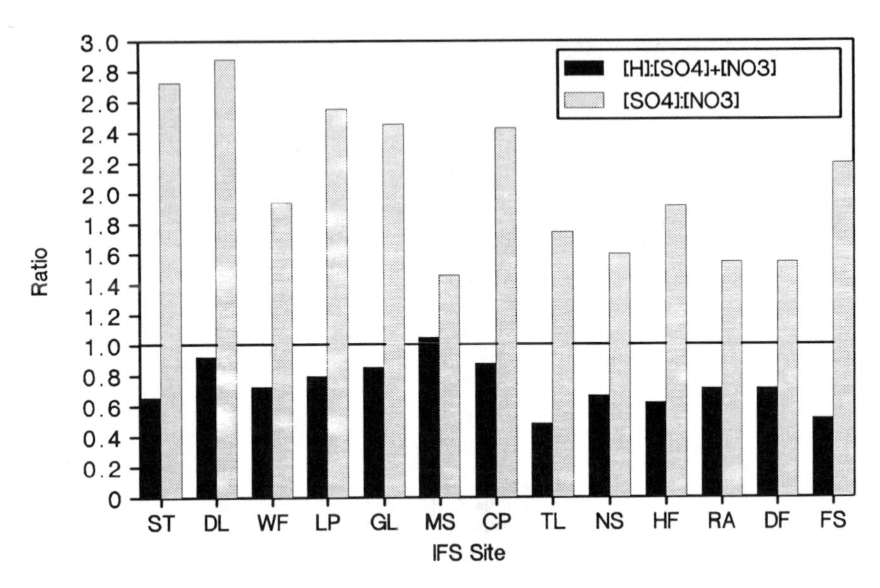

Figure 11.4. Equivalent ratios of strong acidity against [SO_4^{2-} + NO_3^-] (solid bars), and SO_4^{2}:NO_3^- equivalent ratios (stippled bars). Value of 1 for H^+:[SO_4^{2-} + NO_3^-] ratio suggests that all precipitation strong acidity is associated with sulfur and nitrogen oxides. Smaller values suggest either neutralization or nonacidic sources of SO_4^{2}- or NO_3^-. SO_4^{2}-:NO_3^- ratio compares sulfur and nitrogen oxides as sources of precipitation strong acidity.

The different contributions to dry deposition are compared in Figure 11.5. Sites are ordered according to acid deposition totals as in Figure 11.1. Across these IFS sites, HNO_3 contributed an average of 58% of the dry-deposited strong acidity (ranging from 31% to 68%, or 50 to 600 eq ha^{-1} yr^{-1}), and SO_2 an average of 39% of the dry-deposited strong acidity (ranging from 28% to 54%, or 50 to 350 eq ha^{-}1 yr^{-1}).

As was the case with SO_4^{2-} in precipitation, the largest SO_2 contributions to dry-deposited acidity were found at the Southeastern sites ST, DL, LP, and GL (see Figure 11.5), accompanying the higher SO_2 concentrations in that region (see previous chapters). While SO_2 air concentrations generally exceeded those of HNO_3 (see Table 11.1), the greater deposition of HNO_3 reflected the far greater deposition velocity of HNO_3.

Fine (impacting) aerosols contributed only 3% on average to dry-deposited acidity (see Figure 11.5) and might not merit careful attention in further studies of acid deposition in rural environments. Acidic aerosols, however, can be quite important in urban environments and for human health effects. Coarse (sedimenting) particles were not found to be a source of deposited acidity. In fact, an anion deficit was generally observed in analyses of the coarse particle extracts and was treated as HCO_3^- alkalinity (see next section).

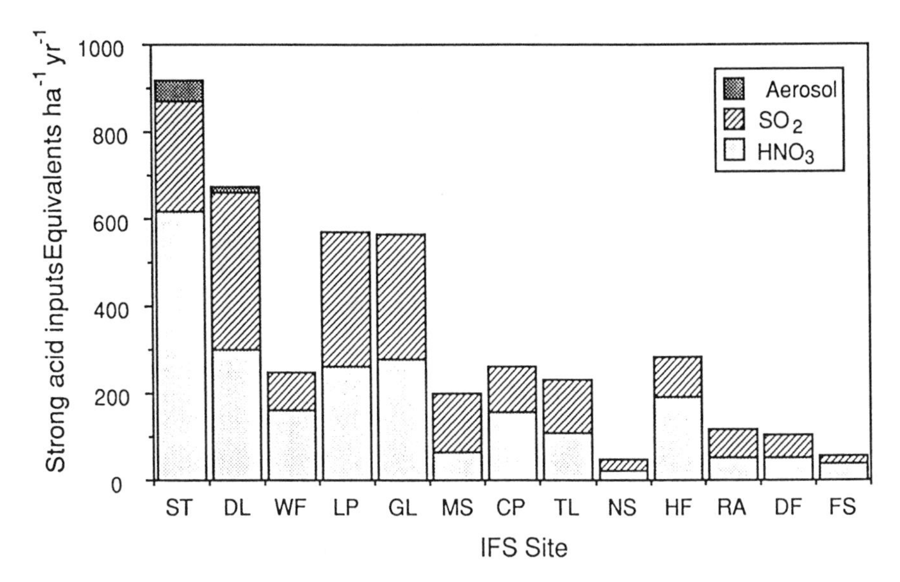

Figure 11.5. Contributions to total dry deposition of impacting (fine) aerosol acidity, H^+ from SO_2 (following hydration), and H^+ from HNO_3 vapor deposition at IFS sites (eq ha^{-1} yr^{-1}).

Dry deposition totals fell in about the same order as total deposition, except for relatively low dry deposition at WF, MS, NS, and possibly FS (see Figure 11.5). Those sites have already been characterized as receiving clean marine or arctic air, at least during dry periods. At sites with the highest acid inputs, dry deposition appears to contribute a greater fraction of the total (Figure 11.6). At low total H^+ deposition, the dry deposition fraction is 20% to 30% of total, increasing to 50% to 55% of the total at high deposition sites.

We attribute this general pattern to two factors. First, there are likely differences in air mass source regions of the IFS sites. The lower latitude southeastern sites are proximate to polluted air mass source regions, and these sites received the greatest dry fraction and the maximum total deposition. The higher latitude northeastern sites and the coastal sites may have had less polluted air mass source regions, particularly during dry periods. The NS and FS sites show particularly low ratios on Figure 11.6.

Second, dry deposition should be a relatively less effective acid input to sites remote from pollutant sources. In remote regions, pollutant concentrations should be vertically well mixed throughout the lower troposphere. Under those conditions, wet deposition, with the potential to extract pollutants from hundreds to thousands of vertical meters, should be much more effective than dry deposition, only extracting pollutants from tens of vertical meters. This is consistent with the role of dry deposition as an effective near-

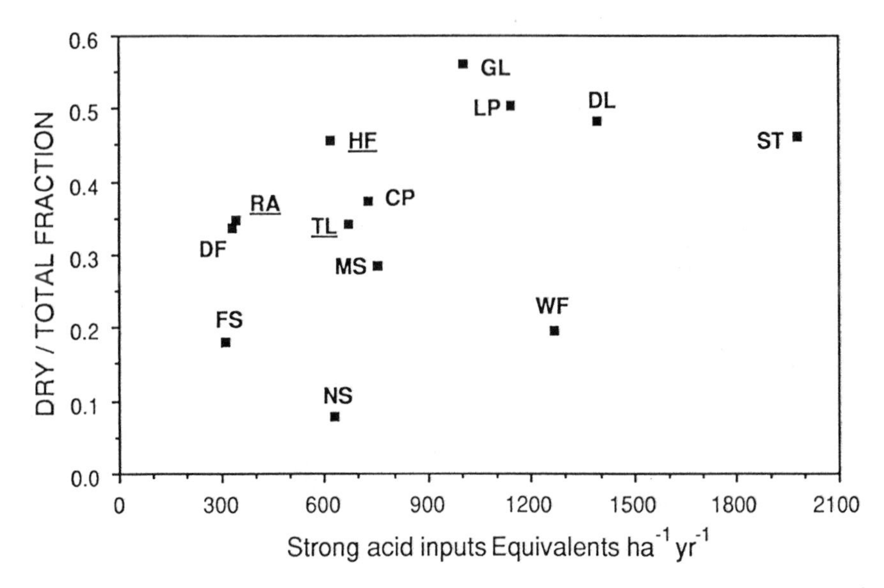

Figure 11.6. Relationship between total atmospheric strong acid deposition (eq ha^{-1} yr^{-1}) and dry deposition fraction of that total at IFS sites. Dry deposition fraction appears to be smaller at lower deposition sites. Underlined points represent deciduous canopies.

field phenomenon (Hicks 1989), occurring before ground-level emissions are mixed far upward in the troposphere. Both because of air mass source region and troposphere mixing, the sites with the least total acidic deposition had the smallest dry to total ratios.

Canopy Net Balance for Strong Acidity

A comparison of the canopy inputs (wet, dry, and cloud deposition) to canopy outputs (throughfall and stemflow; TF and SF) yields a canopy net balance for strong acidity. Unlike other cations, the disappearance of an H^+ ion from the TF and SF flux can result from biological retention or from the protonation of fixed or mobile anions. Here we use the term TF and SF neutralization for the net effect of these removal processes. TF and SF acidification refer to H^+ release from foliage and the dissociation of acids released from the canopy. The next section examines the details of the canopy processing mechanisms.

Figure 11.7 compares wet-only deposition (precipitation plus cloud water flux estimates for WF and ST) with TF and SF outputs. The distance below the 1:1 line represents the degree of TF and SF neutralization by the canopy, while the distance above the line indicates the amount of TF and SF acidification. Forest canopies with low inputs (<600 eq ha^{-1} yr^{-1}) tend to appear balanced or to neutralize a portion of the wet-deposited acidity, while higher

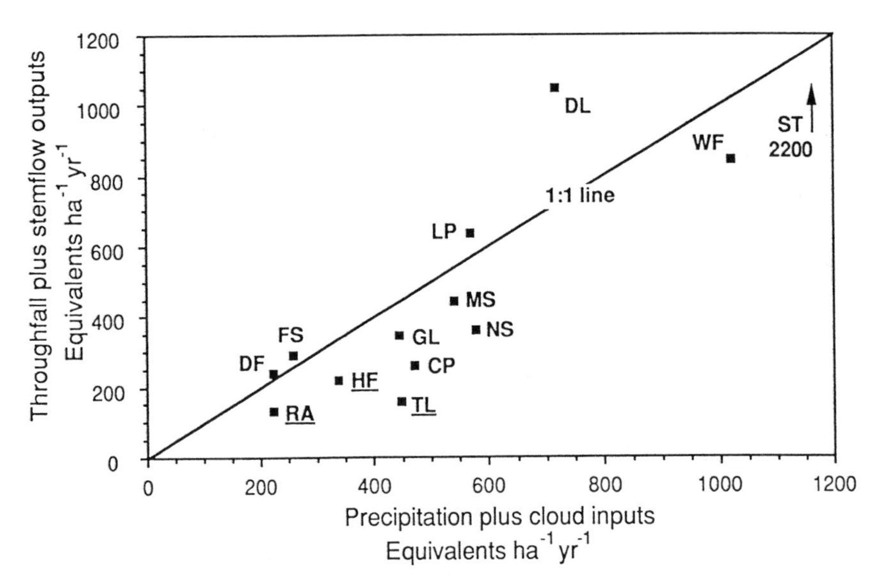

Figure 11.7. Comparison of wet (rain plus cloud) strong acid deposition to stemflow plus throughfall outputs at IFS sites (eq ha^{-1} yr^{-1}). Position of ST on output axis should be well above arrow (2200 eq ha^{-1} yr^{-1}). Sites above 1:1 line acidify wet deposition inputs; those below line neutralize those inputs. Underlined points represent deciduous canopies.

deposition led to more variable results. The deciduous canopies (RA, HF and TL; underlined) show the smallest TF and SF outputs of strong acidity.

By adding dry deposition to the wet deposition considered previously, Figure 11.8 shows the overall effect of the canopy on total strong acid deposition. The vertical distance from the 1:1 line here represents the net canopy exchange of H^+ (NCE, defined as TF plus SF outputs minus total deposition inputs) with positive NCE (acidification) above the 1:1 line and negative NCE (neutralization) below. Net neutralization of deposited acidity occurs in all but the very high input ST canopy. Within the IFS, this canopy was unique in its net release of strong acidity. Possible explanations for this are considered in other portions of this chapter.

The relationship between NCE and acid deposition by vapors in Figure 11.9 suggests that the acid neutralized by the canopy is primarily from vapor deposition. Most sites are close to the 1:1 line, representing total neutralization of deposited vapor. A regression line fitted to all but the ST site shows a significant fraction of the vapor deposition is neutralized in the canopy. Regressions for HNO_3 and SO_2 individually (Figures 11.10 and 11.11; respectively) show similar linear relationships between deposition and neutralization. In Figures 11.10 and 11.11, slopes of the regressions are both more negative than -1, showing that neutralization exceeded the deposition of either SO_2 or HNO_3 individually.

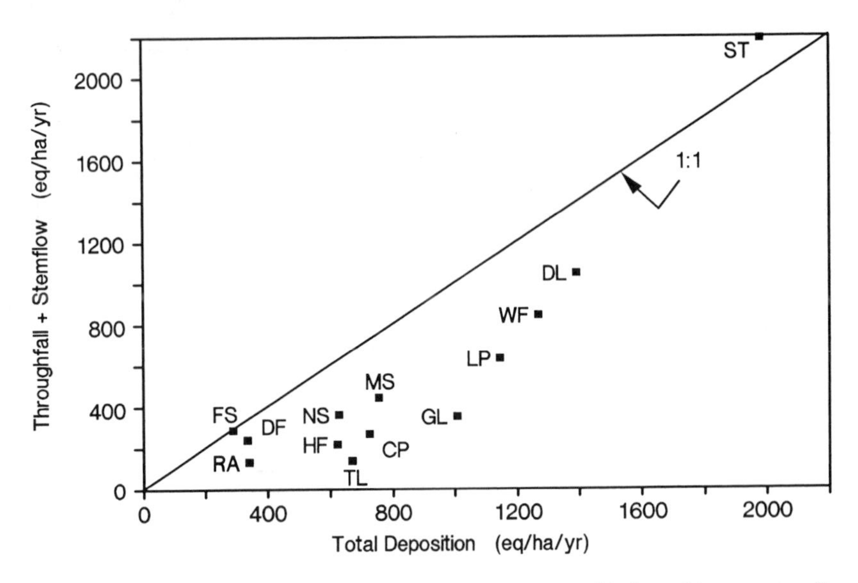

Figure 11.8. Comparison of total (wet + dry) strong acid deposition to stemflow plus throughfall outputs at IFS sites (eq ha^{-1} yr^{-1}). Sites (see Chapter 2) below 1:1 line neutralize total deposition inputs.

Preferential canopy retention of dry-deposited forms of N was suggested by Lindberg et al. (1986). Vapor deposition of acidity is more likely to be neutralized by forest canopies than is wet deposition for three reasons. (1) Dry-deposited acids remain in contact with canopy surfaces for much longer than precipitation-borne acids do, because dry interstorm intervals are much longer than precipitation events. By the same token, dry-deposited alkalinity (primarily from coarse particles) can neutralize acidity during the entire dry period (Lindberg et al. 1990). (2) Substomatal deposition of vapors (the primary pathway for SO_2 and secondary for HNO_3) allow direct contact between living cells and the acidity. Wet deposition is in contact only with cuticularized and suberized canopy surfaces (Schaefer and Reiners 1990). (3) Wet-deposited acidity may not completely contact canopy surfaces, but dry acidity almost certainly does. If crown closure is incomplete, a fraction of precipitation falls through the canopy without contacting any canopy surface. Dry-deposited acidity for the most part contacts canopy surfaces, because dry deposition in forest clearings is generally reduced compared to the same area of forest. Further, waterborne acidity is dissolved in droplets of water with incomplete contact with the canopy, because of contact angles on hydrophobic surfaces (e.g., Cape 1983). If these droplets are not well mixed, then the acid contained there may not have the potential to interact chemically with canopy surfaces before the droplets fall.

Figure 11.12 illustrates the canopy processing of acid inputs as a function of total acid deposition. Again, negative values of net canopy effect imply

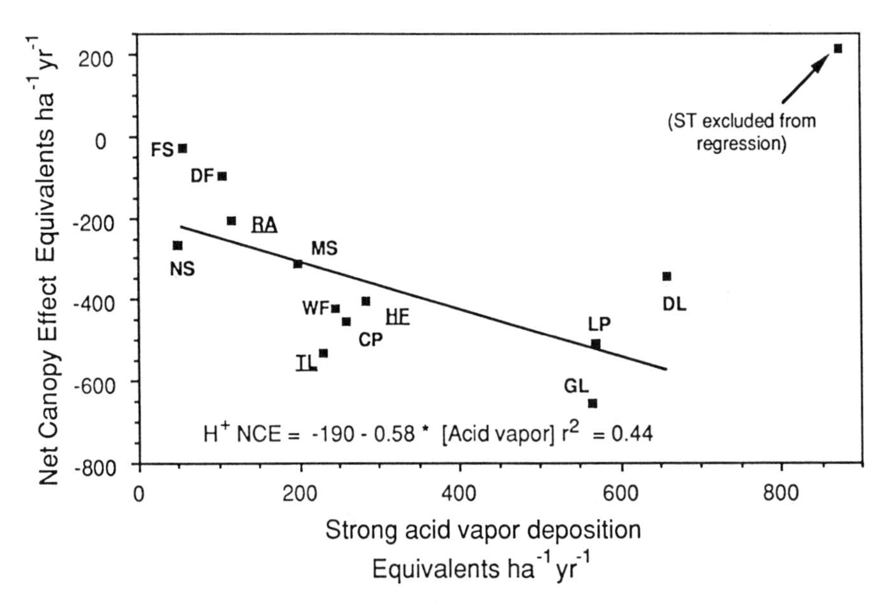

Figure 11.9. Comparison of strong acid vapor (SO_2 and HNO_3) deposition to net canopy effect (NCE; TF + SF outputs minus total deposition inputs) at IFS sites (eq ha^{-1} yr^{-1}). Regression is based on sites other than ST (n = 12). Underlined points represent deciduous canopies, appearing on both sides of regression line.

neutralization of inputs. The IFS sites occupy a broad band of possible canopy responses. With low inputs, there is minimal neutralization, or even slight acidification as found in the IFS high-pH leaf rinsing experiments (described in Chapter 8). As total acid deposition increases, we might expect forest canopies in general to neutralize a constant fraction of the total deposition up to an apparent limit of neutralization. This neutralization limit (negative NCE) appears from our data to be 500 to 700 eq ha^{-1} yr^{-1}. If acid loadings were to increase, it is uncertain whether neutralization could also increase, but it is likely that the dry deposition would be neutralized to a greater extent than the wet deposition.

The range of possible responses for any canopy is set both by the acid inputs and the canopy itself. For instance, at any level of acid deposition the hardwood canopies (underlined in Figure 11.12) neutralize the greatest fraction of deposition (these points are closest to a 1:1 line of complete neutralization). The sites with the lowest dry to wet deposition ratios (overscored in Figure 11.12) neutralize the smallest fraction of acid deposition, again suggesting inefficient neutralization of wet deposition during its relatively short residence on canopy surfaces.

It is important to note the effect of potential errors in acid deposition measurement and modeling on the location of data points on Figure 11.12. Correction of acid deposition overestimates would move the point diagonally

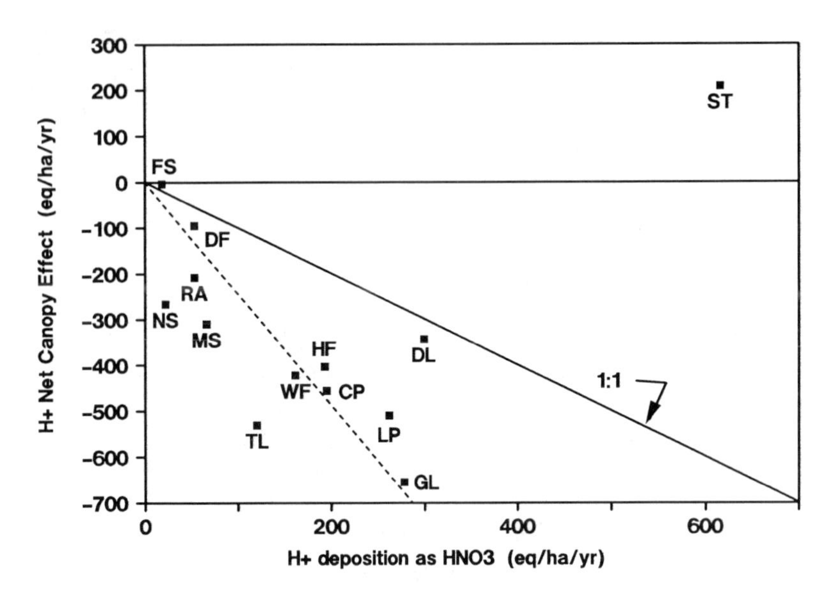

Figure 11.10. Relationship of HNO_3 acid deposition to net canopy effect (NCE; TF + SF outputs minus total deposition inputs) at IFS sites (eq ha^{-1} yr^{-1}). Regression (dotted line) shown is based on all sites except ST (n = 12). All 12 sites fall below 1:1 (solid) line, indicating that strong acid NCE exceeds HNO_3 deposition.

up and to the left, as both inputs and NCE would be overestimated. Correction of acid deposition underestimates would move the point diagonally down and to the right. We regard both dry and cloud deposition to be subject to a greater degree of uncertainty than precipitation acid fluxes (Lindberg and Johnson 1989, and earlier chapters in this volume). Errors in TF + SF acid flux measurements would move a point vertically, but these fluxes are among the most accurately measured in the IFS. Canopy sources of strong acidity would also raise a point vertically. Given these patterns, it does not seem plausible that the observed "saturation" of acid neutralization capacity by acid deposition could be caused by systematic errors in any deposition measurements. DL would move downward and to the right (approaching the asymptotic neutralization discussed earlier) if deposition had been underestimated. Moving ST in the same manner to an NCE of -500 eq ha^{-1} yr^{-1} would require alteration of deposition estimates far in excess of the uncertainty estimated by Lindberg and Johnson (1989).

Deposition of Undissociated Acids

Inputs of undissociated acidity were estimated as well as strong (dissociated) acidity, so that forest canopy contributions of these compounds and total canopy acid balance could be assessed. Canopy cation exchange calculations also utilize net canopy fluxes of undissociated acidity.

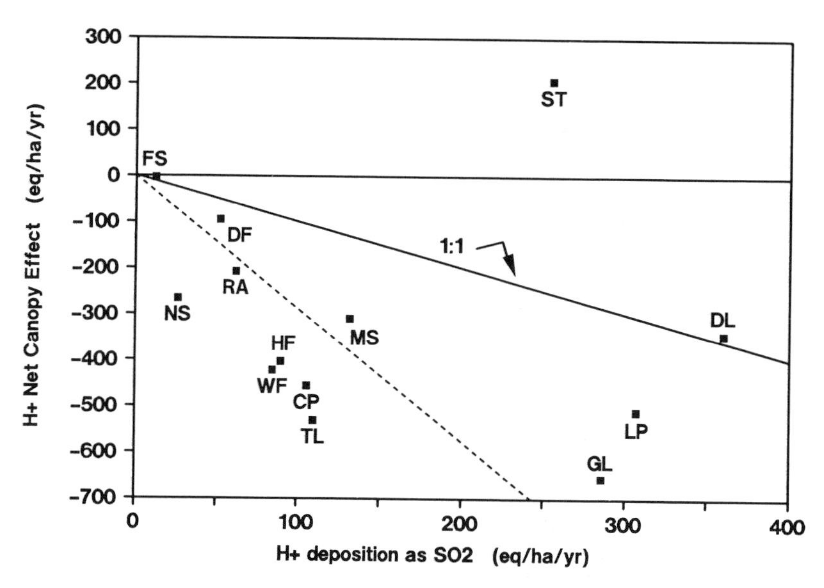

Figure 11.11. Relationship of SO_2 acid deposition to net canopy effect (NCE; TF + SF outputs minus total deposition inputs) at IFS sites (eq ha^{-1} yr^{-1}). Regression (dotted line) shown is based on all sites except ST (n = 12). Most sites fall below 1:1 (solid) line, indicating that strong acid NCE generally exceeds SO_2 deposition.

Estimation of Fluxes

Experimental methods for undissociated acid flux determinations (Schaefer et al. 1989) are not described in Chapter 2 on protocols, so a brief summary is presented here. Precipitation, fog/cloud, SF, and TF fluxes for the IFS sites were estimated as follows. Samples from a number of wet events were preserved with 0.5% $CHCl_3$ and shipped to Oak Ridge National Laboratory (ORNL) for Gran's plot titrations. Ten sites submitted precipitation, TF, SF, and fog/cloud water (ST, WF, and LP only) samples for titration. Samples titrated from the DF, LP, NS, RA, and ST sites represented essentially complete annual fluxes. We expect that the uncertainty in estimating these fluxes should increase with decreasing sample coverage. Samples representing 10% to 50% of the annual fluxes were titrated from the CP, HF and WF sites. Samples representing less than 10% of the annual fluxes were titrated from the DL and GL sites.

For each participating site, undissociated and strong acidities were averaged for each sample type and weighted according to water flux of the individual events. Strong acid fluxes of precipitation, cloud deposition, TF, and SF were reported by individual site operators as described earlier. Ratios of undissociated to strong acidity for each sample type were multiplied by the strong acid fluxes to estimate fluxes of undissociated acidity by each pathway. These data are summarized in Table 11.2. The undissociated to

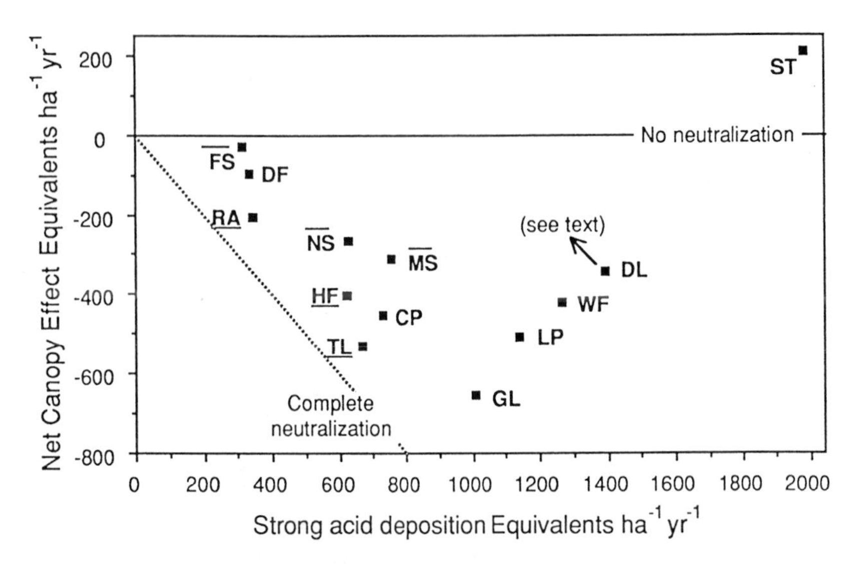

Figure 11.12. Relationship of total strong acid deposition to net canopy effect (NCE; TF + SF outputs minus total deposition inputs) at IFS sites (eq ha^{-1} yr^{-1}). Dotted line indicates complete neutralization of strong acidity by canopy; solid line represents no neutralization. Underscored points represent deciduous canopies (with nearly complete neutralization); overscored points represent sites with low ratios of dry to wet deposition (with little neutralization).

strong acid ratios were used because the subset of events submitted from each site was not chosen to be statistically representative of distributions of either event depth or pH. Studies of LP rain, TF, and SF indicated that undissociated and strong acidity were reasonably well correlated over a wide range of event depths (D.A. Schaefer, unpublished results).

Dry deposition of undissociated acids was anticipated to be a minor input (Galloway et al. 1976) and was not measured independently in the IFS. Because TF and SF fluxes of undissociated acidity cannot be completely attributed to forest canopy sources (Schaefer et al. 1989), atmospheric deposition estimates are required to determine the net canopy effect. Canopy acid balance was estimated for each participating site according to the following equations:

$$\text{Total Input} = \text{Strong}_{\text{dry}} + \text{Strong}_{\text{ppt.}} + \text{Strong}_{\text{cloud}} + \text{Undissociated}_{\text{ppt.}} + \text{Undissociated}_{\text{cloud}}$$

$$\text{Total Output} = \text{Strong}_{\text{TF}} + \text{Strong}_{\text{SF}} + \text{Undissociated}_{\text{TF}} + \text{Undissociated}_{\text{SF}}$$

$$\text{Net Canopy Exchange (NCE)} = \text{Total Output} - \text{Total Input}$$

Table 11.2. Cross-Site Comparison of Undissociated (HA) and Strong (S) Acid Concentrations in Precipitation (PPT) and Throughfall (TF) Titration Samples Collected at 10 IFS Sites and Cloud/Fog Water (C) from 3 Sites[a]

Site	PPT-S	PPT-HA:S	PPT-HA	PPT-TOT	TF-H^+	TF-HA:S	TF-HA	TF-TOT
CP	457	0.45	208	665	244	1.53	372	616
DF	223	0.68	152	375	224	1.54	345	569
DL	720	0.09	66	786	1004	0.14	145	1149
FS[b]	258	—	—	—	279	—	—	—
GL[c]	443	1.36	605	1048	285	2.28	649	934
HF	338	0.53	180	518	218	0.95	206	424
LP	566	0.71	399	965	632	1.43	904	1536
MS[b]	539	—	—	—	444	—	—	—
NS	579	0.49	284	863	362	1.13	410	772
RA	223	0.68	152	375	134	2.64	353	487
ST	541	0.67	363	904	2157	0.67	1443	3600
TL[b]	440	—	—	—	139	—	—	—
WF	560	0.30	169	729	844	0.50	419	1263

Site		CH-A	C-S	C-HA/S
LP		55	36	1.5
ST	130	220	0.59	
WF	45	199	0.43	

[a] Undissociated:strong acid ratios (U:S) from titrated samples were used to estimate undissociated acid fluxes at each site as described in text. Units, $\mu eq\ L^{-1}$, except for ratios. TOT, totals.

[b] Only four precipitation events were titrated.

[c] Only six precipitation events were titrated.

Deposition Pathways (Canopy Inputs)

As mentioned previously, dry deposition of undissociated acidity was presumed to be negligible for this study. Total acid wet deposition fluxes from the atmosphere have previously been reported for the LP and ST sites (Schaefer et al. 1989) and for nearby Tennessee forests (Hoffman et al. 1980). Very few other measurements of undissociated acidity inputs have been reported, and the chemical composition of that acidity remains unclear. Most reports have been based on ion chromatographic analyses of short-chain aliphatic acids [see Keene and Galloway (1988) for review].

The inputs of undissociated acidity were one-third to one-half of the strong acid inputs in precipitation (compare Figures 11.1 and 11.13). In Figure 11.13, the sites are shown in the same order as in Figure 11.1, with FS, MS, and TL deleted because undissociated acidity titrations were not performed for those sites. Undissociated acidity followed the same pattern as the strong acidity fluxes except at the DL and GL sites. For DL and GL, samples from less than 10 precipitation events were titrated for undissociated acidity, so those two undissociated acidity estimates may not accurately reflect the actual fluxes. At the two montane sites (ST and WF), cloud deposition added as much undissociated acidity as it did strong acidity (compare Figures 11.1 and 11.13).

Acid Loss Pathways (Canopy Outputs)

Both strong and undissociated acidity are transferred from the forest canopy to the forest floor in TF and SF. Stemflow did not exceed 5% of the TF

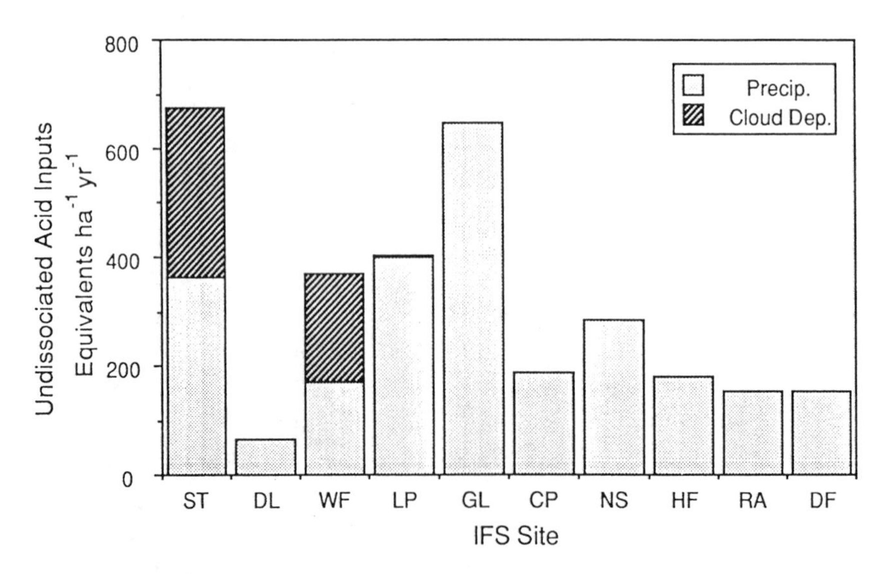

Figure 11.13. Input of undissociated acidity to 10 IFS sites (eq ha^{-1} yr^{-1}). Rain plus snow fluxes (Precip.) and cloud deposition fluxes (Cloud Dep.) were only inputs measured.

flux of either strong or undissociated acidity at any of the IFS sites, and those fluxes are treated together here.

Canopy outputs of undissociated acidity ranged from 14% to 64% of the total acidity outputs, with a median value of 47% (Figure 11.14). The largest canopy outputs of undissociated acidity were measured at ST and LP, where samples were available for titration most rapidly after collection. This raises the possibility that samples from the other sites were not preserved quickly enough to prevent deterioration of the undissociated acidity prior to analyses.

Undissociated acidity in TF and SF substantially exceeded the input fluxes (Figures 11.13 and 11.14). At the same time, total acid input and output fluxes from the canopies were quite similar (Figure 11.15). Here, as reported by Hoffman et al. (1980), forest canopies "exchanged" strong acidity for undissociated acidity, without necessarily changing the magnitude of the total acid flux to the forest floor. The magnitude of this process did not appear to vary with the strong acid input flux across the IFS canopies.

Both strong and undissociated acidity were released by the canopy at ST, possibly as a result of in-canopy decomposition of biological material in this epiphyte-laden forest (Schaefer et al. 1989), or because atmospheric (specifically cloud water) deposition may have been underestimated (Lindberg and Johnson 1989).

Total acidity inputs to some IFS forest canopies exceeded the outputs (see Figure 11.15). This indicates that strong acidity is retained in those canopies in addition to the protonation of acids released by canopies. The exchange

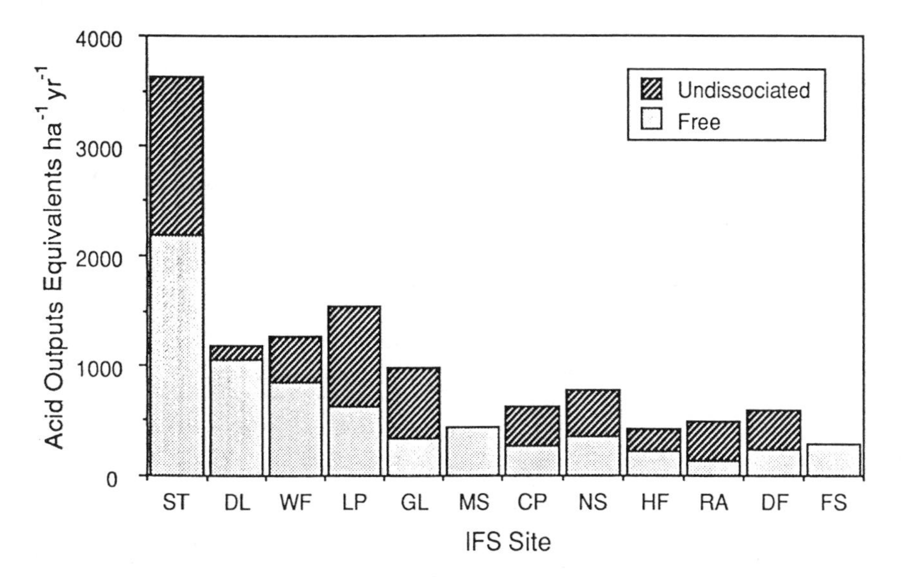

Figure 11.14. Canopy output fluxes of strong and undissociated acidity from IFS sites (eq ha^{-1} yr^{-1}). MS and FS are shown with no undissociated acidity output because samples from those sites were not titrated.

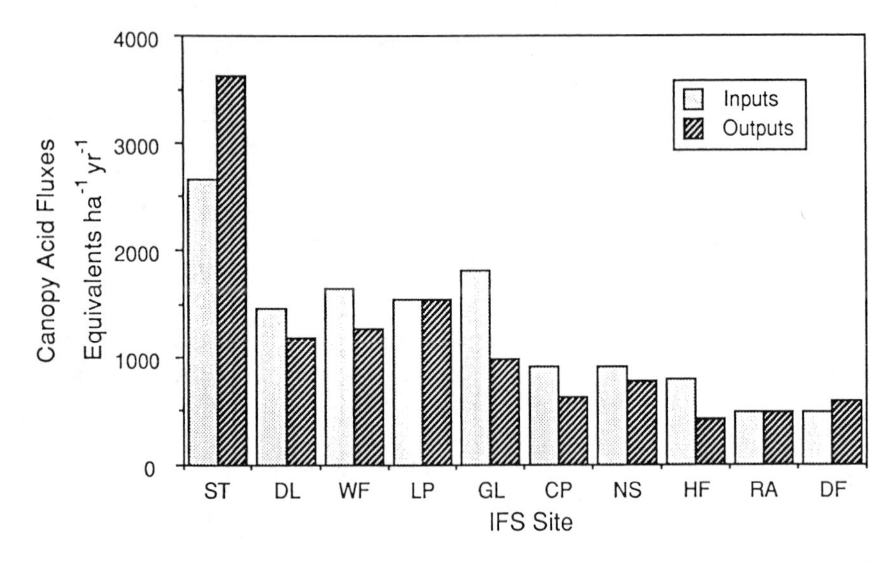

Figure 11.15. Canopy total acid input and output fluxes for 10 IFS sites (eq ha^{-1} yr^{-1}).

of cations for hydrogen ions and other canopy acid balance relationships are considered in detail in the following section.

Canopy Interactions

D.A. Schaefer, S.E. Lindberg, and G.M. Lovett

Strong acidity deposited on a forest canopy from the atmosphere has four possible fates:

1. Protonation of an organic acid anion released from the canopy
2. Proton exchange for another cation at a canopy exchange site
3. Retention in the canopy by some other mechanism
4. Transfer to the forest floor in TF or SF.

Estimating total canopy inputs and outputs of strong and undissociated acidity enables us to examine points 1 and 4, but data on other chemical fluxes are required to examine points 2 and 3.

The presumptive mechanism of cation exchange for protons involves fixed exchange sites at or near foliar surfaces (Mecklenburg et al. 1966). However, this cation exchange per se is not the the only mechanism that could lead to the coupled appearance of cations and the disappearance of strong acidity from throughfall (TF), as indicated in Figure 11.16. The top panel (a) shows the cation exchange mechanism, with strong acidity consumed as cations are released. The bottom panel (b) illustrates inorganic cation release

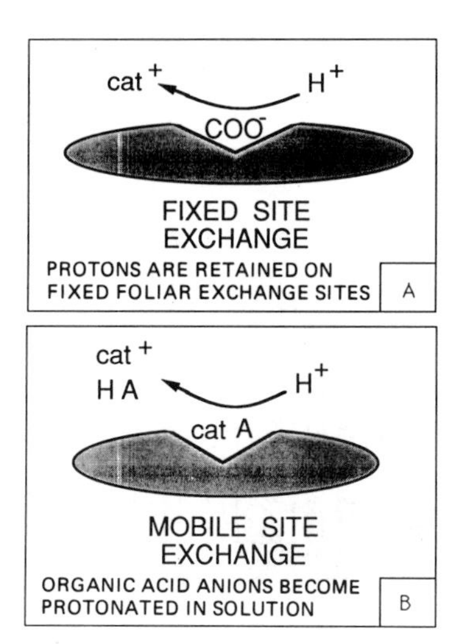

Figure 11.16. Schematic representation of canopy strong acid retention accompanied by cation release from fixed exchange sites (a) and "apparent" retention by protonation of organic anions (b).

coupled with organic acid anions (which become protonated in TF). Both processes release cations from foliage and decrease the strong acidity of water descending through the canopy as TF (Cronan and Reiners 1983), but only in the former would cation exchange necessarily respond positively to increased deposition of acidity (Lovett et al. 1985).

Lovett et al. (1985) illustrated the use of canopy charge balance methods to determine the fraction of the "net canopy exchange" of cations (NCE) that could actually be attributed to canopy cation exchange. They found that for three deciduous forest types in Tennessee, 40% to 60% of the NCE of cations was attributable to cation exchange in the canopy. Here we apply the same methods to make three estimates of canopy cation exchange for the 10 IFS forests for which undissociated acid fluxes and adequate other flux data are available. We hypothesized that canopy cation exchange would be greater at sites with greater atmospheric deposition of acids. Estimates of undissociated acid fluxes were lacking for two other sites, but two of the three methods (described next) could be employed there as well.

This approach can be used for ions that are either released or retained by the canopy. We describe the release method first. By our earlier definition of NCE, if the sum of positive cation NCE values exceeds the sum of positive anion NCE values, charge balance requires canopy retention (negative NCE values) for protons or other cations. In Figure 11.17, the results of these calculations are shown as "net loss." One important anion in deposition, SF, and TF is not quantified explicitly; rather, it is calculated by difference from all other ion values. That ion is bicarbonate, HCO_{3-} (cf.

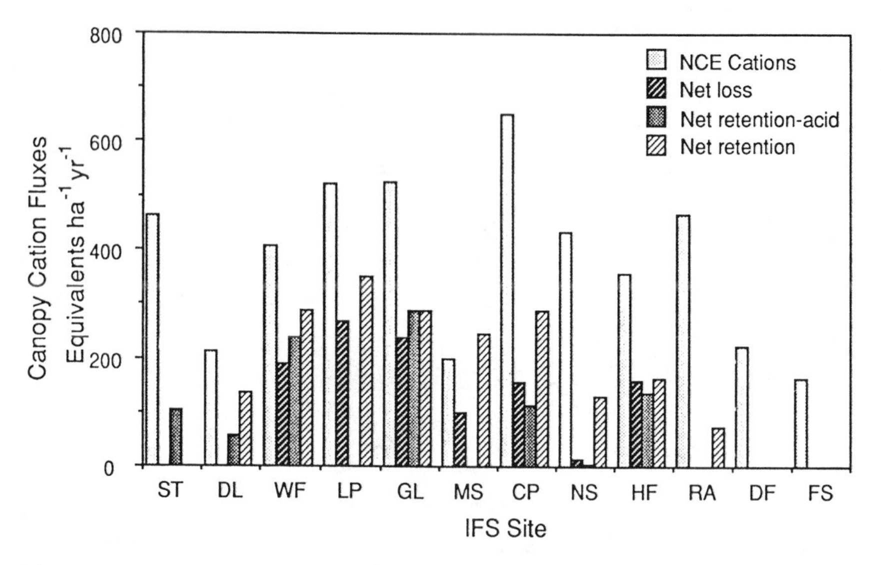

Figure 11.17. Comparison of net canopy effect on cations (NCE Cations) lost by 12 IFS canopies with three estimates of canopy cation exchange: "Net loss," difference between cations lost and anions lost from canopy; "Net retention-acid," difference between cations retained (reduced by undissociated acid flux) and anions retained by canopy; and "Net retention," difference between cations retained (ignoring undissociated acid flux) and anions retained by canopy. If no cation exchange estimate could be calculated or was negative, no bar was plotted (eq ha^{-1} yr^{-1}).

Lindberg et al. 1990). For most IFS sites coarse particle deposition appeared to include a substantial HCO_3^- flux (see Appendix). Where the apparent anion deficit of TF plus SF exceeded that of canopy inputs, there is a canopy release (positive NCE) of HCO_3^-. In TF plus SF, it is recognized that the anions not analyzed probably included dissociated organic acids, but our method of Gran Plot analysis specifically measured undissociated organic acids (p. 439).

The other estimates of canopy cation exchange are based on the difference between cations and anions retained by the canopy (that is, cations and anions whose NCE is negative). This can be calculated in two ways, with contrasting assumptions concerning the NCE of undissociated acidity. Undissociated acidity in TF and SF may appear as organic anions that are protonated by deposited H^+ and so represent strong acid retention in themselves. Conversely, undissociated acidity may be released from the canopy already protonated before any interaction with TF and SF. That release of neutral organics from the canopy has no effect on ion fluxes. Therefore, the "ion retention" method yields the second and third estimates of cation exchange. In the second estimate only deprotonated acids are released from the canopy. Therefore, H^+ retention by the canopy is reduced by the positive

NCE of undissociated acidity, decreasing cation retention overall (shown as "net retention-acid" in Figure 11.17).

The third estimate assumes that only protonated acids are released from the canopy. Therefore, the NCE of undissociated acidity is taken to have no effect on cation retention (shown as "net retention" in Figure 11.17). The difference between the second and third estimates results from the extent to which organic acids are protonated before they are released from canopy surfaces. This uncertainty would be reduced by determining the pKa of the acids and thus their degree of protonation at the pH within the plant tissue. As a first approximation, acids are released from foliage deprotonated, as the pKa values should be less than the plant internal pH of approximately 6. Therefore the "net retention-acid" would seem more likely than the "net retention" estimates in Figure 11.17, but these estimates should provide reasonable bounds on cation exchange.

The foregoing three estimates of canopy cation exchange are presented with the NCE for cations overall (NCE of cations released from the canopy minus NCE of cations retained by the canopy shown as "NCE cations") for average annual fluxes in 12 IFS sites in Figure 11.17. Where the cation exchange estimates could not be calculated or yielded negative values, no bars were plotted. In Figure 11.17, the sites are again arranged according to decreasing strong acid inputs. Note first that all estimates of cation exchange are smaller than (usually much smaller than) the NCE of cations. The average of the three estimates of cation exchange across these forests ranged from 0% to 59% (median 30%) of the net canopy exchange of cations, compared to 40% to 60% obtained by Lovett et al. (1985) for three low-elevation deciduous forests in Tennessee. Interestingly, cation exchange was also 40% of cation NCE at the HF site, the only IFS deciduous forest that received substantial strong acid loading where cation exchange was estimated.

Second, note that the cation exchange estimates varied widely at particular sites (especially at ST, LP, and MS). Where an individual estimate of cation exchange was calculated to be negative, a value of zero is plotted (see Figure 11.17). At sites with low NCE of undissociated acidity, the two estimates based on ion retention converge by definition. Third, there was only a weak relationship betwen strong acid deposition and the individual estimates of cation exchange. Based on these analyses, the highest acid deposition sites (ST and DL) showed some of the lowest canopy cation exchange estimates. These sites in particular failed to support the hypothesis that cation exchange increases with deposition.

Canopy total acid outputs versus total acid inputs (the canopy acid balance; open squares) are plotted in Figure 11.18. Several canopies appear nearly balanced, and others show evidence of canopy retention of acidity (being below the 1:1 line). Is calculated cation exchange greatest where acid retention appears? The average of the three estimates of cation exchange is added to canopy total acid outputs and shown as filled triangles in Figure

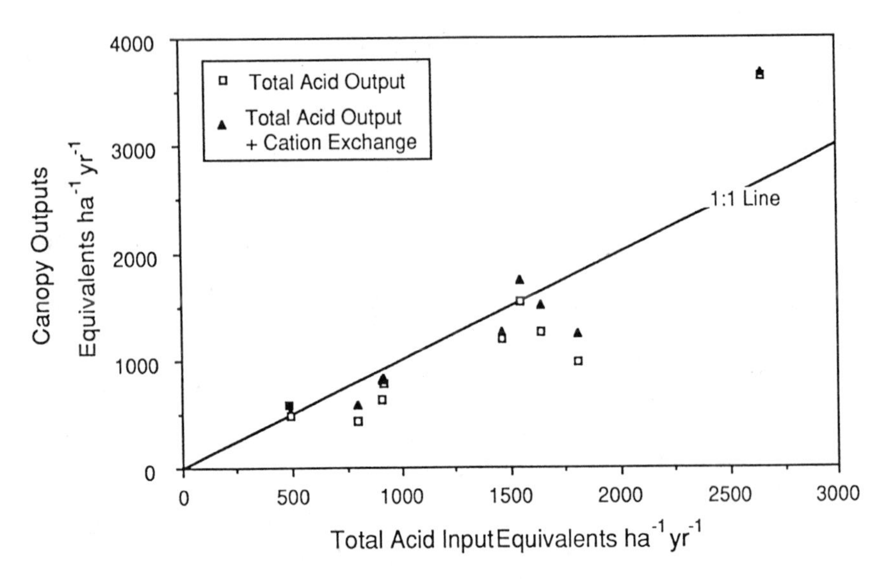

Figure 11.18. Canopy total acid outputs versus total acid inputs for 10 IFS sites (eq ha^{-1} yr^{-1}). Total acid output is represented by open squares; filled triangles represent values increased by averaged cation exchange estimate for each site; 1:1 line represents canopy acid balance.

11.18. These values are shifted closer to the 1:1 line than the open squares, suggesting that some acid neutralization can be attributed to canopy cation exchange. One site (ST) shows much higher output than input (well above the 1:1 line), as both strong and undissociated acid outputs exceeded inputs at that site (see next section).

Figure 11.19 more clearly illustrates the relationship between cation exchange and strong acid neutralization in the forest canopy (ST was excluded because that canopy was a source of strong acidity). The canopies that neutralized acid inputs to the greatest degree also showed the most cation exchange, with approximately 42% of the acid neutralization attributable to cation exchange. The two deciduous forests (northern hardwoods at HF and red alder at RA) do not stand out from the coniferous forests (all pine and spruce) in that relationship (Figure 11.19). As was shown in Figures 11.17 and 11.18, canopies with intermediate strong acid loading appear to exhibit the greatest degree of cation exchange. No nutritional factors (i.e., rhizosphere cation availability, see Chapter 8) have been identified that account for greater cation exchange from the canopies receiving intermediate acid loading.

Because all ionic fluxes are used to develop the canopy cation exchange estimates, the particular cationic species being exchanged cannot be identified. However, the NCE for cations is dominated by potassium and calcium

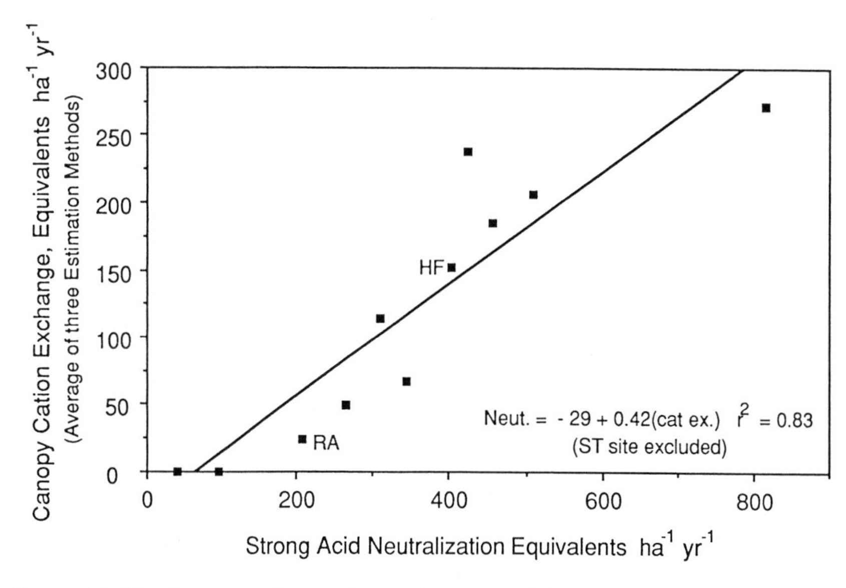

Figure 11.19. Canopy cation exchange (average of three estimates) versus canopy neutralization of strong acid inputs (eq ha^{-1} yr^{-1}). Deciduous canopies RA and HF are labeled. Strong linear relationship is obtained after excluding ST canopy, a strong acid source.

at all sites (see Chapter 8), making these ions the most likely candidates for cation exchange.

Some unexplained patterns remain, especially at ST. It is not known why strong and undissociated acidity are both released in this canopy, but in-canopy decomposition of lichens and other biological material may release organic acids. This release of strong acidity at ST causes calculated cation exchange values to be negative (plotted as zero in Figure 11.18), so canopy cation exchange could not be determined for that site. Retention of acidity in the canopy by some mechanism other than organic acid protonation or cation exchange is the final consideration. Other sections of this book suggest that for vapor-phase deposition of both SO_2 and HNO_3, deposition may take place in the substomatal cavities (Chapters 5 and 6), and lead to irreversible retention of SO_4^{2-} and NO_3^- in some canopy types. The strong acidity so deposited may be irreversibly retained as well, but experiments to test this have not yet been reported in the literature.

Summary of Acid Deposition and Canopy Interactions

D.A. Schaefer

Strong acid deposition to IFS forest canopies ranged from about 300 to 2000 eq ha^{-1} yr^{-1}. Strong acid deposition was higher for the southeastern and montane sites and lower for the northeastern and coastal sites (including NS). Wet deposition ranged from 45% to 90% of the strong acid deposition total, with a greater proportion of wet deposition in the lower acid deposition sites. About 40% of precipitation acidity was attributed to NO_3^- and 60% to SO_4^{2-}.

Dry deposition of acidity was dominated by HNO_3 and SO_2, with N and S contributions reversed in comparison to wet deposition (60% HNO_3 and 40% SO_2). Sulfur dioxide deposition was higher in the southeastern United States because of air concentrations there. There was a general tendency for dry deposition to reflect near-field sources. Cloud and fog deposition was insignificant except at the montane sites (ST and WF), where one-quarter to one-third of the strong acid inputs were attributed to this source.

Comparing strong acid canopy outputs in TF and SF with the inputs shows that neutralization may increase with deposition, but not more than 700 eq ha^{-1} yr^{-1} of strong acidity were neutralized. Dry deposition may be more readily neutralized than is wet deposition, and this neutralization may include a constant fraction of the acid vapors. Both cation exchange and organic acid exchange also acted to partially neutralize the deposition of strong acids across the IFS sites, but neither mechanism appeared to be controlled by strong acid loading to forests at these sites. IFS forest canopies released nearly as much acidity as they received, when both strong and undissociated acids were considered.

The canopies that neutralized strong acid inputs to the greatest extent also exchanged the most cations in their canopies. The responses of spruce, pine, and deciduous canopies all followed that pattern, but the deciduous canopies transferred the smallest strong acid fluxes to the forest floor in TF and SF. One canopy (ST) released large amounts of both strong and weak acidity, perhaps from organic matter decomposition in the canopy or from underestimation of acidic inputs.

H^+ Budgets

D. Binkley

General Description of H^+ Budgets and Soil Acidity

The pH and ion concentrations in soil solutions derive from the equilibrium with the readily exchangeable pools of adsorbed ions. Changes in the soil solution are driven either by changes in the ionic strength of the solution (caused by atmospheric deposition, soil respiration, nitrification, or other processes) or by changes in the exchange complex (see Reuss and Johnson 1986). Three types of changes may occur in the chemistry of the exchange complex (see Binkley et al. 1989a, 1989b):

1. Quantitative replacement of one type of ion (such as 3 Ca^{2+}) by another (such as 2 Al^{3+});

2. Quantitative increase or decrease in the size of the exchange complex from addition or removal of soil organic matter or perhaps from the generation or dissolution of secondary clay minerals through weathering; and
3. Qualitative changes in the affinity of the exchange sites for different ions. Such qualitative changes result from either changes in soil pH (which protonates or deprotonates exchange sites) or the quality of organic exchange sites. In both cases, the selectivity of the exchange complex for various ions would appear to shift.

A soil becomes more acidic if the proportion of the exchange sites occupied by acidic cations (H^+ and Al^{n+}) increases. Such changes are driven by processes that generate H^+, and these processes can be tracked by either accounting for the major processes that produce and consume H^+ (Driscoll and Likens 1980; Sollins et al. 1980; Binkley and Richter 1987), or by following the associated fluxes of base cations and anions (see Reuss and Johnson 1986). Both approaches provide identical estimates of soil acidification.

Budgets of H^+ fluxes account for changes in the quantitative replacement of ions on a constant exchange complex (point 1 in preceding list), and identify the processes responsible for driving the changes. For example, an excess production of H^+ from cation accumulation in biomass may result in H^+ displacing K^+ from the exchange, lowering the dissociation of the exchange complex and therefore lowering soil pH. Quantitative and qualitative changes in the size and selectivity of the exchange complex could also be included in a H^+ budget (e.g., Binkley et al. 1989a; Binkley and Sollins, 1990; Binkley and Valentine, 1991), but long-term data on changes in pools of soil organic matter are usually unavailable.

The H^+ budget approach has several strengths relative to the simpler accounting of the net flux of base cations. At a conceptual level, it provides a framework for integrating the component fluxes for each biogeochemical cycle. For example, nitric acid is a major component of acidic deposition in many polluted regions, but the processing (reduction) of nitrate typically neutralizes a substantial portion of the acidity. Budgets of H^+ fluxes also allow comparison of the relative importance of the major biogeochemical fluxes to ecosystem acidification. If soil pH is above 5.5, the production and dissociation of carbonic acid may drive ecosystem acidification, whereas nitrification may dominate the H^+ budget of N-rich ecosystems. Budgets accounting only for the net flux of base cations would not distinguish between these very different processes.

As with any accounting framework, the line items in a H^+ budget vary with the perpsective of the accountant and the objective of the exercise. The net accounting system used for these H^+ budgets focuses on changes in the H^+ content of the exchange complex. Each flux is written so that a positive value indicates H^+ added to the exchange complex and a negative value represents depletion of H^+ from the exchange complex:

H^+ input $-$ H^+ output
Ammonium input $-$ ammonium output
Aluminum input $-$ aluminum output
Cation increment in biomass
Nitrate output $-$ nitrate input
Sulfate output $-$ sulfate input
Alkalinity output $-$ alkalinity input

It might appear that any estimate of the net H^+ load resulting from the cumulative contribution of so many fluxes would be subject to cumulative uncertainties of the true rate of each flux. However, the necessity of maintaining equivalent charges of anions and cations in all solutions constrains these uncertainties and minimizes the uncertainty in the overall net H^+ loads.

If H^+ input exceeds output, a net increase occurs in the load of H^+ in the ecosystem from this portion of the H^+ budget (summarized from Binkley and Richter 1987). Similarly, if inputs of ammonium exceeded outputs, the storage of ammonium in organic form ($R\text{-}NH_2$) would involve a net addition of H^+ to the ecosystem. Aluminum input is essentially 0 for all sites; Al^{3+} leaving the ecosystem represents the removal of H^+. The IFS protocol did not include determination of the form of aluminum leaching from the soils, so all aluminum is assumed to have a charge of 2^+ in the H^+ budgets (see later discussion).

The increment of cations in biomass results in an efflux of H^+ into the soil. If the cations orginated from decomposing organic matter, then the release of H^+ associated with uptake would precisely balance a consumption of H^+ in decomposition. If the nutrient cation originated from the exchange complex or from atmospheric deposition, then the release of H^+ would represent a net H^+ load to the system. Because soil organic matter usually stays constant or increases in intact forests, we assumed that recycling from decomposition could not supply the nutrient cations accumulated in the standing biomass and forest floor. Therefore, the uptake of cations was assumed to represent a net load of H^+ to the ecosystem. If some of the increment of cations in the forest floor and standing biomass originated from a net depletion of cations in soil organic matter, then our estimated H^+ flux would be too high. The rate of increment was calculated by dividing the cation content of the forest floor and standing biomass by the age of the stand. This approach results in a long-term average, whereas other components of the H^+ budgets represent current annual rates.

The uptake of the anion nutrients, nitrate and sulfate, involves an equivalent uptake of H^+ to balance the reduction reaction that incorporates them into organic molecules. Uptake of nitrate and sulfate deposited from the atmosphere therefore consumes H^+. Uptake of nitrate or sulfate recycled within the ecosystem (from soil organic matter to plant organic matter) has no net effect on the H^+ budget. In the H^+ budgets, therefore, the within-system cycles of N and S were not considered. The effects of sulfate and nitrate inputs and outputs were calculated as a net input of H^+ if outputs exceeded inputs, or a net removal of H^+ if inputs exceeded outputs. The same treatment was used for bicarbonate.

Chloride is often considered to pass through ecosystems with little opportunity for storage or reaction. Nevertheless, the chloride balances were substantially different from 0 for most of the sites. Net retention of chloride was considered to represent a net consumption of H^+ and net loss of chloride a generation of H^+. This convention was used because of the need to maintain electroneutrality.

The input and output of base cations is not included in the calculation of H^+ budgets, as their charge effects are already accounted for by the differences between the anion budgets and the acid cation budgets.

The net H^+ loading in an ecosystem can be aggregated into three fluxes:

1. The increment of cation nutrients in biomass;
2. The net flux of base cations, which results from the integration of the input and output budgets for all elements; and
3. The charge discrepancy in inputs and outputs.

The accumulation of cation nutrients in biomass is a within-ecosystem transfer of cations from electrostatic bonds on the soil exchange complex to largely covalent bonds in plant biomass. This internal ecosystem shift is driven by the generation and dissociation of weak organic acids within the plants and the transfer of the H^+ to the weak-acid exchange complex. The net acidification involves no input or output of ions from the ecosystem, unless biomass harvest occurs.

The net flux of base cations (inputs − outputs) integrates the H^+ fluxes associated with the input and output of the major biogeochemical cycles; net loss of base cations represents the net rate at which the exchange complex may become acidified. In soils with adequate rates of mineral weathering, the net loss of base cations from the exchange complex may be replenished and the rate of acidification could be partially or completely reduced.

The quantity of anion and cation charges must be equivalent for deposition into an ecosystem and for outputs from an ecosystem. The necessity of maintaining charge balance constrains the possible errors that might otherwise accumulate in the summing the many processes that contribute to H^+ budgets. In most biogeochemical studies, the quantity of anions and cations measured in atmospheric deposition and soil leachates do not balance precisely, owing to either unmeasured elements or to analytic imprecision. If the quantity of cations in deposition exceeds anions, then either the quantity of one or more anions is underestimated or the quantity of one or more cations is overestimated.

Interestingly, the effect of the charge imbalance on H^+ budgets is the same regardless of whether the cation sum or the anion sum is in error. Increasing the quantity of anions to account for the imbalance would increase the calculated consumption of H^+, whereas decreasing the quantity of cations deposited would decrease the production of H^+ (see the foregoing equations). The reverse applies to the charge imbalances in soil leachates: increasing the quantity of anions would increase the production of H^+, whereas decreasing the quantity of cations would decrease the consumption of H^+. The calculated net H^+ load can be corrected for these charge imbalances, even without knowledge of the exact source of the error. It is important to note, however, that if inaccurate measurements produced an erroneous estimate of the proportion of acid and base cations, the error would be independent of the charge balance and would constitute a real error in the H^+ budget.

Relative Importance of H^+ Source

The deposition of H^+ was greatest at the high-elevation sites and the southeastern sites (Figure 11.20). The output of H^+ in soil leachates was low in most ecosystems, but exceeded one-third of the deposition rate in the Smokies sites. Deposition of ammonium (Figure 11.21) averaged about one-third of the H^+ deposition rate across sites and greatly exceeded leaching losses of ammonium in all sites. Most sites had little leaching of aluminum (assumed to have a charge of 2^+, based on charge balance in soil leachates for the Smokies sites) from the soil, but this flux was substantial at all three Smokies sites (Figure 11.22). In fact, the removal of H^+ from the forests by aluminum leaching from these sites rivaled the rate of H^+ deposition from the atmosphere.

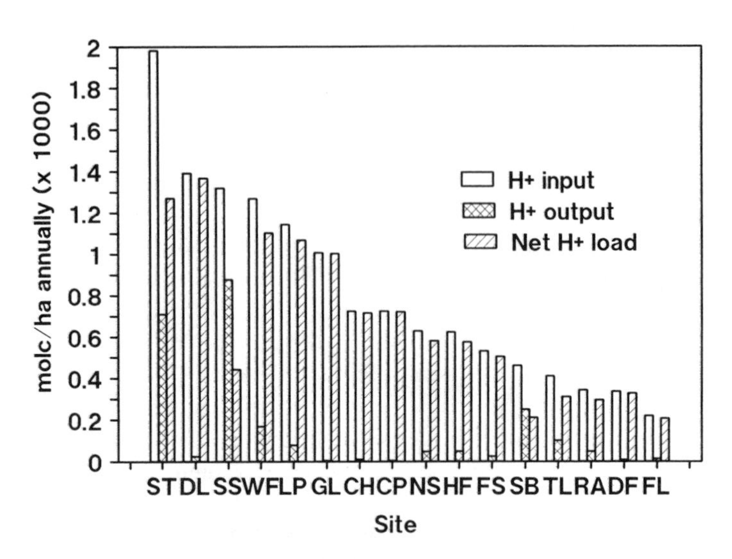

Figure 11.20. Atmospheric deposition of H^+, leaching loss of H^+, and net H^+ input remaining from atmospheric deposition. Positive value for net budget indicates H^+ loaded into ecosystem.

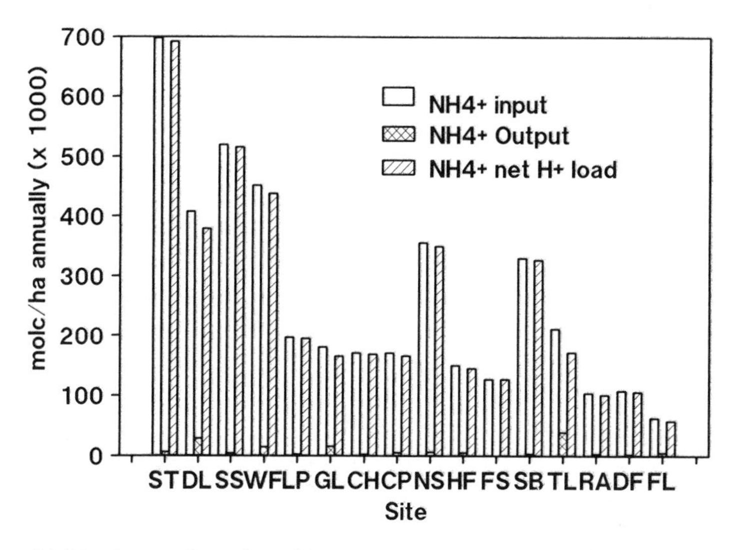

Figure 11.21. Ammonium deposition, ammonium output, and net ammonium retained from atmospheric deposition. Positive value for net budget indicates H^+ loaded into ecosystem.

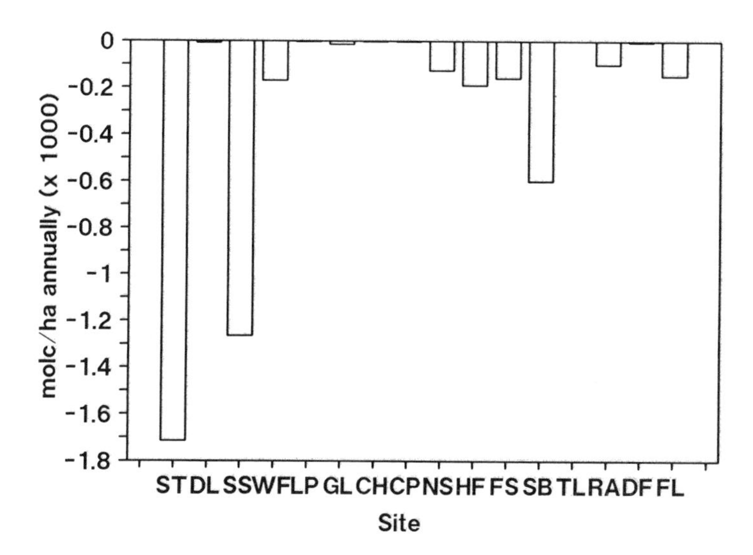

Figure 11.22. Aluminum output represents net removal of H^+ from ecosystem. Aluminum was assumed to have a charge of 2+, based on charge balance in leachates from Smokies sites.

The increment of cations in biomass (averaged over the age of the forests) generated a substantial amount of H^+ in most forests, especially in hardwood forests and in young conifer forests (Figure 11.23).

The nitrate budget produced highly variable H^+ fluxes across the sites. The Smokies Tower site (Figure 11.24) showed a marginal net loss of nitrate (generating about 200 mol(c) ha^{-1} of H^+), whereas nitrate losses from the Smokies Becking site were double the rate of deposition (generating about 800 mol(c) ha^{-1} of H^+). Other major nitrate exporters were the Smokies Beech forest (net 1200 mol(c) ha^{-1} H^+), Turkey Lakes forest (1400 mol(c) ha^{-1} H^+), and the red alder forest (2500 mol(c) ha^{-1} H^+; indirectly resulting from acceleration of N cycling by high rates of N fixation). The other forests retained essentially all deposited nitrate, showing net consumption of H^+ (averaging about -400 mol(c) ha^{-1} H^+).

In general, losses of sulfate from high-elevation sites closely matched atmospheric inputs, resulting in little contribution to the H^+ budget (Figure 11.25). No generalization applied to the other forests; three of the southern pine forests (DL, LP, and FS) showed no net retention of sulfate whereas a fourth (GL) retained almost all the deposited sulfate. The low-elevation hardwood forests also included sites that retained substantial sulfate and others that demonstrated a net loss.

Inputs of alkalinity were not measured in all sites, and rates were assumed to be 0 unless otherwise specified. With this assumption, the alkalinity budget appeared important in the high-elevation sites and at Findley Lake, Duke loblolly pine, and Georgia loblolly pine sites (Figure 11.26).

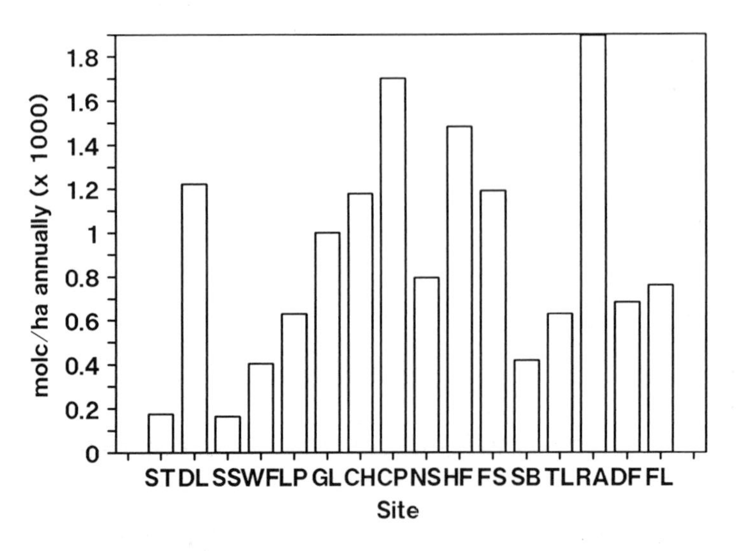

Figure 11.23. Content of nutrient cations derived from current content of biomass divided by stand age. Positive value indicates H^+ loaded into ecosystem.

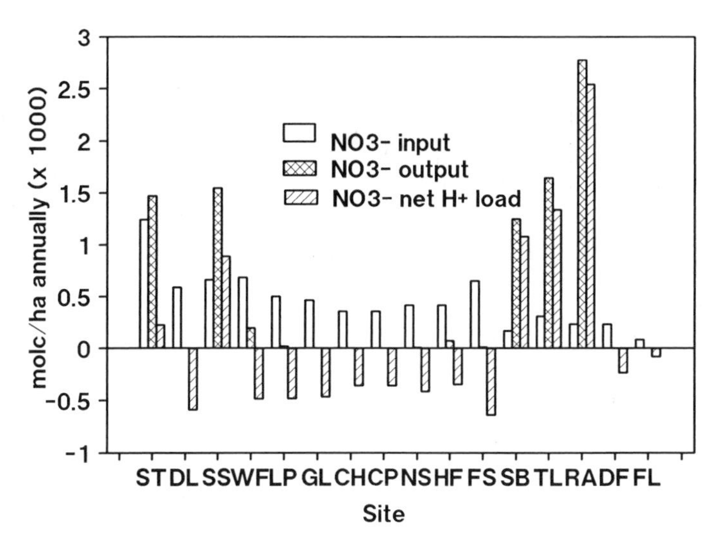

Figure 11.24. Nitrate deposition, nitrate output, and net nitrate retained. Net retention involves consumption of H^+ (negative net values); net loss involves H^+ production (positive net values).

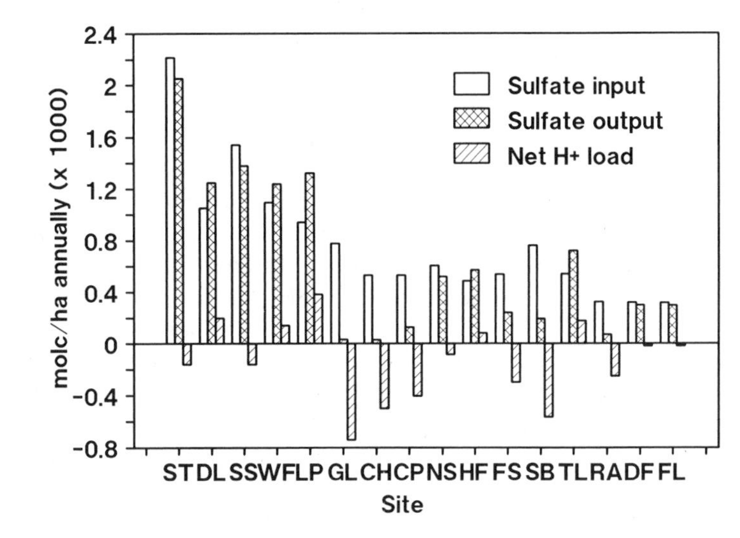

Figure 11.25. Sulfate deposition, sulfate output, and net sulfate retained. Net retention involves consumption of H^+ (negative net values); net loss involves H^+ production (positive net values).

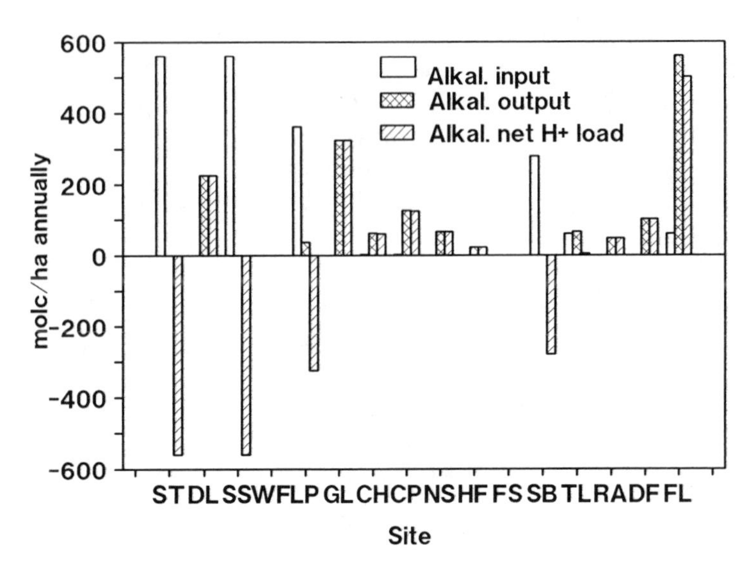

Figure 11.26. Alkalinity input, alkalinity output, and net alkalinity retained. Net retention involves consumption of H^+ (negative net values); net loss involves H^+ production (positive net values). Some sites did not measure alkalinity input, and inputs were assumed to be 0, introducing probable error of <100 mol(c) ha^{-1}.

The net chloride budgets were marginally important (>100 mol ha^{-1} H^+) in consuming H^+ in the Duke loblolly pine, Florida slash pine, and Findley Lake sites, and in generating H^+ in the Georgia loblolly pine site (Figure 11.27).

Net H^+ Balances

High-Elevation Appalachian Sites

The net H^+ budget for the Smokies Beech site indicated a rate of acidification of about 290 mol(c) ha^{-1} annually (Figure 11.28). This net loading of H^+ can be divided into three components. The accumulation of base cations in biomass generated 420 mol(c) ha^{-1}. The net balance of base cation input and output summarizes the rest of the system's biogeochemistry (such as deposition, nitrification, and sulfate retention) and contributed a net negative value of −220 mol(c) ha^{-1}, represented by the net accumulation of base cations within the ecosystem. The final component of the net H^+ loading derives from imbalances in anions and cations in both deposition and leachate outputs. At this site, the sum of cations in deposition was estimated to be 1490 mol(c) ha^{-1}, compared with 1350 mol(c) ha^{-1} of anions. The net cation excess of 140 mol(c) ha^{-1} gives a 140-mol(c) ha^{-1} overestimation of the net H^+ loading. The sum of cations in leachate output was 1660 mol(c) ha^{-1}, and the anion sum was 1610 mol(c) ha^{-1}. The −50-mol(c) ha^{-1}

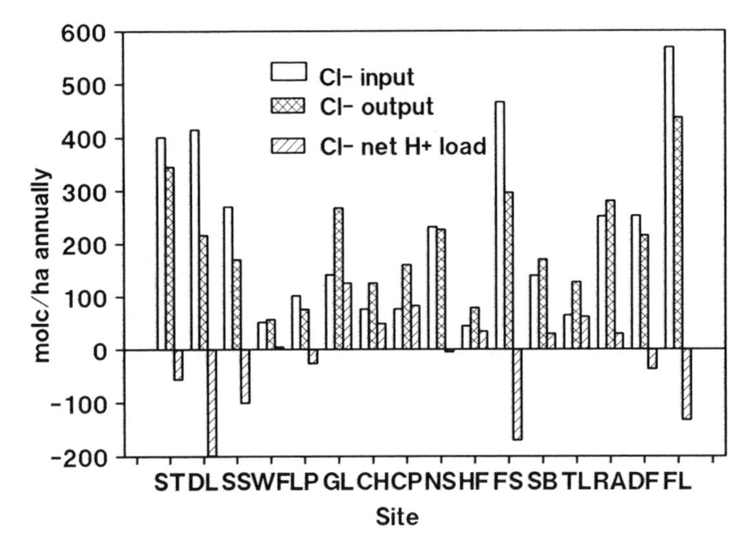

Figure 11.27. Chloride input, chloride output, and net chloride retained. Mechanisms responsible for net release or accumulation of chloride are not clear, but to maintain charge balance, chloride budgets were treated as for other anions: net retention involves consumption of H^+ (negative net values), and net loss involves H^+ production (positive net values).

error in output charge balance partially offset the +140 mol(c) ha^{-1} error in inputs, leaving a net contribution to the estimated H^+ loading of +90 mol(c) ha^{-1}.

The net H^+ budgets for the two spruce stands in the Smokies (see Figure 11.28) indicated essentially no net acidification or alkalization of the ecosystems. This conclusion is based on the assumption that the average charge on aluminum was 2^+. If the charge averaged 1^+, a net acidification of more than 500 mol(c) ha^{-1} would occur, and a charge average of 3^+ would have produced a net consumption of more than 500 mol(c) ha^{-1} of H^+. Note that the absence of acidification currently in the spruce forests is based primarily on the large export of aluminum from the soil. Acidic deposition seems incapable of further acidifying these forests; however, this resistance to further acidification also means the forests provide essentially no buffering to water flowing from the atmosphere to aquatic ecosystems.

The Whiteface site showed a net load of 1340 mol(c) ha^{-1} of H^+, derived almost equally from the accumulation of cations in biomass, the net loss of base cations from the ecosystem, and the error in charge balance.

Coweeta Sites

Both Coweeta forests showed large net loads of H^+, with 1310 mol(c) ha^{-1} acidification in the hardwood forest and 2030 mol(c) ha^{-1} acidification in

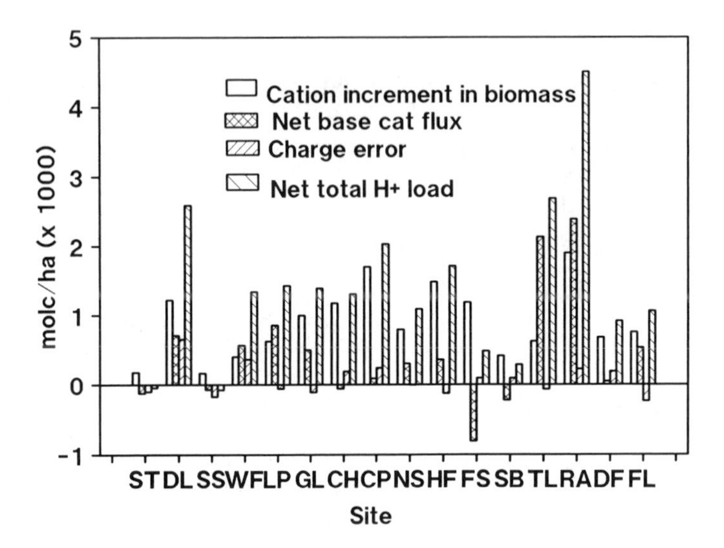

Figure 11.28. Net load of H^+ calculated from H^+ budgets can be summarized with three components: increment of cations in biomass, biogeochemical input and output budgets summarized as net flux of base cations, and error associated with charge imbalances in inputs and outputs.

the white pine forest (see Figure 11.28). The H^+ flux in both sites is dominated by the effects of cation accumulation in biomass; the input and output budgets (as summarized in the net cation flux) are almost balanced. The current rate of deposition in these forests is contributing little to ecosystem acidification, whereas forest growth is having a substantial effect.

Southern Pine Sites

The three loblolly pine sites all had net positive H^+ fluxes. At the Duke loblolly pine site, the cation increment in biomass was the primary factor in acidification. This site displayed the largest error from charge imbalance (660 mol(c) ha^{-1}), which rivaled the estimate of the contribution of biogeochemical cycles to the net H^+ load. The large imbalance in charges derived mostly from a 480-mol(c) ha^{-1} cation excess in deposition. If the input error resulted from overestimating cation inputs, then the net H^+ load is overestimated by 480 mol(c) ha^{-1}. If the deposition of anions were underestimated, an increase in anions would also involve proportional decreases in the net H^+ load.

At the Oak Ridge loblolly pine site, the contribution of cation accumulation in biomass and the overall biogeochemical balance of inputs and outputs contributed similar amounts (about 700 mol(c) ha^{-1}) to the net H^+ loading. The balances of cations and anions were very good, and the charge balance error was negligible.

The net H^+ flux at the Georgia loblolly pine site was also dominated by the accumulation of cations in biomass, with the net role of the biogeochemical input-output budgets contributing only about half as much.

The Florida slash pine site showed a moderate net load of H^+ (+480 mol(c) ha^{-1}), resulting from the offsetting effects of cation accumulation in biomass (1190 mol(c) ha^{-1}) and the net input-output budget (−810 mol(c) ha^{-1}). The charge error was modest (+100 mol(c) ha^{-1}).

Northern Hardwood Sites

The Huntington Forest site showed a net load of 1710 mol(c) ha^{-1} of H^+, again with the contribution from cation accumulation in forest biomass (1480 mol(c) ha^{-1}) far exeeding the importance of the biogeochemical input:output budgets (360 mol(c) ha^{-1}) and the charge error (−130 mol(c) ha^{-1}).

The Turkey Lake site had one of the highest net loads of H^+, 2690 mol(c) ha^{-1}. This was one of the few sites where the biogeochemical input:output budgets (2130 mol(c) ha^{-1}) were more important than the increment of cations in biomass (630 mol(c) ha^{-1}). The high net loss of nitrate (releasing 1340 mol(c) ha^{-1} of H^+) was particularly important.

Washington Sites

The net load of H^+ in the Douglas fir forest was 920 mol(c) ha^{-1}, almost all of which derived from cation accumulation in biomass (680 mol(c) ha^{-1}) rather than from biogeochemical input:output budgets (50 mol(c) ha^{-1}). Errors in charge balance accounted for the rest of the 190 mol(c) ha^{-1} of net H^+ loading.

The red alder forest demonstrated the greatest net H^+ load, 4510 mol(c) ha^{-1}, largely from the net loss of nitrate (generating 2540 mol(c) ha^{-1}). The rapid accumulation of cations in biomass (1900 mol(c) ha^{-1}) also contributed a large portion of the net H^+ loading.

The high-elevation silver fir forest at Findley Lake had a net H^+ load of about 1000 mol(c) ha^{-1}, from roughly equal contributions of cation increment in biomass and the net biogeochemical input and output fluxes (largely resulting from high leaching losses of bicarbonate).

Norway Site

The net load of H^+ of 1090 mol(c) ha^{-1} was again dominated by cation accumulation in biomass (800 mol(c) ha^{-1}), with a moderate contribution from biogeochemical input/output budgets (290 mol(c) ha^{-1}). The charge balances in deposition and output estimates were very good, accounting for none of the net H^+ load.

Comparisons of H^+ Budgets for Polluted and Unpolluted Environments

In contrasting the polluted and unpolluted sites at high elevation, the annual deposition of H^+ at the Smokies Tower site (almost 2000 mol(c) ha^{-1}; see

Figure 11.20) was almost an order of magnitude greater than for the Findley Lake site (about 250 mol(c) ha^{-1}). Ammonium deposition showed the same pattern (see Figure 11.21). Most of the acid load on the Smokies Tower site was neutralized by the leaching of aluminum from the soil (consuming about 1700 mol H^+ ha^{-1} annually, assuming a charge of 2+ for aluminum; see Figure 11.22). Aluminum leaching was much lower at Findley Lake (consuming about 150 mol H^+ ha^{-1} annually). Despite the low rate of aluminum leaching at Findley Lake, this processes accounted for a relatively large proportion of the neutralization of H^+ deposited from the atmosphere. If assumptions about cation increments in biomass are correct, the Smokies Tower site has much lower generation of H^+ by this process (about 200 mol H^+ ha^{-1}; see Figure 11.23) than the Findley Lake site (about 800 mol H^+ ha^{-1}).

The nitrate budgets were relatively unimportant in the H^+ budgets at both locations, but for different reasons. Deposition rates were very high at the Smokies Tower site (about 1500 mol(c) ha^{-1} annually; see Figure 11.24), but leaching losses of nitrate were almost as great. At Findley Lake, deposition rates were so low (and leaching approaches 0) that the net retention of nitrate within the ecosystem accounted for only a slight neutralization of H^+. The input-output budgets for sulfate also balanced closely at both sites, despite great differences in deposition rates (see Figure 11.25). The alkalinity budgets for the two sites were almost mirror images (see Figure 11.26); high deposition and low output for Smokies Tower, and low deposition and high outputs for Findley Lake.

The overall balance of processes generating and consuming H^+ led to little net loading of H^+ in the polluted Smokies Tower site, and a moderate loading at Findley Lake. In both cases, the rate of cation increment in biomass (whether small or large) constituted a substantial component of the net H^+ budget. No net soil acidification should be occurring at the Smokies Tower site; it has reached a "flow-through" stage where export of acidity (in the form of aluminum) matches the generation of acidity. The Findley Lake site might experience soil acidification, but the rate of mineral weathering (see Chapter 10) may be high enough to consume the net loading of H^+.

At low elevation, the Duke Loblolly site received about four times the deposition of H^+ as the Douglas fir site (about 1400 mol H^+ ha^{-1} annually compared to 320 mol H^+ ha^{-1}; see Figure 11.20), with ammonium deposition following the same trend (see Figure 11.21). Consumption of H^+ via aluminum leaching was 0 in both sites (see Figure 11.22). The H^+ flux from the accumulation of cations in biomass was greater at the Duke Lobolly site (about 1200 mol(c) ha^{-1} annually; see Figure 11.23) than at the Douglas fir site (about 700 mol(c) ha^{-1}), resulting from a combination of faster tree growth and more rapid accumulation of forest floor biomass.

The input of nitrate, sulfate, and chloride was greater at the Duke Loblolly site (see Figures 11.24, 11.25, and 11.27), whereas the input of alkalinity was estimated to be negligible at both sites (Figure 11.26).

The net rate of H^+ loading was several times greater in the Duke Lobolly site, owing to both greater rate of biomass increment and greater leaching of base cations by high inputs of anions. The weathering rates of these two sites should show the opposite pattern, with very low (near 0) rates for the Duke Loblolly site (see Chapter 10). Therefore, soil acidification should be occurring at the Duke Loblolly site. The comparison of the Douglas fir site with the red alder site showed that mineral weathering appeared rapid and capable of increasing to consume the extra H^+ generated by nitrification under alder (Van Miegroet and Cole 1988).

Summary and Implications of H^+ Budgets for Soil Acidification

In 12 of the 16 forests included in this analysis, the accumulation of nutrient cations in biomass was a more important source of H^+ than the net input and output budgets of all elements. The exceptions involved major contributions from the biogeochemical input-output budgets in the net acidification of the Whiteface Mountain, Oak Ridge loblolly pine, Turkey Lake, and red alder sites. The production of H^+ through nitrification (without associated uptake of nitrate) played a major role at Turkey Lake and at the N-fixing red alder site.

This pattern has several major implications. The first is that biogeochemical processing of ions deposited from the atmosphere largely mediated the acidifying effects of atmospheric deposition to the point where natural forest growth has a greater acidifying effect on forest soils than does acidic deposition at most sites. As many authors have stated, the long-run ability of the biogeochemical cycles to continue to mediate acidic deposition will depend on whether the output of base cations derives from stripping of the exchange complex or from the weathering of soil minerals. If weathering supplies the base cations, the patterns in Figure 11.28 could be sustained far into the future.

If the net losses of base cations were derived from the exchange complex, the bars in Figure 11.28 would shift. For example, if the supply of base cations at the Georgia loblolly pine site were depleted, the net leaching loss of base cations would decline and the output of aluminum would increase (all else remaining constant). This would result in a reduction in the net H^+ load to the ecosystem, resulting from the export of acidity in the form of aluminum. This feedback effect would tend to limit the acidification of the soil, as was observed for the spruce forests in the Smokies. The good news that soil acidification would slow as base cations became depleted is somewhat offset, however, by the fact that this reduced soil acidification involves increasing export of aluminum with the strong potential for acidification of aquatic ecosystems. The final implication is that forest harvest, which essentially makes the temporary removal of base cations from the soil permanent, may have a greater long-term effect on soil acidification than does acidic deposition at current rates across the IFS sites.

References

Albritton D., Fehsenfeld F., Hicks B., Miller J., Liu S., Hales J., Shannon J., Durham J., Patrinos A. 1988. Atmospheric Processes, Vol. III, Chapter 4. In NAPAP Interim Assessment, Environmental Protection Agency, Washington, D.C.

Binkley D., Richter D. 1987. Nutrient cycles and H^+ budgets in forest ecosystems. Adv. Ecol. Res. 16:1–51.

Binkley D., Sollins P. 1990. Acidification of soils in mixtures of conifers and red alder. Soil Sci. Soc. Am. J. 54:1427–1433.

Binkley D., Valentine D. 1991. Fifty-year biogeochemical effects of green ash, white pine, Norway spruce in a replicated experiment. For. Ecol. Manage. 40:13–25.

Binkley D., Valentine D., Wells C., Valentine U. 1989a. An empirical analysis of the factors contributing to 20-year decrease in soil pH in an old-field plantation of loblolly pine. *Biogeochemistry* 7:139–154.

Binkley D., Driscoll C., Allen H.L., Schoeneberger P., McAvoy D. 1989b. Acidic Deposition and Forest Soils: Context and Case Studies in the Southeastern United States, Springer-Verlag, New York.

Cape J.N. 1983. Contact angles of water droplets on needles of Scots pine (*Pinus sylvestris*) growing in polluted atmospheres. New Phytol. 93:293–299.

Cole D.W., Rapp M. 1981. Elemental cycling in forest ecosystems. In Reichle D.E. (ed.) Dynamic Properties of Forest Ecosystems. Cambridge University Press, London, pp. 341–409.

Cronan C.S., Reiners W.A. 1983. Canopy processing of acidic precipitation by coniferous and deciduous forests in New England. Oecologia 59:216–223.

Driscoll C., Likens G. 1980. Hydrogen ion budget of an aggrading forested watershed. Tellus 34:283–292.

Evans L.S., Curry T.M., Lewin K.F. 1981. Responses of leaves of *Phaesolus vulgaris* L. to simulated acid rain. New Phytol. 88:403–420.

Fairfax J.A.W., Lepp N.W. 1975. Effects of simulated "acid rain" on cation loss from leaves. Nature (London) 255:324–325.

Galloway J.N., Likens G.E., Edgerton E.S. 1976. Acid precipitation in the northeastern United States: pH and acidity. Science 194:722–724.

Galvin P., Sampson P.J., Coffey P.E., Romano D. 1978. Transport of sulfate to New York State. Environ. Sci. Technol. 12:580–584.

Hauhs M., Rost-Siebert K., Raben G., Paces T., Vigerust B. 1989. Summary of European data. In Malanchuk J.L., Nilsson J. (eds.) The Role of Nitrogen in the Acidification of Soils and Surface Waters. Miljorapport 10, Nordic Council of Ministers, Gotab, Stockholm, pp. 5-1-5-37.

Hicks B.B. 1989. Overview of deposition processes. In Malanchuk J.L., Nilsson J. (eds.) The Role of Nitrogen in the Acidification of Soils and Surface Waters. Miljorapport 10, Nordic Council of Ministers, Gotab, Stockholm, pp 3-1-3-21.

Hoffman W.A., Lindberg S.E., Turner R.R. 1980. Precipitation acidity: the role of the forest canopy in acid exchange. J. Environ. Qual. 9:95–100.

Husain L., Samson P.J. 1979. Long range transport of trace elements. J. Geophys. Res. 84:1237–1240.

Husain L., Halstead J.A., Parekh P., Dutkiewicz V.A. 1982. A critical investigation of sulfate transport to the Adirondack Mountains. In Proceedings of the Second Symposium on the Composition of the Nonurban Troposphere, May 25–28, Williamsburg, Virginia, pp. 218–221, American Meteorological Society, Boston, Massachusetts.

Keene W.C., Galloway J.N. 1988. The biogeochemical cycling of formic and acetic acids through the troposphere:an overview of current understanding. Tellus 40B:322–334.

Kelley T.J., Tanner R.L., Newman L., Galvin P.J., Kadlecek J.A. 1984. Trace gas and aerosol measurements at a remote site in the northeastern United States. Atmos. Environ. 18:2565–2576.
King W.J., Vukovich F.M. 1981. Some dynamic aspects of extended pollution episodes. Atmos. Env. 15:1171–1181.
Likens G.E., Bormann F.H., Pierce R.S., Eaton J.S., Johnson N.M. 1977. Biogeochemistry of a Forested Ecosystem. Springer-Verlag, New York.
Lindberg S.E., Lovett G.M., Meiwes K.J. 1986. Deposition and forest canopy interactions of airborne nitrate. In Hutchinson H.J., Meema K.M. (eds.) Effects of Atmospheric Pollutants on Forests, Wetlands, and Agricultural Ecosystems. Springer-Verlag, Berlin, pp. 117–130.
Lindberg S.E., Johnson D.W. 1989. 1988 Annual Report of the Integrated Forest Study. ORNL/TN-11211, Oak Ridge National Laboratory, Oak Ridge, Tennessee.
Lindberg S.E., Bredemeier M., Schaefer D.A., Li Q. 1990. Atmospheric concentrations and deposition during the growing season in conifer forests in the United States and West Germany. Atmos. Environ. 24A:2207–2220.
Lindberg S.E., Johnson D.W., Lovett G.M., Taylor G.E., Van Miegroet H., Owens J.G. 1989. Sampling and Analysis Protocols of the Integrated Forest Study. ORNL/TN-11214, Oak Ridge National Laboratory, Oak Ridge, Tennessee.
Lovett G.M., Lindberg S.E., Richter D.D., Johnson D.W. 1985. The effects of acidic deposition on cation leaching from three deciduous forest canopies. Can J. For. Res. 15:1055–1060.
Matzner E., Khanna P.K., Meiwes K.J., Lindheim M., Prenzel J., Ulrich B. 1982. Elementflusse in Waldokosystemen in Solling-Datendodocumentation. Gottingen Bodenkundliche Berichte 71, Universitat Gottingen, FRG.
Mayer R., Ulrich B. 1982. Input of atmospheric sulfur by dry and wet deposition to two Central European forest ecosystems. Atmos. Environ. 12:375–377.
Mecklenburg R.A., Tukey H.B., Morgan J.V. 1966. A mechanism for leaching of calcium from foliage. Plant Physiol. 41:610–613.
Raynor G.S., Hayes J.V. 1981. Acidity and conductivity of precipitation on Central Long Island, New York in relation to meteorological variables. Water Air Soil Pollut. 15:229–245.
Raynor G.S., Hayes J.V. 1982a. Concentrations of some ionic species in Central Long Island, New York precipitation in relation to meteorological variables. Water Air Soil Pollut. 17:309–335.
Raynor G.S., Hayes J.V. 1982b. Effects of varying air trajectories on spatial and temporal precipitation chemistry patterns. Water Air Soil Pollut. 18:173–189.
Reuss J., Johnson D. 1986. Acid Deposition and the Acidification of Soils and Waters. Springer-Verlag, New York.
Richter D.D., Lindberg S.E. 1988. Wet deposition estimates from long-term bulk and event wet-only samples of incident precipitation and throughfall. J. Environ. Qual. 17:619–622.
Schaefer, D.A., Reiners W.A. 1990. Throughfall chemistry and canopy processing mechanisms. In Lindberg S.E., Page A.L., Norton S.A. (eds.) Advances in Environmental Science, Vol. 3, Acid Precipitation: Sources, Deposition, and Canopy Interactions. Springer-Verlag, New York, pp. 241–284.
Schaefer D.A., Lindberg S.E., Hoffman W.A. Jr. 1989. Fluxes of undissociated acids to terrestrial ecosystems by atmospheric deposition. Tellus 41B:207–218.
Schwartz S.E. 1989. Acid deposition: unraveling a regional phenomenon. Science 243:753–763.
Sollins P., Grier C.C., McCorison F.M., Cromack K. Jr., Fogel R., Fredricksen R.L. 1980. Internal element cycles of an old-growth ecosystem in western Oregon. Ecol. Mongr. 50:261–285.

Summers P.W., Fricke W. 1989. Atmospheric decay distances and times for sulfur and nitrogen oxides estimated for air and precipitation. Tellus 41B:286–295.

Ulrich B. 1983. Soil acidity and its relations to acid deposition. In Ulrich B., Pankrath J. (eds.) Effects of Accumulation of Air Pollutants in Forest Ecosystems. D. Reidel, Dordrecht, Netherlands, pp. 127–146.

van Breeman N., Driscoll C.T., Mulder J. 1984. Acidic deposition and internal proton sources in acidification of soils and surface waters. Nature (London) 307:599–604.

Van Miegroet H.D., Cole W. 1985. Acidification sources in red alder and Douglas fir—importance of nitrification. Soil Sci. of Am. J. 49:1274–1279.

Van Miegroet H., Cole D.W. 1988. Influence of nitrogen-fixing alder on acidification and cation leaching in a forest soil. In Cole D.W., Gessel S.P., (eds.) Forest Site Evaluation and Long-Term Productivity. University of Washington Press, Seattle, pp. 113–124.

Vet R.J., Sirois A., Jeffries D.S., Semkin R.G., Foster N.W., Hazlett P., Chan C.H. 1988. A comparison of bulk, wet-only and wet-plus-dry deposition measurements at the Turkey Lakes watershed. Canadian J. Fisheries Aqua. Sci. 45:26–37.

Vukovich F.M., Fishman J. 1986. The climatology of summertime O_3 and SO_2 (1977–1981). Atmos. Environ. 20:2423–2433.

12. Recovery from Acidification

A.O. Stuanes, H. Van Miegroet, D.W. Cole,
and G. Abrahamsen

Introduction

Acidification and, to a lesser extent, alkalinization occur in all soils. That they do so indicates these reactions are of great importance because they affect such vital ecological processes as the solubility and exchange reactions of inorganic nutrients and toxic metals, the activities of soil animals and microorganisms, and the weathering of soil minerals.

Changes in soil acidification or alkalinization result from various interacting processes that produce and consume hydrogen ions. Some of these processes are discussed in detail in previous chapters. As mentioned, soil acidification is a much more widespread process than the soil alkalinization. During soil formation all soils in areas with precipitation higher than the evapotranspiration tend to become more acid. In areas with surplus precipitation, the natural soil-forming processes that have acidified soils since their formation will continue in the same direction, that is, the process is irreversible. This trend is, of course, not a smooth line; the acidification can be faster or slower because of both natural and anthropogenic factors.

It is easy to find documentation of soil acidification introduced by different acidifying processes. Some tree species are more acidifying than others (e.g., Van Miegroet and Cole 1984, 1985; Liljelund et al. 1986; Stoner 1987; Billet et al. 1988; Holstener-Jørgensen et al. 1988; Sohet et al. 1988; Frank 1989). Increased soil acidification during the past decades in areas exposed for air pollution has been interpreted to be caused by this pollution (Matzner and Ulrich 1984; Tamm and Hallbäcken 1986; Falkengren-Grerup 1987). Application of diluted strong acids in field experiments and lysimeters have been used to examine the effect of the atmospheric input of acidifying substances (Lee and Weber 1982; Brown 1987; Abrahamsen et al. 1989; Tamm and Popovic 1989).

The crucial question is, however, if a terrestrial system that has been under acidification stress can recover when input of acid terminates. In a review of reversibility of soil and water acidification, Hauhs and Wright (1988) referred to several papers indicating recovery from water acidification but none showing recovery of soil acidification. They indirectly linked the water and soil by stating that because the rate of acidification of surface waters depends on the inherent sensitivity of the soil of the catchment, the rate of reversibility following reduction in acidifying stress also depends on the sensitivity and degree to which the system has been affected.

The important issue of recovery of soil and soil water acidification have been addressed under the IFS program. Recovery from internal acidification has been demonstrated by the Douglas fir/red alder conversion experiment in Washington State (Van Miegroet and Cole 1984, 1985, 1988; Van Miegroet et al. 1989a, 1989b). Recovery from external acidification introduced by addition of sulfuric acid has been demonstrated by the Norwegian acidification plots (Stuanes et al. 1988; Abrahamsen et al. 1989).

IFS Case Study

External Acidification

During the period from 1972 to 1975, five field experiments were established in Norwegian forests to examine the effect of artificial acidification on soil properties and growth of trees (Abrahamsen et al. 1976). One of these experiments was included in the IFS program for further studies. This experiment (A-2) was located at Nordmoen (11°6′ E, 60°16′ N) and consisted of 12 field plots of 150 m^2. The experiment was in a homogenous stand of Norway spruce (*Picea abies*) planted in 1956 on a Typic Udipsamment (Abrahamsen et al. 1976; Stuanes and Sveistrup 1979). The average stand height in spring of 1973 was 3.3 m, rising to 8.7 m in the fall of 1988. The plots were watered 27 times, 5 cm each month, in the frost-free period from July 1973 to September 1978 by water applied above the canopy by a sprinkler system. The plots received artificial precipitation in addition to an annual average of 74.2 cm of natural precipitation for the period in question (Abrahamsen 1980a). Since the watering terminated in the fall of 1978 the plots have only received natural precipitation.

Groundwater with the pH adjusted by sulfuric acid was used as artificial rain, giving treatments of pH 6, pH 4, pH 3, and pH 2.5. In addition, areas between the watered plots were used for the not-watered (n.w.) treatment. Each treatment was replicated three times in a randomized design. The pH and lime potential (LP) of bulk precipitation for the watering period are shown in Table 12.1. The high LP of the pH 6 and the pH 4 treatments compared to the not-watered treatment are caused by higher amounts of Ca and Mg in the groundwater than in the precipitation.

Tree height and girth increments were measured annually from fall of 1974. Foliar nutrient concentrations in both current and previous year needles were measured from fall of 1973, except for 1979 and 1980. Needle samples from eight trees per plot were taken from the third or fourth branch whorl from the top of the trees. Soil samples were taken from the O, E, Bs1, and Bs2 horizons in 1975, 1978, 1981, 1984, and 1988. Twenty subsamples were pooled by horizon for each plot. Some of the results from this experiment were reported earlier (Abrahamsen 1980a, 1980b; Stuanes 1980; Tveite 1980a, 1980b; Tveite and Abrahamsen 1980; Abrahamsen 1987; Stuanes et al. 1988; Abrahamsen et al. 1989).

Ten throughfall and litterfall collectors have been in place on each plot (random placement) since 1982. In 1986, four plots were chosen for more intensive studies as part of the Norway spruce site (NS) within the IFS program. Two plots were watered with artificial rain of pH 6 and two with artificial rain of pH 2.5 during the period mentioned. Soil solution has been collected within these plots since October 1986. When discussing the aspect of recovery from acidification, the results from the whole experiment will be taken into account.

Internal Acidification

At the University of Washington Thompson Research Center in the Cedar River Watershed, a naturally established stand of red alder (*Alnus rubra* Bong.) abuts a Douglas fir plantation (*Pseudotsuga menziesii* (Mirb.) Franco). The two stands developed almost simultaneously following logging of the original old-growth forest approximately 55 years ago. The soil at the sites is a Dystric Entic Durochrepts (Van Miegroet and Cole 1984, 1985). Differences in parent material, microclimate, stand age, or prior logging history are excluded as significant sources of variability between the forest ecosystems.

Table 12.1. pH and Lime Potential (LP) of Bulk Precipitation and Total SO_4-S Supplied in Different Treatments (data are for watering period)

Treatment	Not Watered (n.w.)	pH 6	pH 4	pH 3	pH 2.5
Bulk pH	4.36	4.46	4.26	3.61	3.16
Bulk LP	1.8	2.2	2.0	1.4	0.9
SO_4-S (kmol ha^{-1})	1.25	1.72	4.06	10.00	23.44

The soil in the red alder stand has been influenced by high nitric acid inputs originating from internal nitrification for several years. The nitrification under Douglas fir has shown to be practically nonexistent. Similarities in other factors suggest the differences in soil and solution chemistry can be attributed primarily to differences in vegetation composition and associated changes in N status. In 1979 a 20 × 20 m growth plot was established in each of the red alder and Douglas fir stands. Soil pits were dug just outside the growth plots, and tension lysimeters were installed below the forest floor and at 10 and 40 cm below the forest floor–mineral soil boundary. In addition, precipitation and throughfall were collected. After 50 years of alder occupancy, an average of 3500 mol H^+ ha^{-1} is currently added annually via nitrification of the organic N pool compared with annual input by precipitation of about 250 mol (Van Miegroet and Cole 1984, 1985; Van Miegroet 1986).

These two adjacent ecosystems, one with high nitrification and one with almost no nitrification, was an unique opportunity for the IFS project to study the recovery from internal acidification. In 1985 clear-cut/forest conversion plots were established to provide further information on (1) the initial stages of soil acidification from nitrification by planting alder seedlings on N-poor "Douglas fir" soil; and (2) reversal of some of the soil/solution changes after acid input has stopped by removing the alder cover and replacing it by Douglas fir seedlings.

Two 50 × 100 m planting plots were established in each of adjacent red alder and Douglas fir stands. Within each planting plot, eight 15 × 15 m subplots were established. Before clear-cutting and replanting, measurements and sampling of overstory biomass, understory biomass and elemental content, forest floor mass and elemental content, and soil chemical content were carried out. The planting plots were clear-cut in 1984 and the subplots were planted with red alder or Douglas fir seedlings in February 1985, yielding four distinct treatments:

Original	Replanted
Red alder	Red alder
Red alder	Douglas fir
Douglas fir	Red alder
Douglas fir	Douglas fir

Near the center of each planting plot, four tension lysimeter plates were installed in November 1984 at each of three depths: directly beneath the forest floor and 10 and 40 cm below the forest floor–mineral soil boundary. Samples of seedlings, understory, forest floor, and mineral soil have been taken at fixed intervals (Cole 1985).

Acidification

Sources of Acidification

The most dominant internal source of acidification under red alder is nitrification. Nitrification is a microbiological process that converts NH_4^+ into NO_3^+. During this reaction H^+ is released in a quantity of 2 moles of H^+ per mole of NO_3^+ into the soil solution. The H^+ can exchange base cations from the exchange complex, which can be leached down in the soil profile with mobile NO_3^+ as counter ion (Wiklander 1976; Johnson 1981). Nitrification can therefore contribute to soil and soil solution acidification by internal H^+ production and by the production of a mobile anion. Nitrification is highly dependent on the supply of nitrogen in the system. Red alder may fix N at rates ranging between 50 to 200 kg N ha^{-1} per year (Franklin et al. 1968; Zavitkovski and Newton 1968; Cole et al. 1978; DeBell and Radwan 1979; Binkley 1981).

Under spruce forest on acid soils ($pH < 4.5$), carbonic acid contributes very little to the internal acidification because of limited dissociation and thereby a little influence on the cation leaching. The upper horizons of these soils are dominated by organic acids (Johnson et al. 1977; Ugolini et al. 1977; Krug and Frink 1983). Cronan and Aiken (1985) found more DOC (dissolved organic compounds) in the soil solution under the O/A horizons in mixed and coniferous forest than in hardwood forest. The amounts of DOC decreased sharply when passing the B horizons. The concentrations of soluble humic substances varied significantly among different soils, vegetation types, and seasons. They concluded that dissolved humic substances are of great importance for the soil solution acidity of these soils. Mollitor and Raynal (1982) found that organic anion leaching was more important than sulfate leaching in a conifer site, but that they were equal in a hardwood site.

The amounts of DOC can be regulated by changes in ionic strength and anion composition from external sources as acid precipitation (Evans et al. 1988). The amounts of dissolved humic substances decreased with increasing ionic strength, but addition of SO_4^{2-} to the leaching solution resulted in elevated DOC levels. The effect of added NO_3^- was not very clear (Evans et al. 1988). David et al. (1989) found that the solutions from the O horizon, which were dominated by hydrophobic and hydrophilic acids, were altered by acid inputs. The main change was a decease in hydrophobic acids and an increase in hydrophilic acids with increasing acidity. The input of external mineral acids may therefore lead to altered acidity caused by charge differences between hydrophilic and hydrophobic acids.

As already mentioned, external sources of acidification can be mineral acids such as HNO_3 and H_2SO_4. These acids contribute to the acidification of the IFS site to a different degree. However, measurements must be taken over a long span of time to measure any effects of their contribution to the acidity of soil and soil solution. To speed up this acidification effect, groundwater acidified with sulfuric acid was added to part of the Norwegian site. In this case the most acidic treatments were overwhelming the internal sources of acidification.

How Acidification Is Measured

Different measurements can be used to express acidification of the soil solid phase. Soil pH, base saturation, exchangeable acidity, and exchangeable Al have been used in this study. In addition the nitrification potential by use of the buried bag technique was used in the conversion plots (Van Miegroet et al. 1989b). The methods differ slightly between the Douglas fir/red alder conversion plots and the artificial acidification plots, but the methods have been consistent within plots and over time. In the Douglas fir/red alder plots pH is determined in both H_2O and 0.01 *M* $CaCl_2$ solution at a soil to solution ratio of 1:1. Exchangeable bases and Al have been measured after extraction with 1 *M* NH_4Cl, which gave a CEC and base saturation at soil pH. NH_4^+ and NO_3^+ were determined after extraction with 2 *M* KCl (see Cole 1985 for further information).

In the Norwegian acidification plots, pH was determined in water at a soil:water ratio of 1:2.5. Exchangeable bases were measured after extraction with 1 *M* NH_4OAc at pH 7 and potential acidity by titrating the extract back to pH 7 (Ogner et al. 1984). This gives a CEC and base saturation at pH 7, that is, a higher CEC and lower base saturation compared to the determination at soil pH. However, the pH 7 procedure gives a more stable baseline for using changes in base saturation as a measure of changes in soil acidification (Stuanes et al. 1988). Exchangeable Al was determined after extraction with 1 *M* KCl or 1 *M* NH_4NO_3. Results obtained by using 1 *M* KCl, NH_4Cl, or NH_4NO_3 as extractants for determination of Al and exchangeable acidity are quite similar (Stuanes et al. 1984). The amounts of exchangeable Na^+, K^+, Mg^{2+}, and Ca^{2+} after extraction with 1 *M* NH_4Cl, NH_4NO_3, or NH_4OAc are also quite similar.

The chemistry of the solution phase has been measured in the red alder and Douglas fir stands since 1979 and was continued after the conversion plots were established in 1984. In the Norwegian acidification plots the solution phase has been measured only since fall 1986. In all plots the solution phase has been collected by use of tension lysimeters at 10 kPa; the alundum type at the conversion plots (Cole 1968) and the fritted-glass type at the acidification plots. Solution samples were removed from each collection bottle every month. Changes in all parameters in the solution phase can be an indication of an acidification trend. The most widely used are pH, concentration of Al, and alkalinity, but increasing amounts of mobile anions are also a good indicator of leaching and thereby acidification. Each sample was therefore analyzed for pH, specific conductance, and concentrations of cations and anions. The cations, except for NH_4^+, were determined with an atomic absorption spectrophotometer or an inductively coupled plasma spectrophotometer (ICP). NH_4^+ was determined on a Technicon Auto Analyzer. The anions, except for HCO_3^-, were determined on auto analyzer or ion chromatograph. The HCO_3^- concentration was calculated from the alkalinity determined by titration down to pH 4.5 (for the acidification plots and through 1984 for the red alder/Douglas fir) and down to 5.0 for the conversion plots after 1985 (Ogner et al. 1984; Van Miegroet and Cole 1984; Cole 1985).

The response of organisms including trees on acidification can be measured in different ways. Height increment and diameter at breast height have been measured on a certain number of trees every year or every second year. Foliar analysis have been carried out at the same intervals (Cole 1985; Stuanes et al. 1988). Changes in the understory biomass can be followed at the conversion plots (Cole 1985). At the acidification plots changes in the populations of some species of the soil fauna have been used as indicators of soil acidification (Hågvar 1980; Hågvar and Amundsen 1981; Abrahamsen 1983a; Hågvar 1984).

Effects of Acidification

The effects of acidification can be expressed by changes in soil chemical properties in both the solid and solution phase. These effects can introduce changes on the biota living in the soil. The organisms that have been studied in connection with the reported experiments are trees and soil fauna.

Effects of Acidification on the Soil Solid Phase

External Acidification. Before the watering of the Norwegian acidification plots was begun, some random soil samples were taken and pooled by horizon. The first sampling by treatment was carried out in 1975. These results clearly showed that the variation between the replicated plots within the treatments was quite significant, and therefore the results from the random sampling could not be used as a starting point for all plots. The first results for the solid phase of the soil are therefore from 1975 even though the watering started in July 1973. Samples were not collected from the Bs2 layer in 1975. The last watering was performed in September 1978 and the soil samples were taken in October this year. The extra external application of H_2SO_4 had a clear effect on the soil pH and base saturation especially in the upper soil horizons (Figures 12.1 and 12.2, respectively).

In 1975, the difference between highest and lowest pH in the O horizon was 0.3 pH units compared to a difference of 0.5 pH units in 1978 (Figure 12.1). The differences in the lower horizons were of the order of 0.3 pH units. There was a clear effect on the soil pH of the higher lime potential of the pH 6 and pH 4 treatments (see Table 12.1) for the upper horizons. The pH of the not-watered treatment also showed a decreasing trend from 1975 to 1978. The decrease was 0.25 pH units for the O horizon and 0.15 pH units for the E horizon (see Figure 12.1). This is most likely caused by different internal acidification sources, for example, excess cation uptake

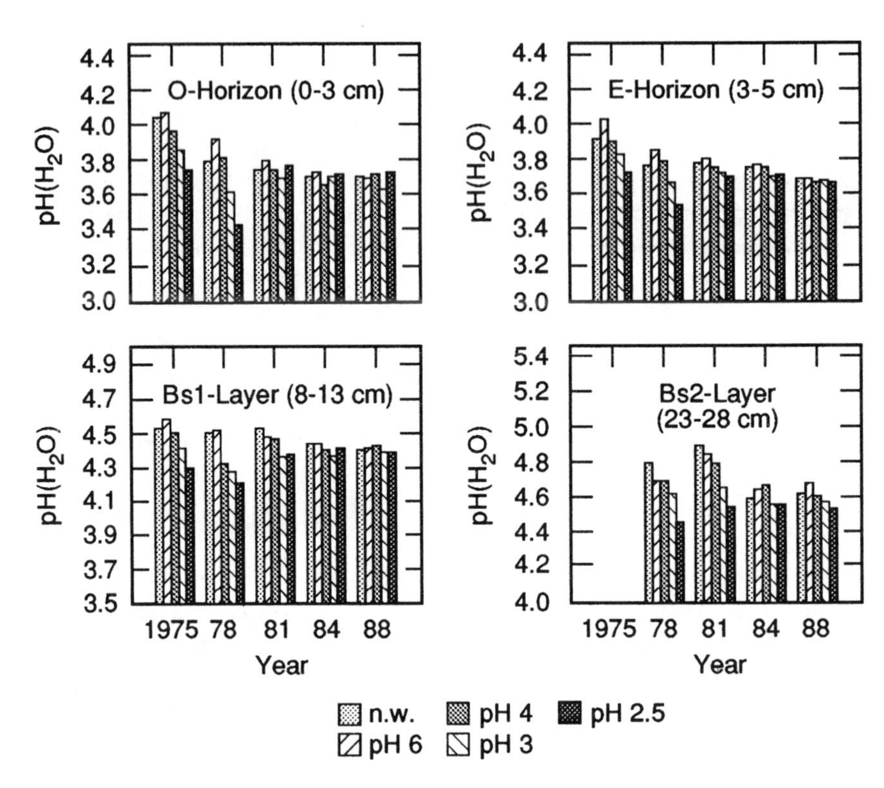

Figure 12.1. Effects of qualities of artificial rain on soil pH with increasing soil depth.

over anion uptake because most of the N is taken up as NH_4^+ (Abrahamsen 1987).

The potential CEC (1 *M* NH_4OAc) did not show changes by treatment and varied from 750 mmol(c) kg^{-1} in the O horizon to about 40 mmol(c) kg^{-1} in the Bs2 layer (Stuanes et al. 1988). Because CEC is not changed, the base saturation is a good measure of alterations in the pool of exchangeable cations in the soil. The higher lime potential of the pH 6 and pH 4 treatments have given a more clear effect on the base saturation than on the soil pH for the years with watering for the upper horizons (see Figure 12.2). The pH 3 and pH 2.5 treatments, however, were overshadowing the positive effect of some additional Ca^{2+} and Mg^{2+} in the water used for watering. For the pH 2.5 treatment the base saturation in the O horizon was 9% in 1975, decreasing to 5% in 1978; the same figures for the not-watered treatment were 18% and 14%. There was also a clear effect of the most acidic treatment in the E horizon but not in the Bs layers (see Figure 12.2). For the Bs layers, it must be mentioned that both CEC and amounts of exchangeable bases was so low that very small differences can give large variations in base saturation. The figures for exchangeable Ca^{2+} and Mg^{2+} followed the

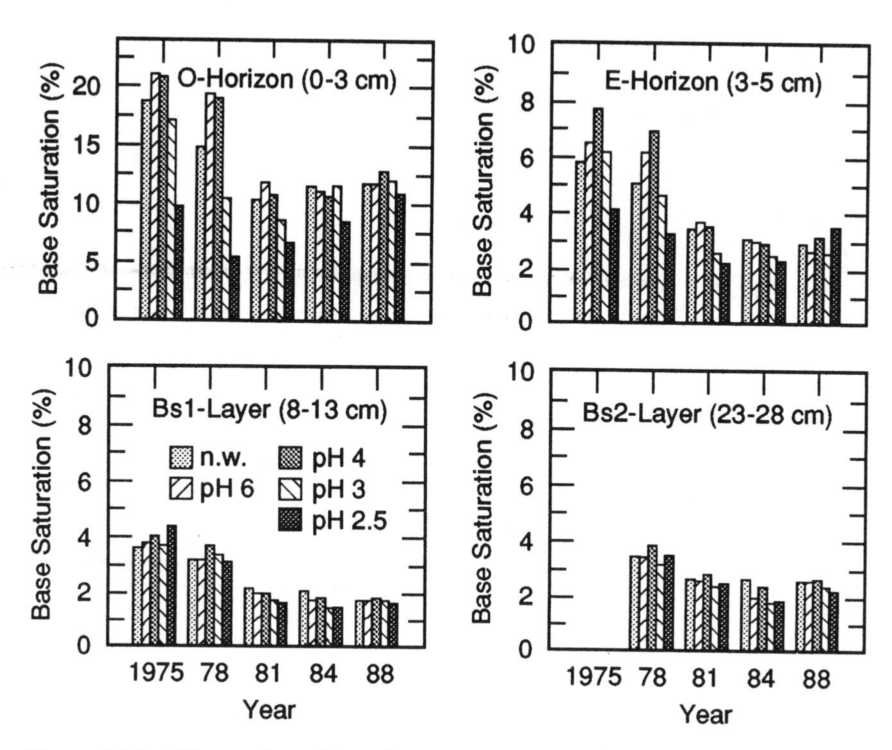

Figure 12.2. Effects of qualities of artificial rain on soil base saturation with increasing soil depth.

same trends as for base saturation, although with greater variations. The amount of exchangeable K^+ appeared to be influenced by treatment only in the O horizon, and even there the influence was relatively small (Abrahamsen 1980a).

The total load of SO_4^{2-} during the watering period is shown in Table 12.1. The highest load is equivalent to 27 years with 100 cm precipitation of pH 4.2 and a concentration of Ca^{2+} + Mg^{2+} equal to 7 μM (Stuanes et al. 1988). The amounts of soil extractable SO_4^{2-} showed only small changes in the upper horizons because of low sorption capacity for SO_4^{2-}. In the Bs1 layer, the amount of extractable SO_4^{2-} for the most acidic treatment was twofold higher than the amount for the not-watered treatment in 1975 and 3.5 fold higher in 1978 (Abrahamsen 1980a).

Bits of pure cellulose and sticks of aspen wood placed in the Norwegian acidification plots showed no reduced decomposition rate except for the pH 2.5 treatment. No effect of acid application alone was found in three other field experiments (Hovland and Abrahamsen 1976).

Internal Acidification. At the end of the first rotation of red alder (after about 50 years), there were some striking differences in the soil chemistry between the red alder and Douglas fir stands (see Table 12.2).

Table 12.2. pH, CEC, Base Saturation, and Exchangeable K^+, Mg^{2+}, and Ca^{2+} for Soil Horizons in Douglas Fir and Red Alder Plots[a]

Horizon and Depth (cm)	Total N (mmol kg^{-1})	pH	CEC (cmol(c) kg^{-1})	Base Saturation (%)	Exchangable K^+ (kmol(c) ha^{-1})	Exchangable Mg^{2+} (kmol(c) ha^{-1})	Exchangable Ca^{2+} (kmol(c) ha^{-1})
Douglas Fir							
A1 (0–7)	136	5.0	12.3	31	1.3	1.2	10.3
A2 (7–15)	93	5.2	9.1	16	0.8	0.7	4.5
B21 (15–30)	71	5.1	8.1	9	1.2	1.0	4.0
B22 (30–45)	57	5.1	5.8	8	1.6	0.7	2.2
Total					4.9	3.6	21.0
Red Alder							
A1 (0–7)	243	4.6	10.1	15	0.8	0.4	2.6
A2 (7–15)	164	4.8	9.8	16	0.7	0.4	3.4
B21 (15–30)	107	5.2	7.4	19	1.3	0.6	6.9
B22 (30–45)	79	5.2	5.7	26	1.7	0.7	6.2
Total					4.5	2.1	19.1

[a] N, pH and CEC are from Johnson et al. (1981); other numbers from Van Miegroet and Cole (1984).

Much higher amounts of N have accumulated in the soil of the red alder stand than in the Douglas fir stand (see Table 12.2). Except for Mg^{2+}, the base cations have apparently not been leached out below the 45-cm depth but only redistributed. This is clearly seen from the total amounts of K^+ and Ca^{2+} down to 45 cm in the two stands. For Mg^{2+} there was no such clear redistribution, and the net loss of Mg^{2+} appear to be about 40% higher in the red alder than in the Douglas fir. The redistribution of especially Ca^{2+} was also clearly expressed in the base saturation. The pH of the A horizon was 0.4 pH unit lower in the red alder, but slightly higher (0.1 pH unit) in the B horizon (see Table 12.2).

Effects of Acidification on the Soil Solution

External Acidification. No soil solution measurements were taken from the Norwegian acidification plots during the watering period, but a lysimeter experiment was performed with the same soil as in the plots. The lysimeters were also watered for 5 years (Abrahamsen 1980a, 1983b). The lysimeters were watered with the same amount and frequency of acidified groundwater as the field plots, but the pH 2.5 treatment was not included, instead there was a pH 2 treatment. The pH of the leachate leaving the lysimeter at the bottom (40 cm depth) had been influenced by the amount of H_2SO_4 in the artificial rain. There were no clear differences between the not-watered and the two least acidic treatments because differences in bulk pH and lime potential were small. The pH of the pH 3 treatment (bulk pH 3.5) dropped from about pH 5.5 at the start of the watering to about 4.5 after 5 years. The pH 2 treatment (bulk pH 2.5) gave a clearly lower pH in the leachate, quite constant around pH 4 with only a slight decreasing trend (Abrahamsen 1980a). The same differences could also be seen in the concentrations of Mg^{2+}, Ca^{2+}, and Al (see Table 12.3), but they were especially pronounced for Al (Abrahamsen 1980a, 1983b).

There was a net loss of K^+, Mg^{2+}, Ca^{2+}, and Al from the lysimeters and a net accumulation of H^+ and SO_4^{2-} (Abrahamsen 1980a). Only 9% and 2% of the input H^+ in the lysimeters was found in the output for the pH 6 treatment and pH 2 treatment, respectively. This high retention of H^+ in the soil reflects the importance of buffering by exchange of bases and by weathering (Abrahamsen 1980a).

Table 12.3. Average Concentrations of Ca^{2+}, Mg^{2+}, and Al (in μmol L^{-1}) in Leachate from 40-cm-Deep Lysimeters for April 1976 to June 1979[a]

Treatment	Not Watered (n.w.)	pH 6	pH 4	pH 3	pH 2
Mg^{2+}	26	19	36	55	112
Ca^{2+}	62	71	88	145	214
Al	13	12	23	83	1288

[a] Data from Abrahamsen (1983b).

Internal Acidification. The high nitrification rate in the red alder is also reflected in the ion fluxes in the two stands as shown in Table 12.4. Most of the NO_3^- in the red alder site is produced in the upper part of the soil, and the amount leaving at the 40-cm depth was about 25 times higher than what was added in precipitation. The net loss of NO_3^--N from the red alder site was about 50 kg ha^{-1} yr^{-1}, compared to about 2 kg from the Douglas fir site. This high annual production of NO_3^- had to be accompanied by the release of an equivalent amount of H^+ (Van Miegroet and Cole 1984). The nitrification process in this soil represented an internal H^+ source 10 times stronger than the input by the ambient precipitation (see Table 12.4).

The pH of the soil solutions in the red alder was also lower than at the Douglas fir. At the 10-cm depth, the difference was about 1 pH unit. This gave a lower pH in the upper soil horizons than in the ambient precipitation for the red alder, while the opposite was true for Douglas fir (Van Miegroet and Cole 1984). As a consequence of the lower pH at the 10-cm depth under red alder, the concentration of Al in the soil solution was 23 μmol L^{-1}, periodically as much as 150 μmol L^{-1}, compared to 5 μmol L^{-1} under Douglas fir (Cole 1985; Van Miegroet and Cole 1988). There was a net leaching of SO_4^{2-} under Douglas fir while a net accumulation was measured under red alder (see Table 12.4). This may result from a higher anion sorption capacity under red alder caused by a lower pH.

The H^+ flux was considerably lower than the NO_3^- flux in the red alder soil, which must result from a strong H^+ buffering of the soil solution. An increase in the flux of cations (see Table 12.4) and a decline in exchangeable base cations in the upper part of the soil (see Table 12.2) indicated exchange of base cations for H^+ as one of the buffering mechanisms involved (Van Miegroet and Cole 1985; Van Miegroet et al. 1989b). Although the net annual leaching losses of K^+, Mg^{2+}, and Ca^{2+} were only 1.7% of the exchangeable pool in the top 45 cm of the soil in the Douglas fir, as much as 14% was lost in the red alder (see Tables 12.2 and 12.4). There were interesting differences in the leaching pattern among K^+, Mg^{2+}, and Ca^{2+}: Mg^{2+} was leached faster than K^+ through the soil under red alder, and the exchangeable pool of Mg^{2+} down to 45 cm in the soil under red alder had almost been reduced by 50% compared to the soil under Douglas fir (see Table 12.2), indicating that weathering has not kept up with the leaching of Mg^{2+}.

The internal H^+ production was dominated by H_2CO_3 in the Douglas fir stand (see Table 12.5). The production of H_2CO_3 was slightly higher than the H^+ input by precipitation. In the red alder stand the internal H^+ production by H_2CO_3 was the same size as the input by precipitation while the internal production by HNO_3 was about 10 fold higher (see Table 12.5). The concentration of organic acids indirectly calculated from anion deficit was negative for Douglas fir and close to 0 for red alder. The leaching output of H^+ below the 40-cm depth was about 7% of the total H^+ input by precipitation and internal sources for the Douglas fir and 1.4% for red alder.

Table 12.4. Annual Flux of H^+, NH_4^+, Mg^{2+}, Ca^{2+}, NO_3^-, SO_4^{2-}, Total Cations, and Total Anions for January 1981 to May 1982[a]

	H^+	NH_4^+	Mg^{2+}	Ca^{2+}	Total Cations	NO_3^-	SO_4^{2-}	Total Anions
Douglas Fir								
Precipitation	332	120	80	183	1312	155	524	1368
Throughfall	124	30	137	316	1376	13	553	1288
Forest floor	82	18	228	743	1570	13	572	1152
10 cm	14	43	229	545	1480	8	536	1610
40 cm	48	40	280	497	1653	10	763	1854
Balance[b]	+284	+80	−200	−314	−341	+145	−239	−486
Red Alder								
Precipitation	332	120	80	183	1312	155	524	1368
Throughfall	57	54	139	239	1271	21	441	1328
Forest floor	275	64	560	1698	3741	2272	843	3857
10 cm	327	26	721	2021	4227	4185	427	5137
40 cm	60	69	803	2765	5035	3791	355	5084
Balance[b]	+272	+51	−723	−2582	−3723	−3636	+169	−3716

[a] Data from Cole (1985) (in mol(c) ha^{-1} yr^{-1}).
[b] +, net accumulation within the ecosystem; −, net loss from ecosystem.

Table 12.5. Mean Annual H^+ Input via Precipitation and Internal H^+ Production by H_2CO_3, HNO_3, and Annual H^+ Leaching Loss[a]

	Precipitation Input	Internal Production by:		Leaching Output of H^+
		H_2CO_3	HNO_3	
Douglas fir	330	400	0	50
Red alder	330	260	3640	60

[a] Data (in mol (c) ha^{-1} yr^{-1}) taken from Table 12.4 and Cole (1985).

The initial stages of soil acidification from nitrification were studied by planting red alder seedlings on N-poor "Douglas fir" soil after clear-cutting the Douglas fir stand in 1984 (see p. 470) and comparing the effect of red alder seedlings with that of Douglas fir seedlings (Cole 1985). The mean height of the red alder seedlings at planting in 1985 was 0.73 m, rising to 6.61 m in 1988, while the comparable mean heights of the Douglas fir seedlings were 0.38 m and 1.55 m, respectively.

Volume-weighted average concentrations of Ca^{2+} and NO_3^- in the soil solution for the years 1985, 1986, 1987, and 1988 did not indicate any clear effects of planting alder on previous Douglas fir soil (see Table 12.6). There was a slight increase in the NO_3^- concentration in the forest floor for the years 1987 and 1988 where alder was planted. This might be an indication of a nitrogen fixation and increased nitrification but will have to be verified by further sampling.

Tree Growth Response to Acidification

External Acidification. Effects of acidification on the tree growth have been studied in the Norwegian acidification experiment. Little effect of the acid application could be observed during the period with watering (see Figure 12.3). In the following years the growth, both height and basal area growth, declined significantly in the most acidified plots and minimum growth was recorded in the years 1980 to 1984. Since then, the height growth in particular, but also to some extent the basal area growth, has improved (see Figure 12.3) (Stuanes et al. 1988; Abrahamsen et al. 1989). A similar effect has been reported from an acidification experiment in a young Scots pine stand (Abrahamsen et al. 1987, 1989; Tveite et al. 1990/91). In three other acidification experiments, however, no significant effect of the acid application on tree growth was found (Abrahamsen et al. 1989).

The effects on tree growth are most probably nutritional effects. The stimulatory effect on tree growth in the first years after the commencement of acid application was probably from a better supply of available nitrogen, and the later negative growth development was probably caused by shortage in Mg. This was more pronounced in the Scots pine experiment (Abrahamsen et al. 1987) than in the Norway spruce experiment (Stuanes et al. 1988). Whether the reduced content resulted from too low concentrations of ex-

Table 12.6. Volume-Weighted Average Concentrations for Ca^{2+} and NO_3^- (in $\mu mol\ L^{-1}$) in Soil Solution at Different Soil Horizons and Treatment

	Ca^{2+}				NO_3^-			
	1985	1986	1987	1988	1985	1986	1987	1988
Douglas fir control								
Forest floor	128	130	191	205	2	3	1	0
A horizon	53	51	102	104	1	0	0	0
B horizon	44	39	45	46	1	0	0	0
Douglas fir to red alder								
Forest floor	75	204	197	235	6	5	28	30
A horizon	92	66	78	89	6	4	0	3
B horizon	51	53	47	55	0	2	0	0
Douglas fir to Douglas fir								
Forest floor	123	—	168	103	35	—	14	2
A horizon	51	36	47	47	3	0	0	0
B horizon	55	45	45	53	0	1	1	0

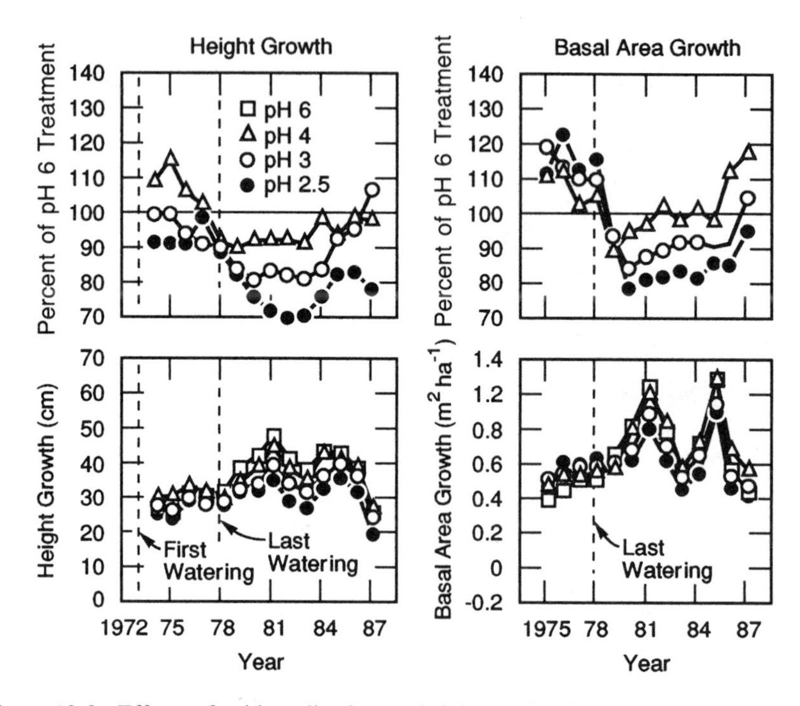

Figure 12.3. Effects of acid application on height and basal area growth of Norway spruce. Two upper figures show growth in percent of control (pH 6); two lower figures show actual growth (Stuanes et al. 1988).

changeable Mg^{2+} in the soil or from antagonism between Mg^{2+} and other elements in the uptake from soil is unclear. Needle analyses also showed increased K^+ concentrations with increasing acid load. At the same time, the concentration of exchangeable Mg^{2+} in the soil was reduced whereas the concentration of exchangeable K^+ was about the same. Other nutrients, except for N, appeared to be in surplus (Abrahamsen et al. 1989). Results from the lysimeter experiment indicated that Al concentrations greater than 0.4 m*M* may have prevailed in the soil solution over long periods, at least in the pH 2.5 treatment (Stuanes et al. 1988). This is quite close to concentrations found to be injurious to spruce, but not to pine (Abrahamsen 1984; Eldhuset 1988).

Internal Acidification. Seedlings of red alder and Douglas fir were planted in 1985 on soil previously occupied by red alder. Height increments since planting (see Figure 12.4) indicated that prior alder occupancy had reduced the growth potential of planted alder (Van Miegroet et al. 1989a). Alder seedlings have grown slower on the site more N enriched for all reported years after planting as compared to the Douglas fir site, which is poorer in N. This suggests that the previous growth of red alder for more than 50

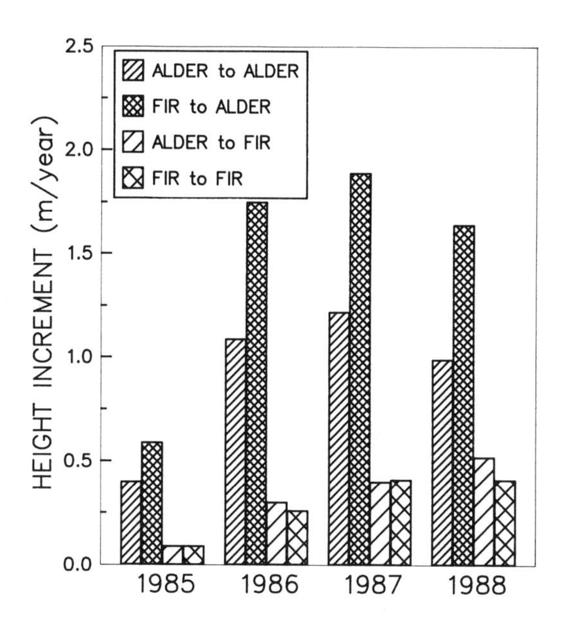

Figure 12.4. Height increment of red alder and Douglas fir seedlings planted on soil previously covered by red alder and Douglas fir, respectively.

years had introduced changes in soil properties leading to a decline in fertility of the soil. An indication of these changes was the decreased growth of red alder (Van Miegroet et al. 1989a). There was no difference in the growth of Douglas fir seedlings between the sites (see Figure 12.4), which might result from a more abundant understory vegetation in the N enriched site. When the trees are fully established and can shade out the competing vegetation, a positive growth response might be seen on the N enriched site (Van Miegroet et al. 1989a). In 1987 there was already a significantly higher foliar N concentrations of Douglas fir seedlings growing on previous red alder sites as compared to those growing on previous Douglas fir sites.

It is difficult to find a single parameter that can explain the slower growth of red alder on soil previously occupied by red alder compared to growth on the same soil previously occupied by Douglas fir. Some selected soil parameters for the two plots are shown in Table 12.7. The data shown in Table 12.7 confirmed the soil data from the control plots reported in Table 12.2. The pH difference is very clearly expressed in Table 12.7, as also are differences in N, Al, and available P, and CEC was measured to be higher in the soil previously covered by red alder than in the soil covered by Douglas fir. Van Miegroet et al. (1989a) discussed the possibility that the acidification in the upper part of the rooting zone might have negatively affected the availability of certain nutrient such as P. They indicated that higher concentrations of Al in soil solution under red alder might have influenced the precipitation of P.

Table 12.7. Selected Soil Parameters for Red Alder Soil Previously Covered by Red Alder or Douglas fir[a]

Depth (cm)	Kjeld.-N (%)	pH (H_2O)	Base sat. NH_4Cl (%)	CEC NH_4Cl (mmol(c) kg^{-1})	Exch. Al (mmol kg^{-1})	Available P (mmol kg^{-1})
Red Alder to Red Alder						
0–15	0.55	4.49	10.4	165.6	6.30	0.63
15–30	0.36	4.91	6.1	140.6	1.16	0.39
30–45	0.36	4.93	11.3	133.3	4.30	0.29
Douglas Fir to Red Alder						
0–15	0.15	5.30	9.2	92.3	0.93	2.05
15–30	0.11	5.34	10.5	86.3	0.03	1.04
30–45	0.11	5.31	7.4	70.0	0.03	0.89

[a] Values are means of eight subplots.

Another possibility is that microbial immobilization in the upper part of the soil with highest contents of N and C might have influenced the retention of P. The C:N ratio in the upper 15 cm was highest in the soil under Douglas fir, 25.4 compared to 20.2 under red alder. As shown in Table 12.7, however, available P was lower in the soil under red alder compared to Douglas fir. This difference could not be seen in foliar P concentrations of alder seedlings growing on previous Douglas fir sites and those growing on previous red alder sites; the figures were 1.42 mg g^{-1} and 1.47 mg g^{-1}, respectively. There was, however, a significant lower P concentration in fireweed (*Epilobium angustifolim* L.) growing on previous red alder sites compared to Douglas fir sites (Van Miegroet et al. 1989a). This difference might be because the root system of the weeds remained largely in the upper soil horizons; it does not prove that the observed slower growth of red alder on previous red alder sites is linked to P availability (Van Miegroet et al. 1989a).

Soil Fauna Response to Acidification. The abundance of soil animals such as collembola, mites, and enchytraeids in the soil of the acidified plots has been reported by Hågvar and Amundsen (1981), Abrahamsen (1983a), and Hågvar (1984). Plots supplied with artificial rain of pH equal to or above 3 generally had a higher abundance of these soil animals. At pH 2.5, the abundance decreased. The most significant decrease was found for the enchytraeids and the least decrease for the mites. An experiment allowing the animals to colonize sterile soil cores pretreated with dilute H_2SO_4 or lime supported the results from the field plots (Hågvar and Abrahamsen 1980). Some species preferred limed soil and others acidified soil, but the total number of enchytraeids and collembola was not significantly affected by any of the treatments. The mites, however, were most abundant in the acidified cores.

Recovery from Acidification

Technique Used to Stop Acidification

The artificial acidification of the Norwegian field plots was terminated in September 1978. Since then, the plots have only been exposed to the ambient wet and dry atmospheric depositions. The internal natural acidification is still active after finishing the watering. The ambient deposition and the natural internal acidification have been the only contributors to the not-watered plots during the whole period. The recovery can be estimated by comparing the changes in the previously acidified plots with the not-watered or pH 6 treatment plots.

The source for internal acidification by increased nitrification from N fixation by red alder was stopped by removing the red alder cover, which was done in September 1984. In the acidified soil, seedlings of red alder and Douglas fir were planted in February 1985 (see p. 470). The recovery from acidification after more than 50 years of N fixation and nitrification can be studied by comparing soil and soil solution under the replanted soil previous covered by alder and Douglas fir.

Measuring Recovery

External Acidification

The pH in the O horizon of the Norwegian acidification plots increased by 0.34 pH unit from 1978 to 1981 for the most acidic treatment (pH 2.5) (see Figure 12.1). A pronounced but moderate increase was also measured for the pH 3 treatment, but for the other treatments there were no changes. The same pattern was also seen for the E horizon. For the Bs1 layer, there was even a small increase for the pH 4 treatment (see Figure 12.1) (Stuanes et al. 1988). The soil pH responded quickly to the termination of the artificial acidification.

In the O horizon, the base saturation has increased gradually for the most acidified treatment from 1978 to 1988. In 1988 the base saturation was about 11% for all treatments (see Figure 12.2). For the pH 3 treatment the base saturation was lowest in 1981. Measured differences in base saturation between treatments for the E horizon leveled out after the watering was stopped. There were no differences between treatments for the Bs1 and the Bs2 layers, but a slight decreasing trend was measured until 1988 (see Figure 12.2).

The amount of extractable SO_4^{2-} in the Bs2 layer decreased from 1978 to 1988 for the pH 2.5 and the pH 3 treatments (see Figure 12.5). The three other treatments have been on almost the same level. The amount for the pH 2.5 treatment was higher than for the other treatments, even 10 years after the artificial acidification stopped (see Figure 12.5).

No soil solution data were available from these acidification plots before October 1986 when two of the pH 6 treatment plots and two of the pH 2.5

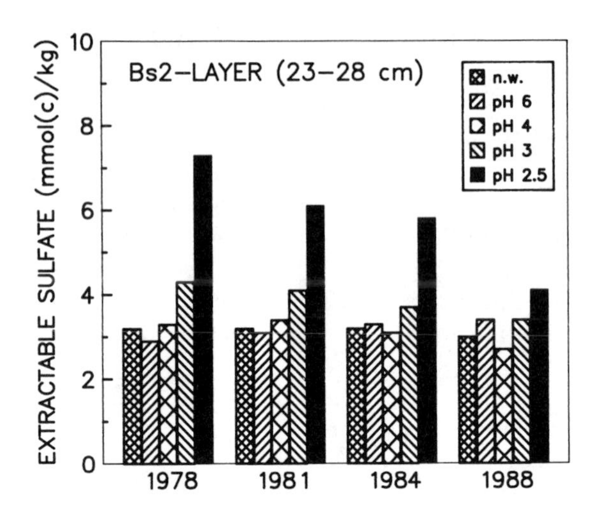

Figure 12.5. Effects of qualities of artificial rain on soil extractable sulfate in Bs2 layer.

treatment plots were included in the IFS study. Three replicated fritted-glass lysimeters for each depth were placed within each plot. For most of the elements there was no difference between the fluxes for the two treatments, but very distinct differences were measured for H^+, SO_4^{2-}, Al, and Mn (see Table 12.8). The soil solution data in Table 12.8 support the differences in extractable SO_4^{2-} shown in Figure 12.5. The pH 6 treatment was accumulating while the pH 2.5 treatment was leaching SO_4^{2-}.

Internal Acidification

The question of recovery from acidification can also be illustrated by studying the plots where previous red alder was replaced with red alder or Douglas fir. The N fixation was effectively stopped by clear-cutting part of the old red alder stand. Soil properties were quite similar for the sites planted with red alder and Douglas fir. Data for red alder to red alder in Table 12.7 can therefore be used as the starting point. There has been no resampling of the soil yet, so the recovery of the soil solid phase cannot be judged.

Young red alder seedlings had not shown to have capacity to increase the NO_3^- leaching after planting on the same soil previous covered by Douglas fir. As shown in Figure 12.6, nitrate leaching below the B horizon (40 cm) decreased sharply after clear-cutting the alder, independent of the replanted species. By the second year after clear-cutting, the NO_3^- concentrations already were lower than in the uncut red alder control plots, and after 2 years the concentrations were down to almost 0 (Van Miegroet and Cole 1988; Van Miegroet et al. 1989a). Leaching of Ca^{2+} followed almost the same trend as for NO_3^- (see Figure 12.7). As already discussed, only low con-

Table 12.8. Two-Year Average Fluxes for H^+, SO_4^{2-}, Al, and Mn[a]

	H^+	SO_4^{2-}	Al	Mn
	(mol(c) $ha^{-1}yr^{-1}$)		(mol $ha^{-1}yr^{-1}$)	
pH 6 Treatment				
Precipitation	535	528	7	2
Throughfall	236	352	7	8
Forest Floor	67	207	47	43
E-horizon (3.2 cm)	50	189	59	23
Bs-horizon (24 cm)	29	311	37	10
BC-horizon (60 cm)	12	324	15	2
Balance[b]	+523	+204	−8	0
pH 2.5 Treatment				
Precipitation	535	528	7	2
Throughfall	278	435	7	14
Forest floor	183	441	132	83
E-horizon (2.9 cm)	133	387	73	79
Bs-horizon (21 cm)	101	568	128	18
BC-horizon (55 cm)	80	713	127	23
Balance[b]	+455	−185	−120	−21

[a] Data for October 1986 to September 1988. Water flux was based on same output of Cl from BC horizon as input to soil.
[b] +, net accumulation within ecosystem; −, net loss from ecosystem.

centrations of H^+ were leaving the B horizon. These concentrations decreased after clear-cutting, giving higher pH values for the soil solution in the B horizon over time (see Figure 12.8).

After 2 years the concentrations in the B horizon of the soil previously covered by red alder were about the same as in the same horizon of the soil previous covered by Douglas fir and now planted with Douglas fir (see Table 12.6 and Figures 12.6 and 12.7). This was also true in the forest floor for Ca^{2+} and NO_3^-, but the H^+ concentration was still higher in the forest floor of the soil previously under red alder cover. Concerning the soil solution, the recovery from acidification has been almost complete 2 to 3 years after removing the acidification source, namely the red alder stand.

Rate of Recovery

Three years after stopping watering with artificial rain, the pH of the O horizon showed no sign of treatment effects, and only small differences could be seen deeper in the soil. All differences in soil pH between treatments have disappeared 6 to 10 years after the watering was stopped. In base saturation, treatment effects could not be seen 10 years after the watering was stopped. After 10 years an effect of the most acidic treatment could still be seen in the amount of extractable SO_4^{2-} in the lower B layer. Comparison of the soil solution between the most acidic treatment and the watered control for the years 1986–1988 still showed an effect of the most acidic treatment by higher flux of H^+, Al, Mn, and SO_4^{2-} from the BC horizon.

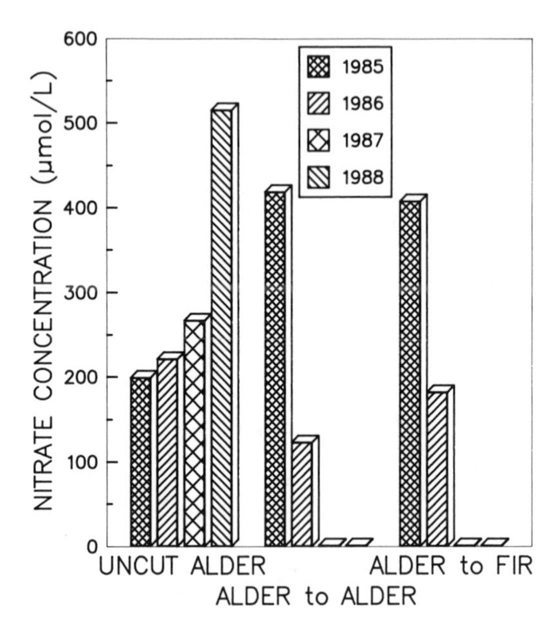

Figure 12.6. Volume-weighted average nitrate concentration in B horizon leachates from uncut alder and clear-cut alder now planted with alder and Douglas fir, for first 4 years after clear-cutting.

Tree growth shows clear signs of recovery, but 10 years after the watering was stopped the forest in the plots that have received the most acidic treatment still grows less than in the other plots (see Figure 12.3). Certain time lags obviously exist between soil and tree growth (Stuanes et al. 1988).

NO_3^- concentrations in the leachate from the B horizon in the soil of the red alder decreased to almost zero 2 years after clear-cutting the alder. Soil solution concentrations in the forest floor also decreased to the same level as in the soil previously covered by Douglas fir. Concentrations of H^+ in the forest floor of the acidified soil were still higher.

Recovery from acidification after the artificial acidification seems to take more time than recovery after internal acidification by nitrification. This is likely to result from a longer term leaching of SO_4^{2-} sorbed during the acidification period compared to the nonsorbed NO_3^-. This difference has to be taken into account when evaluating the recovery from acidification introduced by different acidifying stresses.

Summary and Conclusions

External acidification was studied in a Norwegian field experiment by adding artificial precipitation above the canopy of a Norway spruce stand for 5

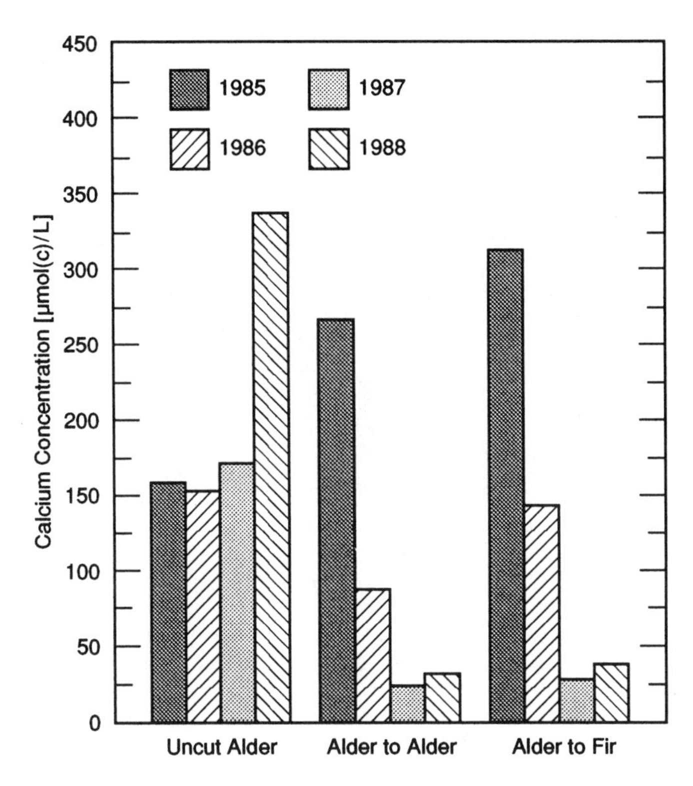

Figure 12.7. Volume-weighted average calcium concentration in B horizon leachates from uncut alder and clear-cut alder now planted with alder and Douglas fir, for first 4 years after clear-cutting.

years. The recovery from acidification was examined during the following 10 years. Internal acidification was studied in a stand of red alder at the University of Washington Thompson Research Center. A neighbouring Douglas fir plantation was used as a control. Recovery was studied after clear-cutting the red alder and planting red alder and Douglas fir in different plots.

Strong external acidification loads can produce marked changes in the soil solid and solution phase. The net leaching from the soil increased with increasing loads, but the changes in the soil solid phase were only found in the upper horizons. Only a few percent of the added H^+ was leached from the soil at 40 cm depth.

More than 50 years with growth of red alder have built up the soil N content by symbiotic N_2 fixation to a level that gives intensive nitrification. In spite of the high internal production of H^+ in this soil, less than 2% of the total external and internal added H^+ were leached below a depth of 40 cm. This indicates that the soil acts as a strong buffer against the acidifi-

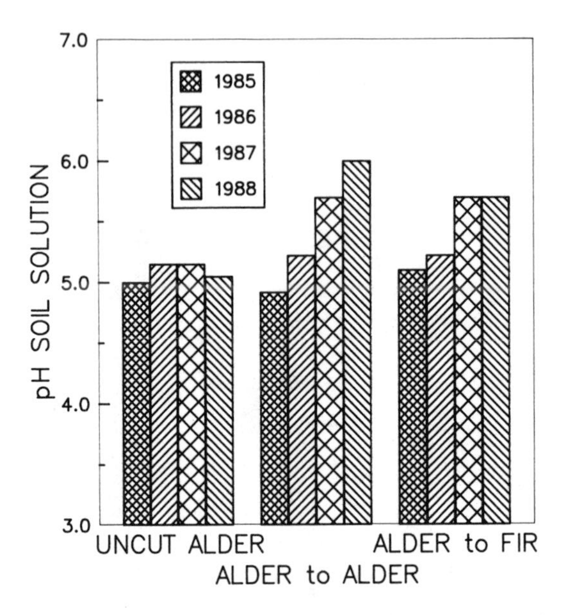

Figure 12.8. Volume-weighted average pH in B horizon leachates from uncut alder and clear-cut alder now planted with alder and Douglas fir, for first 4 years after clear-cutting.

cation. No clear effect of 5 years of red alder growth on soil solution chemistry has been found in soil previously covered by Douglas fir.

The soil can recover from acidification introduced by both external and internal sources. For recovery to be measurable, the acidification must be strong enough to introduce a more severe acidification than the general acidification trend in soils in areas with a surplus of precipitation. This implies that some soil properties might be different from those at the starting point even after a complete recovery from acidification introduced by the defined acidification sources.

Recovery from acidification introduced by increased nitrification seemed to be faster than the recovery from sulfuric acid acidification. Recovery from nitric acid acidification was almost complete after 2 years when using soil solution concentrations as a measure of recovery. Soil solution appears to be more sensitive to chemical changes than the soil solid phase. Soil pH was the first parameter showing recovery after sulfate acidification. Complete recovery of soil pH and base saturation were achieved 3 to 10 years after stopping the artificial acidification. Higher amounts of extractable sulfate was measured in the lower B layer even after 10 years. Differences caused by sulfate load were also found in the soil solution after 10 years.

The soil seemed not to reached a new steady state with respect to acidification under the external acidification pressure. This seemed, however, to be more likely for the soil under internal acidification stress. The external acidification has also introduced a more pronounced loss of cations from the soil than the internal acidification. The redistribution of cations in the upper soil horizons is only noticeable in the first phase of the external acidification. The process of recovery clearly started when the acidifying load of H+ and the mobile cations NO_3^- or SO_4^{2-} was released. This made it possible for the weathering and mineralization to keep up with or even exceed the acidification.

When comparing these two case studies, one must take into account the difference in acid loading. In addition, both the degree of acidification and the degree of recovery are dependent on amount of weatherable minerals and the weathering rate.

References

Abrahamsen G. 1980a. Impact of atmospheric sulphur deposition on forest ecosystems. In Shriner D.S., Richmond C.R., Lindberg S.E. (eds.) Atmospheric Sulfur Deposition, Environmental Impact and Health Effects. Ann Arbor Science, Ann Arbor, Michigan, pp. 397–415.

Abrahamsen G. 1980b. Acid precipitation, plant nutrients and forest growth. In Drabløs D., Tollan A. (eds.) Ecological Impact of Acid Precipitation. SNSF Project, Oslo-Ås, Norway, pp. 58–63.

Abrahamsen G. 1983a. Effects of lime and artificial acid rain on the enchytraeid (Oligochaeta) fauna in coniferous forest. Holarct. Ecol. 6:247–254.

Abrahamsen G. 1983b. Sulphur pollution: Ca, Mg and Al in soil and soil water and possible effects on forest trees. In Ulrich B., Pankrath J. (eds.) Effects of Accumulation of Air Pollutants in Forest Ecosystems. D. Reidel, Dordrecht, pp. 207–218.

Abrahamsen G. 1984. Effects of acidic deposition on forest soil and vegetation. Philos. Trans. R. Soc. Lond. B 305:369–382.

Abrahamsen G. 1987. Air pollution and soil acidification. In Hutchinson T.C., Meema K.M. (eds.) Effects of Atmospheric Pollutants on Forests, Wetlands and Agricultural Ecosystems. Springer-Verlag, Berlin, pp. 321–331.

Abrahamsen G., Bjor K., Teigen O. 1976. Field Experiments with Simulated Acid Rain in Forest Ecosystems. SNSF Project FR 4/76, Oslo-Ås, Norway.

Abrahamsen G., Tveite B., Stuanes A.O. 1987. Wet acid deposition effects on soil properties in relation to forest growth. Exprimental results. In Lavender D.P. (ed.) Woody Plant Growth in a Changing Physical and Chemical Environment, Proceedings of the workshop of IUFRO working party on Shoot Growth Physiology (S2.01–11), Vancouver, Canada, pp. 189–197.

Abrahamsen G., Seip H.M., Semb A. 1989. Long-term acidic precipitation studies in Norway. In Adriano D.C., Havas M. (eds.) Acidic Precipitation, Vol. 1. Case Studies, Springer-Verlag, New York, pp. 138–179.

Billett M.F., FitzPatrick E.A., Cesser M.S. 1988. Long-term changes in the acidity of forest soils in north-east Scotland. Soil Use Manage. 4:102–107.

Binkley D. 1981. Nodule biomass and acetylene reduction rates of red alder and Sitka alder on Vancouver Island, B.C. Can. J. For. Res. 11:281–286.

Brown K.A. 1987. Chemical effects of pH 3 sulphuric acid on a soil profile. Water Air Soil Pollut. 32:201–218.

Cole D.W. 1968. A system for measuring conductivity, acidity and rate flow in a forest soil. Water Resour. Res. 4:1127–1136.

Cole D.W. 1985. Integrated Forest Study of Effects of Atmospheric Deposition. Investigation at Thompson Research Center and Findley Lake. College of Forest Resources, University of Washington, Seattle.

Cole D.W., Gessel S.P., Turner J. 1978. Comparative mineral cycling in red alder and Douglas-fir. In Briggs D.G. (ed.) Utilization and Management of Alder. USDA Forest Service Gen. Tech. Rep. PNW-70, U.S. Government Printing Office, Washington, DC, pp. 327–336.

Cronan C.S., Aiken G.R. 1985. Chemistry and transport of soluble humic substances in forested watersheds of the Adirondack Park, New York. Geochim. Cosmochim. Acta 49:1697–1705.

David M.B., Vance G.F., Rissing J.M., Stevenson F.J. 1989. Organic carbon fractions in extracts of O and B horizons from a New England Spodosol: effects of acid treatment. J. Environ. Qual. 18:212–217.

DeBell D.S., Radwan M.A. 1979. Growth and nitrogen relations of coppiced black cottonwood and red alder in pure and mixed plantings. Bot. Gaz. 140:5102–5107.

Eldhuset T.D. 1988. Effects of aluminium on vascular plants at low pH: a literature survey. Medd. Nor. Inst. Skogforsk. 40.8:1–19 (in Norwegian).

Evans A. Jr., Zelazny L.M., Zipper C.E. 1988. Solution parameters influencing dissolved organic carbon levels in three forest soils. Soil Sci. Soc. Am. J. 52:1789–1792.

Falkengren-Grerup U. 1987. Long-term changes in pH of forest soils in Southern Sweden. Environ. Pollut. 43:79–90.

Frank J. 1989. Acidification of soils caused by the planting of Norway spruce (*Picea abies*) on former birch (*Betula pubescens*) forests in West Norway. In Barth H. (ed.) Effects of Land Use in Catchments on the Acidity and Ecology of Natural Surface Waters. Air Pollution Research Report 13, Commission of the European Communities, Luxembourg, pp. 18–25.

Franklin J.F., Dyrness C.T., Moore D.G., Tarrant R.F. 1968. Chemical soil properties under coastal Oregon stands of alder and conifers. In Trappe J.M. (ed.) Biology of Alder. USDA Pacific Northwest Forest Range Experimental Station, Portland, Oregon, pp. 157–172.

Hauhs M., Wright R.F. 1988. Acid Deposition: Reversibility of Soil and Water Acidification—A Review. Air Pollution Research Report 11, Commission of the European Communities, Luxembourg.

Holstener-Jørgensen H., Krag M., Olsen H.C. 1988. The influence of 12 tree species on the acidification of the upper soil horizons. Forst. Forsøgsvæs. Dan. 42:15–25.

Hovland J., Abrahamsen G. 1976. Acidification Experiments in Conifer Forest. I. Studies on Decomposition of Cellulose and Wood Material. SNSF Project IR 27/76, Oslo-Ås, Norway (in Norwegian).

Hågvar S. 1980. Effects of acid precipitation on soil and forest. 7. Soil animals. In Drabløs D., Tollan A. (eds.) Ecological Impact of acid precipitation. SNSF Project, Oslo-Ås, Norway, pp. 202–203.

Hågvar S. 1984. Effects of liming and artificial acid rain on Collembola and Protura in coniferous forest. Pedobiologia 27:341–354.

Hågvar S., Abrahamsen G. 1980. Colonisation by Enchytraeidae, Collembola and Acari in sterile soil samples with adjusted pH levels. Oikos 34:245–258.

Hågvar S., Amundsen T. 1981. Effects of liming and artificial rain on the mite (Acari) fauna in coniferous forest. Oikos 37:7–20.

Johnson D.W. 1981. Effects of acid precipitation on elemental transport from terrestrial to aquatic ecosystems. In Fozzolare R.A., Smith C.B. (eds.) Beyond the Energy Crisis—Opportunity and Challenge, Pergamon, Oxford, pp. 539–545.

Johnson D.W., Cole D.W., Gessel S.P., Singer M.J., Minden R.V. 1977. Carbonic acid leaching in a tropical, temperate, subalpine and northern forest soil. Arct. Alp. Res. 9:329–343.
Johnson D.W., Cole D.W., Horng F.W., Van Miegroet H., Todd D.E. 1981. Chemical Characteristics of Two Ultisols and Two Forested Inceptisols Relevant to Anion Production and Mobility. Publication 1670, Environmental Sciences Division, Oak Ridge National Laboratory, Oak Ridge, Tennessee.
Krug E.C., Frink C.R. 1983. Acid rain on acid soil: A new perspective. Science 221:520–525.
Lee J.J., Weber D.E. 1982. Effects of sulfuric acid rain on major cation and sulfate concentrations of water percolating through two model hardwood forests. J. Environ. Qual. 11:57–64.
Liljelund L.-E., Nilsson I., Andersson I. 1986. Träslagsvalets betydelse för mark och vatten. En litteraturstudie med speciell referens till luftföroreningar och försurning. Rapport 3182, Statens naturvårdsverk. Solna, Sweden (summary in English).
Matzner E., Ulrich B. 1984. Rates of deposition, internal production, and turnover of protons in two forest ecosystems. Z. Pflanzenernähr. Bodenkd. 147:290–308.
Mollitor A.V., Raynal D.J. 1982. Acid precipitation and ionic movements in Adirondack forest soils. Soil Sci. Soc. Am. J. 46:137–141.
Ogner G., Haugen A., Opem M., Sjøtveit G., Sørlie B. 1984. The chemical analysis program at the Norwegian Forest Research Institute, 1984. Norwegian Forest Research Institute, Ås, Norway.
Sohet K., Herbauts J., Gruber W. 1988. Changes caused by Norway spruce in an ochreous brown earth, assessed by the isoquartz method. J. Soil Sci. 39:549–561.
Stoner J.H. (ed.) 1987. Llyn Brianne acid waters project. An investigation into the effects of afforestation and management on stream acidity. First technical summary report, Welsh Water Authority, Dyfed.
Stuanes A.O. 1980. Effects of acid precipitation on soil and forest. 5. Release and loss of nutrients from a Norwegian forest soil due to artificial rain of varying acidity. In Drabløs D., Tollan A. (eds.) Ecological Impact of Acid Precipitation, SNSF Project, Oslo-Ås, Norway, pp. 198–199.
Stuanes A., Sveistrup T.E. 1979. Field Experiments with Simulated Acid Rain in Forest Ecosystems. 2. Description and Classification of the Soils Used in Field, Lysimeter and Laboratory Experiments. SNSF Project FR 15/79, Oslo-Ås, Norway.
Stuanes A.O., Abrahamsen G., Tveite B. 1988. Effect of artificial rain on soil chemical properties and forest growth. In Mathy P. (ed.) Air Pollution and Ecosystems. D. Reidel, Dordrecht, pp. 248–253, Kluwer Academic Publishers.
Stuanes A.O., Ogner G., Opem M. 1984. Ammonium nitrate as extractant for soil exchangeable cations, exchangeable acidity and aluminum. Commun. Soil Sci. Plant Anal. 15:773–778.
Tamm C.O., Hallbäcken L. 1986. Changes in soil pH over a 50-yr period under different forest canopies in SW Sweden. Water Air Soil Pollut. 31:337–341.
Tamm C.O., Popovic B. 1989. Acidification Experiments in Pine Forest. Report 3589, National Swedish Environmental Protection Board, Solna, Sweden.
Tveite B. 1980a. Effects of acid precipitation on soil and forest. 8. Foliar nutrient concentrations in field experiments. In Drabløs D., Tollan A. (eds.) Ecological Impact of Acid Precipitation, SNSF Project, Oslo-Ås, Norway, pp. 204–205.
Tveite B. 1980b. Effects of acid precipitation on soil and forest. 9. Tree growth in field experiments. In Drabløs D., Tollan A. (eds.) Ecological Impact of Acid Precipitation, SNSF Project, Oslo-Ås, Norway, pp. 206–207.

Tveite B., Abrahamsen G. 1980. Effects of artificial acid rain on the growth and nutrient status of trees. In Hutchinson T.C., Havas M. (eds.) Effects of Acid Precipitation on Terrestrial Ecosystems. Plenum, New York, pp. 305–318.

Tveite B., Abrahamsen G., Stuanes A.O. 1990/91. Liming and wet acid deposition effects on tree growth and nutrition. Experimental results. Water Air Soil Pollut. 54:409–422.

Ugolini F.C., Minden R., Dawson H., Zachara J. 1977. An example of soil processes in the *Abies Amabilis* zone of the central Cascades, Washington. Soil Sci. 124:291–302.

Van Miegroet H. 1986. Role of N Status and N Transformations in H^+ Budget, Cation Loss and S Retention Mechanisms in Adjacent Douglas-fir and Red Alder Forests. Ph.D. Dissertation, University of Washington, Seattle.

Van Miegroet H., Cole D.W. 1984. The impact of nitrification on soil acidification and cation leaching in a red alder ecosystem. J. Environ. Qual. 13:586–590.

Van Miegroet H., Cole D.W. 1985. Acidification sources in red alder and Douglas-fir soils—Importance of nitrification. Soil Sci. Soc. Am. J. 49:1274–1279.

Van Miegroet H., Cole D.W. 1988. Influence of nitrogen-fixing alder on acidification and cation leaching in a forest soil. In Cole D.W., Gessel S.P. (eds.) Forest Site Evaluation and Long-Term Productivity. University of Washington Press, Seattle, pp. 113–124.

Van Miegroet H., Cole D.W., Homann P.S. 1989a. The effect of alder forest cover and alder forest conversion on site fertility and productivity. In Gessel S.P., Lacate D.S., Weetman G.S., Powers R.F. (eds.) Proceedings of the 7th North America Forest Soils Conference. Forestry Publications, Faculty of Forestry, University of British Colombia, Vancouver, BC, Canada, pp. 333–354.

Van Miegroet H., Cole D.W., Binkley D., Sollins P. 1989b. The effect of nitrogen accumulation and nitrification on soil chemical properties in alder forests. In Olsen R.K., Lefohn A.S. (eds.) Effects of Air Pollution on Western Forests. APCA Transaction Series No. 16, Pittsburg, pp. 515–528.

Wiklander L. 1976. The influence of anions on adsorption and leaching of cations in soils. Grundforbättring 27:125–135.

Zavitkovski J., Newton M. 1968. Effect of organic matter and combined nitrogen on nodulation and nitrogen fixation in red alder. In Trappe J.M. (ed.) Biology of Alder. USDA Pacific Northwest Forest Range Experimental Station, Portland, Oregon, pp. 209–221.

13. Regional Evaluations of Acid Deposition Effects on Forests

Introduction

Having reviewed the results of the Integrated Forest Study (IFS) project, we now try to place the results in a larger perspective by very briefly summarizing acid deposition effects and their potential role in forest health in the several forest types represented in the IFS project. This chapter gives brief overviews of the situation in eastern spruce-fir, eastern hardwood, and southern pine forests in North America; and a very brief overview of air pollution in arid forest ecosystems in Europe (with special emphasis on the situation in Norway where the single European IFS site was located). What follows in this chapter is by no means intended to be a comprehensive review of forest health and atmospheric deposition; such an analysis would require a volume of this size for each forest type and is well beyond the scope of this chapter. A comprehensive analysis of forest health and the role of atmospheric deposition has been published for Norway spruce in Europe (Schulze et al. 1989), and one is in preparation for red spruce (Eagar and Adams, in press); the reader is referred to those volumes for a far more detailed discussion than is possible here. The intent of this chapter is merely to highlight those aspects of forest health and atmospheric deposition that have bearing on, or that can be illuminated by, the IFS results.

Eastern Spruce-Fir

A.H. Johnson, A.J. Friedland, E.K. Miller, J.J. Battles,
T.G. Huntington, D.R. Vann, and G.R. Strimbeck

Introduction

The decline of red spruce in the mountain forests of the northern Appalachians has been the focus of research in the past decade because of its possible relationship to acid deposition. The portion of the IFS study carried out at Whiteface Mt., New York, was part of an extensive regionwide, integrated research program that was designed to determine (1) the cause(s) of the decline and (2) the impact of airborne chemicals on the mountain forests of the Northeast. In this chapter we summarize the current understanding of the spruce decline in the mountain forests of the Northeast with emphasis on the studies done at Whiteface Mt.

Red spruce (*Picea rubens* Sarg.) is a canopy dominant in the low-elevation forests of northern New England, southern Quebec, and the Maritime provinces of Canada. In addition, it is a dominant species on the middle and upper slopes of the Adirondack and Appalachian Mountains as far south as North Carolina. On the northern Appalachian Mountains, red spruce importance peaks at about cloudbase (900–1000 m), then decreases. Red spruce is very rare above 1300 m, where natural stresses probably exceed the physiological limits of this species. Key elements in the success of red spruce in the mountain forests are its longevity (it can live 400 years) and shade tolerance. It can remain suppressed for more than 100 years, then occupy the canopy when balsam fir and white birch, its shorter lived competitors, die.

Weiss et al. (1985) reviewed the major causes of mortality of red spruce and balsam fir during the past century, and Hopkins (1901) summarized reports of red spruce mortality for the nineteenth century. Historically, the spruce budworm and spruce beetle have been the most important insect pests. Fire, wind stress, winter injury, and drought are considered the major abiotic factors related to red spruce mortality (Weiss et al. 1985).

Aside from cases where a single cause has been identified, widespread synchronized tree mortality has often occurred from combinations of stress factors. Those multiple stress diseases have been called "declines" (e.g., Houston 1981; Manion 1981). Various combinations of drought, frost, waterlogging, insect defoliation, old age, poor site conditions, insects, and fungal diseases have been designated as key stress factors in declines.

Red spruce has undergone at least one other recorded episode of decline resulting in widespread mortality. There is clear evidence that red spruce died in large numbers across New York and northern New England from about 1871 to 1885 (Hopkins 1901). It was estimated that 50% of the mature spruce in the Adirondacks died during that episode (New York State 1891).

Several accounts written by foresters and entomologists are equivocal as to the causes, although it is clear that the spruce beetle (*Dendroctonus rufipennis* Kirby) was involved in many areas (Hopkins 1901).

Figure 13.1 shows a model for stress-related diseases. The three components of declines suggested by Manion (1981) are portrayed as they affect the carbon that a tree can allocate to defense and repair. Predisposing factors such as old age and poor site conditions can be expected to cause reduced vigor and reduced carbon reserves. An inciting or initiating stress is usually an event of acute injury from drought, defoliation, or frost damage. This causes the mobilization of carbon reserves to repair the injury and defend against pathogens and insects that are present in the stands. Symptoms of stress (dieback of fine branches, adventitious shoots, growth reduction, altered foliar characteristics) appear. If the injury is severe enough, the tree will be unable to effectively repair damage, carry out all of its vital functions, and resist attacks from pests and pathogens. Death, although perhaps years away, is inevitable. Trees that sustain less injury, or those with sufficient reserves to effectively repair damage and resist the secondary assaults, recover.

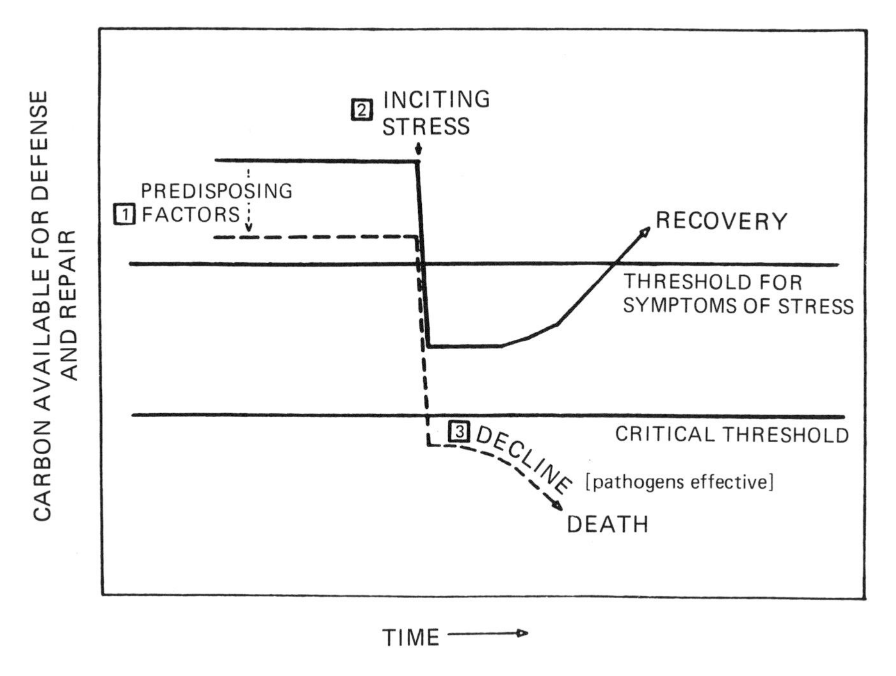

Figure 13.1. Changes in carbon (or energy) available for defense and repair during stages of decline.

Evidence of Recent Spruce Mortality at Whiteface Mountain

Several studies have shown unequivocally that there has been a period of rapid mortality and a marked decline in vigor of red spruce in the Adirondacks and northern Appalachians. Those studies are reviewed in detail by Barnard et al. (1990).

Battles et al. (1988, 1989) measured live and dead red spruce and associated species in 395 permanent plots on Whiteface Mt., New York. Those data extend to elevations 350 m below and 350 m above the IFS site, which is located at about 1020 to 1080 m. Figure 13.2 shows the distribution of live spruce and other species as a function of elevation. Above 999 m, 57% of the spruce with breast height diameter (dbh) greater than 5 cm, were dead in contrast to 28% dead below 1000 m. Using a test of proportions (Fleiss 1981), the difference between high (>999 m) and low (<1000 m) elevations is very highly significant, as is the difference between spruce and the associated species at elevations above 999 m. At Whiteface Mt., the northwest aspect has the greatest percentage of dead spruce, with approximately 70% of the canopy trees dead in 1987. Figure 13.3 shows that the mortality rate was at least 6% per year at elevations above 900 m between 1982 and 1987.

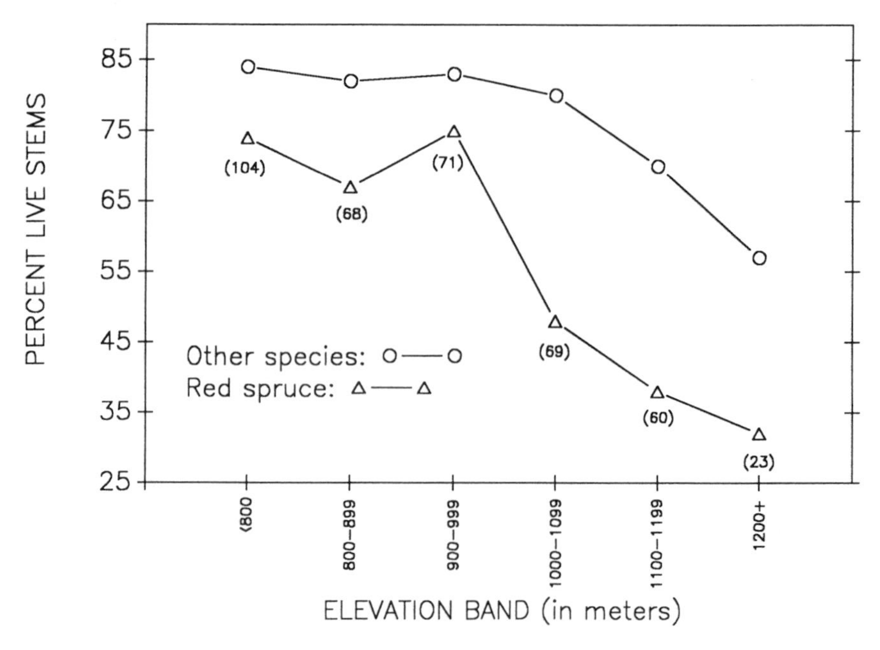

Figure 13.2. Elevational gradients in percent spruce and other species alive in 1987–88. Difference between spruce and other species above 999 m is significant at $p < .01$ using test of proportions. (After Battles et al. 1988).

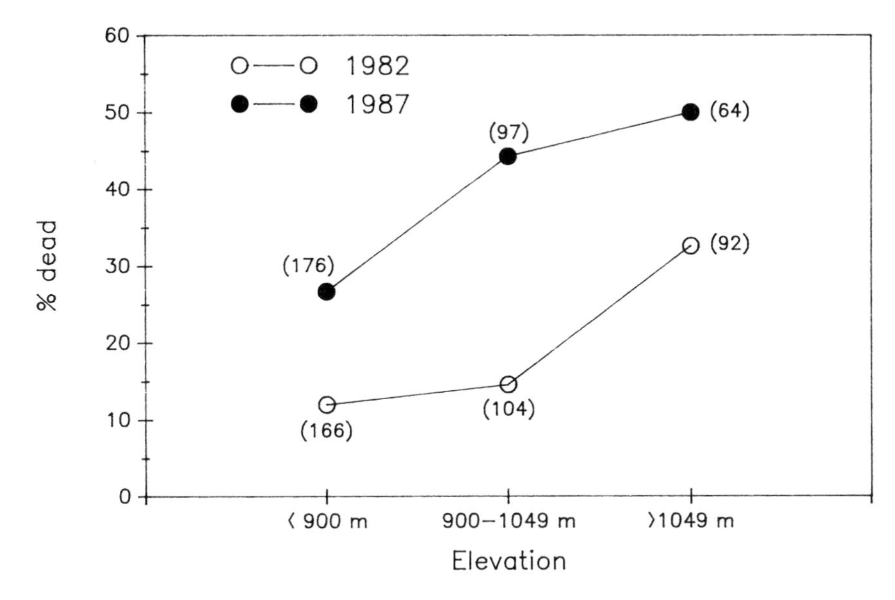

Figure 13.3. Percent standing dead red spruce along twenty 100-m transects on Whiteface Mt., New York in 1982 and 1987. Using test of proportions (Fleiss 1981), differences in percent dead between 2 years are different at $p < .05$ in all elevational bands (data from Silver et al. 1989).

Symptoms of Declining Spruce

Weidensaul et al. (1989) characterized symptoms and the occurrence of pests and pathogens at Whiteface Mt., New York:

1. Of the hundreds of fungi isolated from twigs and foliage, eight or nine were considered potentially pathogenic, but none were judged capable of causing an epidemic.
2. No foliage disorders caused by a biotic pathogen were observed on red spruce.
3. Chlorosis and necrosis of red spruce foliage occurred with the highest frequency and greatest severity above 1000 m. Chlorosis of spruce needles was not observed below 985 m.
4. Neither spruce beetles nor their galleries were frequent.

They concluded that primary pathogens are not causing significant problems, and that the insects and facultative parasites present were not important triggering stresses in the process of decline on Whiteface Mt.

Johnson and Siccama (1983), Johnson and McLaughlin (1986), McLaughlin et al. (1987), Johnson et al. (1988a), Van Duesen (1990), LeBlanc and Raynal (1990), and Leblanc (1990a, 1990b) carried out tree-ring studies using cores from Whiteface Mt. All the studies report a sharp decline in ring width, basal area increment, and volume increment in high-elevation red spruce since the late 1950s or early 1960s. The decreased growth occurred in dominant and codominant spruce of all ages and in stands varying greatly in composition, age, and disturbance history. With one dissension, the consensus is that competition does not have a major influence on the growth trends. In fact, the tallest dominant trees were the earliest trees to be affected, and they have been affected to the greatest degree (LeBlanc and Raynal 1990). Spruce showing the least decline in growth are those codominants that are sheltered by competitors, reinforcing the conclusion that spruce with exposed crowns are most vulnerable (LeBlanc and Raynal 1990).

In dendroclimatogy studies of high-elevation spruce, tree-ring width was consistently associated with late summer and early winter temperatures for more than a century until about 1960, when there was an abrupt change in the ring width–climate relationship (McLaughlin et al. 1987; Johnson et al. 1988).

Hypotheses Advanced to Explain Red Spruce Decline in the High Elevation Forests of the Northeast

Air pollution has been suspected by many as a factor contributing to the spruce decline, largely because: (1) spruce were visibly declining only at the higher elevations where trees are subject to relatively high doses of acid cloud water and ozone; (2) no obvious biological causes explained the regionwide nature of the decline; and (3) a concomitant deterioration in the health of conifers in the Federal Republic of Germany appeared to some to have many similar characteristics (e.g., Liedecker 1988) and most German scientists suspected that air pollution was involved in that "new type of forest decline."

The ideas about air pollution involvement met with skepticism because obvious symptoms of air pollution injury are lacking, the types and levels of air pollution that act as subtle stresses to spruce were unknown, and major, synchronized episodes of mortality in red spruce and other species have occurred at times and in places where air pollution clearly could not have been a factor. At present, there is a substantial amount of data that can be used to test the major hypotheses which have been advanced to explain the observed spruce mortality. The data relevant to natural stresses are reviewed in detail by Barnard et al. (1990). Most important are the data linking winter injury to the initiation of the decline, as outlined next.

Winter Injury

Red spruce is a species known to be particularly prone to winter injury. Both the sensitivity of red spruce to winter injury and the ability of winter injury to cause an abrupt reduction in growth and decline resulting in tree death have been demonstrated by Curry and Church (1952). Wilkinson (1989a, 1989b) studied winter injury to red spruce in a provenance test in the Coleman State Forest (New Hampshire) that had been planted in 1960. Each of 12 seed sources covering the range of the species (New Brunswick to North Carolina) were represented by three plots of 25 trees planted in 1960. Winter injury occurred in each of the last 5 years. Seed source differences in naturally induced winter injury were highly significant, and the degree of winter injury was significantly correlated with tree heights ($p < .5$) and with seed source mortality ($p < .5$). About a third of the individuals in the most sensitive seed sources died. Those associations suggested to the author that the periodic loss of photosynthetic surface area led to reduced growth and greater mortality in sensitive populations. Those findings underscore the importance of winter injury as a primary control of the health and growth of native red spruce populations.

Curry and Church (1952) reported on a winter injury event that occurred in the Adirondacks during the winter of 1947–1948. They noted that red spruce was the most severely damaged species, and that there was considerable bud injury that was roughly proportional to the extent of foliar damage. They reported on the fate of 25 winter-injured red spruce selected in April of 1948 and checked in June 1948 and October 1949. They picked 10 spruce that initially had more than 75% defoliation. At the last examination, 18 months after the initial survey, 6 of those trees had died and 4 were "dying slowly." The other 15 trees selected in April 1948 had initial defoliation of 25% to 75%. At the last examination, 7 were recovering and 8 were "doubtful." Growth in the 2 years after injury was about 50% of that in preceding years. They noted that there was considerable mortality of saplings in that event, and called the effect "permanent winter injury," attributing it to desiccation ("winter drying").

The Department of Forests and Parks of Vermont (1959–1962) reported that there was "severe winter drying" affecting "large areas of natural red spruce" in the winter of 1958–1959. With the flush of new shoots, the damage "cured itself" in the summer of 1959, except in Christmas tree lots. The same agency reported that "severe winter drying of red spruce at higher elevations in northern Vermont occurred again during the winters of 1961 and 1962," indicating that there was extensive damage to red spruce foliage in at least three of four successive years.

The Committee on Insects and Diseases (1959) reported much winter injury in the Adirondacks in 1959 and attributed it to "frost damage" (e.g., freezing damage) associated with an "open winter" (not much snow). They also noted "severe winter drying" on spruce and hemlock in 1966. The Annual Report of the State of New York Conservation Commission reported low snow and widespread winter injury in 1960–1962. The New York State Forest Insect and Disease Control Annual Reports recorded "heavy winter damage to spruce and hemlock in 1966, and "winter drying" and "winter injury" in 1968 and 1969. The New York Forest Pest Report recorded light "winter scorch" of evergreens in Hamilton County in 1971. In New York, foresters filed annual reports as early as the early 1900s, but there were no reports of winter injury to red spruce before 1948.

In New Hampshire, there are fairly detailed reports on the death of high-elevation red spruce from "winterkill" during 1962–1965. Tegethoff (1964) reported on high-elevation spruce mortality in the White Mountain National Forest. Several trained foresters and pathologists observed this "general dying of spruce" where "no primary entomological or pathological organisms could be found." They attributed the mortality to "protracted drouth with the severe winter winds a probable contributing factor." They mention a similar condition in the Middlebury District of the Green Mt. National Forest where no organic vector could be determined. The latter mortality was subsequently studied by Hadfield (1968) in a nearby district (Rochester, Vermont).

Wheeler (1965) reported that in the summer of 1962 in the White Mountains "we started to notice the 'browning' or decadence of upper slope spruce and fir." They noted softwood mortality, but no evidence of insects or disease. The mortality became "more apparent" in 1963, and he notes "sizeable extensions" of mortality in the old-growth softwood stands by 1965. "Winterkill" from combined drought and cold winters with high winds was mentioned as a possible cause. He notes his concern that the possible insect buildup and fire hazard created by the dead trees might be a serious problem.

Kelso (1965) reported on investigations of high-elevation mortality in the White Mountain National Forest, summarizing much of the information cited previously. He reports that mortality on Mt. Kancamagus extended from 3800′ down to 2800′. The lowest mortality noted elsewhere was at about 2300 ft. He also noted a rapid expansion of the damage between 1964 and 1965 and dead spruce from approximately 30 to 60 cm (1–2 ft.) tall. Again "recent winter conditions" and drought were considered to be the causes. The areas he reported on had been visited annually since 1962 with the insects appearing for the first time in 1965.

D. Stark (Maine Department of Conservation, Entomology Laboratory, personal communication) has records of problems in Maine forests for three decades. He notes that the first record of large-scale winter "drying and dying" was in March 1962. It occurred in the west-central part of Maine; "red spruce died along the crests of many mountains . . . at this time." He noted that only red spruce were damaged and that the condition in Maine in 1962 was similar to that in Vermont and in the Presidential Range of New Hampshire; spruce of all ages and competitive status were affected, and the damage he recorded was largely confined to the 1961 foliage, with a variable amount present (5%–70%). He also reported bud death of 0% to 98% during that episode.

"Winter drying" was observed in red spruce (only) in 1968, in 1966 and 1967 with needles missing, and in 1969 with 1968 twigs bare. Mountain sites or areas of exposed ledge and thin soils were most heavily affected.

Stark reported decline and mortality of spruce from 1970–1975 attributed to "winter drying associated with some kind of stress factor" (e.g., exposure from heavy cutting or old age). Throughout the 1962–1975 period, balsam fir remained undamaged. 1975 was the last year for which Stark has detailed reports of winter injury. Importantly, Stark (personal communication) found no record of winter injury to red spruce before 1955 in reports on forest condition in Maine, which date back to the early 1900s.

Widespread winter injury to high-elevation red spruce occurred in 1981, 1984 (Friedland et al. 1984), and 1989 (Herrick and Friedland 1991). Those events have had a sufficient impact to alter the patterns of carbon allocation, as shown by Wargo et al. (1989). Evans (1986) noted winter injury to new shoots on red spruce in both the Northeast and southern Appalachians, and suggested that twig death resulted from a lack of bark formation.

There is not a consensus on the climatic and physiological mechanisms leading to winter injury in red spruce. Many investigators favor desiccation (winter drying), which occurs on warm sunny days when water flow to the needles is blocked, and some scientists favor freezing as the primary mechanism. The data from Whiteface support a major role for freezing injury as explained in a later section, but the exact sequence of physiological and climatic events remains unknown.

Initiation of Long-Term Decline of High-Elevation Spruce

At present, several lines of reasoning lead to the conclusion that the period of winter injury and mortality of the early 1960s not only resulted in the death of red spruce but that it initiated a spiral of decline which is still in evidence in the Northeast. Several independent data sets from several locations show that a number of parameters related to tree vigor show an abrupt change during the 1959–1965 period. Those data are reviewed in detail by Barnard et al. (1990); the studies relevant to Whiteface Mt. are summarized here.

LeBlanc and Raynal (1990) did a stem-analysis study of 46 dominant and codominant red spruce growing at 900 m and 1100 m at Whiteface. They reported that at both sites the annual basal area increment decreases sharply after 1965 in older trees and after 1955 in young trees. The annual basal area increment continues to decline up to the time of sampling. The analysis showed that a very high incidence of terminal death was recorded from the late 1960s to the time of sampling and that there was no evidence of terminal death from 1880 until about 1955.

Johnson et al. (1988a) showed that the annual increment widths of red spruce growing at elevations above 800 m from Whiteface, the Catskills, and Green and White Mountains were closely correlated. As in the studies cited, annual increments decreased sharply after 1960 regardless of stand disturbance history. Using the procedures explained by Cook et al. (1987), Johnson et al. (1988a) showed that for healthy and slightly declining spruce, the high-frequency variation in standardized ring widths were well predicted by a model based on average August and December temperatures. The model was calibrated for the 1885–1940 period and used to predict standardized ring widths in three periods: 1856–1884, 1940–1960, and 1961–1981. After 1960, the relationship between ring widths and climate, which had prevailed for at least 100 years, changed abruptly signifying a rather sudden shift in the way spruce were integrating weather conditions in their production of bole wood. By 1977, ring widths were again predicted by the climate model, suggesting a "shock" that lasted for about 15 years. Those findings suggested that the patterns of carbon allocation were altered for many years, at least through the 1960s and into the late 1970s.

Figure 13.4 shows another analysis of the climatic influence on red spruce growth. In this analysis, a smoothing spline (Cook 1985) was used to identify the lower frequency trends in dominant and codominant spruce in uncut, unburned stands above 850 m on Whiteface Mt., New York, Mt. Mansfield (Vermont), and Mt. Washington (New Hampshire). Low-frequency variations in ring width follow the low-frequency variation in August and December temperatures until about 1960, when ring widths rapidly decrease to values much lower than predicted by the climate-based model.

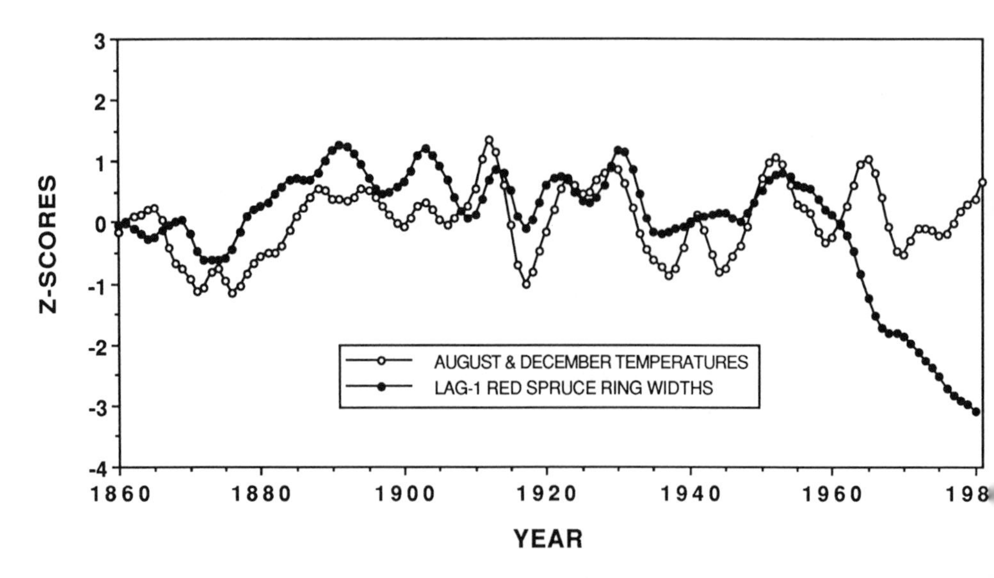

Figure 13.4. Low-frequency variation in December and August temperatures and ring widths of high-elevation red spruce in stands undisturbed by fire or logging in past 130 years. Temperature data are calculated as ($Z_{Dec} - Z_{Aug}$), then smoothed using an 11-year smoothing spline (Cook 1985). Ring widths, shown as Z scores, are pooled averages of trees growing at elevations above 800 m at Mt. Washington, New Hampshire, Mt. Mansfield, Vermont, and Whiteface Mt., New York. Data are from Johnson et al. (1988a).

In summary, the timing of the initiation of the growth decline, recorded red spruce death, and changes in the tree-ring–climate relationships are coincident with a period of several nearly consecutive years when severe winter injury was observed in the Adirondack, Green and White Mountains, and west-central Maine. Red spruce was the species most seriously affected, and in many reports it was the only species affected. Severe winter injury has been shown to lead to loss of growth of red spruce within months, and many trained forest pathologists attributed the death of high-elevation spruce during the early 1960s to the prevalent episodes of winter injury. Other site-related stresses and drought were often considered to be contributing factors. There is conflicting evidence as to the most important mechanism leading to the winter injury observed. Most investigators believed that desiccation was the cause, but the most recent evidence from Whiteface Mt. indicates that freezing injury plays a key role.

Airborne Chemicals in High-Elevation Forests

Air pollution has been regarded by many as a factor involved in the decline of high-elevation spruce, and it is now possible to evaluate those hypotheses using the data from field and laboratory studies done at Whiteface Mt. As a start, two parameters need to be quantified: (1) the length of time forests are exposed to dissolved chemicals in clouds and to gases in potentially injurious concentrations, and (2) the rate of deposition of potentially injurious chemicals to the forests. The data reported here are those from the IFS study at Whiteface and the Mountain Cloud Chemistry Project (MCCP) summarized by Mohnen (1989a,1989b).

Exposure to Airborne Chemicals

Trees in high-elevation forests are subject to prolonged exposure to cloud water, its dissolved constituents, and ozone. Cloud water has considerably higher concentrations of dissolved chemicals than rain water, and many experiments have been carried out to test the effect of acidified mist on red spruce. Thus an understanding of cloud water chemistry, exposure times, and cloud water deposition is necessary to evaluate the applicability of the experimental results. Tables 13.1 and 13.2 summarize the exposure and flux data for dissolved constituents at the IFS site at Whiteface Mt. (elevation, 1020–1050 m) and at two higher elevation MCCP sites, respectively. Most important is the occurrence of cloud water with high acid concentrations. Based on the IFS results and those of the MCCP, the period of cloud immersion at Whiteface ranges from about 15% of the time at 1000 m to 40% of the time at the summit (1485 m).

Mohnen (1989a) indicated that concentrations of SO_2 and NO_x are very low virtually all the time at Whiteface Mt. and at the other MCCP mountain sites. The concentrations of those gaseous pollutants were near the detection limit of the instruments used and were well below the levels known to affect

Table 13.1. Ion Concentration and Flux Data for IFS Site (1020 m) Whiteface Mt., New York for 2 years

Time Period: 1-Year Period (1 June 1986 to 31 May 1987)

SOLUTION CONCENTRATION DATA (Weighted average concentration, μmol L^{-1})

Level	H	Ca	Mg	K	Na	NH_4	NO_3	HCO_3	Cl	SO_4	Orth.-P	Total N	Total Al
Precipitation	50.74	10.08	2.56	0.97	1.40	13.69	21.50	0.00	2.96	48.35	0.15	35.19	0.91
Cloud water	220.04	20.00	6.00	3.00	4.00	130.36	122.58	0.00	7.14	226.07	0.00	252.94	0.00
Throughfall	62.81	28.72	4.63	11.18	3.74	10.77	15.65	0.00	5.79	80.52	2.28	45.81	1.42
Stemflow	0.00	0.00	0.00	0.00	0.00	0.00	0.00	0.00	0.00	0.00	0.00	0.00	0.00
FF	86.06	69.94	37.70	1.71	11.85	1.26	23.78	0.00	9.97	193.28	0.22	82.80	75.53
ABhs	31.78	61.10	54.00	3.82	8.86	2.59	16.98	0.00	7.51	177.90	0.23	40.14	42.21
Bs	26.92	55.98	47.64	2.39	6.82	3.04	15.17	0.00	6.94	167.12	0.22	32.49	39.22

SOLUTION FLUX DATA (kg ha^{-1} yr^{-1})

Level	H_2O (cm)	H	Ca	Mg	K	Na	NH_4-N	NO_3-N	HCO_3	Cl	SO_4-S	Orth.-P	Total N	Total Al
Precipitation	116.95	0.60	2.36	0.36	0.44	0.38	2.24	3.52	0.00	1.23	9.07	0.02	5.76	0.29
Cloud water	12.95	0.29	0.52	0.09	0.15	0.12	2.36	2.22	0.00	0.33	4.69	0.00	4.59	0.00
Dry deposition	0.18	0.16	0.07	0.03	0.08	0.14	2.30	0.00	0.11	0.61	0.00	2.43	0.00	
Throughfall	113.53	0.72	6.53	0.64	4.96	0.98	1.71	2.49	0.00	2.33	14.66	0.27	7.28	0.43
Stemflow	0.00	0.00	0.00	0.00	0.00	0.00	0.00	0.00	0.00	0.00	0.00	0.00	0.00	0.00
FF	91.09	0.79	12.77	4.17	0.61	2.48	0.16	3.03	0.00	3.22	28.22	0.02	10.56	18.56
ABhs	79.75	0.26	9.76	5.23	1.19	1.62	0.29	1.90	0.00	2.12	22.74	0.02	4.48	9.08
Bs	73.30	0.20	8.22	4.24	0.69	1.15	0.31	1.56	0.00	1.80	19.64	0.02	3.34	7.76

Time Period: 1-Year Period (1 June 1987 to 31 May 1988)

SOLUTION CONCENTRATION DATA (Weighted average concentration, μmol L^{-1})

Level	H	Ca	Mg	K	Na	NH_4	NO_3	HCO_3	Cl	SO_4	Orth.-P	Total N	Total Al
Precipitation	46.79	6.00	1.62	0.72	1.28	16.30	24.23	0.00	2.39	40.08	0.15	40.53	0.91
Cloud water	340.79	20.00	6.00	3.00	4.00	198.61	186.75	0.00	10.05	368.40	0.00	385.36	0.00
Throughfall	80.26	32.76	10.38	12.55	4.00	17.49	38.59	0.00	7.18	94.53	2.28	75.47	1.42
Stemflow	0.00	0.00	0.00	0.00	0.00	0.00	0.00	0.00	0.00	0.00	0.00	0.00	0.00
FF	105.70	69.12	30.10	0.58	9.85	1.03	26.06	0.00	13.46	185.72	1.21	97.04	66.14
ABhs	17.16	63.76	32.24	4.30	8.20	0.39	28.79	0.00	10.31	148.12	0.80	40.67	37.73
Bs	15.72	61.54	28.52	3.64	6.85	0.29	41.74	0.00	8.44	159.42	0.59	50.20	39.49

SOLUTION FLUX DATA (kg ha^{-1} yr^{-1})

Level	H_2O (cm)	H	Ca	Mg	K	Na	NH_4-N	NO_3-N	HCO_3	Cl	SO_4-S	Orth.-P	Total N	Total Al
Precipitation	112.41	0.53	1.35	0.22	0.32	0.33	2.57	3.81	0.00	0.95	7.22	0.02	6.38	0.28
Cloud water	19.24	0.66	0.77	0.14	0.23	0.18	5.35	5.03	0.00	0.69	11.36	0.00	10.39	0.00
Dry deposition	0.32	0.02	0.13	0.76	0.62	0.18	2.40	0.00	0.40	2.55	0.00	2.58	0.00	
Throughfall	121.37	0.98	7.97	1.53	5.96	1.12	2.97	6.56	0.00	3.09	18.39	0.29	12.83	0.47
Stemflow	0.00	0.00	0.00	0.00	0.00	0.00	0.00	0.00	0.00	0.00	0.00	0.00	0.00	0.00
FF	97.07	1.03	13.45	3.55	0.22	2.20	0.14	3.54	0.00	4.63	28.90	0.12	13.19	17.32
ABhs	84.79	0.15	10.83	3.32	1.43	1.60	0.05	3.42	0.00	3.10	20.13	0.07	4.83	8.63
Bs	77.81	0.12	9.60	2.70	1.11	1.23	0.03	4.55	0.00	2.33	19.89	0.05	5.47	8.29

Table 13.2. Ion Concentration in Cloud Water and Wet Deposition Data (rain plus cloud water) for MCCP sites on Whiteface Mt., New York[a]

	Samples with Corresponding LWC Measurements				Samples with or without Corresponding LWC Measurements	
Elevation	Id/Name	Weighted	Unweighted	*n*	Unweighted Average	*n*
1475 m	H ion	180.13	174.18	48	172.39	274
	H ion	179.15	168.92	50	158.75	278
	Lab cond	101.18	99.57	50	88.61	278
	Sulfate	117.87	119.72	51	102.68	281
	Nitrate	81.50	92.31	51	82.08	281
	Chloride	4.49	5.24	51	5.16	281
	Ammonium	118.88	128.79	51	104.78	281
	Sodium	1.85	2.24	51	2.66	281
	Potassium	1.56	1.91	51	3.85	281
	Calcium	8.45	14.57	38	9.03	175
	Magnesium	1.45	2.18	38	1.67	175
1220 m	(Site − WF SUBSITE = 2)					
	H ion	487.61	461.13	21	344.95	51
	Lab cond	263.59	252.11	21	189.46	50
	Sulfate	316.83	308.15	21	247.47	51
	Nitrate	152.07	145.88	21	122.11	51
	Chloride	10.16	9.75	21	8.65	51
	Ammonium	216.86	223.95	21	210.26	51
	Sodium	3.39	3.69	21	4.73	51
	Potassium	1.97	2.24	21	2.63	51
	Calcium	18.06	17.28	20	14.15	45
	Magnesium	4.89	4.63	20	3.58	45

Relative deposition of pollutant ions in cloud and rain

		H	NH_4	NO_3	SO_4	NH
Whiteface	Rain	0.16	1.3	3.8	9.1	
	Cloud	0.7	9	19	45	
	Cloud	(0.2–1.2)[b]	(3–15)	(6–33)	(14–76)	

[a] Concentration, $\mu mol\ L^{-1}$; deposition, $kg\ ha^{-1}$, May–October 1987. (After Mohnen 1989a, 1989b).
[b] Note: a range of modal input parameters were used to derive the ranges of cloud water deposition shown in parenthesis for Whiteface Mt.

vegetation. The gases are not treated further in this report, except as they contribute to dry deposition.

At elevations above 1000 m, ozone exposure at Whiteface Mt. increases sharply because of slightly greater peak concentrations but mostly because of a lack of diurnal fluctuation at the higher elevations (e.g., Figure 13.5). Peak values exceed 100 ppb, but those occur only a few times each summer. Figure 13.6 shows the mean monthly ozone concentrations measured at the Whiteface Mt. summit since 1974. Those data indicate summer monthly ozone means in the range of 50 to 60 ppb. Judging from Figure 13.5, the concentrations of ozone, in the spruce-fir zone where the IFS site is located, would have been a few parts per billion lower.

Hypotheses Linking Airborne Chemicals to Spruce Decline

The hypotheses that involve air pollution have focused on the following questions, each of which can be addressed to some degree.

1. Does acidic deposition reduce the base saturation of soils resulting in cation deficiencies in spruce trees?
2. Do the anions supplied in acidic deposition increase the soil solution ratio of Al:Ca or Al:Mg, resulting in an Al-induced Ca or Mg deficiency?
3. Does acidic deposition acidify soils, increasing Al availability, which results in toxicity to fine roots?
4. Do anthropogenically derived acids in cloud water (or rain) increase the leaching of nutrient ions from foliage, resulting in nutrient deficiency?
5. Does ozone or the components of acidic mist alter the response of red spruce to winter stresses?
6. Does ozone or components of acidic mist cause decreased photosynthesis or increased respiration, which reduces growth or the carbon available for defense and repair? Are there other effects on the physiology or biochemistry that add to the natural stresses of the mountain forests?

Each of these questions is addressed next, using the results of field studies and controlled experiments with seedlings, saplings, and branches of mature trees enclosed in chambers.

Long-Term Changes in the Base Cation Pool of Adirondack Forest Soils

As detailed in earlier chapters, the most important mechanism leading to acidification from anthropogenic sources is the leaching of exchangeable base cations by the mobile anions sulfate and (in some forests with excess N) nitrate. As the base cations from exchange sites are depleted by leaching or plant uptake, H^+ added to the soil in precipitation and in the cycles of base cations, N, and S replace the base cations on exchange sites. A clear understanding of the rate at which this will occur (or if it will occur at all) necessitates measuring all the sources and sinks of H^+ and the rates of trans-

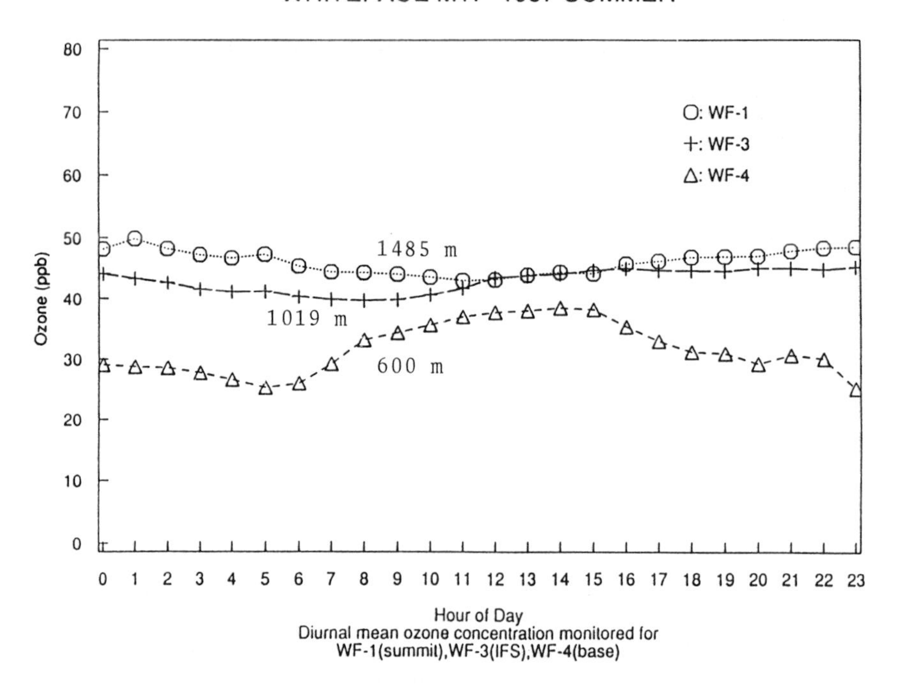

Figure 13.5. Diurnal patterns in ozone concentration at three elevations on Whiteface Mt., New York (Mohnen 1989a).

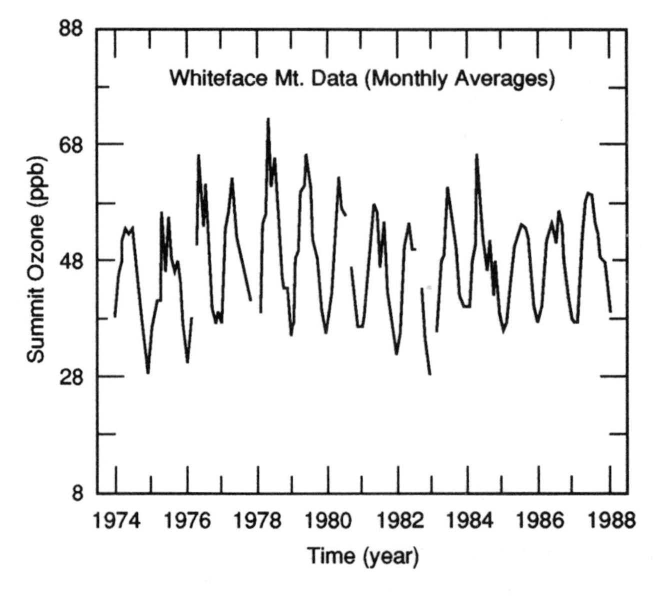

Figure 13.6. Monthly average ozone concentrations at summit of Whiteface Mt., New York (Mohnen 1989a).

fer among the pools. The concepts and data requirements have been thoroughly reviewed elsewhere in this volume.

In Europe, some long-term comparisons of forest soils (e.g., soils at the same site resampled after 30–60 years) have indicated that acidification has taken place, and some authors have attributed at least a portion of the acidification to airborne acids (e.g., Falkengren-Grerup 1986; Hallbacken and Tamm 1986).

Andersen (1988) analyzed soils from sites in the Adirondacks sampled approximately 50 years earlier by C.C. Heimburger (1934). Organic horizon pH and dilute acid-extractable Ca from 48 sites were stratified by confidence in site relocation and compared using student's *t* test. On average, organic horizons did not acidify in highly acid soils, while organic horizons with an original pH > 4.0 did acidify (pH decreased and extractable Ca decreased). In the mineral soil, E horizons acidified over the five dacades (e.g., they lost extractable Ca), but there was on average no change in extractable Ca in B horizons or in strongly acid organic horizons (see Figure 13.7). The same patterns were obtained when the 12 sites dominated by spruce and fir ($>$800 m in elevation) were considered separately.

Andersen (1988) determined the loss of Ca from the solum at 15 mixed hardwood/spruce-fir sites in a low-elevation experimental forest near Newcomb, New York. Table 13.3 shows that Ca was lost from the upper horizons but not from the B horizons, and that the loss of Ca from the soil was more than accounted for by uptake in tree biomass during the 52-year period. Thus any effect of acid deposition on base supply is masked by the other processes of the H^+ cycle.

Overall, Andersen's findings suggested that base cation depletion from acid deposition is unlikely to be a major factor in the high-elevation spruce decline in the Northeastern montane forests.

Aluminum Mobilization

In addition to the fact that Al compounds become more soluble as pH decreases, theoretical considerations and controlled experiments indicate that increased concentrations of strong acid anions (sulfate and nitrate) in the soil solution will result in a greater ratio of Al^{3+} to Ca^{2+} and Mg^{2+} in the soil solution. As a result it has been hypothesized that the increased aluminum availability might result in direct toxicity to roots or indirect effects on Ca or Mg uptake (Shortle and Smith 1988). Several experiments have been designed to assess the sensitivity of red spruce seedlings to acidic solutions containing aluminum.

Hutchinson et al. (1986), Ohno et al. (1988), Thornton et al. (1988), Schier (1985), Joslin and Wolfe (1988), Cronan et al. (1989) and Raynal et al. (1990), have used solution cultures or artificially acidified soils to test for the effect of Al on cation uptake, and in other studies spruce have been subjected to mixtures of ozone and acidic mist to test for effects of airborne

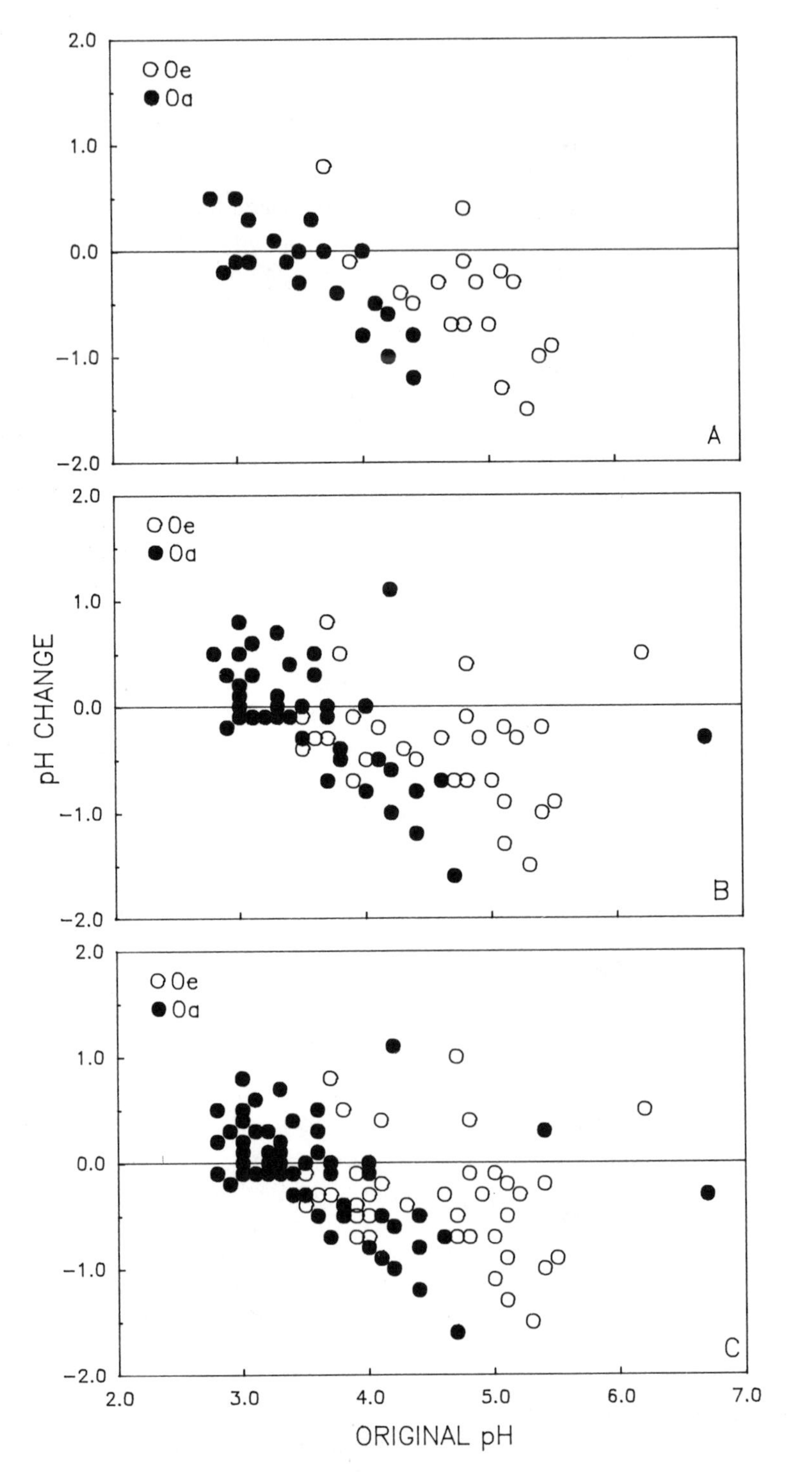

Figure 13.7. (a) pH of Oe and Oa horizons measured in 1930–1932 and 1984. *Pool* with high confidence; *pool C* includes *pool A and B* plus sites relocated with ac- 1932 did not acidify, whereas sites with pH > 4 did acidify ($p < .05$). (b) Com- described in Figure 13.7a (Andersen 1988). E horizons and forest floor samples that

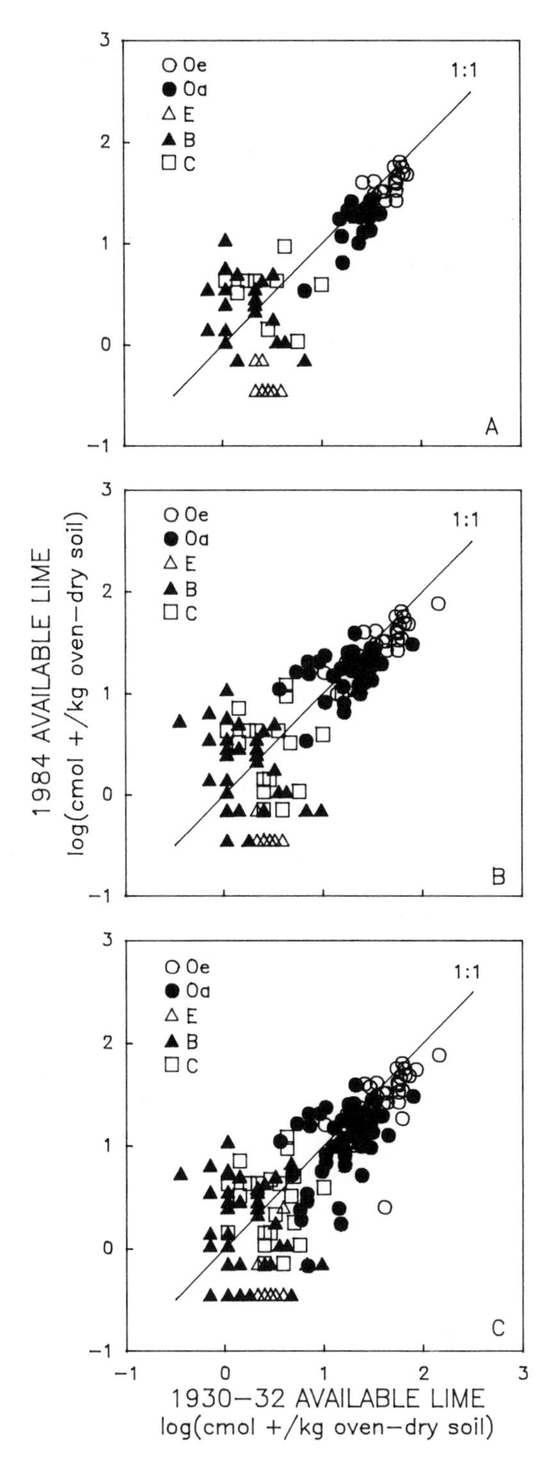

A is sites relocated by survey; *pool B* includes *pool A* and sites that were relocated ceptable confidence (Andersen 1988). On average, sites that had pH < 4.0 in 1930– parison of dilute acid-extractable Ca measured in 1930–1932 and 1984. Pools as were not strongly acid showed significant decrease in Ca ($p < .05$).

Table 13.3. Measured Changes in Dilute Acid-Extractable Ca in Soils and Estimated Uptake of Ca in Biomass for 1932–1984 in Newcomb Experimental Forest, New York[a]

Horizon	Mean ± SE		Mean Difference
n	1930	1984	(1984–1930)
Oe 15	309.18 ± 19.47	242.16 ± 17.25	−67.02
Oa 18	781.56 ± 58.03	661.05 ± 56.55	−120.51
E 10	245.52 ± 14.52	27.81 ± 2.46	−217.71
B 17	572.83 ± 95.90	603.94 ± 79.06	31.11
		Sum (without B)	−405.24

Changes in biomass Ca (kf^{-1} ha)

Horizon	Mean ± SE		Mean Difference
n	1930	1984	(1984–1930)
16	366	892	526

[a] From Andersen (1988).

chemicals on the nutrient status of seedlings and saplings. Negative effects of Al on red spruce seedling root and foliar biomass have been observed when inorganic monomeric Al (MAL) in the soil solution or solution cultures exceeded 185 to 200 μM L^{-1}. Negative effects on Ca and Mg uptake by red spruce seedlings occurred at lower Al concentrations.

Thornton et al. (1988) compared the results of their seedling studies to plant tissue elemental concentrations and total Al in lysimeter solutions at two field sites in the Adirondacks (Big Moose Lake watershed, Huntington Forest). They showed that Al concentrations in the water extracted by lysimeters in the field were considerably lower than the levels associated with root or foliar biomass reductions in the seedling experiments, and that foliar elemental concentrations in the field trees were well within the sufficiency range for Ca and Mg. They pointed out that it is difficult to extrapolate the experimental results to the field and suggested that if Al is a problem for the spruce at their field sites, it is caused by the long-term effects of reduced Ca or Mg uptake rather than acute effects.

McLaughlin et al. (1990b) conducted studies of red spruce in two stands in the Smoky Mountains. At the higher elevation site, diameter increment and height growth were lower during the past 20 years. This corresponded to a lower ratio of net photosynthesis to dark respiration. The trees at the higher site had reduced foliar Ca and Mg, reduced foliar chlorophyll, increased foliar Al, and a low Ca:Al ratio. The authors suggested that the higher site may have been impacted by acidic deposition, leading to some degree of Al toxicity.

Huntington et al. (1989) and Miller et al. (in press) summarized the results of a controlled field study conducted at about 900 m elevation on Whiteface Mt., New York. Their site is a few hundred meters below the

IFS plots. Four blocks of soil that were hydrologically isolated from lateral flow were irrigated with solutions representing the 11-year mean MAP3S ambient precipitation chemistry, $\frac{1}{2}$ the mean values, and the concentrations exceeded in 1% of the MAP3S precipitation samples.

Rapidly reactive aluminum (RRAL) determined using the pyrocatechol violet method (Bartlett 1987) was determined in soil solution samples extracted from three depths at 70 kPa. RRAL was strongly correlated with the total anion equivalents. When the experimental samples were compared to the frequency distribution of anion equivalents measured at the undisturbed IFS plots, the results were as follows:

1. The effect of nitrification caused by disturbance of the soil from the experiment caused nitrate concentration to be unrealistically high in some samples; thus, there were not consistent differences in soil solution chemistry attributable to treatment pH, N, or S.
2. RRAL levels were substantially higher in the forest floor soil solution than in the upper or lower B horizon soil solution.
3. Assuming that RRAL has a charge of 3+, RRAL balanced approximately one-third of the anion charge in the forest floor soil solution and less than one-fourth of the anion charge in the soil solution extracted from the upper and lower B horizon.
4. Assuming that RRAL has a charge of 3+, RRAL was present in a small percentage of the forest floor soil solution samples at concentrations equal to the total inorganic anion charge.
5. In the forest floor lysimeter samples from the IFS plots, total inorganic anion concentrations exceeded 420 μeq L^{-1} less than 5% of the time, so that even if Al^{3+} balanced all the inorganic anion charge, the RRAL levels would exceed 140 μM (= 420 μeq L^{-1}) infrequently. Events exceeding 200 μM Al would be very rare in the forest floor and even rarer in the mineral soil. Figure 13.8 shows a summary of the data for the forest floor soil solution, the zone where Al concentrations were highest.

Effects of Al on Ca and Mg Uptake

Theoretical considerations dictate that increasing the anion concentration in the soil solution will increase the ratio of Al to the divalent and monovalent cations in the soil solution. It has recently been proposed that the increased availability of Al by mobile anion deposition has decreased Ca or Mg uptake by red spruce (e.g., Shortle and Smith 1988; Bondietti et al. 1989). The results of seedling experiments support the possibility, and McLaughlin et al. (1990b) have shown that reduced growth of red spruce saplings in the field is associated with relatively high Al and low Ca and Mg in foliage. Shortle and Smith (1988) have shown that there are adverse ratios of Ca to Al in the tissues of very fine red spruce roots in areas where there is high spruce mortality.

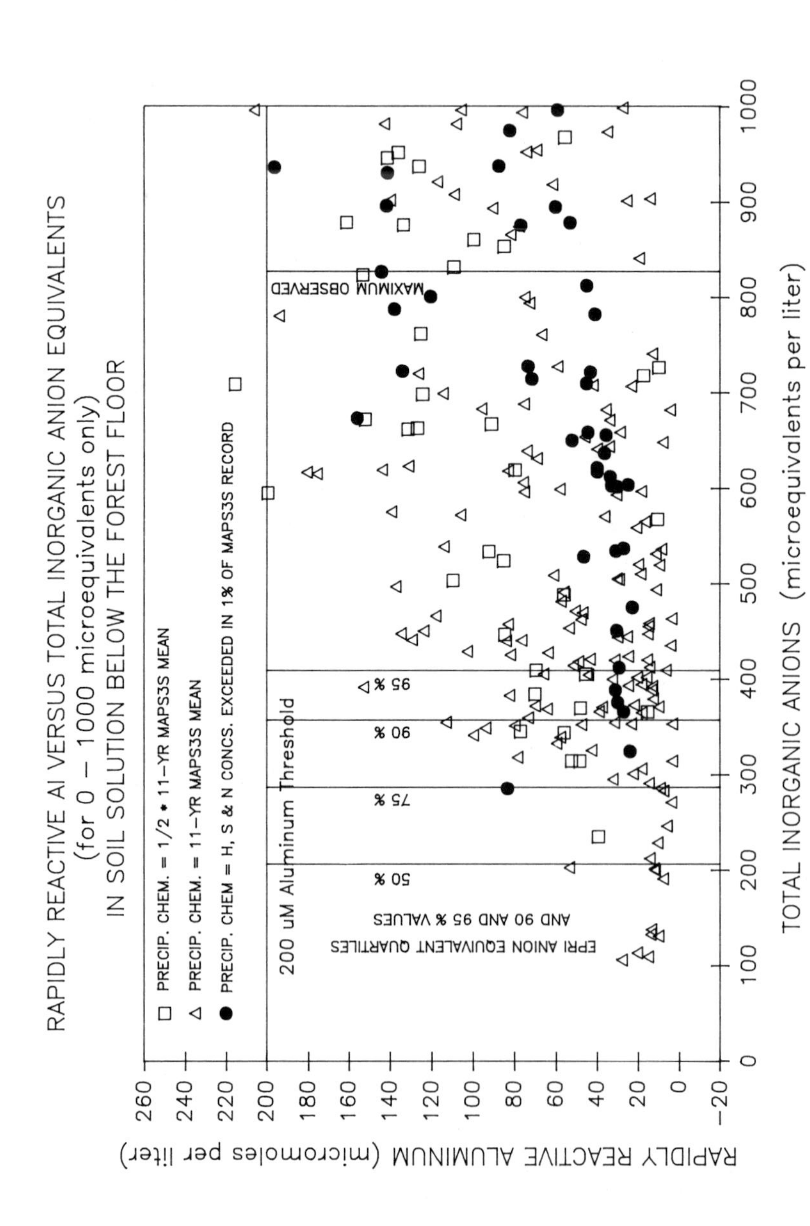

Figure 13.8. Rapidly reactive Al in soil solution in O horizons plotted against total inorganic anion concentration in controlled irrigation experiment of Huntington et al. (1989), carried out at Whiteface Mt., New York. Vertical lines show frequency distribution of inorganic anion concentrations measured in soil solution during 3-year period at IFS site (IFS anion equivalent quartiles). Symbols represent simulated acid rain at three different pHs.

Table 13.4 shows the results of foliar and soil nutrient analyses from Whiteface Mt. Values considered to be sufficient or deficient from the study of Swan (1971) are shown in Table 13.5. Swan considered his standards to be "provisional." From the results, it appears that there are no unequivocal acute deficiencies in the red spruce sampled, but there are moderate deficiencies in Mg and P. If those deficiencies are important as a component of the decline, it is difficult to find unequivocal evidence in the field measurements as described next. Although K in foliage is related to decline class at Whiteface, it is present in the foliage of severely declining trees at levels Swan (1971) considered adequate for good growth. It is equivocal whether the lower K associated with the declining trees is a cause or an effect of the decline.

At Whiteface Mt., the P content of current and year-old foliage was marginally ($p < .07$) or significantly ($p < .001$) correlated with crown condition, but it is unclear whether this is a cause of decline or an effect of it. At present, there is no evidence linking P condition of the foliage to airborne chemicals.

The relatively low levels of Mg are of interest and worthy of further study. Joslin et al. (1988) have shown in their field study on Whitetop Mt. that Mg leaching from red spruce crowns increases with decreasing cloud water pH, thus raising the possibility that foliar nutrition has been adversely affected by acidic clouds. Jacobson et al. (1990a) have shown that in some of their experiments, exposure to acidic mist reduced the foliar Ca and Mg content of red spruce seedlings when they were exposed to acidified mist for periods of 6 to 13 weeks.

Coupled with foliar leaching, a reduced ratio of Mg to Al in the soil solution or on exchange sites could lead to poorer foliar Mg status. Unequivocal field evidence in support of this possibility is currently lacking, however. Mg concentrations in spruce foliage on Whiteface Mt. are *not* correlated with available Mg in the soil nor are they related to neutral salt extractable Al or the ratio of extractable Mg:Al (Figure 13.9). Similarly, there are no correlations among foliar Ca and soil Al or Ca.

Effects of Airborne Chemicals on Winter Hardiness

Because of the evidence linking winter injury to the red spruce decline and the prolonged exposure to acidic cloud water and elevated ozone levels, many experiments have been initiated to determine if airborne chemicals at present and past levels increase sensitivity to winter injury. As reviewed by Barnard et al. (1990), most experiments with seedlings have indicated that acidic mist at ambient and near-ambient levels reduces freezing resistance (Fowler et al. 1990, D. DeHayes, University of Vermont, personal communication). Some (less than half) of the experiments with ozone indicated a significant effect on winter hardiness (Cumming et al. 1988; Fincher et al. 1990), again at ambient levels. To bridge the gap between seedling stud-

Table 13.4. Concentrations of Ca, Mg, K and P in Red Spruce Foliage on Whiteface Mt[a]

		Ca (wt%)				Mg (μg/g)				K (wt%)				P (wt%)			
Vigor Code	Distribution	86	+/−	85	+/−	86	+/−	85	+/−	86	+/−	85	+/−	86	+/−	85	+/−
				0													
1	45	0.22	0.01	.30	0.01	646	21	600	19	0.61	0.02	0.48	0.01	0.12	0.002	0.11	0.002
2	14	0.21	0.01	0.31	0.03	582	36	546	26	0.57	0.03	0.43	0.01	0.12	0.004	0.10	0.003
3	10	0.22	0.02	0.30	0.03	631	27	547	37	0.52	0.03	0.39	0.03	0.11	0.004	0.08	0.004
Statistical Data[b]																	
p		0.64		0.71		0.42		0.12		0.022		5.4×10^{-4}		0.07		0.001	
F ratio		0.22		0.14		0.66		2.48		5.45		13.24		3.50		11.33	
r		−0.057		0.046		−0.10		−0.19		−0.28		−0.41		−0.23		−0.39	

		Ca (wt%)				Mg (μg/g)				K (wt%)				P (wt%)			
Vigor Code	Distribution	86	+/−	85	+/−	86	+/−	85	+/−	86	+/−	85	+/−	86	+/−	85	+/−
<800	14	0.24	0.02	0.33	0.02	638	44	606	32	0.61	0.03	0.48	0.02	0.12	0.004	0.10	0.002
800–899	12	0.24	0.02	0.29	0.02	720	33	643	37	0.55	0.04	0.46	0.03	0.12	0.005	0.11	0.025
900–999	13	0.19	0.01	0.26	0.02	636	40	566	36	0.60	0.04	0.45	0.03	0.12	0.004	0.11	0.005
1000–1099	9	0.21	0.02	0.30	0.03	616	41	565	46	0.65	0.03	0.47	0.01	0.12	0.003	0.10	0.002
1100–1199	17	0.21	0.02	0.31	0.02	574	22	542	16	0.57	0.02	0.44	0.02	0.12	0.003	0.11	0.003
>1200	4	0.24	0.04	0.32	0.05	589	39	575	83	0.51	0.04	0.38	0.02	0.10	0.008	0.08	0.006
Statistical Data[b]																	
p		0.24		0.99		0.033		0.062		0.40		0.058		0.80		0.05	
F ratio		1.36		0.0025		4.73		3.60		0.71		3.73		0.07		3.84	
r		−0.14		−0.019		−0.26		−0.23		−0.12		−0.23		−0.03		−0.24	

[a] Samples were collected from the Northwest face, at elevations above and below the IFS site, in October 1986. The data are analyzed for trends related to elevation and to crown condition. Crown class 1 trees had less than 10% foliage loss, class 2 had 11%–50% loss, and class 3 had 50%–99% loss.

[b] Statistical data show the probability (p), F ratio, and correlation coefficient (r) for each element as related to vigor code. Distributions are not always true for all variables because of missing data.

Table 13.5. Suggested Provisional Standards for Evaluation of Results of Foliar Analyses Species[a]

Element	Range of Acute Deficiency[b]	Range of Moderate Deficiency	Transition Zone from Deficiency to Sufficiency	Range of Sufficiency for Good to Very Good Growth	Range of Luxury to Excess (Toxic) Consumption
N	Below 1.10	1.10–1.30	1.30–1.60	1.60–2.80	2.80 and up
P	Below 0.10	0.10–0.14	0.14–0.18	0.18–0.28	0.28 and up
K	Below 0.19	0.19–0.30	0.30–0.40	0.40–1.10	1.10 and up
Mg	Below 0.04	0.04–0.06	0.06–0.08	0.08–0.17	0.17 and up
Ca	Below 0.05	0.05–0.08	0.08–0.12	0.12–0.30	0.30 and up

[a] Modified from Swan (1971). Red spruce foliar concentration is expressed as percent dry matter.
[b] These suggested standards are essentially judgments; they are based both on results of greenhouse studies and on experience gained from use of foliar analysis in field studies.

ies and mature trees growing under field conditions, we designed and used branch chambers at Whiteface Mt. to determine what effect excluding airborne chemicals had on foliar properties, especially those related to winter injury.

The experimental site was located 300 m uphill from the IFS monitoring site. Vann and Johnson (1988), Vann et al. (1990), and Berlyn et al. (1989) have described the materials and methods in detail. The 1988 experiment yielded the most important data for evaluating the effect of airborne chemicals on winter injury because a regionwide winter injury event occurred in late winter 1989 that damaged some of the branches used in the experiment.

In the 1988–1989 experiment, we selected four red spruce more than 75 years old growing at 1170 m on Whiteface Mt. The branches that were enclosed in chambers were selected to be as similar as possible using the following criteria: position on the northerly or westerly aspects on a fully exposed portion of the upper crown (whorls 7–13); branch age; visual estimation of foliage color (subsequently checked by chlorophyll analysis); and ease of installation (to maximize technician safety).

The chambers were half-ovoids with a volume of approximately 85 L. They were constructed of polyethylene and Teflon to minimize chemical reactivity, and contained 187 ± 40 1988 shoots.

One open branch on each tree was used as a reference for chamber effects (designated "open"), and the following treatments were randomly assigned: (1) charcoal-filtered air from which cloud water and charcoal-reactive gases were removed (designated "filtered"); (2) ambient air with cloud water removed by a series of teflon mesh filters (designated "dry"); (3) a charcoal-filtered treatment like number (1) with deionized water mist added to the chambers during naturally occurring cloud events (designated "misted"); and (4) a control treatment in which ambient air and clouds were drawn through the chambers.

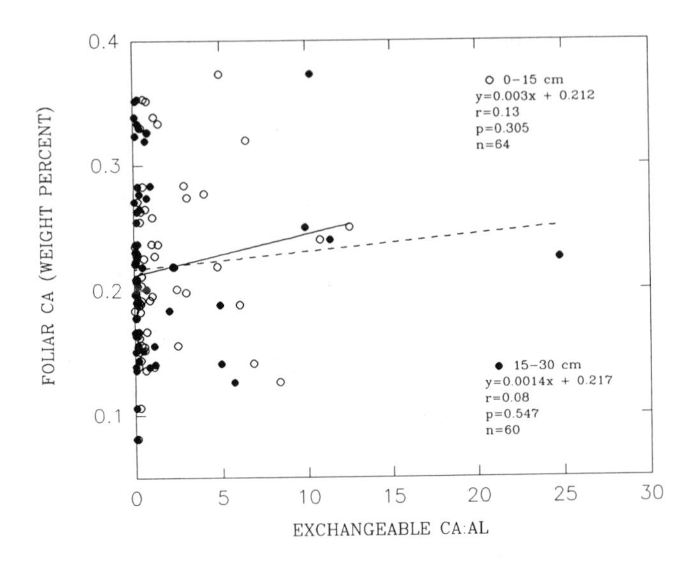

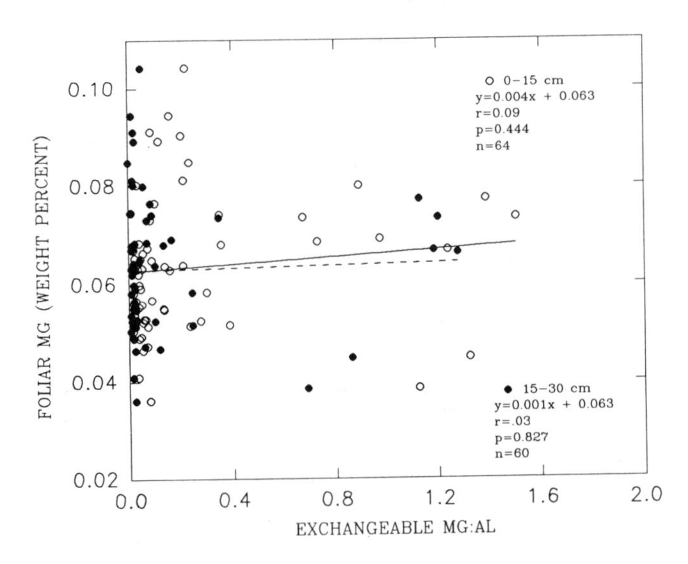

Figure 13.9. Foliar Ca versus exchangeable Ca:Al and foliar Mg versus exchangeable Mg:Al in root zone of trees sampled for foliar analysis. Trees were located on northwest face of Whiteface Mt. between 700 and 1200 m. (After Johnson et al. 1989a).

Ozone measurements were made inside and outside the chambers in August 1988. In chambers from which we wished to exclude ozone, the levels inside the chambers were 31% of ambient; inside the control and "dry" treatments, ozone levels were 91% of ambient. Temperatures were recorded inside the chambers using thermocouples. Average temperature over the 3 months of exclusions averaged 1–2°C warmer than ambient, with peak values in full sun at midday several additional degrees warmer.

After the 3 months of treatments (early July–September 30), there were no treatment effects on needle weight, but there were significant effects on chlorophyll content, carotenoid content, cuticle thickness, and stomatal wax plug thickness (Berlyn et al. 1989; Vann et al. 1990). Analysis of temperature, aspect, and previous-year needle chlorophyll showed that the treatment effects were not caused by artifacts, but that they can be reasonably attributed to differences in airborne chemicals. Figure 13.10 shows the data for total chlorophyll (chlorophyll a + chlorophyll b) as an example of the treatment effects.

We tested samples for freezing tolerance in January 1989 using the ion leakage method of DeHayes and Williams (1989). When sampled, none of the trees showed signs of winter injury; there were no visible symptoms, and ion leakage values were characteristic of uninjured needles. The trees differed in their sensitivity to freezing as shown in Figure 13.11. Our related

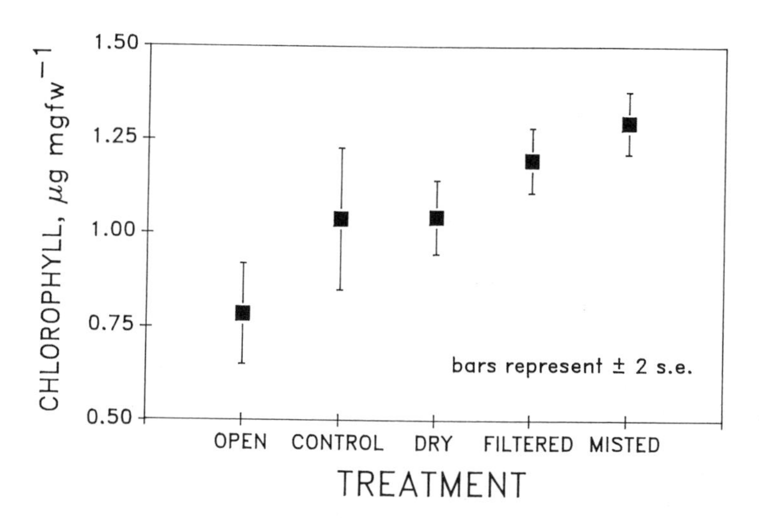

Figure 13.10. Total chlorophyll measured on samples collected on September 1988 as function of branch chamber treatment. Chambers had been in place for 3 months. Error bars are ±2 SE. (After Vann et al. 1990.)

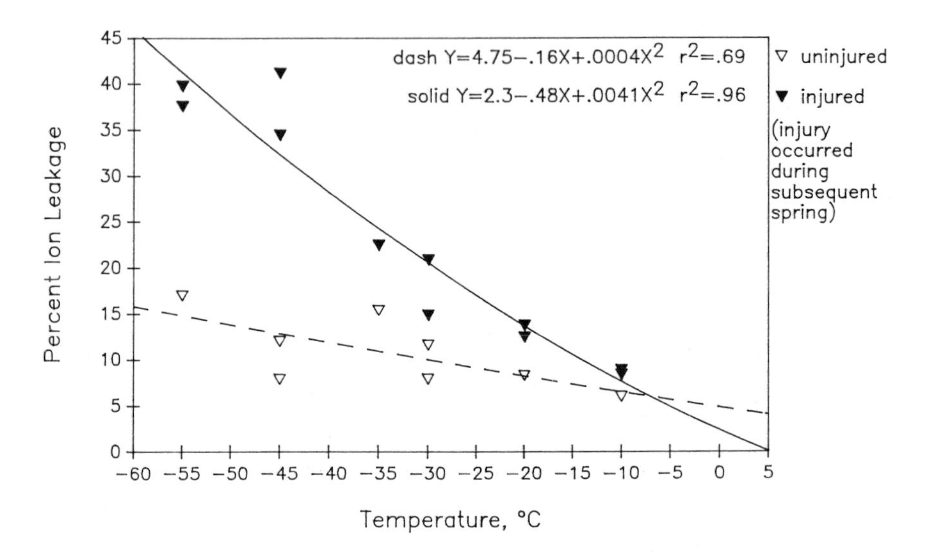

Figure 13.11. Effect of experimental freezing on ion leakage in January 1989. Needles tested were from control chambers on four trees used by Vann et al. (1990) for summer 1988 branch chamber experiment. None of four trees showed visible injury when sampled in January. Subsequently, two trees showed injury in following spring (solid triangles) and two trees were uninjured. Higher levels of ion leakage indicate more freezing injury. Methods used were those of DeHayes and Williams (1989).

experiments have shown that visible winter injury (red needles) occurs when the relative ion leakage exceeds 15%–20%. Based on this threshold, we would expect that trees designated 4 and 6 would show visible injury if temperatures dropped below −25°C, whereas trees 3 and 5 would show visible injury only if temperatures dropped below −45°C.

In late March, visible winter injury was observed on trees 4 and 6 but not on trees 3 and 5. Air temperatures measured at the IFS site showed that after mid-January, temperatures dropped below −25°C on four nights, with a low of −31°C. On a branch-by-branch basis, there were statistically significant relationships between ion leakage measured in January before the trees were injured, and the degree of winter injury scored by three observers in the spring (Figures 13.12 and 13.13). Interestingly, none of the cuticle or wax parameters measured in the previous fall were correlated with the degree of injury. Accordingly, our data suggest that freezing is a key factor that produces winter injury in subalpine red spruce, while the role of desiccation is uncertain.

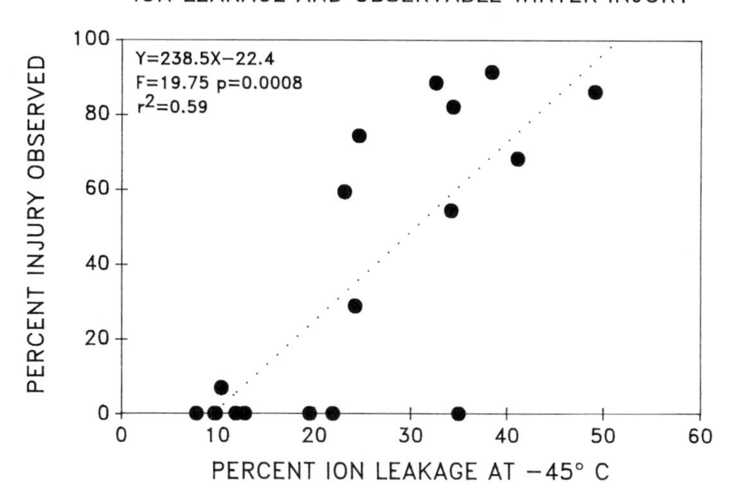

Figure 13.12. Winter injury at Whiteface Mt., New York, scored in May–June 1989 compared to susceptibility to freezing injury measured in January 1989 before winter injury event. (After Vann et al. 1990.)

Figure 13.14 shows the effect of the treatments on the freezing susceptibility of January foliage samples from the winter-injured trees (trees 4 and 6). Except for foliage from the filtered treatments, ion leakage at all temperatures was consistent across the trees. When linear models are fitted to the ion leakage versus temperature data, the curve for the control treatments (ambient air and cloud water) differs from that for the misted treatment (deionized water mist and charcoal-filtered air) at $.05 < p < .10$. Overall, the winter injury data from the branch chamber experiment are consistent with the growing body of results that indicate that exposure to airborne chemicals reduced winter hardiness of high-elevation red spruce. At present, the mechanisms involved are unknown and constitute a topic for continuing research.

Summary

The decline of red spruce has provided one focus for research in the effort designed to better delineate the effects of air pollution on North American forests. Whiteface Mt. is a site representative of the Adirondack Mountains, a region where more than half of the canopy spruce have died during the past 25 years. The IFS project was a key source of data on atmospheric chemistry and elemental deposition and cycling. Along with several other projects, it has allowed the testing of several hypotheses regarding the role

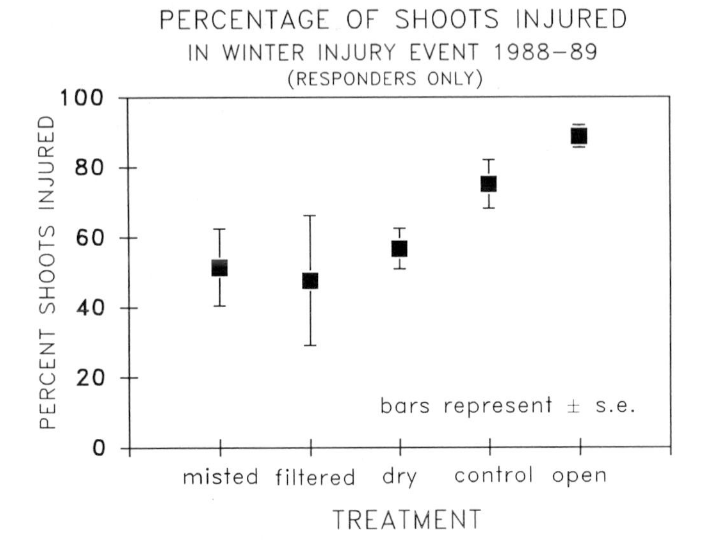

Figure 13.13. Degree of winter injury versus treatment for 1988 branch chamber experiment of Vann et al. (1990). Winter injury was scored as percentage of 1988 shoots in each chamber, which were read in spring. Score is average of three independent observers. Error bars are ±1 SE.

of airborne chemicals in the mortality of red spruce. Our current understanding is summarized here:

1. The recent period of spruce mortality in the northern Appalachians was triggered by repeated, severe winter injury that occurred in the late 1950s and extended to the mid-1960s. Mortality has been especially pronounced at the top of the red spruce elevational range where, we believe, it has been living near its physiological limits.
2. Along with pathogens and insects, winter injury in the 1970s and 1980s sustained the decline.
3. At current levels, airborne chemicals, especially acid mist, appear to be an important factor in reducing the winter hardiness of red spruce. Experimental results from several research groups support this conclusion. At present we lack a precise understanding of which airborne chemicals are involved or how they act to interfere with winter hardiness. The evidence in hand suggests that sulfuric acid in cloud water may be the most important constituent.
4. We have not found evidence at Whiteface Mt. or nearby Adirondack sites that implicates loss of bases from the soil, aluminum mobilization, or unfavorable base cation:aluminum ratios as important stresses contributing to decline.

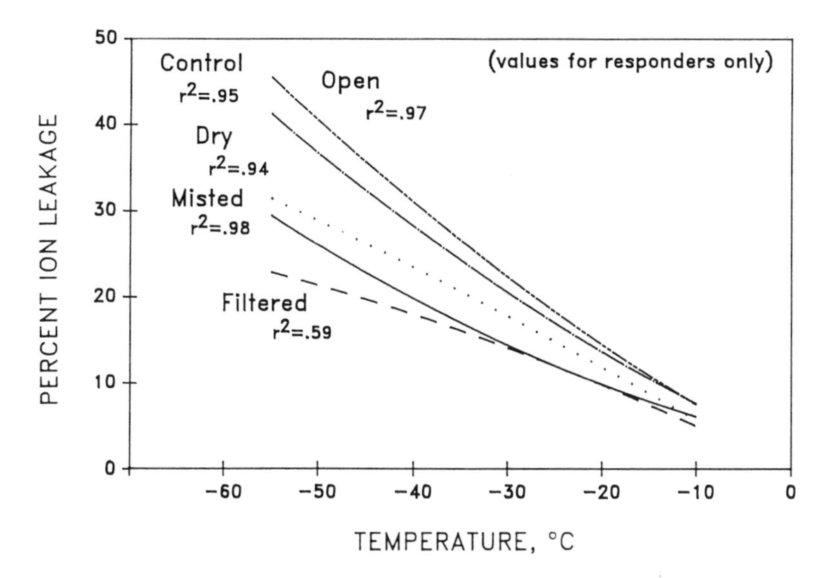

Figure 13.14. Effect of summertime treatments on experimentally induced freezing injury in January 1989 (Vann et al. 1990). Needles were experimentally frozen in January 1989, 3.5 months after treatments ceased. Ion leakage is shown for two trees that later showed naturally induced injury.

In addition to the results reported here, several other studies have shown that ozone or acidic mist has detrimental effects on red spruce seedlings. Direct injury from mist at pH 3–3.5 has been observed (Jacobson et al. 1990b,1990c) with sulfuric acid mist applied with wetting and drying cycles. Alterations of the patterns of antioxidant levels have been observed by Hausladen et al. (1990), and field measurements suggest that ozone exposure may add to the relatively high levels of oxidant stress observed in spruce growing at high-elevation on Whiteface Mt. (Madamanchi et al. 1990). The extent to which these are occurring and their roles in the decline, if any, are undetermined at present.

It appears that exposure to ozone or acidic mist for one growing season does not substantially alter biomass production in seedlings and saplings grown under controlled conditions; however, most of the experiments are scheduled to run for several years. A sound judgment on direct effects on photosynthesis, respiration, and growth must await those results, but the tendency for airborne chemicals to promote winter injury suggests that periodic foliage loss will cause reduced growth.

Eastern Hardwoods

D.J. Raynal, N.W. Foster, M.J. Mitchell, and D.W. Johnson

Introduction

Although dieback in eastern deciduous forests of North America is not as dramatic as that in some subalpine Appalachian coniferous forests, concern about the long-term productivity of hardwoods has heightened as awareness of the quantitative chemical nature of atmospheric deposition has increased. Although individual tree species such as sugar maple may be declining in parts of their distributional ranges, there is currently no evidence of regional forest decline in North American deciduous forests. Nonetheless, because it is widely acknowledged that atmospheric deposition can cause soil acidification, leading to nutrient deficiency and metal toxicity in forest soils and reduction in tree growth, linkages between atmospheric deposition and forest productivity should be determined. Forest-atmosphere relationships are complex, however. Possible adverse influences of atmospheric deposition may be offset by stimulatory effects such as fertilization influences. Studies addressing the effects of air pollutants on tree physiology and forest soils ultimately are necessary if the consequences of atmospheric deposition on forests are to be understood.

Because there are very few hardwood forests in which actual soil changes associated with atmospheric deposition can be measured, research must concentrate on an elemental budget approach to determine the degree to which atmospheric deposition has or will cause acidification and nutrient loss in soils. This methodology, described in detail by Binkley and Richter (1987), involves measuring hydrogen ion input to forests from the atmosphere and determining internal ecosystem H^+ generation from (1) natural acid production, (2) uptake of cations by vegetation, and (3) formation of humus. This measurement allows assessment of the extent at which current acidification rates are accelerated by atmospheric deposition (Johnson and Taylor 1989). Cation budgets can be compared with exchangeable cation pools or total soil pools to estimate which soils may change with further ion inputs. Rates of mineral weathering release of cations, generally poorly quantified, must be determined if the rate of ion change is to be evaluated.

Coupled with nutrient cycling investigations, studies of toxicity of metals, particularly Al, are needed. The Aluminum in the Biosphere (ALBIOS) project evaluated patterns of Al biogeochemistry and phytotoxicity in North American and northern European forests (Cronan and Goldstein 1989). The ALBIOS findings indicated that relationships between inputs of H_2SO_4 and HNO_3 are not simple. In general, however, watersheds with the highest concentrations of labile Al were those characterized by soils with base saturation of less than 10% to 15% and elevated solution concentrations of strong acid anions. Under such conditions, there is a strong positive relationship between acidic deposition and Al mobilization (Cronan and Schofield 1990).

Phytotoxicity screening of tree seedlings indicated that deciduous trees ranged from "sensitive" to "moderately insensitive" to "insensitive" to soluble Al and that toxicity thresholds for sensitive species were sometimes within the range of highest soluble Al found in forest watersheds (Kelly et al. 1990; Sucoff et al. 1990).

The nutrient budget analysis approach provides a means of determining whether cations such as Ca, K, or Mg are or will become limiting for tree growth. Johnson and Taylor (1989) reported that recent nutrient budget analyses of the influences of whole-tree harvesting and atmospheric deposition have predicted greater loss of base cations, especially Ca, than of N or P. Such budgets seem to accurately reflect actual elemental losses when compared to long-term analyses of nutrients in studies where forest soils have been resampled.

One aspect of cation cycling and nutrition in eastern hardwoods that deserves special mention here is the propensity for many of these species to take up and accumulate high concentrations of base cations, especially Ca (Marion 1979). This pattern of high Ca uptake and accumulation was also observed in the IFS network. As noted in Chapter 8, Ca uptake in the CW site (which contains a significant oak and hickory component) and beech sites rivals that of the faster growing DL pine site because of higher tissue concentrations in the hardwood sites.

The tendency for certain hardwoods to accumulate Ca, which is especially pronounced in oak and hickory species, leads to inordinately high rates of soil Ca depletion with harvesting in many of these forests and the possibility of significant soil acidification from Ca uptake even without harvesting. Both literature reviews and collaborative field studies have indicated that, while Ca limitations until now have been virtually unknown in North American forests, nutrient budget data most often point toward impending Ca depletion with intensive harvesting (Johnson 1983; Johnson et al. 1988a; Federer et al. 1989). The recent analysis by Federer et al. (1989) is especially convincing in that they showed that significant depletion of total ecosystem Ca was likely to occur with intensive harvesting in several forests of the eastern United States. Johnson and Todd (1990) concluded that uptake was primarily responsible for large observed decreases in exchangeable Ca in subsoils. Several studies on the effects of Ca uptake on and near Walker Branch Watershed, Tennessee, showed that Ca uptake by oak and hickory species has led to significant depletion of subsoil exchangeable Ca^{2+} in as little as a decade (Johnson et al. 1988; Johnson and Todd 1990).

Thus, while there is currently no direct evidence that Ca depletion is leading to Ca deficiencies or forest decline in eastern hardwood forests, there are reasons to closely monitor the situation to see if these nutrient budget analyses are, in fact, pointing in that direction. The results from Walker Branch Watershed clearly indicate that soils have become depleted of exchangeable Ca^{2+} over a very short time in sites with high rates of Ca uptake (Johnson et al. 1988b; Johnson and Todd 1990). The question is, will deficiencies develop before soils stop changing? Will succession in these mixed forests favor species that are less Ca demanding in the future? These and many other interesting questions related to the role of Ca in tree nutrition arise from the analyses of nutrient budget data from eastern hardwood forests.

Three IFS sites are classified as eastern hardwoods. The two northern sites are Huntington Forest (Adirondack Mountains of New York, HF) and Turkey Lakes Watershed (Ontario, Canada, TL); both have sugar maple and yellow birch as major constituents of the overstory vegetation, but only the former site has a major component of American beech. The TL site is also significantly older with major overstory trees being about 140 years old versus 100 years old at HF. The only southern hardwood site is located at Coweeta Watershed (CH) in North Carolina, which has been left undisturbed since 1923. Oaks constitute about one-half of the overstory of this mixed southern hardwood.

Sugar Maple Decline

The isolated declines in sugar maple and other tree species in the eastern deciduous forests during recent decades are widely recognized (Houston 1987). A current, severe, and widespread dieback of deciduous forests in southwestern Quebec has been intensively examined since 1982. An aerial survey of approximately 2 million ha of forest by the Quebec Ministere de l'Energie et des Resources indicated that approximately 50% of the trees of that area experienced 11% to 25% defoliation, and crown damage greater than 26% was observed on less than 3% of the area (Gagnon and Roy 1989). This survey also revealed that between 1986 and 1988 the percentage of trees affected increased from 76% to 89% but the severity of the damage did not change.

The health of the deciduous forest in Ontario has been recently summarized by McLaughlin et al. (1990). In the 1980s maple decline was an isolated phenomenon, with the geographic distribution of maple decline correlating only superficially with atmospheric pollution. The prime inciting factor appears to have defoliation by forest tent caterpillar in 1976 to 1978, concurrent with spring droughts in 1976 and 1977. Tree conditions, stable in 1984 to 1986, deteriorated in 1987 and 1988 following severe defoliation and two consecutive unusually warm and dry growing seasons.

In the United States, in contrast to parts of Quebec and Ontario, widespread recent reductions in sugar maple health have not been observed. In the northeastern states the population and growing stock volume of sugar maple has been increasing in recent decades, although isolated incidences of dieback have occurred (Gansner et al. 1987). Sugar maple growth between 1950 and 1980 in this region was considered normal (Hornbeck et al. 1988). However, in the northeastern Adirondack Mountains of New York, Bauce (1989) demonstrated a steady 30-year growth decline of sugar maple in a 85-year-old-stand. There, growth reduction was attributable to the occurrence of warm winter and dry summer conditions and the role of secondary agents including defoliating insects.

Maple Decline and Acid Deposition

Much of the area experiencing maple decline in southern Quebec receives the highest rate of atmospheric pollutants in Canada (12 kg ha^{-1} yr^{-1} wet S, 0_3 maxima exceeding 80 ppb) (McLaughlin et al. 1990a). Whether atmospheric acidity is a major stress contributing to deciduous forest decline, as suggested by Gagnon et al. (1986), has been vigorously debated. One hypothesis suggests that acid deposition contributes to nutrient leaching and unbalanced nutrient availability on sites of marginal soil fertility (Bernier and Brazeau 1989). Another suggests that acid rain-induced soil acidification could increase Al levels in the soil solution to levels that may interfere with root growth and nutrient assimilation (e.g., Kinch 1989). Thornton et al. (1986) found sugar maple seedlings moderately sensitive to Al in solution culture. A third hypothesis suggests that the N in acid deposition is likely to stimulate tree growth and increase demand for other nutrients that are in short supply (Nihlgard 1985). However, it has been found that there is little growth response to N addition for some sugar maple stands, and in some cases fertilization actually decreases growth (Stone 1986; Stanturf et al. 1989). This may be caused in part by the elevated N capital in conjunction with the elevated inputs that are common for these stands. Some of these stands may be exhibiting NO_3^- leaching, and this may have potential consequences associated with the loss of base cations and potential acidification processes including the mobilization of Al^{+3}. However, Burke and Raynal (in manuscript) found that N fertilization of seedlings resulted in greater production of foliage than fine roots, in part because of an extended period of foliage development. Extended production and smaller partitioning of photosynthate to reserve carbohydrate may lead to increased susceptibility to natural stresses including defoliation and drought.

In comparing two northern IFS sites (see Appendix for details on data for each site), TL and HF, it is apparent that the TL site has higher N capital (11,641 and 9,567 kg N ha^{-1}, respectively) and much higher NO_3^- leaching (1,146 and 71 eq ha^{-1} yr^{-1}, respectively). This was attributed to the greater maturity of the former site and concomitant low demand of N by the forest vegetation. The third hardwood site (CH) is southern and has lower N capital (5,347 kg ha^{-1}) and neglible NO_3^- leaching (1.3 eq ha^{-1} yr^{-1}). Further details on the N budgets of IFS sites including the three eastern hardwoods sites are given in Chapter 7.

Potassium Deficiency

Very low potassium (K) concentrations in sugar maple foliage have been reported in declining stands in the Quebec Appalachians since 1983 (Bernier and Brazeau 1988a). Seasonal declines in foliar K concentrations from late August to October are cited by Bernier et al. (1989) as evidence that foliar leaching may not be compensated by uptake when soil K supplies are low. However, Hogan and Foster (in manuscript), have deduced that the amount of K, Ca and Mg lost from sugar maple canopies exposed to acid deposition (pH 4.4, wet SO_4^{2-} and NO_3^- at 25 and 20 kg ha^{-1}, respectively) was too small to reduce the foliar concentrations of bases during the growing season. Further, they reported that K loss from sugar maple foliage, under laboratory conditions, was not influenced by increasing acidity from pH 5.0 to 3.0. Likewise, Robitaille and Boutin (1989) observed that the leaching of K from a sugar maple canopy was not augmented by precipitation of increased acidity.

Translocation of K from foliage before abscission might be an alternate explanation for late season declines in foliar K concentrations; a mechanism whereby the trees are able to compensate for low availability of K in the soil. Considerable translocation of K from foliage during senescence was estimated at the two tolerant hardwood IFS sites at HF and at TL.

The concentration of K in sugar maple foliage at HF (0.65%) is less than that for TL (1%) and for other maples growing on granitic material (Foster et al., in manuscript). The level of K at HF is close to that found in Quebec for declining maples (0.55%; Bernier and Brazeau 1988a), but greater than that associated with acute deficiencies (0.2% to 0.3%; Bernier and Brazeau 1988b). In earlier work at HF when the stand was 70 years old, fertilization with K resulted increased foliar concentration but did not increase tree growth (Lea et al. 1980). At the CH site, the foliage, which has a very different vegetation composition (50% of basal area is oak), had a mean K concentration (0.7%) that was intermediate between the two northern sites.

Phosphorus Deficiency

Bernier et al. (1989) determined that some declining sugar maple stands, growing on soils with moderately acid mull layers with low available P had deficient or very low foliar P concentrations. They have hypothesized that, in response to increased H^+ inputs to soil, additional P is fixed in Ah horizons that contain Fe and Al hydrous oxides (Pare and Bernier 1989). Mineral acids are likely to increase the activity of Al in soil solution. Kinch (1989) considered that the expression of decline in Ontario sugar maple stands is related to elevated soil Al concentrations. Declining maple stands examined by Kinch exhibited P (and Mg) concentrations in foliage that were near critical levels (0.11%) at which visual deficiency symptoms have been shown to appear (Bernier and Brazeau 1988a). Kinch observed lower growth of bioassay seedlings and elevated Al in seedling roots in soil from beneath declining trees than from healthy ones. Seedlings in surface (F, H, Ah) soils from declining trees contained suppressed P in foliage and responded to P additions with improved growth. Mineral horizons, however, were capable of meeting the P requirements of the seedlings. Perhaps when trees are exposed to severe drought they may be forced to obtain a greater proportion of their P requirements from surface horizons that retain all the water when precipitation is light and infrequent.

Acid Deposition, Nutrient Deficiency, and Other Stresses

There is considerable evidence from biogeochemical studies that acid deposition alone is unlikely to produce the interference in hardwood nutrition associated with maple decline. The causes of the observed nutritional disturbances, however, have yet to be determined. Fertilization has proven successful in improving vigor (Hendershot et al. 1989) and in raising foliar levels of K (Ouimet and Fortin 1989) and P for at least 3 years after treatment with these nutrients (Bernier et al. 1989). It remains to be seen whether improved tree nutrition improves either the levels or rate of recovery in declining stands.

Because of high rates of NO_3^- leaching, TL is losing more base cations than HF. The HF site has larger reserves of base cations and is potentially less sensitive to acidification (Foster et al., in manuscript). The HF site has higher levels of Al leaching (93 moles ha^{-1} yr^{-1}; data from Appendix) than either TL (0.11 mol ha^{-1} yr^{-1}; Foster et al. 1986) or CH (2.1 moles ha^{-1} yr^{-1}; data from Appendix). In the two northern sites (Foster et al., in manuscript) as well as the CH site, because of the high levels of native soil acidity as well as weathering and atmospheric inputs of base cations it is expected that base saturation and soil acidity will change very slowly (also see Chapter 8 for detailed discussion of cation cycling).

Natural stresses can explain some but not all the recent decline in sugar maple health and vigor that has been observed in parts of North America. Maple dieback in Quebec and New York has been related in part at least to insect defoliation, drought, winter damage, soil infertility, and stand density (Bauce 1989; Bernier et al. 1989). For example, a 1980–1981 winter freeze-thaw capable of inducing cavitation injury in trees was followed by dieback of sugar maple the following summer (Auclair 1989). Some areas experienced severe defoliation by forest tent caterpillar in 1980 and 1981 and growing season droughts in 1982 and 1983. For a more detailed discussion of these and other possible causes of maple declind in Quebec, see Hendershot and Jones (1989).

Tree-ring growth studies also suggest that sugar maple has been largely unaffected by current levels of acid deposition in eastern North America (Bauce 1989). Witter et al. (1990) reported no change in sugar maple productivity across a pollution gradient of S deposition from 12 kg ha^{-1} yr^{-1} in Ohio to 3 kg ha^{-1} yr^{-1} in northern Minnesota. Likewise, Frelich et al. (1989) found no relation between sugar maple growth and S deposition (4 to 7 kg ha^{-1}) in Wisconsin. Considerable uncertainty exits, however, on whether air pollutants may react to exacerbate the effects of natural stresses or whether the cumulative impacts of air pollutants may impact future forest productivity.

Al Phytotoxicity

Soluble Al is expected to increase markedly in sites with soils of 10% to 15% effective base saturation or less (Reuss 1983), pH below 4.9, and solutions with SO_4^{2-} concentrations greater than 80 μmol L^{-1} (Cronan and Goldstein 1989). Soils associated with many shade-tolerant hardwoods have effective base saturations of 18% to 78% (Table 13.6). [The effective base saturation of soils with 6% to 13% (NH_4OAc) would be 12% to 53% using conversions suggested by Kalisz and Stone (1980)]. Most of the tolerant hardwood soils in Table 13.6 would have pH and solution SO_4^{2-} levels conducive to Al release but exhibit base saturation levels greater than suggested critical levels. Many soils under tolerant hardwoods, thereby, would appear more likely to leach bases than Al when exposed to acid deposition.

The HF site has greater concentrations of Al and SO_4^{2-} in leachates from the B horizon (21.3 μmol L^{-1} and 124.3 μeq L^{-1}, respectively) than Turkey Lake (14 μmol L^{-1} and 90.8 μeq SO_4^{2-} L^{-1}). The CH site (0.32 μmol L^{-1} and 8.23 μeq SO_4^{2-} L^{-1}) has much lower concentrations of both constituents leaching from the B horizon than the northern sites; the low concentration of SO_4^{2-} results from the retention of S in this site (see Chapter 6). It has been previously been found that about one-third of the Al leached from the B horizon of the HF site is monomeric, and this is the form which is of greatest concern regarding biological toxicity (David and Driscoll 1984).

Further, organic matter in surface horizons is likely to buffer acidity by exchange with Ca and by protonation of weak acid functional groups, without increasing Al concentrations in solution (James and Riha 1986). Many soils below 44° latitude (see Table 13.6) have pH values less than that of ambient rainfall, so acidification of these soils by atmospheric deposition is unlikely. Soils beneath tolerant hardwoods at higher latitudes are exposed to precipitation with a mean pH of 4.0 to 4.4 and therefore appear at risk to further acidification by acid deposition. However, Roberge (1987) observed that the humus layers and surface mineral horizon from stands of sugar maple, within the region of high dieback and acid precipitation, have not become more acidic in recent years irrespective of their initial pH.

ALBIOS studies indicated that growth of sugar maple can be reduced at solution Al concentrations only above 600 μm (Thornton et al. 1986). However, Al reduced tissue contents of Ca and Mg at much lower levels (100 μM) and therefore may impair growth of sugar maple by causing nutrient stress.

Implications of IFS Findings and Forest Decline

Witter (1990) observed that S and N cycling in tolerant hardwood stands was related to wet deposition of SO_4^{2-} and NO_3^-. Both anions are contrib-

Table 13.6. Soil Properties[a] from Selected Tolerant Hardwood Forests[b]

Lat. (N)	Long (W)	Elevation	Vegetation	pH (H_2O)	C ($g\ kg^{-1}$)	C/N	CEC[c]	Exchangeable Cations K	Ca	Mg	Na	Base Sat. (%)	Solution SO_4^{2-} (μ mol L^{-1})
								(c mol(c) kg^{-2})					
47°03′	84°24′	350	Sugar maple	4.5	45	16	15(7)	0.12	1.20	0.25	0.22	12(23)	55
46°	74°	380	Sugar maple	5.1	71	—	(4)	0.12	1.66	0.20	0.05	(53)	60–70
45°38′	71°42′	300	Sugar maple	4.6	51	19	37	0.10	3.23	0.41	0.08	13	—
45°25′	79°10′	—	Sugar/red maple	4.4	48	19	(4)	0.18	2.40	0.43	—	(78)	80
44°	74°	700	Beech/spruce/maple	3.9	—	—	11	0.06	0.66	0.10	0.01	7	105
43°59′	74°14′	530	Maple/beech/birch	4.1	95	—	25	0.13	1.14	0.19	0.01	6	105
43°59′	74°14′	530	Maple/beech/birch	4.0	95	25	(14)	0.12	2.18	0.22	0.09	(18)	65
43°57′	71°43′	1900	Maple/birch/beech	4.0	45	22	26	0.10	1.68	0.15	0.15	10	—

[a]Average values from surface horizons to 30 cm depth; organic horizons included (if availabie).
[b]Cited from Cronan (1985); Foster et al. (1986); Hendershot (pers. comm.): Lazerte (pers. comm.); Lozano et al. (1987); Mollitor and Raynal (1982); Shepard et al. (1990); Pilgrim (pers. comm.); Robarge (pers. comm).
[c] CEC by buffered NH_4OA_c Cac or, in brackets, unbuffered chloride salts (effective base saturation).

uting to the leaching of cations from IFS sites at HF (Shepard et al. 1990) and TL (Foster 1985), although atmospheric NO_3^- inputs at the latter are overshadowed by the considerable mineralization and nitrification of organic N in soil (Foster et al. 1986). In general SO_4^{2-} is contributing significantly to the leaching of cations from soils supporting tolerant hardwood forests (see Table 13.5; see also Chapters 6 and 8). However, the suggestion that acid deposition is contributing to K leaching from soils and thereby exacerbating inherently low K availability is not supported by nutrient budget analyses in tolerant hardwood forests. For example, leaching of K as a percentage of that in solution in surface soil horizons was 17% at TL and only 5% in the more K-limited HF soil (Foster et al., in manuscript). Low concentrations of K in percolating water at 30 cm depth in both fertilized (281 kg ha^{-1} of K as K_2SO_4) and unfertilized Spodosols, beneath a Quebec Appalachian mature sugar maple forest, were also reported by Boutin et al. (1989).

Some soils in the Quebec Appalachian region contain abundant Mg but are low in K, foliage from damaged trees is enriched in Mg but deficient in K (Ouimet and Fortin 1989). Trees may accumulate Mg in excess of their needs and in the process restrict K uptake. The uptake of K by the trees might be favored if Mg availability in the soil could be reduced (Roy and Gagnon 1989). Because Ca and Mg are more likely to be leached from tolerant hardwoods soils than K (Foster 1985), acid deposition is more likely to favor improved K nutrition than not. Foliar deficiency has also been reported for Mg (Bernier and Brazeau 1988b) in sugar maple-dominated stands in the Quebec Laurentians, on sites with low reserves of available Mg (Bernier et al. 1989).

Southern Pines

D. Binkley and D.W. Johnson

For centuries, foresters have known that the productivity of a forest stand reaches a maximum at a relatively young age and then declines. The growth of individual trees within stands also demonstrate a wide range of patterns, with the growth of suppressed trees declining while the growth of dominant trees accelerates. A variety of interacting ecological processes underlie these patterns, including changes in leaf area that are driven by competition among trees and by changes in site fertility. An understanding of these normal changes in rates of growth in forests provides the necessary foundation for considering possible growth declines driven by stress from pollutants.

Growth Patterns with Stand Development

The accumulation of biomass in forest stands is typically expected to follow a sigmoidal pattern, with growth rates accelerating to a maximum then declining with further stand development (Figure 13.15; stem biomass and production curves for Piedmont loblolly pine; site index, 20 m at 25 years). The derivative of the biomass accumulation curve represents the annual rate increment. The common explanation for this widespread pattern is that productivity increases as stand leaf area increases (Möller 1945; Daniel et al. 1979; Waring and Schlesinger 1985). A peak in productivity is reached when leaf area reaches a maximum. The decline in productivity is often expected to result from increasing respiration losses of carbon as the quantity of respiring tissue accumulates in the forest (Waring and Schlesinger 1985).

In reality, the pattern of biomass accumulation in forests does tend to follow these expectations, but variations around these smooth expectations are pronounced. For example, the rate of biomass accumulation in the Duke Forest site was greater than the rate expected based on site index curves developed for similar stands on the Piedmont (Figure 13.16). The annual rate of stemwood production was also variable from year to year, peaking below the expected maximum. Perhaps the most important point to note in Figure 13.16 is that the productivity of this stand should be expected to decline rather dramatically during the next 5 years but that the appearance of the decline may be obscured by year-to-year variability.

The causes behind the changes in growth rates have not been well characterized, and they vary from site to site. For example, the initial accumulation of pine leaf area depends on nutrient availability and competition with hardwoods. Fertilization and herbicide treatments can accelerate the

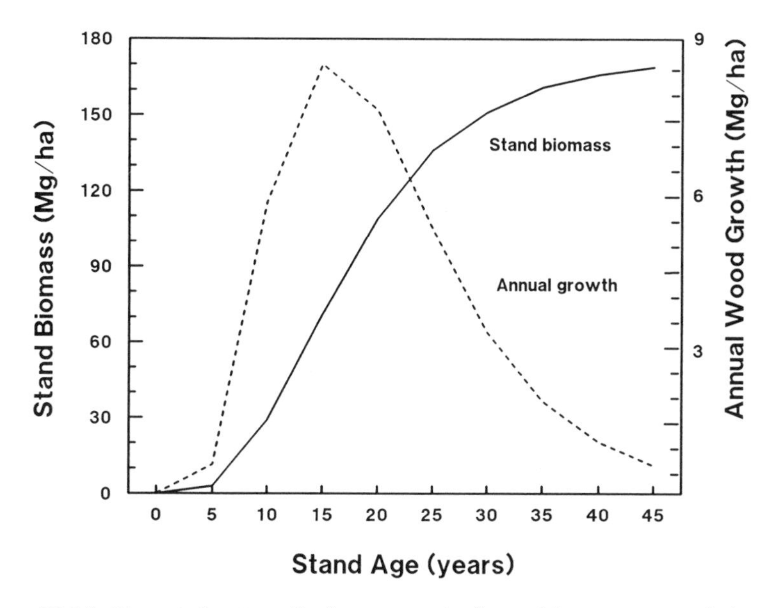

Figure 13.15. Expected pattern (in 5-year steps) of stem biomass accumulation and production for site index 20 m at 25 years in Piedmont forests of North Carolina.

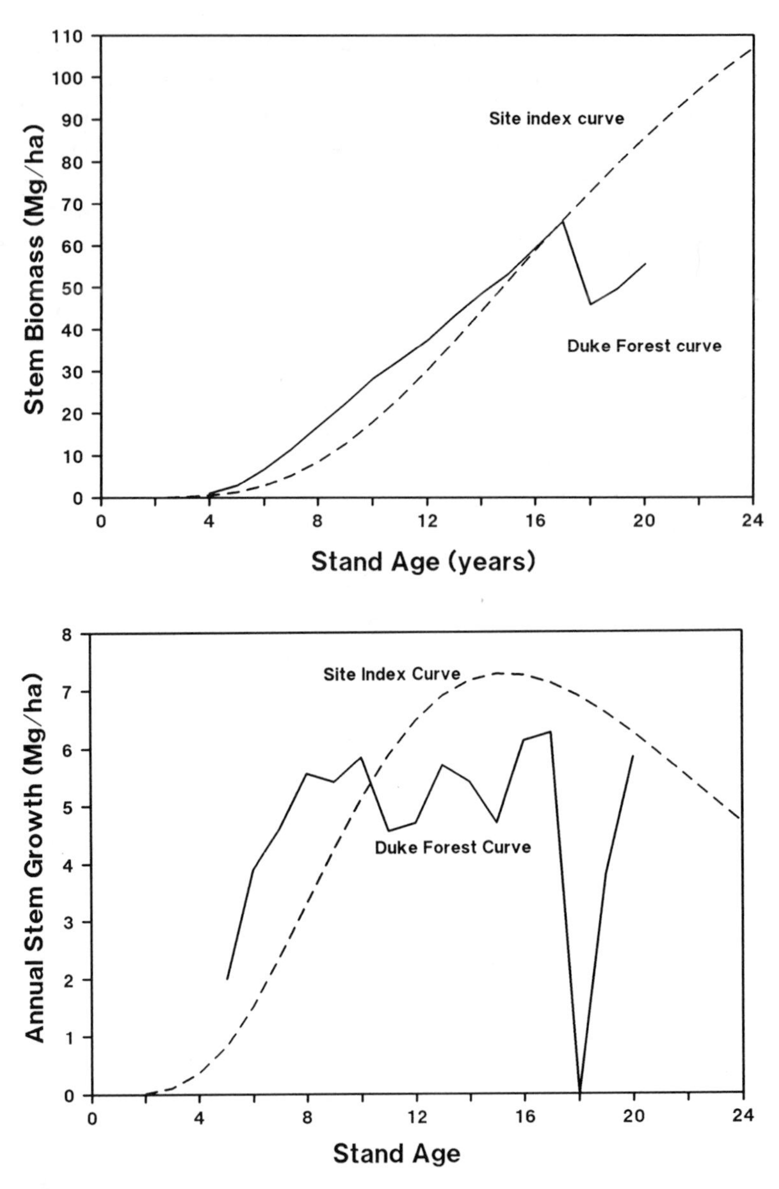

Figures 13.16. Expected patterns for site index 18 m at 25 years, contrasted with observed patterns reconstructed for Duke Forest IFS site [the drop at age 17 resulted (bottom) from commercial thinning of stand].

achievement of maximum stand leaf area. The maximum leaf area obtained in a pine stand may also be constrained by nutrient availability. For example, Vose and Allen (1987) found that fertilization of loblolly pine forests increased the leaf area of some stands from 2.7 to 4.3 m^2 of projected leaf area per meter squared of ground area. Stands with high leaf area indexes did not increase leaf area after fertilization.

The cause of the decline in productivity in later stand development has rarely been examined. Fertilization experiments in older conifer stands have demonstrated increased rates of stem growth (Allen 1987) and leaf area (Gholz 1986; Vose and Allen 1987). Some of the decline in productivity may relate to increasing respiration costs, but this expectation should not be assumed to be a sufficient explanation until experimental manipulations (such as fertilization and competition control) have examined direct limitations on stand leaf area and growth.

The most complete picture available of ecosystem dynamics with stand development in the South comes from the work of H. Gholz and colleagues (Gholz and Fisher 1984; Gholz et al. 1985; Gholz 1986) on a chronosequence of stands. Their work documented an increase of pine leaf area up through 10 years, remaining constant through about 25 years, and then declining by 25% by age 35 (Figure 13.17). The rate of aboveground productivity reached a peak about the same time as stand leaf area, but productivity declined sooner than did leaf area (Figure 13.17). The decline in wood production per unit of leaf area (Figure 13.18) indicates increasing

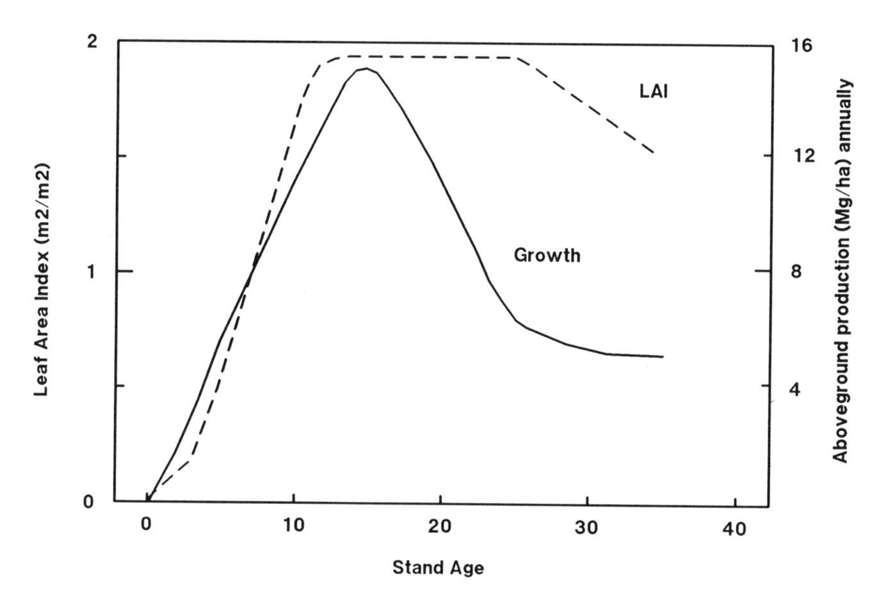

Figure 13.17. Pattern of development of stand leaf area and aboveground net primary productivity for chronosequence of slash pine forests. (From Gholz 1986.)

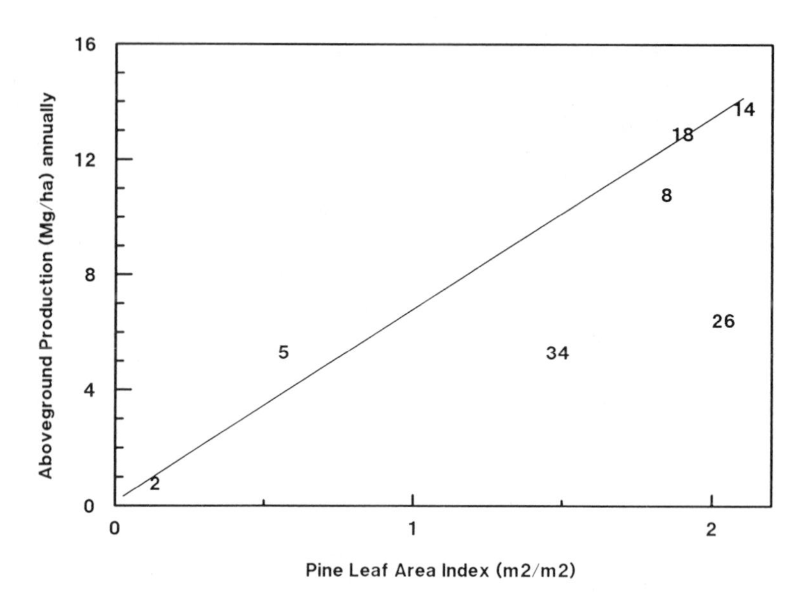

Figure 13.18. Efficiency of wood production per unit of pine leaf area was relatively constant through age 18, but dropped sharply for ages 26 and 34 indicating increased stress (probably from increased nutrient limitation). (From Gholz 1986.)

stress in stands older than 25 years. Fertilization of a 26-year-old stand increased leaf area by about 10%, demonstrating that at least part of the decline in leaf area and productivity after age 25 may be caused by nutrient limitations.

The relative degree of nutrient limitation on the growth of pine stands at various stages of stand development has not been determined in the South. Most researchers expect that nutrient availability (at least for N) is greatest early in stand development and probably declines with stand age (e.g., Binkley et al. 1989b). This expectation is based on observations of high rates of N availability after harvest (e.g., Vitousek and Matson 1985), and of the rapid accumulation of a high proportion of ecosystem nutrient totals into pine biomass (Table 13.7). Pine plantations, especially on old-field sites, probably "mine" the more readily available supplies of nutrients and experience declining nutrient availability and increasing nutrient limitation. Information is limited, but the rapid accumulation of 30% of an ecosystem's N capital in biomass (vegetation + forest floor) is certainly consistent with these expectations.

Growth Trends across the Region

Concern over possible pollution-induced declines in the growth of pine forests in the South was ignited by an analysis of USDA Forest Service forest inventory data (Sheffield et al. 1985). The average annual increment of

Table 13.7. Nitrogen Accumulation (kg^{-1} ha) in Ecosystem Biomass (vegetation + forest floor)[a]

Location	Stand Age	Soil (depth)	N contained in Vegetation	N contained in Forest Floor	Proportion Biomass
Loblolly Pine					
Duke Forest (IFS)	23	1,680 (0–0.8 m)	235	450	0.29
Oak Ridge (IFS)	33	5,260 (0–0.6 m)	195	110	0.05
Duke Forest (Wells and Jorgensen 1975)	16	1,750 (0–0.7 m)	260	300	0.24
Duke Forest (Wells and Jorgensen 1978)	32	1,425 (0–0.7 m)	430	360	0.36
North Carolina (Tew et al. 1986)	22	3,940 (0–0.6 m)	210	360	0.13
Tennessee (Johnson and Todd 1987)	31	3,050 (0–0.45 m)	195	445	0.17
Slash Pine					
Florida (IFS)	20	1,670 (0–0.8 m)	460	285	0.31
Florida (Gholz and Fisher 1985)	36	2,300 (0–1.0 m)	220	400	0.21
Florida	40	2,960 (0–1.0 m)	140	270	0.12

[a]From Morris et al. (1983).

southern pines in the Piedmont and mountains across the South declined by 30% to 50% between 1972 and 1982 relative to trees of the same diameter in the period from 1957% to 1966. Pine stands in the Coastal Plain showed much smaller declines or no declines in growth at all. Sheffield et al. (1985) listed eight potential causes of the growth declines:

Atmospheric deposition
Increased stand density
Increasing stand age
Increasing competition from hardwoods
Drought
Reductions in the water table
Loss of old-field conditions
Increased impacts of diseases.

These authors had no evidence regarding atmospheric deposition. Detailed analyses of their database did reveal that changes in stand structure (density, average age of plots sampled, and hardwood competition) all changed in a direction that would tend to cause reductions in radial growth. Similarly, the late 1970s and early 1980s experienced more frequent droughts than in the earlier reference periods. Many pine stands in the Piedmont originated from the abandonment of agricultural fields between 1945 and 1965, and the growth of these stands likely benefited from prior soil management (fertilization, plowing, and associated minimization of hardwood competition).

Burkhart (1988) assessed the state of knowledge of declines in pine growth in the South. He concluded that no evidence linking acidic deposition to growth declines existed and that any observed declines likely resulted from natural causes or from ozone exposure.

Zahner et al. (1989) performed a more rigorous analysis of growth trends in 131 well-stocked, even-aged stands of loblolly pine in Georgia, South Carolina, and North Carolina. They concluded that growth of trees between the ages of 20 and 45 years has declined by about 1% annually since 1950 in the Piedmont; they attributed the decline to increased competition, regional drought, and unidentified factors that could include atmospheric deposition (suggested to result from ozone exposure rather than acidic deposition).

Implications from the Southern Pine IFS Sites

The southern pine sites in the IFS provided no evidence supporting a role of acidic deposition in any growth declines. Soil solution concentrations of cation nutrients were not low, and concentrations of aluminum were orders of magnitude below the likely threshold of toxicity to pines. Our data are consistent with the "loss of old-field fertility" pattern, in that the rapid accumulation of nutrients in vegetation and undecomposed forest floor might indicate increasing nutrient limitations with time.

The monitoring of ozone concentrations at IFS sites across the South demonstrated frequent episodes of concentrations in excess of the toxicity values reported for southern pines from experimental studies (see Chapter 4). The southern pine IFS sites were generally medium to high in exchangeable soil cations and low in nonexchangeable cations (see Chapter 8). The Duke, Georgia, and Oak Ridge loblolly pine sites have the highest exchangeable base cation pools and lowest total cation pools in the IFS network (Chapter 8). As indicated in Chapter 10, weathering at the Duke site is essentially zero and weathering at the other southern pine sites is very low. The Oak Ridge sites represent an extreme case with respect to Ca, where nearly all soil Ca is on exchange sites. The turnover calculation in Chapter 8 indicated that the three loblolly pine sites have lower than average potential for soil acidification when only exchangeable pools are considered, but much greater than average potential for soil acidification and base cation depletion when total soil cation pools are considered.

The Florida slash pine site is rather unique in that it has one of the lowest pools of both exchangeable and total base cations in soils, yet turnover rates generally do not indicate great potential for change (where data are available) because of net Ca and Mg gains from atmospheric deposition (i.e., deposition minus leaching). From this perspective, these sites would appear to be more susceptible to depletion of soil cations than those from glaciated regions with much greater quantities of weatherable primary minerals. Binkley et al. (1989b) estimated that about 10% to 15% of commercial pine forests in the Southeast would be limited by the availability of cation nutrients if limitations by nitrogen and phosphorus were removed by fertilization. They forecasted that about 10% to 30% of forest soils in the region will experience significant acidification as a result of the combined effects of acid deposition and other management activities (especially harvesting).

Because leaching losses of base cations decline markedly as base saturation declines (Chapter 8), acid deposition alone may be incapable of causing such deficiencies. In addition, Lucier (1988) noted that current growth declines reported by Sheffield et al. (1985) seemed to be restricted primarily to nonindustrial lands of the Piedmont and to not be observed on intensively managed sites. Depletion of soil cations by forest growth and harvest must be greater on intensively managed sites, so any current declines in growth on less intensively managed sites are unlikely to be related to cation deficiencies. The reported declines in forest growth for some regions of the Southeast are reminiscent of declines in the growth of second rotation stands of *Pinus radiata* in Australia: although the specific causes remain poorly quantified, the effects can be reversed by good management practices (including fertilization with Ca, K, or Mg if necessary) (Keeves 1966; Squires 1982).

Trends in Soil Chemistry across the Region

Observed Changes

Despite the importance of soil fertility in maintaining the forest industries in the South, surprisingly little research has focused on rates of changes in soil chemistry in the region (Binkley et al. 1989b). Changes in soil chemistry in southern pine stands with time have been examined in only two locations.

The chronosequence study of slash pine forests by Gholz and Fisher (1985) found no indication of significant changes in soil acidity with stand age. The A1 horizon of three 2-year-old stands averaged pH 3.8, with Bt horizons averaging pH 4.8. The same values for three 34-year-old stands were 0.1 units lower, a nonsignificant difference.

In the Calhoun Experimental Forest in South Carolina, C. Wells sampled soils in a loblolly pine plantation 5 years after planting in an abandoned agricultural field and again 20 years later. Analysis of these samples showed that pH in the 0- to 75-mm depth declined from 5.0 to 4.1 and from 4.7 to 4.3 in the B horizon (Binkley et al. 1989a). The process responsible for soil acidification appeared to be increased saturation of the exchange complex with aluminum. The increase in aluminum saturation balanced the depletion of base cations. Simulations with the MAGIC (Cosby et al. 1985) model suggested that the decrease in soil pH combined with the increased aluminum saturation of the exchange complex may have increased the concentrations of Al^{3+} in soil solutions from 0.3 μmol(c) L^{-1} at age 5 to 3 μmol(c)

L^{-1} at age 25 (Binkley et al. 1989b). Over the same period, alkalinity of the soil solution may have declined from 125 to 40 μmol(c) L^{-1}. The rate of base cation depletion was about 2200 mol(c) ha^{-1} annually, which was probably close to the unmeasured rate of accumulation of base cations in vegetation and forest floor biomass. No connection with acidic deposition was apparent.

Soil Sensitivity Criteria

A variety of studies have suggested criteria for classifying the sensitivity of soils to acidification in the South (Table 13.8). By the criteria used for cation exchange capacity, all the southern pine sites would classify as sensitive to acidification. By the base saturation criteria, none of the sites would be sensitive by any of the classification schemes except for the Florida slash pine site, where low base saturation would indicate sensitive soils by two of the four suggested schemes. Interestingly, the chronosequence study by Gholz and Fisher (1984) indicated no change in soil acidity during 35 years of stand development; this pattern suggests that the critical base saturation values may not be appropriate for these soils. The classification schemes showed divergent opinions about critical levels of soil pH for indicating sensitivity to acidification. Some included high pH as an indicator of sensitivity (based on likely change in pH), whereas others focused on low pH

Table 13.8. Sensitivity Criteria for Southern Forest Soils[a]

Reference	Class	CEC (mmol (c) kg^{-2})	Base Saturation (%)	pH[b]
McFee (1980)	Sensitive	<60		
	Slightly sensitive	60–150		
	Nonsensitive	>150		
Klopatek et al. (1980)	Sensitive	<120	30–50	>5
	Slightly sensitive	120–200	30–40	>5
	Insensitive	>120	<30	<5
	Very insensitive	>200	>50	>6
Olson et al. (1982)	Highly sensitive	<62	>5.5	
	Moderately sensitive	62–90	>5.5	
	Insensitive	>90	—	
Turner et al. (1986)	Sensitive:			
	Base cation depletion	<150	20–60	>4.5
	Al toxicity	<150	<20	<4.5
Binkley et al. (1989b)	Sensitive	<100	<10	<5.0
	Moderately sensitive	100–150	10–20	<5.5
	Insensitive	>150	>20	>5.5

[a] From Binkley et al. (1989b).
[b] pH is soil pH in laboratory measurements, except for Binkley et al. (1989b), where pH is pH of soil solutions.

as a critical indicator (based on response of soil solution chemistry). Accordingly, the southern pine sites were classified as sensitive by some schemes (particularly that of Turner et al. 1986) and as insensitive by other schemes.

Implications from Southern Pine IFS Sites

The soils from the southern pine sites showed substantial differences in pH and exchange chemistry; the only unifying feature across all sites was the very low quantities of weatherable minerals (see Chapter 10). The loblolly pine sites all had greater quantities of exchangeable base cations than the site in the Calhoun Forest (Binkley et al. 1989a), which showed rapid acidification with stand development, so we expect that any acidification during the next few decades will likely be less than was observed at the Calhoun site. The slash pine forest in Florida had particularly low concentrations of exchangeable base cations, but the very low pH falls within the range where buffering by aluminum oxides will likely prevent further declines in soil pH.

Past soil management practices have been important in modifying soil properties. Fertilization during agricultural management at the Duke Forest was probably fundamental in providing the relatively high concentrations of exchangeable base cations in the soil.

The southern pine sites in the IFS project showed no immediate cause for concern over acidification. However, the low cation supplies (relative to plant uptake and biomass harvest) indicate that nutrient management may need attention to sustain long-term site productivity. Management practices in the future will likely have major impacts as well.

Europe

D. Aamlid, K. Venn, and A.O. Stevens

Forest Decline in Europe

Some phenomena were noticed in the Federal Republic of Germany in the late 1970s that were later named "neuartige Waldschaden," the novel forest disease or forest decline. Reports on forest decline of this new type soon appeared extensively in public media as well as in scientific journals. Air pollution was immediately blamed as a causal factor. However, very few scientific facts could support this statement. Early in the 1980s it became obvious that this new forest decline was of a very complex nature. Somewhat similar symptoms were found in many of the European countries. However, differences in decline symptoms occurred (Figure 13.19). In the most severely affected parts of central Europe, extensive forest dieback was recorded over tens of thousands of hectares, while in western and northern forests only minor defoliation of living trees was apparent.

Figure 13.19. Severely defoliated Norway spruce.

Classification of Forest Decline in Europe

Forest damage is classified today in two major groups.

1. Traditional types as fungal and insect attacks are well documented: emission damage caused by different gases, climatic conditions, and deficiency of nutrient elements such as potassium, magnesium, nitrogen, and phosphorus.
2. The novel forest disease, which is characterized by several diffuse symptoms of decline resulting in a disease syndrome earlier not known (Schutt et al. 1983).

The syndrome differs from ordinary tree diseases because many affected forest ecosystems appear likely to be destroyed. Symptoms are of two general types, according to Schutt and Cowling (1985):

1. Growth-reducing symptoms including loss of foliar biomass, loss of feeder root biomass, decrease in height and diameter increment, chlorosis, premature senescence and death of older needles and leaves, increased susceptibility to secondary root and foliar pathogens, death of affected trees, and change of ground vegetation.

2. Abnormal growth symptoms, including abscission of green leaves and green shoots, altered branching habits (lametta syndrome), and abundant production of adventitious (secondary) shoots, altered morphology of leaves, repeated abnormal crops of seeds and cones, and changed allocation of photosynthetic products.

German scientists have recognized five different types of forest decline (FBW 1986):

1. Needle yellowing at higher elevations in the German "Mittelgebirge." This decline type is characterized by chlorosis of needles and loss of older needles. Magnesium deficiency may be an associated factor (Zöttl and Hüttl 1985; Rehfuess 1987).
2. Thinning of tree crowns at medium to high altitudes in the German "Mittelgebirge." This decline type is also characterized by loss of needles, but without discoloration.
3. Needle reddening in older stands in southern Germany. This decline type is associated with the needle fungi *Lophodermium piceae, L. macrosporum,* and *Rhizosphaera kalkhoffii* (Rehfuess and Rodenkirchen 1984).
4. Chlorosis of needles and loss of younger needles at higher elevations in the Bavarian alps. The chlorosis is mainly caused by potassium and magnesium deficiencies (Rehfuess 1983).
5. Thinning of tree crowns in northern coastal areas. Loss of needles and reduced growth characterize this decline type.

One or more of the German types are found over nearly all Europe. The ultimate symptoms for several of these types are needle discoloration and needle loss; therefore, further investigations have concentrated on the study of causal factors leading to such symptoms. These symptoms have also been chosen as the two major criteria in nationwide surveys. Yield and death rates of stands are additional criteria of increasing interest. Most countries in Europe were alerted by the German findings and are today seriously concerned for the future of their forests, looking for symptoms that may indicate early warning signals of any type of forest decline.

European activities on forest damage assessment and monitoring are organized under the United Nations Economic Commission for Europe (UNECE) under the Convention on Long-Range Transboundary Air Pollution, International Co-ordinated Programme on Assessment and Monitoring of Air Pollution Effects on Forests (ICP-Forests). A more detailed description is given in Aamlid et al. (1990).

In addition, many investigations are being carried out individually by university units and research institutes. Summarized information and discussions of results from a wide selection of these investigations that are concerned with problems linked to forest decline are presented by Schmidt-Vogt (1989).

Methods of Assessment

The UNECE ICP–Forests Programme has three main levels of activities in monitoring forest vitality. Details of levels 1 and 2 are further described in Aamlid et al. (1990).

Level 1. Large-scale representative surveys and assessment
Level 2. Intensive studies on permanent plots
Level 3. Special forest ecosystem analysis.

Special Forest Ecosystem Analysis

Despite the fact that many forest ecosystem studies are in operation in Europe, very few appear to have a national representative perspective. The present chapter explains the Norwegian national system of permanent plots for special forest ecosystem analysis, which has a national representative perspective, to demonstrate types of data that can be extracted from such a system. The locations of the intensive study plots for special forest ecosystem analysis in Norway are shown in Figure 13.20.

The Norwegian system is probably one of few European systems for special forest ecosystem analysis operating today that can be linked to both the UNECE monitoring program and a larger research program on forest ecosystem nutrient cycling, for example, the Integrated Forest Study. Thus far it has not been possible to apply all desirable variables in the monitoring program. The most important deficiency is probably the dry deposition analysis. Physical and chemical analyses of shrubs and of whole trees are also lacking. The most important additions are assessment of trees according to the ECE Manual (1986) (crown density, crown color, forest increment and volume) and recording of epiphytic lichen flora on branches and trunks. A mycorrhiza and root biology study is also in progress.

Forest Decline Extent and Trends in Europe: Large-Scale Representative Surveys

Crown Assessment–Defoliation

The major species observed are Norway spruce, Scots pine, white fir, beech and oak. The results on defoliation from the survey year 1988 (GEMS 1990) are given in Figure 13.21.

From the 1988 European forest damage survey, it is possible to conclude that:

- Despite considerable differences in forest structure, species composition, and levels and types of air pollution, surveys of forest health based on commonly agreed methods are now being conducted in 25 countries across Europe.

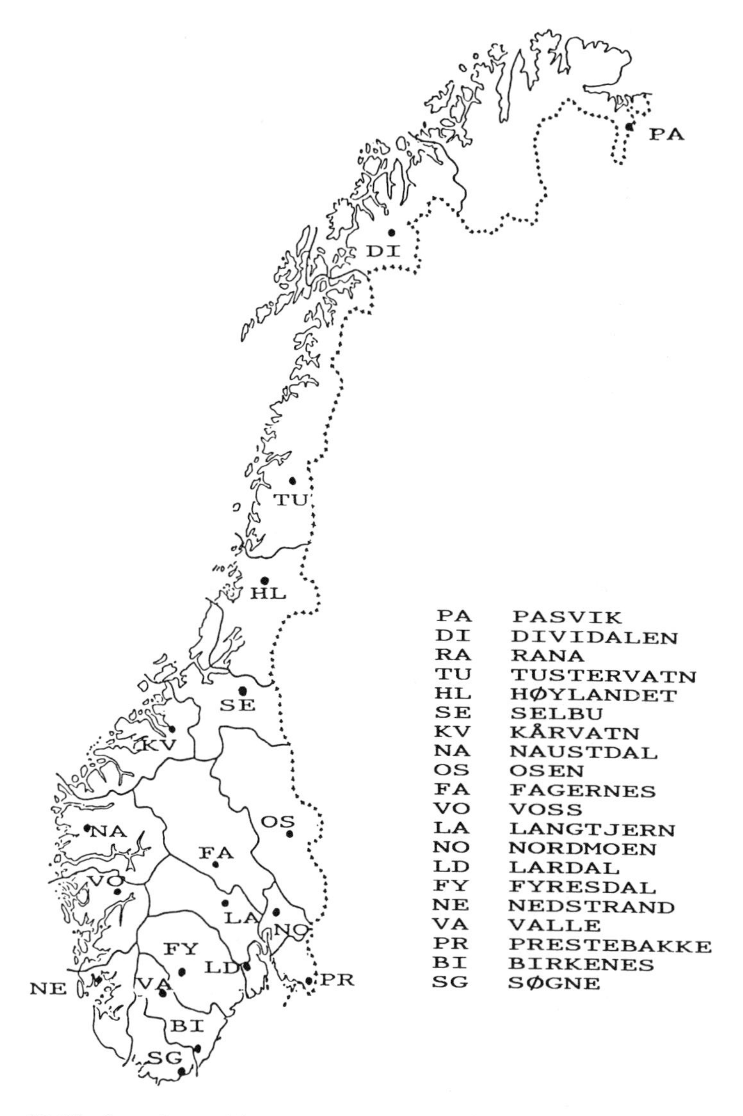

Figure 13.20. Locations of intensive study plots for special forest ecosystem analysis in Norway. Plots with codes PR, NO, BI, NA, and HL are treated further in this report.

- Forest damage, expressed as a loss of needles or leaves, has been observed in all the 25 countries.
- In many regions, forests at higher elevations and forests older than 60 years are considerably more defoliated than younger stands and forests at lower elevation.

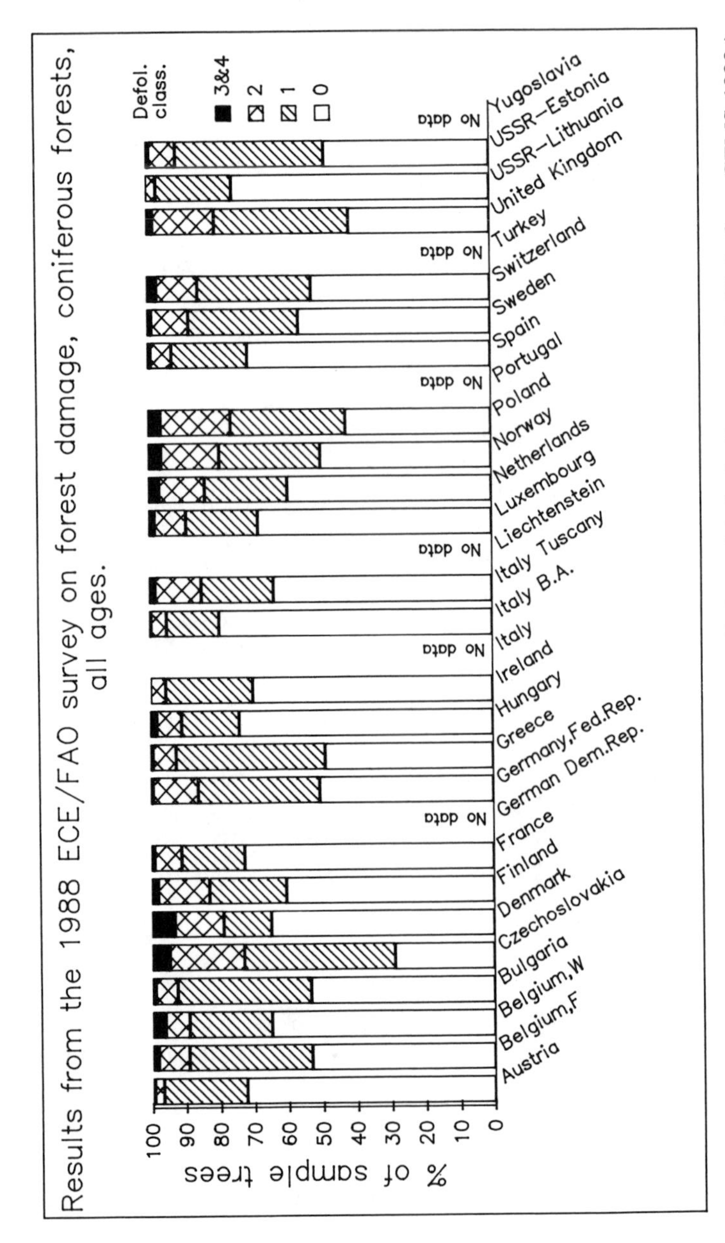

Figure 13.21. Results from survey year 1988 on defoliation, alphabetically arranged. (Data from GEMS 1990.)

- Old spruce, fir, and oak are presently the most affected species.
- Nordic coniferous forests have shown to be as defoliated as many Central European forests.

Trends

In 1984, annual and standardized nationwide surveys began in the Federal Republic of Germany and a few other European countries. The methodology used in these surveys later became the basis for the methodology adopted by the UNECE system, as described in the ECE Manual. The UNECE coordinated, official monitoring of forest health (mostly conifers) in Europe began in 1986. Figure 13.22 gives the trends in defoliation in the Federal Republic of Germany for Norway spruce, Scots pine, and white fir (Anonymous 1989). Table 13.9 summarizes the changes in defoliation for a selected group of countries up to 1988. There seems to have been a slight increase in defoliation during these years.

Yield and Volume

Forest growth and increment are compiled in some countries by tentatively applying standardized methodology (e.g., Kramer 1986). The ECE Manual has prescribed some preliminary methods. However, the investigation periods are still too short to produce adequate yield data. However, crown density has shown to be correlated to growth parameters in some regions (Horntvedt and Tveite 1986, Bjorkdal and Eriksson 1989).

Areas with totally dead forest (damage class 4) have not increased during this period. In the Federal Republic of Germany dead forest covered 7360 hectares in 1988, which is 0.1%–0.2% of the total forest area (Kandler 1989).

Intensive Studies on Permanent Plots and Special Forest Ecosystem Analysis in Norway

General Site Data and Air Pollution

The monitoring of air and precipitation quality in Norway is performed by the Norwegian Institute for Air Research (NILU). The results of 1988 (SFT 1989) demonstrate quite well the pollution levels of the country.

The pH of precipitation is given in Figure 13.23a, which shows a pattern with clear gradients of precipitation acidity. The most acid precipitation (pH 4.3) is found in the southern regions. However, the pH is not directly correlated to the deposition of strong acids (H) because of great differences in the amount of precipitation. This aspect is reflected in Figure 13.23b.

The pattern for sulfate-sulfur and nitrogen compounds in 1988 is similar. The deposition of sulfate-sulfur varies from 1.6 g S m^{-2} in the southern regions to less than 0.2 g S m^{-2} in northern regions. The deposition of nitrogen compounds varies from 2.3 g N m^{-2} in the southern regions to less than 0.2 g N m^{-2} in the northern regions (SFT 1989).

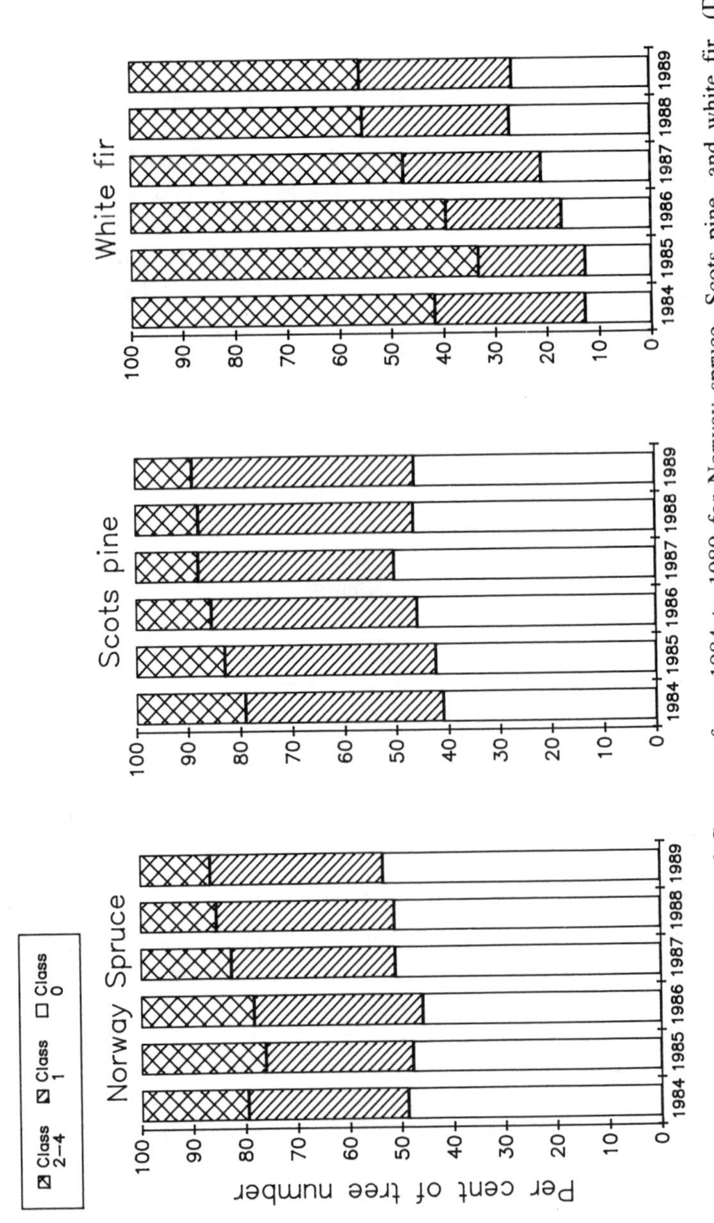

Figure 13.22. Trends in Federal Republic of Germany from 1984 to 1989 for Norway spruce, Scots pine, and white fir. (Data from Anonymous 1989.)

Table 13.9. Changes in Degree of Defoliation for Some Countries During 1986 to 1988[a]

Country	Change Defoliation, Classes 2–4 (%)			Past 2 years (±)
	1986	1987	1988	
Austria	4.5	3.5	3.3	−0.2
Belgium–Flanders	—	4.7	10.8	+6.1
Bulgaria	4.7	3.8	7.6	+3.8
Czechoslovakia	16.4	15.6	27.0	+11.4
Denmark	—	24.0	21.0	−3.0
Finland	—	13.5	17.0	+3.5
France	12.5	12.0	9.1	−2.9
Germany, (FRG)	19.5	15.9	14.0	−1.9
Italy–Bolzano	—	3.1	5.2	+2.1
Liechtenstein	22.0	27.0	23.0	−4.0
Luxembourg	4.2	3.8	11.1	+7.3
Netherlands	28.9	18.7	14.5	−4.2
Norway	7.0[b]	—	20.8	+13.8[c]
Spain	18.2	10.7	7.3	−3.4
Sweden	11.1	5.6	12.3	+6.7
Switerland	16.0	14.0	15.0	+1.0
United Kingdom	—	23.0	20.0	−3.0
Yugosalvia[d]	23.0	16.1	17.5	+1.4
USSR–Lithuania	—	14.8	3.0	−11.8

[a] Data from GEMS (1990).
[b] Regional survey in 1984–85.
[c] Change in 1984/85–1988.
[d] Regional survey in 1988.

The crown density has been monitored since 1986 on nine of the 19 permanent plots (Figure 13.24). The results indicate that mean crown density has been quite stable during this period (OPS 1990) (Figure 13.24). Although variable results, this is probably within the range of errors of the method itself.

Selected Plots along a Pollution Gradient

From the 19 permanent plots (see Figure 13.20), five plots (coded PR, NO, BI, NA, and HL) are further discussed. These are selected plots located along a decreasing pollution gradient from south to north. This feature can be observed when comparing Figures 13.20, 13.23a, and 13.23b. General site information concerning these five monitoring plots is given in Table 13.10.

Soil Chemistry

The general conifer forest soil type in all the Norwegian monitoring plots is Gleyed Humo-Ferric Podzol, according to the Canadian soil classification

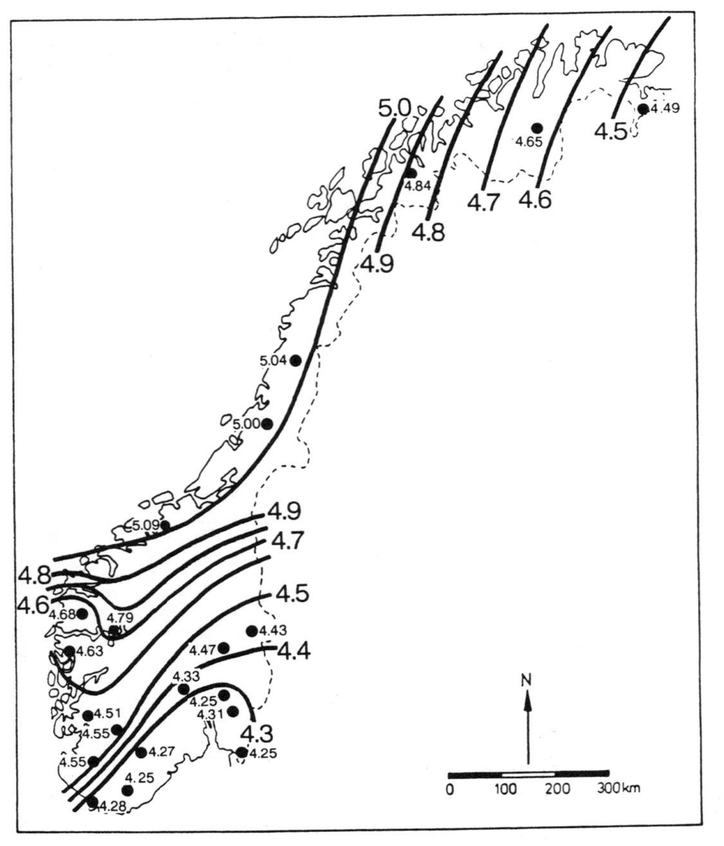

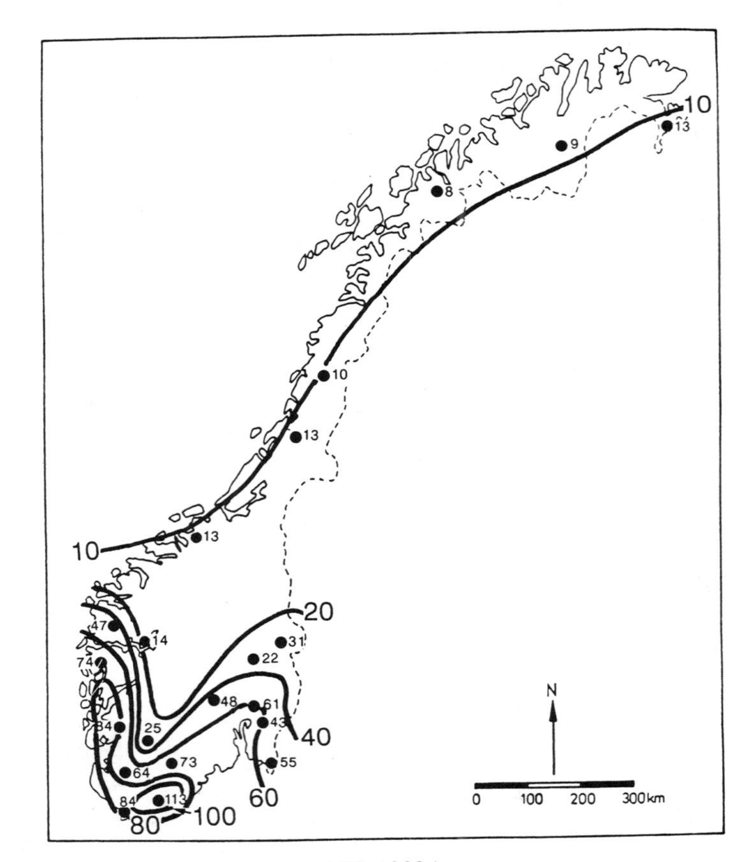

Figure 13.23. (a and b) pH and H^+ depositions in Norway 1988. (From SFT 1989.)

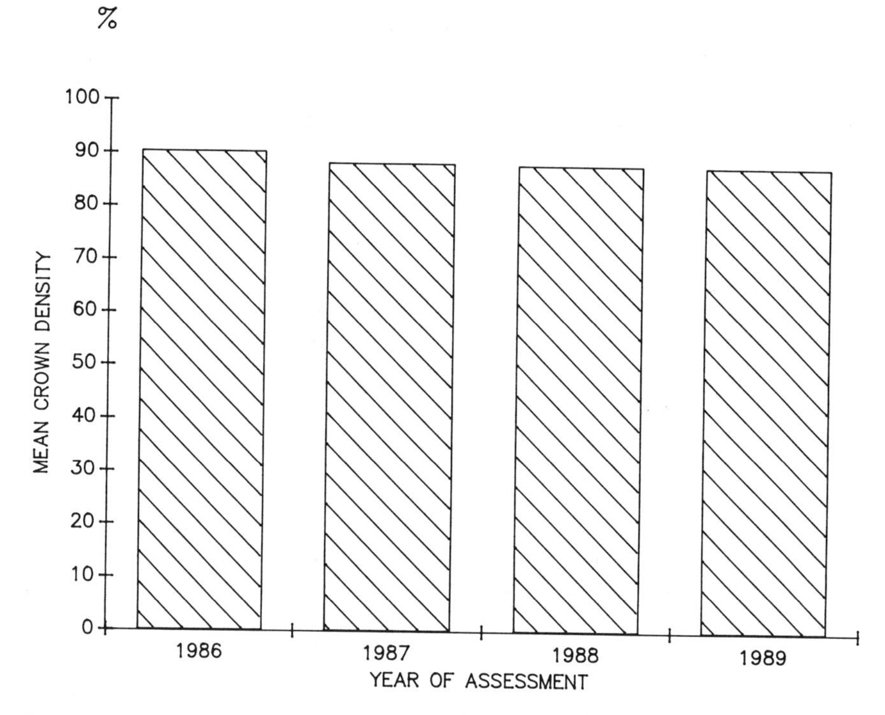

Figure 13.24. Mean crown density of nine permanent plots for special forest ecosystem analysis followed from 1986 in Norway. (From OPS 1990.)

Table 13.10. General Site Information for Five Monitoring Plots Located along a Gradient from South to North in Norway[a]

Plot Code	Latitude (N)	Location Longitude (E)	Altitude (m.a.s.1.)	Main Tree Species	Stand Age	Stand Density[c]
BI	58°23′	8°15′	210	Norway spruce	95	1333
PR	59°00′	11°32′	135	Norway spruce	77	924
NO	60°16′	11°06′	205	Norway spruce	76	727
NA	61°34′	5°53′	70	Norway spruce	37	2166
HL	64°39′	12°19′	120	Norway spruce	142	1210

[a] Data from OPS data bank.
[b] Mean age at 1.3 m above ground.
[c] Tree number per hectare.

system (OPS 1988a, 1988b, 1989). The nutrient capacity varies from plot to plot depending on age, relief, vegetation, climate, soil parent material, and other conditions (Table 13.11) (OPS 1988a, 1988b, 1989). The soil is acid in these plots. The magnesium and calcium content of the soil has not reached any deficit levels, but increasing nitrogen deposition might accelerate the demand for these two elements. These results probably reflect quite well the chemical and physical properties of Norwegian forest soils in spruce forests of blueberry type.

Soil solution is collected from the plots by use of alundum tension lysimeter plates at a constant suction of 10 kPa. No unusual trends are seen in the results for the first 3 years, although the most southeastern plot (PR) has the highest concentrations of total Al found (80 μmol L^{-1}).

Precipitation within the Stand and Free-Falling Precipitation

Typical examples of the flux and concentrations of some elements in the throughfall and free-falling precipitation for 1988 are given in Table 13.12. Volume-weighted mean pH in bulk and throughfall precipitation for the growing season for the years 1986–1988 are given in Table 13.13.

The main result from these data is a verification of the pattern of air pollution distribution in Norway shown earlier. It is also an important fact that some of the nitrogen input is held back in the tree crowns. This is recognized as a certain uptake of nitrogen compounds by the foliage, and it might express a nonsaturated situation of nitrogen in the forests.

Perspectives for Special Forest Ecosystem Analysis

Some data are available from level 2 and 3, mostly published in separate reports from various national programs. Examples are the "Sanasilva" program in Switzerland (Sanasilva 1989), "Monitoring Programme for Forest Damage" ("Overvåkingsprogram for skogskader") in Norway (OPS 1988a, 1988b, 1989), the "Höhenprofil Zillertal" in Austria [e.g., articles in *Phyton*, Vol. 29 (1989)] and studies in the Federal Republic of Germany, such as "Das Forschungsprogramm Waldschaden am Standort Poststurm" (Bauch and Michaelis 1988) and the ecosystem study in Lange Bramke (Hauhs 1989).

However, data are lacking for many countries. The results presented in this chapter are examples from the Norwegian system of plots for special forest ecosystem analysis (level 3), which we hope demonstrate that it is possible to run a national network of stations where the recorded parameters are important inputs to a model based on nutrient cycling.

Causes of Forest Decline in Europe: Accepted Explanations

The forest decline in Europe has led to the formulation of several hypotheses, and numerous experiments have been carried out to test them. Most of the published information is well summarized in Schmidt-Vogt (1989).

Table 13.11. Chemical Properties of Soil Samples in Five Monitoring Plots Located along a Gradient from South to North in Norway[a]

Plot Code	Horizon[b]	pH (H_2O)	CEC[a]	BS (%)	Ca	K	Mg	N	C:N	Soil Classification
					(mmol kg^{-1})					
BI	LFH	3.6	334	53	58	17	19	1033	32	
	Ae/Ahe	3.9	28	15	1.0	0.9	0.4	51	28	
	AB	4.0	34	9	0.6	0.6	0.2	40	30	Gleyed eluviated dystric
	B	4.4	41	9	0.6	0.8	0.2	116	32	brunisol
PR	LFH	3.7	315	49	54	18	13	988	32	Gleyed humoferric podzol
	Ae/Ahe	3.8	50	10	1.1	1.3	0.4	96	28	
	Bh	4.2	68	7	0.9	1.2	0.4	164	25	
	Bf (20 cm)	4.5	30	7	0.5	0.6	0.1	120	25	
	Bf (15 cm)	4.5	20	7	0.3	0.2	0.2	87	26	
NO	LFH	3.7	305	51	56	22	10	1045	29	Eluviated dystric brunisol
	Ae	3.8	100	4	1.2	0.7	0.5	77	31	
	B (10 cm)	4.4	52	3	0.3	0.5	0.2	49	29	
	B (10–20 cm)	4.6	22	5	0.2	0.5	0.04	35	31	
	B (20–35 cm)	4.7	12	8	0.1	0.4	0.04	26	33	
	B/BC (35–50 cm)	4.7	8	11	0.1	0.3	0.04	19	36	
NA	LFH	4.0	326	55	41	19	35	1270	25	Gleyed humoferric podzol
	Ahe	4.0	79	18	1.5	3.3	3.1	446	16	
	AB	4.1	58	12	0.7	1.7	1.2	265	17	
	Bf (15 cm)	4.4	48	8	0.6	0.7	0.5	169	23	
	Bf (15 cm)	4.5	40	8	0.6	0.7	0.3	120	27	
HL	LFH	4.0	421	74	84	28	50	1104	32	Gleyed ferrohumic podzol
	Ae/Ahe	4.2	37	29	2.2	1.0	1.7	54	30	
	Bhf/Bf (10 cm)	4.6	98	16	3.6	1.6	2.2	198	34	
	Bf/Bm (10 cm)	4.7	53	18	2.2	1.3	0.9	151	36	

[a] Data from OPS data bank.
[b] After the Canadian System of Soil Classification.
[c] In mmol(c) kg^{-1}.

Table 13.12. Depositions and Concentration (in parentheses) of Sulfate-Sulfur and Nitrate-Nitrogen and Ammonium-Nitrogen in Bulk and Throughfall Precipitation in June–October 1988 in Five Monitoring Plots Located along a Gradient from South to North in Norway[a]

	SO_4-S		N (total)		NH_4-N		NO_3-N	
Plot Code	Bulk	Throughfall	Bulk	Throughfall	Bulk	Throughfall	Bulk	Throughfall
BI	28	35	67	45	31	21	37	33
	(58)	(93)			(32)	(29)	(39)	(31)
PR	11	16	15	18	6	10	9	7
	(51)	(118)			(14)	(36)	(19)	(26)
NO	12	19	18	17	7	9	10	8
	(48)	(88)			(15)	(19)	(21)	(19)
NA	7	5	9	6	3	3	6	2
	(29)	(34)			(6)	(11)	(12)	(7)
HL	4	6	5	4	2	1	3	2
	(24)	(34)			(5)	(4)	(9)	(7)

[a] Deposition, mmol(c) m^{-2}; concentration, weighted means, μmol(c) L^{-1}. Data from OPS (1990) OPS data bank.

Table 13.13. Volume-Weighted Mean pH from Collecting Season 1986–1988 for Bulk and Throughfall Precipitation in Five Monitoring Plots Located along a Gradient from South to North in Norway[a]

Plot	Bulk	Throughfall	Bulk	Throughfall	Bulk	Throughfall
Code	1986		1987		1988	
BI	4.37	4.20	4.40	4.30	4.30	4.27
PR	4.33	4.04	4.39	4.23	4.37	4.23
NO	4.30	4.17	4.39	4.31	4.36	4.33
NA	4.81	4.90	4.60	5.00	4.70	5.00
HL	—	—	4.68	4.72	4.92	4.75

[a] Data from OPS (1990) data bank.

The report of the 1988 forest damage survey in Europe (GEMS 1990) pointed out that differences in the spatial and temporal development of forest damage have become particularly evident in 1988. These features support the opinion of many scientists that forest decline can best be described as a process of high causal complexity involving both abiotic and biotic factors. The novel forest decline damages in Europe are being intensively studied in light of new and old observations and experimental results. Most scientist agree that no single hypothesis is able to explain all recorded facts. However, the following presentation of the complex relationships involved demonstrates how a monocausal (partial) hypothesis may be integrated into a basic multiple stress model.

Air pollutants may produce acid precipitation, which leads to soil acidification. Sulfate and nitrate ions passing through the soil horizons cause leaching of base cations, thus progressively decreasing the buffering capacity and depleting the soil of nutrient elements such as magnesium, calcium, potassium, and others. With a decrease in pH, metal ions are set free, some of which may be toxic to tree roots (e.g., aluminum species) or to mycorrhizal fungi (Ulrich 1984, 1986). To what degree these processes are counteracted by weathering partly depends on edaphic and climatic conditions at the site. If net leaching takes place, the result over time may be reduced availability and reduced uptake of nutrient elements important for tree growth, which leads to deficiency symptoms, reduced crown and root growth, and finally to decline symptoms (Schulze 1989). At the same time the acid precipitation may accelerate the natural leaching of ions from the foliage, thus exacerbating a possible deficiency in leaves (Ulrich and Matzner 1983).

Increased nitrogen deposition may affect the natural development and physiology of trees, rendering them more liable to frost, drought, parasites, and insects, and may exert a negative influence on the mycorrhizae (Nihlgård 1985; Göbl 1986). Air pollutants such as ozone and other photooxidants cause damage to cell membranes in living tissues. They may also interfere with physiological processes, rendering trees susceptible to frost, and accelerate senescense of needles and leaves. These effects increase the degree of multiple stresses accumulated in the forest ecosystem (Arndt 1985a, 1985b;

Krause and Prinz 1986). Depositions of heavy metals may also cause inhibition of nutrient uptake mainly by negative effects on mycorrhizae and the soil microbiology (Glatzel 1985; Gobl and Mutsch 1985).

Hypotheses

A few basic hypotheses have governed the type of approach taken when searching for causal explanations of the new type of forest decline. The discussion has left the stage of monocausal explanations, although Koch's postulate (Last et al. 1984) is still most useful and also necessary in elucidating etiological relationships. Today, a range of multiple stress hypotheses have gained solid support. They are being extensively accepted as the basis for explaining how interactions between predisposing, inciting, and contributing factors lead to the decline and death of a tree (or a forest). This is well described as the decline disease spiral by Manion (1981).

A second hypothesis assumes that the sum of stresses loaded onto a forest ecosystem is decisive for its response to a new additional stress impact (Schütt et al. 1983; Schütt 1988). This idea is visualized by the situation in which a barrel will overflow by the addition of only a single drop of water—if the barrel is already filled to the brim. The consequence of this hypothesis is that even a slight reduction in air pollution loads may bring the added sum of stresses below a certain damaging or lethal threshold. If this is true, one will have to admit that removal of this stress, regardless of which category of factors it belongs to, is of utmost importance, even though it may quantitatively be of minor magnitude compared to the other (natural) stresses involved. This view must be kept in mind when evaluating statments made by various countries on the significance of air pollutants in the forest decline etiology.

A third hypothesis seems to be needed to explain more of the forest decline phenomena that have been reported from Europe. If predisposing factors cause a latent susceptibility to inciting factors, this latent susceptibility increases under sustained stress leading to a higher degree of predisposition, which is only revealed by the impact of inciting factors.

The forest ecosystem may be compared to an exposed but undeveloped photographic film. Nobody can see the picture until it is developed. The developer incites a sudden and profound change in the film. The information hidden in the exposed picture is revealed, and strong relationships become evident between the intensity of the exposing light and the resulting colors of the film.

When predisposing, inciting, and contributing factors are acting in sequence, it is commonly presumed that the resulting responses in a forest ecosystem will develop more or less gradually. The expected response may be explained as a continuous, linear or curved relationship expressing a direct correlation to the impact of factors in action. However, it can be claimed that most of the existing reports of forest decline in central Europe and elsewhere indicate discontinuity in damage development. Sudden changes in forest health status have been observed in several countries. In the Federal Republic of Germany, the German Democratic Republic, Poland, and Czechoslovakia, vast areas of coniferous forest, mainly Norway spruce, have died rather suddenly or started to decline rapidly after experiencing abrupt temperature variations in midwinter or spring, severe insect attacks during summer, summer frost, or early frosts during autumn, severe droughts, winds, or winter storms (Bosch and Rehfuess 1988). These are all typical inciting factors. Their impact on the forest ecosystem depends on the degree of predisposition already present in the system, that is, the accumulated effect of the predisposing factors in action at earlier stages. This level of predisposition may be linked in a continuous way to the level of these predisposing factors. However, to date there is no direct method for displaying and recording this level of predisposition.

In a forest ecosystem, various predisposing factors may be acting for periods of many years; for example, in cases of displaced species or ecotypes of trees, severe climate during certain periods or constant emissions of air pollutants at subtoxic levels. Over time, increasing levels of predisposition may develop. It is presumed that the accumulating degree of predisposition is a function of the total predisposing load. After development of symptoms or damage due to some inciting factor(s), this level of predisposition is revealed and may then show strong correlations to levels of impact of the predisposing factors, as well as to inciting ones.

If this hypothesis is true, it must lead to a change in the concept of the present forest decline situation and its progress in various regions. Primarily, one should not interpret stability in levels of symptoms or damages as an absence of threatening factors. Only after action of inciting factors the true situation will become evident. Second, predisposing factors should be focused to mitigate the impact of inciting factors (which are often inevitable occurrences of natural phenomena).

Areas approaching natural tree limits (boreal, alpine, or oceanic) constitute a substantial part of the forested land in northern Europe, as in Scandinavia. Severe natural climatic stresses prevail in these regions. According to our hypothesis, such a situation renders only slight margins for additional (anthropogenic) stresses to the forest ecosystem before the maximum thresholds are reached.

Summary and Conclusions

It is widely accepted that forest decline is a multiple syndrome and that there is less synchrony in its incidence than was previously thought. There are also some doubts as to the "novelty" of some kinds of damage, for example, magnesium deficiency and damages associated with needle fungi. Nevertheless, the appearance of so many ailments in the forest at roughly the same time suggests presence of linking and triggering factors.

The large-scale surveys of forest damage are useful and very important for identifying and quantifying the problems with novel forest decline, but scientific searching for the causes of the novel forest decline can probably be carried out only through detailed work in long-term studies of forest ecosystems. Models of nutrient cycling are necessary tools for revealing the true status of health in our forests. A continuation of the present pollution loads for extended periods of time or an increase in pollution levels may threaten the vitality of forests over large areas of Europe. A reduction of air pollution loads would improve the condition of forests and postpone a possible expansion of forest decline. The important role of forests and the threat of global warming from the possible greenhouse effect support the need for further measures to reduce air pollution.

References

Aamlid D., Venn K., Stuanes A.O. 1990. Forest decline in Norway, monitoring results, international links and hypotheses. Norw. J. Agr. Sci. supplement No. 4, 27 pp.

Allen H.L. 1987. Forest fertilizers. J. For. 85:37–46.

Andersen S.B. 1988. Long-Term Changes (1930–32 to 1984) in the Acid-Base Status of Forest Soils in the Adirondacks of New York. Ph.D. Thesis, University of Pennsylvania, Philadelphia.

Anonymous. 1989. Zustand des Waldes 1989 in der Bundesrepublik Deutchland. Allg. Forstztg. 44:1296.

Arndt U. 1985a. Ozon und seine Bedeutung für das Waldsterben in Mitteleuropa. In Niesslein E., Voss G. (eds.) Was wir über das Waldsterben Wissen. Köln, Deutscher Instituts–Verlag, pp. 160–174.

Arndt U. 1985b. Ozon als möglicher Verursacher von Waldschaden. In Kortzfleisch G von (Ed.) Waldchäden: Theorie und Praxis auf der Suche nach Antworten. München, Wien, Oldenbourg, pp. 195–212.

Auclair A.N.D. 1989. Cavitation as a mechanism of dieback in northern hardwoods. In L'atelier sur Dépérissement dan les Érablières. Ministèrede l'Agriculture, des Pêcheries et de lálimentation du Quebec. Centre de Recherche Acéricole. St.-Hyacinthe (QC). pp. 34–39.

Bartlett R.J. 1987. Simple kinetic fractionations of reactive aluminum in soil "solutions." Soil Sci. Soc. Am. J. 51:1479–1482.

Barnard J.E., Lucier A.A., Johnson A.H., Brooks R.T., Karnosky D.F., Dunn P.H. 1990. Changes in Forest Health and Productivity in the United States and Canada. SOS/T Report No. 16. National Acid Deposition Assessment Program, Washington, D.C.

Battles J.J., Johnson A.H., Siccama T.G. 1988. Recent changes in spruce-fir forests of the New York, Vermont and New Hampshire. In Proceedings of the UA/FRG Research Symposium: Effects of Atmospheric Pollutants on the Spruce-Fir Forests of the Eastern United States and the Federal Republic of Germany, Burlington, Vermont. Gen. Tech. Report NE-120, U.S. Forest Service, Radnor, Pennsylvania.

Battles J.J., Johnson A.H., Siccama T.G., Friedland A.J., Miller E.K. 1989. Spatial Patterns of Red Spruce Decline on Whiteface Mt., New York. Report SF08-

1, U.S. Forest Service, Forest Response Program. Available from the Northeastern Forest Experiment Station, Radnor, Pennsylvania.
Bauce E.S. 1989. Sugar Maple, *Acer saccharum* Marsh., Decline Associated with Past Disturbances. Ph.D. Thesis. SUNY College of Environmental Science and Forestry, Syracuse, New York.
Bauch J., Michaelis W. (eds.) 1988. Das Forschungsprogramm Waldschäden am Standort "Poststurm," Forstamt Farchau/Ratzeburg. GKSS-Forschungszentrum, Geesthacht.
Berlyn G.P., Anoruo A.O., Boyce R.L., Silver W.L., Letourneau A.L., Johnson A.H., Vann D.R. 1989. Effect of branch chamber and ambient environments on six foliar properties of high elevation red spruce on Whiteface Mt., New York. U.S. Forest Service, Forest Response Program, Northeastern Forest Experiment Station, Radnor, Pennsylvania.
Bernier B., Brazeau M. 1988a. Foliar nutrient status in relation to sugar maple dieback in the Quebec Appalachians. Can. J. For. Res. 18:754–761.
Bernier B., Brazeau M. 1988b. Magnesium deficiency symptoms associated with sugar maple dieback in a lower Laurentians site in southeastern Quebec. Can. J. For. Res. 18:1265–1269.
Bernier B., Pare D., Brazeau M. 1989. Natural stresses, nutrient imbalances and forest decline in southeastern Quebec. Water Air Soil Pollut. 48:239–250.
Binkley D., Valentine D., Wells C., Valentine U. 1989a. An empirical analysis of the factors contributing to 20-year decrease in soil pH in an old-field plantation of loblolly pine. Biogeochemistry 8:39–54.
Binkley D., Richter D. 1987. Nutrient cycles and H^+ budgets of forest ecosystems. Adv. Ecol. Res. 16:1–51.
Binkley D., Driscoll C., Allen H.L., Schoeneberger P., McAvoy D. 1989b. Acidic Deposition and Forest Soils: Context and Case Studies in the Southeastern U.S. Ecological Studies #72, Springer-Verlag, New York.
Bjorkdal G., Eriksson H. 1989. Effects of crown decline on increment in Norway spruce (*Picea abies* (L.) Karst) in Southern Sweden. Medd. Nor. Inst. Skogforsk. 42:19–36.
Bondietti E.A., Baes C.F., McLaughlin S.B. 1989. The potential of trees to record aluminum mobilization and changes in alkaline earth availability. In Biologic Markers of Air-Pollution Stress and Damage in Forests. National Academy Press, Washington, D.C.
Bosch C., Rehfuess K.E. 1988. Über die Rolle von Frostereignissen bei den "neuartigen" Waldschäden. Forstwiss. Cbl. 107:123–130.
Boutin R., Robitaille G., Meyer L. 1989. Experience de fertilisation dans l'érabliere de Norbertville. Impact sur la concentration des elements nutrifs dans la solution du soil. In L'Atelier sur le Dépérissement dans les Erablières. Centre de Recherche Acericole, St.-Hyacinthe (QC). Ministèrede l'Agriculture, des Pêcheries et de lâlimentation du Quebec, pp. 120–125.
Burkhart H. 1988. Southern Pines. Assessment prepared for Utility Air Resources Group, Washington, D.C.
Committee on Insects and Diseases (New York) Society of American Foresters. 1959. M. Bermingham, New York DEP, Albany, New York.
Cook E.R. 1985. A Time Series Approach to Tree Ring Analysis. Ph.D. Thesis, University of Arizona, Tucson.
Cook E.R., Johnson A.H., Blasing T.J. 1987. Modelling the climate effect in tree rings for forest decline studies. Tree Physiol. 3:27–40.
Cosby B.J., Hornberger G.M., Galloway J.N., Wright R.F. 1985. Modeling the effects of acid deposition: Assessment of a lumped parameter model of soil water and streamwater chemistry. Water Resour. Res. 21:51–63.

Cronan C.S., April R., Bartlett R.J., Bloom P.R., Driscoll C.T., Gherini S.A., Henderson G.S., Joslin J.D., Kelly J.M., Newton R.M., Parnell R.A., Patterson H.H., Raynal D.J., Schaedle M., Schofield C.L., Sucoff E.I., Tepper H.B., Thornton F.C. 1989. Aluminum toxicity in forests exposed to acidic deposition: the ALBIOS results. Water Air Soil Pollut. 48:181–192.

Cronan C.S. 1985. Biochemical influence of vegetation and soils in the ILWAS watersheds. Water Air Soil Pollut. 26:355–371.

Cronan C.S., Goldstein R.A. 1989. ALBIOS: a comparison of aluminum biogeochemistry in forested watersheds exposed to acidic deposition. In Adriano D.C., Havas M. (eds.) Acidic Precipitation, Vol. 1. Case Studies, Springer-Verlag, New York, pp. 113–135.

Cronan C.S, Schofield C.L. 1990. Relationships between aqueous aluminum and acidic deposition in forested watersheds of North America and northern Europe. Environ. Sci. Technol. 24:1100–1105.

Cumming J.R., Alscher R.G., Chabot J., Weinstein L.H. 1988. Effects of ozone on the physiology of red spruce seedlings. In Proceedings of the US/FRG Research Symposium: Effects of Atmospheric Pollutants on the Spruce-Fir Forests of the Eastern U.S. and the Federal Republic of Germany. Gen Tech Rep. NE-120, USDA Forest Service, Northeastern Forest Experiment Station, Radnor, Pennsylvania. pp. 355–364.

Curry J.R., Church T.W. 1952. Observations on winter drying of conifers in the Adirondacks. J. For. 50:114–116.

Daniel T.W., Helms J.A., Baker F.S. 1979. Principles of Silviculture. McGraw-Hill, New York.

David M.B., Driscoll C.T. 1984. Aluminum speciation and equilibria in soil solutions of a haplorthod in the Adirondack mountains (New York, U.S.A.). Geoderma 33:297–318.

DeHayes D.H., Williams M.W. 1989. Critical Temperature: A Quantitative Method of Assessing Cold Tolerance. USDA Forest Service. Northeastern Forest Experiment Station. General Technical Report NE-134. 6p.

Department of Forests and Parks, State of Vermont. 1959–1960, 1961–62. Biennial Reports of the Department of Forests and Parks. Montpelier, Vermont.

Eager C., Adams M.B. Decline of Red Spruce in the Eastern United States, Springer-Verlag, New York, (in press).

ECE Manual. 1986. Manual on Methodologies and Criteria for Harmonized Sampling, Assessment, Monitoring and Analysis of the Effects of Air Pollution on Forests. Programme Co-Ordinating Centres/UN-ECE, Hamburg/Geneva.

Evans L. 1986. Proposed mechanisms of initial injury causing apical dieback in red spruce at high elevation in eastern North America. Can. J. For. Res. 16:1113–1116.

Falkengren-Grerup U. 1986. Soil acidification and vegetation changes in deciduous forests in southern Sweden. Oecologia 70:339–347.

FBW (Forschungsbeirat Waldschäden/Luftverunreinigungen der Bundesregierung und der Länder). 1986. 2.Bericht. Krnforschungszentrum, Karlsruhe.

Federer C.A., Hornbeck J.W., Tritton L.M., Martin C.W., Pierce R.S., Smith C.T. 1989. Long-term depletion of calcium and other nutrients in eastern U.S. forests. Environ. Manage. 13:593–601.

Fincher J., Cumming J.R., Alscher R.G., Ruben G., Weinstein L. 1990. Long-term ozone exposure affects winter hardiness of red spruce. New Phytol. 113:321–355.

Fleiss J.L. 1981. Statistical Methods for Rates and Proportions. Wiley, New York.

Foster N.W. 1985. Acid precipitation and soil solutions chemistry within a maple-birch forest in Canada. For. Ecol. Manage. 12:215–231.

Foster N.W., Morrison I.K., Nicolson J.A. 1986. Acid deposition and ion leaching from a podzolic soil under hardwood forest. Water Air Soil Pollut. 31:879–890.
Fowler D.J.N., Cape J.D., Deans I.D., Leith M.B., Murray R.I., Smith L.H., Sheppard A.N., Unsworth M.H. 1990. Effects of acid mist on the frost hardiness of red spruce seedlings. New Phytol. 113:321–355.
Frelich LE, Bockheim J.G., Leide J.E. 1989. Historical trends in tree-ring growth and chemistry across an air-quality gradient in Wisconsin. Can. J. For. Res. 19:113–121.
Friedland A.J., Gregory R.A., Karenlampi L., Johnson A.H. 1984. Winter damage to foliage as a factor in red spruce decline. Can. J. For. Res. 14:963–965.
Gagnon G., Roy G., Gravel C., Gagné J. 1986. State of dieback research at the Min. de l'Énergie et des Ressources. In Maple Producers Information Session, May 8, 1986, Quebec City, pp. 43–80. Conseil des Productions Végétales du Québec, Quebec City.
Gagnon G., Roy G. 1989. Etat du dépèrissement des forets au Quebec. In L'Atelier sur le Dépérisement dans les Érablières. Ministèrede l'Agriculture, des Pêcheries et de Lálimentation du Quebec. Centre de Recherche Acericole, St.-Hyacinthe (QC), pp.14–17.
Gansner D.A., Birch T.W., Frieswyk T.S. 1987. What's up with sugar maple? Nat. For., pp. 5–6.
GEMS (Global Environment Monitoring System). 1990. International Co-operative programme on assessment and monitoring of air pollution effects on forests. Forest damage and air pollution. Report on the 1988 Forest Damage Survey in Europe, Convention on Long-Range Transboundary Air Pollution.
Gholz H.L. 1986. Canopy development and dynamics in relation to primary production. In Crown and Canopy Structure in Relation to Productivity. Fujimori T., Whitehead D. (eds.) Forestry and Forest Products Research Laboratory, Ibaraki, Japan, pp. 224–242.
Gholz H.L., Fisher R.F.L. 1984. The limits to productivity: fertilization and nutrient cycling in Coastal plain slash pine forests. In Stone E. (ed.) Forest Soils and Treatment Impacts. University of Tennessee, Knoxville, pp. 105–120.
Gholz H.L., Fisher R.F., Pritchett W.L. 1985. Nutrient dynamics in slash pine plantation ecosystems. Ecology 66:647–659.
Glatzel G. 1985. Schwermatallbelastung von Wäldern in der Umgebung eines Hüttenwerkes in Brixlegg/Tirol II. Wachstum und Mineralstoffernährung von Fichtenplanzen (*Picea abies*) auf Schwermetallbelasteten Waldbodenhumus. Forstwiss. Centralbl (Hamb.) 102:42–51.
Göbl F. 1986. Wirkung simulierter saurer Niedersläge auf Böden und Fichtenjungplanzen im Gefässversuch. III. Mykorrhiza—untersuchungen. Forstwiss. Centralbl (Hamb.) 103:89–107..
Göbl F., Mutsch F. 1985. Schwermatallbelastung von Wäldern in der Umgebung eines Hüttenwerkes in Brixlegg/Tirol I. Untersuchung der Mykorrhiza und Humusauflage. Forstwiss. Centralbl. (Hamb.) 102:28–40.
Hadfield J.S. 1968. Evaluation of diseases of red spruce on the Chamberlain Hill sale, Rochester Ranger District, Green Mt. Nat. Forest. File Report A-68–8 5230. USDA-Forest Service Northeastern Area, State and Private Forestry, Amherst FPC Field Office, Amherst, Maine.
Hallbäcken L., Tamm C.O. 1986. Changes in soil acidity from 1927 to 1982–1984 in a forest area of south-west Sweden. Scand. J. For. Res. 1:219–232.
Hauhs M. 1989. Lange Bramke: an ecosystem study of a forested catchment. In Adriano D. C., Havas M. (eds.) Acid Precipitation, Vol. 1. Case Studies, Springer-Verlag, New York, pp. 275–305.
Hausladen A., Madamanchi N.R., Fellows S., Alscher R.G., Amundson R.G. Seasonal changes in antioxidants in red spruce as affected by ozone. New Phytologist (in press).

Heimburger C.C. 1934. Forest Type Studies in the Adirondacks. Ph.D. Thesis, Cornell, University, Ithaca, New York.

Hendershot W.H., Jones A.R.C. 1989. Maple decline in Quebec: a discussion of possible causes and the use of fertilizers to limit damage. For. Chron. 65:280–287

Hendershot W.H., Lalonde H., Champagne M. 1989. Response of sugar maple in Beauce-Megantic region to fertilization. In L'atelier sur Dépérissement dan les Érablières. Ministèrede l'Agriculture, des Pêcheries et de lálimentation du Quebec. Centre de Recherche Acéricole. St.-Hyacinthe (QC), pp. 94–97.

Herrick G.T., Friedland A.J. 1991. Winter desiccation and injury of subalpine red spruce. Tree Physiol. 8:23–36.

Hopkins A.D. 1901. Insect enemies of the spruce in the Northeast. Bull. No. 28 (new series), U.S. Dept. of Agriculture, Div. of Entomology, Washington, D.C. pp. 15–29.

Hornbeck J.W., Smith R.B., Federer C.A. 1988. Growth trends in 10 species of trees in New England. 1950–1989. Can. J. For. Res. 18:1337–1340.

Horntvedt R., Tveite B. 1986. Overvåking av skogskader. Norsk institutt forkogforskning. Årbok 85:38–44 (in norwegian).

Houston D.R. 1981. Stress-triggered tree diseases, the diebacks and declines. NE-INF-41–81, USDA Forest Service, Washington, D.C.

Houston D.R. 1987. Forest tree decline of past and present: current understanding. Can. J. Plant Pathol. 9:349–360.

Huntington T.G., Miller E.K., Johnson A.H. 1989. Rapidly reactive aluminum levels in the soil solution in field experiments at Whiteface Mt., New York. Forest Response Program, U.S. Forest Service, Northeastern Forest Experiment Station, Radnor, Pennsylvania.

Hutchinson T.C., Bozic L., Munoz-Vega G. 1986. Responses of five species of conifer seedlings to aluminum stress. Water Air Soil Pollut. 31:283–294.

Jacobson J.S., Lassoie J.P., Osmeloski J.F., Yamada K.E. 1990a. Changes in foliar elements in red spruce seedlings after exposure to sulfate and nitrate acidic mist. Water Air Soil Pollut. 48:141–160.

Jacobson J.S., Heller L.I., Yamada K.E., Osmeloski J.F., Bethard T., Lassoie J.P. 1990b. Foliar injury and growth response of red spruce to sulfate and nitrate acidic mist. Can. J. For. Res. 20:58–65.

Jacobson J.S., Bethard T., Heller L.I., Lassoie J.P. 1990c. Response of *Picea rubens* seedlings to varying concentrations of acidity, sulfur- and nitrogen-containing pollutants in mist. Unpublished manuscript available from the USDA Forest Service Northeastern Forest Experiment Station, Radnor, Pennsylvania.

James B.R., Riha S.J. 1986. pH buffering in forest soil organic horizons: relevance to acid precipitation. J. Environ. Qual. 15:229–234.

Johnson A.H., McLaughlin S.B. 1986. The nature and timing of the deterioration of red spruce in the northern Appalachians. Report of the Committee on Monitoring and Trends in Acidic Deposition. National Research Council, National Academy Press, Washington, D.C.

Johnson A.H., Siccama T.G. 1983. Acid deposition and forest decline. Environ. Sci. Technol. 17:294–305.

Johnson A.H., Cook E.R., Siccama T.G. 1988a. Relationships between climate and red spruce growth and decline. Proc. Natl. Acad. Sci. USA 00:000–000.

Johnson A.H., Schwartzman T.N., Miller R.W. 1989. Relationships between foliar chemistry, soil chemistry and forest condition at Whiteface Mt., New York. Forest Response Program, U.S. Forest Service, Northeastern Forest Experiment Station, Radnor, Pennsylvania.

Johnson D.W. 1983. The effects of harvesting schedules on nutrient depletion from forests. In Ballard R., Gessel S.P. (eds.) Proceedings of the IUFRO Symposium on Forest Site and Continuous Productivity. Gen. Tech. Report PNW-163, U.S.D.A. Forest Service, Portland, Oregon, pp. 157–166.

Johnson D.W., Taylor G.W., 1989. Role of air pollution in forest decline in eastern North America. Water Air Soil Pollut. 48:21–44.

Johnson D.W., Todd D.E. 1987. Nutrient export by leaching and whole-tree harvesting in a loblolly pine and mixed oak forest. Plant Soil 102:99–109.

Johnson D.W., Todd D.E. 1990. Nutrient cycling in forests of Walker Branch Watershed Tennessee: roles of uptake and leaching in causing soil changes. J. Environ. Qual. 19:97–104.

Johnson D.W., Henderson G.S., Todd D.E. 1988c. Changes in nutrient distribution in forests and soils of Walker Branch Watershed, Tennessee, over an eleven-year period. Biogeochemistry 5:275–293.

Johnson D.W., Kelly J.M., Swank W.T., Cole D.W., Miegroet H., Hornbeck J.W., Pierce R.S., van Lear D. 1988b. The effects of leaching and whole-tree harvesting on cation budgets of several forests. J. Environ. Qual. 17:418–424.

Joslin J.D., Wolfe M.H. 1988. Response of red spruce seedlings to change in soil solution aluminum in six amended forest horizons. Can. J. For. Res. 18:1614–1623.

Joslin J.D., McDuffie C., Brewer P.F. 1988. Acidic cloud water and cation loss from red spruce foliage. Water Air Soil Pollut. 39:355–363.

Kalisz P.J., Stone E.L. 1980. Cation exchange capacity of acid forest humus layers. Soil Sci. Soc. Am. J. 44:407–413.

Kandler O. 1989. Epidemiological evaluation of the course of "Waldsterben" from 1983 to 1987. In Buchner J.B., Buchner-Wallin I., (eds.) Air Pollution and Forest Decline, Proceedings, 14th International Meeting for Specialists in Air Pollution Effects on Forest Ecosystems, Interlaken, Switzerland. IUFRO P2.05, Birmsdorf, pp. 297–302.

Keeves A. 1966. Some evidence of loss of productivity with successive rotations of *Pinus radiata* in the south-east of South Australia. Aust. For. 30:51–63.

Kelly J.M., Schaedle M., Thornton F.C., Joslin J.D. 1990. Sensitivity of tree seedlings to aluminum II. Red oak, sugar maple, European beech. J. Environ. Qual. 19:172–179.

Kelso E.G. 1965. Memorandum 5220,2480, July 23, 1965. U.S. Forest Service, Northern FPC Zone, Amherst, Massachusetts.

Kinch JG. 1989. The Relationships between Soil and Foliar Chemistry and the Decline of *Acer saccharum* in Ontario. M.Sc. Thesis, University of Toronto, Toronto, Ontario.

Klopatek J., Harris W.F., Olson R.J. 1980. A regional ecological assessment approach to atmospheric deposition: effects on soil systems. In Shriner D., Richmond C., Lindberg S., (eds.) Atmospheric Sulfur Deposition: Environmental Impacts and Health Effects. Ann Arbor Science, Michigan, pp. 539–553.

Kramer H. 1986. Relation between crown parameters and volume increment of *Picea abies* stands damaged by environmental pollution. Scand. J. For. Res. 1:251–263.

Krause G.H.M., Prinz B. 1986. Zur Wirkung von Ozon und saurem Nebel (eizeln und in Kombination auf phänomenologische und physiologische Parameter an Nadel- und Laubgehölzen im kombinierten Begasungsexperiment. Statusseminar KFA Jülich. Jül-Spez 369:208–221.

Last F.T., Fowler D., Freer-Smith P.-H. 1984. Die Postulate von Koch und die Luftverschmutzung. Forstwiss. Centralbl. (Hamb.) 103:28–48.

Lea R., Tierson W.C., Bickelhaupt D.H., Leaf A.L. 1980. Differential foliar responses of northern hardwoods to fertilization. Plant Soil 54:419–439.

LeBlanc D.C. 1990a. Red spruce decline on Whiteface Mt., New York: I. Relationships with elevation, tree age and competition. Can J. For. Res. 20:1408–1414.

LeBlanc D.C. 1990b. Relationships between breast-height and whole-stem growth indices for red spruce on Whiteface Mt., New York. Can. J. For. Res. 20:1399–1407.

LeBlanc D.C., Raynal D.J. 1990. Red spruce decline on Whiteface Mt., New York: relationships between apical and radial growth decline. Can. J. For. Res. 20:1415–1421

Leidecker H. 1988. Fichtensterben and spruce decline—a diagnostic comparison in Europe and North America. In Proceedings US/FRG Research Symposium: Effects of Atmospheric Pollutants on the Spruce-Fir Forests of the Eastern U.S. and the Federal Republic of Germany. Gen. Tech. Rept. NE-120, Northeastern Forest Experiment Station, Radnor, Pennsylvania. pp. 245–256.

Lozano F.C., Parton W.J., Lau J.K.M., Vanderstar L. 1987. Physical and chemical properties of the soils at the southern biogeochemical study site. Tech. Rep. BGC-018, Faculty of Forestry, University of Toronto. Ontario Ministry of Environment. Dorset Research Centre, Ontario.

Lucier A.A. 1988. Pine growth-rate changes in the Southeast: a summary of key issues for forest managers. South. J. Appl. For. 12:84–89.

Madamanchi N.R., Hausladen A., Alscher R.G., Amundson R.G., Fellows S. 1990. Seasonal changes in antioxidant in red spruce (*Picea rubens* Sarg. from three field sites in the Northeastern U.S. Unpublished manuscript, available from the USDA Forest Service, Northeastern Forest Experiment Station, Radnor, Pennsylvania.

Manion P.D. 1981. Tree Disease Concepts. Prentice-Hall, Englewood Cliffs, New Jersey.

Marion G.M. 1979. Biomass and nutrient removal in long-rotation stands. In Leaf A.L. (ed.) Impact of Intensive Harvesting or Forest Nutrient Cycling. State University of New York, Syracuse, pp. 98–110.

McLaughlin D.L., Gagnon D.L., Cox R.M. 1990a. Canadian forest decline problems. pp. 38–54. In Pearson R.G., Percy K.E. (eds.) The 1990 Canadian Long-Range Transport of Air Pollutants and Acid Deposition Assessment Report. Part 5. Terrestrial Effects Subgroup, Federal/Provincial Research Monitoring and Coordinating Committee, pp. 38–54.

McLaughlin S.B., Anderson C.P., Edwards N.T., Roy W.K., Layton P.A. 1990b. Seasonal patterns of photosynthesis and respiration of red spruce saplings from two elevations in declining southern Appalachian stands. Can. J. For. Res. 20:485–495.

McLaughlin S.B., Downing D.J., Blasing T.J., Cook E.R., Adams H.S. 1987. An analysis of climate and competition as contributors to the decline of red spruce in high elevation Appalachian forests. Oecologia 72:487–501.

Miller E.K., Huntington T.G., Johnson A.H., Friedland A.J. Aluminum release from soils in a fir-spruce forest at Whiteface Mountain, New York: implications for red spruce mortality. J. Environ. Qual. (in press).

Mohnen V.A. 1989a. Exposure of Forests to Air Pollutants, Clouds, Precipitation, Climatic Variables. EPA/60/53-89-003, U.S. Environmental Protection Agency, Atmospheric Research and Exposure Assessment Laboratory, Raleigh, North Carolina.

Mohnen V.A. 1989b. Mountain Cloud Chemistry Project—Wet, Dry and Cloud Water Deposition. EPA/60/53–89/009, U.S. Environmental Protection Agency, Atmospheric Research and Exposure Assessment Laboratory, Raleigh, North Carolina.

Möller C.M. 1945. Untersuchungen uber Laubmenge, Stoffverlust und Stoffproduktion des Waldes. Forstl. Forsogsvaes. Dan. 17:1–287.

Mollitor A.V., Raynal D.J. 1982. Acid precipitation and ionic movements in Adirondack forest soils. Soil Sci. Soc. Am. J. 46:137–141.

Morris L.A., Pritchett W.L., Swindel B.F. 1983. Displacement of nutrients into windrows during site preparation of a flatwood forest. Soil Sci. Soc. Am. J. 47:591–594.

New York State. Forest Pest Reports, 1971–1977. Forest Insect and Disease Control Ann. Reports 1968–1976. Annual Report of the New York State Conservation Commission 1911–1963. Albany, New York.

New York State. 1891. Seventh Report of the Forest Commission. Albany, New York. (Available from M. Bermingham, New York DEP, Albany, New York.)

Nihlgård B. 1985. The ammonium hypothesis—an additional explanation to the forest dieback in Europe. Ambio 14:2–8.

Ohno T., Sucoff E.I., Erich M.S., Bloom P.R., Buschena C.A., Dixon R.K. 1988. Effect of soil aluminum on the growth and chemical composition of red spruce. J. Environ. Qual. 17:666–672.

Olson R.J., Johnson D.W., Shriner D.S. 1982. Regional assessment of potential sensitivity of soils in the eastern United States to acid precipitation. ORNL/TM-8374. Oak Ridge National Laboratory, Oak Ridge, TN. 50 p.

OPS (Monitoring Programme for Forest Damage). 1988a. Monitoring Programme for Forest Damage. Annual Report 1986. Ås, Norway (in Norwegian; summary in English).

OPS (Monitoring Programme for Forest Damage). 1989. Monitoring Programme for Forest Damage, Annual Report 1988. Ås, Norway (in Norwegian; summary in English).

OPS (Monitoring Programme for Forest Damage). 1990. Monitoring Programme for Forest Damage. Annual Report 1989. Ås, Norway (in Norwegian; summary in English).

OPS (Monitoring Programme for Forest Damage). 1988b. Monitoring Programme for Forest Damage. Annual Report 1987. Ås, Norway (in Norwegian; summary in English).

Ouimet R., Fortin J.M. 1989. Resultat du projet semi-operationnel de fertilisation des érablères dépérissantes dans la region de l'Estrie-Beauce. In L'Atelier sur Dépérissement dan les Érablières. Ministèrede l'Agriculture, des Pêcheries et de lálimentation du Quebec. Centre de Recherche Acéricole. St.-Hyacinthe (QC), pp. 102–107.

Paré D., Bernier B. 1989. Origin of the phosphorus deficiency observed in declining sugar maple stands in the Quebec Appalachians. Can. J. For. Res. 19:24–34.

Raynal D.J., Joslin J.D., Thornton P.C., Schaedle M., Henderson G.S. 1990. Sensitivity of tree seedlings to aluminum: III. Red spruce and loblolly pine. J. Environ. Qual. 19:180–187.

Rehfuess K.E., 1983. Walderkrankungen und Immissionen—eine Zwischenbilanz. Allg. Forstz. 38:601–610.

Rehfuess K.E., 1987. Perceptions on forest diseases in central Europe. Forestry 60:1–11.

Rehfuess K.E., Rodenkirchen H. 1984. Uber die Nadelrote-Erkrankung der Fichte (*Picea abies* Karst). in Suddeutschland. Forstwiss. Centrabl. (Hamb.) 103:248–262.

Reuss J.O. 1983. Implications of the calcium-aluminum exchange system for the effect of acid precipitation on soils. J. Environ. Qual. 12:591–595.

Roberge M.R. 1987. Evolution of the pH of various forest soils over the past 20 years. Inf. Rep. LAU-X77, Laurentian Forestry Centre, Sainte-Foy, Quebec.

Robitaille G., Boutin R. 1989. Influence de la cime de l'erable à sucre surles précipitations acides et non acides. In L'Atelier sur Dépérissement dan les Érablières. Ministèrede l'Agriculture, des Pêcheries et de lálimentation du Quebec. Centre de Recherche Acéricole. St.-Hyacinthe (QC), pp. 155–159.

Roy G., Gagnon G. 1989. Resultats préliminaires du chaulage et de la fertilisation dans des érablières, trois ans apres traitement. In L'Atelier sur Dépérissement dan les Érablières. Ministèrede l'Agriculture, des Pêcheries et de lálimentation du Quebec. Centre de Recherche Acéricole. St.-Hyacinthe (QC), pp. 115–119.

Sanasilva. 1989. Sanasilva-Waldschadenbericht 1988. Bundesamt für Forstwesen und Landschaftsschutz, Bern.

Schier G.A. 1985. Response of red spruce and balsam fir seedlings to aluminum toxicity in nutrient solutions. Can J. Forest Res. 15:29–33.

Schmidt-Vogt H. 1989 Die Fichte. Paul Parey, Hamburg and Berlin.

Schulze E.D. 1989. Air pollution and forest decline in a spruce (*Picea abies*) forest. Science 244:776–783.

Schulze E.D., Lange O.L., Oren R. 1989. Forest decline and air pollution. Ecological Studies No. 77, Springer-Verlag, Berlin. 475 pp.

Schütt P. 1988. Waldsterben—Wichtung der Ursachenhypothesen. Fortschr. Bot. 59:17–18.

Schütt P., Cowling E.B. 1985. Waldsterben, a general decline of forests in central Europe: symptoms, development and possible causes. Plant Dis. 69:548–558.

Schütt P.W., Blaschke H., Lang K.J., Schuck H.J., Summerer H. 1983. So stirbt der Wald. BLV, Verlagsgesellschaft, München.

SFT (Statens Forurensningstilsyn). 1989. Statlig program for Forurensningsovervaking. Overvaking av langtransportert forurenset luft og nedbør. Årsrapport, 1988, Statens Forurensningstilsyn, Oslo (in Norwegian).

Sheffield R., Cost N., Bechtold W., McClure J. 1985. Pine growth reductions in the Southeast. Resource Bulletin SE-83, U.S.D.A. Forest Service, Asheville, North Carolina.

Shepard J.P., Mitchell M.J., Scott T.J., Driscoll C.T. 1990. Soil solution chemistry of an Adirondack spodosol: lysimetry and N dynamics. Can. J. For. Res. 20:818–824.

Shortle W.C., Smith K.T. 1988. Aluminum-induced calcium deficiency syndrome in declining red spruce. Science 240:1017–1018.

Silver W.L., Siccama T.G., Johnson C., Johnson A.H. 1991. Changes in red spruce populations in montane forests of the Appalachians: 1982–1987. Amer. Midlands Naturalist 125:340–347.

Squires R.O. 1982. Review of second rotation silviculture of *P. radiata* plantations in southern Australia: establishment practice and expectations. In Ballard R., Gessel S.P. (eds.) IUFRO Symposium on Forest Site and Continuous Productivity, pp. 136–137. Gen. Tech. Rep. PNW-163, USDA Forest Service, Portland, Oregon.

Stanturf J.A., Stone E.L. Jr., McKittrick R.C. 1989. Effects of added nitrogen on growth of hardwood trees in southern New York. Can. J. For. Res. 19:279–284

Stone D.M. 1986. Effect of thinning and nitrogen fertilization on diameter growth of pole-sized sugar maple. Can. J. For. Res. 16:1245–1249

Sucoff E.I., Thornton F.C., Joslin J.D. 1990. Sensitivity of tree seedlings to aluminum. I. Honeylocust. J. Environ. Qual. 19:163–171.

Swan H.S.D. 1971. Relationships between nutrient supply, growth and nutrient concentrations in the foliage of white and red spruce. Woodlands Rep. WR/34, Feb. 1971. Pulp and Paper Research Institute of Canada.

Tegethoff A.C. 1964. High Elevation Spruce Mortality. Memorandum 5220, September 25, 1964. U.S. Forest Service, Northern FPC Zone, Amherst, Massachusetts.

Tew D.T., Morris L.A., Allen H.L., Wells C.G. 1986. Estimates of nutrient removal, displacement and loss resulting from harvest and site preparation of a *Pinus taeda* plantation in the Piedmont of North Carolina. For. Ecol. Manage. 15:257–267.

Thornton F.C., Joslin J.D., Raynal D.J. 1988. The possible role of aluminum in red spruce decline. Paper presented at the 1st Annual Meetings of APCA, Dallas, Texas, June 19–24, 1988.
Thornton F.C., Schaedle M., Raynal D.J. 1986. Effect of aluminum on the growth of sugar maple in solution culture. Can. J. For. Res. 16:892–896.
Turner R.S., Olson R.J., Brandt C.C. 1986. Areas Having Soil Characteristics That May Indicate Sensitivity to Acidic Deposition under Alternative Forest Damage Hypotheses. Environmental Sciences Division Publication #2720, Oak Ridge National Laboratory, Tennessee.
Ulrich B. 1984. Waldsterben durch saure Niederschläge. Umschau. Wissen schaft. 11:348–355.
Ulrich B. 1986. Die rolle der Bodenversaurung beim Waldsterben: Langfristige Konsequenzen und forstliche Moglichkeiten. Forstwiss. Centralbl. (Hamb.) 105:421–435.
Ulrich B., Matzner E. 1983. OkosystemareWirkungsketten beim Wald- und Baumsterben. Allg. Forst. Holzwirtsch. 38:468–474.
Van Deusen P.C. 1990. Stand dynamics and red spruce decline. Can. J. For. Res. 20:743–749.
Vann D.R., Johnson A.H. 1988. Design and testing of a field branch enclosure for the exclusion of atmospheric components. In Proceedings of the US/FRG Research Research Symposium: Effects of Atmospheric Pollutants on the Spruce-Fir Forests of the Eastern U.S. and the Federal Republic of Germany, pp. 431–440.
Vann D.R., Johnson A.H., Strimbeck G.R., Dranoff M.M. 1990. Effects of ambient levels of airborne chemicals on the foliage of mature red spruce. Forest Response Program, U.S. Forest Service, Northeastern Forest Experiment Station, Radnor, Pennsylvania.
Vitousek P., Matson P. 1985. Disturbance, nitrogen availability, nitrogen losses in an intensively managed loblolly pine plantation. Ecology 66:1360–1376.
Vose J., Allen H.L. 1987. Leaf area, stemwood growth, nutrition relationships in loblolly pine. For. Sci. 34:547–563.
Wargo P.M., Bergdahl D.R., Olson C.W., Tobi D.R. 1989. Root vitality and decline of red spruce. U.S. Forest Service, Forest Response Program, Northeastern Forest Experiment Station, Radnor, Pennsylvania.
Waring R.H., Schlesinger W.H. 1985. Forest Ecosystems: Concepts and Management. Academic Press, New York.
Weidensaul T.C., Fleck A.M., Hartzler D.M., Capek C.L. 1989. Survey and assessment of forest insect and disease relationships in red spruce and balsam fir. Forest Response Program, U.S. Forest Service, Northeastern Forest Experiment Station, Radnor, Pennsylvania.
Weiss M.J., McCreery L.R., Millers I.R., O'Brien J.T., Miller-Weeks M. 1985. Cooperative survey of red spruce and balsam fir decline in New York, New Hampshire and Vermont—1984. Interim Report. USDA Forest Service, Forest Pest Management. Durham, New Hampshire.
Wells C.G., Jorgensen J.R. 1975. Nutrient cycling in loblolly pine plantations. In Bernier B., Winget C.H. (eds.) Forest Soils and Forest Land Management. University of Laval Press, Quebec, pp. 137–158.
Wells C.G., Jorgensen J.R. 1978. Nutrient cycling in loblolly pine—silvicultural implications. TAPPI 61:29–32.
Wheeler G.S. 1965. Memorandum 2400, 5100 July 1, 1965. U.S. Forest Service Northern FPC Zone, Laconia, New Hampshire.
Wilkinson R.C. 1989a. Geographic variation in needle morphology of red spruce in relation to winter injury and decline. Forest Response Program, U.S. Forest Service, Northeastern Forest Experiment Station, Radnor, Pennsylvania.

Wilkinson R.C. 1989b. The effects of winter injury on growth of 30- year-old red spruce from twelve provenances growing in New Hampshire. Forest Response Program, U.S. Forest Service, Northeastern Forest Experiment Station, Radnor, Pennsylvania.

Witter J. (Comp.). 1990. Effect of an air pollution gradient on northern hardwood forests in the northern Great Lakes region. Poster presented in the International Congress on Forest Decline, Research: State of Knowledge and Perspectives, Friedrichshafen, Federal Republic of Germany. October 1989.

Zahner R., Saucier J., Myers R. 1989. Tree-ring model interprets growth decline in natural stands of loblolly pine in the Southeastern United States. Can. J. For. Res. 19:612–621.

Zöttl H.W., Hüttl R. 1985. Schadsymptome und Ernährungszustand von Fichtenbeständen im südwestdeutschen Alpenvorland. Allg. Forstztg. 40:197–199.

14. Synthesis and Modeling of the Results of the Integrated Forest Study

Introduction and Acknowledgments

D.W. Johnson and S.E. Lindberg

The opportunity to summarize and synthesize the results of a major study such as the Integrated Forest Study (IFS) ultimately falls to those individuals who coordinated the study and edited this volume. Because summaries and syntheses are not effectively produced by committee, they necessarily reflect the opinions of the editors. This synthesis is no exception. We are pleased to have this opportunity to summarize the findings of all the IFS scientists and to express our personal interpretations and opinions of the project results. Similarly, the modeling activities of the IFS have been largely conducted and coordinated by a small group. However, they incorporate the results of several workshops, numerous laboratory studies, and the field data that were collected by the IFS scientists. None of the material in this chapter could have been prepared without the dedicated help and cooperation of all the IFS participants, which is gratefully acknowledged.

Summary and Synthesis of the Integrated Forest Study

Atmospheric Deposition and Its Interactions with the Forest Canopy

S.E. Lindberg

To achieve the primary objective of the IFS, that is, to determine the effect of atmospheric deposition in forest nutrient cycles, required establishment of common deposition sampling and data analysis protocols at all sites. These were developed from published state-of-the-art methods and models and used to measure wet deposition, meteorological parameters, hydrologic fluxes, and atmospheric concentrations, and to estimate dry deposition and cloudwater interception from these physical measurements. The IFS represents the first long-term effort to measure and compile "as-accurate-as-possible" estimates of total atmospheric deposition to a wide variety of forest stands and to synthesize these data in a nutrient cycling context.

The IFS deposition data presented in the preceding chapters illustrate several important aspects of the atmospheric exposure characteristics of forest ecosystems across a wide elevational gradient and over an extensive spatial scale. Atmospheric deposition clearly plays a significant role in the biogeochemical cycles at all IFS forests, generally being most important at the eastern U.S. high-elevation sites. The flux of sulfate ions in throughfall at all sites is dominated by atmospheric deposition, with internal foliar leaching sources generally accounting for less than about 10% of the throughfall plus stemflow flux. A similar condition exists for N compounds and H^+. The export of sulfur in soil solution from many sites can be accounted for almost totally by atmospheric sources, while the export of N is determined by many factors, including deposition (especially in the Smoky Mountains). The base cation budgets indicate that deposition also provides important nutrient cations such as Ca in the base cation-poor forests such those at the Smoky Mt., Coweeta, and Florida sites.

In general, atmospheric deposition increases from the "background" sites in Washington to the northeastern sites but is highest at the southeastern IFS sites. However, elevational effects tend to dominate the spatial patterns. The fluxes of the major ions are highest at the high-elevation (1740 m) Smoky Mountains site, and second or third highest at Whiteface Mt. (1000 m). Only Na^+ and Mg^{2+} consistently deviate from this trend, with the highest fluxes measured at the near-coastal sites in Florida and Washington.

Many of these trends could not have been predicted from existing wet deposition network data, and were, in fact, contrary to expectations based on these networks (NADP 1990). The U.S National Atmospheric Deposition Program (NADP) operates nearly 200 precipitation chemistry sites, and these data have been used in the absence of similar data on total deposition to estimate spatial and temporal trends in atmospheric loading to ecosystems (e.g., Fox et al. 1989). Figures 14.1 and 14.2 illustrate isopleths of ion wet deposition from the NADP network for the period corresponding to the IFS study. These are overlain on maps of the annual mean fluxes (total wet plus dry) of the major ions measured at each IFS forest. The NADP network data have consistently proved highly useful for analysis of trends in precipitation chemistry and wet deposition (Sisterson et al., in press). However, it is clear that they cannot replace site-specific measurements of total atmospheric deposition because of limitations in spatial interpolation and lack of total deposition measurements. Analysis of the role of deposition in forest nutrient

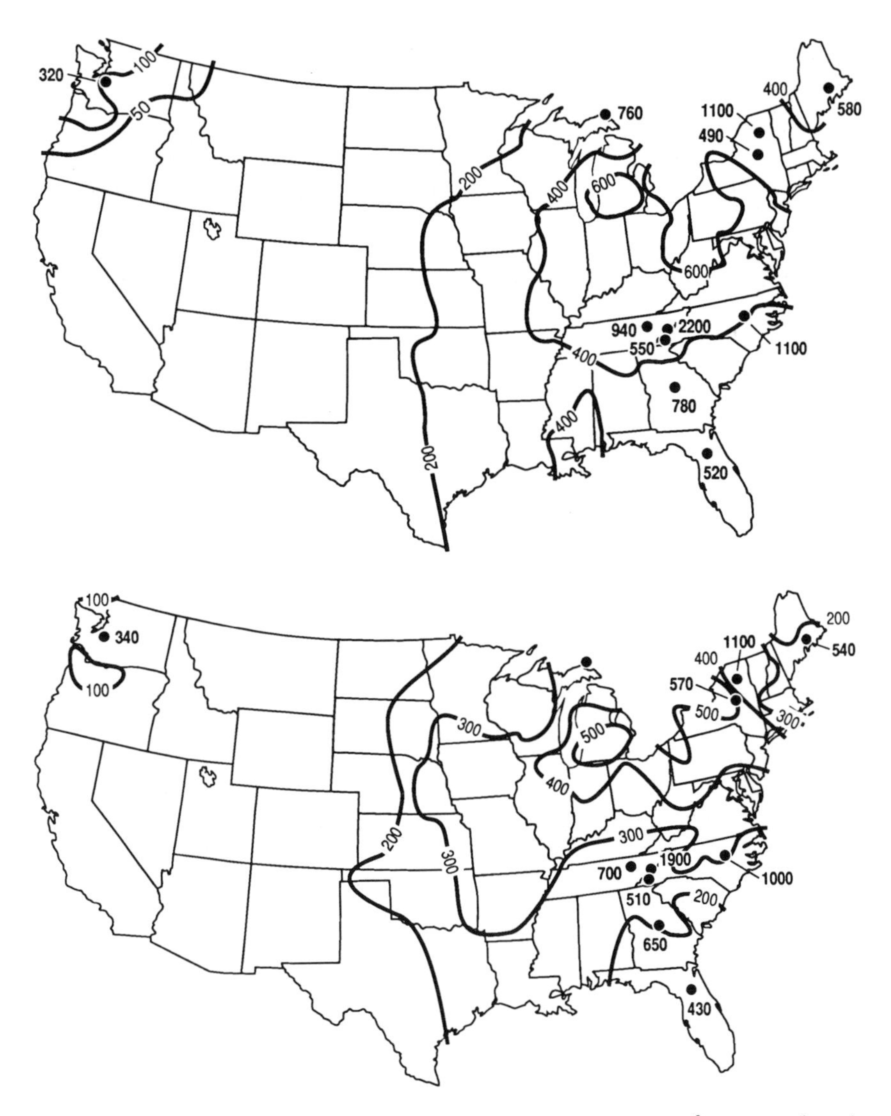

Figure 14.1. Isopleths of mean annual wet deposition rates of SO_4^{2-} (Eq ha^{-1} yr^{-1}) (top) and of N as NO_3^- + NH_4^+ ions (mol ha^{-1} yr^{-1}) (bottom) determined from data collected by U.S. National Atmospheric Deposition Program (NADP) for April 1986–March 1989. Isolines were drawn based on those sites (150 of possible 206) whose data met standard NADP sampling completeness criteria for entire period (NADP 1990; source of maps, G. Scott and C. Simmons, NADP Coordination Office, Colorado State University). Mean total annual deposition of these ions determined for IFS sites is indicated in same units at each site location for comparison.

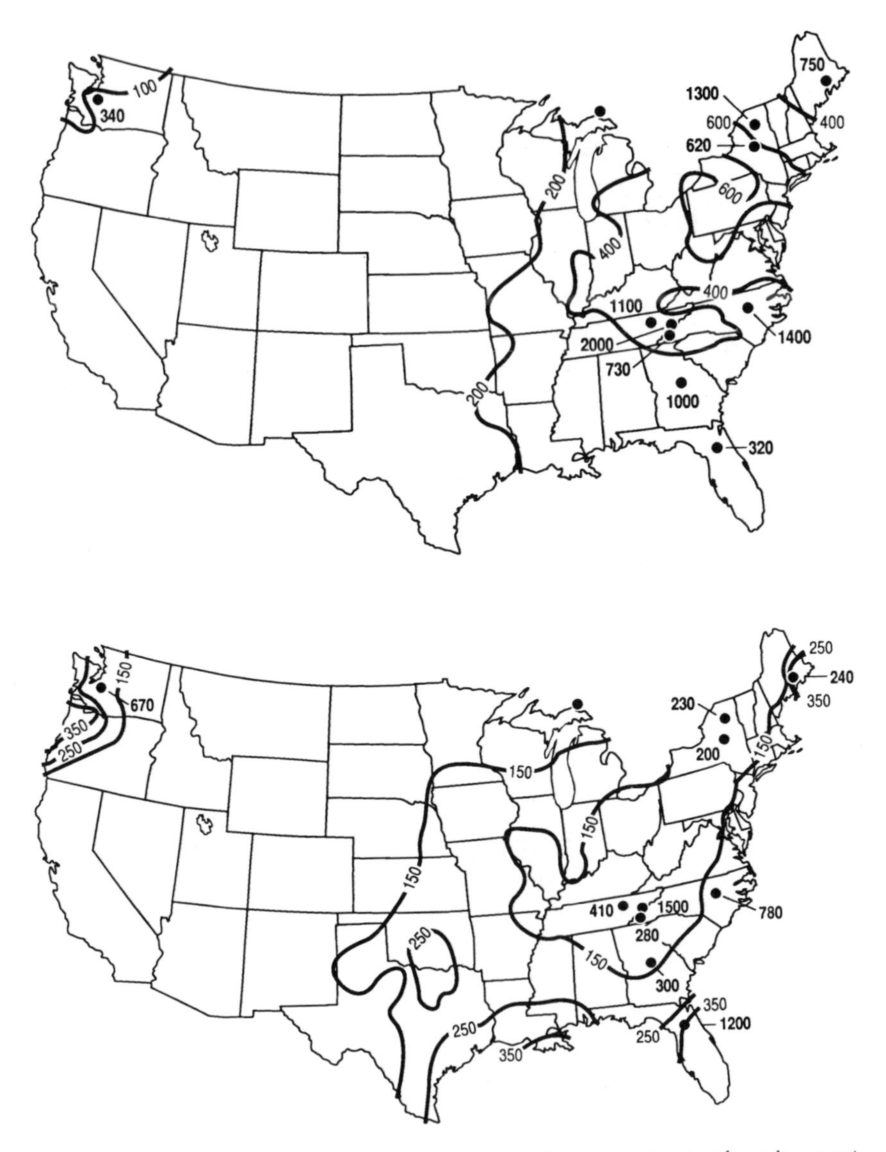

Figure 14.2. Isopleths of mean annual wet deposition rates (Eq ha^{-1} yr^{-1}) of H^+ (top) and sum of base cations (K^+ + Na^+ + Ca^{2+} + Mg^{2+}) (bottom) determined from data collected by U.S. National Atmospheric Deposition Program (NADP) for April 1986–March 1989. Isolines were drawn based on those sites (150 of possible 206) whose data met standard NADP sampling completeness criteria for entire period (NADP 1990; source of maps, G. Scott and C. Simmons, NADP Coordination Office, Colorado State University). Mean total annual deposition (Eq ha^{-1} yr^{-1}) of these ions determined for IFS sites is indicated at each site location for comparison.

cycles based solely on wet deposition network data would be subject to considerable uncertainty.

The IFS deposition data differ so drastically from the fluxes that might be predicted based on wet deposition network data because such network data do not include the often large contributions of dry and cloudwater deposition. Dry deposition is roughly comparable to wet for most ions, contributing about 30% to 50% of the total input at most IFS sites. Nitric acid vapor is the most important component of the dry deposition of N species at all sites because of its high reactivity with surfaces (median contribution, 60%), while dry deposition of SO_2 and aerosol SO_4^{2-} are comparable at most IFS sites. At all IFS sites the dry deposition of H^+ is dominated by gaseous oxides of S and N (>90% of input), while dry deposition of base cations is dominated by coarse aerosols (>90% of input). Cloud water interception contributes significantly to the total wet plus dry deposition of all ions at the Whiteface and Smoky Mountains forests, ranging from about 20% for the base cations at Whiteface Mt. to about 50% for SO_4^{2-} at the Smoky Mt. site, and certainly cannot be ignored in studies of biogeochemical cycles in many high-elevation forests in the eastern United States.

The IFS high-elevation forests not only receive significantly higher ion loadings but are also exposed to highly concentrated cloud water for several hundred hours each year (the mean ion concentrations in cloud water are generally 10 to 20 fold greater than those in rain, with pH values less than 3.0 having been recorded at each site). In addition, these sites are exposed to elevated levels of ozone for extended periods of time. Unlike the low-elevation sites where ozone patterns are characterized by strong diurnal cycles with nighttime minima, the mountain sites exhibit little diurnal variation in ozone but had midday concentrations similar to those at the low-elevation sites, leading to higher 24-hour and 12-hour mean levels overall. These sites also experience among the highest 1-hour maximum ozone levels measured in the IFS. At comparable elevations, the ozone levels are generally highest at the southeastern IFS sites (Smoky Mt. > Whiteface Mt.; Tennessee, North Carolina, and Georgia sites > New York and Maine sites). In addition, the mean ozone levels across the sites are positively correlated with dry deposition of both S and N compounds, suggesting the possibility of multiple pollutant interactions in these forests.

A commonly hypothesized, and often reported, effect of atmospheric deposition in the canopy is the leaching of base cations (Sherbatskoy and Klein 1983). Despite the breadth of data collected by the IFS, regression models (using 14 canopy, soil, atmospheric, and climatic characteristics) failed to consistently explain the large differences in annual leaching rates of these cations from the various IFS forest canopies. Either the extensive IFS data are still too limited or are not sufficiently accurate to explain such trends, or unmeasured factors are involved (such as leaf wax chemistry or local peroxide levels). Two potential factors, precipitation acidity and ozone levels, were investigated in detail in controlled studies on several IFS forest species (white pine, sugar maple, red spruce, and loblolly pine). Five-hour ozone exposures at levels up to 140 ppb combined with water misting had no effect on the leaching of Ca^{2+} and Mg^{2+} from branches on mature trees. However, increasing the leaching water acidity to pH levels at 3.8 and less significantly increased the leaching of these cations. Charge balance calculations indicated that these losses are caused by simple cation exchange reactions and that extended exposures of the canopy to highly acidic cloud water (pH < 3.3) would result in much greater leaching of cations from forest canopies than would exposure to rain only. The field data generally confirmed these findings.

All the IFS forest canopies neutralize deposited H^+ to some extent, with the deciduous forests consistently showing the greatest ability to consume deposited H^+. Dry deposition appears to be more efficiently neutralized than wet, and mechanisms of neutralization include organic acid protonization and cation exchange. Total acidity fluxes above and below the canopy are often nearly in balance, but net acid uptake occurs in several canopies. The canopies that neutralize strong acid inputs to the greatest extent also exchange the most cations in their canopies, and H^+ exchange accounts for a median of approximately 30% of the total leaching of base cations from these canopies. However, relative to the foliar requirement and content of cations, the current foliar leaching at most IFS sites is low. In addition, controlled studies showed that foliar leaching rates under moderate pH conditions were small compared to root uptake rates for pine seedlings. Thus, for most IFS forests, with the possible exception of the high-elevation spruce-fir sites, deposition of acidity appears to have little potential to influence total canopy cation pools. However, we do not know whether a smaller, exchangeable pool of cations in the canopy might be more seriously affected.

This does not imply the lack of important canopy interactions of deposited ions, however. Among the most significant results of the IFS is quantification of the degree to which atmospheric N is retained by the forest canopy. Throughfall fluxes indicated that the canopies of all IFS forests consistently act as sinks for deposited inorganic N and that the amount of retention is positively correlated with the amount of N deposited. The dominant form retained in the canopy is inorganic, and this uptake is partially offset by release of organic N, perhaps indicating in-canopy transformation. On average, the IFS forest canopies retain or transform 40% of the deposited inorganic N, and the amount of retention is positively correlated with the amount of N deposited. Significant canopy N uptake is not often considered in traditional interpretation of N cycles in forests.

The behavior of deposited S and Na^+ in the IFS forest canopies is notably different from that of the other ions. Within the error of most measurements, these elements behave nearly conservatively in the canopy, with the estimated deposition and measured fluxes in throughfall plus stemflow generally within 10% to 15% of each other. If one assumes that Na^+, being nonessential to plants, should behave conservatively in the forest canopy, then the fact that conservative behavior was observed tends to confirm the validity and utility of the methods developed in the IFS for estimating atmospheric inputs, especially dry and cloud water deposition. This also lends credibility to the canopy budgets of the other base cations. For deposited S compounds, only dry-deposited SO_2 appears to exhibit a significant canopy interaction, being retained in the canopies with efficiencies ranging from 30% to 70%, the remainder being washed back into throughfall. Because SO_2 is a relatively small fraction of the total input, and because some SO_2 uptake is offset by minor foliar leaching, throughfall measurements of S have a strong potential for prediction of deposition fluxes (r^2 = .97). This finding has proved very useful in the IFS and also has wide-ranging implications for future monitoring of temporal and spatial trends in S fluxes to forests.

Finally, it is clear to this author that the IFS data firmly establish the inadequacy of the bulk deposition collectors so commonly used in many previous studies of ecosystem nutrient cycling (e.g., Cole and Rapp 1981). While this has been debated for some time, it has generally been accepted as fact for systems in highly polluted environments where dry deposition is significant (NRC 1983). However, use of bulk collectors has been advocated for systems in less impacted environments and to estimate inputs of some base cations under all circumstances. The IFS data suggest that even these applications are subject to considerable uncertainty. Figure 14.3 illustrates the differences in bulk deposition and total wet + dry fluxes of several ions across the IFS sites. The underestimates of the actual input provided by bulk collectors are clearly shown, ranging from a 30% underestimate of SO_4^{2-} at the "background" red alder site in Washington to a factor of approximately 6 for N at the Smoky Mt. site. It is clear that bulk methods are questionable even at the "background" IFS sites, especially for N. The IFS results have clarified the implications of using incomplete and inaccurate results from bulk deposition measurements to attempt to quantify foliar leaching or uptake rates and to understand the role of the atmosphere in forest biogeochemical cycles.

Relationships among Atmospheric Deposition, Forest Nutrient Status, and Forest Decline

D.W. Johnson

The principal aim of the IFS project was to determine the effects of atmospheric depositon on nutrient status of a variety of forest ecosystems and to determine if these effects are in any way related to current or potential forest decline. In summarizing and synthesizing the IFS results in the previous chapters, it is this author's opinion that acid deposition is having a significant effect on nutrient cycling in most of the forest ecosystems in this project (the exceptions being the relatively unpolluted Douglas fir, red alder,

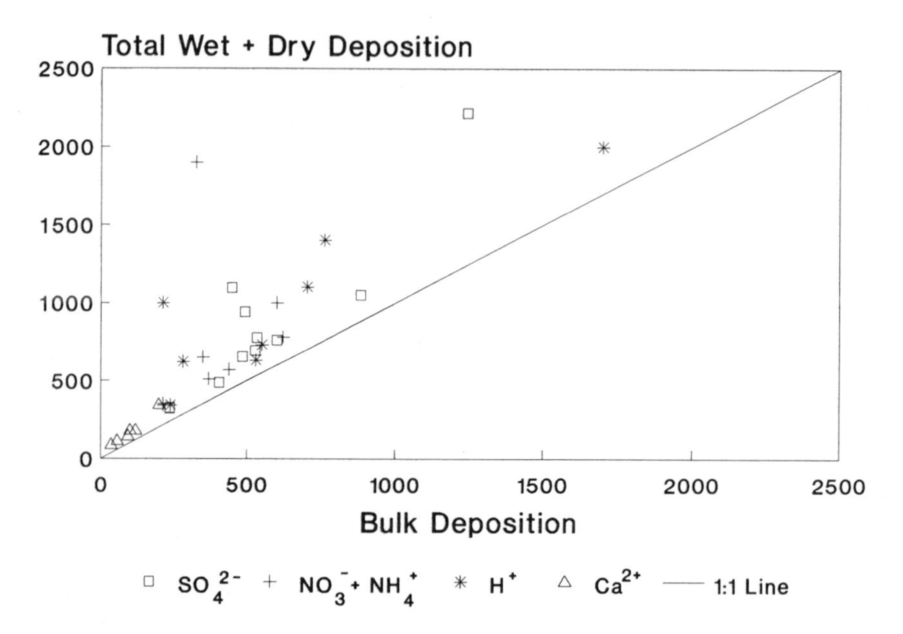

Figure 14.3. Relationship between measured annual mean bulk deposition and total atmospheric deposition (wet plus dry) determined at each IFS site for various ions (eq ha^{-1} yr^{-1}). Total deposition was determined from measured and modeled results based on methods described in Chapter 2.

and Findley Lake sites in Washington State). The nature of these effects varies from one location to another. In all but the relatively unpolluted Washington sites, atmospheric sulfur depositon is having a significant, often overwhelming effect on cation leaching rates from soils. None of these sites is currently deficient and it is open to question as to whether they will ever become deficient, even with accelerated cation leaching rates: soils may stop changing before deficiencies develop, or weathering rates may be adequate to maintain adequate cation nutrient supplies, especially if weathering is also accelerated by increased cation leaching.

Some of the soils in the southeastern sites are extremely low in weatherable minerals and are therefore susceptible to base cation depletion. Soils with low weatherable minerals are often protected from rapid base cation depletion by large exchangeable base cation pools. However, cases of base cation depletion over one to two decades in southeastern soils not blessed with such high levels of exchangeable bases have been documented (Binkley et al. 1989; Johnson et al. 1988). Whether such changes in base cation levels will result in base cation deficiencies is not clear. Cation exchange equations predict that those cations being most significantly depleted from exchange sites will be increasingly resistant to leaching. As deficiency thresholds are approached, leaching losses of the potentially deficient cation will decrease, even to the point where leaching losses are less than atmospheric deposition. At that stage, only harvesting can result in a net loss of the potentially deficient cation from the ecosystem, and even harvesting removals may be substantially offset by deposition in certain cases.

In general, nutrient budget data from the IFS study and from the literature suggest that the susceptibility of southeastern sites to base cation depletion from soils and the development of cation deficiencies by that mechanism appears to be greater than in northern sites. Having said this, it is noteworthy that the only known, documented, base cation deficiencies in North America are actually in the Northeast (Heiberg and White 1953; Stone and Kszystyniak 1977). The existence of these deficiencies in the Northeast does not necessarily negate the statement made earlier, however, in that these K- and Mg-deficient sites are limited to soils with inherently low K or Mg supplies because of the nature of soil parent material. Thus, it seems that the size of the weatherable mineral pool is critical to whether cation deficiencies will develop or not. In that southeastern soils generally have very low weatherable mineral pools, the statement that they are susceptible to base cation depletion and development of deficiencies still stands. The actual extent to which base cation depletion and cation deficiencies will or will not occur depends very much on site-specific conditions such as exchangeable base cation reserves, leaching rate, and, most importantly, harvesting intensity and frequency.

Atmospheric deposition may have significantly affected the nutrient status of some of the IFS sites through the mobilization of Al. Specifically, soil solution Al levels in the Smokies sites approach and sometimes exceed levels noted to impede base cation uptake in solution culture studies (Raynal et al 1990). It is therefore possible that the rates of base cation uptake and cycling in these sites have been reduced because of soil solution Al. To the extent that atmospheric deposition has contributed to these elevated soil solution Al levels, it has likely caused a reduction in base cation uptake and cycling rates at these sites. Nitrate and sulfate are the dominant anions in the Smokies sites, and nitrate pulses are the major cause of Al pulses in soil solution.

At first glance, it would seem that atmospheric N deposition is the major cause of high nitrate leaching rates in the Smokies sites, given the fact that atmospheric N deposition approximately equals nitrate-N leaching (see Appendix). However, there are several lines of evidence that soil nitrogen mineralization and nitrification contribute substantially to nitrate leaching at these sites as well: (1) there is a substantial increase in nitrogen flux (primarily as nitrate) in the forest floor and A horizons of these sites as compared to throughfall and stemflow; (2) in situ soil incubations indicate periodically high rates of N mineralization and nitrification in these horizons; and (3) the nitrate pulses noted in these horizons cannot be accounted for by pulses of nitrogen input from atmospheric deposition and throughfall and therefore must be a result of N mineralization and nitrification. Atmospheric deposition is most likely the only significant historical source of N to these systems, in that no known N fixers are present. However, it is not clear as to what extent current atmospheric N deposition as compared to internal N mineralization is causing current nitrate leaching. This uncertainty can be best stated in the form of a thought experiment: What would nitrate leaching rates be if N deposition were an order of magnitude lower than current estimates? If, leaching rates remained initially the same (which is the author's hypothesis), how long would these rates be maintained under a lower deposition regime?

The connection between Al mobilization and forest decline is not clear. The decline in red spruce has certainly been more severe in the Northeast than in the Southeast, yet all evidence (from the IFS project as well as previous studies; see A. Johnson 1983 and Joslin et al. 1988) indicates that Al mobilization is most pronounced in the southern Appalachians. The Whiteface site was selected because it was in a state of decline, yet soil solution Al levels there are lower than in the Smokies sites, which are not in an visually obvious state of decline (e.g., no dieback other than fir killed by the balsam woolly adelgid, no needle yellowing, etc.). Thus, as was the case with soil change in the Southeast, the Al mobilization situation does not constitute a "smoking gun," but certainly does constitute a situation worthy of further research and monitoring.

Implications of the Integrated Forest Study Project for Forest Nutrient Cycling

D.W. Johnson and S.E. Lindberg

The IFS project has, for the first time, accurately quantified atmospheric inputs to nutrient cycles using state-of-the-art techniques to measure wet and dry deposition. Without exception, these new estimates of atmospheric deposition are larger, often by nearly an order of magnitude, than estimates obtained by traditional methods (e.g., bulk collectors). These new, larger estimates of input create significant changes in how nutrient cycles are interpreted. For instance, the amounts of atmospheric nutrient input to most of the IFS sites are considerably greater than previously reported in the nutrient cycling literature (Cole and Rapp 1981). It is probable that nutrient inputs have always been greater than previously reported, given the importance of dry and sometimes cloud water deposition. These new, higher estimates of atmospheric nutrient inputs have a very significant effect on calculations of potential changes in forest nutrient status. If atmospheric nutrient inputs are twice what was previously known (a typical situation when bulk and total deposition rates are compared), forecasts of nutrient depletion from intensive harvesting, for example (Boyle et al. 1973; Federer et al. 1989; Johnson et al. 1988), will require substantial revision.

The IFS project has added substantial insight into the cycling of nitrogen and sulfur as well as that of base cations, as already discussed. Until very recently, it was widely assumed that North American forests were almost uniformly nitrogen deficient and always accumulated nitrogen from atmospheric deposition (even with the low estimates of N deposition from bulk collectors) (Cole and Rapp 1981; Gessel et al. 1973). In recent years, however, several case studies of so-called nitrogen-saturated forests have been found (Van Breemen et al. 1987), some of which are within the IFS network (Foster et al. 1989; Johnson et al., in press; Van Miegroet and Cole 1984). These studies have revealed that N saturation may be caused by a variety of factors, including excessive N inputs (via either deposition or N fixation), low uptake, lack of fire history, or simply naturally rich site conditions. The emergence of these studies requires us to rethink traditional views that North American forests are uniformly conservative with nitrogen. Much of the controversy as to the effects of harvesting on nitrate leaching (see Vitousek and Melillo 1979) now proves to be moot; harvesting or any other factor that reduces uptake (such as insect attack; Foster et al. 1989; Swank et al. 1981) can clearly cause increased nitrate leaching in sites where soils are naturally nitrogen rich or deposition rates are high. Harvesting can actually cause a decrease in nitrate leaching when it results in the removal of a source of nitrogen input via fixation (see Chapter 11; Van Miegroet et al. 1989).

Until a little more than a decade ago, sulfur was among the least studied of the macronutrients. The advent of the acid deposition issue changed that situation quickly, and many studies of sulfur cycling and soil sulfur dynamics have appeared in the literature since 1980 (David et al. 1982; Johnson and Cole 1977; Johnson and Henderson 1981; Johnson 1984; Johnson et al. 1981 1982b; Lindberg and Garten 1988). An interesting and significant pattern of ecosystem sulfur retention was noted early in the acid depositon era: systems of the southeastern United States uniformly showed greater S input than output, whereas ecosystems of the Northeast showed an approximate net balance (Johnson et al. 1980; Rochelle et al. 1987). This regional pattern was often roughly related to the limit of the Wisconsin glaciation. Two alternative hypotheses were set forth to explain the apparent net retention of sulfur in southeastern soils: (1) sulfate adsorption to Fe and Al hydrous oxides, and (2) organic sulfur accumulation. With regard to the former, regional patterns of soil sulfate adsorption were found to match regional patterns of apparent net ecosystem S retention (Johnson et al. 1980; Johnson and Todd 1983). Southeastern subsoils were found to be enriched in the Fe and Al hydrous oxides, which adsorb sulfate, and relatively free of organic matter, which blocked adsorption sites. Northeastern soils were found to be occasionally enriched in Fe and Al hydrous oxides, but these were often coated with organic matter, which blocked adsorption sites.

The matching patterns of soil sulfate adsorption and ecosystem S retention on a broadly regional scale gave circumstantial evidence that one resulted in the other. There were exceptions, to the regional pattern, however: in glaciated subsoils with relatively high Fe and Al hydrous oxides and relatively low organic matter, laboratory-determined sulfate adsorption capacities were quite high even though ecosystem S retention was zero (e.g., Johnson et al. 1986). Those who posed the organic retention hypothesis noted that organic sulfur pools are always the largest in forest ecosystems, and that it was therefore only logical to assume that organic S retention was greater than inorganic retention (e.g., David et al. 1982; Swank et al. 1984).

One of the specific objectives of the IFS project was to determine the relative importance of organic versus inorganic S retention in the forest ecosystems under study. Laboratory studies of both processes were undertaken, and the results indicated that inorganic retention (adsorption) was of greater significance than organic retention for short-term sulfur retention (see Chapter 5). From the size of the inorganic versus organic soil S pools, however, it was concluded that incorporation of adsorbed sulfate into organic matter was the major long-term mechanism of ecosystem S retention. These results have significant implications for ecosystem recovery from sulfur deposition: long-term incorporation of S into organic matter will prevent desorption of sulfate from adsorption sites, the net result being that ecosystem S retention constitutes a permanent mitigation against base cation depletion by atmospheric sulfur deposition.

One of the greatest impediments to predicting soil change caused by either acid deposition or natural causes is a lack of quantitative knowledge on soil weathering rates. The IFS project attacked this problem in two ways: (1) characterization of primary minerals at each of the sites, and (2) process-level studies on weathering, including rhizosphere effects, at specific sites. This combination of approaches has allowed us to rank the IFS sites with respect to importance of weathering and to make some quantitative statements about specific sites (i.e., weathering is near zero at the Duke and Florida sites, whereas weathering is the main soure of Na and Ca flux from the soil at Huntington Forest; see Chapter 10). However, the goal of obtaining quantitative data on soil weathering remains elusive for most of the sites.

Thus, the IFS study has contributed to our knowledge of nutrient cycling in forests in many significant respects. In addition, the IFS project has considerably expanded the basic data base on nutrient cycling, much of which was summarized nearly a decade ago by Cole and Rapp (1981). Unlike other large scale studies of nutrient cycling, the IFS data was collected with standard protocols so that intersite comparisons can be made with confidence. Even though we have drawn many interpretations from the IFS data set in this volume, there are surely many more interpretations that could be drawn and uses to which this data could be put. Thus, we have presented summaries of the IFS data in the Appendix in the hope that the scientific community will find them useful in assessing emerging environmental problems as well as the basic patterns of forest nutrient cycling.

The Nutrient Cycling Model (NuCM): Overview and Application

S. Liu, R. Munson, D.W. Johnson, S. Gherini, K. Summers,
R. Hudson, K. Wilkinson, and L.F. Pitelka

Introduction

A general model of forest nutrient cycling was well established by the time the acid rain issue emerged (e.g., Cole et al. 1968; Curlin 1970), but this model tended to concentrate on aboveground processes such as litterfall, uptake, translocation, and was not widely used in early assessments of acid deposition effects. Early acid deposition studies tended to concentrate on selected processes (primarily foliar and soil leaching) rather than ecosystem-level responses (Johnson and Cole 1977; Wood and Bormann 1974). There was an unstated tendency to treat the soil as a homogenous, unvegetated block with certain exchangeable cation reserves that would become more acid, given sufficiently large acid inputs over a sufficient length of time (Engstrom et al. 1971).

Reuss (1976), Sollins et al. 1980), and Ulrich (1980) began to introduce an ecosystem-level perspective to the acid deposition–soil acidification issue by noting that acid deposition is merely an increment to a variety of internal acidification processes (carbonic, organic, and nitric acid production within the soil; humus formation; and plant uptake). These authors proposed that internal acid generating processes could be described and quantified by constructing "hydrogen ion budgets," a concept later elaborated by Binkley and Richter (1987). Hydrogen ion budgets have been used to evaluate the role of acid deposition in causing soil acidification, but they have some serious limitations and flaws, including the inability to describe or predict soil solution properties (an important need when considering the aluminum mobilization/toxicity issue) and an inability to predict the H^+ generation or consumption from plant uptake of nitrogen.

There are some interesting interactions between uptake and leaching of base cations that produce the kind of nonintuitive results which call for simulation modeling. These interactions are illustrated in the results of a study of harvesting and leaching in loblolly pine and mixed oak forests in eastern Tennessee (Johnson and Todd 1987). In comparing these two sites, we found that high rates of Ca^{2+} uptake in the mixed oak forest had depleted subsoil exchangeable Ca^{2+} and reduced base saturation relative to the loblolly pine site. This depletion by the mixed-oak vegetation caused large differences in cation leaching: Ca^{2+} and total cation leaching were lower in the mixed-oak site than in the loblolly pine site, whereas Mg^{2+} leaching in the mixed-oak site was greater than in the loblolly pine site. The differences in Mg^{2+} leaching rates were especially interesting and nonintuitive because soil exchangeable Mg^{2+} was actually lower in the mixed-oak than in the loblolly pine site. This apparent anomaly was in reality predictable because of the ratios of exchangeable Ca^{2+} to Mg^{2+}, which were greater in the loblolly pine than in the mixed-oak site. These results imply that (1) tree uptake can result in the conservation of cations in short supply by depleting soil exchangeable reserves, and (2) whole-tree harvesting will take a greater toll on potentially limiting cations than will leaching.

To test this and related hypotheses, the Nutrient Cycling Model (NuCM) has been developed by Tetra Tech, (Lafayette, California) as a part of the Electric Power Research Institute's Integrated Forest Study (IFS). The model simulates vegetation growth, litterfall, and decay, soil biogeochemical processes, and moisture routing (Gherini et al. 1989). The model was developed for use on IBM PCs or compatibles and incorporates a graphical user interface.

The response of soil nutrient status to acidic deposition has been simulated using the model for three forested sites. These include the Huntington Forest in the Adirondack mountains of New York, the Smokies Tower site in eastern Tennessee, and the Duke Forest in central North Carolina. This section presents a summary of the NuCM model and then describes the results of its application to these three sites.

Summary of Nutrient Cycling Model (NuCM)

NuCM model theory derives from ideas of the IFS principal investigators and the literature (e.g., Johnson and Cole 1977; Johnson et al. 1982; Melillo et al. 1982; Paster and Post 1986; Reuss and Johnson 1986), the ILWAS model (Gherini et al. 1985; Goldstein et al. 1984), and the OR-NATURE model (Ungs et al. 1985). The available nutrients in soil strata and vegetation pools and the fluxes between them are explicitly tracked and provided as model output, as shown in Figure 14.4. The interaction of processes occurring in these pools ultimately determines nutrient status and thus forest health and productivity (Johnson et al. 1986). The model can be used to simulate the response of forests to atmospheric deposition and to various management practices (e.g., application of fertilizers). Factors are included in the model that allow the user to easily increase or decrease atmospheric deposition loads. Sulfur and nitrogen deposition can be adjusted either alone or in conjunction with changes in the deposition of each base cation. The model can be used to simulate seasonal effects as well as the longer term response. The model can also be used to test hypotheses relating to tree growth, conservation of nutrients, and forest floor chemical behavior.

The forested ecosystem is represented as a series of vegetation and soil components. The model provides for both an overstory and understory, each of which can be divided into canopy, bole, and roots. The key processes occurring in each vegetation component are noted in Figure 14.5. Tree growth in the model is a function of user-defined developmental stage and the availability of nutrients and moisture. Translocation of nutrients before senescence is included. The understory is simulated in a similar manner to the

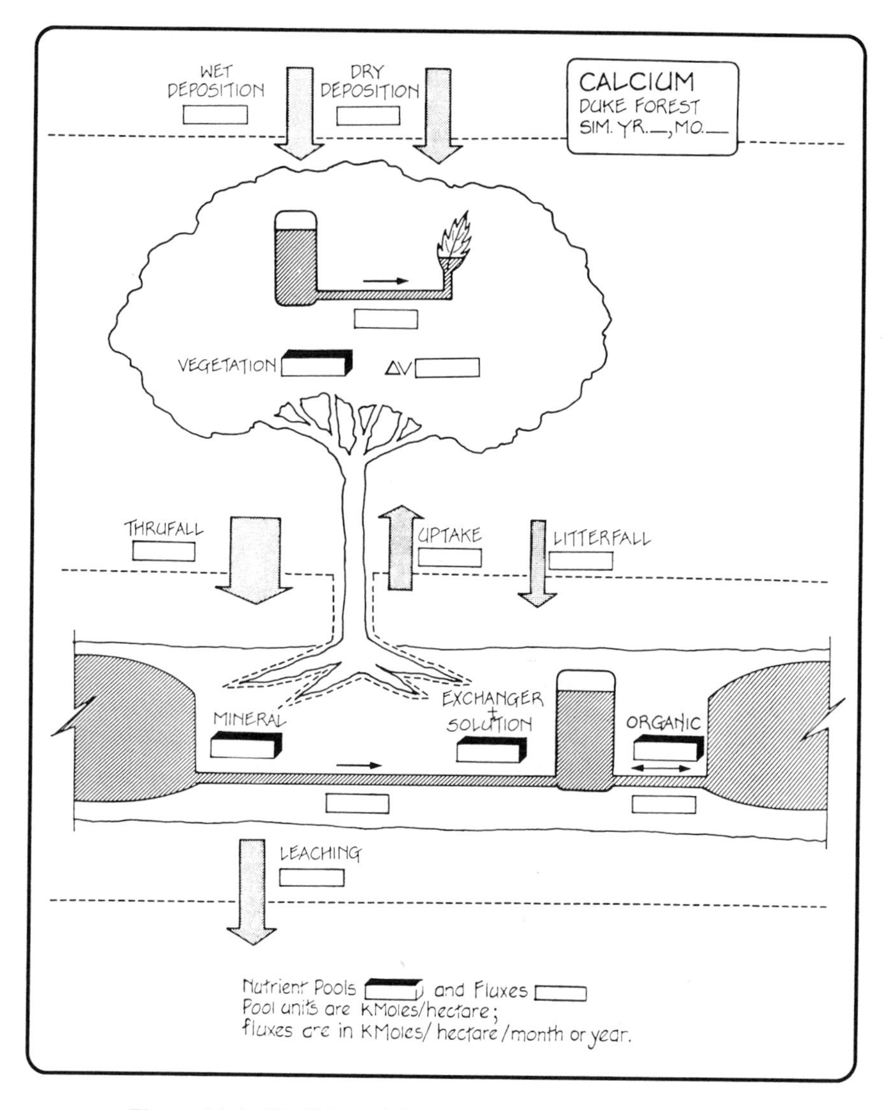

Figure 14.4. NuCM model: example pool and flux diagram.

overstory, except that its "incident precipitation" is the overstory's throughfall and its biomass stoichiometries are allowed to be different. For both the overstory and the understory, variable stoichiometries are allowed to accommodate multiple species and availability of nutrients.

Using mass balance and transport formulations, the model tracks 16 solution-phase components including the major cations and anions (analytical totals), ANC (acid-neutralizing capacity), an organic acid analogue, and total aluminum (upper portion of Table 14.1). The concentrations of hydrogen

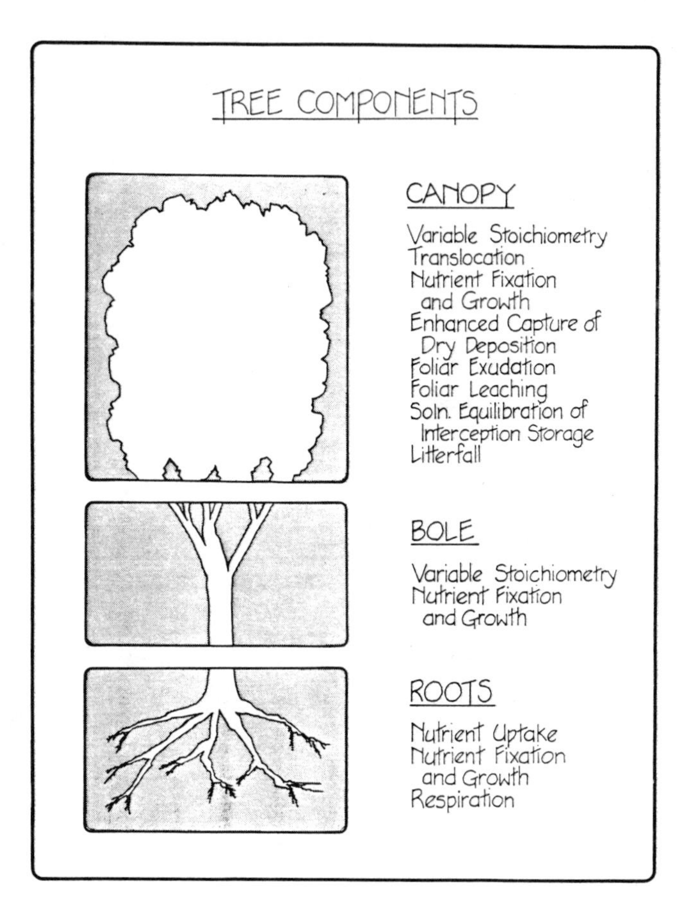

Figure 14.5. Tree components and processes associated with those components.

Table 14.1. Solution-Phase Chemical Constituents Simulated in NuCM

	Cations	Anions	Neutral Species	Analytical Totals	Gases (Input Only)
Solutes tracked by mass balance	Ca_T	SO_{4T}	H_4SiO_4	Alk(ANC)	(SO_x)
	Mg_T	NO_{3T}	$CO_2(aq)$	Org Acid (triprotic, R)	(NO_X)
	K_T	Cl_T	AlR	Al_T	
	Na_T	PO_{4T}	$Al(OH)_3^0$	C_T(TIC)	
	NH_{4T}	$Al(OH)_4^-$	H_3R		
Solutes which are calculated from those above	H^+	HCO_3^-			
	Al^{3+}	CO_3^{2-}			
	$Al(OH)^{2+}$	R^{3-}			
	$Al(OH)_2^+$	HR^{2-}			
	$Al(SO_4)^+$	H_2R^-			
		$Al(SO_4)_2^-$			

ion, aluminum, and carbonate species, and organic acid ligands and complexes are then calculated on the basis of the 16 components. The acid–base characteristics of the forest soil solution are computed by the model to properly account for the influence of hydrogen ion concentration on cation exchange and mineral weathering.

The major processes included in the model are noted in Figure 14.6. A brief description of selected processes and their representation in the model is included below.

Hydrologic Processes

The model routes precipitation through the canopy and soil layers and simulates evapotranspiration and deep seepage as well as lateral flow out of the soil system. The soil includes multiple layers (as many as 10), and each layer can have different physical and chemical characteristics. The hydrologic processes simulated by the model are depicted schematically in Figure 14.7.

The movement of water through the system is simulated using the continuity equation, Darcy's equation for permeable media flow, and Manning's equation for free surface flow. Moisture content can vary with depth in the soil; percolation occurs between layers as a function of layer permeabilities and differences in moisture content; and lateral flow occurs in those layers with saturated moisture contents. The lower portion of a layer with average

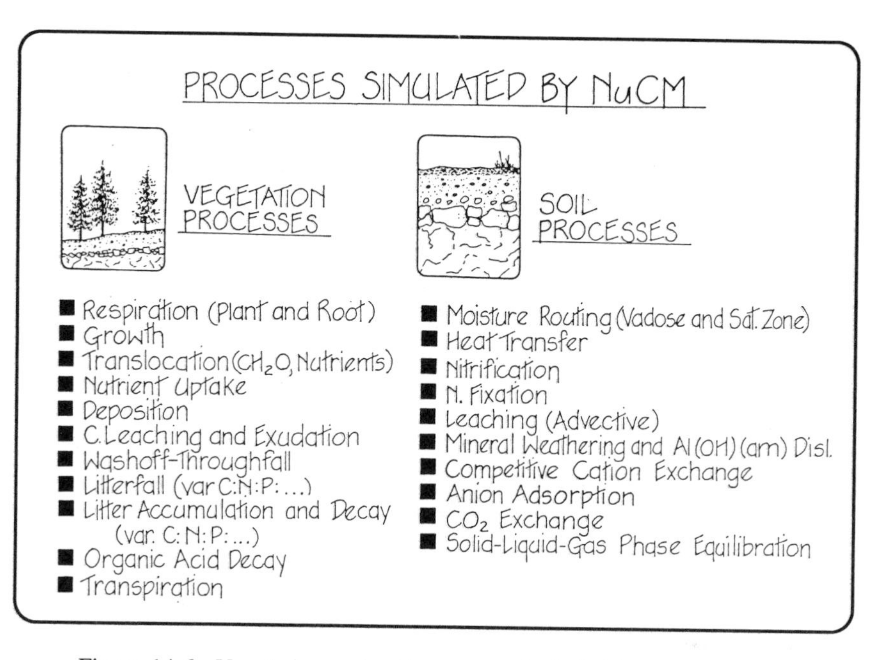

Figure 14.6. Vegetation and soil processes included in NuCM model.

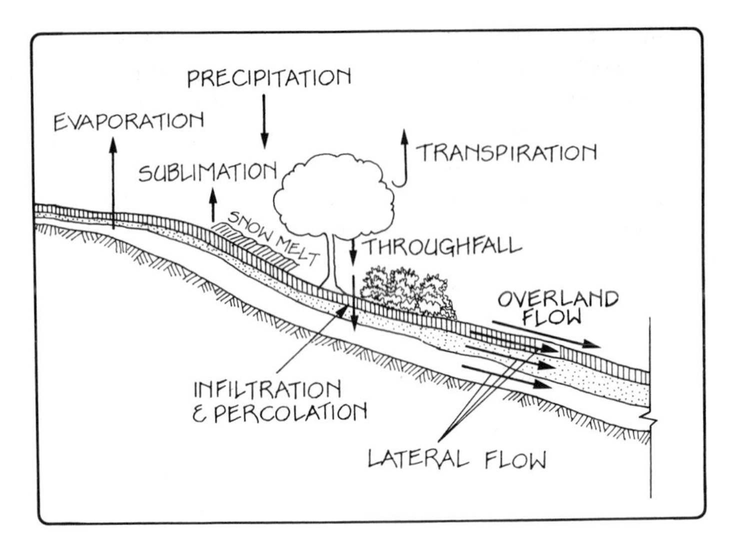

Figure 14.7. Terrestrial hydrologic processes influencing nutrient cycling (Munson and Gherini 1991).

moisture level approaching saturation is also allowed to produce lateral flow. This is achieved using a linearization of Darcy's equation to account for variation in moisture content within that horizon.

Canopy Processes

As indicated in Figure 14.5, the model simulates processes that occur in the canopy, bole, and roots. Although all these processes can influence nutrient cycling, this section will focus on three occurring in the canopy.

The canopy is the first component that interacts with deposition. It intercepts wet and dry deposition and stores a fraction of the precipitation volume on the leaf surfaces, up to a user-defined interception storage capacity. The solution stored on the canopy undergoes chemical reaction with constituents on the leaf surface (e.g., foliar exudates), provides for direct nutrient uptake by the leaf, and is subject to concentration by evaporation. The dry deposition (gaseous, aerosol, and particulate matter) is enhanced by the canopy to a greater capacity than would occur in an open area because the leaves provide additional surface area (Chen et al. 1983; Lindberg et al. 1986). This enhanced collection is proportional to the leaf area index (LAI).

The constituents on leaf surfaces with which precipitation reacts include dry deposition and the products of foliar leaching and exudation. Both foliar leaching and exudation result in the flux of nutrients from within the leaf to the leaf surface (Lovett, this volume; Schaefer, this volume). Exudation is thought to be driven by internal processes taking place within the leaf, whereas leaching is driven by deposition. Leaching is simulated as a rate process dependent on the hydrogen ion concentration in the moisture on the leaf surface. When mixed with precipitation, the constituents from all three processes (dry deposition, leaching, and exudation) produce the markedly different chemical characteristics observed in throughfall relative to incident precipitation.

Nutrient translocation is simulated as the partial withdrawal of nutrients from the leaves prior to leaf fall. This results in a smaller total input of nutrients to the soil system via litterfall and its subsequent decay. If translocation were neglected, excess litterfall-associated nutrients would be available for leaching after mineralization.

Soil Biogeochemical Processes

The nutrient pools associated with soil solution, the ion exchange complex, minerals, and soil organic matter are all tracked explicitly. The processes that govern interactions among these pools include decay, nitrification, anion adsorption, cation exchange, and mineral weathering. These are described briefly next.

Decay Processes

When litter falls to the forest floor, insects and microbes mediate its decay. The decay is represented as a series reaction (Figure 14.8) with first-order dependencies on the reactant concentrations and C:N ratios. The decay products include nutrients and organic matter, both solid (e.g., humus) and solution phase (e.g., organic acids). The nutrients produced enter the solution

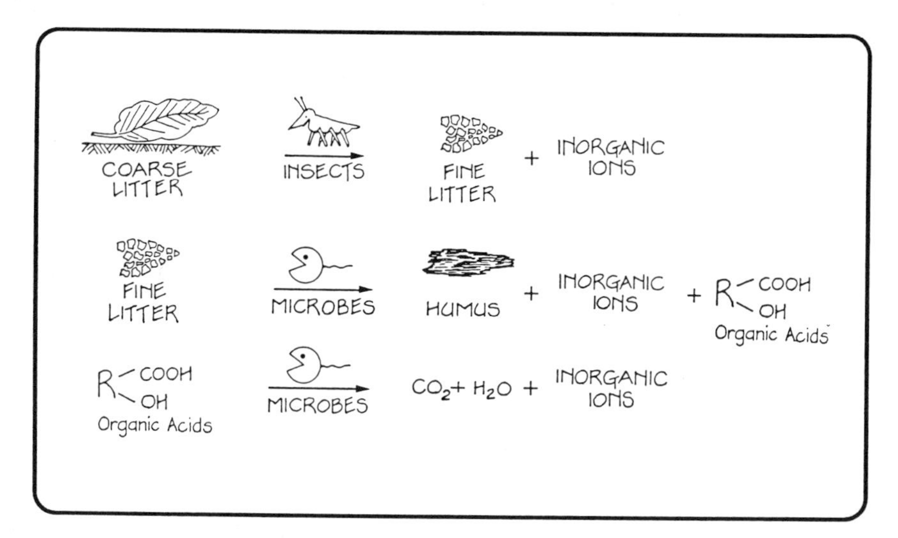

Figure 14.8. Sequential decay of litter and organic matter (Munson and Gherini 1991).

phase where they are available for uptake by vegetation or the exchange complex and for transport from the forest floor by percolation or lateral flow.

The dissolved organic matter, having a preponderance of acidic functional groups (carboxyl, phenolic and enolic hydroxyls), generally depresses solution pH but also serves as a buffer by minimizing subsequent changes in pH. Measurements show that organic acid concentrations [as indicated by dissolved organic compounds (DOC)] generally decrease with depth in the soil profile (Cronan 1985). This implies that the dissolved organic matter is being adsorbed/precipitated and mineralized as it moves from the upper to the lower soil horizons. The model simulates this behavior using a pH-dependent isotherm for adsorption and a first-order reaction rate expression for decay.

Nitrification

One of the major products of litter decay is ammonium. In the soil system, ammonium can be readily converted to nitrate (Figure 14.9). This reaction results in the loss of ANC and the production of the mobile nitrate anion. The resulting strong acid can lead to exchange reactions that accelerate cation leaching. However, nitrogen, as a major plant nutrient, is largely taken up by the trees and understory vegetation during the growing season. During the dormant season, nitrate concentrations in upper soil horizons can build up to significant levels in forests where percolation is limited by the formation of snowpack. In fact, nitrification reactions continue during the win-

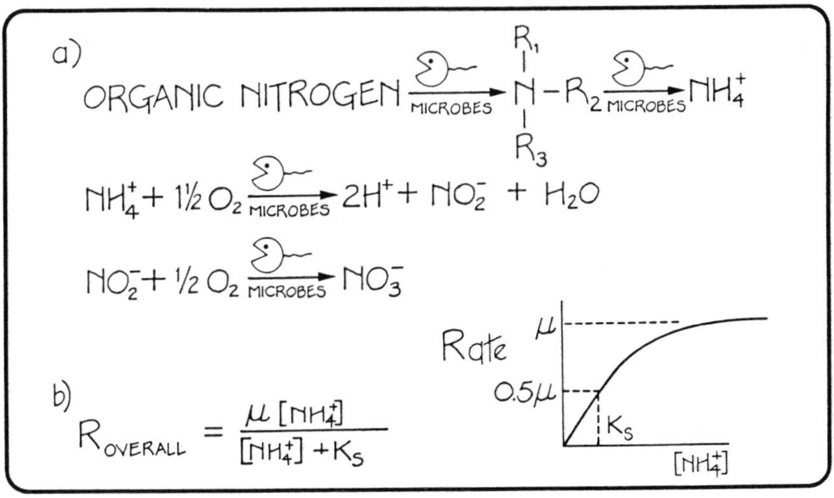

a)

$$\text{ORGANIC NITROGEN} \xrightarrow{\text{MICROBES}} R_1{-}N(R_3){-}R_2 \xrightarrow{\text{MICROBES}} NH_4^+$$

$$NH_4^+ + 1\tfrac{1}{2}\,O_2 \xrightarrow{\text{MICROBES}} 2H^+ + NO_2^- + H_2O$$

$$NO_2^- + \tfrac{1}{2}\,O_2 \xrightarrow{\text{MICROBES}} NO_3^-$$

b)

$$R_{OVERALL} = \frac{\mu\,[NH_4^+]}{[NH_4^+] + K_S}$$

Rate

μ

0.5μ

K_S

$[NH_4^+]$

Figure 14.9. Mineralization and oxidation of organic nitrogen: (a) general stoichiometry; (b) Rate expression, with NH_4^+ oxidation-limiting reaction (Munson and Gherini 1991).

ter even though soil temperatures are significantly reduced (Peters and Driscoll 1987). Reaction rates are lower, but the process proceeds, and without flushing by rain or snowmelt, soil solution nitrate concentrations increase. Snowmelt flushes some of the nitrate from the system. Later in the spring, concentrations can be drastically reduced by uptake by vegetation.

Anion Adsorption

The model simulates the noncompetitive adsorption of sulfate, phosphate, and organic acid. Although phosphate and organic acids are adsorbed, sulfate adsorption often has a more significant effect on acid–base chemistry in systems where this process is active. Adsorption of sulfate increases the acid-neutralizing capacity (ANC) of the soil solution and thereby reduces hydrogen ion concentration and thus cation exchange and weathering reactions. Sulfate adsorption can be especially significant in old, highly weathered soils such as those found in the southeastern United States. These soils can have high concentrations of aluminum and iron oxides and hydroxides, which serve as effective adsorption sites (Figure 14.10). Sulfate adsorption is simulated in NuCM using both linear and Langmuir (saturation) adsorption isotherms.

Even though phosphate is strongly sorbed, its concentration is generally much lower on an equivalent basis and therefore has a minimal direct acid–base effect. Phosphate adsorption in the model is represented by a linear isotherm. As noted, organic acid adsorption is represented by a pH-dependent isotherm.

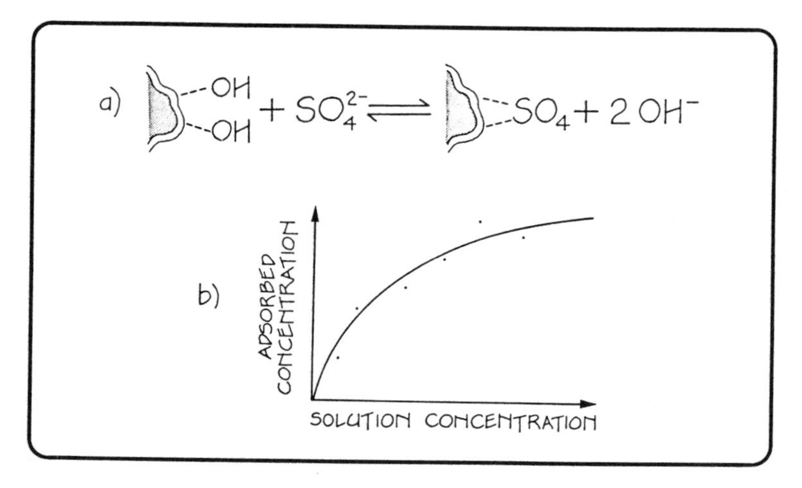

Figure 14.10. Anion adsorption: (a) idealized representation; (b) Langmuir adsorption isotherm (Munson and Gherini 1991).

Cation Exchange

The cation exchange complex represents a major accessible pool of cationic nutrients. Cation exchange is also an effective solution-phase buffering process. It limits changes in soil solution pH by exchanging hydrogen and aluminum ions for sorbed bases (Figure 14.11). If excess base cations are introduced to a system, for example by liming, the base cations are sorbed and hydrogen and aluminum ions are released. The hydrogen ions are neutralized by reaction, in the case of liming, with bicarbonate and carbonate, and the aluminum is precipitated, most likely as an amorphous aluminum hydroxide. The cation exchange process proceeds rapidly and thus provides effective short-term buffering. The sorbed cation pool is, however, limited. For example, Schnoor and Stumm (1986) reported that forest soils in the northeastern United States have enough exchangeable bases (without weathering) to buffer acid inputs for 50 to 200 years. This is in contrast to the potential base supply from mineral weathering, which is very large (e.g., millions of years), but is relatively unresponsive to changes in solution composition as described next.

Mineral Weathering

Primary minerals (usually aluminum silicates) react with hydrogen ions to form secondary minerals and release base cations (and occasionally small quantities of strong acid anions) and silica to solution (Figure 14.12). These reactions, which are normally slow, are described in the model using rate expressions with dependencies on the mass of mineral present and solution-

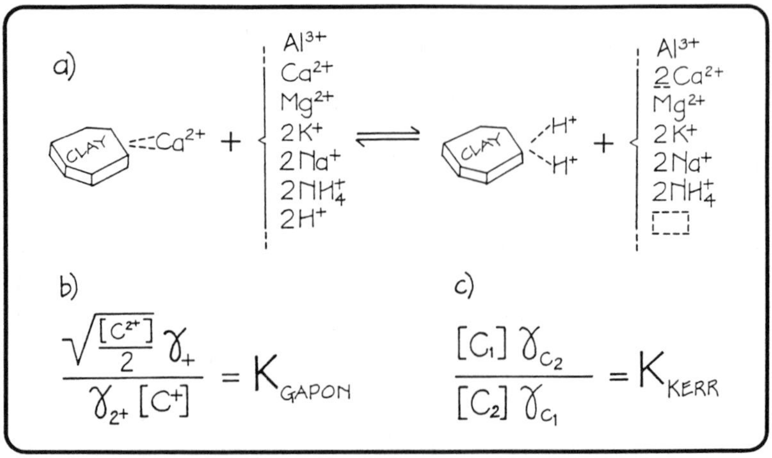

a)

$$\text{CLAY}\cdots Ca^{2+} + \left\{ \begin{array}{l} Al^{3+} \\ Ca^{2+} \\ Mg^{2+} \\ 2K^{+} \\ 2Na^{+} \\ 2NH_4^{+} \\ 2H^{+} \end{array} \right. \rightleftharpoons \text{CLAY} \begin{array}{l} \cdots H^{+} \\ \cdots H^{+} \end{array} + \left\{ \begin{array}{l} Al^{3+} \\ 2Ca^{2+} \\ Mg^{2+} \\ 2K^{+} \\ 2Na^{+} \\ 2NH_4^{+} \\ [\;\;] \end{array} \right.$$

b)

$$\frac{\sqrt{\frac{[C^{2+}]}{2}}\,\gamma_{+}}{\gamma_{2+}\,[C^{+}]} = K_{GAPON}$$

c)

$$\frac{[C_1]\,\gamma_{C_2}}{[C_2]\,\gamma_{C_1}} = K_{KERR}$$

Figure 14.11. Competitive cation exchange: (a) idealized representation; (b) heterovalent exchange equation; (c) homovalent exchange equation (Munson and Gherini 1991).

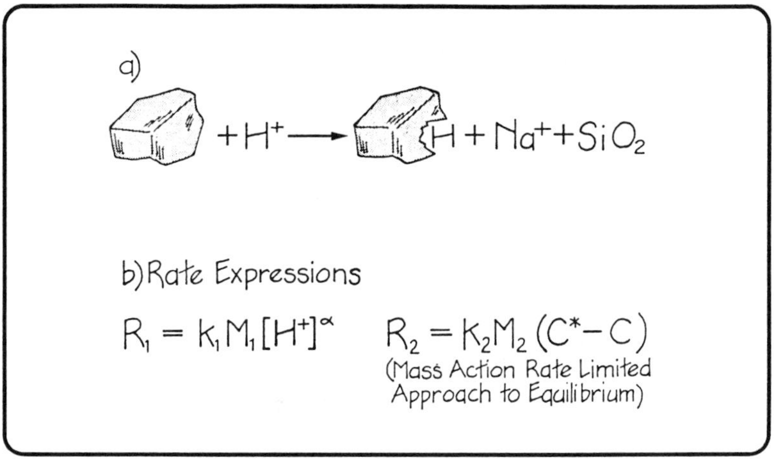

a)

$+ H^+ \longrightarrow H + Na^+ + SiO_2$

b) Rate Expressions

$R_1 = k_1 M_1 [H^+]^\alpha \qquad R_2 = k_2 M_2 (C^* - C)$

(Mass Action Rate Limited Approach to Equilibrium)

Figure 14.12. Mineral weathering: (a) Idealized representation; (b) rate expressions − C^* = equilibrium concentration of reaction product (Munson and Gherini 1991).

phase hydrogen ion concentration taken to a fractional power. This hydrogen ion dependence has been reported to range from 0.0 to 0.8 (e.g., Drever et al. 1985; Gherini et al. 1988). This dependence is weak; changes in weathering rate from changes in solution-phase hydrogen ion concentration are less than proportional. Further, the buffering resulting from cation exchange limits changes in soil solution pH, thereby minimizing the response of weathering rates to changes in atmospheric deposition acidity.

Because weathering is the ultimate source of cationic nutrients, its role in nutrient cycling can be extremely important. To the extent that weathering supplies the required nutrients to soil solution, the demand on the exchange pool is reduced. This can have important implications for long-term nutrient status, particularly in soils with low cation exchange capacities.

Model Input and Output

Model input is based on measurable parameters. The categories of input data used are noted in Figure 14.13. All input is accomplished using series of menus or automatic transfer of data from previously entered files. For example, Figure 14.14 shows the menu for entering adsorbed and solution cation concentrations for initializing the cation exchange complex for a given soil layer. The model uses such data to compute selectivity coefficients for each soil layer simulated. The use of previously entered files can be particularly convenient for large data sets such as those associated with meteorological observations. Function keys have been programmed to enable the user to move quickly from one parameter or menu to another. "Help" menus

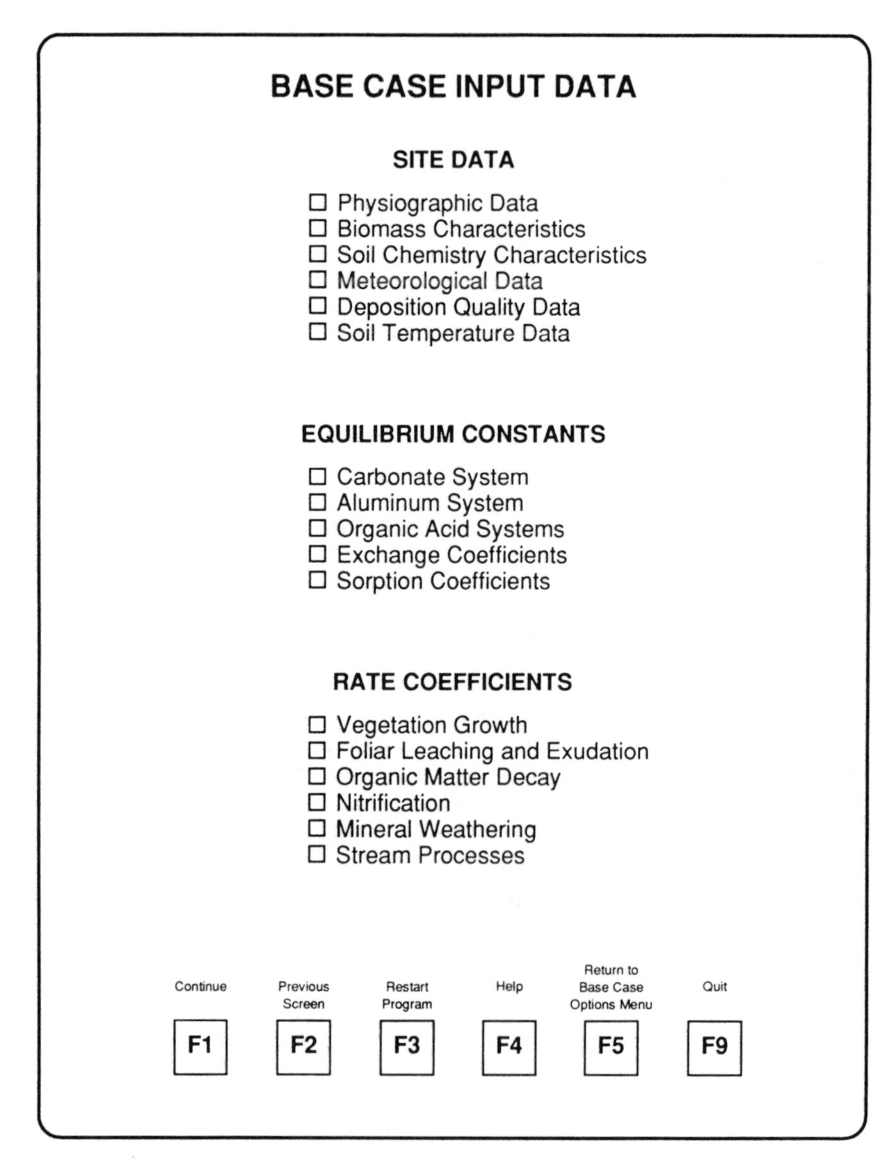

Figure 14.13. Control menu that allows NuCM users access to data input menus.

to define the variables and provide typical ranges of parameters have been included to make the model easier to use.

Model output options (Figure 14.15) include nutrient pool sizes, fluxes between components, the relative contribution or loss by process, and soil solution and adsorbed concentrations versus time (Figure 14.16). Long-term

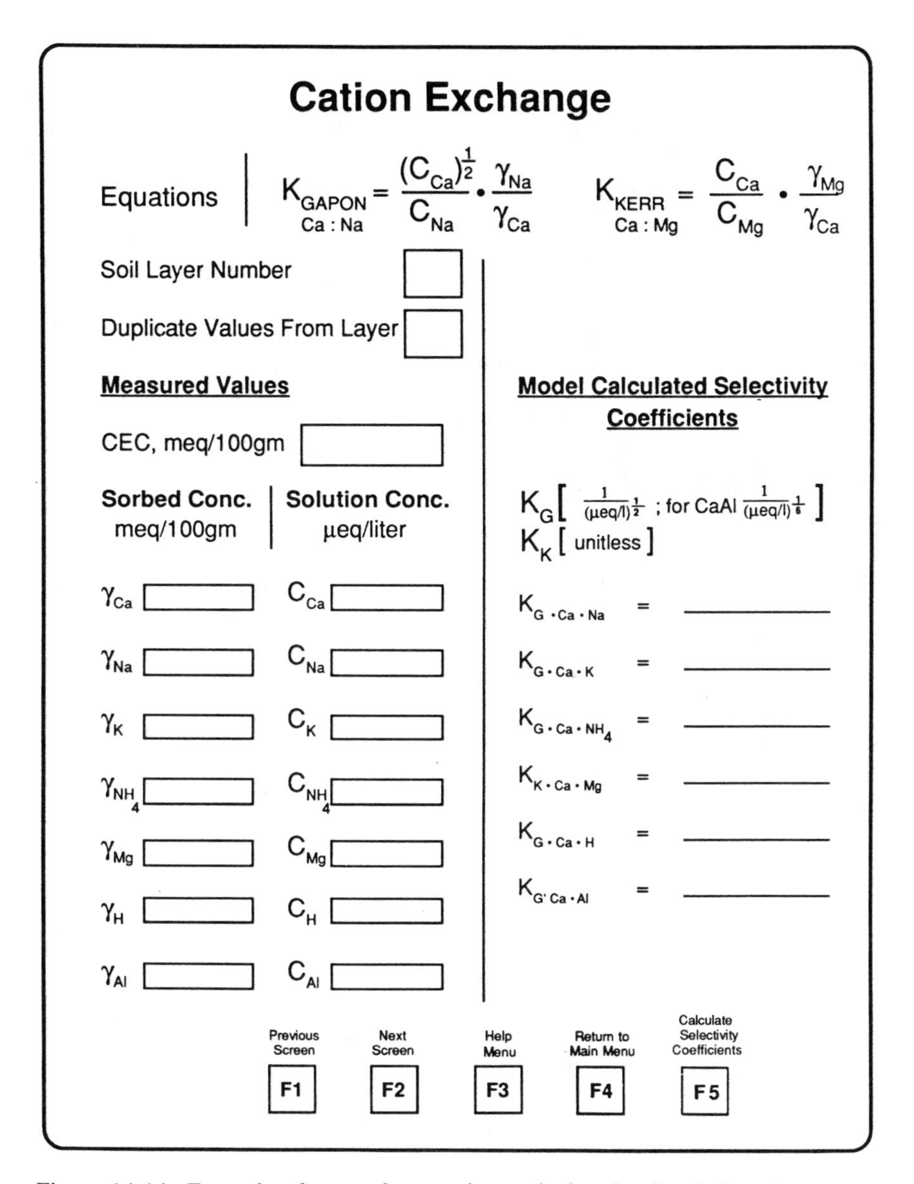

Figure 14.14. Example of menu for entering sorbed and soil solution-phase concentrations (model then uses values and calculates selectivity coefficients).

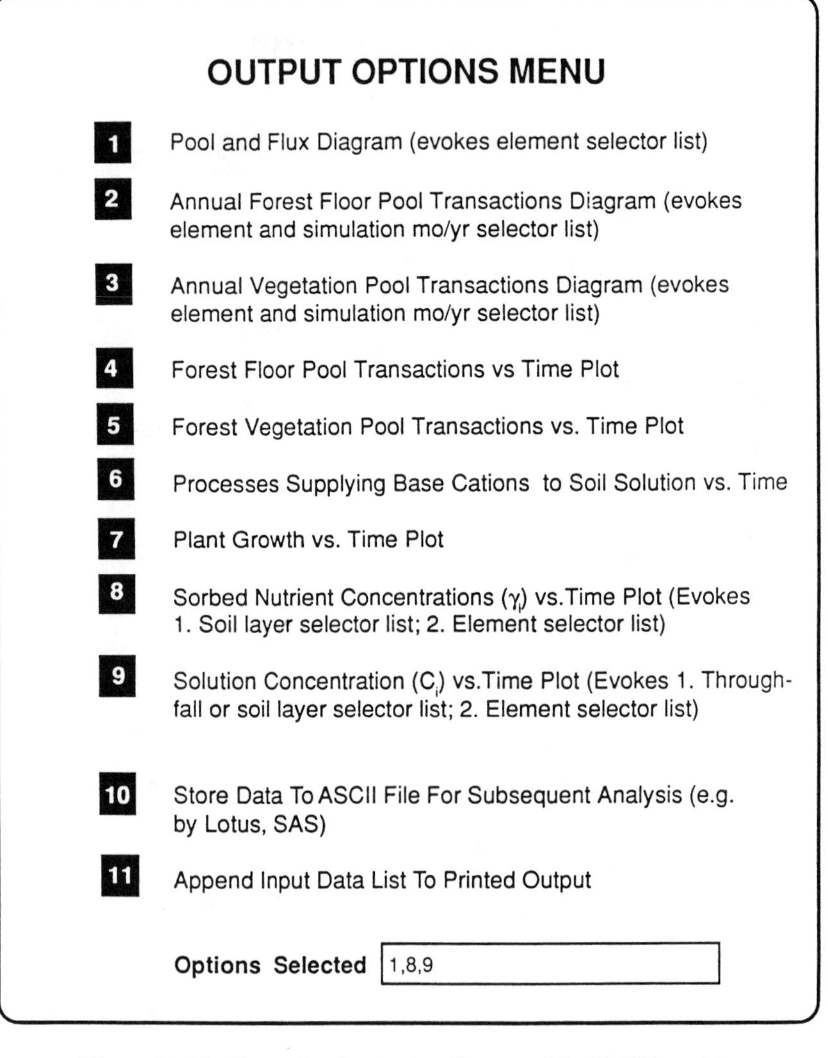

Figure 14.15. Example of output options used in NuCM model.

nutrient loss or accumulation can be tracked by following annual pool and flux charts (see Figure 14.4).

Computer Requirements

The model operates on IBM or compatible PCs with a math coprocessor, EGA or VGA card, color monitor, and at least 500K RAM. The model operates under DOS 3.00 or more recent versions. The code is written in "C," Fortran, and Assembly languages. The model has a user-selectable

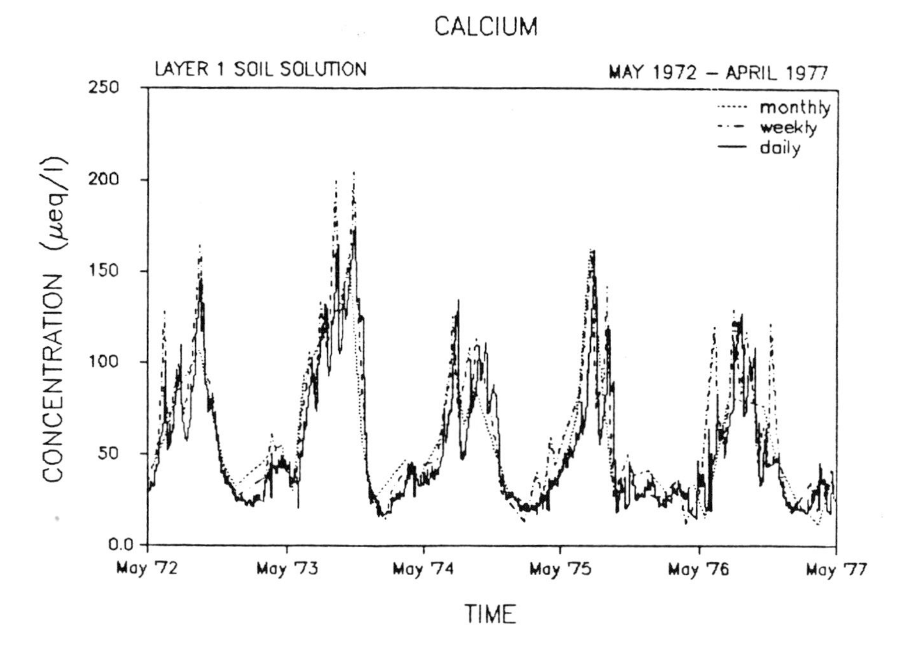

Figure 14.16. Example NuCM Output: soil solution concentration versus time plot for calcium for uppermost soil layer. Plot shows minimal loss of resolution with increasing time step size from daily to weekly to monthly.

time step (daily, weekly, monthly), which facilitates making long-term simulations.

Model Simulation Results

The calibrated data sets for the Huntington Forest, Smokies Tower, and Duke Forest were used to compare the effects of different deposition scenarios. Long-term simulations were made for the base case with constant deposition and two alternative deposition scenarios. The scenarios called for staged reductions in sulfur deposition over the next 65 years (Figure 14.17). The total sulfur reduction was similar for both scenarios, but the timing of deposition reductions varied considerably. The scenario with the more rapid decrease in sulfur deposition has been designated as the retrofit scenario while the slower deposition reduction is called the replacement scenario. These scenarios represent estimates of changes in deposition resulting from the application of scrubber technology (retrofit) versus clean coal technology (replacement) at United States power plants (EPRI 1990). For the base case and each scenario, graphs of the readily available soil nutrients (Ca, Mg, K) and sulfate were produced. The readily available nutrient pools include nutrients in solution and adsorbed to the cation exchange complex. Nutrients

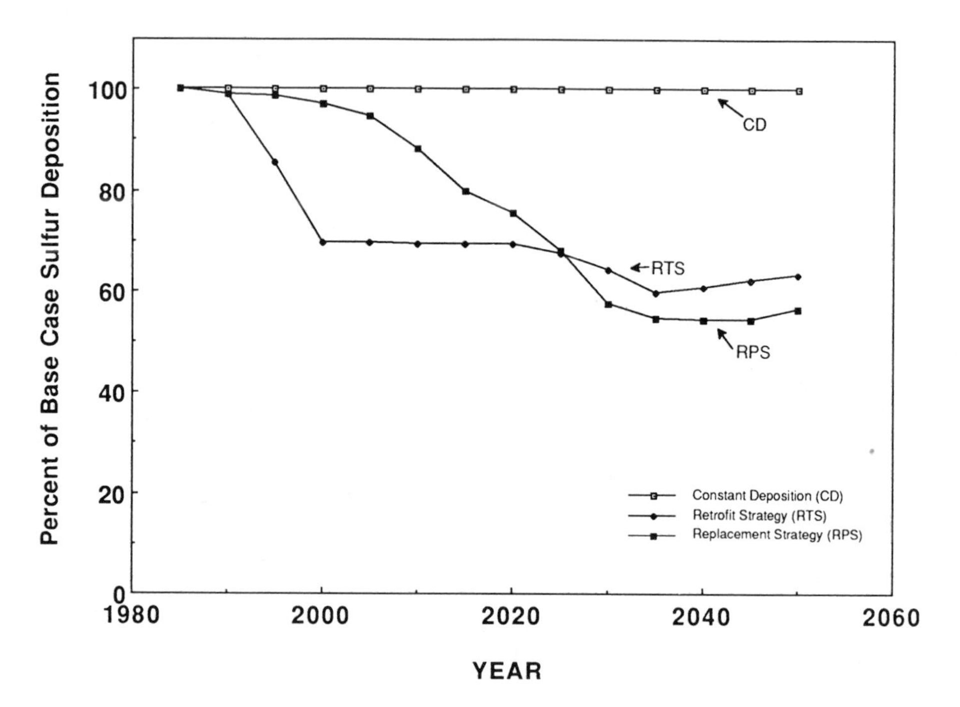

Figure 14.17. Sulfur deposition for base case and reductions for alternative scenarios.

tied up in organic matter or in mineralogical phases are simulated but are not considered a part of the readily available pools.

As shown in Figure 14.18 for Huntington Forest, the adsorbed plus solution-phase sulfate concentrations are low for the base case and for both alternative strategies during the 65-year simulation period. The differences at 65 years between the two alternative scenarios are small (<0.1 kmol ha^{-1}) with the replacement scenarios producing a slightly smaller final pool. Changes between the available soil nutrient pools for Ca, Mg, and K were small relative to the base case (Figure 14.18b, 14.18c, and 14.18d, respectively). The differences between the base case and the alternative scenarios at the end of the 65-year simulations were about 0.3 kmol ha^{-1} for Ca, 1.5 kmol ha^{-1} for Mg, and 0.1 kmole ha^{-1} for K. Differences between the replacement and retrofit scenarios were negligible for all three nutrients.

The Smokies Tower site showed more change over time for sulfate (Figure 14.19a) for the alternative strategies. The larger changes in the solution plus adsorbed sulfate result mostly from desorption of sulfate from the soils at this site. At the end of the 65-year simulation period, the solution plus

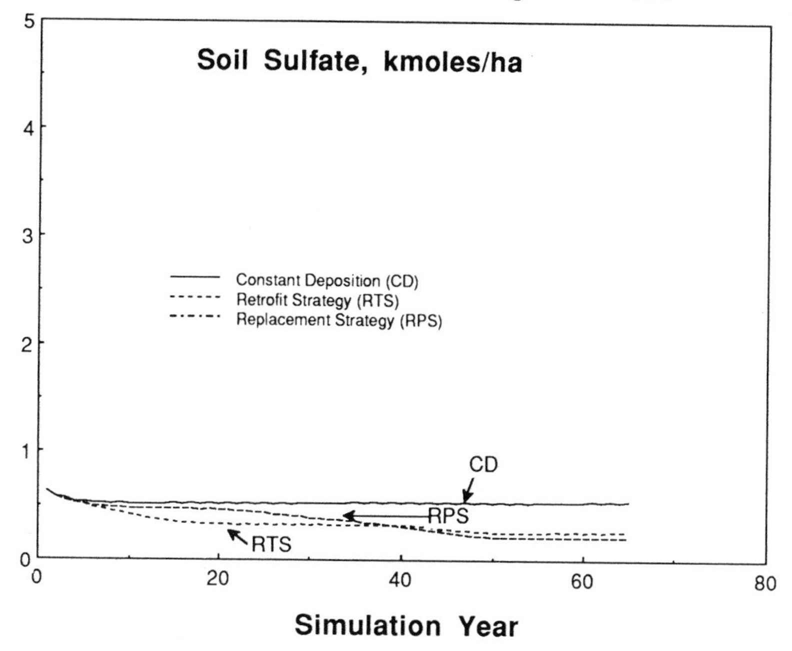

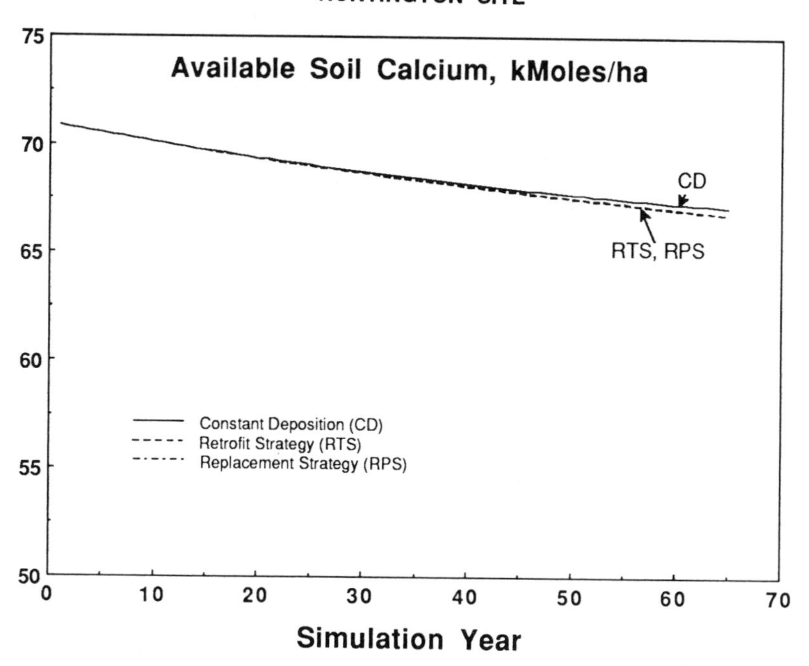

Figure 14.18. Comparison of (a) soil sulfate, (b) available soil calcium, (c) soil magnesium, and (d) soil potassium at Huntington Forest for base case and alternative strategies (see following page).

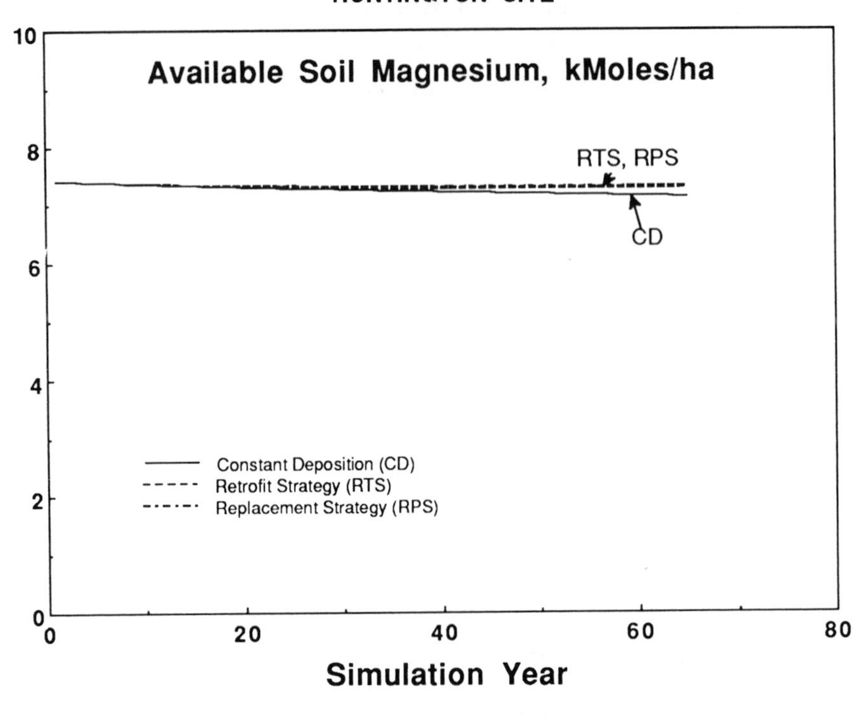

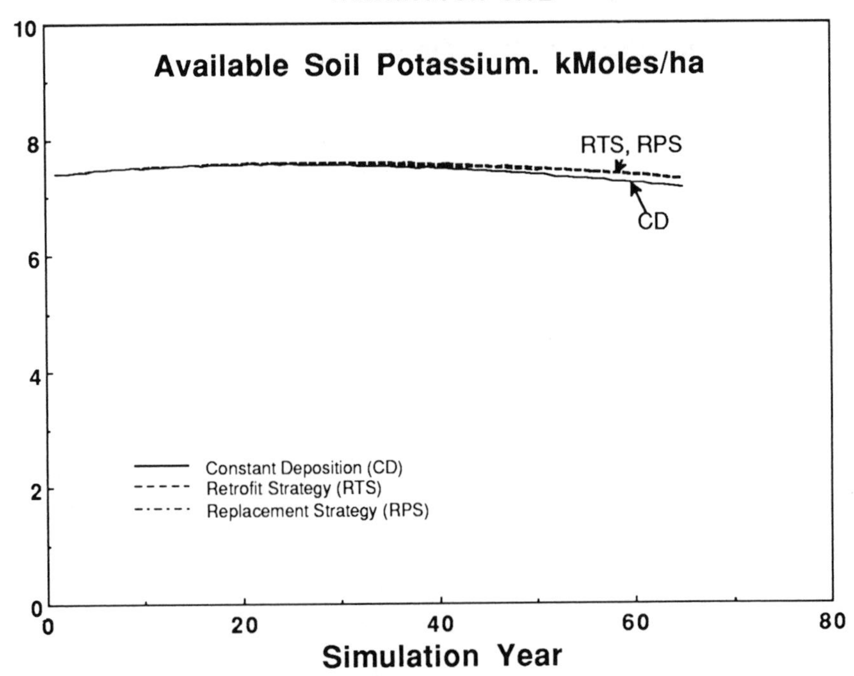

Figure 14.18. *Continued.*

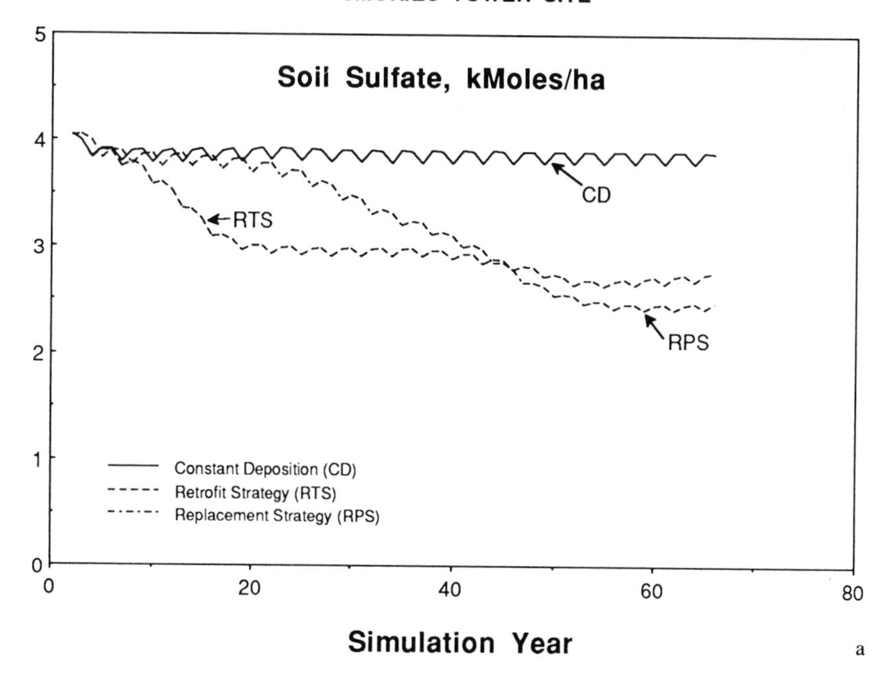

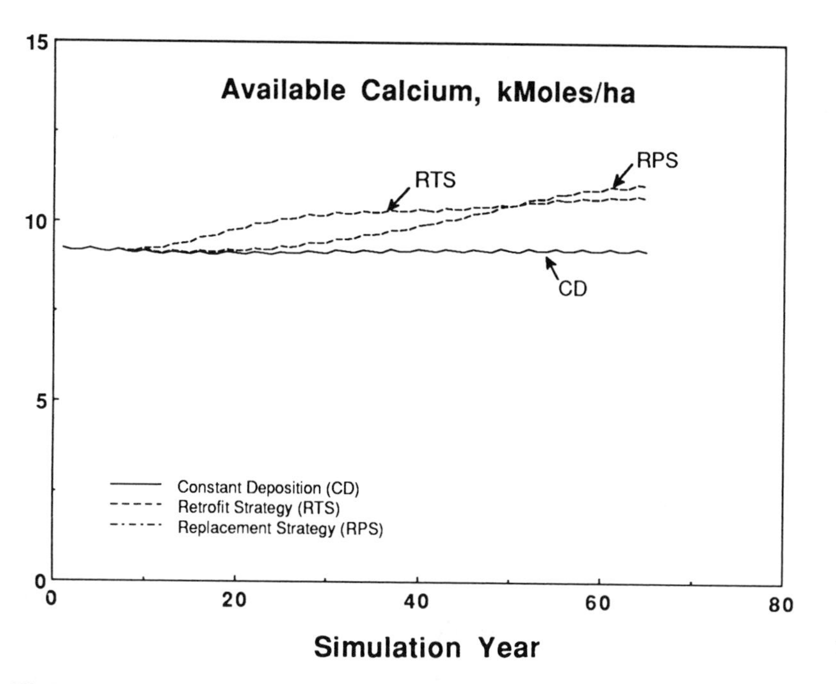

Figure 14.19. Comparison of (a) soil sulfate, (b) available soil calcium, (c) soil magnesium, and (d) soil potassium at Smokies Tower for base case and alternative strategies (see following page).

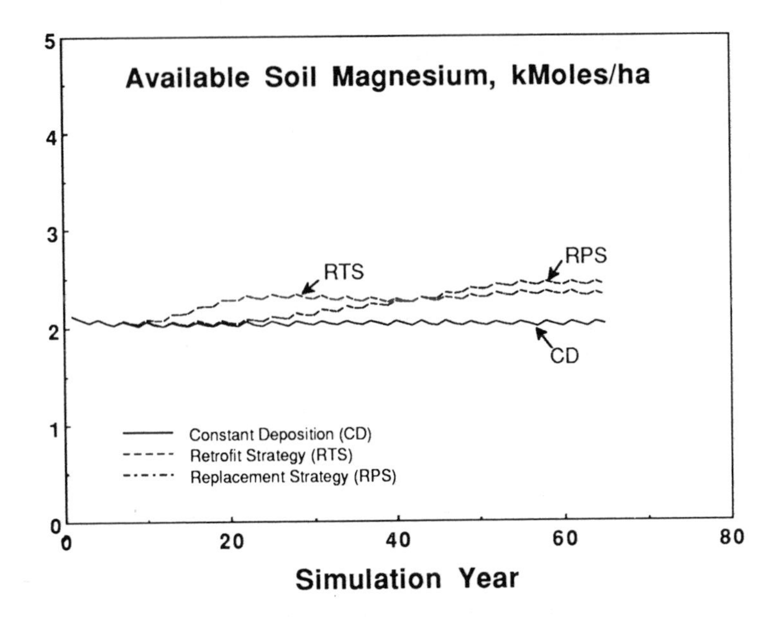

c

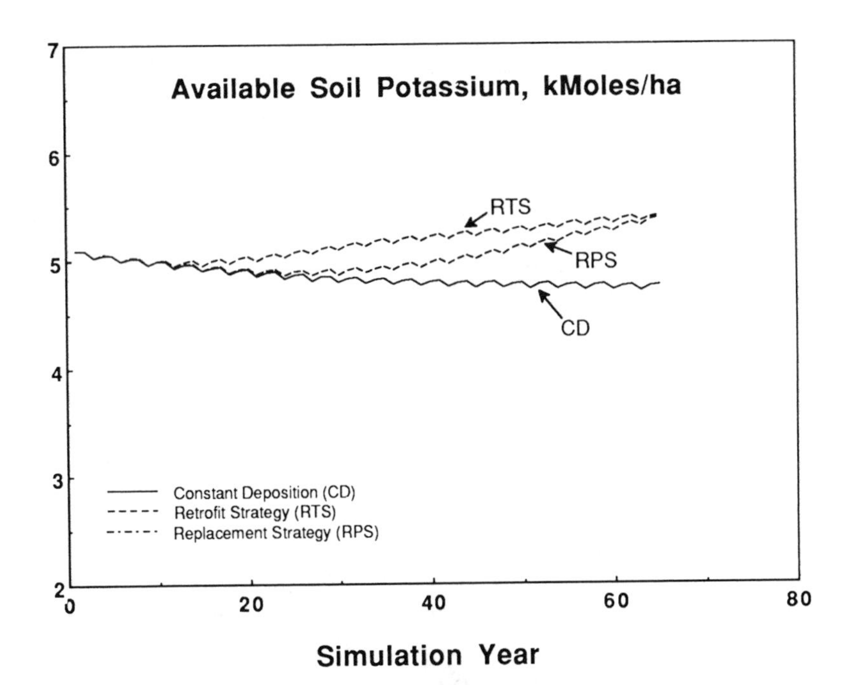

d

Figure 14.19. *Continued.*

sorbed sulfate pool was lower with the replacement than the retrofit scenario. The available soil nutrient pools for Ca, Mg, and K are shown in Figures 14.19b, 14.19c, and 14.19d, respectively. Larger nutrients pools were projected after 65 years for the alternative scenarios than the constant deposition case. Differences between the two alternative scenarios were small for all the cations (<1 kmol ha^{-1}). The available Ca and Mg were slightly higher for the replacement scenario than the retrofit scenario; pool sizes were about the same for these two alternatives for K. Available nutrients were higher for both scenarios than the base case by 0.5 kmol ha^{-1} for K, by about 1 kmol ha^{-1} for Mg, and by about 2 kmol ha^{-1} for Ca.

The response of Duke Forest sulfate pool to the various alternative scenarios was similar to the response for the Smokies Tower site. Figure 14.20a shows that the difference in adsorbed plus solution-phase sulfate concentration for the Duke Forest site between the base case and the two alternatives was about 1.7 kmol ha^{-1}. The replacement scenario was about 0.3 kmol ha^{-1} less than the retrofit scenario. Changes in the cationic nutrients Ca, Mg, and K are shown in Figures 14.17b, 14.17c, and 14.17d, respectively. The soil calcium concentrations for the two alternative scenarios were about 1.2 kmol ha^{-1} higher than the constant deposition case at the end of the 65-year simulation. The difference between the constant deposition case and the alternative scenarios was about 0.4 kmol ha^{-1} for soil Mg and 0.2 kmol ha^{-1} for soil K. The differences between the retrofit and replacement scenarios were negligible for all the cationic nutrients. Silica measurements are planned for the Duke site to further constrain the model's weathering rate estimates. These rates are currently being constrained by sodium concentration differentials and ion ratios.

The projected change in soil solution pH, after 65 years of simulation, for either the constant deposition case or the alternative scenarios at any of the three sites was less than 0.15 pH units.

Summary

The nutrient cycling model, NuCM, was developed as part of the Electric Power Research Institute's Integrated Forest Study. The NuCM model incorporates state-of-the-art understanding of the biogeochemical and transport processes controlling nutrient cycles. The model has been developed to facilitate application to complex cycling problems.

The model simulates vegetation growth, litterfall and decay, soil biogeochemical processes, and moisture routing. Output of the model includes the available nutrients in soil strata and vegetation pools and the fluxes between the pools on a weekly, monthly, or annual basis. Solution and adsorbed concentrations in the various soil layers can be plotted versus time. The model can be used to test hypotheses related to forest growth and to simulate response to changes in atmospheric deposition.

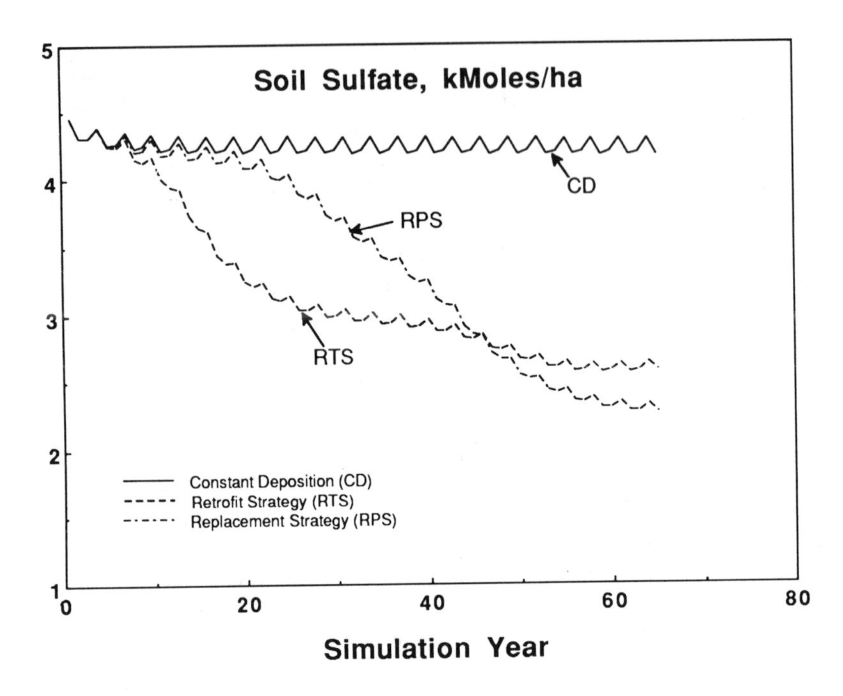

a

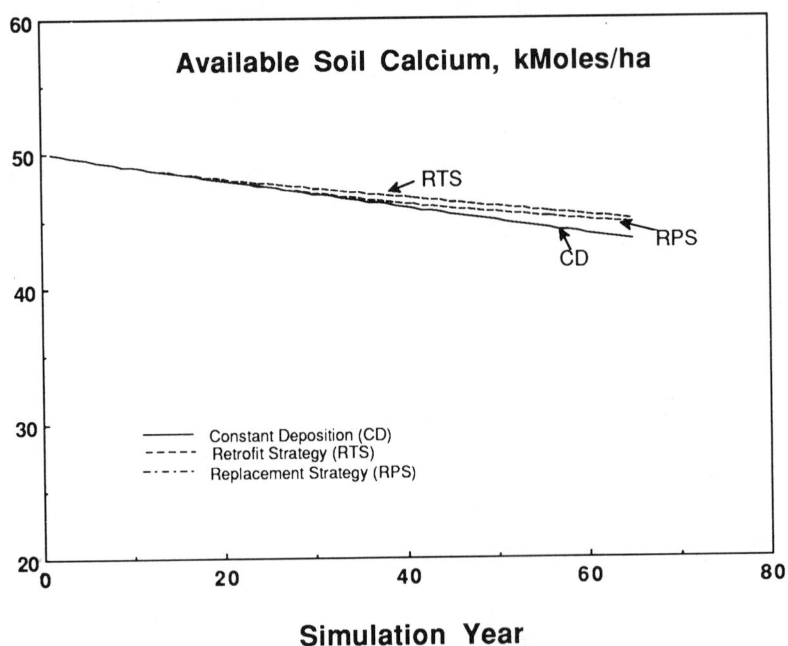

b

Figure 14.20. Comparison of (a) soil sulfate, (b) available soil calcium, (c) soil magnesium, and (d) soil potassium at Duke Forest for base case and alternative strategies (see following page).

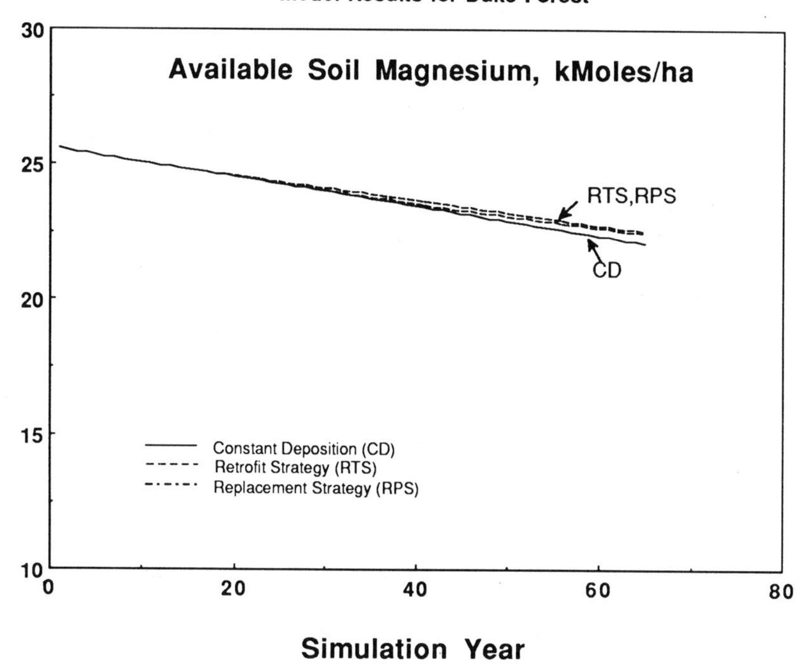

c

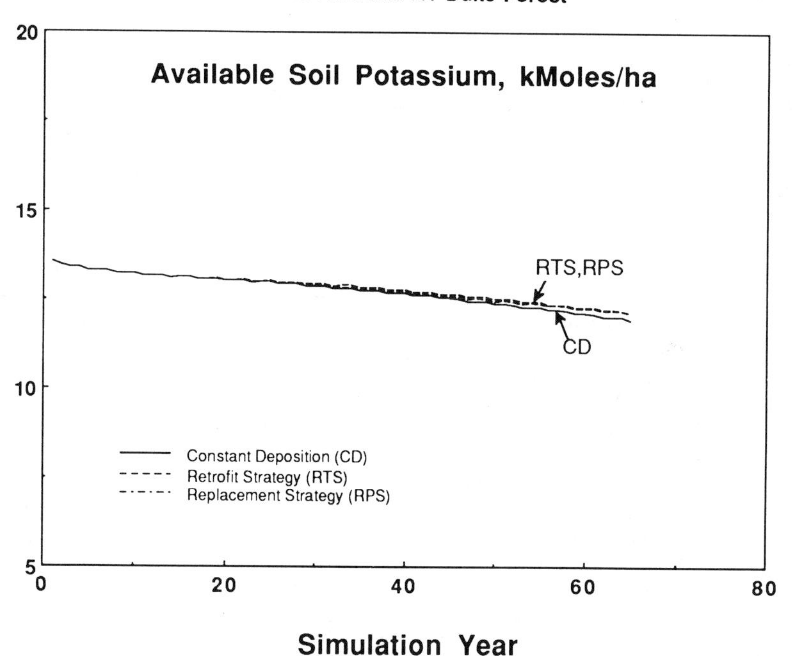

d

Figure 14.20. *Continued.*

The model has been used to simulate effects on nutrient status of acidic deposition at three sites: Huntington Forest, New York; Smokies Tower, Tennessee, and Duke Forest, North Carolina. Model results show only minor changes in nutrient status at the four sites during the next 65 years at current deposition levels. The results also show only small differences in soil nutrient status between the two alternative scenarios for reduction of SO_x emissions. Finally, neither "threshold effects" nor abrupt changes in nutrient pool sizes occurred in any of the simulations.

References

Binkley D., Valentine D., Wells C., Valentine, U. 1989. An empirical model of the factors contributing to 20-yr decrease in soil pH in an old-field plantation of loblolly pine. Biogeochemistry 8:39–54.

Binkley D., Richter. D. 1987. Nutrient cycles and H+ budgets of forest ecosystems. Adv. Ecol. Res. 16:1–51

Boyle J.R., Phillips J.J., Ek A.R. 1973. "Whole-tree" harvesting: nutrient budget evaluation. J. For. 71:760–762.

Chen C.W., Hudson R.J.M., Gherini S.A., Dean J.D., Goldstein R.A. 1983. Acid Rain Model: Canopy Module. J. Environ. Eng. *109*:585–603.

Church M.R., Thornton K.W., Shaffer P.W., Stevens D.L., Rochelle B.P., Holdren G.R., Johnson M.G., Lee J.J., Turner R.S., nCassell D.L., Lammers D.A., Campbell W.G., Liff C.I., Brandt C.C., Liegel L.H., Bishop G.D., Mortenson D.C., Pierson S.S., Schmoyer D.D. Pierson, S.S. D.D. Schmoyer. 1989. Direct/Delayed Response Project: Future Effects of Long-Term Sulfur Deposition on Surface Water Chemistry in the Northeast and Southern Blue Ridge Province. EPA Report EPA/600/3–89/061a, U.S. EPA ORD, Washington, D.C., ERL, Corvallis, Oregon. U.S.

Cole D.W., Rapp M. 1981. Elemental cycling in forest ecosystems. In Reichle D.E. (ed.) Dynamic Properties of Forest Ecosystems. Cambridge University Press, London, pp. 341–409.

Cole D.W., Gessel S.P., Dice S.F. 1968. Distribution and cycling of nitrogen, phosphorus, potassium, calcium in a second-growth Douglas-fir forest. In Young H.E. (ed.) Primary Production and Mineral Cycling in Natural Ecosystems. University of Maine Press, Orono, Maine, pp. 197–213.

Cronan C.S. 1985. Biogeochemical Influence of vegetation and soils in the ILWAS watersheds. Water Air Soil Pollut. 26:354–371.

Curlin J.W. 1970. Nutrient cycling as a factor in site productivity and forest fertilization. In Youngberg C.T., Davey C.R. (eds.) Tree Growth and Forest Soils. Oregon State University Press, Corvallis, pp. 313–326.

David M.B., Mitchell M.J., Nakas J.P. 1982. Organic and inorganic sulfur constituents of a forest soil and their relationship to microbial activity. Soil Sci. Soc. Am. J. 46:847–852.

Engstrom A., Backstrand G., Stenram H. (eds.) 1971. Air Pollution across National Boundaries: The Impact on the Environment of Sulfur in Air and Precipitation. Rep. No. 93, Ministry for Foreign Affairs/Ministry for Agriculture, Stockholm, Sweden.

Federer C.A., Hornbeck J.W., Tritton L.M., Martin C.W., Pierce R.S., Smith, C.T. 1989. Long-term depletion of calcium and other nutrients in eastern U.S. Forests. Environ. Manage. 13:593–601.

Foster N. W., Nicholson J.A., Hazlett P.W. 1989. Temporal variation in nitrate and nutrient cations in drainage waters from a deciduous forest. J. Environ. Qual. 18:238–244.

Fox D.G., Bartuska A., Byrne J., Cowling E., Fisher R., Likens G., Lindberg S., Linthurst R., Messer J., Nichols D. 1989. A Screening Procedure to Evaluate Air Pollution Effects on Class I Wilderness Areas. General Technical Report RM-168, U.S. Department of Agriculture Forest Service, Rocky Mt. Forest and Range Experiment Station. RM-168.

Gessel S.P., Cole D.W., Steinbrenner E.C. 1973. Nitrogen balances in forest ecosystems of the Pacific Northwest. Soil Biol. Biochem. 5:19–34.

Gherini S.A., Mok L., Hudson R.J., Davis G.F., Chen C.W., Goldstein R.A. 1985. The ILWAS Model: formulation and application. Water Air Soil Pollut. 26:425–459.

Gherini S., Munson R., Altwicker E., April R., Chen C., Clesceri N., Cronan C., Driscoll C., Johannes A., Newton R., Peters N., Schofield C. 1989. Regional Integrated Lake-Watershed Acidification Study (RILWAS): Summary of Major Findings. EPRI Report EN-6641, Electric Power Research Institute, Palo Alto, California.

Goldstein et al. 1984. Integrated lake-watershed acidification study (ILWAS): a mechanistic ecosystem analysis. Trans. R. Soc. Lond. B 305:409–425.

Heiberg S.O., White D.P. 1953. Potassium deficiency of reforested pine and spruce stands in northern New York. Soil Sci. Soc. Am. Proc. 15:369–376.

Johnson D.W. 1984. Sulfur cycling in forests. Biogeochemistry 1:29–43.

Johnson D.W., Cole D.W. 1977. Sulfate mobility in an outwashsoil in western Washington. Water Air Soil Pollut. 7:489–495.

Johnson D.W., Henderson G.S. 1979. Sulfate adsorption and sulfur fractions in a highly weathered soil under a mixed deciduous forest. Soil Sci. 128:34–40.

Johnson D. W., Todd D. E. 1983. Some relationships among aluminum, carbon, sulfate in a variety of forest soils. Soil Sci. Soc. Am. J. 47:792–800.

Johnson D.W., Todd D.E. 1987. Nutrient export by leaching and whole-tree harvesting in a loblolly pine and mixed oak forest. Plant Soil 102:99–109.

Johnson D.W., Henderson G.S., Todd D.E. 1988. Changes in nutrient distribution in forests and soils of Walker Branch Watershed over an eleven-year period. Biogeochemistry 5:275–293.

Johnson D.W., Henderson G.S., Todd D.E. 1981. Evidence of modern accumulation of sulfate in an east Tennessee forested Ultisol. Soil Sci. 132:422–426.

Johnson D.W., West D.C., Todd D.E., Mann L.K. 1982a. Effects of sawlog versus whole-tree harvesting on the nitrogen, phosphorus, potassium, calcium budgets of an upland mixed oak forest. Soil Sci. Soc. Am. J. 46:1304–1309.

Johnson D. W., Hornbeck J.W., Kelly J.M, Swank W.T., Todd D.E. 1980. Regional patterns of soil sulfate accumulation: relevance to ecosystem sulfur budgets. In Shriner D.S., Richmond C.R., Lindberg S.E. (eds.) Atmospheric Sulfur Deposition: Environmental Impact and Health Effects. Ann Arbor Science, Ann Arbor, Michigan, pp. 507–520.

Johnson D.W., Richter D.D., Van Miegroet H., Cole D.W., Kelly J.M. 1986c. Sulfur cycling in five forest ecosystems. Water Air Soil Pollut. 30:965–979.

Johnson D.W., Van Miegroet H., Lindberg S.E., Harrison R.B., Todd D.E. Nutrient cycling in red spruce forests of the Great Smoky Mountains. Can. J. For. Res. (in press).

Johnson D.W., Henderson G.S., Huff D.D., Lindberg S.E., Richter D.D., Shriner D.S., Todd D.E., Turner J. 1982b. Cycling of organic and inorganic sulphur in a chestnut oak forest Oecologia 54:141–148.

Lindberg S.E., Garten C.T. 1988. Sources of sulfur in forest canopy throughfall. Nature (London) 336:148–151.

Lindberg S.E., Lovett G.M., Richter D.D., Johnson D.W. 1986. Atmospheric deposition and canopy interactions of major ions in a forest. Science 231:141–145.

Melillo J.M., Aber J.D., Murstore J.F. 1982. Nitrogen and lignin control of hardwood leaf litter decomposition dynamics. Ecology 63:621–623.

Munson R.K., Gherini S.A. 1991. Processes influencing the acid-base chemistry of surface waters. pp. 9–34. In: Acid Deposition and Aquatic Ecosystems: Regional Case Studies. Charles D.F. (ed.) Springer-Verlag, New York.

NADP. 1990. NADP/NTN Annual Data Summary of Precipitation Chemistry in the United States for 1989. United States National Atmospheric Deposition Program, Coordination Office, Natural Resource Ecology Lab, Colorado State University, Fort Collins, Colorado.

National Research Council (NRC). 1983. Atmospheric Processes in Eastern North America. National Academy of Sciences Press, Washington, D.C.

Pastor J., Post W.M. 1986. Influence of climate, soil moisture, succession on forest carbon and nitrogen cycles. Biogeochemistry 2:3–27.

Peters N.E., Driscoll C.T. 1987. Hydrogeologic controls on surface water chemistry in the adirondack region of New York State. Biogeochemistry 3:163–180.

Raynal D.J., Joslin J.D., Thornton F.C., Schadedel M., Henderson G.S. 1990. Sensitivity of tree seedlings to aluminum: III. Red spruce and loblolly pine. J. Environ. Qual. 19:180–187

Reuss J.O. 1976. Chemical/biological relationships relevant to ecological effects of acid rainfall. Water Air Soil Pollut. 7:461–478.

Reuss J.O., Johnson D.W. 1986. Acid Deposition and the Acidification of Soils and Waters. Springer-Verlag, New York.

Rochelle B.P., Church M.R., David M.B. 1987. Sulfur retention at intensively-studied watersheds in the U.S. Canada. Water Air Soil Pollut. 33:73–83.

Scherbatskoy T., Klein R. 1983. Response of spruce and birch foliage to leaching by acid mists. J. Environ. Qual. 12:189–193.

Sisterson D.L., Bowersox V., Olsen A., Meyers T., Vong R. 1990. Deposition Monitoring: Methods and Results. NAPAP Report 6, In Acidic Deposition: State of Science and Technology. National Acid Precipitation Assessment Program, Washington, D.C.

Sollins P., Grier C.C., McCorison F.M., Cromack K., Jr., Fogel R., Fredriksen R.L. 1980. The internal element cycles of an old-growth Douglas-fir ecosystem in western Oregon. Ecol. Monogr. 50:261–285.

Stone E.L., Kszystyniak. R. 1977. Conservation of potassium in the *Pinus resinosa* ecosystem. Science 198:192–194.

Swank W.T., Fitzgerald J.W., Ash J.T. 1984. Microbial transformation of sulfate in forest soils. Science 223:182–184.

Swank W.T., Waide J.B., Crossley D.A. Jr., Todd R.L. 1981. Insect defoliation enhances nitrate export from forest ecosystems. Oecoligia 51:297–299.

Ulrich B. 1980. Production and consumption of hydrogen ions in the ecosphere. In Hutchinson T.C., Havas M. (eds.) Effects of Acid Precipitation on Terrestrial Ecosystems. Plenum Press, New York, pp. 255–282.

Ungs M.J., Boersma L., Yingjajaval S., Klock G.O. 1985. Users Manual for OR-NATURE, the Numerical Analysis of Transport of Water and Solutes Through Soil and Plants.

Van Breemen N., Mulder J., Van Grinsven J.J.M. 1987. Impacts of atmospheric deposition on woodland soils in the netherlands: II. Nitrogen transformations. Soil Sci. Soc. Am. J. 51:1634–1640.

Van Miegroet H., Cole D.W. 1984. The impact of nitrification on soil acidification and cation leaching in red alder ecosystem. J. Environ. Qual. 13:586–590.

Van Miegroet, H., Cole D.W., Homann P.S. 1989. The effect of alder forest cover and alder forest conversion on site fertility and productivity. In Gessel S.P., Lacate D.S., Weetman G.F., Powers R.F. (eds.) Sustained Productivity of Forest Soils, Proceedings of the 7th North American Forest Soils Conference. University of British Columbia, Faculty of Forestry Publication, Vancouver, B.C., Canada, pp. 333–354.

Vitousek P.M., Melillo J.M. 1979. Nitrate losses from disturbed forests: patterns and mechanisms. For. Sci. 25:605–619.

Wood T., Bormann F.H. 1974. The effects of an artificial acid mist upon the growth of *Betula alleghaniensis*. Environ. Pollut. 7:259–268.

Appendix: Nutrient Flux and Content Data from the Integrated Forest Study (IFS) Sites

The following appendix pages provide a summary of the nutrient and flux data from the (IFS) sites. There are four tables for each site: biomass and nutrient content (set 1); atmospheric deposition and soil solution fluxes (set 2); organic matter and nutrient fluxes (set 3); and atmospheric concentrations in precipitation solutions and in air (set 4). We have decided to vary the units, primarily for traditional reasons: for example, those interested in the details of solution fluxes typically use data expressed on an equivalent basis to compare cation and anion fluxes. On the other hand, nutrient content data are not conveniently expressed on an equivalent basis (e.g., what charge does one assign to nitrogen or sulfur in plant tissue?), and are typically expressed on a weight basis. An alternative was to express all data on a molar basis, but we judged that this would burden every user with the tedious chore of converting to more familiar units.

A strong word of caution is in order in regard to some of the biomass flux data. Our confidence in the quality of the data in the biomass flux tables decreases substantially with the position in the table, with the lowest rows of data deserving the least confidence. As specified in the footnotes, we have taken only one of many possible approaches to calculating uptake and translocation; this approach obviously does not always provide a realistic estimate. Specifically, net canopy exchange (NCE, which is equal to throughfall plus stemflow minus deposition) is sometimes very large at those sites with only bulk deposition collectors, in turn causing estimates of uptake to be very large and translocation to become negative (see sulfur at the Smokies Becking site, for example). This is obviously a sampling artifact caused by the inefficiency of bulk collectors in collecting dry deposition, and the true uptake values in these cases are much lower than the values calculated according to the formula used here. Perhaps a more realistic estimate of deposition would be to use throughfall + stemflow (TF + SF), assuming NCE to be (or to approximate) zero in these particular cases. It was not deemed either feasible or desirable to calculate uptake using all possible formulations in these tables; however, the basic data for the use of other formulations is provided, and the users of these data are encouraged to use whichever alternative approach they see fit.

Table CH-1. Biomass and Nutrient Content of the Coweeta Hardwood (CH) Site
Principal Investigators: Wayne T. Swank, Lee J. Reynolds, James M. Vose

Component	Organic Matter	Ca	Mg	K	P	N	S	Na
				(kg ha^{-1})				
OVERSTORY								
FOLIAGE	5,364	28	11	38	5	93	6	—
BRANCH	62,503	206	20	80	11	141	29	—
BOLE	144,461	590	32	159	14	289	66	—
STUMP	—	—	—	—	—	—	—	—
ROOTS	52,432	278	89	167	40	434	—	—
TOTAL	264,760	1,102	152	444	70	957	101	—
UNDERSTORY	24,722	57	11	44	4	68	18	—
FOREST FLOOR								
O HORIZONS	25,291	298	56	76	29	235	25	—
WOOD	2,011	15	0	1	2	10	3.7	—
TOTAL	27,302	312	56	77	31	245	28.7	—
SOIL, EXTRACTABLE								
A (0–13 cm)	—	1,259	116	275	10	—	16	—
BA (13–22 cm)	—	16	30	120	5	—	24	—
Bt (22–67 cm)	—	69	549	576	9	—	364	—
BC (67–89 cm)	—	23	321	289	6	—	236	—
TOTAL	—	1,367	1,016	1,260	30	—	640	—
SOIL, TOTAL[a]								
A (0–13 cm)	62,314	1,609	7,539	16,188	286	1,320	173	—
BA (13–22 cm)	23,584	1,228	7,962	16,260	255	715	134	—
Bt (22–67 cm)	44,262	5,318	40,702	80,175	1,522	1,474	835	—
BC (67–89 cm)	11,388	3,300	29,254	53,995	794	568	282	—
TOTAL	141,548	11,455	85,457	166,618	2,857	4,077	1,424	—

[a] Soil organic matter, soil carbon × 2.

Table CP-1. Biomass, Nutrient Content of the Coweeta Pine (CP) Site
Principal Investigators: Wayne T. Swank, James M. Vose

Component	Organic Matter	Ca	Mg	K	P	N	S	Na
				(kg ha^{-1})				
OVERSTORY								
FOLIAGE	9,500	30	12	43	13	114	14	—
BRANCH	38,028	117	16	53	13	147	25	—
BOLE	145,350	97	31	111	36	190	104	—
STUMP	—	—	—	—	—	—	—	—
ROOTS	60,296	204	57	163	35	422	N.D.	—
TOTAL	253,174	448	116	370	97	873	143	—
UNDERSTORY								
FOREST FLOOR								
O HORIZONS	23,384	180	28	48	28	263	26	—
WOOD	4,739	13	0	12	4	19	1.8	—
TOTAL	28,123	193	28	60	32	282	27.8	—
SOIL, EXTRACTABLE								
A (0–13 cm)	—	289	36	76	8	—	5	—
BA (13–22 cm)	—	57	59	80	7	—	18	—
Bt (22–67 cm)	—	287	55	110	6	—	138	—
BC (67–89 cm)	—	34	163	123	13	—	146	—
TOTAL	—	667	313	389	34	—	307	—
SOIL, TOTAL[a]								
A (0–9 cm)	27,254	767	5,204	15,828	211	776	98	—
BA (9–28 cm)	19,552	412	12,028	35,654	668	643	235	—
Bt (28–62 cm)	22,060	1,371	20,070	66,721	629	869	299	—
BC (62–91 cm)	7,872	1,221	28,375	60,989	755	210	88	—
TOTAL	76,738	3,771	65,677	179,192	2,263	2,498	720	—

[a] Soil organic matter, soil carbon × 2.

Table DF-1. Biomass, Nutrient Content of the Douglas Fir (DF) Site
Principal Investigators: Dale W. Cole, Helga Van Miegroet, Peter Homann

Component	Organic Matter	Ca	Mg	K	P	N	S	Na
				(kg ha^{-1})				
OVERSTORY								
FOLIAGE	6,180	38	7	38	19	92	9	0.2
BRANCH	22,900	86	8	39	10	68	5	1
BOLE	242,200	100	8	23	10	149	11	5
STUMP								
ROOTS	44,500	68	5	31	8	43	7	3
TOTAL	315,780	292	28	131	47	352	32	9
UNDERSTORY	3,050	25	6	29	4	30	5	0.5
FOREST FLOOR								
O HORIZONS	23,500	182	23	28	24	295	29	12
WOOD[a]	13,200	45	8	9	7	84	9	6
TOTAL	36,700	227	31	37	31	379	38	18
SOIL, EXTRACTABLE								
A (0–7 cm)	—	429	36	74	60	—	8	17
A (7–15 cm)	—	210	18	52	4	—	16	16
B (15–30 cm)	—	201	21	70	5	—	61	8
BC (30–45 cm)	—	253	25	67	3	—	121	76
TOTAL	—	1,093	100	263	72	—	206	117
SOIL, TOTAL[b]								
A (0–7 cm)	61,800	6,000	3,200	2,800	623	1,200	123	7,600
A (7–15 cm)	40,400	7,000	4,100	3,600	596	860	92	9,300
B (15–30 cm)	55,800	13,800	8,400	6,900	1,168	1,450	184	17,900
BC (30–45 cm)	60,400	19,600	13,100	10,500	1,315	1,570	259	24,400
TOTAL	218,400	46,400	28,800	23,800	3,702	5,080	658	59,200

[a] Includes only wood <6 cm diameter.
[b] Soil organic matter, soil carbon × 2.

Table DL-1. Biomass, Nutrient Content of the Duke (DL) Site
Principal Investigator: D. Binkley

Component	Organic Matter	Ca	Mg	K (kg ha^{-1})	P	N	S	Na
OVERSTORY								
FOLIAGE	7,624	21	7	46	11	83	8	—
BRANCH	15,590	48	7	23	5	36	3	—
BOLE[a]	78,180	59	17	61	10	86	27	—
STUMP	—	—	—	—	—	—	—	—
ROOTS	25,670	19	8	20	6	29	10	—
TOTAL	127,064	147	39	150	32	234	48	—
UNDERSTORY	54	0.3	0.1	0.8	0.1	0.4	0.1	—
FOREST FLOOR	—	—	—	—	—	—	—	—
O HORIZONS	63,000	145	18	50	71	450	52	—
WOOD	0	0	0	0	0	0	0	—
TOTAL	63,000	145	18	50	71	450	52	—
SOIL, EXTRACTABLE								
Ap (0–20 cm)	—	28	8	44	115	—	N.D.	10
E (20–35 cm)	—	377	81	37	33	—	N.D.	5
Bt1 (35–60 cm)	—	766	174	162	25	—	N.D.	11
Bt2 (60–80 cm)	—	860	278	191	10	—	N.D.	16
TOTAL	—	2,125	541	434	42	—	N.D.	42
SOIL, TOTAL[b]								
Ap (0–20 cm)	27,786	1,178	657	4,228	604	829	38	544
E (20–35 cm)	11,988	1,147	1,761	7,770	233	849	111	984
Bt1 (35–60 cm)	18,400	1,464	3,944	10,880	206	0	712	3,968
Bt2 (60–80 cm)	11,522	1,152	4,400	12,320	211	0	944	5,232
TOTAL	69,696	4,941	10,762	35,198	1,254	1,678	1,805	10,728

[a] Included in bole and roots.
[b] Organic matter, soil carbon × 2.

Table FL-1. Biomass, Nutrient Content of the Findley Lake (FL) Site
Principal Investigator: Dale W. Cole

Component	Organic Matter	Ca	Mg	K	P	N	S	Na
				kg ha^{-1}				
OVERSTORY								
FOLIAGE	24,400	130	22	157	37	271	26	1
BRANCH	75,700	268	24	108	25	156	10	4
BOLE	463,900	892	63	326	47	410	37	23
STUMP	—	—	—	—	—	—	—	—
ROOTS	149,000	496	44	230	58	228	26	10
TOTAL	713,000	1,786	153	821	167	1,065	99	38
UNDERSTORY	660	2	1	6	1	8	1	0
FOREST FLOOR								
O HORIZONS	—	—	—	—	—	—	—	—
WOOD	—	—	—	—	—	—	—	—
TOTAL	54,800	160	47	84	60	575	59	80
SOIL, EXTRACTABLE								
E (0–12 cm)	—	45	9	42	8	—	7	0
Bhs (12–17 cm)	—	37	9	44	2	—	4	9
Bs (17–41 cm)	—	86	18	96	4	—	100	38
BC (41–71 cm)	—	31	6	37	4	—	85	0
TOTAL	—	199	42	219	18	—	196	47
SOIL, TOTAL[a]								
E (0–12 cm)	33,400	37,200	12,000	5,400	133	790	83	35,400
Bhs (12–17 cm)	40,000	5,700	2,100	1,700	145	690	79	7,200
Bs (17–41 cm)	357,600	15,200	15,400	12,400	1,486	4,090	786	31,700
BC (41–71 cm)	127,800	15,800	17,700	19,900	1,152	2,050	454	45,600
TOTAL	558,800	73,900	47,200	39,400	2,916	7,620	1,402	119,900

[a] Soil organic matter, soil carbon × 2.

Table FS-1. Biomass, Nutrient Content of the Florida (FS) Site
Principal Investigator: H.L. Gholz

Component	Organic Matter	Ca	Mg	K	P	N	S	Na
				(kg ha^{-1})				
OVERSTORY								
FOLIAGE	4,801	9	6	11	4	37	3	—
BRANCH	7,414	17	5	6	2	18	2	—
BOLE	94,126	91	26	27	11	119	13	—
STUMP	—	—	—	—	—	—	—	—
ROOTS	29,789	47	16	11	8	122	13	—
TOTAL	136,130	164	53	55	25	296	31	—
UNDERSTORY	16,011	37	17	13	10	166	21	—
FOREST FLOOR								
O HORIZONS	31,700	98	22	9	10	263	37	—
WOOD	5,700	24	3	1	1	23	3	—
TOTAL	37,400	122	25	10	11	286	40	—
SOIL, EXTRACTABLE								
A (0–14 cm)	—	76	22	17	5	—	0.1	5
E (14–24 cm)	—	16	4	0	2	—	0.2	4
Bh (24–36 cm)	—	38	7	0	4	—	0.1	13
B/E (36–60 cm)	—	51	9	0	7	—	0.1	18
E′ (60–95)	—	35	7	0	9	—	0.1	13
Bt (95–120+)	—	27	132	17	16	—	5	40
TOTAL	—	243	181	34	43	—	5.6	93
SOIL, TOTAL[a]								
A (0–14 cm)	49,100	221	184	328	184	92	79	240
E (14–24 cm)	5,276	37	34	42	19	69	75	29
Bh (24–36 cm)	36,348	68	105	117	99	236	48	47
B/E (36–60 cm)	40,470	117	184	213	108	501	87	85
E′ (60–95)	8,186	141	227	274	18	770	112	108
Bt (95–120+)	10,450	174	2,316	1,531	108	1,182	98	270
TOTAL	149,830	758	3,050	2,505	536	2,850	499	779

[a] Soil organic matter, soil carbon × 2.

Table GL-1. Biomass, Nutrient Content of the Georgia (GL) Site
Principal Investigators: H.L. Ragsdale, J. Dowd

Component	Organic Matter	Ca	Mg	K	P	N	S	Na
				(kg ha^{-1})				
OVERSTORY								
FOLIAGE	5,428	11	6	19	6	72	4	—
BRANCH	11,270	24	5	13	3	43	4	—
BOLE	129,418	240	45	110	12	360	41	—
STUMP	—	—	—	—	—	—	—	—
ROOTS	28,103	50	26	50	11	78	18	—
TOTAL	174,219	325	82	192	32	553	67	—
UNDERSTORY	18,392	129	19	39	6	94	4	—
FOREST FLOOR								
O HORIZONS	—	—	—	—	—	—	—	—
WOOD	—	—	—	—	—	—	—	—
TOTAL	32,713	125	39	30	21	310	22	—
SOIL, EXTRACTABLE								
A (0–20 cm)	—	677	234	145	8	—	34	12
A/B (20–40 cm)	—	481	157	199	3	—	267	49
Bt (40–80 cm)	—	602	685	736	6	—	837	125
BC (80+ cm)	—	597	635	738	2	—	N.D.	125
TOTAL	—	2,357	1,711	1,818	311	—	1,138	311
SOIL, TOTAL[a]								
A (0–20 cm)	118,634	5,600	2,225	22,259	520	3,821	242	2,266
A/B (20–40 cm)	75,616	4,489	2,091	32,061	328	2,160	578	2,889
Bt (40–80 cm)	57,216	15,497	9,533	53,939	1,121	3,612	1,063	9,105
BC (80+ cm)	52,096	7,288	6,398	46,257	1,185	2,996	N.D.	7,472
TOTAL	303,562	32,874	20,247	154,516	3,154	12,589	1,883	21,732

[a] Soil organic matter, soil carbon × 2.

Table HF-1. Biomass, Nutrient Content of the Huntington Forest (HF) Site
Principal Investigators: M.J. Mitchell, D.J. Raynal, E.H. White, R. Briggs, J. Shepard, T. Scott, M. Burke, J. Porter

Component	Organic Matter	Ca	Mg	K (kg ha^{-1})	P	N	S	Na
OVERSTORY								
FOLIAGE	3,545	22	6	27	4	77	6	0.3
BRANCH	42,537	311	20	67	14	170	15	3
BOLE	146,155	521	38	127	12	218	25	13
STUMP	—	—	—	—	—	—	—	—
ROOTS[a]	7,659	35	5	14	4	66	5	1
TOTAL	—	—	—	—	—	—	—	—
UNDERSTORY	847	3	1	6	1	9	1	0.1
FOREST FLOOR								
O HORIZONS	—	—	—	—	—	—	—	—
WOOD	—	—	—	—	—	—	—	—
TOTAL	44,923	301	48	131	68	1,152	70	6
SOIL, EXTRACTABLE								
A (0–4 cm)	—	261	17	25	N.D.[b]	—	1	5
E (4–8 cm)	—	52	4	7	N.D.	—	1	3

continued

Table HF-1. *Continued*

Component	Organic Matter	Ca	Mg	K (kg ha^{-1})	P	N	S	Na
Bh (8–12 cm)	—	78	6	11	N.D.	—	2	5
Bhs (12–20 cm)	—	78	5	10	N.D.	—	16	8
Bs1 (20–31)	—	54	3	8	N.D.	—	16	8
Bs2 (31–44)	—	43	2	7	N.D.	—	17	8
Bs3 (44–58)	—	40	1	5	N.D.	—	9	11
TOTAL	—	606	38	73	N.D.	—	62	48
SOIL, TOTAL[c]								
A (0–4 cm)	68,506	1,894	597	3,305	173	1,424	108	5,120
E (4–8 cm)	14,754	4,345	1,767	18,341	148	340	49	15,359
Bh (8–12 cm)	36,272	4,005	1,509	9,934	306	853	96	11,110
Bhs (12–20 cm)	86,124	9,579	3,700	17,205	892	1,522	209	16,832
Bs1 (20–31)	84,954	23,808	7,921	32,770	1,455	1,601	283	41,502
Bs2 (31–44)	73,278	30,433	9,958	43,761	1,528	1,192	246	42,849
Bs3 (44–58)	175,694	49,984	14,937	64,736	2,168	1,053	188	77,201
TOTAL	539,582	124,048	40,389	190,052	6,670	7,985	1,179	48,421

[a] Fine roots only.
[b] N.D., not determined.
[c] Soil organic matter, soil carbon × 2.

Table LP-1. Biomass, Nutrient Content of the Oak Ridge Loblolly Pine (LP) Site
Principal Investigators: D.W. Johnson, S.E. Lindberg

Component	Organic Matter	Ca	Mg	K (kg ha^{-1})	P	N	S	Na
PLOT 1								
OVERSTORY								
FOLIAGE	2,950	5	2	18	4	30	4	—
BRANCH	15,380	21	5	12	2	22	9	—
BOLE	98,800	70	19	66	10	89	59	—
STUMP	—	—	—	—	—	—	—	—
ROOTS	—	—	—	—	—	—	—	—
TOTAL	117,130	97	26	96	16	140	72	—
UNDERSTORY	17,950	65	8	44	5	52	14	—
FOREST FLOOR								
O HORIZONS	9,754	157	24	43	22	273	13	—
WOOD	4,380	3	1	3	1	4	2	—
TOTAL	14,134	160	25	46	23	277	15	—
SOIL, EXTRACTABLE								
Ap (0–13 cm)	—	2,227	361	852	83	75	75	39
Bw1 (13–28 cm)	—	2,574	504	1,112	60	93	93	31
Bw2 (28–41 cm)	—	1,617	338	668	16	64	64	15
Bw3 (41–45 cm)	—	479	100	223	2	19	19	6
TOTAL	—	6,897	1,303	2,856	161	251	250	91
SOIL, TOTAL[a]								
Ap (0–4 cm)	57,668	2,663	9,302	40,514	1,816	2,428	422	13,222
Bw1 (4–7 cm)	29,104	3,138	14,786	60,520	1,813	1,824	479	13,062
Bw2 (7–15 cm)	10,764	2,210	10,676	44,298	1,010	847	186	8,899
Bw3 (15–57 cm)	4,842	554	3,485	17,671	346	156	68	3,522
TOTAL	102,378	8,565	38,249	163,003	4,985	5,257	1,156	38,705

continued

Table LP-1. *Continued*

Component	Organic Matter	Ca	Mg	K (kg ha^{-1})	P	N	S	Na
PLOT 2								
OVERSTORY								
FOLIAGE	3,020	5	2	19	4	31	4	—
BRANCH	15,250	21	5	12	2	21	9	—
BOLE	101,460	72	19	68	10	91	61	—
STUMP	—	—	—	—	—	—	—	—
ROOTS	—	—	—	—	—	—	—	—
TOTAL	119,730	99	26	99	16	143	74	—
UNDERSTORY	17,770	90	10	50	6	61	25	—
FOREST FLOOR								
O HORIZONS	9,754	157	24	43	22	273	13	—
WOOD	4,380	3	1	3	1	4	3	—
TOTAL	14,134	160	25	46	23	277	16	—
SOIL, EXTRACTABLE								
Ap (0–17 cm)	—	928	91	174	184	—	20	6
Bw1 (17–33 cm)	—	761	98	157	171	—	13	17
Bw2 (33–83 cm)	—	3,771	292	455	254	—	201	36
TOTAL	—	5,460	481	786	609	—	234	59
SOIL, TOTAL[a]								
Ap (0–17 cm)	51,648	938	3,304	21,999	1,453	1,739	255	6,060
Bw1 (17–33 cm)	39,942	1,040	3,562	21,999	1,319	1,423	197	5,499
Bw2 (33–83 cm)	45,814	4,966	13,200	69,785	6,736	2,607	621	23,524
TOTAL	137,404	6,944	20,066	113,783	9,508	5,769	1,073	35,083

[a] Soil organic matter, soil carbon × 2.

Table MS-1. Biomass, Nutrient Content of the Maine (MS) Site
Principal Investigator: I. Fernandez

Component	Organic Matter	Ca	Mg	K	P	N	S	Na
				(kg ha^{-1})				
OVERSTORY								
FOLIAGE	17923	80	17	123	27	207	N.D.	—
BRANCH	22044	88	10	29	11	68	N.D.	—
BOLE	124730	292	38	97	28	191	N.D.	—
ROOT/STUMP	42319	127	80	51	22	95	N.D.	—
TOTAL	207016	587	145	300	88	561	—	—
UNDERSTORY N.D.[a]	N.D.	N.D.	N.D.	N.D.	N.D.	N.D.	—	—
FOREST FLOOR								
O HORIZONS	146940	368	88	175	87	1121	148	—
WOOD	3752	6	1	2	1	2	N.D.	—
TOTAL	150692	374	89	177	88	1123	N.D.	—
SOIL, EXTRACTABLE								
O (11–0 cm)	—	225	40	68	2	—	14	12
E (0–8 cm)	—	13	5	9	2	—	12	49
B (8–13 cm)	—	8	2	6	1	—	35	125
B (13–28 cm)	—	15	5	17	2	—	296	125
B (28–52)	—	27	6	20	7	—	348	N.D.
TOTAL	—	288	58	120	14	—	705	311
SOIL, TOTAL[b]								
O (11–0 cm)	146940	368	88	175	87	1121	148	28
E (0–8 cm)	16952	3878	938	3065	400	209	25	4317
B (8–13 cm)	21049	4554	2360	4006	510	283	76	6421
B (13–28 cm)	80437	25152	13031	22122	2818	961	424	35456
B (28–52)	40158	25662	13274	22535	2871	484	246	36118
TOTAL	305536	59614	29691	51903	6686	3058	919	82340

[a] N.D., not determined
[b] Soil organic matter, soil carbon × 2.

Table NS-1. Biomass, Nutrient Content of the Norway (NS) Site
Principal Investigators: Ingvald Rösberg, Arne O. Stuanes

Component	Organic Matter	Ca	Mg	K	P	N	S	Na
				(kg ha^{-1})				
PLOT R1								
OVERSTORY								
FOLIAGE	16,230	78	14	74	29	167	13	—
BRANCH	14,110	45	5	43	12	94	7	—
BOLE	83,287	111	17	124	13	94	8	—
STUMP	9,150	16	1	10	1	14	1	—
ROOTS	31,020	31	6	29	9	61	5	—
TOTAL	153,797	281	43	280	64	430	34	—
UNDERSTORY	2,973	10	2	16	6	46	4	—
FOREST FLOOR								
O HORIZONS	63,285	175	53	112	70	842	82	—
WOOD	0	0	0	0	0	0	0	—
TOTAL	63,285	175	53	112	70	842	82	—
SOIL, EXTRACTABLE								
O (5–0 cm)	—	121	14	55	5	—	3	2
E (0–1 cm)	—	15	3	7	2	—	0.5	1
Bs (1–18 cm)	—	35	5	46	86	—	22	8
BC (18–32 cm)	—	11	3	27	258	—	74	8
C (32–41 cm)	—	5	1	10	113	—	16	3
IIC (41+)	—	7	2	30	161	—	12	3
TOTAL	—	194	28	175	625	—	128	25
SOIL, TOTAL[a]								
E (0–1 cm)	10,652	316	267	774	N.D.[b]	207	—	667
Bs (1–18 cm)	60,868	7,527	9,291	16,768	—	1,482	—	13,404
BC (18–32 cm)	27,020	10,115	17,387	25,822	—	1,103	—	20,243
C (32–41 cm)	6,732	3,676	7,684	12,279	—	366	—	8,897
IIC (41+)	8,234	7,406	13,846	23,577	—	577	—	16,569
TOTAL	113,506	29,040	48,475	79,220	—	3,735	—	59,780

PLOT R2								
OVERSTORY								
FOLIAGE	14,950	85	12	67	23	146	11	—
BRANCH	11,902	42	4	35	9	73	5	—
BOLE	34,895	49	8	54	6	45	4	—
STUMP	3,118	7	1	4	1	6	1	—
ROOTS	28,930	25	5	25	8	49	4	—
TOTAL	93,795	208	30	185	47	319	25	—
UNDERSTORY	2,657	13	2	18	4	38	3	—
FOREST FLOOR								
O HORIZONS	57,750	180	49	100	54	605	62	—
WOOD	0	0	0	0	0	0	0	—
TOTAL	57,750	180	49	100	54	605	62	—
SOIL, EXTRACTABLE								
O (7–0 cm)	—	112	13	58	5	—	4	2
E (0–1 cm)	—	12	3	10	3	—	1	1
Bs (1–26 cm)	—	30	4	66	83	—	46	10
BC (26–46 cm)	—	12	2	31	263	—	49	8
C (46+ cm)	—	6	1	10	77	—	7	3
TOTAL	—	172	23	175	431	—	107	24
SOIL, TOTAL[a]								
E (0–1 cm)	13,490	519	438	1,270	N.D.	223	—	1,093
Bs (1–26 cm)	77,140	10,804	13,335	24,069	—	1,797	—	19,241
BC (26–46 cm)	21,148	9,593	16,491	24,490	—	878	—	19,200
C (46+ cm)	4,460	3,032	6,337	10,127	—	238	—	7,338
TOTAL	116,238	23,948	36,601	59,956	—	3,136	—	46,872

[a] Soil organic matter, soil carbon × 2.
[b] N.D., not determined.

Table RA-1. Biomass, Nutrient Content of the Red Alder (RA) Site.
Principal Investigators: Dale W. Cole, Helga Van Miegroet, Peter Homann

Component	Organic Matter	Ca	Mg	K ($kg\ ha^{-1}$)	P	N	S	Na
OVERSTORY								
FOLIAGE	3,490	23	6	43	10	116	7	0.2
BRANCH	29,400	96	11	91	15	169	10	0.7
BOLE	193,000	403	40	189	21	545	27	9
STUMP	—	—	—	—	—	—	—	—
ROOTS	45,800	134	12	74	3	167	14	4
TOTAL	262,690	656	69	397	49	997	58	14
UNDERSTORY	6,010	24	11	109	8	87	9	1
FOREST FLOOR								
O HORIZONS	79,600	666	86	112	80	1,528	140	38
WOOD[a]	18,900	165	11	8	5	101	8	2
TOTAL	98,500	831	97	120	85	1,629	148	40
SOIL, EXTRACTABLE								
A (0–7 cm)	—	151	13	39	1	—	6	16
A (7–15 cm)	—	91	7	27	1	—	8	13
B (15–30 cm)	—	242	18	35	1	—	15	12
BC (30–45 cm)	—	420	36	44	1	—	15	8
TOTAL	—	904	74	145	4	—	44	49
SOIL, TOTAL[b]								
A (0–7 cm)	60,400	3,800	2,300	1,900	290	1,740	165	4,900
A (7–15 cm)	36,400	4,400	3,100	2,200	275	1,150	115	6,200
B (15–30 cm)	68,800	9,900	6,600	4,900	602	2,160	194	13,500
BC (30–45 cm)	64,000	11,300	8,300	6,400	690	2,000	192	16,900
TOTAL	229,600	29,400	20,300	15,400	1,857	7,050	666	41,500

[a] Includes only wood <6 cm diameter.
[b] Soil organic matter, soil carbon × 2.

Table SB-1. Biomass, Nutrient Content of the Smokies Beech (SB) Site
Principal Investigators: D.W. Johnson, S.E. Lindberg

Component	Organic Matter	Ca	Mg	K ($kg\ ha^{-1}$)	P	N	S	Na
PLOT 1								
OVERSTORY								
FOLIAGE	2,440	12	3	20	4	59	5	—
BRANCH	77,820	113	14	75	5	106	39	—
BOLE	99,990	180	21	101	8	141	44	—
STUMP	—	—	—	—	—	—	—	—
ROOTS	23,050	68	20	65	36	185	8	—
TOTAL	203,300	372	59	261	54	491	—	
UNDERSTORY	10,510	31	8	66	7	71	6	—
FOREST FLOOR								
O HORIZONS	3,568	9	1	4	3	28	3	—
WOOD	25,500	113	12	17	20	204	20	—
TOTAL	29,068	122	13	21	23	232	23	—
SOIL, EXTRACTABLE								
A (0–13 cm)	—	58	19	160	20	—	11	2
A/B (13–21 cm)	—	10	6	94	9	—	19	1
Bw (21–46 cm)	—	25	17	386	6	—	206	4
Bw2 (46–65 cm)	—	11	8	309	4	—	319	3
TOTAL	—	104	50	949	39	—	555	10
SOIL, TOTAL								
A (0–13 cm)	33,922	428	1,572	7,202	774	2,566	151	3,050
A/B (13–21 cm)	21,918	195	1,568	6,666	629	1,125	179	3,472
Bw (21–46 cm)	44,944	1,005	7,592	35,102	2,338	3,318	804	14,978
Bw2 (46–65 cm)	24,626	928	7,129	23,209	1,481	2,044	437	11,384
TOTAL	125,410	2,556	17,861	72,179	5,222	9,053	1,571	32,884

continued

Table SB-1. *Continued*

Component	Organic Matter	Ca	Mg	K (kg ha^{-1})	P	N	S	Na
PLOT 2								
OVERSTORY1								
FOLIAGE	2,120	10	3	17	4	52	4	—
BRANCH	68,020	101	13	65	5	93	37	—
BOLE	81,850	147	18	83	6	116	36	—
STUMP	—	—	—	—	—	—	—	—
ROOTS	8,490	55	11	22	7	72	8	—
TOTAL	—	—	—	—	—	—	—	—
UNDERSTORY	1,540	10	4	36	4	33	4	—
FOREST FLOOR								
O HORIZONS	—	—	—	—	—	—	—	—
WOOD	—	—	—	—	—	—	—	—
TOTAL	28,810	128	15	23	15	247	25	—
SOIL, EXTRACTABLE								
A (0–10 cm)	—	139	26	66	20	—	24	4
Bw1 (10–20 cm)	—	17	9	28	12	—	35	2
BC (10–40 cm)	—	17	12	62	31	—	225	0
TOTAL	—	173	47	156	63	—	284	6
SOIL, TOTAL[a]	—	346	94	312	126	—	568	12
Ap (0–17 cm)	73,460	192	992	4,802	553	2,545	244	2,104
Bw1 (17–33 cm)	34,622	223	1,825	8,630	661	1,477	271	3,673
Bw2 (33–83 cm)	60,570	689	6,178	27,295	1,513	3,036	634	10,025
TOTAL	168,652	1,104	8,995	40,727	2,727	7,058	1,149	15,802

[a] Soil organic matter, soil carbon × 2.

Table SS-1. Biomass, Nutrient Content of the Smokies Becking (SS) Site
Principal Investigators: D.W. Johnson, S.E. Lindberg

Component	Organic Matter	Ca	Mg	K	P	N	S	Na
				(kg ha^{-1})				
OVERSTORY								
FOLIAGE	8,130	20	4	35	7	82	9	—
BRANCH	24,860	69	6	14	3	32	12	—
BOLE	190,330	245	24	62	8	126	110	—
STUMP	—	—	—	—	—	—	—	—
ROOTS	34,570	150	12	42	9	95	22	—
TOTAL	257,890	484	46	154	27	336	154	—
UNDERSTORY	6,000	21	5	28	7	50	7	—
FOREST FLOOR								
O HORIZONS	108,790	125	24	26	213	1,738	90	—
WOOD	57,300	44	1	17	25	178	9	—
TOTAL	166,090	169	25	43	238	1,916	99	—
SOIL, EXTRACTABLE								
Oa (5–0 cm)	—	4	11	15	—	—	3	2
A (0–3 cm)	—	13	5	15	1	—	8	1
A/B (3–13 cm)	—	24	12	33	2	—	17	3
Bw1 (13–29 cm)	—	10	7	27	3	—	14	2
Bw2 (29–46 cm)	—	6	4	29	4	—	15	2
TOTAL	—	57	39	119	10	—	57	10
SOIL, TOTAL[a]								
Oa (5–0 cm)	61,780	35	68	396	58	978	119	203
A (0–3 cm)	13,668	85	272	1,420	79	878	85	951
A/B (3–13 cm)	33,410	418	1,638	8,982	284	2,006	209	6,952
Bw1 (13–29 cm)	16,710	670	3,642	15,501	300	967	189	8,761
Bw2 (29–46 cm)	12,222	1,270	9,116	21,838	793	528	670	9,308
TOTAL	137,790	2,478	14,737	48,137	1,514	5,357	1,272	26,175

[a] Soil organic matter, soil carbon × 2.

Table ST-1. Biomass, Nutrient Content of the Smokies Tower (ST) Site
Principal Investigators: D.W. Johnson, S.E. Lindberg

Component	Organic Matter	Ca	Mg	K	P	N	S	Na
				$(kg\ ha^{-1})$				
PLOT 1								
OVERSTORY								
FOLIAGE	9,750	32	4	60	9	120	12	—
BRANCH	33,150	90	8	17	4	55	24	—
BOLE	209,470	284	29	65	9	154	108	—
STUMP	—	—	—	—	—	—	—	—
ROOTS	12,270	30	6	16	5	110	9	—
TOTAL	264,640	436	47	158	28	439	152	—
UNDERSTORY	8,700	28	8	45	8	85	9	—
FOREST FLOOR								
O HORIZONS	134,800	274	54	103	99	2080	292	—
WOOD	16,300	32	3	4	4	41	13	—
TOTAL	151,100	306	57	107	103	2165	305	—
SOIL, EXTRACTABLE								
O (cm)	—	—	—	—	—	—	—	—
A (0–9 cm)	—	31	21	61	4	—	32	2
A/B (9–21 cm)	—	31	18	61	4	—	38	2
Bw1 (21–38 cm)	—	34	15	66	5	—	69	5
Bw3 (49–55 cm)	—	20	7	39	1	—	48	2
Bw2 (38–49 cm)	—	9	1	18	2	—	12	1
TOTAL	—	125	62	245	16	—	199	12
SOIL, TOTAL								
O (cm)	—	—	—	—	—	—	—	—
A (0–9 cm)	46,702	456	2,094	—	359	2,900	166	4,749
A/B (9–21 cm)	51,694	747	3,901	—	533	2,751	362	7,545
Bw1 (21–38 cm)	47,936	952	5,886	—	590	1,911	542	8,701
Bw3 (49–55 cm)	26,604	751	4,826	—	337	1,184	288	9,589
Bw2 (38–49 cm)	8,418	339	3,568	—	157	296	135	5,230
TOTAL	181,354	3,245	20,275	—	1,976	9,042	1,493	35,814

PLOT 2								
OVERSTORY								
FOLIAGE	8,650	27	3	40	7	83	10	—
BRANCH	28,950	93	7	15	4	47	18	—
BOLE	223,730	303	29	63	9	138	110	—
STUMP	—	—	—	—	—	—	—	—
ROOTS	47,070	208	20	46	13	220	17	—
TOTAL	308,400	631	60	164	33	488	155	—
UNDERSTORY	5,800	21	5	36	5	48	5	—
FOREST FLOOR								
O HORIZONS	88,900	206	27	198	73	1,440	196	—
WOOD	20,000	28	3	4	3	77	11	—
TOTAL	108,900	234	30	202	76	1,517	207	—
SOIL, EXTRACTABLE								
OA (cm)	—	66	7	0	3	—	3	1
A11 (0–3 cm)	—	31	5	12	3	—	4	2
A12 (3–8 cm)	—	8	4	11	3	—	6	2
AB (8–13 cm)	—	10	6	11	2	—	10	1
Bw1 (13–23 cm)	—	12	8	17	2	—	31	3
Bw2 (23–64 cm)	—	23	25	79	40	—	185	14
TOTAL	—	150	55	130	53	—	239	23
SOIL, TOTAL[a]								
OA (cm)	62,200	32	30	204	46	1,063	118	137
A11 (0–3 cm)	39,186	101	269	1,845	83	1,054	130	1,105
A12 (3–8 cm)	22,768	191	920	5,804	180	540	59	3,697
AB (8–13 cm)	25,118	259	1,194	5,804	201	721	73	4,114
Bw1 (13–23 cm)	35,352	472	2,979	10,584	386	985	112	7,312
Bw2 (23–64 cm)	81,258	2,429	18,478	49,177	1,524	2,874	500	31,935
TOTAL	265,882	3,484	23,870	73,418	2,420	7,237	992	48,300

[a] Soil organic matter, soil carbon × 2.

Table TL-1. Biomass, Nutrient Content of the Turkey Lakes (TL) Site
Principal Investigators: I.K. Morrison, N.W. Foster

Component	Organic Matter	Ca	Mg	K	P	N	S	Na
				(kg ha^{-1})				
OVERSTORY								
FOLIAGE	4,176	33	5	46	4	94	10	—
BRANCH	50,410	337	15	141	11	174	18	—
BOLE	154,407	590	28	148	14	246	19	—
STUMP	—	—	—	—	—	—	—	—
ROOTS	35,067	119	8	30	3	52	4	—
TOTAL	244,060	1,079	56	365	32	566	51	—
UNDERSTORY	1,280	9	1	6	1	14	1	—
FOREST FLOOR								
O HORIZONS	—	—	—	—	—	—	—	—
WOOD	—	—	—	—	—	—	—	—
TOTAL	33,004	282	49	55	32	927	90	—
SOIL, EXTRACTABLE								
E (0–4 cm)	—	101	10	17	N.D.[a]	—	N.D.	2
Bhs1 (4–7 cm)	—	108	11	19	—	—	11	2
BhS2 (7–15 cm)	—	121	11	28	—	—	34	3
Bs (15–57 cm)	—	341	24	61	—	—	258	16
TOTAL	—	671	56	125	—	—	303	23
SOIL, TOTAL[b]								
E (0–4 cm)	19,514	2,447	734	3,553	47	645	43	1,817
Bhs1 (4–7 cm)	29,238	2,554	1,056	2,646	103	1,013	103	1,968
BhS2 (7–15 cm)	37,490	6,001	2,725	5,145	244	1,802	197	4,915
Bs (15–57 cm)	232,324	47,846	20,739	38,887	1,427	6,520	753	33,264
TOTAL	318,566	58,848	25,254	50,231	1,821	9,980	1,096	41,964

[a] N.D., Not Determined.
[b] Soil organic matter, soil carbon × 2.

Table WF-1. Biomass, Nutrient Content of the Whiteface (WF) Site
Principal Investigators: A. Friedland, E. Miller

Component	Organic Matter	Ca	Mg	K	P	N	S	Na
				(kg ha^{-1})				
OVERSTORY								
FOLIAGE	7,250	39	5	29	8	102	10	0
BRANCH	32,480	122	20	92	23	245	24	0.4
BOLE	77,790	126	15	49	6	93	62	1.6
STUMP	—	—	—	—	—	—	—	—
ROOTS	35,260	123	14	72	17	188	18	0.7
TOTAL	—	—	—	—	—	—	—	—
UNDERSTORY		Included in Overstory						
FOREST FLOOR								
O HORIZONS	141,400	387	39	89	136	2,642	N.D.[a]	12
WOOD	—	—	—	—	—	—	—	—
TOTAL	141,400	387	39	89	136	2,642	N.D.[a]	12
SOIL, EXTRACTABLE								
ABhs (0–19 cm)	—	85	13	41	5	—	12	49
Bs (19–41 cm)	—	80	0	78	6	—	47	46
TOTAL	—	165	13	119	11	—	59	95
SOIL, TOTAL[b]								
ABhs (0–19 cm)	234,300	20,880	5,268	13,680	655	5,152	471	13,330
Bs (19–41 cm)	209,500	47,120	15,387	27,850	1,091	4,669	617	27,940
TOTAL	443,800	68,000	20,655	41,530	1,746	9,821	1,088	41,270

[a] N.D., Not Determined
[b] Soil organic matter, soil carbon × 2.

Table CH-2. Annual Mean Atmospheric Deposition, Throughfall, Stemflow, and Soil Solution Fluxes for Major Ions at Coweeta Hardwood (CH) Site[a]

Principal Investigators: Wayne T. Swank, Lee J. Reynolds, James M. Vose

SITE = COWEETA HARDWOOD (CH)
Site Code: CH
Vegetation: Mixed Hardwood
Elevation = 725 m
Geology: Tallulah Falls Formation[b]
Mean Total Annual Precipitation = 138 cm
Mean Total Annual Throughfall + Stemflow = 123
Fraction of Total Precipitation Sampled = 100%
Fraction of Total Dry Period Sampled = 0
Mean Annual Cloud/Fog Water Input = 0

SAMPLING PERIOD: 9/20/85–3/31/88
Location: Coweeta Hydrologic Laboratory, North Carolina
Latitude = 35°03′N
Longitude = 83°27′W
Soils: Fannin; fine-loamy, micaceous, mesic Typic Hapludults
Number of Replicate Throughfall Samplers Used = 12 (n = 0 for intensives)

	H_2O	SO_4^{2-}	NO_3^-	Cl^-	$H_2PO_4^-$	H^+	NH_4^+	K^+	Na^+	Ca^{2+}	Mg^{2+}	Al[b]	OrgN[c]	WKACID[d]	Alk[e]
	(cm)	(eq ha^{-1} yr^{-1})													
BULK DEPOSITION[g]	138	493	209	84	8	518	130	27	82	85	32	N/A	161	—	2
BULK THROUGHFALL[g]	115	447	109	138	12	188	59	375	74	266	166	7.9	225	—	48
STEMFLOW	8	107	3.5	19	1	16	5.7	203	7.7	181	47	1.2	68	—	121
FOREST FLOOR AND SOIL SOLUTION FLUXES:															
Forest Floor	118	445	69	194	19	110	69	766	66	1418	812	88	1005	—	74
A HORIZON (14 cm)	99	652	2	247	5	17	11	286	211	193	579	19	142	—	137
BA HORIZON (25 cm)	81	478	1	157	24	12	7	119	162	73	500	3	34	—	123
BC HORIZON (72 cm)	67	55	1	128	3	9	6	41	82	31	177	2	14	—	62

[a] Compiled by averaging measured fluxes for each year and computing sums and differences from these averages. Values in tables should be considered to include only two significant figures, despite up to five digits being expressed (because of program used to print data). Some columns may not total to the exact sums shown for the same reason.
[b] Tallulah Falls Formation: Predominately biotite paragneiss and biotite schist.
[c] Total Al in solution at sample pH, expressed in mole ha^{-1} yr^{-1}.
[d] OrgN = organic nitrogen = total Kjeldahl nitrogen NH_4^+, expressed in mole ha^{-1} yr^{-1}.
[e] Alkalinity.
[f] Weak acidity by Gran's plot titration.
[g] At this site, only bulk deposition and throughfall were sampled.

Table CP-2. Annual Mean Atmospheric Deposition, Throughfall, Stemflow, and Soil Solution Fluxes for Major Ions at Coweeta Pine (CP) Site[a]

Principal Investigators: W.T. Swank, L.J. Reynolds, J.M. Vose

SITE = COWEETA PINE (CP)	SAMPLING PERIOD: 4/1/86 to 3/31/89 (36 months)
Site Code: CP	Location: Coweeta Hydrologic Laboratory, NC
Vegetation: White Pine	Latitude = 35°03′ N
Elevation = 725 m	Longitude = 83°27′ W
Geology: Tallulah Falls Formation	Soils: Fannin; fine-loamy, micaceous, mesic Typic Hapludults
Mean Total Annual Precipitation = 143.5 cm	Number of Replicate Throughfall Samplers Used = 2 (n = 12 for intensives)
Mean Total Annual Throughfall + Stemflow = 102.3 cm	Mean Coarse Particle (Deposition Plate) Extrapolation Factor = 5.8
Fraction of Total Precipitation Sampled = 48%	Approximate Mean Dry Deposition Velocity (V_d) for Fine Particles ≈.092 cm/s
Fraction of Total Dry Period Sampled = 24%	Approximate Mean V_d for SO_2 ≈ .37, for HNO_3 ≈ 2.0 (for Comparison)
Mean Annual Cloud/Fog Water Input = 0 cm	Range of Cloud/Fog Water Input = 0 cm/y

	H_2O (cm)	SO_4^{2-}	NO_3^-	Cl^-	$H_2PO_4^-$	H^+	NH_4^+	K^+	Na^+	Ca^{2+}	Mg^{2+}	Al[b]	OrgN[c]	WKACID[d]	Alk[e]
		(eq ha^{-1} yr^{-1})													
ATMOSPHERIC DEPOSITION															
VAPORS	—	105	155	—	—	260	—	—	—	—	—	—	—	—	—
FINE PARTICLES	—	33	1	1	—	11	22	.5	1	2	.4	—	—	—	—
COARSE PARTICLES	—	55	23	8	5	—	23	26	14	88	17	—	—	—	—
TOTAL DRY DEPOSITION	—	192	179	9	5	271	45	27	15	90	17	—	—	—	—
PRECIPITATION	144	359	151	65	3	457	135	8	47	53	26	—	93	—	—
FOG/CLOUD	—	—	—	—	—	—	—	—	—	—	—	—	—	—	—
TOTAL DEPOSITION	144	552	330	74	8	728	180	34	62	143	43	—	—	—	2
BELOW CANOPY FLUXES:															
THROUGHFALL FLUX	92.4	409	180	118	23	244	78	319	61	285	147	—	—	—	8
SPATIAL VAR. TF (SE)[f]	—	51	16	12	7	45	18	46	8	118	74	—	—	—	—
STEMFLOW FLUX	9.9	66	2	20	5	31	2	85	11	40	25	—	—	—	<1
THROUGHFALL + STEMFLOW	102.3	475	182	138	28	275	80	403	72	325	172	—	279	—	8
NET CANOPY EXCHANGE[g]	—	−76	−148	65	20	−454	−100	369	10	182	129	—	—	—	—
FOREST FLOOR AND SOIL SOLUTION FLUXES:															
Forest Floor	97	728	147	280	73	56	142	911	188	1376	612	94	746	—	59
A HORIZON (10 cm)	87	930	3	223	10	21	9	216	154	971	472	56	350	—	190
BA HORIZON (29 cm)	77	771	3	230	4	11	8	161	185	246	593	3	60	—	92
BC HORIZON (72 cm)	53	129	1	160	3	4	5	53	103	44	279	2	16	—	111

[a] Compiled by averaging measured fluxes for each year and computing sums and differences from these averages. Values in tables should be considered to include only two significant figures, despite up to four digits being expressed (because of program used to print data). Some columns may not total to exact sums shown for same reason.
[b] Total Al in solution at sample pH, expressed in mole ha^{-1} yr^{-1}.
[c] OrgN = organic nitrogen = total Kjeldahl nitrogen $-NH_4^+$, expressed in mole ha^{-1} yr^{-1}.
[d] Weak acidity.
[e] Alkalinity.
[f] Spatial variability of throughfall fluxes expressed as the standard error of the mean flux determined from replicate samplers.
[g] Net canopy exchange = throughfall + stemflow flux − total deposition.

Table DF-2. Annual Mean Atmospheric Deposition, Throughfall, Stemflow, and Soil Solution Fluxes for Major Ions at Douglas Fir (DF) Site[a]

Principal Investigators: Dale W. Cole, Helga Van Miegroet, Peter Homann

SITE = DOUGLAS FIR (DF)
Site Code: DF
Vegetation: Douglas Fir
Elevation = 220 m
Geology: Glacial till
Mean Total Annual Precipitation = 114 cm
Mean Total Annual Throughfall + Stemflow = 97 cm
Fraction of Total Precipitation Sampled = 53%
Fraction of Total Dry Period Sampled = 76% for fine aerosols and gases, 20% for coarse particles
Mean Annual Cloud/Fog Water Input = 0 cm

SAMPLING PERIOD: 9/1/86–9/1/88 (24 months)
Location: Cedar River, Washington
Latitude = 47°23′ N
Longitude = 121°56′ W
Soils: Durocrept
Number of Replicate Throughfall Samplers Used = 2
Mean Coarse Particle (Deposition Plate) Extrapolation Factor = 2.8(GS) 1.0(DS)
Approximate Mean Dry Deposition Velocity (V_d) for Fine Particles ≈ .14 cm/s
Approximate Mean V_d for SO_2 ≈ .34, for HNO_3 ≈ 1.3
Range of Cloud/Fog Water Input = 0 cm/y

	H_2O	SO_4^{2-}	NO_3^-	Cl^-	$H_2PO_4^-$	H^+	NH_4^+	K^+	Na^+	Ca^{2+}	Mg^{2+}	Al[b]	OrgN[c]	WKACID[d]	Alk[e]
	(cm)	(eq ha^{-1} yr^{-1})													
ATMOSPHERIC DEPOSITION:															
VAPORS	—	52	54	—	—	105	—	—	—	—	—	—	—	—	—
FINE PARTICLES	—	12	2	1	—	7	9	1.7	5.2	1.3	1.2	—	—	—	—
COARSE PARTICLES	—	62	58	34	—	—	27	24	94	44	17	—	—	—	—
TOTAL DRY DEPOSITION	—	126	114	36	—	112	36	26	99	45	19	—	—	—	—
PRECIPITATION	114	194	123	217	1	223	72	38	259	131	50	—	7	—	—
FOG/CLOUD	—	—	—	—	—	—	—	—	—	—	—	—	—	—	—
TOTAL DEPOSITION	114	320	237	252	—	335	108	64	358	176	69	—	—	—	6
BELOW CANOPY FLUXES:															
THROUGHFALL FLUX	95	276	62	313	6	224	22	184	345	218	88	—	—	—	19
SPATIAL VAR. TF (SE)[f]	—	—	—	—	—	—	—	—	—	—	—	—	—	—	—
STEMFLOW FLUX	2	15	—	16	0	16	—	18	11	18	5	—	—	—	—
THROUGHFALL + STEMFLOW	97	291	62	329	6	240	23	202	356	236	93	—	57	—	—
NET CANOPY EXCHANGE[g]	—	−29	−174	77	—	−95	−86	138	−3	60	24	—	50	—	—
FOREST FLOOR AND SOIL SOLUTION FLUXES:															
Forest Floor	90	210	4	449	10	36	12	218	391	1780	607	432	458	—	296
A HORIZON (10 cm)	78	274	2	496	1	24	5	221	364	799	273	151	193	—	230
B HORIZON (40 cm)	58	301	1	215	0	8	2	33	292	263	127	2	28	—	172

[a] Compiled by averaging measured fluxes for each year and computing sums and differences from these averages. Values in tables should be considered to include only two significant figures, despite up to five digits being expressed (because of program used to print data). Some columns may not total to the exact sums shown for the same reason.
[b] Total Al in solution at sample pH, expressed in mole ha^{-1} yr^{-1}.
[c] OrgN = organic nitrogen = total Kjeldahl nitrogen -NH_4^+, expressed in mole ha^{-1} yr^{-1}.
[d] Weak acidity.
[e] Alkalinity.
[f] Spatial variability of throughfall fluxes expressed as the standard error of the mean flux determined from replicate samplers.
[g] Net canopy exchange = throughfall + stemflow flux − total deposition.

Table DL-2. Annual Mean Atmospheric Deposition, Throughfall, Stemflow, and Soil Solution Fluxes for Major Ions at Duke Loblolly Pine (DL) Site[a]

Principal Investigators: K. Knoerr, D. Binkley

SITE = DUKE LOBLOLLY PINE (DL) SAMPLING PERIOD: 11/1/86 to 10/31/88 (24 months) deposition; 3/24/86 to 4/9/89 (36 months) for soil solutions.

Site Code: DL
Vegetation: Loblolly Pine
Elevation = 213 m
Geology: highly weathered residual
Felsic igneous parent material

Location: Duke Forest, Durham, NC
Latitude = 36°12′ N
Longitude = 79°17′ W
Soils: Typic Hapludult

Mean Total Annual Precipitation = 113.4 cm
Mean Total Annual Throughfall + Stemflow = 101.4 cm
Fraction of Total Precipitation Sampled = 68%
Fraction of Total Dry Period Sampled = 46%
Mean Annual Cloud/Fog Water Input = 0 cm

Number of Replicate Throughfall Samplers Used = 5 (n = 12 for intensives)
Mean Coarse Particle (Deposition Plate) Extrapolation Factor = 2.3
Approximate Mean Dry Deposition Velocity (V_d) for Fine Particles ≈.12 cm/s
Approximate Mean V_d for SO_2 ≈ 0.29, for HNO_3 ≈ 3.2 (for Comparison)
Range of Cloud/Fog Water Input = 0 cm/y

	H_2O	SO_4^{2-}	NO_3^-	Cl^-	$H_2PO_4^-$	H^+	NH_4^+	K^+	Na^+	Ca^{2+}	Mg^{2+}	Al[b]	OrgN[c]	WKACID[d]	Alk[e]
	(cm)	(eq ha^{-1} yr^{-1})													
ATMOSPHERIC DEPOSITION															
VAPORS	—	360	299	—	—	659	—	—	—	—	—	—	—	—	—
FINE PARTICLES	—	57	3	11	—	13	35	15	5	8	2	—	—	—	—
COARSE PARTICLES	—	59	89	26[f]	1	—	66[f]	78	94	134	35	—	—	—	—
TOTAL DRY DEPOSITION	—	475	391	36	1	671	101	93	99	142	37	—	—	—	—
PRECIPITATION	113	575	198	379[e]	—	720	306	30[f]	143	202	38	—	—	—	—
FOG/CLOUD	0	—	—	—	—	—	—	—	—	—	—	—	—	—	—
TOTAL DEPOSITION	113	1050	589	415	1	1392	408	122	242	344	75	—	—	—	—

continued

Table DL-2. *Continued*

	H_2O	SO_4^{2-}	NO_3^-	Cl^-	$H_2PO_4^-$	H^+	NH_4^+	K^+	Na^+	Ca^{2+}	Mg^{2+}	Al[b]	OrgN[c]	WKACID[d]	Alk[e]
	(cm)	(eq ha^{-1} yr^{-1})													
BELOW CANOPY FLUXES:															
THROUGHFALL FLUX	99	786	413	456[f]	—	1004	241	199[f]	215	407	121	—	—	—	—
SPATIAL VAR. TF (SE)[g]	—	27	25	27	—	43	4	15	8	18	8	—	—	—	—
STEMFLOW FLUX	2	41	6	12[f]	—	45	3	7	4	14	5	—	—	—	—
THROUGHFALL + STEMFLOW	101	827	419	469	—	1050	244	206	220	422	126	—	—	—	—
NET CANOPY EFFECT[h]	12	−223	−170	54	−1	−342	−164	84	−23	78	52	—	—	—	—
FOREST FLOOR AND SOIL SOLUTION FLUXES:															
Forest Floor (0 cm)	40	416	7	222	33	252	22	193	207	443	185	660	471	—	37
Ap HORIZON (20 cm)	40	696	1	189	.6	12	20	146	258	517	161	10	283	—	322
Bt1 HORIZON (40 cm)	40	774	1	180	.5	9	17	153	171	628	186	4	268	—	323
Bt2 HORIZON (70 cm)	40	1249	1	216	.04	25	29	181	291	658	320	4	50	—	225
Subsoil (2 m)	40	316	2	403	1	19	43	63	260	163	144	1	127	—	274

[a] Compiled by averaging measured fluxes for each year and computing sums and differences from these averages. Values in tables should be considered to include only two significant figures, despite up to four digits being expressed (because of program used to print data). Some columns may not total to the exact sums shown for the same reason.
[b] Total Al in solution at sample pH, expressed in mole ha^{-1} yr^{-1}.
[c] OrgN = organic nitrogen = total Kjeldahl nitrogen -NH_4^+, expressed in mole ha^{-1} yr^{-1}.
[d] Weak acidity.
[e] Alkalinity.
[f] Year 2 data only.
[g] Spatial variability of throughfall fluxes expressed as the standard error of the mean flux determined from replicate samplers.
[h] Net canopy effect = throughfall + stemflow flux − total deposition.

Table FL-2. Annual Mean Atmospheric Deposition, Throughfall, Stemflow, and Soil Solution Fluxes for Major Ions at Findley Lake (FL) Site[a]

Principal Investigators: Dale W. Cole, Helga Van Miegroet, Peter Homann

SITE = FINDLEY LAKE (FL)
Site Code: FL
Vegetation: Pacific silver fir
Elevation = 1100 m
Geology: Tephra over fractured andesite
Mean Total Annual Precipitation = 174
Mean Total Annual Throughfall + Stemflow = 147
Fraction of Total Precipitation Sampled = 100
Fraction of Total Dry Period Sampled = 0
Mean Annual Cloud/Fog Water Input = unknown

SAMPLING PERIOD: 8/1/86–8/1/88
Location: Findley Lake, Cascades, Washington
Latitude = 47°04′
Longitude = 121°25′
Soils: Cryohumod
Number of Replicate Throughfall Samplers Used = 4

	H_2O	SO_4^{2-}	NO_3^-	Cl^-	$H_2PO_4^-$	H^+	NH_4^+	K^+	Na^+	Ca^{2+}	Mg^{2+}	Al[b]	OrgN[c]	WKACID[d]	Alk[e]
	(cm)							(eq ha^{-1} yr^{-1})							
BULK DEPOSITION[f]	174	319	84	569	17	219	63	58	648	249	171	—	30	—	60
BULK THROUGHFALL[f]	146	437	19	530	1	285	24	243	512	304	173	—	122	—	32
STEMFLOW	1	5	0	11	2	5	0	19	6	18	4	—	3	—	0
FOREST FLOOR AND SOIL SOLUTION FLUXES:															
Forest Floor	113	208	2	542	17	556	10	256	533	430	224	188	169	—	0
E HORIZON (10 cm)	101	190	2	423	10	318	12	109	795	478	241	544	253	—	28
Bs HORIZON (40 cm)	97	299	3	437	1	13	5	39	1155	292	177	77	85	—	561

[a] Compiled by averaging measured fluxes for each year and computing sums and differences from these averages. Values in tables should be considered to include only two significant figures, despite up to five digits being expressed (because of program used to print data). Some columns may not total to the exact sums shown for the same reason.
[b] Total Al in solution at sample pH, expressed in mole ha^{-1} yr^{-1}.
[c] OrgN = organic nitrogen = total Kjeldahl nitrogen -NH_4^+, expressed in mole ha^{-1} yr^{-1}.
[d] Weak acidity.
[e] Alkalinity.
[f] At this site, only bulk deposition and throughfall were sampled.

Table FS-2. Annual Mean Atmospheric Deposition, Throughfall, Stemflow, and Soil Solution Fluxes for Major Ions at Florida Slash Pine (FS) Site[a]

Principal Investigator: H.L. Gholz

SITE = FLORIDA SLASH PINE (FS)
Site Code: FS
Vegetation: Slash Pine
Elevation = 38.1 m
Geology: Marine sands
Mean Total Annual Precipitation = 112.12 cm
Mean Total Annual Throughfall + Stemflow = 99.57 cm
Fraction of Total Precipitation Sampled = 99%
Fraction of Total Dry Period Sampled = 65%
Mean Annual Cloud/Fog Water Input = 0 cm

SAMPLING PERIOD: 5/15/88 to 5/15/89 (12 months)
Location: Gainesville, FL
Latitude = 29°44′ N
Longitude = 82°30′ W
Soils: Ultic Haplaquod
Number of Replicate Throughfall Samplers Used = 8 (n = 12 for intensives)
Mean Coarse Particle (Deposition Plate) Extrapolation Factor = 2.5
Approximate Mean Dry Deposition Velocity (V_d) for Fine Particles ≈.11 cm/s
Approximate Mean V_d for SO_2 ≈ 0.19, for HNO_3 ≈ 0.85 (for Comparison)
Range of Cloud/Fog Water Input = 0 cm/y

	H_2O	SO_4^{2-}	NO_3^-	Cl^-	$H_2PO_4^-$	H^+	NH_4^+	K^+	Na^+	Ca^{2+}	Mg^{2+}	Al[b]	OrgN[c]	WKACID[d]	Alk[e]
	(cm)	(eq ha^{-1} yr^{-1})													
ATMOSPHERIC DEPOSITION:															
VAPORS	—	17	38	—	—	55	—	—	—	—	—	—	—	—	—
FINE PARTICLES	—	25	6	8	.1	2	16	4	13	—	—	—	—	—	—
COARSE PARTICLES	—	129	105	211	5	—	4	10	352	159	74	—	—	—	—
TOTAL DRY DEPOSITION	—	171	149	219	6	47	20	14	365	159	74	—	—	—	—
PRECIPITATION	112	344	156	198	9	258	106	12	242	243	131	—	—	—	—
FOG/CLOUD	—	—	—	—	—	—	—	—	—	—	—	—	—	—	—
TOTAL DEPOSITION	112	514	305	417	15	315	126	26	608	402	205	—	—	—	—

BELOW CANOPY FLUXES:															
THROUGHFALL FLUX	99	444	181	446	13	279	47	57	558	483	248	—	—	—	—
SPATIAL VAR. TF (SE)[f]	—	—	—	—	—	—	—	—	—	—	—	—	—	—	—
STEMFLOW FLUX	.4	8	.5	9	.1	9	.1	.7	6	7	3	—	—	—	—
THROUGHFALL + STEMFLOW	—	451	181	455	13	288	47	58	564	489	252	—	—	—	—
NET CANOPY EXCHANGE[g]	−13	−63	−124	37	−3	−37	−79	32	−44	87	47	—	—	—	—
FOREST FLOOR AND SOIL SOLUTION FLUXES:															
Forest Floor	86	484	129	695	105	330	N/A	86	650	298	213	1259	—	—	
A HORIZON (10 cm)	74	1007	279	1615	<59	260	N/A	N/A	2840	803	179	355	—	—	—
E HORIZON (20 cm)	55	1011	162	1243	<59	58	N/A	N/A	1961	576	188	370	—	—	—
Bh HORIZON (40 cm)	37	579	102	858	<59	50	N/A	N/A	963	256	123	1854	—	—	—
E' HORIZON (70 cm)	27	390	53	415	<59	23	N/A	N/A	555	135	51	318	—	—	—
Bt HORIZON (120 cm)	13	242	10	296	<59	25	N/A	N/A	370	107	94	79	—	—	—

[a] Compiled by averaging measured fluxes for each year and computing sums and differences from these averages. Values in tables should be considered to include only two significant figures, despite up to four digits being expressed (because of program used to print data). Some columns may not total to the exact sums shown for the same reason.

[b] Total Al in solution at sample pH, expressed in mole ha^{-1} yr^{-1}.

[c] OrgN = organic nitrogen = total Kjeldahl nitrogen -NH_4^+, expressed in mole ha^{-1} yr^{-1}.

[d] Weak acidity.

[e] Alkalinity.

[f] Spatial variability of throughfall fluxes expressed as the standard error of the mean flux determined from replicate samplers.

[g] Net canopy exchange = throughfall + stemflow flux − total deposition.

Table GL-2. Annual mean atmospheric deposition, stemflow, litterfall, Soil Solution Fluxes for Major Ions at the Georgia Loblolly Pine (GL) Site.[a]

Principal Investigators: H.L. Ragsdale, J. Dowd

SITE = GEORGIA LOBLOLLY PINE (GL)
Site Code: GL
Vegetation: Loblolly Pine
Elevation = 145 m
Geology: Gniess
Mean Total Annual Precipitation = 86.5 cm
Mean Total Annual Throughfall + Stemflow = 74.6 cm
Fraction of Total Precipitation Sampled = 58%
Fraction of Total Dry Period Sampled = 78%
Mean Annual Cloud/Fog Water Input = 0 cm

SAMPLING PERIOD: 4/1/87–3/31/89 (24 months)
Location: B.F. Grant Forest, Georgia
Latitude = 33°22′ N
Longitude = 83°27′ W
Soils: Typic Hapludult
Number of Replicate Throughfall Samplers Used = 6 (n = 12 for intensives)
Mean Coarse Particle (Deposition Plate) Extrapolation Factor = 1.4
Approximate Mean Dry Deposition Velocity (V_d) for Fine Particles ≈ .15 cm/s
Approximate Mean V_d for SO_2 ≈ .65, for HNO_3 ≈ 2.0 (for Comparison)
Range of Cloud/Fog Water Input = 0 cm/y

	H_2O	SO_4^{2-}	NO_3^-	Cl^-	$H_2PO_4^-$	H^+	NH_4^+	K^+	Na^+	Ca^{2+}	Mg^{2+}	Al[b]	OrgN[c]	WKACID[d]	Alk[e]
	(cm)							(eq ha^{-1} yr^{-1})							
ATMOSPHERIC DEPOSITION:															
VAPORS	—	286	278	—	—	565	—	—	—	—	—	—	—	—	—
FINE PARTICLES	—	47	2	—	—	—	27	5	7	4	1	—	—	—	—
COARSE PARTICLES	—	74	36	32	5	—	37	15	44	51	17	—	—	—	—
TOTAL DRY DEPOSITION	—	408	316	32	5	565	64	20	51	56	18	—	—	—	—
PRECIPITATION	87	368	150	109	2	443	117	13	73	45	23	—	—	—	—
FOG/CLOUD	—	—	—	—	—	—	—	—	—	—	—	—	—	—	—
TOTAL DEPOSITION	87	776	466	141	7	1007	181	33	124	100	41	—	—	—	—
BELOW CANOPY FLUXES:															
THROUGHFALL FLUX	—	463	370	199	2	285	150	251	93	158	121	—	—	—	—
SPATIAL VAR. TF (SE)[f]	—	26	25	21	—	21	8	29	3	25	18	—	—	—	—
STEMFLOW FLUX	—	97	40	40	1	66	19	52	17	38	29	—	—	—	—
THROUGHFALL + STEMFLOW	75	561	410	239	3	352	169	304	110	196	150	—	—	—	—
NET CANOPY EXCHANGE[g]	—	−315	−57	98	−4	−656	−12	271	−14	96	109	—	—	—	—
FOREST FLOOR AND SOIL SOLUTION FLUXES:															
A HORIZON (10 cm)	43	313	0.7	497	31	10	6	335	805	273	192	211	626	563	560
A/B HORIZON (20 cm)	33	186	0.7	423	15	6	5	210	862	184	92	10	485	529	510
Bt HORIZON (40 cm)	24	38	0.6	287	12	5	17	86	662	68	43	7	410	349	410
BC+ HORIZON (80 cm)	22	35	0.6	267	11	5	16	80	615	63	40	6	380	324	381

[a] Compiled by averaging measured fluxes for each year and computing sums and differences from these averages. Values in tables should be considered to include only two significant figures, despite up to five digits being expressed (because of program used to print data). Some columns may not total to the exact sums shown for the same reason.
[b] Total Al in solution at sample pH, expressed in mole ha^{-1} yr^{-1}.
[c] OrgN = organic nitrogen = total Kjeldahl nitrogen -NH_4^+, expressed in mole ha^{-1} yr^{-1}.
[d] Weak acidity.
[e] Alkalinity.
[f] Spatial variability of throughfall fluxes expressed as the standard error of the mean flux determined from replicate samplers.
[g] Net canopy exchange = throughfall + stemflow flux − total deposition.

Table HF-2. Annual Mean Atmospheric Deposition, Throughfall, Stemflow, and Soil Solution Fluxes for Major Ions at Huntington Forest (HF) Site.[a]

Principal Investigators: J. Shepard, M.J. Mitchell, T.J. Scott, Y. Zhang

SITE = HUNTINGTON FOREST (HF)	SAMPLING PERIOD: 5/1/86–4/30/88 (24 months)
Site Code: HF	Location: Huntington Forest, New York
Vegetation: Mixed Northern Hardwood	Latitude = 45°59′ N
Elevation = 530 m	Longitude = 74°14′ W
Geology: Precambrian hornblende − granitic gneiss bedrock	Soils: coarse-loamy, mixed, frigid, Typic Haplorthod
Mean Total Annual Precipitation = 96.9 cm	Number of Replicate Throughfall Samplers Used = 2 (n = 12 for intensives)
Mean Total Annual Throughfall + Stemflow = 80.6 cm	Mean Coarse Particle (Deposition Plate) Extrapolation Factor = 2.6(GS) 1.0(DS)
Fraction of Total Precipitation Sampled = 75%	Approximate Mean Dry Deposition Velocity (V_d) for Fine Particles ≈.02 cm/s
Fraction of Total Dry Period Sampled = 50%	Approximate Mean V_d for SO_2 ≈ .21, for HNO_3 ≈ 2.1 (for Comparison)
Mean Annual Cloud/Fog Water Input = 0 cm	Range of Cloud/Fog Water Input = 0 cm/y

	H_2O	SO_4^{2-}	NO_3^-	Cl^-	$H_2PO_4^-$	H^+	NH_4^+	K^+	Na^+	Ca^{2+}	Mg^{2+}	Al[b]	OrgN[c]	WKACID[d]	Alk[e]
	(cm)	(eq ha^{-1} yr^{-1})													
VAPORS	—	90	193	—	—	283	—	—	—	—	—	—	—	—	—
FINE PARTICLES	—	7	.3	.1	—	2	5	.1	.2	.6	.2	—	—	—	—
COARSE PARTICLES	—	32	37	12	—	—	13	9	8	58	10	—	—	—	—
TOTAL DRY DEPOSITION	—	129	230	12	—	284	18	9	8	59	10	—	—	—	—
PRECIPITATION	97	358	187	32.0	—	338.1	132.4	31.6	24.3	43.0	11.7	—	—	—	—
FOG/CLOUD	—	—	—	—	—	—	—	—	—	—	—	—	—	—	—
TOTAL DEPOSITION	97	488	417	44	—	622	150	40	33	102	21	—	—	—	—
BELOW CANOPY FLUXES:															
THROUGHFALL FLUX	76	397	208	25	—	218	119	164	44	172	53	—	—	—	—
SPATIAL VAR. TF (SE)[f]	—	40	11	2	—	7	12	11	6	29	7	—	—	—	—
STEMFLOW FLUX	5	47	8	7	—	1	14	95	10	13	4	—	—	—	—
THROUGHFALL + STEMFLOW	—	443	216	32	—	219	133	259	53	185	56	—	260	—	—
NET CANOPY EXCHANGE[g]	—	−44	−201	−12	—	−403	−17	218	21	83	35	—	—	—	—
FOREST FLOOR AND SOIL SOLUTION FLUXES:															
Forest Floor (0 cm)	68	620	307	202	—	256	39	187	207	787	206	113	589	—	—
E HORIZON (8 cm)	63	701	149	154	—	122	11	38	254	557	129	155	331	—	—
B HORIZON (58 cm)	44	572	71	79	—	48	5	12	168	306	67	95	103	—	22

[a] Compiled by averaging measured fluxes for each year and computing sums and differences from these averages. Values in tables should be considered to include only two significant figures, despite up to five digits being expressed (because of program used to print data). Some columns may not total to the exact sums shown for the same reason.
[b] Total Al in solution at sample pH, expressed in mole ha^{-1} yr^{-1}.
[c] OrgN = organic nitrogen = total nitrogen $-NHy^+ + NO_3^-$, expressed in mole ha^{-1} yr^{-1}.
[d] Weak acidity.
[e] Alkalinity.
[f] Spatial variability of throughfall fluxes expressed as the standard error of the mean flux determined from replicate samplers.
[g] Net canopy exchange = throughfall + stemflow flux − total deposition.

Table LP-2. Annual Mean Atmospheric Deposition, Throughfall, and Stemflow, Soil Solution Fluxes for Major Ions at Oak Ridge Loblolly Pine (LP) Site[a]

Principal Investigators: S.E. Lindberg, D.W. Johnson

SITE = OAK RIDGE LOBLOLLY PINE (LP)	SAMPLING PERIOD: 4/1/86 to 3/31/89 (36 months)
Site Code: LP	Location: Oak Ridge, Tennessee, USA
Vegetation: Loblolly Pine	Latitude = 35°54′ N
Elevation = 300 m	Longitude = 84°20′ W
Geology: River alluvium derived from shale	Soils: Fluventic Dystrochrept
Mean Total Annual Precipitation = 113.6 cm	Number of Replicate Throughfall Samplers Used = 6 (n = 12 for intensives)
Mean Total Annual Throughfall + Stemflow = 108.7 cm	Mean Coarse Particle (Deposition Plate) Extrapolation Factor = 1.4
Fraction of Total Precipitation Sampled = 99%	Approximate Mean Dry Deposition Velocity (V_d) for Fine Particles ≈.05 cm/s
Fraction of Total Dry Period Sampled = 40%	Approximate Mean V_d for SO_2 ≈ 0.3, for HNO_3 ≈ 1.9 (for Comparison)
Mean Annual Cloud/Fog Water Input = 1.0 cm	Range of Cloud/Fog Water Input = 0.63–1.8 cm/y

	H_2O	SO_4^{2-}	NO_3^-	Cl^-	$H_2PO_4^-$	H^+	NH_4^+	K^+	Na^+	Ca^{2+}	Mg^{2+}	Al[b]	OrgN[c]	WKACID[d]	Alk[e]
	(cm)	(eq ha^{-1} yr^{-1})													
ATMOSPHERIC DEPOSITION:															
VAPORS	—	307	262	—	—	570	—	—	—	—	—	—	—	—	—
FINE PARTICLES	—	27	.83	.58	—	5	18	.38	.67	4	.86	—	—	—	—
COARSE PARTICLES	—	69	27	8	.45	—	7	9	79	180	24	—	—	—	—
TOTAL DRY DEPOSITION	—	403	290	9	.45	575	26	9	7	184	25	—	—	—	—
PRECIPITATION	114	514	205	89	.87	566	147	7	41	91	26	3	10	362	—
FOG/CLOUD	1	24	7	5	.18	3	25	3	5	9	1	—	—	—	—
TOTAL DEPOSITION	115	941	502	103	1.5	1143	197	20	55	284	53	—	—	—	—

BELOW CANOPY FLUXES:															
THROUGHFALL FLUX	109	855	349	150	4	632	121	295	61	424	152	8	—	965	—
SPATIAL VAR. TF (SE)[f]	—	43	18	18	1	64	13	77	2	96	34	1	—	85	—
STEMFLOW FLUX	—	3	.11	.39	.01	2	.1	.41	.15	.90	.38	.07	—	3	34
THROUGHFALL + STEMFLOW	—	858	349	150	4	634	121	295	61	425	153	8	161	968	34
NET CANOPY EXCHANGE[g]	109	−83	−154	47	2	−509	−77	275	7	141	99	—	—	—	—
FOREST FLOOR AND SOIL SOLUTION FLUXES:															
Plot 1															
Forest Floor	43	753	111	137	16	47	4	672	60	1083	458	133	113	—	41
Ap HORIZON (17 cm)	43	836	53	139	5	138	3	125	128	1078	390	29	59	—	19
Bw2 HORIZON (40 cm)	43	1324	18	76	6	77	2	113	101	958	411	6	20	—	37
Plot 2															
Bw2 HORIZON (80 cm)	43	1208	127	139	6	21	2	121	154	1327	263	8	138	—	75

[a] Compiled by averaging measured fluxes for each year and computing sums and differences from these averages. Values in tables should be considered to include only two significant figures, despite up to five digits being expressed (because of program used to print data). Some columns may not total to the exact sums shown for the same reason.
[b] Total Al in solution at sample pH, expressed in mole ha^{-1} yr^{-1}.
[c] OrgN = organic nitrogen = total Kjeldahl nitrogen $-NH_4^+$, expressed in mole ha^{-1} yr^{-1}.
[d] Weak acidity.
[e] Alkalinity.
[f] Spatial variability of throughfall fluxes expressed as the standard error of the mean flux determined from replicate samplers.
[g] Net canopy exchange = throughfall + stemflow flux − total deposition.

Table MS-2. Annual Mean Atmospheric Deposition, Throughfall, Stemflow, Soil Solution Fluxes for Major Ions at the Maine Spruce (MS) Site.[a]

Principal Investigator: I. Fernandez

SITE = MAINE SPRUCE (MS)	SAMPLING PERIOD: 4/1/88–3/31/89 (12 months)
Site Code: MS	Location: Howland, Maine
Vegetation: Red Spruce	Latitude = 45°10′ N
Elevation = 65 m	Longitude = 64°40′ W
Geology: Basal till	Soils: Aquic Haplorthods
Mean Total Annual Precipitation = 78.7 cm	Number of Replicate Throughfall Samplers Used = 2 (n = 12 for intensives)
Mean Total Annual Throughfall + Stemflow = 57.9 cm	Mean Coarse Particle (Deposition Plate) Extrapolation Factor = 3.2
Fraction of Total Precipitation Sampled = 82.5%	Approximate Mean Dry Deposition Velocity (V_d) for Fine Particles ≈.14 cm/s
Fraction of Total Dry Period Sampled = 57%	Approximate Mean V_d for SO_2 ≈ 0.34, for HNO_3 ≈ 1.8 (for Comparison)
Mean Annual Cloud/Fog Water Input = 0 cm	Range of Cloud/Fog Water Input = 0 cm/y

	H_2O	SO_4^{2-}	NO_3^-	Cl^-	$H_2PO_4^-$	H^+	NH_4^+	K^+	Na^+	Ca^{2+}	Mg^{2+}	Al[b]	OrgN[c]	WKACID[d]	Alk[e]
	(cm)	(eq ha^{-1} yr^{-1})													
ATMOSPHERIC DEPOSITION:															
VAPORS	—	132	67	—	—	199	—	—	—	—	—	—	—	—	—
FINE PARTICLES	—	34	2	4	—	16	14	.5	3	1	1	—	—	—	—
COARSE PARTICLES	—	115	64	35	—	—	102	13	27	79	27	—	—	—	—
TOTAL DRY DEPOSITION	—	281	133	38	—	215	116	13	30	81	28	—	—	—	—
PRECIPITATION	79	304	208	61	—	539	88	10	66	11	4	—	—	—	—
FOG/CLOUD	—	—	—	—	—	—	—	—	—	—	—	—	—	—	—
TOTAL DEPOSITION	79	585	341	99	—	754	204	23.2	95	92	31	—	—	—	—
BELOW CANOPY FLUXES:															
THROUGHFALL FLUX	58	602	86	147	—	444	14	119	109	151	61	—	—	—	—
SPATIAL VAR. TF (SE)[g]	—	44	7	11	—	3	4	10	8	14	6	—	—	—	—
STEMFLOW FLUX[h]	0.1	7	0.01	2	—	2	0.1	8	2	6	2	—	—	—	—
THROUGHFALL + STEMFLOW	58	602	86	147	—	444	14	119	109	151	61	—	—	—	—
NET CANOPY EXCHANGE[g]	—	17	−354	48	—	−310	−190	96	14	59	29	—	—	—	—
FOREST FLOOR AND SOIL SOLUTION FLUXES:															
Forest Floor	—	—	—	—	—	—	—	—	—	—	—	—	—	—	—
Bs HORIZON (75 cm)[h]	61	596	<12	226	—	62	12	76	356	160	168	23	—	—	—

[a] Compiled by averaging measured fluxes for each year and computing sums and differences from these averages. Values in tables should be considered to include only two significant figures, despite up to four digits being expressed (because of program used to print data). Some columns may not total to the exact sums shown for the same reason.

[b] Total Al in solution at sample pH, expressed in mole ha^{-1} yr^{-1}.

[c] OrgN = organic nitrogen = total Kjeldahl nitrogen $-NH_4^+$, expressed in mole ha^{-1} yr^{-1}.

[d] Weak acidity.

[e] Alkalinity

[f] Spatial variability of throughfall fluxes expressed as the standard error of the mean flux determined from replicate samplers.

[g] Net canopy exchange = throughfall + stemflow flux − total deposition.

[h] Stemflow and soil solution fluxes from collections between 1/1/89 and 12/31/89, and are not calculated into the rest of the data. Total precipitation for the period was 107 cm for the 1989 calendar year.

Table NS-2. Annual Mean Atmospheric Deposition, Throughfall, Stemflow, Soil Solution Fluxes for Major Ions at the Norway Spruce (NS) Site[a]

Principal Investigators: Eïnar Joranger, Ingvald Røsberg, Arne O. Stuanes

SITE = NORWAY SPRUCE (NS)	SAMPLING PERIOD: 10/1/86 to 9/30/88 (24 months)
Site Code: NS	Location: Nordmoen, Norway
Vegetation: Norway Spruce	Latitude = 60°16′ N
Elevation = 200 m	Longitude = 11°06′ E
Geology: 60 m deep glacifluvial deposits over Precambrian and Permian crystalline bedrock	Soils: Typic Udipsamment
Mean Total Annual Precipitation = 111 cm	Number of Replicate Throughfall Samplers Used = 2 (n = 20 for intensives)
Mean Total Annual Throughfall + Stemflow = 52 cm	Mean Coarse Particle (Deposition Plate) Extrapolation Factor = 4.4
Fraction of Total Precipitation Sampled = 90%	Approximate Mean Dry Deposition Velocity (V_d) for Fine Particles ≈.25 cm/s
Fraction of Total Dry Period Sampled = 48%	Approximate Mean V_d for SO_2 ≈ 0.6, for HNO_3 ≈ 2.2 (for Comparison)
Mean Annual Cloud/Fog Water Input = 0 cm	Range of Cloud/Fog Water Input = 0 cm/y

	H_2O	SO_4^{2-}	NO_3^-	Cl^-	$H_2PO_4^-$	H^+	NH_4^+	K^+	Na^+	Ca^{2+}	Mg^{2+}	Al^b	$OrgN^c$	$WKACID^d$	Alk^e
	(cm)	(eq ha^{-1} yr^{-1})													
ATMOSPHERIC DEPOSITION:															
VAPORS		27	23	—	—	49	—	—	—	—	—	—	—	—	—
FINE PARTICLES	—	19	8	2	—	—	21	3	—	2	1	—	—	—	—
COARSE PARTICLES	—	79	52	95	—	—	29	40	73	51	19	—	—	—	—
TOTAL DRY DEPOSITION	—	125	82	97	—	49	50	42	73	53	20	—	—	—	—
PRECIPITATION	111	538	337	134	—	579	305	20	108	55	31	—	94	—	8
FOG/CLOUD	—	—	—	—	—	—	—	—	—	—	—	—	—	—	—
TOTAL DEPOSITION	111	663	419	231	—	628	356	63	181	109	51	—	—	—	—

continued

Table NS-2. *Continued*

	H_2O	SO_4^{2-}	NO_3^-	Cl^-	$H_2PO_4^-$	H^+	NH_4^+	K^+	Na^+	Ca^{2+}	Mg^{2+}	Al[b]	OrgN[c]	WKACID[d]	Alk[e]
	(cm)	(eq ha^{-1} yr^{-1})													
BELOW CANOPY FLUXES:															
THROUGHFALL FLUX	78	659	81	251	—	362	119	323	174	248	132	—	146	—	10
SPATIAL VAR. TF (SE)[e]	—	—	—	—	—	—	—	—	—	—	—	—	—	—	—
STEMFLOW FLUX	.1	—	—	—	—	—	—	—	—	—	—	—	—	—	—
THROUGHFALL + STEMFLOW	—	659	81	251	—	362	119	323	174	248	132	—	146	—	—
NET CANOPY EXCHANGE[g]	—	−2.4	−338	19	—	−366	−337	261	−6	140	101	—	—	—	—
FOREST FLOOR AND SOIL SOLUTION FLUXES:															
A-4 R1															
Forest Floor	41	538	9	309	63	317	52	471	299	558	262	132	357	—	—
E HORIZON (4.5 cm)	41	484	5	219	30	266	8	150	209	339	198	287	423	—	—
Bs HORIZON (24 cm)	41	894	5	404	3	128	7	34	341	174	185	237	68	—	10
BC HORIZON (60 cm)	41	522	6	226	3	47	6	41	349	160	148	65	47	—	66
A-4 R2															
Forest Floor	38	297	7	141	23	263	40	102	128	564	123	71	256	—	2
E HORIZON (6.2 cm)	38	312	6	159	20	218	8	108	186	411	117	129	226	—	—
Bs HORIZON (29 cm)	38	495	5	144	3	73	6	35	169	209	82	86	58	—	58
BC HORIZON (60 cm)	38	404	4	228	3	20	7	29	297	203	140	21	28	—	103

[a] Compiled by averaging measured fluxes for each year and computing sums and differences from these averages. Values in tables should be considered to include only two significant figures, despite up to five digits being expressed (because of program used to print data). Some columns may not total to the exact sums shown for the same reason.
[b] Total Al in solution at sample pH, expressed in mole ha^{-1} yr^{-1}.
[c] OrgN = organic nitrogen = total persulfate digested N − NH_4^+ − NO_3^-
[d] Weak acidity.
[e] Alkalinity
[f] Spatial variability of throughfall fluxes expressed as the standard error of the mean flux determined from replicate samplers.
[g] Net canopy exchange = throughfall + stemflow flux − total deposition.

Table RA-2. Annual Mean Atmospheric Deposition, Throughfall, Stemflow, Soil Solution Fluxes for Major Ions at the Red Alder (RA) Site.[a]

Principal Investigators: Dale W. Cole, Helga Van Miegroet, Peter Homann

SITE = RED ALDER (RA)	SAMPLING PERIOD: 9/1/86–9/1/88 (24 months)
Site Code: RA	Location: Cedar River, Washington
Vegetation: Red Alder	Latitude = 47°23′ N
Elevation = 220 m	Longitude = 121°56′ W
Geology: Glacial till	Soils: Durocrept
Mean Total Annual Precipitation = 114 cm	Number of Replicate Throughfall Samplers Used = 2
Mean Total Annual Throughfall + Stemflow = 113 cm	Mean Coarse Particle (Deposition Plate) Extrapolation Factor = 2.8(GS) 1.0(DS)
Fraction of Total Precipitation Sampled = 53%	Approximate Mean Dry Deposition Velocity (V_d) for Fine Particles ≈.06 cm/s
Fraction of Total Dry Period Sampled = 76% for fine aerosols and gases, 20% for coarse particles	Approximate Mean V_d for SO_2 ≈ .35, for HNO_3 ≈ 1.3
Mean Annual Cloud/Fog Water Input = 0 cm	Range of Cloud/Fog Water Input = 0 cm/y

	H_2O	SO_4^{2-}	NO_3^-	Cl^-	$H_2PO_4^-$	H^+	NH_4^+	K^+	Na^+	Ca^{2+}	Mg^{2+}	Al^b	OrgN[c]	WKACID[d]	Alk[e]
	(cm)	(eq ha^{-1} yr^{-1})													
VAPORS	—	62	54	—	—	116	—	—	—	—	—	—	—	—	—
FINE PARTICLES	—	6	1	—	—	3	5	0.5	2.2	0.7	0.5	—	—	—	—
COARSE PARTICLES	—	62	58	34	—	—	27	24	94	44	17	—	—	—	—
TOTAL DRY DEPOSITION	—	131	113	35	—	119	31	24	97	44	18	—	—	—	—
PRECIPITATION	114	194	123	217	1	223	72	38	259	131	50	—	7	—	6
FOG/CLOUD	—	—	—	—	—	—	—	—	—	—	—	—	—	—	—
TOTAL DEPOSITION	114	325	235	251	—	342	104	62	355	176	68	—	—	—	—
BELOW CANOPY FLUXES:															
THROUGHFALL FLUX	112	295	64	350	10	134	38	312	407	280	144	—	—	—	200
SPATIAL VAR. TF (SE)[f]	—	—	—	—	—	—	—	—	—	—	—	—	—	—	—
STEMFLOW FLUX	1	1	—	3	0	—	1	7	3	2	1	—	—	—	—
THROUGHFALL + STEMFLOW	113	296	65	353	10	134	38	319	410	282	145	—	137	—	—
NET CANOPY EXCHANGE[g]	—	−28	−171	102	—	−208	−65	256	55	106	77	—	130	—	—
FOREST FLOOR AND SOIL SOLUTION FLUXES:															
Forest Floor	97	193	4879	725	24	524	85	1032	491	2850	1254	955	740	—	19
A HORIZON (10 cm)	85	104	4871	362	0	257	10	271	466	1783	745	1027	139	—	0
B HORIZON (40 cm)	67	73	2776	281	0	48	3	204	371	1871	602	47	23	—	48

[a] Compiled by averaging measured fluxes for each year and computing sums and differences from these averages. Values in tables should be considered to include only two significant figures, despite up to five digits being expressed (because of program used to print data). Some columns may not total to the exact sums shown for the same reason.
[b] Total Al in solution at sample pH, expressed in mole ha^{-1} yr^{-1}.
[c] OrgN = organic nitrogen = total Kjeldahl nitrogen $-NH_4^+$, expressed in mole ha^{-1} yr^{-1}.
[d] Weak acidity.
[e] Alkalinity.
[f] Spatial variability of throughfall fluxes expressed as the standard error of the mean flux determined from replicate samplers.
[g] Net canopy exchange = throughfall + stemflow flux − total deposition.

Table SB-2. Annual Mean Atmospheric Deposition, Throughfall, Stemflow, Soil Solution Fluxes for Major Ions at the Smokies Beech (SB) Site.[a]
Principal Investigators: D.W. Johnson, S.E. Lindberg

SITE = SMOKIES BEECH (SB)
Site Code: SB
Vegetation: American beech
Elevation = 1600 m
Geology: Anakeesta shale
Mean Total Annual Precipitation = 151 cm
Mean Total Annual Throughfall + Stemflow = 144 cm
Fraction of Total Precipitation Sampled = 100%
Fraction of Total Dry Period Sampled = —
Mean Annual Cloud/Fog Water Input = —

SAMPLING PERIOD: 1/1/86–12/88
Location: Great Smoky Mountain National Park, Tennessee
Latitude = 35°37′ N
Longitude = 85°26′ W
Soils: Umbric Dystrochrepts
Number of Replicate Throughfall Samplers Used = 4 (n = 0 for intensives)

	H_2O	SO_4^{2-}	NO_3^-	Cl^-	$H_2PO_4^-$	H^+	NH_4^+	K^+	Na^+	Ca^{2+}	Mg^{2+}	Al[b]	OrgN[c]	WKACID[d]	Alk[e]
	(cm)	(eq ha^{-1} yr^{-1})													
Plot 1															
BULK DEPOSITION[f]	151	573	29	99	4	1007	8	67	40	139	40	6	13	—	—
BULK THROUGHFALL[f]	144	771	113	119	14	682	10	311	118	374	125	13	124	—	—
STEMFLOW															
FOREST FLOOR AND SOIL SOLUTION FLUXES:															
A HORIZON (13 cm)	84	668	764	124	8	693	3	102	107	495	186	199	15	—	—
Bw1 HORIZON (46 cm)	84	428	967	156	5	265	3	53	106	415	155	350	1	—	—
Bw2 HORIZON (64 cm)	84	195	1248	149	4	250	3	11	97	383	150	315	<1	—	—
Plot 2															
Throughfall	141	780	88	170	10	364	10	366	128	499	128	11	12	—	—
Bw2 HORIZON (60 cm)	81	389	207	172	4	184	4	53	83	246	77	57	<1	—	—

[a] Compiled by averaging measured fluxes for each year and computing sums and differences from these averages. Values in tables should be considered to include only two significant figures, despite up to five digits being expressed (because of program used to print data). Some columns may not total to the exact sums shown for the same reason.
[b] Total Al in solution at sample pH, expressed in mole ha^{-1} yr^{-1}.
[c] OrgN = organic nitrogen = total Kjeldahl nitrogen $-NH_4^+$, expressed in mole ha^{-1} yr^{-1}.
[d] Weak acidity.
[e] Alkalinity
[f] At this site, only bulk deposition and throughfall were sampled.
[g] Note: There was a large pulse of NO_3^- and associated cations during the first year of monitoring at this Site.

Table SS-2. Annual Mean Atmospheric Deposition, Throughfall, Stemflow, and Soil Solution Fluxes for Major Ions at Smokies Becking Spruce (SS) Site.[a]

Principal Investigators: D.W. Johnson, S.E. Lindberg

SITE = SMOKIES BECKING SPRUCE (SS)
Site Code: SS
Vegetation: Red Spruce
Elevation = 1800 m
Geology: Thunderhead Sandstone
Mean Total Annual Precipitation = 151
Mean Total Annual Throughfall + Stemflow = 159
Fraction of Total Precipitation Sampled = 100
Fraction of Total Dry Period Sampled = —
Mean Annual Cloud/Fog Water Input = unknown

SAMPLING PERIOD: 1/1/85–12/31/88
Location: Great Smoky Mountain National Park, North Carolina
Latitude = 35°34′ N
Longitude = 83°29′ W
Soils: Umbric dystrochrepts
Number of Replicate Throughfall Samplers Used = 8

	H_2O	SO_4^{2-}	NO_3^-	Cl^-	$H_2PO_4^-$	H^+	NH_4^+	K^+	Na^+	Ca^{2+}	Mg^{2+}	Al[b]	OrgN[c]	WKACID[d]	Alk[e]
	(cm)	(eq ha^{-1} yr^{-1})													
Plot 1															
BULK DEPOSITION[f]	151	697	125	115	3	45	9	45	65	240	60	9	11	—	0
BULK THROUGHFALL[f]	168	1609	396	263	8	272	12	272	179	938	218	25	19	—	0
STEMFLOW	3	90	4	7	0.1	54	0.2	15	7	50	12	3	2	—	0
FOREST FLOOR AND SOIL SOLUTION FLUXES:															
A HORIZON (20 cm)	108	1732	1084	232	9	1740	4	167	175	670	258	489	30	—	0
Bw1 HORIZON (32 cm)	108	1334	1470	151	8	1146	4	133	134	658	222	516	15	—	0
Bw2 HORIZON (53 cm)	108	1379	1547	170	10	878	4	133	141	620	228	640	11	—	0
Plot 2															
Throughfall	150	1530	398	288	7	1426	19	244	202	794	212	36	11	—	0
Bw2 HORIZON (58 cm)	90	1350	1262	184	8	812	5	127	184	463	265	551	13	—	0

[a] Compiled by averaging measured fluxes for each year and computing sums and differences from these averages. Values in tables should be considered to include only two significant figures, despite up to five digits being expressed (because of program used to print data). Some columns may not total to the exact sums shown for the same reason.
[b] Total Al in solution at sample pH, expressed in mole ha^{-1} yr^{-1}.
[c] OrgN = organic nitrogen = total Kjeldahl nitrogen $-NH_4^+$, expressed in mole ha^{-1} yr^{-1}.
[d] Weak acidity .
[e] Alkalinity.
[f] At this site, only bulk deposition and throughfall were sampled.

Table ST-2. Annual Mean Atmospheric Deposition, Throughfall, Stemflow, and Soil Solution Fluxes for Major Ions at Smokies Tower Spruce (ST) Site.[a]
Principal Investigators: S.E. Lindberg and D.W. Johnson

SITE = SMOKIES TOWER SPRUCE (ST)
Site Code: ST
Vegetation: Red Spruce
Elevation = 1740 m
Geology: Thunderhead Sandstone
Mean Total Annual Precipitation = 203.4 cm
Mean Total Annual Throughfall + Stemflow = 215.6 cm
Fraction of Total Precipitation Sampled = 93%
Fraction of Total Dry Period Sampled = 24%
Mean Annual Cloud/Fog Water Input = 41 cm

SAMPLING PERIOD: 4/1/86–3/31/89 (36 months)
Location: Great Smoky Mountain National Park, NC
Latitude = 35°34′ N
Longitude = 83°28′ W
Soils: Umbrid dystrochrepts
Number of Replicate Throughfall Samplers Used = 2 (n = 12–18 for intensives)
Mean Coarse Particle (Deposition Plate) Extrapolation Factor = 4.4
Approximate Mean Dry Deposition Velocity (V_d) for Fine Particles ≈.4 cm/s
Approximate Mean V_d for SO_2 ≈ 0.3, for HNO_3 ≈ 6 (for Comparison)
Range of Cloud/Fog Water Input = 14–64 cm/y

	H_2O	SO_4^{2-}	NO_3^-	Cl^-	$H_2PO_4^-$	H^+	NH_4^+	K^+	Na^+	Ca^{2+}	Mg^{2+}	Al[b]	OrgN[c]	WKACID[d]	Alk[e]
	(cm)	(eq ha^{-1} yr^{-1})													
ATMOSPHERIC DEPOSITION:															
VAPORS		255	616	—	—	871	—	—	—	—	—	—	—	—	—
FINE PARTICLES	—	135	3	2	—	45	74	2	4	16	3	—	—	—	—
COARSE PARTICLES	—	161	133	19	.85	—	43	10	31	353	47	—	—	—	—
TOTAL DRY DEPOSITION	—	550	752	22	.85	916	118	12	35	369	50	—	—	—	—
PRECIPITATION	203	596	228	90	1	541	218	19	48	132	47	5	42	560	—
FOG/CLOUD	41	1067	261	290	1	524	362	137	244	299	95	—	—	—	—
TOTAL DEPOSITION	244	2214	1241	401	3	1981	698	169	327	800	192	—	—	560	—

BELOW CANOPY FLUXES:															
THROUGHFALL FLUX	215	2465	866	457	3	2157	220	403	294	895	283	20	—	1254	—
SPATIAL VAR. TF (SE)[f]	—	185	171	34	.8	88	27	35	19	79	29	0.2	—	59	—
STEMFLOW FLUX	1	51	2	6.8	.04	32	2	12	3	25	7	.4	—	46	—
THROUGHFALL + STEMFLOW	216	2516	869	464	3	2189	222	415	297	921	290	20	598	1300	—
NET CANOPY EXCHANGE[g]	—	303	−372	63	−.01	208	−476	246	−30	121	98	—	—	—	—
FOREST FLOOR AND SOIL SOLUTION FLUXES:															
Plot 1															
A HORIZON (20 m)	156	2275	1997	396	11	3006	29	427	344	612	409	869	25	—	0
Bw1 HORIZON (38 cm)	156	2194	1274	349	12	824	6	88	309	544	419	822	8	—	0
Bw2 HORIZON (55 cm)	156	2052	1469	345	4	710	6	111	319	744	434	867	10	—	0
Plot 2															
Bw2 HORIZON (65 cm)	156	2349	806	310	18	846	8	136	428	698	346	813	6	—	0

[a] Compiled by averaging measured fluxes for each year and computing sums and differences from these averages. Values in tables should be considered to include only two significant figures, despite up to four digits being expressed (because of program used to print data). Some columns may not total to the exact sums shown for the same reason.
[b] Total Al in solution at sample pH, expressed in mole ha^{-1} yr^{-1}.
[c] OrgN = organic nitrogen = total Kjeldahl nitrogen $-NH_4^+$, expressed in mole ha^{-1} yr^{-1}.
[d] Weak acidity.
[e] Alkalinity
[f] Spatial variability of throughfall fluxes expressed as the standard error of the mean flux determined from replicate samplers.
[g] Net canopy exchange = throughfall + stemflow flux − total deposition.

Table TL-2. Annual Mean Atmospheric Deposition, Throughfall, Stemflow, and Soil Solution Fluxes for Major Ions at the Turkey Lakes (TL) Site[a]

Principal Investigators: N.W. Foster, P.W. Hazlett, J.A. Nicolson

SITE = TURKEY LAKES (TL)	SAMPLING PERIOD: 1/1/86–12/31/86
Site Code: TL	Location: Ontario, Canada
Vegetation: Mixed Northern Hardwood	Latitude = 47°03′ N
Elevation = 350 m	Longitude = 84°25′ W
Geology: basalt-Granite	Soils: Spodosols
Mean Total Annual Precipitation = 120	Number of Replicate Throughfall Samplers Used = 12
Mean Total Annual Throughfall + Stemflow = 107	Mean Coarse Particle (Deposition Plate) Extrapolation Factor = N/A
Fraction of Total Precipitation Sampled = 100%	Approximate Mean Dry Deposition Velocity (V_d) for Fine Particles ≈ 0.1 cm/s
Fraction of Total Dry Period Sampled = 100%	Approximate Mean V_d for SO_2 ≈ 0.17 for HNO_3 ≈ 3.2 (for Comparison)
Mean Annual Cloud/Fog Water Input = 0 cm	Range of Cloud/Fog Water Input = 0 cm/y

	H_2O	SO_4^{2-}	NO_3^-	Cl^-	$H_2PO_4^-$	H^+	NH_4^+	K^+	Na^+	Ca^{2+}	Mg^{2+}	Al[b]	OrgN[c]	WKACID[d]	Alk[e]
	(cm)	(eq ha^{-1} yr^{-1})													
ATMOSPHERIC DEPOSITION:															
VAPORS	—	—	—	—	—	—	—	—	—	—	—	—	—	—	—
FINE PARTICLES	—	—	—	—	—	—	—	—	—	—	—	—	—	—	—
COARSE PARTICLES	—	—	—	—	—	—	—	—	—	—	—	—	—	—	—
TOTAL DRY DEPOSITION[a]	—	110	120	—	—	—	—	—	—	—	—	—	—	—	—
PRECIPITATION	120	650	340	—	—	—	—	—	—	—	—	—	—	—	—
TOTAL DEPOSITION[a]	—	760	460	—	—	—	—	—	—	—	—	—	—	—	—
BULK DEPOSITION[a]	—	542	311	65	2	411	211	18	64	186	52	—	114	—	61
BELOW CANOPY FLUXES:															
BULK THROUGHFALL[a]	107	638	286	119	2	161	215	392	84	360	138	—	457	—	208
FOREST FLOOR AND SOIL SOLUTION FLUXES:															
Forest Floor	92	588	508	120	3	217	219	376	90	1234	324	—	618	—	nd
B HORIZON (10 cm)	82	768	1381	238	1	60	36	265	305	1592	673	334	298	—	171
B HORIZON (65 cm)	79	720	1645	127	1	102	39	38	258	1760	454	314	275	—	66

[a] For the Turkey Lakes site, data were made available from other ongoing studies which were not directly a part of the IFS. Annual deposition fluxes for 1984 were based on measurements reported by Sirois and Vet (1988, Can. J. Fish. Aquat. Sci. 45(Suppl. 1):14–35). Separate measurements of wet and dry deposition were available only for sulfate and nitrate, and only bulk deposition data were available for the other ions. Note that the bulk throughfall fluxes include stemflow only for sulfate.

[b] Total Al in solution at sample pH, expressed in mole ha^{-1} yr^{-1}.

[c] OrgN = organic nitrogen = total Kjeldahl nitrogen $-NH_4^+$, expressed in mole ha^{-1} yr^{-1}.

[d] Weak acidity.

[e] Alkalinity.

Table WF-2. Annual Mean Atmospheric Deposition, Throughfall, Stemflow, and Soil Solution Fluxes for Major Ions at Whiteface Mountain (WF) Site[a]

Principal Investigators: A. Friedland, E. Miller

SITE = WHITEFACE MOUNTAIN (WF)
Site Code: WF
Vegetation: Red Spruce, Balsam Fir
Elevation = 950–1100 m
Geology: Precambrian Anorthsite
Mean Total Annual Precipitation = 114.7 cm
Mean Total Annual Throughfall + Stemflow = 117.5 cm
Fraction of Total Precipitation Sampled = 92.5%
Fraction of Total Dry Period Sampled = 8.1%
Mean Annual Cloud/Fog Water Input = 16 cm

SAMPLING PERIOD: 6/1/86–5/31/88 (24 months)
Location: Whiteface Mountain, New York
Latitude = 44°22′ N
Longitude = 73°54′ W
Soils: Typic Cryohumods
Number of Replicate Throughfall Samplers Used = 20 (n = 25 for intensives)
Mean Coarse Particle (Deposition Plate) Extrapolation Factor = 3
Approximate Mean Dry Deposition Velocity (V_d) for Fine Particles ≈0.047 cm/s
Approximate Mean V_d for SO_2 ≈ 0.14 for HNO_3 ≈ 4.0 (for Comparison)
Range of Cloud/Fog Water Input = 13–19 cm/y

	H_2O	SO_4^{2-}	NO_3^-	Cl^-	$H_2PO_4^-$	H^+	NH_4^+	K^+	Na^+	Ca^{2+}	Mg^{2+}	Al[b]	OrgN[c]	WKACID[d]	Alk[e]
	(cm)	(eq ha^{-1} yr^{-1})													
ATMOSPHERIC DEPOSITION:															
VAPORS	—	85	162	—	—	247	—	—	—	—	—	—	—	—	—
FINE PARTICLES	—	10	.7	.1	—	1	9	.2	1	.7	.3	—	—	—	—
COARSE PARTICLES	—	4	5	7	—	2	2	10	14	4	8	—	—	—	—
TOTAL DRY DEPOSITION	—	99	168	7	—	250	11	10	15	5	8	—	—	—	—
PRECIPITATION	115	508	262	31	2	560	172	10	15	93	24	10	—	—	—
FOG/CLOUD	16	489	254	14	—	460	270	5	6	32	10	—	—	—	—
TOTAL DEPOSITION	131	1096	683	52	2	1270	453	25	37	129	42	10	0	—	—
BELOW CANOPY FLUXES:															
THROUGHFALL FLUX	118	1030	320	76	27	844	167	140	45	362	89	17	228	—	—
SPATIAL VAR. TF (SE)[f]	—	—	—	—	—	—	—	—	—	—	—	—	—	—	—
STEMFLOW FLUX	.1	4	0	.1	0	2	.1	1	.1	1	.1	0	1.3	—	—
THROUGHFALL + STEMFLOW	—	1034	320	76	27	846	167	141	46	363	89	17	229	—	—
NET CANOPY EXCHANGE[g]	—	−62	−363	24	25	−424	−286	116	9	234	47	7	229	—	—
FOREST FLOOR AND SOIL SOLUTION FLUXES:															
Forest Floor (12 cm)	94	1799	230	103	5	860	11	13	106	656	335	686	574	—	—
AB HORIZON (30 cm)	82	1358	181	72	4	207	14	33	71	512	367	332	139	—	—
Bs HORIZON (60 cm)	76	1239	198	57	3	168	14	22	52	440	300	297	89	—	—

[a] Compiled by averaging measured fluxes for each year and computing sums and differences from these averages. Values in tables should be considered to include only two significant figures, despite up to five digits being expressed (because of program used to print data). Some columns may not total to the exact sums shown for the same reason.
[b] Total Al in solution at sample pH, expressed in mole ha^{-1} yr^{-1}.
[c] OrgN = organic nitrogen = total Kjeldahl nitrogen $-NH_4^+$, expressed in mole ha^{-1} yr^{-1}.
[d] Weak acidity.
[e] Alkalinity
[f] Spatial variability of throughfall fluxes expressed as the standard error of the mean flux determined from replicate samplers.
[g] Net canopy exchange = throughfall + stemflow flux − total deposition.

Table CH-3. Organic Matter and Nutrient Fluxes in Coweeta Hardwood (CH) Site
Principal Investigators: Wayne T. Swank, Lee J. Reynolds, James M. Vose

	Organic Matter	Ca	Mg	K	P	N	S
				(kg ha^{-1} yr^{-1})			
Litterfall	—	—	—	—	—	—	—
Needles	0	0	0	0	0	0	0
Leaves	3889	28.3	9.1	15.7	1.5	27.4	3
Roots	—	—	—	—	—	—	—
Other	1093	9.2	1.2	2.1	0.6	9	0.9
Total	4982	37.5	10.3	17.8	2.1	36.4	3.9
Net Canopy Exchange	—	7.2	2.2	21.5	0.2	−0.4	−1.0
Requirement[a]							
Foliage	6351	38	12.5	43.7	5.5	102.4	9
Wood	3571	12.2	0.9	4.1	0.46	7.1	1.6
Roots	2195	10.1	3.3	6.1	1.5	15.8	—
Total	12117	60.3	16.7	53.9	7.46	125.3	10.6
Mortality	—	—	—	—	—	—	—
Stand Increment[b]	3571	12.2	0.9	4.1	0.46	7.1	1.6
Uptake[c]	—	—	—	—	—	—	—
Foliar	—	0	0	0	0	0.4	0
Soil	—	47.7	12.2	41.3	2.2	34.1	5.6
Total	—	47.7	12.2	41.3	2.2	34.5	5.6
Translocation	—	12.6	4.5	12.6	5.3	75	5

[a] Requirement = biomass and nutrients in new growth.
[b] Stand increment = requirement for wood − mortality.
[c] Foliar uptake (FU) = −(Net Canopy Exchange, or NCE) if NCE $<$ 0; FU = 0 if NCE = or $>$ 0. Soil Uptake = Foliar Litterfall + Requirement for Wood + NCE. ROOTS NOT INCLUDED.

Table CP-3. Organic Matter and Nutrient Fluxes in Coweeta Pine (CP) Site
Principal Investigators: Wayne T. Swank, Lee J. Reynolds, James M. Vose

	Organic Matter	Ca	Mg	K	P	N	S
			(kg ha^{-1} yr^{-1})				
Litterfall	—	—	—	—	—	—	—
Needles	3992	25.4	5.5	9	2.4	30	3.3
Leaves	—	—	—	—	—	—	—
Roots	—	—	—	—	—	—	—
Other	2480	7.6	1.2	2.6	0.8	13.1	1.7
Total	6472	33	6.7	11.6	3.2	43.1	5
Net Canopy Exchange	—	3.6	1.6	14.4	0.6	0.4	−1.2
Requirement[a]	—	—	—	—	—	—	—
Foliage	5700	17.8	7	26.1	7.7	68.7	8.2
Wood	3871	3.9	1	3.6	1.1	6.6	2.8
Roots	3322	11.3	3	9	1.7	23.3	—
Total	12893	33	11	38.7	10.5	98.6	11
Mortality	—	—	—	—	—	—	—
Stand Increment[b]	3871	3.9	1	3.6	1.1	6.6	2.8
Uptake[c]	—	—	—	—	—	—	—
Foliar	—	0	0	0	0	0	1.2
Soil	—	32.9	8.1	27.3	4.1	37	4.9
Total	—	32.9	8.1	27.3	4.1	37	6.1
Translocation	—	−11.2	−0.1	2.4	4.7	38.3	4.9

[a] Requirement = biomass and nutrients in new growth.
[b] Stand increment = requirement for wood − mortality.
[c] Foliar uptake (FU) = −(Net Canopy Exchange, or NCE) if NCE < 0; FU = 0 if NCE = or > 0. Soil Uptake = Foliar Litterfall + Requirement for Wood + NCE. ROOTS NOT INCLUDED.

Table DF-3. Organic Matter and Nutrient Fluxes in Douglas-fir (DF) Site.
Principal Investigators: Dale W. Cole, Helga Van Miegroet, Peter Homann

	Organic Matter	Ca	Mg	K	P	N	S
			$(kg\ ha^{-1}\ yr^{-1})$				
Litterfall	—	—	—	—	—	—	—
Needles	1050	10.9	0.9	2.6	1.5	7.8	1.0
Leaves	—	—	—	—	—	—	—
Roots	—	—	—	—	—	—	—
Other	730	4.5	0.6	1.6	0.6	7.0	0.5
Total	1780	15.4	1.5	4.2	2.1	14.8	1.5
Net Canopy Exchange	—	1.2	0.3	5.4	0.2	−2.8	−0.5
Requirement[a]	—	—	—	—	—	—	—
Foliage	1050	6.4	1.2	6.4	3.2	15.6	1.5
Wood	6800	5.1	0.4	1.9	0.6	5.4	0.5
Roots	—	—	—	—	—	—	—
Total	7850	11.5	1.6	8.3	3.8	21.0	2.0
Mortality	4220	3.6	0.3	1.3	0.4	3.7	0.3
Stand Increment[b]	2580	1.5	0.1	0.6	0.2	1.7	0.2
Uptake[c]	—	—	—	—	—	—	—
Foliar	—	0.0	0.0	0.0	0.0	2.8	0.5
Soil	—	17.2	1.6	9.9	2.3	10.4	1.0
Total	—	17.2	1.6	9.9	2.3	13.2	1.5
Translocation	—	−5.7	0.0	−1.6	1.5	7.8	0.5

[a] Requirement = biomass and nutrients in new growth.; wood includes large roots
[b] Stand increment = requirement for wood − mortality.
[c] Foliar uptake (FU) = −(Net Canopy Exchange, or NCE) if NCE $<$ 0; FU = 0 if NCE = or $>$ 0. Soil Uptake = Foliar Litterfall + Requirement for Wood + NCE. FINE ROOTS NOT INCLUDED.

Table DL-3. Organic Matter and Nutrient Fluxes in Duke Loblolly (DL) Site
Principal Investigators: D. Binkley, K. Knoerr

	Organic Matter	Ca	Mg	K	P	N	S
				($kg\ ha^{-1}\ yr^{-1}$)			
Litterfall	—	—	—	—	—	—	—
Needles	3774	12.8	3	7	5.5	14.8	2.6
Leaves	—	—	—	—	—	—	—
Roots	—	—	—	—	—	—	—
Other	1141	2.5	0.3	1	0.7	3.9	0.8
Total	4915	15.3	3.3	8	6.2	18.7	3.4
Net Canopy Exchange	—	1.6	0.6	3.3	0.03	−4.7	−3.6
Requirement[a]	—	—	—	—	—	—	—
Foliage	5390	14.6	4.9	32.3	8.1	58.2	5.4
Wood	4590	13.6	2.6	8.9	1.8	15.4	1.6
Roots	3900	6	5.3	5	5.6	15.2	3.1
Total	13880	34.2	12.8	46.2	15.5	88.8	10.1
Mortality	—	—	—	—	—	—	—
Stand Increment[b]	4590	13.6	2.6	8.9	1.8	15.4	1.6
Uptake[c]	—	—	—	—	—	—	—
Foliar	—	0	0	0	0	4.7	3.6
Soil	—	26.4	5.6	15.9	7.3	25.5	0.6
Total	—	26.4	5.6	12.9	7.3	30.2	4.3
Translocation	—	7.8	7.2	33.3	8.2	58.6	5.8

[a] Requirement = biomass and nutrients in new growth.
[b] Stand increment = requirement for wood − mortality.
[c] Foliar uptake (FU) = −(Net Canopy Exchange, or NCE) if NCE $<$ 0; FU = 0 if NCE = or $>$ 0. Soil Uptake = Foliar Litterfall + Requirement for Wood + NCE. ROOTS NOT INCLUDED.

Table FL-3. Organic Matter and Nutrient Fluxes in Findley Lake (FL) Site.
Principal Investigator: Dale W. Cole

	Organic Matter	Ca	Mg	K	P	N	S
				(kg ha^{-1} yr^{-1})			
Litterfall	—	—	—	—	—	—	—
Needles	1060	9.9	0.7	1.6	0.8	5.9	0.8
Leaves	—	—	—	—	—	—	—
Roots	—	—	—	—	—	—	—
Other	1670	5.4	0.7	1.3	1.1	10.6	1.0
Total	2730	15.3	1.4	2.9	1.9	16.5	1.8
Net Canopy Exchange	—	1.5	0.1	8.0	−0.4	−0.1	2.0
Requirement[a]	—	—	—	—	—	—	—
Foliage	1060	5.6	0.9	6.8	1.6	11.8	1.1
Wood	3020	7.4	0.6	3.0	0.6	3.6	0.3
Roots	—	—	—	—	—	—	—
Total	4080	13.0	1.5	9.8	2.2	15.4	1.4
Mortality	—	—	—	—	—	—	—
Stand Increment[b]	3020	7.4	0.6	3.0	0.6	3.6	0.3
Uptake[c]	—	—	—	—	—	—	—
Foliar	—	0.0	0.0	0.0	0.4	0.1	0.0
Soil	—	18.8	1.4	12.6	1.0	9.4	3.1
Total	—	18.8	1.4	12.6	1.4	9.5	3.1
Translocation	—	−5.8	0.1	−2.8	0.8	5.9	−1.7

[a] Requirement = biomass and nutrients in new growth.; wood includes large roots
[b] Stand increment = requirement for wood − mortality.
[c] Foliar uptake (FU) = −(Net Canopy Exchange, or NCE) if NCE $<$ 0; FU = 0 if NCE = or $>$ 0. Soil Uptake = Foliar Litterfall + Requirement for Wood + NCE. FINE ROOTS NOT INCLUDED.

Table FS-3. Organic Matter and Nutrient Fluxes in Florida (FS) Site
Principal Investigator: H.L. Gholz

	Organic Matter	Ca	Mg	K	P	N	S
	(kg ha^{-1} yr^{-1})						
Litterfall	—	—	—	—	—	—	—
Needles	3624	10.8	3.6	2.7	1.3	18	2.4
Leaves	197	0.6	0.2	0.2	0.1	1	0.2
Roots	—	—	—	—	—	—	—
Other	632	1.7	0.2	0.2	0.1	2.5	0.3
Total	4453	13.1	4	3.1	1.5	21.5	2.9
Net Canopy Exchange	—	1.7	0.6	1.2	0.06	−2.8	−1
Requirement[a]	—	—	—	—	—	—	—
Foliage	1869	2.5	2.5	6.5	1.7	14.9	1.2
Wood	6402	5.9	1.8	1.8	0.8	7.9	0.8
Roots	1462	1.2	0.4	0.4	0.2	1.6	0.2
Total	9733	9.6	4.7	8.7	2.7	24.4	2.2
Mortality	—	—	—	—	—	—	—
Stand Increment[b]	6402	5.9	1.8	1.8	0.8	7.9	0.8
Uptake[c]	—	—	—	—	—	—	—
Foliar	—	0	0	0	0	2.8	1
Soil	—	19	6.2	5.9	2.2	24.1	2.4
Total	—	19	6.2	5.9	2.2	26.9	3.4
Translocation	—	−9.4	−1.5	2.8	0.49	−2.5	−1.2

[a] Requirement = biomass and nutrients in new growth.
[b] Stand increment = requirement for wood − mortality.
[c] Foliar uptake (FU) = −(Net Canopy Exchange, or NCE) if NCE $<$ 0; FU = 0 if NCE = or $>$ 0. Soil Uptake = Foliar Litterfall + Requirement for Wood + NCE. ROOTS NOT INCLUDED.

Table GL-3. Organic Matter and Nutrient Fluxes in Georgia Loblolly (GL) Site
Principal Investigators: H.L. Ragsdale, J. Dowd

	Organic Matter	Ca	Mg	K	P	N	S
				(kg ha^{-1} yr^{-1})			
Litterfall	—	—	—	—	—	—	—
Needles	—	—	—	—	—	—	—
Leaves	—	—	—	—	—	—	—
Roots	—	—	—	—	—	—	—
Other	—	—	—	—	—	—	—
Total	10315	38	13	28	6	76	6
Net Canopy Exchange	—	1.9	1.3	10.6	−0.1	−0.1	−0.9
Requirement[a]	—	—	—	—	—	—	—
Foliage	—	—	—	—	—	—	—
Wood	—	—	—	—	—	—	—
Roots	—	—	—	—	—	—	—
Total	14974	29	10	29	7	99	7
Mortality	—	—	—	—	—	—	—
Stand Increment[b]	—	—	—	—	—	—	—
Uptake[c]	—	—	—	—	—	—	—
Foliar	0	0	0	3.4	—	—	—
Soil	—	—	—	—	—	—	—
Total	—	—	—	—	—	—	—
Translocation	—	—	—	—	—	—	—

[a] Requirement = biomass and nutrients in new growth.
[b] Stand increment = requirement for wood − mortality.
[c] Foliar uptake (FU) = −(Net Canopy Exchange, or NCE) if NCE < 0; FU = 0 if NCE = or > 0. Soil Uptake = Foliar Litterfall + Requirement for Wood + NCE. ROOTS NOT INCLUDED.

Table HF-3. Organic Matter and Nutrient Fluxes in Huntington Forest (HF) Site
Principal Investigators: M.J. Mitchell, D.J. Raynal, E.H. White, R. Briggs, J. Shepard, J. Scott, M. Burke, J. Porter

	Organic Matter	Ca	Mg	K	P	N	S	Na
				($kg\ ha^{-1}\ yr^{-1}$)				
Litterfall	—	—	—	—	—	—	—	—
Needles	—	—	—	—	—	—	—	—
Leaves	3131	29.5	3.7	5.5	1.6	35	3.4	0.1
Roots	1317	3.3	0.8	0.2	0.9	14.8	1.3	0.1
Other	816	6.4	0.5	1.4	0.6	9.8	0.6	0.06
Total	5264	39.2	5	7.1	3.1	59.6	5.3	0.26
Net Canopy Exchange	—	1.7	.4	8.5	—	−2.8	−1.4	0.5
Requirement[a]	—	—	—	—	—	—	—	
Foliage	3545	22.2	6.3	27.4	4.5	77	6.4	0.3
Wood	725	2.7	0.2	0.7	0.06	1.1	0.1	0.1
Roots	1317	3.3	0.8	0.2	0.9	14.8	1.3	0.1
Total	5587	28.2	7.3	28.3	5.46	92.9	7.8	0.5
Mortality	—	—	—	—	—	—	—	
Stand Increment[b]	725	2.7	0.2	0.7	0.06	1.1	0.1	0.1
Uptake[c]	—	—	—	—	—	—	—	
Foliar	—	0	0	0	0	2.8	1.4	
Soil	—	33.9	4.3	14.7	1.66	33.3	2.1	0.7
Total	—	33.9	4.3	14.7	1.66	36.1	3.5	0.7
Translocation	—	−5.7	3	13.6	3.8	56.8	4.3	−0.2

[a] Requirement = biomass and nutrients in new growth.
[b] Stand increment = requirement for wood − mortality.
[c] Foliar uptake (FU) = −(Net Canopy Exchange, or NCE) if NCE $<$ 0; FU = 0 if NCE = or $>$ 0. Soil Uptake = Foliar Litterfall + Requirement for Wood + NCE. ROOTS NOT INCLUDED.

Table LP-3. Organic Matter and Nutrient Fluxes in Oak Ridge Loblolly (LP) Site
Principal Investigator: D.W. Johnson, S.E. Lindberg

	Organic Matter	Ca	Mg	K	P	N	S
			(kg ha^{-1} yr^{-1})				
PLOT 1							
Litterfall	—	—	—	—	—	—	—
Needles	2085	5.7	1.6	1.5	1	9.4	1.6
Leaves	297	8	1.3	0.7	0.3	2.4	0.3
Roots	—	—	—	—	—	—	—
Other	213	1.2	0.3	0.2	0.04	1.9	0.1
Total	2595	14.9	3.2	2.4	1.34	13.7	2.0
Net Canopy Exchange	—	2.8	1.2	10.8	0.6	−1.1	−1.3
Requirement[a]	—	—	—	—	—	—	—
Foliage	2290	4.1	1.8	14.2	3	23.2	2.8
Wood	6510	5.5	1.4	4.7	0.7	6.6	4.1
Roots	—	—	—	—	—	—	—
Total	8800	9.6	3.2	18.9	3.7	29.8	6.9
Mortality	—	—	—	—	—	—	—
Stand Increment[b]	6510	5.5	1.4	4.7	0.7	6.6	4.1
Uptake[c]	—	—	—	—	—	—	—
Foliar	—	0	0	0	1.1	0	1.3
Soil	—	22	4.5	17.7	2.6	17.3	4.7
Total	—	22	4.5	17.7	3.7	17.3	6.0
Translocation	—	−12.4	−1.3	1.2	1.1	11.4	0.9

PLOT 2							
Litterfall	—	—	—	—	—	—	—
Needles	2594	7.4	1.9	1.7	1.1	12	2.2
Leaves	586	11.1	2.6	1.7	0.7	4.5	0.6
Roots	—	—	—	—	—	—	—
Other	320	1.3	1.1	0.3	0.4	2.9	0.2
Total	3500	19.8	5.6	3.7	2.2	19.4	3.0
Net Canopy Exchange	—	3.4	2.1	18.4	1.2	7.8	1
Requirement[a]	—	—	—	—	—	—	—
Foliage	2850	5.1	2.3	17.7	3.6	28.7	3.4
Wood	1165	0.9	0.2	0.8	0.1	1.1	0.7
Roots	—	—	—	—	—	—	—
Total	4015	6	2.5	18.5	3.7	29.8	4.1
Mortality	—	—	—	—	—	—	—
Stand Increment[b]	1165	0.9	0.2	0.8	0.1	1.1	0.7
Uptake[c]	—	—	—	—	—	—	—
Foliar	—	0	0	0	0	0	0
Soil	—	22.8	6.8	22.6	3.1	25.4	4.5
Total	—	22.8	6.8	22.6	3.1	25.4	4.5
Translocation	—	−16.8	−4.3	−4.1	0.6	4.4	−0.5

Note: Leaf litterfall, uptake of Ca are greatly influenced by Ca-rich dogwood litter.

[a] Requirement = biomass and nutrients in new growth.

[b] Stand increment = requirement for wood − mortality.

[c] Foliar uptake (FU) = −(Net Canopy Exchange, or NCE) if NCE $<$ 0; FU = 0 if NCE = or $>$ 0. Soil Uptake = Foliar Litterfall + Requirement for Wood + NCE. ROOTS NOT INCLUDED.

Table MS-3. Organic Matter and Nutrient Fluxes in Maine (MS) Site
Principal Investigator: I. Fernandez

	Organic Matter	Ca	Mg	K	P	N	S
				(kg ha^{-1} yr^{-1})			
Litterfall	—	—	—	—	—	—	—
Needles	—	—	—	—	—	—	—
Leaves	—	—	—	—	—	—	—
Roots	—	—	—	—	—	—	—
Other	—	—	—	—	—	—	—
Total	2105	14.8	1.8	1.8	1.2	12.2	—
Net Canopy Exchange	—	1.2	0.3	3.7	−0.2	−6.2	−0.3
Requirement[a]	—	—	—	—	—	—	—
Foliage	—	1.1	0.2	1.6	0.2	2.2	N.D.
Wood	—	5.1	0.6	3.5	0.5	3.8	N.D.
Roots	—	0.9	0.1	0.6	0.1	0.7	N.D.
Total	—	7.1	0.9	5.7	0.8	6.7	N.D.
Mortality	—	0.1	<0.1	<0.1	<0.1	<0.1	N.D.
Stand Increment[b]	—	—	—	—	—	—	—
Uptake[c]	—	—	—	—	—	—	—
Foliar	—	0	0	0	0.2	1.4	1.4
Soil	—	—	—	—	—	—	—
Total	—	—	—	—	—	—	—
Translocation	—	—	—	—	—	—	—

[a] Requirement = biomass and nutrients in new growth.
[b] Stand increment = requirement for wood − mortality.
[c] Foliar uptake (FU) = −(Net Canopy Exchange, or NCE) if NCE < 0; FU = 0 if NCE = or > 0. Soil Uptake = Foliar Litterfall + Requirement for Wood + NCE. ROOTS NOT INCLUDED.

Table NS-3. Organic Matter and Nutrient Fluxes in Norway (NS) Site
Principal Investigators: Ingvald Røsberg, Arne O. Stuanes

	Organic Matter	Ca	Mg	K	P	N	S
				(kg ha^{-1} yr^{-1})			
PLOT R1							
Litterfall	—	—	—	—	—	—	—
Needles	1992	12.4	1.1	3.2	2.1	16.7	1.5
Leaves	—	—	—	—	—	—	—
Roots	—	—	—	—	—	—	—
Other	1294	2.3	0.8	3.1	1.3	12	1.1
Total	3286	14.7	1.9	6.3	3.4	28.7	2.6
Net Canopy Exchange[d]	—	2.8	0.5	8.1	−0.01	−6.0	−0.2
Requirement[a]	—	—	—	—	—	—	—
Foliage	1830	6	2.2	13.4	4.0	25.4	1.8
Wood	5043	6.1	1.3	10.5	1.9	10.9	0.6
Roots	—	—	—	—	—	—	—
Total	6873	12.1	3.5	23.9	5.9	36.3	2.4
Mortality	—	—	—	—	—	—	—
Stand Increment[b]	5043	6.1	1.3	10.5	1.9	10.9	0.6
Uptake[c]	—	—	—	—	—	—	—
Foliar	—	0	0	0	0.0	6.0	0.2
Soil	—	21.3	2.9	21.8	4.0	21.6	1.9
Total	—	21.3	2.9	21.8	4.0	27.6	1.9
Translocation	—	−9.2	−0.1	0.5	1.9	8.7	−0.2

continued

Table NS-3. *Continued*

	Organic Matter	Ca	Mg	K	P	N	S
			(kg ha^{-1} yr^{-1})				
PLOT R2							
Litterfall	—	—	—	—	—	—	—
Needles	1433	12.1	0.9	1.9	1.3	12.5	1.0
Leaves	—	—	—	—	—	—	—
Roots	—	—	—	—	—	—	—
Other	857	2.1	0.5	2.1	0.8	7.8	0.6
Total	2290	14.2	1.4	4.0	2.1	20.3	1.6
Net Canopy Exchange[d]	—	2.5	0.5	6.4	−0.01	−5.8	−1.2
Requirement[a]	—	—	—	—	—	—	—
Foliage	1882	6.3	2.1	12.9	3.6	25.1	1.8
Wood	3849	5.3	1.4	8.1	1.0	8.4	0.7
Roots	—	—	—	—	—	—	—
Total	5731	11.6	3.5	21.0	4.6	33.5	2.5
Mortality	—	—	—	—	—	—	—
Stand Increment[b]	3849	5.3	1.4	8.1	1.0	8.4	0.7
Uptake[c]	—	—	—	—	—	—	—
Foliar	—	0.0	0.0	0.0	0.0	5.8	1.2
Soil	—	19.9	2.8	16.4	2.3	15.1	0.5
Total	—	19.9	2.8	16.4	2.3	20.9	1.7
Translocation	—	−8.3	0.7	4.6	2.3	12.6	0.8

[a] Requirement = biomass and nutrients in new growth.
[b] Stand increment = requirement for wood − mortality.
[c] Foliar uptake (FU) = −(Net Canopy Exchange, or NCE) if NCE < 0; FU = 0 if NCE = or > 0. Soil Uptake = Foliar Litterfall + Requirement for Wood + NCE. ROOTS NOT INCLUDED.
[d] Based on total deposition, bulk throughfall, stemflow.

Table RA-3. Organic Matter and Nutrient Fluxes in Red Alder (RA) Site.
Principal Investigator: Dale W. Cole

	Organic Matter	Ca	Mg	K	P	N	S
				(kg ha^{-1} yr^{-1})			
Litterfall	—	—	—	—	—	—	—
Needles	—	—	—	—	—	—	—
Leaves	3118	34.2	4.7	13.4	2.1	58.8	3.4
Roots	—	—	—	—	—	—	—
Other	1363	10.8	1.6	5.2	1.6	21.2	1.3
Total	4481	45.0	6.3	18.6	3.7	80.0	4.7
Net Canopy Exchange	—	2.1	0.9	10.0	0.3	−1.4	−0.4
Requirement[a]	—	—	—	—	—	—	—
Foliage	3492	22.7	6.4	43.3	10.4	115.9	6.6
Wood	6850	15.6	1.6	8.3	0.9	21.5	1.2
Roots	—	—	—	—	—	—	—
Total	10342	38.3	8.0	51.6	11.3	137.4	7.8
Mortality	3840	11.2	1.0	6.2	0.2	14.0	1.2
Stand Increment[b]	3010	4.4	0.6	2.1	0.7	7.5	0.0
Uptake[c]	—	—	—	—	—	—	—
Foliar	—	0.0	0.0	0.0	0.0	1.4	0.4
Soil	—	51.9	7.2	31.7	3.3	78.9	4.2
Total	—	51.9	7.2	31.7	3.3	80.3	4.6
Translocation	—	−13.6	0.8	19.9	8.0	57.1	3.2

[a] Requirement = biomass and nutrients in new growth.; wood includes large roots
[b] Stand increment = requirement for wood − mortality.
[c] Foliar uptake (FU) = −(Net Canopy Exchange, or NCE) if NCE < 0; FU = 0 if NCE = or > 0. Soil Uptake = Foliar Litterfall + Requirement for Wood + NCE. FINE ROOTS NOT INCLUDED.

Table SB-3. Organic Matter and Nutrient Fluxes in Smokies Beech (SB)Site
Principal Investigator: D.W. Johnson

	Organic Matter	Ca	Mg	K	P	N	S
			$(kg\ ha^{-1}\ yr^{-1})$				
PLOT 1							
Litterfall	—	—	—	—	—	—	—
Needles	0	0	0	0	0	0	0
Leaves	1660	9.3	1.5	2.9	1.6	22	2.2
Roots	—	—	—	—	—	—	—
Other	584	1.3	0.4	1.8	0.4	4.2	0.4
Total	2244	10.6	1.9	4.7	2	26.2	2.6
Net Canopy Exchange	—	4.7	1.0	9.5	0.3	2.8	3.5
Requirement[a]	—	—	—	—	—	—	—
Foliage	2440	11.7	3	19.8	4.5	58.9	4.9
Wood	8470	13.9	1.7	8.5	0.6	11.7	3.9
Roots	—	—	—	—	—	—	—
Total	10910	25.6	4.7	28.3	5.1	70.6	8.8
Mortality	—	—	—	—	—	—	—
Stand Increment[b]	8470	13.9	1.7	8.5	0.6	11.7	3.9
Uptake[c]	—	—	—	—	—	—	—
Foliar	—	0	0	0	0	0	0
Soil	—	27.9	4.2	20.9	2.5	36.5	9.6
Total	—	27.9	4.2	20.9	2.5	36.5	9.6
Translocation	—	2.3	−0.5	7.4	2.6	34.1	0.8

PLOT 2							
Litterfall	—	—	—	—	—	—	—
Needles	0	0	0	0	0	0	0
Leaves	1510	8.1	1.3	2.5	1.4	20.5	2.1
Roots	—	—	—	—	—	—	—
Other	318	1.4	0.2	0.7	0.3	2.9	0.2
Total	1828	9.5	1.5	3.2	1.7	23.4	2.3
Net Canopy Exchange	—	7.2	1.1	11.7	0.2	0.8	3.6
Requirement[a]	—	—	—	—	—	—	—
Foliage	2120	10	2.6	17	3.9	51.6	4.2
Wood	4720	7.8	0.9	4.7	0.4	6.5	2.2
Roots	—	—	—	—	—	—	—
Total	6840	17.8	3.5	21.7	4.3	58.1	6.4
Mortality	—	—	—	—	—	—	—
Stand Increment[b]	4720	7.8	0.9	4.7	0.4	6.5	2.2
Uptake[c]	—	—	—	—	—	—	—
Foliar	—	0	0	0	0	0	0
Soil	—	23.1	3.3	18.9	2.0	27.8	7.9
Total	—	23.1	3.3	18.9	2.0	27.8	7.9
Translocation	—	−5.3	0.2	2.8	2.3	30.3	−1.5

[a] Requirement = biomass and nutrients in new growth.
[b] Stand increment = requirement for wood − mortality.
[c] Foliar uptake (FU) = −(Net Canopy Exchange, or NCE) if NCE $<$ 0; FU = 0 if NCE = or $>$ 0. Soil Uptake = Foliar Litterfall + Requirement for Wood + NCE. ROOTS NOT INCLUDED.

Table SS-3. Organic Matter and Nutrient Fluxes in Smokies Becking(SS) Site
Principal Investigator: D.W. Johnson

	Organic Matter	Ca	Mg	K	P	N	S
				($kg\ ha^{-1}\ yr^{-1}$)			
Litterfall	—	—	—	—	—	—	—
Needles	1173	5.3	0.3	1.1	0.4	5.9	0.8
Leaves	69	0.6	0.1	0.3	0.1	0.8	0.1
Roots	—	—	—	—	—	—	—
Other	667	1.8	0.1	0.3	0.3	4.5	0.5
Total	1909	7.7	0.5	1.7	0.8	11.2	1.4
Net Canopy Exchange	—	14.0	1.9	8.9	0.2	−3.9	−14.6
Requirement[a]	—	—	—	—	—	—	—
Foliage	2033	4.9	1	8.8	1.8	20.5	2.4
Wood	790	1.2	0.3	2.8	0.2	0.6	0.5
Roots	—	—	—	—	—	—	—
Total	2823	6.1	1.3	11.6	2	21.1	2.9
Mortality	1210	1.8	0.4	4.3	0.3	0.9	0.7
Stand Increment[b]	−420	−0.6	−0.1	−1.5	−0.1	−0.3	−0.2
Uptake[c]	—	—	—	—	—	—	—
Foliar	—	0	0	0	0	3.9	0
Soil	—	21.1	2.6	13.1	0.9	3.4	16
Total	—	21.1	2.6	13.1	0.9	7.3	16
Translocation	—	−15.0	−1.1	−1.5	1.1	13.8	13.1

[a] Requirement = biomass and nutrients in new growth.
[b] Stand increment = requirement for wood − mortality.
[c] Foliar uptake (FU) = −(Net Canopy Exchange, or NCE) if NCE $<$ 0; FU = 0 if NCE = or $>$ 0. Soil Uptake = Foliar Litterfall + Requirement for Wood + NCE. ROOTS NOT INCLUDED.

Table ST-3. Organic Matter and Nutrient Fluxes in Smokies Tower (ST) Site
Principal Investigator: D.W. Johnson

	Organic Matter	Ca	Mg	K	P	N	S
				($kg\ ha^{-1}\ yr^{-1}$)			
PLOT 1							
Litterfall	—	—	—	—	—	—	—
Needles	1196	5.3	0.4	1.2	0.4	6.5	0.9
Leaves	340	2	0.4	0.6	0.2	4.7	0.4
Roots	—	—	—	—	—	—	—
Other	234	0.5	0.1	0.2	0.1	1.8	0.2
Total	1770	7.8	0.9	2	0.7	13	1.5
Net Canopy Exchange	—	−2.4	−1.2	9.6	−0.3	−4.1	4.8
Requirement[a]	—	—	—	—	—	—	—
Foliage	2440	8	1	15	2.4	29.9	3
Wood	2090	3.2	0.3	0.7	0.1	1.8	1.1
Roots	—	—	—	—	—	—	—
Total	4530	11.2	1.3	15.7	2.5	31.7	4.1
Mortality	—	—	—	—	—	—	—
Stand Increment[b]	2090	3.2	0.3	0.7	0.1	1.8	1.1
Uptake[c]	—	—	—	—	—	—	—
Foliar	—	0	0	0	0.3	4.1	0
Soil	—	12.9	2.3	12.1	0.4	8.9	7.2
Total	—	12.9	2.3	12.1	0.7	13	7.2
Translocation	—	−1.7	1.0	3.6	1.8	18.7	−3.1

continued

Table ST-3. *Continued*

	Organic Matter	Ca	Mg	K	P	N	S
	(kg ha^{-1} yr^{-1})						
PLOT 2							
Litterfall	—	—	—	—	—	—	—
Needles	1254	5.4	0.4	1.5	0.5	6.9	0.9
Leaves	124	0.9	0.2	0.3	0.1	1.2	0.1
Roots	—	—	—	—	—	—	—
Other	4037	9.8	0.8	2.4	1.6	24.2	2.3
Total	5415	16.1	1.4	4.2	2.2	32.3	3.3
Net Canopy Exchange	—	−0.4	0.2	3.5	0.4	−16.4	−13
Requirement[a]	—	—	—	—	—	—	—
Foliage	2160	6.7	0.9	10.1	1.8	20.7	2.4
Wood	−800	−1.2	−0.1	−0.3	−0.1	−0.7	−0.4
Roots	—	—	—	—	—	—	—
Total	1360	5.5	0.8	9.8	1.7	20	2
Mortality	—	—	—	—	—	—	—
Stand Increment[b]	−800	−1.2	−0.1	−0.3	−0.1	−0.7	−0.4
Uptake[c]	—	—	—	—	—	—	—
Foliar	—	0.4	0	0	0	16.4	13
Soil	—	4.7	0.7	5	0.9	−9	−12.4
Total	—	5.1	0.7	5	0.9	7.4	0.6
Translocation	—	0.4	0.1	4.8	0.8	12.6	1.4

[a] Requirement = biomass and nutrients in new growth.
[b] Stand increment = requirement for wood − mortality.
[c] Foliar uptake (FU) = −(Net Canopy Exchange, or NCE) if NCE $<$ 0; FU = 0 if NCE = or $>$ 0. Soil Uptake = Foliar Litterfall + Requirement for Wood + NCE. ROOTS NOT INCLUDED.

Table TL-3. Organic Matter and Nutrient Fluxes in Turkey Lakes (TL) Site
Principal Investigators: I.K. Morrison, N.W. Foster

	Organic Matter	Ca	Mg	K	P	N	S
				(kg ha^{-1} yr^{-1})			
Litterfall	—	—	—	—	—	—	—
Needles	—	—	—	—	—	—	—
Leaves	2924	29	3	7	1	29	2
Roots	—	—	—	—	—	—	—
Other	806	9	1	2	1	11	1
Total	3730	38	4	9	2	40	3
Net Canopy Exchange	—	3.6	1.1	15.2	0.4	4.9	1.5
Requirement[a]	—	—	—	—	—	—	—
Foliage	4176	33	5	46	4	94	10
Wood	10239	1.3	0.2	0.8	0.1	1.3	0.1
Roots	1753	1.2	0.1	0.3	<.01	0.1	<.01
Total	16168	35.5	5.3	47.1	4.1	95.4	10.1
Mortality	—	—	—	—	—	—	—
Stand Increment[b]	—	—	—	—	—	—	—
Uptake[c]	—	—	—	—	—	—	—
Foliar	0	0	0	0	0	0	—
Soil	—	33.9	4.3	23	1.5	35.2	3.6
Total	—	33.9	4.3	23	1.5	35.2	3.6
Translocation	—	−0.9	0.7	23	2.5	58.8	6.4

[a] Requirement = biomass and nutrients in new growth.
[b] Stand increment = requirement for wood − mortality.
[c] Foliar uptake (FU) = −(Net Canopy Exchange, or NCE) if NCE < 0; FU = 0 if NCE = or > 0. Soil Uptake = Foliar Litterfall + Requirement for Wood + NCE. ROOTS NOT INCLUDED.

Table WF-3. Organic Matter and Nutrient Fluxes in Whiteface (WF) Site
Principal Investigators: A. Friedland, E. Miller

	Organic Matter	Ca	Mg	K	P	N	S
				(kg ha^{-1} yr^{-1})			
Litterfall							
Needles	—	—	—	—	—	—	—
Leaves	—	—	—	—	—	—	—
Roots	—	—	—	—	—	—	—
Other	—	—	—	—	—	—	—
Total	2597	15	0.7	1.3	1.4	23.7	—
Net Canopy Exchange	—	4.7	0.6	4.5	0.8	−5.9	−1.0
Requirement[a]							
Foliage	2209	10.6	1.5	10	2.5	34.5	2.9
Wood	3640	9.2	1.4	5.6	1	10.8	2.7
Roots	1530	5.1	0.7	3.4	0.8	8.3	0.8
Total	7379	24.9	3.6	19	4.3	53.6	6.4
Mortality	—	—	—	—	—	—	—
Stand Increment[b]	3640	9.2	1.4	5.6	1	10.8	2.7
Uptake[c]							
Foliar	—	0	0	0	0	+5.9	+1.0
Soil[d]	—	34	3.4	14.8	4.0	36.9	—
Total[d]	—	34	3.4	14.8	4.0	42.8	—
Translocation	—	9.1	0.2	4.2	0.3	10.8	—

[a] Requirement = biomass and nutrients in new growth.
[b] Stand increment = requirement for wood − mortality.
[c] Foliar uptake (FU) = −(Net Canopy Exchange, or NCE) if NCE < 0; FU = 0 if NCE = or > 0.
[d] Soil Uptake = Litterfall + Requirement for Wood + Requirement for roots + NCE.
[e] Translocation = Requirement − Uptake.

Table CP-4. Mean Atmospheric Concentrations at Coweeta Pine (CP) Site[a].
SITE = COWEETA PINE SITE (CP)
SAMPLING PERIOD: 4/86 to 3/89

A. Volume-weighted mean ion concentrations in wet deposition ($\mu eq\ L^{-1}$)

	SO_4^{2-}	NO_3^-	Cl^-	PO_4^{3-}	H^+	NH_4^+	K^+	Na^+	Ca^{2+}	Mg^{2+}	Al[b]	WKACID[c]
Cloud water	—	—	—	—	—	—	—	—	—	—	—	—
Rain	25.0	10.5	4.5	.21	32	9.4	.53	3.3	3.7	1.9	—	—
Throughfall	44.1	20.0	12.7	2.4	27	8.8	35	6.6	30	16	—	—
Stemflow	66.6	1.9	20.3	5.4	32	1.9	85	10.7	40	25	—	—

B. Time-weighted mean concentrations in atmospheric samples ($\mu g\ m^{-3}$)

Gas/Vapors		Aerosols									
SO_2	HNO_3	SO_4^{2-}	NO_3^-	Cl^-	PO_4^{3-}	H^+	NH_4^+	K^+	Na^+	Ca^{2+}	Mg^{2+}
3.4	1.52	5.5	.27	.117	.0046	0.39	1.40	.065	.078	.128	.016

C. Time-weighted mean fluxes of coarse particles to inert deposition plates ($\mu eq\ m^{-2}\ d^{-1}$)

SO_4^{2-}	NO_3^-	Cl^-	PO_4^{3-}	H^+	NH_4^+	K^+	Na^+	Ca^{2+}	Mg^{2+}
3.0	1.14	.43	.304	—	1.25	1.48	.76	4.73	.93

[a] These data represent weighted means for complete sampling period. Values in tables should be considered to include only two significant figures, despite up to five digits being expressed (because of nature of the program used to print data).
[b] Total Al in solution at sample pH, expressed in mol L^{-1}.
[c] Weak acidity by Gran's plot titration.

Table DF-4. Mean Atmospheric Concentrations at Douglas Fir (DF) Site[a].
SITE = DOUGLAS FIR SITE (DF)
SAMPLING PERIOD: 9/86 to 8/88

A. Volume-weighted mean ion concentrations in wet deposition ($\mu eq\ L^{-1}$)

	SO_4^{2-}	NO_3^-	Cl^-	PO_4^{3-}	H^+	NH_4^+	K^+	Na^+	Ca^{2+}	Mg^{2+}	Al[b]	WKACID[c]
Cloud water	—	—	—	—	—	—	—	—	—	—	—	—
Rain	17	11	19	—	20	6	3	22	12	4	—	—
Throughfall	29	7	33	—	24	2	20	37	23	9	—	—
Stemflow	88	2	97	—	100	3	107	69	110	30	—	—

B. Time-weighted mean concentrations in atmospheric samples ($\mu g\ m^{-3}$)

Gas/Vapors		Aerosols									
SO_2	HNO_3	SO_4^{2-}	NO_3^-	Cl^-	PO_4^-	H^+	NH_4^+	K^+	Na^+	Ca^{2+}	Mg^{2+}
1.67	0.94	1.49	0.38	0.15	—	0.018	0.42	0.099	0.30	0.066	0.035

C. Time-weighted mean fluxes coarse of particles to inert deposition plates ($\mu eq\ m^{-2}\ d^{-1}$)

SO_4^{2-}	NO_3^-	Cl^-	PO_4^{3-}	H^+	NH_4^+	K^+	Na^+	Ca^{2+}	Mg^{2+}
11.4	10.6	6.3	—	—	4.9	4.4	17.3	8.0	3.2

[a] These data represent weighted means for complete sampling period. Values in tables should be considered to include only two significant figures, despite up to five digits being expressed (because of nature of the program used to print data).
[b] Total Al in solution at sample pH, expressed in mol L^{-1}.
[c] Weak acidity by Gran's plot titration.

Table DL-4. Mean Atmospheric Concentrations at Duke Loblolly (DL) Site[a].
SITE = DUKE LOBLOLLY SITE (DL)
SAMPLING PERIOD: 11/86 to 10/88

A. Volume-weighted mean ion concentrations in wet deposition (μeq L^{-1})

	SO_4^{2-}	NO_3^-	Cl^-	PO_4^{3-}	H^+	NH_4^+	K^+	Na^+	Ca^{2+}	Mg^{2+}	Al[b]	WKACID[c]
Cloud water	—	—	—	—	—	—	—	—	—	—	—	—
Rain	49	17	32	0	61	26	2	13	18	3	—	<0.01
Throughfall	78	41	44	0	100	24	20	21	42	12	—	<0.01
Stemflow	241	36	79	1	275	14	39	27	82	30	—	<0.01

B. Time-weighted mean concentrations in atmospheric samples (μg m^{-3})

Gas/Vapors		Aerosols									
SO_2	HNO_3	SO_4^{2-}	NO_3^-	Cl^-	PO_4^{3-}	H^+	NH_4^+	K^+	Na^+	Ca^{2+}	Mg^{2+}
8.90	2.77	7.53	0.59	1.22	—	0.04	1.75	0.17	0.30	0.43	0.05

C. Time-weighted mean fluxes of coarse particles to inert deposition plates (μeq m^{-2} d^{-1})

SO_4^{2-}	NO_3^-	Cl^-	PO_4^{3-}	H^+	NH_4^+	K^+	Na^+	Ca^{2+}	Mg^{2+}
6.99	10.95	2.44	0.12	—	6.70	12.74	10.59	17.93	4.28

[a] These data represent weighted means for complete sampling period. Values in tables should be considered to include only two significant figures, despite up to five digits being expressed (because of nature of the program used to print data).
[b] Total Al in solution at sample pH, expressed in mol L^{-1}.
[c] Weak acidity by Gran's plot titration.

Table FS-4. Mean Atmospheric Concentrations at Florida Pine (FS) Site[a].
SITE = FLORIDA PINE SITE (FS)
SAMPLING PERIOD: 5/88 to 4/89

A. Volume-weighted mean ion concentrations in wet deposition (μeq L^{-1})

	SO_4^{2-}	NO_3^-	Cl^-	PO_4^{3-}	H^+	NH_4^+	K^+	Na^+	Ca^{2+}	Mg^{2+}	Al[b]	WKACID[c]
Cloud water	—	—	—	—	—	—	—	—	—	—	—	—
Rain	30.66	13.95	17.70	0.83	22.99	9.49	1.10	21.61	21.67	11.70	—	—
Throughfall	44.72	18.24	44.96	1.30	28.12	4.71	5.77	56.26	48.68	25.02	—	—
Stemflow	180.70	11.24	210.45	2.39	211.40	2.38	16.65	136.24	161.82	80.53	—	—

B. Time-weighted mean concentrations in atmospheric samples (μg m^{-3})

Gas/Vapors		Aerosols									
SO_2	HNO_3	SO_4^{2-}	NO_3^-	Cl^-	PO_4^{3-}	H^+	NH_4^+	K^+	Na^+	Ca^{2+}	Mg^{2+}
1.29	1.18	3.32	0.95	0.80	0.01	5.18	0.80	0.40	0.84	—	—

C. Time-weighted mean fluxes of coarse particles to inert deposition plates (μeq m^{-2} d^{-1})

SO_4^{2-}	NO_3^-	Cl^-	PO_4^{3-}	H^+	NH_4^+	K^+	Na^+	Ca^{2+}	Mg^{2+}
19.55	16.01	32.04	1.66	—	1.55	1.63	55.91	26.13	9.03

[a] These data represent weighted means for complete sampling period. Values in tables should be considered to include only two significant figures, despite up to five digits being expressed (because of nature of the program used to print data).
[b] Total Al in solution at sample pH, expressed in mol L^{-1}.
[c] Weak acidity by Gran's plot titration.

Table GL-4. Mean Atmospheric Concentrations at Georgia Loblolly (GL) Site[a].
SITE = GEORGIA LOBLOLLY SITE (GL)
SAMPLING PERIOD: 4/87 to 3/88

A. Volume-weighted mean ion concentrations in wet deposition (μeq L^{-1})

	SO_4^{2-}	NO_3^-	Cl^-	PO_4^{3-}	H^+	NH_4^+	K^+	Na^+	Ca^{2+}	Mg^{2+}	Al[b]	WKACID[c]
Cloud water	—	—	—	—	—	—	—	—	—	—	—	—
Rain	35	11	11	0.163	46	7.7	1.5	7.4	2.9	1.6	—	—
Throughfall	58	25	27	0.37	36	12	33	14	20	15	—	—
Stemflow	—	—	—	—	—	—	—	—	—	—	—	—

B. Time-weighted mean concentrations in atmospheric samples (μg m^{-3})

	Gas/Vapors		Aerosols									
	SO_2	HNO_3	SO_4^{2-}	NO_3^-	Cl^-	PO_4^{3-}	H^+	NH_4^+	K^+	Na^+	Ca^{2+}	Mg^{2+}
	7.56	2.25	3.19	0.45	0.022	0.046	—	0.85	0.23	0.26	0.1	0.003

C. Time-weighted mean fluxes of coarse particles to inert deposition plates (μeq m^{-2} d^{-1})

	SO_4^{2-}	NO_3^-	Cl^-	PO_4^{3-}	H^+	NH_4^+	K^+	Na^+	Ca^{2+}	Mg^{2+}
	11.3	9.4	10.3	0.77	—	3.47	1.13	24.2	10.15	3.74

[a] These data represent weighted means for complete sampling period. Values in tables should be considered to include only two significant figures, despite up to five digits being expressed (because of nature of the program used to print data).
[b] Total Al in solution at sample pH, expressed in mol L^{-1}.
[c] Weak acidity by Gran's plot titration.

Table HF-4. Mean Atmospheric Concentrations at Huntington Forest (HF) Site[a].
SITE = HUNTINGTON FOREST SITE (HF)
SAMPLING PERIOD: 5/86 to 4/88

A. Volume-weighted mean ion concentrations in wet deposition (μeq L^{-1})

	SO_4^{2-}	NO_3^-	Cl^-	PO_4^{3-}	H^+	NH_4^+	K^+	Na^+	Ca^{2+}	Mg^{2+}	Al[b]	WKACID[c]
Cloud water	—	—	—	—	—	—	—	—	—	—	—	—
Rain	36.95	19.30	3.30	—	34.90	13.65	3.55	2.45	4.20	1.30	—	—
Throughfall	49.25	25.90	3.05	—	27.10	14.85	18.90	6.55	22.85	5.50	—	—
Stemflow	98.60	17.45	15.20	—	2.70	30.40	201.75	20.25	26.85	8.20	—	—

B. Time-weighted mean concentrations in atmospheric samples (μg m^{-3})

Gas/Vapors		Aerosols									
SO_2	HNO_3	SO_4^{2-}	NO_3^-	Cl^-	PO_4^{3-}	H^+	NH_4^+	K^+	Na^+	Ca^{2+}	Mg^{2+}
5.8	1.8	4.89	1.09	0.05	—	0.01	1.25	0.08	0.07	0.20	0.06

C. Time-weighted mean fluxes of coarse particles to inert deposition plates (μeq m^{-2} d^{-1})

SO_4^{2-}	NO_3^-	Cl^-	PO_4^{3-}	H^+	NH_4^+	K^+	Na^+	Ca^{2+}	Mg^{2+}
9.68	11.19	3.70	—	—	3.98	2.64	2.44	17.56	2.91

[a] These data represent weighted means for complete sampling period. Values in tables should be considered to include only two significant figures, despite up to five digits being expressed (because of nature of the program used to print data).
[b] Total Al in solution at sample pH, expressed in mol L^{-1}.
[c] Weak acidity by Gran's plot titration.

Table LP-4. Mean Atmospheric Concentrations at Oak Ridge Loblolly Pine (LP) Site[a].
SITE = OAK RIDGE LOBLOLLY PINE (LP)
SAMPLING PERIOD: 4/86 to 3/89

A. Volume-weighted mean ion concentrations in wet deposition ($\mu eq\ L^{-1}$)

	SO_4^{2-}	NO_3^-	Cl^-	PO_4^{3-}	H^+	NH_4^+	K^+	Na^+	Ca^{2+}	Mg^{2+}	Al^b	WKACID[c]
Fog water	230	69	57	1.7	22	240	39	62	96	15	—	—
Rain	46	18	8.0	0.094	51	13	0.65	3.5	8.0	2.3	.26	33
Throughfall	81	33	14	0.39	59	11	28	5.4	41	15	.70	87
Stemflow	570	19	68	1	460	19	77	12	160	68	16	620

B. Time-weighted mean concentrations in atmospheric samples ($\mu g\ m^{-3}$)

Gas/Vapors		Aerosols									
SO_2	HNO_3	SO_4^{2-}	NO_3^-	Cl^-	PO_4^{3-}	H^+	NH_4^+	K^+	Na^+	Ca^{2+}	Mg^{2+}
12	2.8	7.7	0.36	0.072	0.017	0.034	2.0	0.088	0.095	0.52	0.062

C. Time-weighted mean fluxes of coarse particles to inert deposition plates ($\mu eq\ m^{-2}\ d^{-1}$)

SO_4^{2-}	NO_3^-	Cl^-	PO_4^{3-}	H^+	NH_4^+	K^+	Na^+	Ca^{2+}	Mg^{2+}
15	5.5	1.9	0.14	—	1.5	1.9	1.4	39	5.3

[a] These data represent weighted means for complete sampling period. Values in tables should be considered to include only two significant figures, despite up to five digits being expressed (because of nature of the program used to print data).
[b] Total Al in solution at sample pH, expressed in mol L^{-1}.
[c] Weak acidity by Gran's plot titration.

Table MS-4. Mean Atmospheric Concentrations at Main Spruce (MS) Site[a].
SITE = MAINE SPRUCE SITE (MS)
SAMPLING PERIOD: 4/88 to 3/89

A. Volume-weighted mean ion concentrations in wet deposition (μeq L^{-1})

	SO_4^{2-}	NO_3^-	Cl^-	PO_4^{3-}	H^+	NH_4^+	K^+	Na^+	Ca^{2+}	Mg^{2+}	Al[b]	WKACID[c]
Cloud water	—	—	—	—	—	—	—	—	—	—	—	—
Rain	25.05	28	8.56	—	—	14.95	1.38	8.37	1.36	0.68	—	—
Throughfall	49.47	13.92	24.27	—	—	—	22.56	17.92	21.23	8.68	—	—
Stemflow	—	—	—	—	—	—	—	—	—	—	—	—

B. Time-weighted mean concentrations in atmospheric samples (μg m^{-3})

Gas/Vapors		Aerosols									
SO_2	HNO_3	SO_4^{2-}	NO_3^-	Cl^-	PO_4^{3-}	H^+	NH_4^+	K^+	Na^+	Ca^{2+}	Mg^{2+}
4.896	0.796	3.594	0.394	0.326	—	0.025	0.543	0.045	0.157	0.059	0.022

C. Time-weighted mean fluxes of coarse particles to inert deposition plates (μeq m^{-2} d^{-1})

SO_4^{2-}	NO_3^-	Cl^-	PO_4^{3-}	H^+	NH_4^+	K^+	Na^+	Ca^{2+}	Mg^{2+}
12.6	4.7	3.3	—	—	9.82	1.40	2.65	8.60	2.46

[a] These data represent weighted means for complete sampling period. Values in tables should be considered to include only two significant figures, despite up to five digits being expressed (because of nature of the program used to print data).
[b] Total Al in solution at sample pH, expressed in mol L^{-1}.
[c] Weak acidity by Gran's plot titration.

Table NS-4. Mean Atmospheric Concentrations at Norway Spruce (NS) Site[a].
SITE = NORWAY SPRUCE SITE (NS)
SAMPLING PERIOD: 10/86 to 9/88

A. Volume-weighted mean ion concentrations in wet deposition (μeq L^{-1})

	SO_4^{2-}	NO_3^-	Cl^-	PO_4^{3-}	H^+	NH_4^+	K^+	Na^+	Ca^{2+}	Mg^{2+}	Al[b]	WKACID[c]
Cloud water	—	—	—	—	—	—	—	—	—	—	—	—
Rain	48	30	12	—	52	27	1.9	10	5	2.8	—	—
Throughfall	83	11	31	—	46	16	46	20	33	17	—	—
Stemflow	—	—	—	—	—	—	—	—	—	—	—	—

B. Time-weighted mean concentrations in atmospheric samples (μg m^{-3})

Gas/Vapors		Aerosols									
SO_2	HNO_3	SO_4^{2-}	NO_3^-	Cl^-	PO_4^{3-}	H^+	NH_4^+	K^+	Na^+	Ca^{2+}	Mg^{2+}
1.3	0.4	2.5	1.0	0.14	—	—	0.89	0.12	0.25	0.073	0.024

C. Time-weighted mean fluxes of coarse particles to inert deposition plates (μeq m^{-2} d^{-1})

SO_4^{2-}	NO_3^-	Cl^-	PO_4^{3-}	H^+	NH_4^+	K^+	Na^+	Ca^{2+}	Mg^{2+}
7.1	5.0	8.0	—	—	3.0	3.3	6.4	4.5	1.6

[a] These data represent weighted means for complete sampling period. Values in tables should be considered to include only two significant figures, despite up to five digits being expressed (because of nature of the program used to print data).
[b] Total Al in solution at sample pH, expressed in mol L^{-1}.
[c] Weak acidity by Gran's plot titration.

Table RA-4. Mean Atmospheric Concentrations at Red Alder (RA) Site[a].
SITE = RED ALDER SITE (RA)
SAMPLING PERIOD: 9/86 to 8/88

A. Volume-weighted mean ion concentrations in wet deposition ($\mu eq\ L^{-1}$)

	SO_4^{2-}	NO_3^-	Cl^-	PO_4^{3-}	H^+	NH_4^+	K^+	Na^+	Ca^{2+}	Mg^{2+}	Al[b]	WKACID[c]
Cloud water	—	—	—	—	—	—	—	—	—	—	—	—
Rain	17	11	19	—	20	6	3	23	12	4	—	—
Throughfall	27	6	31	—	12	3	28	37	25	13	—	—
Stemflow	23	6	48	—	3	8	110	46	25	11	—	—

B. Time-weighted mean concentrations in atmospheric samples ($\mu g\ m^{-3}$)

Gas/Vapors		Aerosols									
SO_2	HNO_3	SO_4^{2-}	NO_3^-	Cl^-	PO_4^{3-}	H^+	NH_4^+	K^+	Na^+	Ca^{2+}	Mg^{2+}
1.67	0.94	1.49	0.38	0.15	—	0.018	0.42	0.099	0.30	0.066	0.035

C. Time-weighted mean fluxes of coarse particles to inert deposition plates ($\mu eq\ m^{-2}\ d^{-1}$)

SO_4^{2-}	NO_3^-	Cl^-	PO_4^{3-}	H^+	NH_4^+	K^+	Na^+	Ca^{2+}	Mg^{2+}
11.4	10.6	6.3	—	—	4.9	4.4	17.3	8.0	3.2

[a] These data represent weighted means for complete sampling period. Values in tables should be considered to include only two significant figures, despite up to five digits being expressed (because of nature of the program used to print data).
[b] Total Al in solution at sample pH, expressed in mol L^{-1}.
[c] Weak acidity by Gran's plot titration.

Table ST-4. Mean Atmospheric Concentrations at Smokies Tower (ST) Site[a].
SITE = SMOKIES TOWER SPRUCE (ST)
SAMPLING PERIOD: 4/86 to 3/89

A. Volume-weighted mean ion concentrations in wet deposition (μeq L^{-1})

	SO_4^{2-}	NO_3^-	Cl^-	PO_4^{3-}	H^+	NH_4^+	K^+	Na^+	Ca^{2+}	Mg^{2+}	Al[b]	WKACID[c]
Cloud water	250	64	84	0.34	110	94	36	72	74	25	—	—
Rain	30	11	4.4	0.091	27	11	0.97	2.4	6.6	2.3	0.23	18
Throughfall	115	41	21	0.13	100	10	19	13	42	13	1.0	69
Stemflow	310	20	45	0.21	190	14	71	20	150	44	2.6	300

B. Time-weighted mean concentrations in atmospheric samples (μg m^{-3})

Gas/Vapors		Aerosols									
SO_2	HNO_3	SO_4^{2-}	NO_3^-	Cl^-	PO_4^{3-}	H^+	NH_4^+	K^+	Na^+	Ca^{2+}	Mg^{2+}
4.0	2.3	6.2	0.20	0.077	0.015	0.043	1.3	0.05	0.082	0.27	0.034

C. Time-weighted mean fluxes of coarse particles to inert deposition plates (μeq m^{-2} d^{-1})

SO_4^{2-}	NO_3^-	Cl^-	PO_4^{3-}	H^+	NH_4^+	K^+	Na^+	Ca^{2+}	Mg^{2+}
13	10	1.4	0.056	—	3.3	0.82	2.3	26	3.5

[a] These data represent weighted means for complete sampling period. Values in tables should be considered to include only two significant figures, despite up to five digits being expressed (because of nature of the program used to print data).
[b] Total Al in solution at sample pH, expressed in mol L^{-1}.
[c] Weak acidity by Gran's plot titration.

Table WF-4. Mean Atmospheric Concentrations at Whiteface Mountain (WF) Site[a].
SITE = WHITEFACE MOUNTAIN SITE (WF)
SAMPLING PERIOD: 6/86 to 5/88

A. Volume-weighted mean ion concentrations in wet deposition (μeq L^{-1})

	SO_4^{2-}	NO_3^-	Cl^-	PO_4^{3-}	H^+	NH_4^+	K^+	Na^+	Ca^{2+}	Mg^{2+}	Al[b]	WKACID[c]
Cloud water	303	158	8.7	0	286	168	3.0	4.0	20	6.0	—	—
Rain	44.30	22.84	2.68	0.15	48.80	14.97	0.84	1.34	8.08	2.10	0.91	—
Throughfall	87.76	27.50	6.51	2.28	71.82	14.24	11.89	3.88	30.81	7.60	1.42	—
Stemflow	164.55	0.84	8.26	0.15	194.46	8.09	117.55	12.05	85.3	8.5	3.47	—

B. Time-weighted mean concentrations in atmospheric samples (μg m^{-3})

	Gas/Vapors		Aerosols									
	SO_2	HNO_3	SO_4^{2-}	NO_3^-	Cl^-	PO_4^{3-}	H^+	NH_4^+	K^+	Na^+	Ca^{2+}	Mg^{2+}
	3.163	1.205	4.553	0.207	0.001	—	0.006	0.816	0.049	0.116	0.148	0.043

C. Time-weighted mean fluxes of coarse particles to inert deposition plates (μeq m^{-2} d^{-1})

	SO_4^{2-}	NO_3^-	Cl^-	PO_4^{3-}	H^+	NH_4^+	K^+	Na^+	Ca^{2+}	Mg^{2+}
	0.563	0.818	1.064	—	0.512	0.337	1.478	2.060	1.115	0.327

[a] These data represent weighted means for complete sampling period. Values in tables should be considered to include only two significant figures, despite up to five digits being expressed (because of nature of the program used to print data).
[b] Total Al in solution at sample pH, expressed in μmol L^{-1}.
[c] Weak acidity by Gran's plot titration.

Index